SUSTAINABLE WIRELESS SENSOR NETWORKS

Edited by **Dr. Winston Seah and**
Dr. Yen Kheng Tan (Editor-in-Chief)

Sustainable Wireless Sensor Networks
Edited by Dr. Winston Seah and Dr. Yen Kheng Tan (Editor-in-Chief)

Published by InTech
Janeza Trdine 9, 51000 Rijeka, Croatia

Publishing Process Manager Jelena Marusic

Technical Editor Teodora Smiljanic

Cover Designer InTech Design Team

Additional hard copies can be obtained from orders@intechopen.com

Sustainable Wireless Sensor Networks, Edited by Dr. Winston Seah and Dr. Yen Kheng Tan (Editor-in-Chief)
p. cm.
ISBN 978-953-307-297-5

For EU product safety concerns:
IN TECH d.o.o., Prolaz Marije Krucifikse Kozulić 3, 51000 Rijeka, Croatia,
info@intechopen.com or visit our website at intechopen.com.

Meet the editors

Yen Kheng Tan received his B.Eng degree in Electrical and Computer Engineering (ECE) from National University of Singapore (NUS) in 2003 and the Master of Technological Design (Mechatronics Engineering) degree jointly offered by NUS and the Eindhoven University of Technology (TU/e) in 2006. Since 2006, he has been working towards his Ph.D degree in the Department of ECE of NUS. Presently, he is leading a team of research engineers at the Energy Research Institute @NTU (ERI@N). His research interests are micro-power generations from solar, wind, thermal and vibration energy sources, energy harvesting wireless sensor network, nonlinear control and motor drives, electric vehicles, wireless power transfer, rehabilitation engineering and assistive technology. Yen Kheng is on the Technical Program Committee of numerous conferences and reviewer of papers for a number of prestigious international journals. He is also the book editor of several books as well. [http://www.ece.nus.edu.sg/emdl/yenkheng.htm]

Winston K.G. Seah received the Dr.Eng. degree from Kyoto University, Kyoto, Japan, in 1997. He is currently Professor of Network Engineering in the School of Engineering and Computer Science, Victoria University of Wellington, New Zealand. Prior to this, he has worked for more than 16 years in mission-oriented research, taking ideas from theory to prototypes, most recently, as a Senior Scientist (Networking Protocols) in the Institute for Infocomm Research (I2R), Singapore. He is actively involved in research in the areas of mobile ad hoc and sensor networks, and co-developed one of the first Quality of Service models for mobile ad hoc networks. He is a Senior Member of the IEEE and Professional Member of the ACM.

Contents

Preface

Wireless Sensor Networks came into prominence around the start of this millennium motivated by the omnipresent scenario of small-sized sensors with limited power deployed in large numbers over an area to monitor different phenomenon. The sole motivation of a large portion of research efforts has been to maximize the lifetime of the network, where network lifetime is typically measured from the instant of deployment to the point when one of the nodes has expended its limited power source and becomes in-operational – commonly referred as first node failure. Over the years, research has increasingly adopted ideas from wireless communications as well as embedded systems development in order to move this technology closer to realistic deployment scenarios. In such a rich research area as wireless sensor networks, it is difficult if not impossible to provide a comprehensive coverage of all relevant aspects.

In this book, we hope to give the reader with a snapshot of some aspects of wireless sensor networks research that provides both a high level overview as well as detailed discussion on specific areas. The chapters in this book can be generally divided into the following areas: review of technology, wireless communications and networking, optimization for WSN applications, and systems implementation with a brief mention on security.

Dr. Winston Seah and
Dr. Yen Kheng Tan (Editor-in-Chief)

Part 1

Review of Technology

1

A Survey of Routing Protocols of Wireless Sensor Networks

Zhe Yang and Abbas Mohammed
Blekinge Institute of Technology
Sweden

1. Introduction

Networked micro-sensor technology is a key technology for the future. It has been identified as one of the most important technologies for the 21st century and is regarded to revolutionize information gathering and processing in applications (Heinzelman et al., 1999). Advances in Micro-Electro-Mechanical Systems (MEMS) and low-power integrated digital electronics have inspired the development of micro-sensors (Sohrabi, 2000). Such sensors are generally equipped with date processing, communication, and information collecting capabilities. They can detect the variation of ambient conditions in the environment surrounding the sensors and transform them into electric signals. Interests in sensor networks have motivated intensive research in the past few years emphasizing the potential of cooperations among sensors in date collecting and processing, coordination and management of the sensing activity, and date flow to the sink. Wireless Sensor Networks (WSNs) is one of the architecture of sensor networks. WSN can be formed by sensors in an ad-hoc manner. It belongs to the general family of sensor networks that use multiple distributed sensors to collect information on entities of interest.

As in many technologies, research on sensor networks was originally motivated by military applications. Early research was done by military using sensor networks for defence dealing with events at contiguous levels. Around 1980 modern research on sensor networks started with the distributed sensor networks program at the US Defense Advanced Research Projects Agency (DARPA). During this period universities and institutes did an intensive research in technology components for sensor networks about designing acoustic sensors, protocols to link processes of working on a common application in a network, self-location algorithms, distributed software and developing test beds. In 1990s there was an important shift of sensor network research due to advances in computing and communications. Small size, low cost sensors are designed to be based upon MEMS technology, wireless networking and low power processors, which make sensors possible to be deployed in a wireless fashion. The shift has led and influenced the latest research on networking and information processing techniques of sensor networks.

1.1 Communication Architecture and Applications of WSNs

A typical WSN contains a large number of sensor nodes, which send collected data via radio transmitter to a sink. The decrease in both the size and the cost due to the development of MEMS has led to smart disposable micro sensors, which can be networked through wireless connections to the Internet. Fig. 1 shows an architecture of communications in a WSN. Sensor nodes are capable of organizing themselves, and collect information about the phenomenon and route data via neighbouring sensors to the sink. The gateway in Fig. 1 could be a fixed or mobile node with an ability of connecting sensor networks to the outer existing communication infrastructure, such as Internet, cellular and satellite networks.

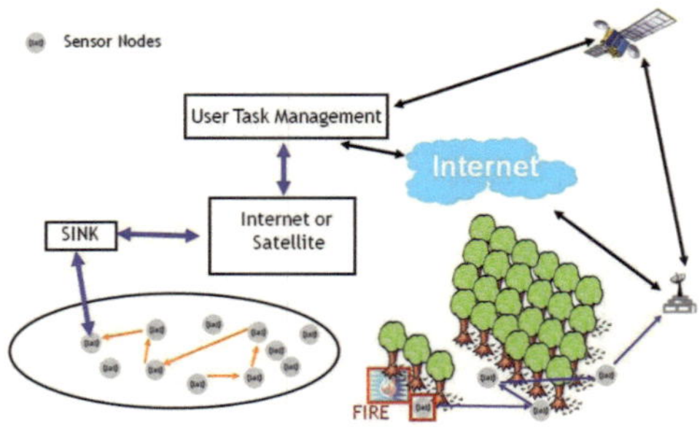

Fig. 1. Communication architecture of a WSN

Depending on applications to reveal some characteristics about phenomena in the area, sensor nodes can be deployed on the ground, in the air, under water, on bodies, in vehicles and inside buildings (Akyildiz et al., 2002; Xu, 2002). Publications and current applications have shown these connected sensor nodes have the potential in both consumer and military applications, e.g. target field imaging, intrusion detection, weather monitoring, security and tactical surveillance, distributed computing, traffic and inventory control, detecting ambient conditions such as temperature, humidity, movement, sound and light. Deployment of these sensor nodes can be in a random fashion like dropping from a helicopter in a disaster management applicaiton for environment survelience, or manually(Akyildiz et al., 2002).

1.2 Network Layer in WSNs

Sensor nodes are constrained by energy supply and bandwidth. Such constraints combined with the deployment of a large number of nodes are challenges to the design and maintenance of the network. Energy-awareness needs to be considered at all layers of a protocol stack. Physical and data link layers, which are generally common for all kind of applications, also need to consider these limitations. Thus research on these layers have focused on radio communication hardware, energy-aware medium access control (MAC) protocols (Demirkol et al., 2006; Intel, 2004). Table 1 gives a full view of protocol stack for communications in sensor networks.

The main aim at the networking layer is to find the route to transmit data from sensor nodes to the sink in an energy-efficient and reliable manner in order to maximally extend the lifetime of the network. Routings in sensor networks are challenging due to several characteristics distinguishing from established wireless communication networks in following areas.

1. It is not possible to build a global addressing scheme for a large number of sensors deployed. The addressing scheme, e.g. Internet Protocol (IP), needs to maintain routing tables for the global topology. Updating in a dynamic environment of a typical sensor network's application leads to heavy overhead in terms of time, memory and energy. Therefore a classical IP-based protocol is not applicable to sensor networks (Akyildiz et al., 2002).

<table>
<tr><td colspan="5">Task management</td></tr>
<tr><td colspan="5">Power management</td></tr>
<tr><td colspan="5">Mobility management</td></tr>
<tr><td>Application</td><td>Transport</td><td>Network</td><td>MAC/DATA link</td><td>Physical layer</td></tr>
</table>

Table 1. Protocol stack for communications in sensor network

2. Compared to a typical communication network, e.g. mobile communication networks, almost all applications of sensor networks require the flow of sensed data from different sources to the same sink (Akyildiz et al., 2002). Most prevalent wireless networks today, e.g. cellular network, are based on cells which are regions divided geographically. A mobile terminal in a cell only communicates with the base station serving the cell. A peer-to-peer communication between two mobile terminals doesn't exist. Communications are established through different base stations. However sensor nodes in WSNs send data to the sink based on a multiple hop routing composed by distributed networking and control functions in sensor nodes.
3. Data traffic generated by sensor nodes have significant redundancy because multiple sensors with a similar distance to the phenomena may generate the same data. Such redundancy needs to be eliminated by using proper routing protocols to improve energy and bandwidth utilization.
4. Different resource management protocols in the stack have to consider constrains of sensor node in terms of its transmission power, residual energy, processing and storage capacity.

Many specific algorithms have been proposed to solve these problems of routing data in sensor networks (Niculescu, 2005). These routing mechanisms have to consider characteristics of sensor nodes and application requirements. Classically most routing protocols could be classified as *data-centric, hierarchical* and *location* based protocols depending on the network structure and applications. A few distinct protocols are based on *network flow and quality of service (QoS)* awareness.

Date-centric protocols are query-based and depend on the naming of data of interest, which could help to reduce repeated transmissions.

Hierarchical protocols use the cluster concept in the network to divide sensors into different clusters and choose cluster heads to aggregate and reduce transmission of data in order to save energy.

Location-based protocols utilize the position information to relay data to the destination.

The chapter is organized as follows: In the rest of section 1, we will briefly summarize the design issues for sensor networks on data routing. In section 2, 3, 4 and 5, different routing approaches of current reserach will be presented. In section 6 information of research platforms, simulation, and development tools of WSNs will be introduced. Conclusions and future work will be given in section 7.

1.3 Design Factors of WSNs

Due to the large number of sensor nodes and the dynamics of their operating environment, it poses unique challenges in the architecture design of sensor networks. Routing design is closely related to the system architecture mode. In this section, we will summarize architectural issues of sensor networks and how they affect routing process in WSNs.

Dynamic Network: Basically a WSN consists of three components: sensor node, sink and event. Sensor nodes and sink are assumed to be fixed or mobile. Currently sensor nodes in most applications are assumed to be stationary, but it is still necessary to support the mobility of sinks or gateways in the network. Thus the stability of routing data is an important design factor, in addition to energy consumptions and bandwidth utilizations (Akyildiz et al., 2002).

Node Deployment: The topology of node deployment is application dependent and affects the performance of the routing protocol. If the nodes are deployed randomly, they need to create an infrastructure in an ad-hoc manner and organize themselves to establish paths to route the data . If nodes are deployed manually with pre-arranged locations, pre-determined path could be built to route data (Akyildiz et al., 2002). In addition, the position of the sink or gateway is also important to optimize routing paths.

Energy Constrains: The process of setting up the routes in the network is greatly affected by energy considerations. Since radio transmission degrades with distance much faster than transmission in free space, it implies that communication distance and energy consumption must be well managed. Directed routing would perform well enough if all sensor nodes are close to the sink. However most of the time it is necessary to use multiple hop routing to consume less power than directive routing, because sensors are usually randomly scattered in the area. However this introduces significant overhead for topology management and MAC protocols.

Data Transmission and Dissemination Models: Based on applications of sensor networks, the data delivery to the sink can be continuous, event-driven, query-driven or hybrid (Akyildiz et al., 2002; Demirkol et al., 2006). In the continuous model, each sensor node sends data periodically. While in the event-driven and a query-driven applications, sensor node only start to transmit data when the event occurs or a query is generated by the sink. Some applications combine continuous, even-driven and query-driven data delivery. Corresponding to the transmission model of data delivery mentioned above, data flow transmitted between sensor nodes can be classified as: broadcast, unicast and multicast subject to different routing protocols. Performance of using protocols is application dependent. For example, broadcasting can generate high overhead, but it is suitable for dynamic changes in the topology of the network. A major advantage of broadcast is the lack of a complex network layer protocol for routing, address and location management. Existing sensor network efforts have mostly relied on this approach.

MAC Protocol Design: To operate a wireless sensor network successfully, MAC protocols are important networking issues, which need to consider energy consumption and complexity of implementation. Numerous energy-efficient MAC protocols have been developed for sensor networks, such as S-MAC (Ye et al., 2002) and T-MAC (Dan & Langendoen, 2003), but these protocols still operate in an address-based fashion, which rely on message passing to individual neighbours. A data-centric based MAC protocol is investigated in (Ditzel & Langendoen, 2005) to follow a data-centric routing. Nodes access the communication resources following an active-sleep regime, alleviating the problem of idle-listening.

There are other issues, such as coverage area, scalability, transmission media, which could also affect the design and performance of routing protocol.

2. Date-centric Protocols

Due to the dense deployment and dynamic environment of sensor nodes in many applications, it is not possible to assign global identifiers to each node (Akyildiz et al., 2002). Random deployment and dynamics make it hard to select a specific set of sensor nodes to be queried. Thus routing in the system should operate autonomously, changing its configuration as required. This means protocols are able to select a set of sensor nodes and can employ data aggregation during the delivery while considering energy consumption. Connections can not be planned in advance but should emerge on-demand. However, in traditional address-based networks, routings are created between nodes and managed in the network layer of the communication stack.
Users would be more interested in querying a specific area rather than a single node because of a large population of sensor nodes. In data-centric routing, the sink requests information from nodes in certain area and waits for responses from sensors located in the selected area. To facilitate date-centric characteristics of sensor queries, an attribute-based naming scheme is used to specify the properties of data. Each node involved in the transmission plays the same role.

2.1 Flooding and Gossiping

Flooding and Gossiping: These two classical mechanisms to deliver data in sensor networks don't need any routing algorithms. In a flooding mechanism, each sensor receives a data packet and then broadcasts to all neighbouring nodes. When the packet arrives at the destination or the maximum number of hops is reached, the process of broadcasting is stopped. It is easy to implement flooding but with several drawbacks like implosion, which could be caused by sending duplicated messages to the same node, overlapping when two nodes sense in the same region and send similar data to its neighbours (Heinzelman et al., 1999). Energy-awareness is not considered in these mechanisms. Gossiping solves the problem of implosion by sending information to a random neighbour instead of a classical broadcasting mechanism sending packets to all neighbours. However gossiping causes another problem - delay in a propagation of data among sensor nodes.

2.2 SPIN

Sensor Protocols for Information via Negotiation (SPIN): SPIN is an outcome of an early research in a data-centric routing mechanism. The main idea in SPIN is to use meta data instead of a full data packet transmitted at each node to all nodes (Heinzelman et al., 1999). It assumes that nodes in close proximity have similar data. Before transmission, data collected by nodes are exchanged among sensors via data advertisement mechanism, which enable nodes to distribute data which other nodes don't pose. Negotiation process between neighboured nodes are performed by naming the data using high-level descriptors before any data start to be transmitted (Akyildiz et al., 2002). SPIN ensures that low redundant data sent throughout the network and solve problems, such as wasting energy and bandwidth to send extra copies of data by sensors in the same area (Akyildiz et al., 2002), of a broadcasting mechanism, e.g. flooding. Each sensor node can operate more efficiently and conserve energy by sending data after negotiation instead of sending all data. However data collecting mechanism cannot guarantee the delivery of data. For example, if the node of interest is far away from the sensing source and nodes between the source and destination are not interested in that data, such data can not be be routed to the destination. Two types of SPIN have been developed. *SPIN-1* doesn't deal with energy efficiency and *SPIN-2* is energy aware.

SPIN is more efficient than the protocol based on flooding and has relatively quick convergence in terms of latency. The type of protocols can be used for both mobile or stationary events. The negotiation process is fairly simple. The main drawback of the protocol is the energy consumption caused by idle nodes being always active.

2.3 Directed Diffusion

Directed Diffusion The protocol is an important breakthrough in a data-centric routing research of sensor networks. The idea behind diffusing data through sensor nodes aims to use a scheme of naming data for all communications. It uses attribute-value pairs for the data and queries the sensors on demand. By defining an interest through a list of attribute-value pairs, such as name of objects, interval, duration, geographical area etc., it can create a query to communicate with nodes. Data is cached at intermediate nodes for aggregation and loop prevention. Interests are propagated by unicast, multicast or broadcast from the sink to nodes. Once sensor nodes collect information, they compare with the data in the interest pre-stored in the cache and respond to specified interest. The local gradient is set by propagating interest from sink to source, where a path reinforcement between a source and sink can be realized by resending interest messages frequently. The data sent back to a sink by unicast and multicast consists of the data rate, duration and expiration time derived from the received interest. Path repairs in directed diffusion could also be possible by employing multiple paths in advance (Sohrabi, 2000). Thus if one path fails, an alternative is chosen to replace it.

A directed diffusion protocol consumes much less energy by having less traffic compared to flooding. It uses the best path based on local gradient to have a good latency bound. A retransmission of interest makes the protocol robust. The drawback of the protocol is that directed diffusion is application dependent, because it is based on a query-driven data delivery model. If the application, like environment monitoring, requires continuous transmission to the sink, it will not work effectively with a query-driven on demand data model (Akyildiz et al., 2002). Comparing the data with the pre-stored interests will also generate redundant overhead at the sensors. A retramission or an alternative path maintenance is needed.

2.4 Other Data-centric Protocols

SPIN and direct diffusion have motivated designs of other data-centric protocols. *Energy-aware routing, rumor routing, gradient-based routing* and *Constrained Anisotropic Diffusion Routing (CADR)* follow a similar concept of using queries to certain regions to get response (Akyildiz et al., 2002; Braninsky & Estrin, 2002; Dan & Langendoen, 2003; Heinzelman et al., 1999; Niculescu, 2005).

3. Hierarchical Protocols

The dense and random deployment of nodes causes the scalability to be one of the design issues in sensor networks. Usually a network with a single gateway is not scalable for a larger set of sensors since sensors are not capable of extended communication period. Networking different clusters is proposed to allow system to cope with an additional load and a large coverage area with a long-haul communication.

The main aim of a hierarchical routing is to maintain the energy consumption of sensor nodes. A cluster formation in a multi-hop sensor network is typically based on the energy reserve of sensors and their distances to the cluster head (Akyildiz et al., 2002). The cluster head with a higher-energy node, can be used to process and send the information. In addition, the rest of sensors in that cluster can perform tasks of sensing.

3.1 LEACH

Low-Energy Adaptive Clustering Hierarchy (LEACH): The idea of LEACH algorithm is to form clusters of the sensor nodes based on the received signal strength, and use local cluster heads as routers to the sink (Heinzelman et al., 2000). This routing mechanism saves energy since the transmissions are mainly managed by cluster heads. Initially cluster heads are randomly selected and changed over time in order to spread load and balance the energy dispersions of nodes. A cluster head compresses data arriving from nodes belonging to its cluster and sends an aggregated packet to the sink. Adaptive clustering is employed to increase the lifetime of the system. LEACH assumes that each node has enough power to transmit signals to reach cluster head and has equal computational power to work in different MAC protocols. Thus it is not applicable to deploy in large regions due to the variation of distances between sensors and head of clusters (Akyildiz et al., 2002). Moreover the idea of dynamic clustering brings extra overhead, such as rotation of cluster head, advertisement etc., and accordingly consumes energy.

LEACH uses Code Division Multiple Access (CDMA) or Time Division Multiple Access (TDMA) MAC to share channels (Akyildiz et al., 2002; Heinzelman et al., 2000). These MAC protocols are widely used in modern cellular communication systems. By scheduling nodes into different sub-channels by codes or time slots, they can avoid interference between each other. So traffic in sensor network is largely collision-free, which saves energy compared to MAC protocols such as Carrier Sense Multiple Access (CSMA) (Ye & Heidemann, 2003). However adopting these scheduled MAC schemes causes idle listening, which happens when the radio is listening to the channel before transmitting possible data. Due to constantly listening to the channel, the cost of energy is especially high in many sensor network applications where no data is transmitted during the period of no reported event. Idle listening is a dominant factor of radio energy consumption (Demirkol et al., 2006).

LEACH optimizes energy by shutting down radios of sensor nodes and load balancing. The scalability can be reached by a distributed hierarchical approach. It is easy to aggregate data at the head of a cluster and send to a user or sink. However disadvantages of LEACH are related to its hierarchical formation, where the failure and selection of the cluster head is a problem and difficult to optimize. It is also expensive to assume that all nodes are capable of communicating over an extended distance.

3.2 GPSR

Greedy Perimeter Stateless Routing (GPSR): The GPSR is a routing protocol to transfer the data packets in wireless datagram networks. GPSR is based on an algorithm which combines Greedy Packet Forwarding and Perimeter Forwarding methods. Greedy Packet forwarding is a strategy enabling the source to know the geographic position of the destination integrated in the route request. GPSR provides each node of the network with a local table to list identifies and positions of neighbouring nodes. A proactive broadcast refreshes the table of each node in a regular time interval. The source node gives the packet a destination address. This address will not be changed by any node which forwards the packet.

GPSR performs better than dynamic source routing and only needs local information for packet forwarding by using greedy forwarding strategy. The protocol can be operated by using energy-efficient MAC protocols to increase the energy efficiency.

3.3 Other Hierarchical Protocols

LEACH inspires many hierarchical protocols such as Power-Efficient Gathering in Sensor Information System (PEGASIS), Threshold sensitive Energy Efficient sensor Network protocol (TEEN), Adaptive TEEN (APTEEN) etc..

PEGASIS is an enhanced protocol using CDMA capable nodes over LEACH to extend network lifetime by using only one node in a chain to transmit to the sink instead of multiple nodes. *TEEN* and *APTEEN* (Zhao et al., 2006) are based on time-critical applications using TDMA schedule. They are developed to be responsive to sudden changes, which require the precision of time in non-period and periodic reports such as the temperature in the sensed area in a disaster management application.

4. Location-based Protocols

Since sensor nodes are mostly spatially and randomly scattered in an area, there is no addressing scheme, e.g. IP-addresses, for sensor network. In most applications, location information is needed in order to know the separating distance between particular nodes and optimize routing in an energy efficient way. Relative coordinates can be built by exchanging distances between neighbour nodes in order to approximate the strength of incoming signals. Alternatively equipping low-power global positioning system (GPS) devices into hardware of sensor nodes can obtain location information directly through communications with satellites. While global coordinates and compatibility are desirable, the GPS may not always be used because of line-of-sight conditions, power requirement and other limits (Niculescu, 2005). In some applications information of the sensor area is known, so using locations of sensors can build a query directly diffused only to region of interests, and decrease the number of transmission significantly.

4.1 MECN and SMECN

Minimum Energy Communication Network (MECN): By using low power GPS devices, sensor nodes in MECN can setup and maintain a minimum energy network (Xu, 2002). MECN assumes a master-node as the information sink and develops a minimum power topology for each node. MECN identifies a relay region for each node, which is consisted of nodes in a surrounding area where transmitting through those nodes is more energy-efficient than direct transmission. A sub-network build in MECN is based on having less number of nodes which can consume less power for transmission between any two specific nodes. In this way, global minimum power paths are found without considering all the nodes in the network. Optimal links are calculated based on the position coordinates updated by using GPS. Moreover it can dynamically adapt to elimination of nodes or the deployment of new sensors since it is capable of self-reconfiguring.

Small MECN (SMECN): It is assumed that in MECH each node can transmit to others, which are not possible in all cases if there are obstacles between any pair of considered nodes. SMECN is proposed to cope with obstacles. Although it is still assumed in SMECN that node could be fully connected, the sub-network established in SMECN for minimum energy is smaller in terms of the number of edges compared with the one in MECH if broadcasts are able to reach all nodes in a circular region around the broadcaster. Thus the number of hops in SMECH will decrease, therefore energy can be saved. More overhead happens when finding a sub-network with a smaller number of edges in SMECN (Xu, 2002).

4.2 GAF and GEAR

Geographic Adaptive Fidelity (GAF): It is a GPS location-based routing algorithm designed primarily for mobile ad-hoc network (Akyildiz et al., 2002). The idea of the protocol is to setup a virtual grid based on location information and conserve energy by turning off some nodes depending on the redundancy in the network without affecting the system fidelity to extend the network lifetime. GAF may also be considered as a hierarchical protocol, where the clusters are based on geographic locations . A representative node is selected in each particular cluster to transmit the data to other nodes. GAF performs at least as well as a normal ad-hoc routing protocol, e.g. dynamic source routing, but with substantial conservation of energy, which is realized by the protocol to tune for parameters like estimated node active time, time for node discovery and status being active and sleep.
Geographic and Energy Aware Routing (GEAR): Estimating separation distance is an alternative to use location information from GPS. GEAR uses of geographic information and relays queries to certain regions because data queries contain geographic attributes. The main idea is to restrict the number of interests in directed diffusion by only considering a certain region rather than sending the interests to the whole network, thus it conserves energy and improves the lifetime of network.

5. Network Flow and QoS-based Protocols

There are some effective routing protocols proposed in different approaches which don't fit the above classification. In network flow, route is modeled and solved in a network flow. QoS-based protocols consider end-to-end delay requirements and establish paths in sensor networks. A few examples of these are discussed in this section.
Maximum Life Energy Routing (MLER): It is proposed in (Chang & Tassiulas, 2000) as a solution to the problem of routing in sensor networks based on a network flow approach. The main idea of this approach is to maximally extend the network lifetime by defining link cost as a function of residual energy of node, and the require transmission energy using that link. By maximizing the lifetime of the network, the protocol leads to establish traffic distribution, which is a possible solution to the routing problem in sensor networks.
Sequential Assignment Routing (SAR): SAR is the first protocol for WSN that includes a notion of QoS. The objective of the SAR algorithms is to minimize the average weighted QoS metric throughout the lifetime of the network (Akyildiz et al., 2002). It creates trees rooted at one-hop neighbours of the sink by taking QoS metric, energy resource on each path and priority level of each packet into consideration. Through creating trees, SAR built multiple paths from sink to sensors. It ensures the fault-tolerance and easy recovery. However when huge sensors are deployed, SAR will suffer from the overhead of maintaining the table and states at each sensor nodes.
Different QoS-aware MAC protocols have also been proposed for WSNs. Reinforcement Learning based MAC (RL-MAC) (Liu & Elhanany, 2006) is an novel adaptive MAC for WSNs, which employs a reinforcement learning framework. Nodes actively infer the state of other nodes, using a reinforcement learning based control mechanism. QoS is easily implemented in this proposed framework. QoS-aware MAC (Q-MAC) in (Liu et al., 2005) is another innovative idea, which minimizes the energy consumption in multi-hop WSNs and provides QoS by differentiating network services based on priority level. It allows sensor networks to reflect on the criticality of data packets originating from the different sensor nodes.

6. Research Platforms and Tools

Great interests have motivated intensive research to realize the vision of WSN in the past few years. Research prototype sensor nodes (UCB motes, WINS, Smart Dust, PC104, etc.) are designed and manufactured. Simulation and development tools are also being developed.

6.1 Sensor platforms

1. MICA motes

   MICA mote is a commercially available product that has been used widely by researchers and developers. MICA motes use an Atmel Atmega 128L microcontroller to provide bidirectional communication at 50 kbps and a pair of *AA* batteries to provide energy. The operation system (OS) cooperating with the MICA is called the TinyOS , which is currently widely used.

2. Rockwell WINS

   Rockwell WINS uses a StrongARM 1100 CPU running at 133 MHz, 1 MB of FLASH memory, 1 MB of RAM, a 100 kbps radio, and has to operate on two 9V batteries. This is considered to be towards the high end of sensor network devices.

3. Smart Dust

   Tiny nodes, called Smart Dust, are densely deployed to float in the air and organize themselves into a sensor network to achieve a surveillance task. It has more strict constraint with energy consumption and a simply undivided architecture. Currently sensor networks are considered to evolve toward this small dust if technological advance permits such miniaturization and copes with other existing limits (Hollick et al., 2004).

4. PC-104 based nodes

   Nodes based on PC-104 are much larger than Mica motes. They are widely used as parent nodes in hierarchical sensor networks. The PC-104 based testbed is mainly funded by DARPA SenseIT program. It is built upon off-the-shelf PC-104 based products. Each node has an AMD ElanSC400 CPU,16MB RAM and 16MB IDE Flash Disk.

6.2 Simulation and development tools

1. UCB tools chain for development in Motes

   The tool chains are composed of four parts: TinyOS kernel, NesC compiler, TOSSIM simulator and TinyDB processing system. TinyOS has a component-based programming model, codified by the NesC language. TinyOS is not an OS in the traditional sense; it is a programming framework for embedded systems and set of components that enable building an application-specific OS into each application.

2. NS-2 and Nam NS-2 is developed since 1985 by collaborations of USC Berkeley, USC/ISI, MIT, C-Mellon, Sun, DARPA and NSF (Shih et al., 2001). It is suitable for simulating wireless sensor network operations. Nam is a package for visualization using for NS-2.

3. Sensor Sim is a simulation framework for modeling sensor networks built upon the NS-2 simulator and provides additional feature for modeling sensor networks.

7. Conclusions and Future Work

Routing technologies in sensor networks have attracted considerable attention in recent years. They are subject to challenges which are different to traditional networks. In general, a routing protocol needs to deal with scalability, energy efficiency, robustness, latency, low computation and memory usage. In this survey, we have summarized typical research results on routings protocols in sensor networks and classified them into different classes as data-centric, hierarchical and location-based. Examples of network flow and QoS modeling methods, which follows other approaches, have also been discussed.

Data-centric protocols name the data and query the nodes based on attributes of the data. The most important aspect of this paradigm is the content of the sensor-generated data, which drives most implementations of the upper layers: discovery, routing, and querying (Niculescu, 2005). Research follow this paradigm in order to avoid the overhead of forming clusters. On the other hand, the naming scheme is not sufficient for complex queries and is not easily extended to cover a larger area.

Cluster-based routing protocols divide sensor nodes into different groups to efficiently relay the sensed data to the sink. A cluster head performs aggregation and fusion of data and sends data directly to the sink on behalf of other member nodes. It gives solutions to the problem of the formation of clusters and optimization of the energy consumption. The process of the communication between the head sensor is an open issue for reserach.

Protocols employing location information and topological deployment of sensor nodes are classified as location based ones. There is no need for routing tables in this network since each node can decide how to relay packets based on the destination to the packet and some local information about its immediate neighbours. However since it is the source must know the position of the destination, it is still an implicit requirement in many applications. Moreover how to aid energy efficient routing by using the local geometrical information is still a problem.

Most research protocols pay main attentions to the energy efficiency without addressing many issues like QoS. QoS-aware routings in sensor networks have many applications like real time targeting, emergent event triggering in monitoring applications etc. Current research is aiming at controlling QoS requirement in an energy-efficient application environment. Common issues like routing around obstacles, scalabilities, adaptive applications etc. are open for designs of protocols.

Sensor network is a popular research area and has applications in the real world. Protocols present today have their own set of problems which need to be improved. Most protocols dealing with energy efficiency can be significantly improved with robustness and scalability. Most results are empirical nowadays and more theoretical work can be done to incorporate game theory for modeling. Simulation tools can also be improved by focusing on sensor network in mind.

8. References

Akyildiz, I. F., Su, W., Sankarasubramaniam, Y. & Cayirci, E. (2002). A survey on sensor network, *IEEE Communications Magazine* .

Braninsky, D. & Estrin, D. (2002). Rumor routing algorithm for sensor networks, *First Workshop on Sensor Networks and Applications (WSNA)*, Atlanta,GA.

Chang, J. H. & Tassiulas, L. (2000). Maximum lifetime routing in wireless sensor networks, *The Advanced Telecommunications and Information Distriution Research Program (ATIRP'2000)*, College Park,MD.

Dan, T. V. & Langendoen, K. (2003). An adaptive energy-efficient mac protocol for wireless sensor networks, *the first international conference on Embedded Networked Sensor Systems, ACM press*, pp. 171–180.

Demirkol, I., Ersoy, C. & Alagoz, F. (2006). Mac protocols for wireless sensor networks: a survey, *Communications Magazine, IEEE* **44**(4): 115–121.

Ditzel, M. & Langendoen, K. (2005). D3: Data-centric data dissemination in wireless sensor networks, *Wireless Technology 2005* pp. Page(s):185 – 188.

Heinzelman, W., Chandrakasan, A. & Balakrishnan, H. (2000). Energy-efficient communication protocol for wireless sensor networks, *the Hawaii International Conference System Sciences*, Hawaii.

Heinzelman, W., Kulik, J. & Balakrishnan, H. (1999). Adaptive protocols for information dissemination in wireless sensor networks, *the 5th Annual ACM/IEEE International of the First Workshop on Sensor Networks and Applications (WSNA)*, Altlanta, GA, USA.

Hollick, M., Martinovic, I., Krop, T. & Rimac, I. (2004). A survey on dependable routing in sensor networks, ad hoc networks, and cellular networks, *the 30th EUROMICRO Conference*.

Intel (2004). Instrumenting the word-an introduction to wireless sensor networks.

Liu, Y., Elhanany, I. & Qi, H. (2005). An energy-efficient qos-aware media access control protocol for wireless sensor networks, *IEEE MASS 2005* .

Liu, Z. & Elhanany, I. (2006). Rl-mac: A qos-aware reinforcement learning based mac protocol for wireless sensor networks, *2006 IEEE International Conference on Networking, Sensing and Control 2006, ICNSC '06*, pp. 768 – 773.

Niculescu, D. (2005). Communication paradigms for sensor networks, *IEEE Communication Magazine* .

Shih, E., Calhoun, B. H., Cho, S. H. & Chandrakasan, A. P. (2001). Energy-efficient link layer for wireless microsensor networks, *IEEE 2001 Computer Society Workshop on VLSI*, pp. 16–21.

Sohrabi, K. (2000). Protocols for self-organization of a wireless sensor network, *IEEE Personal Communications* **7**(5): 16–27.

Xu, N. (2002). A survey of sensor network applications, *IEEE Communications Magazine* **40**.

Ye, W. & Heidemann, J. (2003). Medium access control in wireless sensor networks, *Technical Report ISI-TR-580*.

Ye, W., Heidemann, J. & Estrin, D. (2002). An energy efficient mac protocol for wireless sensor networks, *21st conference of the IEEE Computer and Communications Societies (INFOCOM)* **11**(1): 1567–1576.

Zhao, L., Xu, C. N., Xu, Y. J. & Li, X. W. (2006). Energy-aware qos control for wireless sensor network, *2006 1st IEEE Conference on Industrial Electronics and Applications, Vols 1-3*, pp. 1663–1668.

2

Review of Energy Harvesting Technologies for Sustainable Wireless Sensor Network

Yen Kheng Tan and Sanjib Kumar Panda
National University of Singapore
Singapore

1. Introduction

The rapid growth in demands for computing everywhere has made computer a pivotal component of human mankind daily lives. Whether we use the computers to gather information from the Web, to utilize them for entertainment purposes or to use them for running businesses, computers are noticeably becoming more widespread, mobile and smaller in size. What we often overlook and did not notice is the presence of those billions of small pervasive computing devices around us which provide the intelligence being integrated into the real world. These pervasive computing devices can help to solve some crucial problems in the activities of our daily lives. Take for examples, in the military application, a large quantity of the pervasive computing devices could be deployed over a battlefield to detect enemy intrusion instead of manually deploying the landmines for battlefield surveillance and intrusion detection Chong et al. (2003). Additionally, in structural health monitoring, these pervasive computing devices are also used to detect for any damage in buildings, bridges, ships and aircraft Kurata et al. (2006).

To achieve this vision of *pervasive computing*, also known as *ubiquitous computing*, many computational devices are integrated in everyday objects and activities to enable better human-computer interaction. These computational devices are generally equipped with sensing, processing and communicating abilities and these devices are known as *wireless sensor nodes*. When several wireless sensor nodes are meshed together, they form a network called the *Wireless Sensor Network (WSN)*. Sensor nodes arranged in network form will definitely exhibit more and better characteristics than individual sensor nodes. WSN is one of the popular examples of ubiquitous computing as it represents a new generation of real-time embedded system which offers distinctly attractive enabling technologies for pervasive computing environments. Unlike the conventional networked systems like Wireless Local Area Network (WLAN) and Global System for Mobile communications (GSM), WSN promise to couple end users directly to sensor measurements and provide information that is precisely localized in time and/or space, according to the users' needs or demands. In the Massachusetts Institute of Technology (MIT) technology review magazine of innovation published in February 2003 MIT (2003), the editors have identified Wireless Sensor Networks as the first of the top ten emerging technologies that will change the world. This explains why WSN has swiftly become a hot research topic in both academic and industry.

2. Smart Environment with Pervasive Computing

Pervasive computing is the trend towards increasingly ubiquitous and connected computing devices in the environment. These pervasive computing devices are not personal computers as we tend to think of them, but they are very tiny computing devices, either mobile or embedded in almost any type of object imaginable, including cars, tools, appliances, clothing and various consumer goods. According to Dan Russell, director of the User Sciences and Experience Group at IBM's Almaden Research Center, by the near future, computing will have become so naturalized within the environment that people will not even realize that they are using computers Kumar (2005). Russell and other researchers expect that in the future smart devices all around us will maintain current information about their locations, the contexts in which they are being used, and relevant data about the users. The goal of the researchers is to create a system that is pervasively and unobtrusively embedded in the environment, completely connected, intuitive, effortlessly portable and constantly available. *Smart environment* is among the emerging technologies expected to prevail in the pervasive computing environment of the future.

The notion of smart environment is becoming a reality with pervasive computing as well as advancements of various related technologies such as wireless networking, micro-fabrication and integration using micro-electromechanical system (MEMS) technology and embedded intelligent with microprocessors. Smart environments represent the next evolutionary development step in various application areas such as building, utilities, industrial, home, marine, animal habitat, traffic, etc. Like any sentient organism, the smart environment relies first and foremost on sensory data from the real world. Sensory data comes from multiple sensors of different modalities in distributed locations. Similarly for the smart environment, information about its surroundings is also needed just like what is captured by the receptors in the biological systems. The information needed by the smart environments is provided by the distributed WSN which has its pervasive sensor nodes for sensing, processing and communicating the information to the base station. To facilitate smart environments in various application areas, a general architecture of the data acquisition and distribution network is provided in Figure.1. The data acquisition network is designed to gather real-world information as well as to monitor the condition of the targeted application. Data are collected at the base station in a wireless manner, preprocessed and then distributed to the end users via different communicating devices.

Referring to Figure.1, it can be seen that the entire data network is a very large and complex system that is made up of many different subsystems i.e. sensor nodes, base station, management center, wireland and wireless communication systems. The sensor nodes and the base station are part of the data acquisition network and the wireland and wireless communication systems belongs to the data distribution network. Once the sensor nodes are deployed in the application areas, the nodes would sense and collect data from the environment and the collected data are then sent to the base station in a wireless manner. The base station consolidates the collected data and preprocesses the data so that it can be delivered quickly and safely over the data distribution network to the end users. Most importantly, the end users must be able to access the information at anywhere and at any time. In between the data acquisition network and the data distribution network, a management center is incorporated so as to better coordinate, monitor and control the data flow between the two networks. When data is transferred within the entire network, there are two important factors that need to be well considered namely data integrity and data security.

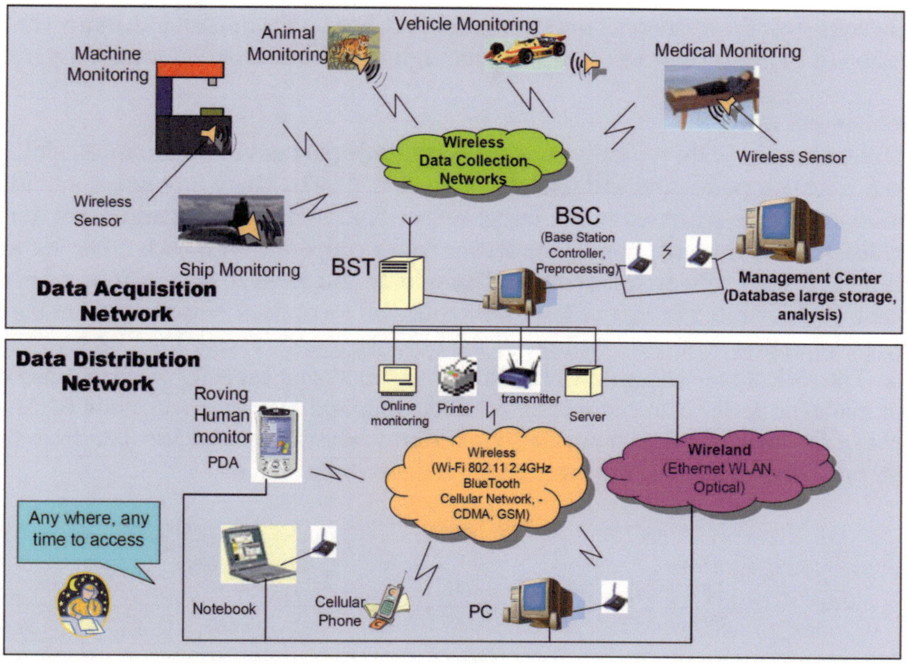

Fig. 1. A general architecture of the data acquisition and distribution network to facilitate smart environments Cook et al. (2004)

The framework of a WSN is similar to the architecture of the general data acquisition and distribution network described in Figure.1. Likewise, the main objective of WSN is to provide the end user with intelligence and a better understanding of the environment so as to facilitate a smart environment. WSN is considerably a new research field and it has a widespread of research problems for both academic scholars and industrial researchers to resolve. WSN itself has also several attractive advantages as discussed in Kuorilehto et al. (2005) Callaway (2003) that make it suitable for many potential implementation areas. These implementation areas include environmental monitoring, health monitoring, miliary surveillance and many others listed in Table.1. The challenging part of the WSN research work is that WSN requires an enormous breadth of knowledge from a vast variety of disciplines such as embedded microprocessor, networking, power, wireless communication and microelectronic to be able to optimize WSN for specific application.

3. Overview of Wireless Sensor Networks

The original motivation of WSN can be traced back to the design of military applications such as battlefield surveillance and intrusion detection mentioned by Chong et al. in Chong et al. (2003). Based on the previous endeavors to build efficient military sensor networks as well as the fast developments in microelectronic design and wireless communication, WSN are gradually introduced to many civil application areas. With the continuous dedications of academic scholars and industrial researchers, people are getting closer and closer to the essential points to understand WSN technology. The unique characteristics of WSN make it advantageous over the former networks on one hand, but on the other hand, many challenges are inevitable. Hence further research and thorough reflections on WSN are greatly needed.

3.1 Architecture of WSN

The architecture of a WSN typically consists of multiple pervasive sensor nodes, sink, public networks, manager nodes and end user Akyildiz et al. (2002). Many tiny, smart and inexpensive sensor nodes are scattered in the target sensor field to collect data and route the useful information back to the end user. These sensor nodes cooperate with each other via wireless connection to form a network and collect, disseminate and analyze data coming from the environment. As illustrated in Figure.2, the data collected by node A is routed within the sensor field by other nodes. The data will reach the boundary node E and then be transferred to the sink. The sink serves as a gateway with higher processing capacity to communicate with the task manager node. The connection between sink and task manager node is the public networks in the form of Internet or satellite. The end user will receive the data from the task manager node and perform some processing on these received data.

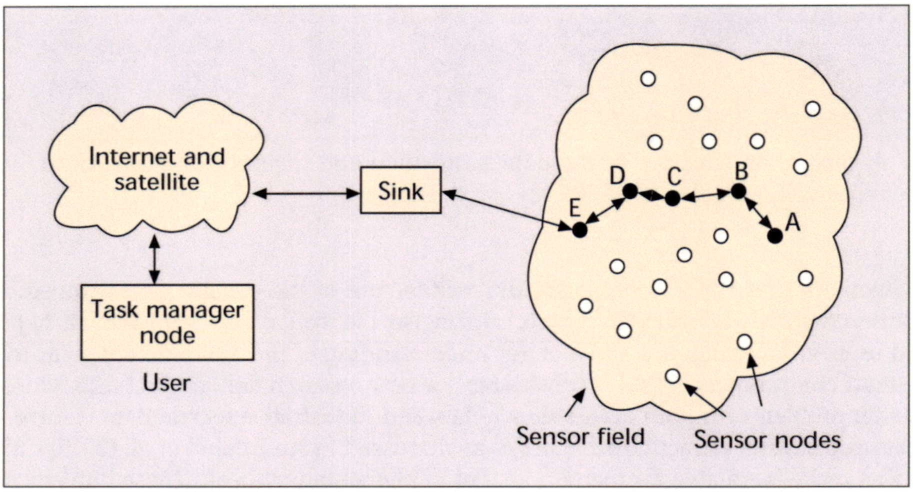

Fig. 2. Architecture of WSN to facilitate smart environments Akyildiz et al. (2002)

In Figure.2, the sink is essentially a coordinator between the deployed sensor nodes and the end user and it can be treated like a gateway node. The need of a sink in WSN architecture is due to limited power and computing capacity of the wireless sensor nodes. The gateway node is equipped with better processor and sufficient memory space because the node can provide the need for extra information processing before data is transferred to the final destination. The gateway node can therefore share the loadings posed on the wireless sensor nodes and hence prolong their working lifetime. The communication means amongst the sensor nodes is through wireless media because in most application scenarios, the physical contacts among the sensor nodes for configuration, maintenance and replacement are rather difficult or even impossible. The design of the wireless sensor network as described by Figure.2 is influenced by many factors, including fault tolerance, scalability, production costs, operating environment, sensor network topology, hardware constraints, transmission media and power consumption. These design factors are important because they serve as a guideline to design a protocol or an algorithm for sensor networks Akyildiz et al. (2002). In addition, these influencing factors can be used to design the WSN to meet the specific requirement of various real-life applications.

Let us examine how the design factors can be used as a guideline to design a suitable protocol for the WSN in the military surveillance and intrusion detection application mentioned earlier. During each deployment of the sensor nodes either by dropping from a plane or delivering in an artillery shell or rocket or missile, hundreds to several thousands of sensor nodes are deployed throughout the battlefield for sensing. Since the WSN consists of a large number of sensor nodes, the cost of a single node is very important to justify the overall cost of the network. If the cost of the network is more expensive than deploying traditional sensors, the sensor network is not cost-justified. Next, to achieve good coverage of the whole deployment ground, the sensor nodes in the WSN are desired to be deployed closely to each other. However, deploying a high number of nodes densely requires careful handling of the WSN topology. The topology of the WSN is first considered during the predeployment phases when the sensor nodes are deployed into the battlefield by a plane or an artillery shell. After deployment which is the post-deployment phase, the topology of the WSN may need to be changed due to change in sensor nodesŠ position, reachability (due to issues like jamming and moving obstacles), available energy, malfunctioning and task details. In some cases, redeployment of additional sensor nodes are carried out at any time to replace malfunctioning nodes or to cater for changes in the task dynamics in the WSN. In the mist of the sensing operations, some sensor nodes in the WSN may fail due to the lack of power or experiencing some physical damage or encountering environmental interference. This would interrupt the WSN functionalities. As such, the fault tolerance level of the WSN must be high enough to ensure that the failure of sensor nodes should not affect the overall task of the sensor network. Despite that the fault tolerance of the WSN can be designed to be as high as possible, there is bound to have some limits to where the fault tolerance level of the WSN can achieve. Hence to sustain the WSN functionalities without any interruption, many researchers have been focusing on power conservation and power management for the sensor nodes Sinha et al. (2001) Merrett et al. (2005) and design of energy-aware protocols and algorithms for the WSNs Sohrabi et al. (2000) Lattanzi et al. (2007) in order to reduce the power consumption of the overall wireless sensor network. By doing so, the lifetime of the WSN can be extended.

To understand how data are communicated within the sensor nodes in a WSN, the protocol stack model of a WSN as shown in Figure.3 is investigated. With this understanding, the energy hungry portions of the WSN can be identified and then the WSN redesigned accordingly for lower power consumption. To start with the basic communication process, it consists of sending data from the source to the destination. Primarily, it is the case of two wireless sensor nodes wanting to communicate with each other. Hence, the sensor node at source generates information, which is encoded and transmitted to destination, and the destination sensor node decodes the information for the user. This entire process is logically partitioned into a definite sequence of events or actions, and individual entities then form layers of a communication stack. Traditionally, the process of communication is formally organized as the ISO-OSI reference model stack which consists of seven layers with application layer as the highest layer and physical layer as the lowest layer of the OSI model. However, the protocol stack model adopted by WSN is different from the conventional OSI model. As illustrated in Figure.3, the protocol stack of WSN introduces extra features such as power awareness, mobility control and task management. It offers flexibility for WSN applications which are built on the stack and promotes the cooperativeness among sensor nodes.

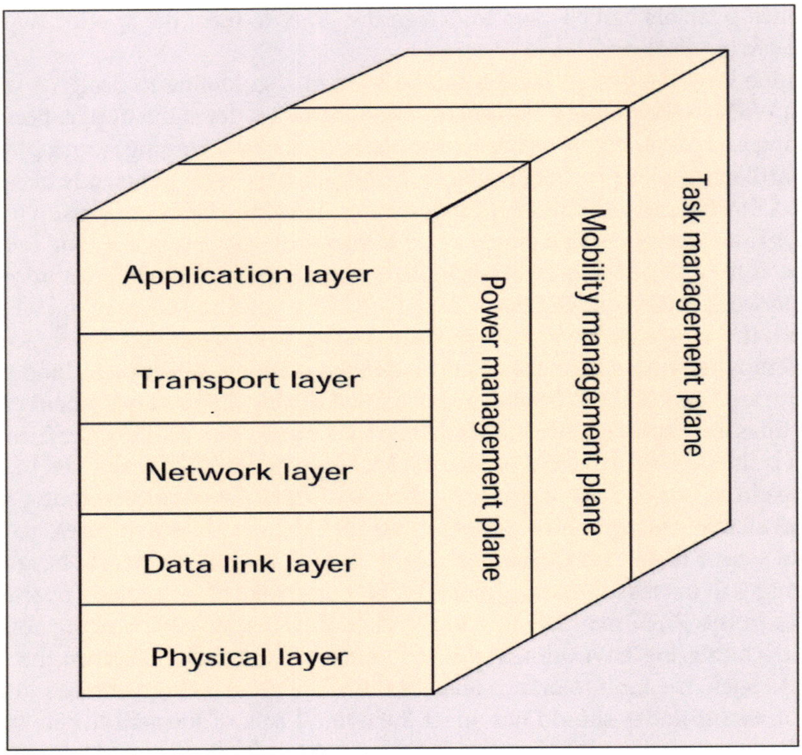

Fig. 3. Sensor networks protocol stack Akyildiz et al. (2002)

The WSN protocol stack shown in Figure.3 consists of five network layers namely physical (lowest), data link, network, transport and application (highest) layers and three new elements: power management plane, mobility management plane and task management plane Akyildiz et al. (2002). Starting from the lowest level, the physical layer is to meet the needs of receiving and transferring data collected from the hardware. It is well known that long distance wireless communication can be expensive, in terms of both energy and implementation complexity. While designing the physical layer for WSNs, energy minimization is considered significantly more important over and above the other factors like propagation and fading effects. Energy-efficient physical layer solutions are currently being pursued by researchers to design for tiny, low-power, low-cost transceiver, sensing and processing units. The next higher layer is the data link layer which ensures reliable point-to-point and point-to-multipoint connections for the multiplexing of data streams, data frame detection, medium access and error control in the WSN. The data link layer should be power-aware and at the same time to minimize the collisions between neighbors' signals because the environment is noisy and sensor nodes themselves are highly mobile. This is also one of the layers in the WSN whereby power saving modes of operation can be implemented. The most obvious means of power conservation is to turn the transceiver off when it is not required. By using a random wake-up schedule during the connection phase and by turning the radio off during idle time slots, power conservation can be achieved. A dynamic power management scheme for WSNs has been discussed in Sinha et al. (2001) where five power-saving modes are proposed and intermode transition policies are investigated. The network layer takes care of routing the data supplied by the transport layer. In WSN deployment, the routing protocols in the network layer are important because an efficient routing protocol can help to serve various applications and save energy. By setting appropriate energy and time delay thresholds for data relay, the protocol can help prolong the lifetime of sensor nodes. Hence the network layer is another layer in the WSN to reduce power consumption. The transport layer helps to maintain the flow of data if the sensor networks application requires it. Depending on the sensing tasks, different types of application software can be built and used on the application layer.

The three special planes in the stack help the sensor nodes to coordinate tasks and keep the power consumption low Akyildiz et al. (2002). The power management plane is designed to control the power usage of each node. For example, when the power level is low, the sensor node will broadcast to the neighbors telling that its remaining power is low and can only be reserved for sensing rather than participating in routing. The mobility management plane will detect and record the movement of sensor nodes to keep track of the route as well as the neighbors. By having the knowledge of neighbors, each sensor node in the network can balance power usage and task processing. The task management plane will schedule the sensing tasks and balance the work loads. As a result, sensor nodes can perform the task depending on current power level and situation of their neighbors. In summary, the three management planes help the sensor nodes to work together in a power efficient way and share resources more wisely.

3.2 WSNs Applications

WSNs can be used in virtually any environment, even where wired connections are not possible or the terrain inhospitable or physical placement of the sensors are difficult. Besides that, WSNs also enable unattended monitoring of physical quantities over large areas on a scale that would be prohibitively expensive to accomplish with human beings. These attractive features promote the potential of WSNs for more application areas. To ensure full connectivity, fault tolerance and long operational life, wireless sensor networks are deployed in ad hoc manner and the networks use multi-hop networking protocols to make real-world information and control ubiquitously available Sohrabi et al. (2000). There have been many applications suggested for WSNs and they are listed in Table.1.

These wide range of applications described in Table.1 for WSNs can be roughly classified into three categories suggested in Culler et al. (2004):

- monitoring space
- monitoring things
- monitoring the interactions of things with each other and the encompassing space

The first classification includes environmental monitoring, indoor climate control, military and space surveillance. The second classification includes structural monitoring, condition-based equipment maintenance, medical health diagnostics, vehicle safety and urban terrain mapping. The most dramatic applications fall under the third classification which involve monitoring complex interactions, including wildlife habitats, disaster management, emergency response, asset tracking and manufacturing process flow. Based on the collaborative efforts of a large number of sensor nodes, WSNs have become proven by many researchers as good candidates to provide economically viable solutions for a wide range of applications such as environmental monitoring, scientific data collection, health monitoring and military operations as tabulated in Table.1. These sensor nodes are coordinated based on some network topologies to cooperate with one another within the WSNs to satisfy the applications

Application	**Type**	**Requirements**	**Scale and density**
Great Duck Island Mainwaring et al. (2002)	Environmental monitoring	Data archiving, Internet access, long lifetime	32 nodes in 1 km^2
Flood detection Boulis et al. (2003)	Disaster management	Current condition evaluation	200 nodes 50m apart
SSIM (artificial retina) Schwiebert et al. (2001)	Health	Image identification, realtime, complex processing	100 sensors per retina
Human monitoring Thomas (2006)	Health	Quality of data, security, alerts	Several nodes per human
WINS for military Marcy et al. (1999)	Military	Target identification, realtime, security, quality of data	Several distant nodes
Object tracking Romer (2004)	Military	Collaborative processing, realtime, location-awareness	7 nodes in proximity
WINS condition monitoring Marcy et al. (1999)	Machinery monitoring	Data aggregation, machinery lifetime projection	Few nodes per machinery
Smart kindergarten Srivastava et al. (2001)	Space surveillance	Video streaming, identification, location-awareness	Tens of sensors, indoor
Pressure in automobile tires Roundy (2003)	Vehicle safety	Real time data, improve the safety and performance of the vehicle	Volume constraint of 1 cm^3

Table 1. Examples of prototyped applications for WSNs Kuorilehto et al. (2005)

requirement stated in Table.1. Because of the great potential in WSN, many groups around the world have invested lots of research efforts and time in the design of sensor nodes for their specific applications. These include Berkeley's Mica motes Hill et al. (2002), PicoRadio projects Rabaey et al. (2000), MIT's μAmps MIT (2008) as well as many others. In addition, the TinyOS project TinyOS (2008) provides a framework for designing flexible distributed applications for data collection and processing across the sensor network. All of these sensor nodes have similar goals which are small physical size, low power consumption and rich sensing capabilities.

3.3 Challenges on WSNs

The unique features of the WSNs pose challenging requirements to the design of the underlying algorithms and protocols. Several ongoing research projects in academia as well as in industry aim at designing protocols that satisfy these requirements for sensor networks Chong et al. (2003), Kuorilehto et al. (2005), Akyildiz et al. (2002) and Tubaishat et al. (2003). Some of the important challenges are presented as shown below.

- *Sensor nodes are limited in energy, computational capacities and memory:*

  Sensor nodes are small-scale devices with volumes approaching a cubic millimeter in the near future. Such small volumetric devices are very limited in the amount of energy that the storage element such as batteries can store. Hence the batteries with finite energy supply must be optimally used for both processing and communication tasks. The communication task tends to dominate over the processing task in terms of energy consumption. Thus, in order to make optimal use of energy, the amount of communication task should be minimized as much as possible. In practical real-life applications, the wireless sensor nodes are usually deployed in hostile or unreachable terrains, they cannot be easily retrieved for the purpose of replacing or recharging the batteries, therefore the lifetime of the network is usually limited. There must be some kind of compromise between the communication and processing tasks in order to balance the duration of the WSN lifetime and the *energy density* of the storage element. In summary, limitation in the device size and energy supply typically means restricted amount of resources i.e. CPU performance, memory, wireless communication bandwidth used for data forwarding and range allowed.

- *Sensor nodes in the WSN are ad hoc deployed and distributed for processing and sensing:*

  Sensor nodes are either placed one by one in the vicinity of the phenomenon or deployed in an ad hoc fashion by air or by some other means. Once the sensor nodes are deployed, the WSNs would not have any human intervention to interrupt their operations. The sensor nodes must be able to configure themselves to form connections to set up the network to meet the application requirement. In case of any changes in the operating conditions or environmental stress on the sensor nodes that causes them to fail leading to connectivity changes, this requires reconfiguration of the network and re-computation of routing paths. Another point to take note is that using a WSN, many more data can be collected as compared to just one sensor. Even deploying a sensor with great line of sight, it could still have obstructions. Thus, distributed sensing provides robustness to environmental obstacles. Because there are many sensor nodes densely deployed, neighbor nodes may be very close to each other. Hence, multihop communication in WSNs is expected to consume less power than the traditional single hop broadcast communication because the transmission power levels can be kept low. Additionally, multihop communication can also effectively overcome some of the signal propagation effects experienced in long-distance wireless communication.

- *Network and communication topology of a WSN changes frequently:*

  When the sensor nodes are deployed, the position of sensor nodes is not predetermined. This means that the sensor nodes must be able to configure themselves after deployment. They must possess some means to identify their location either globally or with respect to some locally determined position. Once the network is set up, it is required that the WSN be adaptable to the changing connectivity (for e.g., due to addition of more nodes, failure of nodes, etc.) as well as the changing environmental conditions.

Unlike traditional networks, where the focus is on maximizing channel throughput or minimizing node deployment, the major consideration in a sensor network is to extend the system lifetime as well as the system robustness.

In contrast to the traditional networks which focus mainly on how to achieve high quality of service (QoS) provisions, WSN protocols tend to focus primarily on power conservation and power management. However, there must be some embedded trade-off mechanisms that give the end user the option of prolonging the WSN lifetime but at the cost of lower throughput or higher transmission delay. Conversely, the power consumption of the WSN can be reduced by sacrificing the QoS provisions i.e. lowering the data throughput or having higher transmission delay. Among the several challenging requirements posed on the design of the underlying algorithms and protocols of the WSNs, it is well-known among the academia as well as industry that energy constraint is one of the most significant challenges in the WSN research field Callaway (2003). The functionalities of the WSN are highly dependent on the amount of energy that is available to be expended by each of the sensor node in the network. That is why the energy constraint challenge is substantial enough to be chosen for further investigations and discussions in my research work.

4. Wireless Sensor Nodes in WSN

A wireless sensor network consists of many energy-hungry wireless sensor nodes distributed throughout an area of interest. Each sensor node monitors its local environment, locally processing and storing the collected data so that other sensor nodes in the network can use it. Network nodes share these information via a wireless communication link. The block diagram of an energy-hungry wireless sensor node residing in the WSN is shown in Figure.4. The sensor node typically consists of four sub-units namely the sensor itself, data acquisition system, local microcontroller and radio communication block. The sensor, data acquisition, microprocessor and radio communications are all power sink modules because they need to consume electrical energy from the power source in order to operate. These sub-units of the sensor node are all energy hungry. Since the power source is driven by batteries, the energy-hungry sub-units would use up all the energy in the batteries after some times and the sensor node would then go into an idle state.

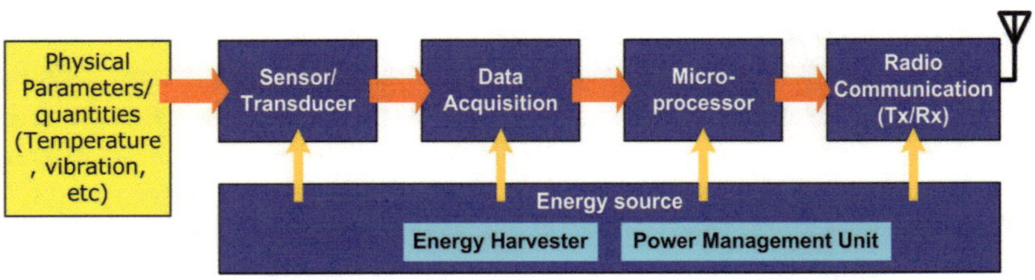

Fig. 4. Block diagram of energy-hungry wireless sensor node

The sensor/transducer converts an environmental sensing parameter such as temperature, vibration, humidity, etc to an electrical signal. A data acquisition unit is incorporated in the sensor node to realize amplification and pre-processing of the output signals from sensors, for example conversion from analog to digital form and filtering. To encompass some level of intelligences like data processing and time scheduling in the sensor node, a microprocessor has been incorporated in the sensor node. A radio communication block is included in the sensor node to enable the node to communicate with its neighbor node or the base station in a wireless manner. If one of the sensor nodes fails, the other sensor nodes in the network would take over the responsibility of the failed node. This provides redundancy and therefore reliability of the wireless sensor network. However, in order to optimize the WSN in practical situations, there must be some considerations to be taken into account i.e. how many sensor nodes to be deployed; should all of them be active at all times; or the nodes communicate with each other and collectively gather and transmit data such that energy consumption of the sensor is minimized at the same time the reliability is not sacrificed. All these requirements are application specific and need to be addressed appropriately.

Other than the above mentioned considerations for optimizing the WSN in practical situations, the information about how much electrical power a sensor node consume during operation also plays an important part. Hence the power consumed by each individual components i.e. processor, radio, logger memory and sensor board in a sensor node has been tabulated in Table.2. It can been observed that all the components in the sensor node consume mW level of power during the active mode of operation and then drop to μW of power when in sleep or idle mode.

In most sensor nodes applications, the processor and radio need to run only for a brief period of time, followed by a sleep cycle. During sleep, current consumption is in the μA range as opposed to mA range. This results in the sensor node drawing very little amount of current for the majority of the time and short duration of current spikes while processing, receiving and transmitting data. This method is known as *duty cycling* which helps to extend the lifetime of the battery. However, due to the current surges during the active mode of operation, the power density of the battery must be high enough to support the current surge. Based on the high energy capacity battery i.e. 3000 mAh, the life of the battery powering the sensor node can last at most 1.5 years as shown in Table.2. After which, without battery replacement, the sensor node can be considered as an idle node. The higher the battery capacity, the bigger will be the size of the battery. Take for an example an AA alkaline battery of 2850 mAh, the size of the battery is 14.5 mm x 23 mm x 50.5 mm but the size of a coin type of alkaline battery of 290mAh is 24.5 mm x 24.5 mm x 3mm. So for the case of 250 mAh battery which is smaller in size, the battery can only sustain the operation of the sensor node for at most 2 months. This time duration is really too short for the WSN to be useful in the practical situations.

5. Problems in Powering the Sensor Nodes

As the network becomes dense with many wireless sensor nodes, the problem of powering the nodes becomes critical, even worst when one considers the prohibitive cost of providing power through wired cables to them or replacing batteries. In order for the sensor nodes to be conveniently placed and used, these nodes must be extremely small, as tiny as several cubic centimeter. When the sensor nodes are small, there would be severe limits imposed on the nodes' lifetime if powered by a battery that is meant to last the entire life of the device.

5.1 High Power consumption of Sensor Nodes

Compared with conventional computers, the low-cost and battery-powered miniaturize sensor nodes have limited energy supply from very small batteries as well as stringent processing and communications capabilities plus memory is scarce. State of the art, non-rechargeable lithium batteries can provide energy up to 800 WH/L (watt hours per liter) or 2880 J/cm^3 Doherty et al. (2001). If an electronic device with a 1 cm^3 coin-size battery is to consume 200 μW of power on average (this is a challenging average power consumption by the load), the device could last for 4000 hours or 167 days which is equivalent to half a year. In fact, the

SYSTEM SPECIFICATIONS		
Currents **Processor**		**Example Duty Cycle**
Current (full operation)	8 **mA**	1
Current sleep	8 μA	99
Radio		
Current in receive	8 mA	0.75
Current transmit	12 mA	0.25
Current sleep	2 μA	99
Logger Memory		
Write	15 mA	0
Read	4 mA	0
Sleep	2 μA	100
Sensor Board		
Current (full operation)	5 mA	1
Current sleep	5 μA	99
Computed mA-hr used each hour		
Processor		0.0879
Radio		0.0920
Logger Memory		0.0020
Sensor Board		0.0550
Total current (mA-hr) used		**0.2369**
Computed battery life vs. battery size		
Battery Capacity (mA-hr)		Battery Life (months)
250		1.45
1000		5.78
3000		17.35

Table 2. Battery life estimation for a senor node operating at 1% duty cycle Crossbow (2007)

calculation is a very optimistic estimate as the entire capacity of the battery usually cannot be completely used up depending on the voltage drop. Additionally, it is also worth mentioning that the sensors and electronics of a wireless sensor node could be far smaller than 1 cm^3.

Hence in this case, the battery would dominate the system space usage. Clearly, a lifetime of half a year for the electronic device to operate is far from sufficient because the duration of the device's operation could last for several years. This implies that the battery supply of the electronic device has to be regularly maintained. The need to develop alternative method for powering the wireless sensor and actuator nodes is acute. Hence the research direction is targeted to resolve the energy supply problems faced by the energy hungry wireless sensor nodes.

5.2 Limitation of Power Sources for Sensor Nodes

Like any other electronic devices, the sensor nodes in the WSN need to be powered by energy sources in order to operate. If a wired power cable is used, many of the advantages such as self-autonomous and mobility of the sensor nodes enabled by the wireless communications are sacrificed. In many applications, a power cable is not a preferable option to power the sensor nodes knowing the advantages of wireless option. Instead, there are many types of portable energy systems listed in Figure.5 that are suitable for powering sensor nodes in the wireless sensor networks. Among these energy systems or sources, the rechargeable/alkaline battery is one of the most popular method so far. Although batteries have been widely used in powering sensor nodes in WSN presently, the problem is that the energy density of batteries are limited and they may not be able to sustain the operation of the sensor nodes for a long period of time. In many application scenarios, the lifetime of the sensor node typically ranges from two to ten years depending on the requirement of the specific application. Take for the case of deploying sensor nodes on the ice mountain to detect the thickness level of the ice on the mountain, it will take years for the melting process to be measurable. Hence the lifetime of the sensor nodes must be to last for several years before they go into idle state. If that is the case, the lifetime of one or several sensor nodes, depending on the size of the WSN, would affect the performance of the WSN.

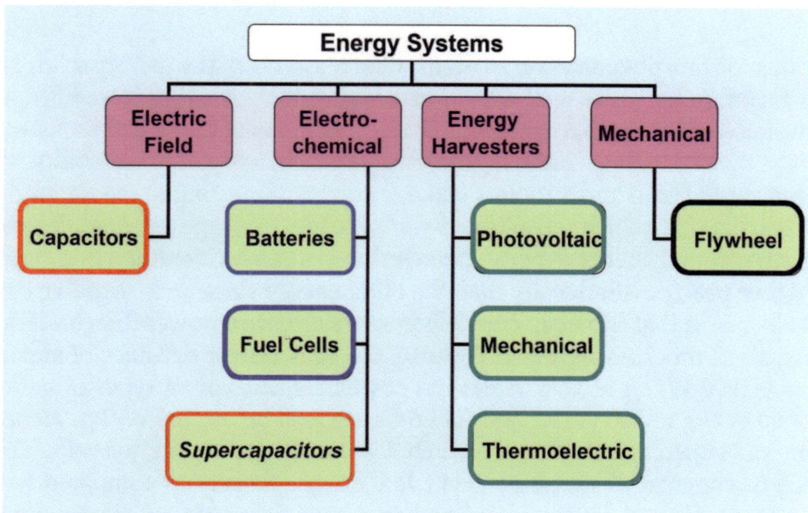

Fig. 5. General types of portable energy systems

Supercapacitor, in short supercap, is another electrochemical energy system other than batteries that has been gaining its presence in powering the wireless sensor nodes. There are

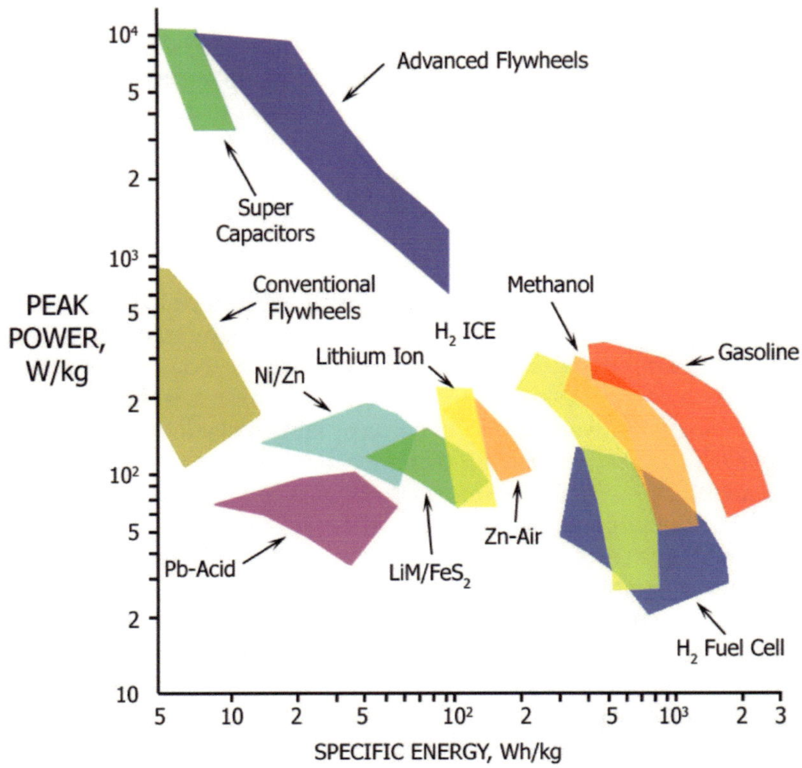

Fig. 6. Ragone plot for comparing the energy storage technologies and their power density versus energy density characteristics Tester (2005)

several reasons for this phenomenon to occur. One reason is that supercapacitor is very scalable and its performance scales well with its size and weight. Another reason is that supercap has many desirable characteristics that favour the operations of the sensor nodes such as high power density, rapid charging times, high cycling stability, temperature stability, low equivalent series resistance (ESR) and very low leakage current. Referring to the Ragone plot Tester (2005) shown in Figure.6 which consolidates various energy storage technologies and compare their power density and energy density characteristics, it can be identified that supercapacitor has much higher peak power density than the other energy storage devices like batteries and fuel cells. This means that supercap can deliver more electrical power than batteries and fuel cells within a short time. As shown in Figure.6, the peak power densities of supercapacitors are well above 1000 W/kg level whereas the power densities of all types of batteries are in the range of 60 W/kg to 200 W/kg and fuel cells are well below 100 W/kg. Hence for burst power operation, supercapacitors are better choice than batteries and fuel cells. The only major drawback of supercap is that it has very low energy density as compared to the rest of the energy storage devices. Batteries and fuel cells have much higher energy storage capacities than the supercapacitors, therefore they are more suitable for those energy-hungry sensor nodes that need to operate for a long time.

The electrical characteristics of a battery define how it would perform in the circuit and the physical properties of the battery have a large impact on the overall size and weight of the sensor node. Batteries convert stored chemical energy directly into electrical energy. They are generally classified into two groups namely 1) single-use/primary and 2) rechargeable/secondary batteries. The distinction between the two groups is based on the nature of the chemical reactions. Primary batteries are discarded when sufficient electrical energy can no longer be obtained from them. Secondary batteries, on the other hand, convert chemical energy into electrical energy by chemical reactions that are essentially reversible. Thus, by passing the electrical current in the reverse direction to that during discharge, the chemicals are restored to their original state and the batteries are restored to full charge again. Some important parameters of the batteries that help to determine the performances of the battery are listed as follows: -

- Energy density by weight (Wh/kg) and volume (Wh/cm^3) determines how much energy a battery contains in comparison to its weight and volume respectively
- Power density by weight (W/kg) determines the specific power available per use
- Self-charge rate determines the shelf life of a battery
- Cost of battery

The performances of the wireless sensor nodes meshed together in a network form are largely constrained by some limitations in the electrochemical type of energy system. One significant limitation is the limited energy storage capabilities of the batteries or supercapacitors. The energy stored in the storage elements would definitely be drained off by the connected loads after some time. If this is the case, the distance range and data transmission frequency of the communication device in the sensor nodes are highly dependent upon the available electrical power supply and the electrical energy stored in the storage elements. Usually, the wireless sensor networks are preferred to be left unattended once deployed in inaccessible environments where maintenance would be inconvenient or impossible, therefore replacement of the batteries in the wireless sensor nodes is out of the question. The lifetime of the wireless sensor network is therefore determined by the characteristics of the batteries used. In order to overcome the energy constraint of the WSN due to the energy hungry sensor nodes and the limited energy density of the storage elements, some solutions have been proposed in the next section. The proposed solutions are suggested to extend and sustain the operation of the WSNs.

6. Proposed Solutions for WSN problems

Often WSNs are deployed in regions that are difficult to access and so the sensor nodes should not require any maintenance at all under ideal condition. They must be energetically autonomous and independent. This implies that once the batteries/supercapacitors are installed for the sensor nodes, they do not need to be replaced or recharged for a long period of time and really operate in an autonomous manner for life-long operation. In many application scenarios, the lifetime of the sensor node typically ranges from two to ten years depending on the requirement of the specific application. For that, the stringent condition imposes drastic constraints on the power consumption of the sensor node. Take for an example, a single 1.5 V good AA alkaline battery is used to power a wireless sensor node for two to ten years, it can be roughly estimated that the average power consumption of the sensor node ranges from 250 μW to 50 μW. Given that today's commercially available low power radio transceivers typically consume several tens of milliwatts, keeping the transceiver constantly active is clearly impossible. Several possible solutions to address these problems related to powering the emerging wireless technologies have been suggested in the below list and these solutions will be further elaborated in the following sections.

- Improve the performance of the finite power sources for e.g. by increasing the energy density of the power sources
- Reduce the power consumption at different levels of the sensor nodes hierarchy i.e. signal processing algorithms, operating system, network protocols and integrated circuits
- Develop energy harvesting techniques that enable a sensor node to generate its own power by harvesting energy from the ambient

6.1 Improvements on Finite Power Sources

Research to increase the energy storage density of both rechargeable and primary batteries has been conducted for many years and continues to receive substantial focus Blomgren (2002). The past few years have also seen many efforts to miniaturize fuel cells which promise several times the energy density of batteries. While these technologies promise to extend the lifetime of wireless sensor nodes, they cannot extend their lifetime indefinitely. Other than that, there are many disadvantages such as risk of fire, short shelf life of typically 2-3 years, limited energy density, low power density, etc in the existing rechargeable or alkaline batteries that are not only impacting on the operation of the sensor nodes but also causing problems to the environmental conditions.

6.2 Reduce Power Consumption of Sensor Nodes

Low power consumption by each individual sensor node is paramount to enable a long operating lifetime for a wireless sensor network. A long sensor node lifetime under diverse operating conditions demands power-aware system design. In a power-aware design, the energy consumption of the sensor node at different levels of the system hierarchy, including the signal processing algorithms, operating system, network protocols and even the integrated circuits themselves have to be considered. Computation and communication are partitioned and balanced for minimum energy consumption. This is facilitated by low duty cycle operation typically of the order of 0.1 % to 1 % (most of the time the sensor nodes are sleeping), local signal processing, multi-hop networking among sensor nodes can also be introduced to reduce the communication link range for each node in the sensor network. Since the loss in the communication path increases with the communication range, this reduction in the nodes linkage range would result in massive reductions in power requirements. Compared with characteristics of conventional long-range wireless systems, the reduced link range and data bandwidth yield a significant link budget advantage for typical wireless sensor applications. However, the severely limited energy sources (compact battery systems) for wireless sensor nodes create profound design challenges.

6.3 Proposed Sustainable Power Source for WSN

The wireless sensor node harvests its own power to sustain its operation instead of relying on finite energy sources such as alkaline/rechargeable batteries. This is an alternative energy system for the WSN. The idea is that a node would convert *renewable energy* abundantly available in the environment into *electrical energy* using various conversion schemes and materials for use by the sensor nodes. This method is also known as "*energy harvesting*" because the node is harvesting or scavenging unused freely available ambient energy. Energy harvesting is a very attractive option for powering the sensor nodes because the lifetime of the nodes would only be limited by failure of theirs own components. However, it is potentially the most difficult method to exploit because the renewable energy sources are made up of different forms of ambient energy and therefore there is no one solution that would fit all of applications. However, this option would be able to extend the lifetime of the sensor node to a larger extent compared to the other two possibilities i.e. improvements on the existing finite energy sources and reduce the power consumption of sensor nodes.

7. Overview of Energy Harvesting

Energy harvesting is a technique that capture, harvest or scavenge unused ambient energy (such as vibrational, thermal, wind, solar, etc.) and convert the harvested energy into usable electrical energy which is stored and used for performing sensing or actuation. The harvested energy is generally very small (of the order of mJ) as compared to those large-scale energy harvesting using renewable energy sources such as solar farms and wind farms. Unlike the large-scale power stations which are fixed at a given location, the small-scale energy sources are portable and readily available for usage. Energy harvested from the ambient are used to power small autonomous sensors that are deployed in remote locations for sensing or even to endure long-term exposure to hostile environments. The operations of these small autonomous sensors are often restricted by the reliance on battery energy. Hence the driving force behind the search for energy harvesting technique is the desire to power wireless sensor networks and mobile devices for extended operation with the supplement of the energy storage elements if not completely eliminating the storage elements such as batteries.

7.1 Concept of Energy Harvesting

Energy harvesting systems generally consist of: energy collection elements, conversion hardware and power conditioning and storage devices as shown in Figure.7. Power output per unit mass or volume i.e. power/energy density is a key performance unit for the collection elements. The harvested power must be converted to electricity and conditioned to an appropriate form for either charging the system batteries or powering the connected load directly. Load impedance matching between the energy collectors/energy sources and storage elements/connected to the load is necessary to maximize the usage of the scavenged energy. Appropriate electronic circuitry for power conditioning and load impedance matching may be available commercially or may require custom design and fabrication.
Various scavengable energy sources, excluding the biological type, that can be converted into electrical energy for use by low power electronic devices are shown in Figure.7. Our environment is full of waste and unused ambient energy and these energy sources like solar, wind, vibration, ocean wave, ambient radio frequency waves, etc are ample and readily available in the environment. Since these renewable energy sources are already available, it is not necessary to deliberately expend efforts to create these energy sources like the example of burning the non-renewable fossil fuels to create steam which in turn would cause the steam turbine to rotate to create electrical energy. Unlike fossil fuels which are exhaustible, the environmental energies are renewable and sustainable for almost infinite long period. The energy harvesting process can be easily accomplished. As long as the conversion hardware are chosen correctly in relation to the energy sources, the environmental energy can then be harvested and converted into electrical energy. The energy conversion hardware are designed in different forms to harvest various types of renewable energies. Take for an example, the material of the photovoltaic cell in the solar panel is doped in such a way that when the solar radiation is absorbed by the cell, the solar energy from the sun would be harvested and converted into electrical

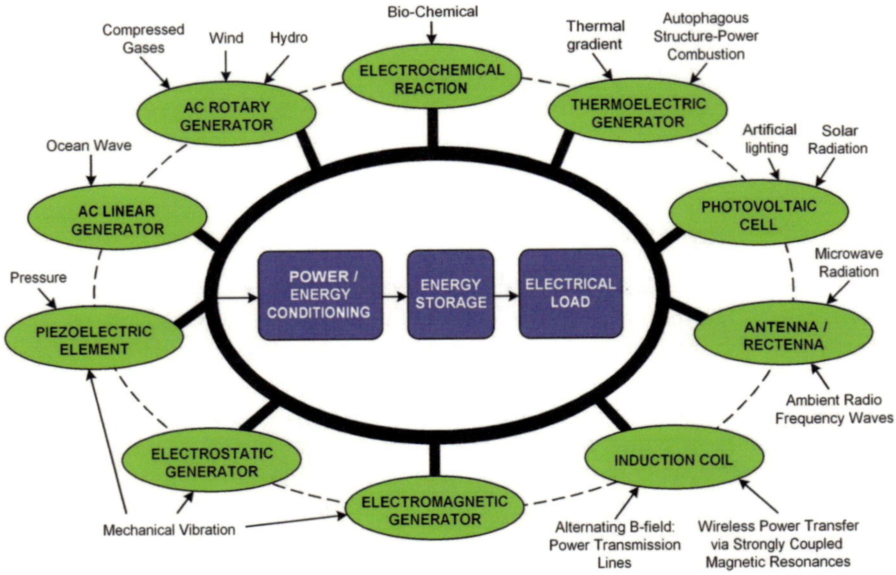

Fig. 7. Energy sources and respective transducers to power autonomous sensor nodes. Adapted from Thomas (2006) with additional power sources.

energy. The whole energy harvesting process involves energy conversion hardware that converts the environmental energy into electrical energy, electrical energy conditioning by the power management circuit and then store in energy storage elements and finally supply to the electrical load.

7.2 Benefits of Energy Harvesting

Energy harvesting provides numerous benefits to the end user and some of the major benefits about EH suitable for WSN are stated and elaborated in the following list. Energy harvesting solutions can:

1. Reduce the dependency on battery power. With the advancement of microelectronics technology, the power consumption of the sensor nodes are getting lesser and lesser, hence harvested ambient/environmental energy may be sufficient to eliminate battery completely.
2. Reduce installation cost. Self-powered wireless sensor nodes do not require power cables wiring and conduits, hence they are very easy to install and they also reduce the heavy installation cost.
3. Reduce maintenance cost. Energy harvesting allows for the sensor nodes to function unattended once deployed and eliminates service visits to replace batteries.
4. Provide sensing and actuation capabilities in hard-to-access hazardous environments on a continuous basis.
5. Provide long-term solutions. A reliable self-powered sensor node will remain functional virtually as long as the ambient energy is available. Self-powered sensor nodes are perfectly suited for long-term applications looking at decades of monitoring.

6. Reduce environmental impact. Energy harvesting can eliminate the need for millions on batteries and energy costs of battery replacements.

7.3 Various Energy Harvesting Techniques

In both academic research works and industry applications, there are many research and development works being carried out on harnessing large-scale energy from various renewable energy sources such as solar, wind and water/hydro NREL (2010). Little attention has been paid to small-scale energy harvesting methods and devices in the past as there are hardly any need. Having said that, it does not mean that there is no research activity being conducted on small-scale energy harvesting. In fact, there are quite a significant amount of research works recorded in the literature that discuss about scavenging or harvesting small-scale environmental energy for low powered mobile electronic devices especially wireless sensor nodes. Figure.8 shows various types of ambient energy forms suitable for energy harvesting along with examples of the energy sources. The energy types are thermal energy, radiant energy and mechanical energy.

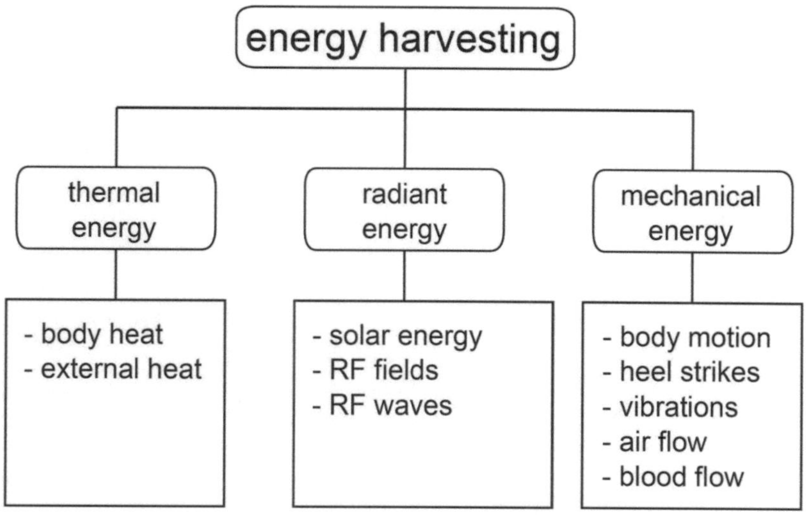

Fig. 8. Types of ambient energy sources suitable for energy harvesting

Some energy harvesting research prototypes for harvesting various energy sources have been discussed. A substantial piece of the research work done by Roundy et al. in Roundy et al. (2004) describes the extraction of energy from kinetic motion. Roundy gave a comprehensive examination on vibration energy scavenging for wireless sensor network. There are other vibration based energy harvesting research works being reported for instances piezoelectric generators in shoes Schenck et al. (2001), wearable electronic textiles Emdison et al. (2002) and electromagnetic vibration-based microgenerator devices for intelligent sensor systems Glynne et al. (2004). In the research area of thermal energy harvesting, both Stevens Stevens (1999) and Lawrence et al. Lawrence et al. (2002) consider the system design aspects for thermal energy scavenging via thermoelectric conversion that exploits the natural temperature difference between the ground and air. Similarly, Leonov et al. Leonov et al. (2007) have considered thermal energy harvesting through thermoelectric power generation from body heat to power wireless sensor nodes. Research on small-scale wind energy harvesting have also been performed by several group of researchers like Weimer et al. Weimer et al. (2006), Myers et al. Myers et al. (2007) and the author himself Tan et al. (2007) and Ang et al. (2007). Heliomote is a sensor node prototype developed by Aman Kansal et al. Raghunathan et al. (2005) that utilizes solar energy harvesting to supplement batteries to power the wireless embedded systems.

7.4 Comparison of Energy Harvesting Sources

To make the sensor node truly autonomous and self-sustainable in the WSN, the energy consumption of the sensor node must be entirely scavenged from the environment. The choice of the energy harvesting technique is crucial. Numerous studies and experiments have been conducted to investigate the levels of energy that could be harnessed from the ambient environment. A compilation of various power densities derived from various energy harvesting sources has been listed in Table.3.

Energy Source	Performance (Power Density)	Notes
Solar (direct sunlight)	100 mW/cm^3	Common polycrystalline solar cells are 16 %-17 % efficient, while standard mono-crystalline cells approach 20 %
Solar (illuminated office)	100 $\mu W/cm^3$	
Thermoelectric	$^{a)}60\mu W/cm^2$ at 5^oC gradient $^{b)}135$ $\mu W/cm^2$ at 10^oC gradient	Typical efficiency of thermoelectric generators are $\leq 1\%$ for $\Delta T < 40^oC$ $^{a)}$Seiko Thermic wristwatch at 5^oC body heat, $^{b)}$Quoted for a Thermo Life generator at $\Delta T = 10\ ^oC$
Blood Pressure	0.93W at 100mmHg	When coupled with piezoelectric generators, the power that can be generated is order of μW when loaded continuously and mW when loaded intermittently
Proposed Ambient airflow Harvester	177 $\mu W/cm^3$	Typical average wind speed of 3 m/s in the ambient.
Vibrational Micro-Generators	4 $\mu W/cm^3$ (human motion-Hz) $800\mu W/cm^3$ (machines-kHz)	Predictions for 1 cm^3 generators. Highly dependent on excitation (power tends to be proportional to ω, the driving frequency and y_o, the input displacement
Piezoelectric Push Buttons	50 $\mu J/N$	Quoted at 3 V DC for the MIT Media Lab Device

Table 3. Power density comparison on various energy harvesting sources

From Table.3, it can be clearly seen that the solar energy source which is abundant outdoors during the daytime provides the best performance in terms of power density, measuring up to 100 mW/cm^3. The power density of the solar radiation on the earth's surface indicates that in a small volume of 1 cm^3, 100 mW of power can be harvested from the sun by using the solar panel. To achieve this high power density, the solar panel has to be exposed in outdoor condition which has direct bright sunlight. When the solar panel is brought into indoor conditions such as illuminated office, the light intensity is reduced tremendously and the power density of the solar energy source drops to almost 100 μW/cm^3. This shows that the available solar power in indoors is drastically lower than that available in outdoors. For design of embedded wireless sensor nodes to be deployed indoors or overcast areas such as buildings and installations, and forestry terrains, where access to direct sunlight is often not available, solar energy source may not be a suitable choice. Hence there is a need to search for alternative energy sources either to replace the solar energy source as a whole or to supplement the solar energy source when the intensity of the light is low. Thermal energy is an example of the alternative energy sources. To harvest the thermal energy, the *thermoelectric generator (TEG)* has been developed and it harvests the heat energy based on Seebeck effect. One commercial application example of TEG is illustrated by the Seiko Thermic wristwatch. The thermoelectric module in the wristwatch is recorded to yield 60 μW/cm^2 at 5°C temperature gradient with 10 thermoelectric generators coupled together Kanesaka (1999). However typical efficiency for thermoelectric generators is less than 1% for temperature gradient less than 40°C and it is hard to find such temperature gradient in the normal ambient environment. Hence thermal energy harvesting is more suitable for low power applications that consume power less than a few mW or hundreds of μW.

Other than solar and thermal energy sources, there is another type of energy source that is available in human blood pressure. Assuming an average blood pressure of 100 mmHg (normal desired blood pressure is 120/80 above atmospheric pressure), a resting heart rate of 60 beats per minute and a heart stroke volume of 70 milliliters (ml) passing through the aorta per beat Braunwald (1980), then the power generated is about 0.93 W. Ramsay and Clark Ramsay et al. (2001) found that when the blood pressure is exposed to a piezoelectric generator, the generator can generate power of the order of μW when the load applied changes continuously and mW as the load applied changes intermittently. However harnessing power from blood pressure would only limit the application domains to wearable micro-sensors. Taking an interesting turn, Shenck and Paradiso Schenck et al. (2001) have built shoe inserts capable of generating 8.4 mW of power under normal walking conditions. This shows that mechanical vibration is another promising energy source worth investing effort to investigate. Chandrakasan and Amirtharajah Meninger et al. (2001) have demonstrated an electromagnetic vibration-to-electricity converter that produces 2.5 $\mu W/cm^3$. Similarly, another piece of research work discussed by Mitcheson et. al in Mitcheson et al. (2004) has made an analysis indicated that up to 4 μW/cm^3 can be achieved from vibrational microgenerators (of order 1 cm^3 in volume) that typical human motion (5 mm motion at 1 Hz) stimulates and up to 800 μW/cm^3 from machine-induced stimuli (2 nm motion at 2.5 kHz). Additionally, Joe Paradiso and Mark Feldmeir in Paradiso et al. (2002) have successfully demonstrated a piezoelectric element with a resonantly matched transformer and conditioning electronics that, when struck by a button, generate 1 mJ at 3V per 15N push, enough power to run a digital encoder and a radio that can transmit over 50 feet. The mechanical vibration energy harvesting is restricted to specific applications where vibration energy source is available.

In summary, the comparison table has shown the performance of each energy harvesting source in terms of the power density factor. Although the table shows that the solar energy source yields the highest power density, this may not be always the case. Under illuminated indoor condition, the solar energy harvested by the solar panel drops tremendously. The other energy harvesting sources would provide higher power density. Depending on the renewable energy sources available at the specific application areas like outdoor bright sunny day with rich amount of solar energy, along coastal area with a lot of wind energy, bridge structure with vehicles travelling has strong vibrations, etc, a suitable energy harvesting source should be selected to power the load for the specific application. Additionally, there is also a possibility that two or more energy sources are available for harvesting, so hybrid energy harvesting could also be an interesting option for energy-hunger load.

8. Energy Harvesting for Wireless Sensor Network

The concept of energy harvesting in relation to wireless sensor network (WSN) entails the idea of scavenging energy from mechanical, vibrational, rotational, solar or thermal means rather than relying on mains power or alkaline/rechargeable batteries to power the sensor nodes in the WSN. For instance, power can be harvested from the mechanical force of a conventional mechanical ON and OFF switch being turned on or off. Alternately, power can be derived from the difference in temperature between the human body and the surrounding ambient environment. Energy harvesting is increasingly gaining notice in the WSN research as well as industry market because it is a very potential solution to extend the lifetime of the sensor node's operation.

8.1 Architecture of Self-Powered Wireless Sensor Nodes

Figure.9 shows an overview functional diagram of a self-powered wireless sensor node in a WSN which contains the four key units namely

- Energy harvesting unit i.e. power supply, power management/conditioning and energy storage
- Microcontroller unit i.e. signal processing, data manipulation and networking
- Sensor unit for parameters such as temperature, humidity, light and speed sensing
- Wireless communication i.e. transmitter and receiver pair or transceiver unit

The energy harvesting system consists of three main components namely energy harvester, power management/conditioning and energy storage. Figure.10 shows the general block diagram representation of a typical energy harvesting unit. The function of the energy harvester is to convert energy harnessed from ambient energy sources into electrical energy. For examples, the Lead Zirconate Titanate (PZT) ceramic material converts mechanical (strain or stress) energy into electrical energy due to the piezoelectric effect, the photovoltaic cell converts solar energy into electrical energy, the thermoelectric generator output electrical voltage when there is a thermal gradient across it and the wind turbine converts kinetic energy from wind flow into electrical energy. The generated electrical energy from the energy harvester needs to be conditioned by the power management unit before supplying to the load. The main objective of the power electronics technology in the power management unit is to process and control the flow of electrical energy from the source to the load in such a way that energy is used efficiently. This matching process is a crucial step in ensuring that maximum power is transferred from the source to the load. Another function of the power conditioning unit involves the conversion and control of electrical voltage at higher levels into suitable levels for the loads.

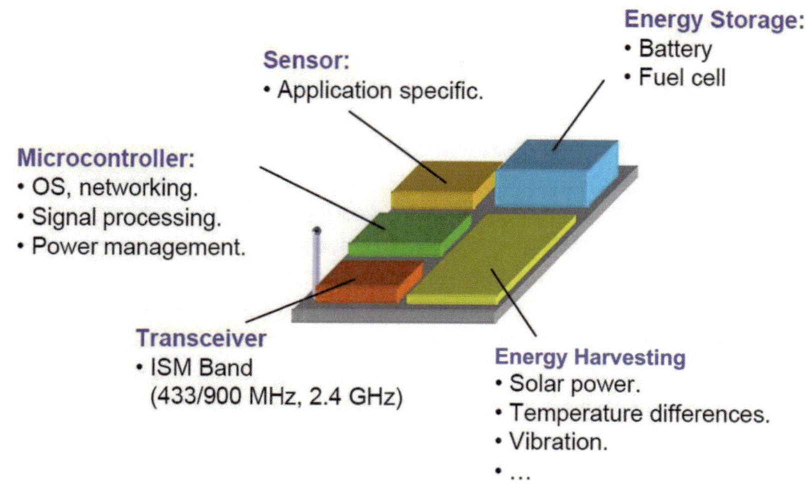

Fig. 9. Key components of a self-powered wireless sensor node

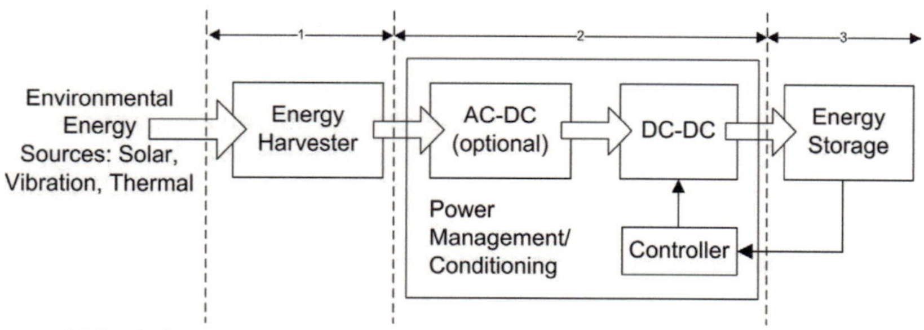

Fig. 10. General block diagram representation of energy harvesting system unit

To ensure continuity in the load operation even when the external power source is temporarily unavailable, the excess energy harnessed has to be stored either in a rechargeable battery or electrochemical double layer capacitors, also known as supercapacitors/ultracapacitors. As mentioned before, batteries have higher energy density (more capacity for a given volume/weight) but lower power density compared to supercapacitors. Recently, such capacitors have been explored for energy storage because they are more efficient and suitable to handle short duration power surges than batteries. Supercapacitors also offer higher lifetime in terms of charge-discharge cycles. However they involve leakage (intrinsic and due to parasitic paths in the external circuitry), which precludes their use for long term energy storage. The overall coverage of this research work involves the investigations on several potential renewable energy harvesting sources and applying these energy sources on some technically feasible application areas to verify that energy harvesting is indeed applicable for real-life applications. The power conditioning electronic circuits in the energy harvesting system are designed based on the energy harvesting input energy sources and the connected output loads, hence different types of power conditioning circuit designs have been proposed to bridge between the source and the load. It is worth noting that the design of the energy harvesting system to power the sensor node in the WSN may differ from one application to another application because of the variations in the load requirements and the differences in

the condition of the deployment area. This would be covered in greater detail in the next few chapters.
The other units of the self-powered wireless sensor node are treated as loads to the energy harvesting system unit. They consume electrical power from the energy sources i.e. energy harvester and/or energy storage to perform their respective operations. Sensors are devices that responds to a physical stimulus (such as thermal energy, electromagnetic energy, acoustic energy, pressure, magnetism or motion) to produce an electrical sensed signal. These sensor devices generally consume relatively low power as compared to the processing and communication units. Hence they are normally not regarded as the major bottleneck in the electronic circuitry. Microcontroller (MCU), which includes an integrated CPU, memory (a small amount of RAM, ROM, or both) and other peripherals such as counters, timer and Analog-Digital Converter (ADC) on the same chip, is a highly integrated single purpose processing unit capable of executing small control programs such as signal processing, power management and networking. The processing power of the MCU is a function of the electrical power consumed i.e. the higher the processing speed, the higher the electrical power consumed by the MCU. Microcontroller is one of the energy hungry units in the wireless sensor node which typically consumes few tens of mW to hundreds of mW during processing and very little power in the order of μW is needed to keep in standby mode. Another energy hunger unit in the sensor node is the communication unit. The function of the communications unit is to transmit or receive data in a wireless manner. A transmitter or receiver has only one function in the communication unit whereas a transceiver has both transmit and receive functions. Some sensor nodes might have only the transmitter to perform uni-directional data transmission whereas others may need to have a transceiver for bi-directional communication.

8.2 Sensor nodes operation with Energy Harvesting Principle

The energy harvester of the energy harvesting system described in Figure.10 converts the environmental energy into electrical energy, at a certain efficiency. The harvested energy is then either stored in the energy storage element or supplied to the load. Energy storage is a very essential element of the energy harvesting system because it acts like a stable bridge between the source and load that provides a constant power flow to the load from a variable environmental source. In short, the power conditioning unit is used to condition the harvested energy so as to properly charge the storage unit and also to provide the appropriate power supply to the load. For a perpetual sensor node operation, it must be such that

$$\overline{P}_g \geq \overline{P}_c \tag{1}$$

where $\overline{P}_g$ and $\overline{P}_c$ are the generated and consumed average/mean powers respectively. As illustrated in Section.4, the power consumed by the sensor node is typically few tens to hundreds of mW and the power generated by the various energy sources of the same area/volume space as the sensor node are much smaller, in the range of units or tens of μW. This is very obvious that energy harvesting is not able to power the operation of the wireless sensor node continuously. One of the possible approach is to reduce the power consumption of the sensor node by duty cycling the node's operation into intermittent form. However, the intermittent mode of operation of the sensor node should not affect the monitoring process of the WSN. In duty cycling type of approach, autonomous sensor nodes are often designed to operate in a very low duty cycle, D, with moderate power consumption in active mode, P_{active} (tens or hundreds of mW), and very low power consumption while idle (sleep mode), P_{sleep} (units or tens of μW), in order to minimize the average power consumed by the sensor node. By doing so, the operation of the sensor node in the WSN can then be sustained by the energy harvesting source. This is one of the methods to sustain the operational lifetime of the wireless sensor node with aid of energy harvesting principle. Let's investigate the amount of power consumed by sensor node when duty-cycling operation is implemented. The consumed average power can be approximated as follows: -

$$\overline{P}_c = P_{sleep} + DP_{active} \tag{2}$$

From Equations (1) and (2), it is observed that when D is large which means the sensor node is active for a long period of time, the average power consumed by the node would be high. Hence the generated power may not be sufficient to power the sensor node's operation. Conversely, if D is small, the sensor node is put to idle state for most of the time and it wakes up to perform sensing and communicating when needed, the average power consumption of the node would be reduced tremendously. If this is the case, there is a higher possibility that the generated power is either able to power the sensor node directly or able to accumulate enough energy in the energy storage and then release to sensor node. Based on the above two equations, it can be worked out that the maximal duty cycle to maintain the operation of the sensor node in continuous mode is given as: -

$$D_{max} = \frac{\overline{P}_g - P_{sleep}}{P_{active}} \tag{3}$$

D_{max} must selected such that it is neither too small until the WSN operation is affected due to the lack in communicating or sensing time nor too large until the average power consumed by the sensor node is much larger than the average power generated by the energy harvesting source. Often, it is hard to have such a situation whereby the generated average power is more than the consumed average power. This is because the power consumed by the sensor node is much more than what the energy harvester of the similar size can provide. Another reason is the variability of environmental energy sources. As the environmental energy sources change their characteristics from time to time, the harvested electrical energy would change accordingly and so there would be times where $P_g < P_c$. To overcome that, a storage unit is needed. This energy reservoir must be able to supply power to the load whenever $P_g < P_c$. For any arbitrary long period of time, T, a long-term storage ($E_{storage}$) unit must be designed to fulfill the condition of: -

$$E_{storage} \geq max \int (P_c - P_g)dt \tag{4}$$

The burst power operation of the sensor node is another condition to be considered for energy harvesting source. Even if generated power, P_g, is constant, for example solar power coming from permanent indoor lights, a short-term storage is still needed to withstand the burst power consumption profile of an autonomous sensor node. Figure.11 illustrates this situation when $P_{active} > P_g$. The capacity of the energy storage should not be selected to be too high because the physical size of the storage would become too large. Depending on the operational requirement of the application, the characteristic of the energy harvesting source and the energy consumption of the sensor node in the WSN, the energy storage and the duty cycle, D, of the sensor node can be determined accordingly.

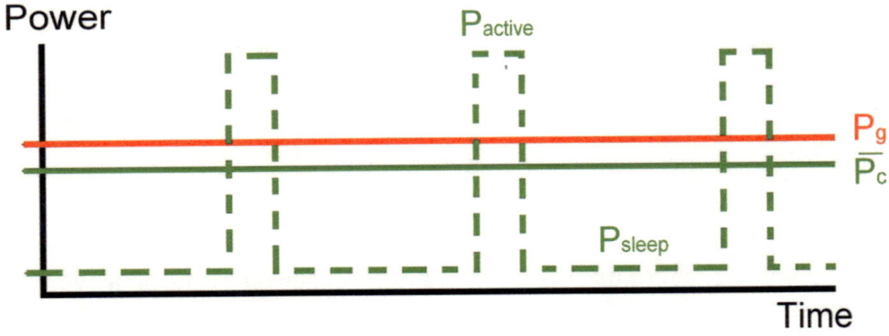

Fig. 11. Burst power consumption by the sensor node when $P_{active} > P_g$

9. Conclusions

The major hindrances of the "deploy and forget" nature of the wireless sensor networks (WSNs) are the limited energy capacity and unpredictable lifetime performance of the battery. In order to overcome these problems, energy harvesting/scavenging, which harvests/scavenges energy from a variety of ambient energy sources and converts into electrical energy to recharge the batteries, has emerged as a promising technology. With the significant advancement in microelectronics, the energy and therefore the power requirement for sensor nodes continues to decrease from few mWs to few tens of μW level. This paves the way for a paradigm shift from the battery-operated conventional WSN, that solely relies on batteries, towards a truly self-autonomous and sustainable energy harvesting wireless sensor network (EH-WSN).

10. References

Chong C. & Kumar S.P. (2003). Sensor Networks: Evolution, Opportunities, and Challenges, *Proceeding of the IEEE, Sensor Networks and Applications*, vol.91, no.8, pp.1247-1256, 2003.

N. Kurata, S. Saruwatari & H. Morikawa (2006). Ubiquitous Structural Monitoring using Wireless Sensor Networks, *International Symposium on Intelligent Signal Processing and Communications*, pp.99-102, 2006.

Technology Review (2003). 10 Emerging Technologies That Will Change the World, *February 2003 Issue of Technology Review*, >http://www.technologyreview.com/Infotech/13060/?a=f<.

Rakesh Kumar (2005). Shaping Ubiquitous Technology for developing countries, *International Telecommunications Union (ITU) Workshop on Ubiquitous Network Societies*, >itu.int/osg/spu/ni/.../Papers/Paper_Ubiquity_and_developing_world.pdf<.

D.J. Cook & S.K. Das (2004). Wireless Sensor Networks, *Smart Environments: Technologies, Protocols and Applications*, John Wiley, New York, 2004.

M. Kuorilehto, M. Hannikainen & T.D. Hamalainen (2005). A survey of application in wireless sensor networks, *EURASIP Journal on Wireless Communications and Networking*, vol.2005, no.5, pp.774-788, 2005.

Edgar H. Callaway (2003). Wireless sensor networks: architectures and protocols, *Auerbach Publications*, Boca Raton, 2003.

I.F. Akyildiz, W.L. Su, S. Yogesh & C. Erdal (2002). A Survey on Sensor Networks, *IEEE Communications Magazine*, vol.40, no.8, pp.102-114, 2002.

D. Culler, D. Estrin & M. Srivastava (2004). Overview of sensor networks, *IEEE Computer*, vol.37, no.8, pp.41-49, 2004.

A. Mainwaring, J. Polastre, R. Szewczyk, D. Culler & J. Anderson (2002). Wireless sensor networks for habitat monitoring, *Proceedings of the ACM International Workshop on Wireless Sensor Networks and Applications*, pp.88-97, 2002.

A. Boulis, C.C. Han & M.B. Srivastava (2003). Design and implementation of a framework for efficient and programmable sensor networks, *Proc. 1st International Conference on Mobile Systems, Applications and Services (MobiSys'03)*, San Francisco, Calif, USA, 2003.

L. Schwiebert, S. K. S. Gupta & J. Weinmann (2001). Research challenges in wireless networks of biomedical sensors, *Proc. 7th ACM International Conference on Mobile Computing and Networking (MobiCom '01)*, pp.151-165, Rome, Italy, 2001.

Thomas von Buren (2006). Body-Worn Inertial Electromagnetic Micro-Generators, *Ph.D Thesis*, Swiss Federal Institute of Technology Zurich, 2006.

H.O. Marcy, J.R. Agre, C. Chien, L.P. Clare, N. Romanov & A. Twarowski (1999). Wireless sensor networks for area monitoring and integrated vehicle health management applications, *Proc. AIAA Guidance, Navigation, and Control Conference and Exhibit*, Portland, Ore, USA, 1999.

K. Romer (2004). Tracking real-world phenomena with smart dust, *Proc. 1st European Workshop on Wireless Sensor Networks (EWSN'04)*, pp.28-43, Berlin, Germany, 2004.

M.B. Srivastava, R.R. Muntz & M. Potkonjak (2001). Smart kindergarten: sensor-based wireless networks for smart developmental problem-solving enviroments, *Proc. 7th ACM International Conference on Mobile Computing and Networking (MobiCom'01)*, pp.132-138, Rome, Italy, 2001.

S. Roundy (2003). Energy Scavenging for Wireless Sensor Nodes with a Focus on Vibration to Electricity Conversion, *Ph.D Thesis*, University of California, Berkeley, 2003.

TinyOS (2008). The TinyOS Project, *TinyOS Community Forum*, >http://www.tinyos.net<.

M. Tubaishat & S. Madria (2003). Sensor networks: an overview, *IEEE Potentials*, vol.22, pp.20-23, 2003.

A. Sinha & A. Chandrakasan (2001). Dynamic Power Management in Wireless Sensor Networks, *IEEE Design Test Comp.*, 2001.

G.V. Merrett, B.M. Al-Hashimi, N.M. White & N.R. Harris (2005). Resource aware sensor nodes in wireless sensor networks, *Journal of Physics*, vol.15, no.1, pp.137-42, 2005.

K. Sohrabi, J. Gao, V. Ailawadhi & G. Pottie (2000). Protocols for self-organization of a wireless sensor network, *IEEE Personal Communications*, vol.7, no.5, pp.16-27, 2000.

E. Lattanzi, E. Regini, A. Acquaviva & A. Bogliolo (2007). Energetic sustainability of routing algorithms for energy-harvesting wireless sensor networks, *Computer Communications*, vol.30, no.14-15, pp.2976-2986, 2007.

Crossbow Technology Inc.(2007). MPR-MIB Users Manual, *Crossbow Resources*, Revision A, 2007.

L. Doherty, B.A. Warneke, B.E. Boser & K.S.J. Pister (2001). Energy and performance considerations for smart dust", *International Journal of Parallel and Distributed Systems and Networks*, vol.4, no.3, pp.121-133, 2001.

J. L. Hill & D. E. Culler (2002). Mica: A wireless platform for deeply embedded networks, *IEEE Micro*, vol.22, pp.12-24, 2002.

J. M. Rabaey, M. J. Ammer, J. L. da Silva, Jr. D. Patel & S. Roundy (2000). PicoRadio supports ad hoc ultra-low power wireless networking, *IEEE Computer*, vol.33, pp.42-48, 2000.

Massachusetts Institute of Technology (2008). μAmps projects, *Microsystems Technology Laboratories*, >http://www-mtl.mit.edu/researchgroups/icsystems/uamps/<.

J.W.Tester (2005). Energy Transfer and Conversion Methods, *Sustainable Energy Lecture Notes*, Topic on Energy Storage Modes, MIT, 2005.

Crossbow Technology Inc. (2007). MPR-MIB Users Manual, *Crossbow Resources*, Revision A, 2007.

G.E. Blomgren (2002). Perspectives on portable lithium ion batteries liquid and polymer electrolyte types, *Seventeenth Annual Battery Conference on Applications and Advances*, pp.141-144, 2002.

J. P. Thomas, M. A. Qidwai, and J. C. Kellogg (2006). Energy scavenging for small-scale unmanned systems, *Journal of Power Sources*, vol.159, pp.1494-1509, 2006.

Renewable Resource Data Center (RReDC). *National Renewable Energy Laboratory*, >http://www.nrel.gov/rredc/< accessed on 14-06-2010.

S. Roundy, P.K. Wright and J.M. Rabaey (2004). Energy Scavenging for Wireless Sensor Networks with Special Focus on Vibrations, *Kluwer Academic Press*, Boston, MA, 2004.

N.S. Schenck and J.A. Paradiso (2001). Energy Scavenging with Shoe-Mounted Piezoelectrics, *IEEE Micro*, vol.21, pp.30-41, 2001.

J. Edmison, M. Jones, Z. Nakad and T. Martin (2002). Using piezoelectric materials for wearable electronic textiles, *Proceedings of Sixth International Symposium on Wearable Computers (ISWC)*, 2002.

P. Glynne-Jones, M.J. Tudor, S.P. Beeby and N.M. White (2004). An electromagnetic, vibration-powered generator for intelligent sensor systems, *Sensors and Actuators*, vol.110, no.1-3, pp.344-349, 2004.

J.W. Stevens (1999). Heat transfer and thermoelectric design considerations for a ground-source thermo generator, *Proceedings of 18th International Conference on Thermoelectrics*, 1999.

E.E. Lawrence and G.J. Snyder (2002). A study of heat sink performance in air and soil for use in a thermoelectric energy harvesting device, *Proceedings of 21st International Conference on Thermoelectrics (ICT Š02)*, 2002.

V. Leonov, T. Torfs, P. Fiorini, C. Van Hoof (2007). Thermoelectric Converters of Human Warmth for Self-Powered Wireless Sensor Nodes, *IEEE Sensors Journal*, vol.7, issue.5, pp.650-657, 2007.

Michael A. Weimer, Thurein S. Paing, and Regan A. Zane (2006). Remote area wind energy harvesting for low-power autonomous sensors", *37th IEEE Power Electronics Specialists Conference*, pp.2911-2915, 2006.

R. Myers, M. Vickers, Kim Hyeoungwoo and S. Priya (2007). Small scale windmill, *Applied Physics Letters*, vol.90, no.5, p.54106-1-3, 2007.

Y.K. Tan and S.K. Panda (2007). A Novel Piezoelectric Based Wind Energy Harvester for Low-power Autonomous Wind Speed Sensor, *33th Annual IEEE Conference of Industrial Electronics Society (IECON'07)*, pp.2175-2180, 2007.

R.J. Ang, Y.K. Tan and S.K. Panda (2007). Energy harvesting for autonomous wind sensor in remote area, *33th Annual IEEE Conference of Industrial Electronics Society (IECON'07)*, pp.2104-2109, 2007.

V. Raghunathan, A. Kansal, J. Hsu, J. Friedman and M. Srivastava (2005). Design considerations for solar energy harvesting wireless embedded systems, *Fourth International Symposium on Information Processing in Sensor Networks (IPSN 2005)*, pp.457-462, 2005.
T. Kanesaka (1999). Development of a Thermal Energy Watch", *Proc. 64th Conference on Chronometry (Société Suisse de Chronométrie)*, Le Sentier, Switzerland, pp.19-22, 1999.
E. Braunwald (1980). Heart Disease: A Textbook of Cardiovascular Medicine, *W. B. Saunders Company*, Philadelphia, 1980.
M. Ramsay and W. Clark (2001). Piezoelectric energy harvesting for bio-MEMs applications, *Proceedings of the SPIE - The International Society for Optical Engineering*, vol.4332, pp.429-439, 2001.
S. Meninger, A.P. Amirtharajan and R.Chandrakasan (2001). Vibration-to-Electric Energy Conversion, *IEEE Transaction on VLSI System*, vol.9, pp.64-71, 2001.
P.D. Mitcheson, T.C. Green, E.M. Yeatman and A.S. Holmes (2004). Architectures for Vibration-Driven Micropower Generators, *Journal of Microelectromechanical Systems*, vol.13, no.3, pp.429-440, 2004.
J. A. Paradiso and Mark Feldmeier (2002). A compact, wireless, self-powered pushbutton controller, *MIT Media Laboratory*, 2002.

3

Monitoring of Wireless Sensor Networks

Benahmed Khelifa[1], Haffaf Hafid[2] and Merabti Madjid[3]
[1]*Béchar University, Algeria, benahmed_khelifa@yahoo.fr*
[2]*University of Es-Sénia Oran, Algeria, haffaf_hafid@yahoo.fr*
[3]*School of Computing & Mathematical Sciences, Liverpool John Moores University, U.K, M.Merabti@ljmu.ac.uk*

1. Introduction

A Wireless Sensor Network consists of small battery powered wireless devices, that are capable of monitoring environmental conditions such as humidity, temperature, noise, etc. Sensor networks do not have a fixed infrastructure but form an ad hoc topology. Wireless sensor networks are emerging as a promising platform that enable a wide range of applications in both military and civilian domains such as battlefield surveillance, medical monitoring, biological detection, home security, smart spaces, inventory tracking, etc. Such networks consist of small, low-cost, resource limited (battery, bandwidth, CPU, memory) nodes that communicate wirelessly and cooperate to forward data in a multi-hop fashion. Thus, they are especially attractive in scenarios where it is infeasible or expensive to deploy a significant networking infrastructure.

In wireless sensor networks, services may fail due to various reasons, including radio interference, de-synchronization, battery exhaustion, or dislocation. Such failures are caused by software and hardware faults, environmental conditions, malicious behavior, or bad timing of a legitimate action. In general, the consequence of such an event is that a node becomes unreachable or violates certain conditions that are essential for providing a service, for example by moving to a different location, the node can no further provide sensor data about its former location. In some cases, a failure caused by a simple software bug can be propagated to become a massive failure of the sensor network. This results in application trials failing completely and is not acceptable in safety critical applications.

The open nature of the wireless communication, the lack of infrastructure, the fast deployment practices, and the hostile deployment environments, make them vulnerable to a wide range of intrusions and security attacks. The motivation for attacking a sensor networks could be, for example, to gain an undeserved and exclusive access to the collected data. There has been a multitude of attacks described in the literature: probabilistic data packet dropping, topology manipulation, routing table manipulation, prioritized data and control packet forwarding, identity falsification, medium access selfishness etc. The protection system of a sensor networks usually relies on the following two mechanisms: (i) authentication and secure protocols and (ii) intrusion and attack (misbehavior) detection. As the experience from the Internet shows, the weaknesses in authentication and secure protocols are frequently exploited. These protocols alone are in general considered being insufficient to provide the necessary level of protection. Therefore, there has been a lot of effort invested in providing networks with means for a timely detection of an attack or intrusion. Such detection is often based on methods and algorithms known from the field of machine learning.

Additionally, After sensors get deployed in the monitored area, the access to them can be difficult. For example, a sensor network, with the goal to monitor conditions in the sewer system of a large city, might be inaccessible for maintenance, software updates or battery exchange. Therefore, a special focus has been put on designing energy efficient protocols at all layers of the OSI (Open Systems Interconnection protocol stack. Additionally to addressing energy constraints, these protocols should impose a high degree of robustness in order to minimize the need for human intervention.

The use of these sensor networks in hostile environments means that providing quality of service is essential and requires the implementation of fault-tolerant mechanisms that can ensure availability and continuity of service. For example, the maximum coverage of the regions monitored by the network and connectivity of the various nodes of the network must be maintained. However in an environment where each node can fail unexpectedly resulting in the isolation of some parts of the network, this guarantee is neither automatic nor easy to achieve.

For all this problems, the integration of mechanisms for monitoring wireless sensor networks, for the reason of topology control, fault tolerance and security are crucial for the effective use of wireless sensor networks. There are many current management approaches, but each provides only partial solutions to the problems of monitoring and fault tolerance, and they do not adapt to the properties and constraints of many wireless sensor networks.

In summarize there are many papers tried to tackle monitoring methodologies in wireless sensor networks. In this chapter we will try to give an overview on the use of monitoring mechanisms to supervise wireless sensor networks. Then we detailed the description of some research using monitoring mechanisms for reasons of security, topology control or fault tolerance in wireless sensor network, and we will describe our contribution in this field.

2. A survey of monitoring mechanism in WSN

To address these problems , many researchers have used the concept of centralized monitoring, where a control center is responsible for monitoring all network nodes (such as base station, the central controller or manager, and sink) . Other researchers have used a decentralized approach to monitor network nodes as fault detection, security, connectivity and coverage control.

2.1 Monitoring of Connectivity and Coverage in WSN

Connectivity is particularly important for wireless sensor networks. In a wireless sensor network, the deployment strategy often involves using more nodes then necessary and turning off the ones that are not being used for communication or sensing. When the network becomes disconnected, one or more of the redundant nodes can be turned on to repair connectivity [1]. The main problem with this technique is the requirement for extra nodes, and when several nodes in a limited region fail it may no longer be possible to repair the network. Li and Hou study the problem of adding as few nodes as possible to a disconnected static network so that the network remains connected [2]. They show that the problem is NP-Complete and propose some heuristic solutions. These algorithms require global knowledge of the graph and they are time-consuming to apply. Consequently they are typically not applicable in real-time with dynamic networks.

Using mobility to maintain connectivity has attracted many researchers. The general approach has been the use of mobility to carry data between disconnected components of the network [3]. Another approach is the storage of data when connectivity is disrupted, and sending the data when connectivity is subsequently repaired [4, 5]. A significant problem with these approaches is the latency in data transfer for time critical applications.

K. Benahmed and al. used graph theory and self-organisation mechanism for monitoring connectivity in wirless sensor networks [6].

There are also approaches that can be used to maintain uninterrupted connectivity with dynamic networks. Spanos and Murray propose a technique for providing radio connectivity while moving a group of robots from one configuration to another [7]. However, this analysis is not valid when there are obstacles.

Several other solutions for fault tolerance are based on the nature of redundant sensor networks. Fusion techniques [7, 8] may merge or aggregate the different readings of the sensors. Multi routing paths [9, 21] and techniques to ensure k-connectivity between nodes [10, 11, 12, 13, 14] can be applied to increase the reliability of the transmission of messages in wireless sensors networks.

First, most existing solutions have treated the problems of sensing coverage and network connectivity separately. The problem of sensing coverage has been investigated extensively. Several algorithms aim to find close-to-optimal solution based on global information. Both [22] and [23] apply linear programming techniques to select the minimal set of active nodes for maintaining coverage. More sophisticated coverage model is used to address exposure-based coverage problems in [24][25]. The maximal breach path and maximal support path in a sensor network are computed using voronoi diagram and delaunay triangulation techniques in [24]. The problem of finding the minimal exposure path is addressed in [25]. In [26] , sensor deployment strategies were investigated to provide sufficient coverage for distributed detection. Provided scalability and fault-tolerance, the localized algorithms can be more suitable and robust for large-scale wireless sensor network that operate in dynamic environments. The protocol proposed in [27] uses a local geometric calculation of sponsored sectors to preserve the sensing coverage. However, these protocols do not address the problem of maintaining network connectivity. Several other protocols (e.g., ASCENT [28], SPAN [29], AFECA [30], and GAF [31]) aim to maintain network connectivity, but do not guarantee sensing coverage. Unfortunately, satisfying only coverage or connectivity alone is not sufficient for a sensor network to provide sufficient service. Without sufficient sensing coverage, the network cannot monitor the environment with sufficient accuracy or may even suffer from "sensor voids" where no sensing can occur. Without sufficient connectivity, nodes may not be able to coordinate effectively or transmit data back to base stations. The combination of coverage and connectivity is a special requirement introduced by sensor networks that integrate multi-hop wireless communication and sensing capabilities into a single platform. In contrast, traditional mobile ad hoc networks comprised of laptops only need to maintain network connectivity.

A second limitation of the aforementioned coverage protocols (except for the global algorithm in [22]) is that they can only provide a fixed degree of coverage. They cannot dynamically reconfigure to meet the requirements of different applications and environments, or a same application with varying operational conditions. Finally, while the PEAS [32] protocol was designed to address both coverage and connectivity in a configurable fashion, it does not provide analytical guarantees on the degree of coverage and connectivity. For many critical sensor network applications (e.g., surveillance and structural monitoring) guaranteed degrees of coverage and connectivity are required, and a best effort approach is not sufficient.

2.2 Monitoring Wireless sensor Networks for Security Reason

Wireless sensors networks are vulnerable to many types of attacks. In recent years there have been many proposals using cryptography to ensure secure communication such as SPINS [33], etc. Nevertheless, cryptography alone is not sufficient for node compromise attacks and novel misbehaviors in sensor networks [34].

A protocol called DICAS using local monitoring is proposed by Khalil et al. [35], for secure routing, which mitigates the control and data traffic attacks in sensor networks. They propose a countermeasure for wormhole attacks, called LITEWORP [37], which uses guard nodes to attest the source of each transmission. Neighbor watch [36] is employed by a hop-by-hop resilient packet-forwarding scheme. For reputation and trust based systems, neighbor watch is used as a component to monitor neighborhoods and collect information to build trust relationships among nodes in the network, such as RFSN[38], CONFIDANT[39], CORE[40], etc. For intrusion detection systems, local monitoring is used to build decentralized protocols [41, 42]. Khalil et al. [43] propose a on-demand sleep-wake protocol to shorten the time a node needs to be awake for the purpose of monitoring. They do not, however, consider the optimized selection of monitoring nodes in the network, but focusing on how to schedule nodes to meet the monitoring requirement for given communication links. Hsin et al. [44] propose self-monitoring mechanism, this proposition pay more attention on the system-level fault diagnosis of the network, especially detecting node failures. They do not deal with malicious behaviors as what are considered in the works [36,45,46]. On the other hand, our study emphasizes the optimized node selection for the local monitoring scheme. In [47], the authors present DAMON, a distributed system for monitoring multi-hop mobile networks. DAMON uses agents within the network to monitor network behavior and send collected measurements to data repositories. Zhao et al. [48] propose to scan the residual energy and monitor parameter aggregates including link loss rate and packet count. Such information is collected locally at each node and transmitted back to the sink for analysis. In [49], the authors propose Sympathy tool to actively collect run-time status from sensor nodes like routing table and flow information and detects possible faults by analyzing node status together with observed network exceptions. In [50], an IDS model for ad-hoc networks is presented following the behavioral paradigm. The IDS is decentralized and detection is made by clusters. A technique to safely elect the responsible node for monitoring each cycle was developed. This solution is expensive, thus being inadequate to a WSN. In [51], Marti et al. used Watchdog technique or local monitoring for ad-hoc networks in order to improve the detection of mischievous nodes. It uses a technique called pathrater to help routing protocols to avoid those nodes. In this work, the monitor node watches its neighbors to know what each one of them will do with the messages it receives from another neighbor. If the neighbour of the monitor nodes changes, delays, replicates, or simply keeps the message that should be retransmitted, the monitor counts a failure. This technique is also used to detect other types of attacks. This approach is not efficient because watchdog needs more memory. If watchdog's neighbor sensors send large number of messages, the watchdog will run out of its memory quickly. However, none of these previous works has sought to give more importance to the election criteria of nodes responsible for monitoring the network. In addition, the audit data used in monitoring and detecting abnormal behavior in the network, does the flow of traffic, but nobody has taken the resources consumed in a sensor as an index of screening abnormalities. The highlight of our work is summarized in a comprehensive strategy for monitoring the network, in order to detect and remove nodes to abnormal behavior. Our work therefore focuses around a strategy of distributed resolution on the algorithmic level, that is to say an implementation of the distributed algorithm throughout the network, in which each sensor involved through local pre-treatment. On the other hand in most of the work, monitoring keys entire population of the sensor network at the same time, this poses a problem of congestion at the communication channel and overloads the sensors responsible for network monitoring. In our case the sensor responsible for monitoring selects and analysis a single sample from the population to monitor.

3. Hybrid Approach for Monitoring the Connectivity and Coverage in Wireless Sensor Networks

The use of sensor networks in hostile environments means that providing quality of service is essential and requires the implementation of fault-tolerant mechanisms that can ensure availability and continuity of service. For example, the maximum coverage of the regions monitored by the network and connectivity of the various nodes of the network must be maintained. However in an environment where each node can fail unexpectedly resulting in the isolation of some parts of the network, this guarantee is neither automatic nor easy to achieve.

The integration of mechanisms for surveillance, topology control and fault tolerance are crucial for the effective use of wireless sensor networks. There are many current management approaches, but each provides only partial solutions to the problems of monitoring and fault tolerance, and they do not adapt to the properties and constraints of many wireless sensor networks. Therefore, the work presented in this paper gives a new approach for monitoring connectivity in wireless sensors networks. We provide a rigorous analysis for the development of fault-tolerance to ensure both ongoing monitoring of network connectivity and self organization, mainly to enhance the degree of connectivity at critical nodes presenting articulation points in the network.

The rest of this sub-chapter is organized as follows. The following sections 3.1 and 3.2 introduce the concepts of connectivity, monitoring and fault tolerance . We model our problem in Section 3.3. Sections 3.4 and 3.5 describe our solution. In Section 3.6, we present our simulation results.

3.1. Connectivity

A network of sensors is considered to be connected only if there is at least one path between each pair of nodes in the network. Connectivity depends primarily on the existence of paths.

It is affected by changes in topology due to mobility, the failure of nodes, attacks and so on. The consequences of such occurrences include the loss of links, the isolation of nodes, the partitioning of the network, the upgrading of paths and re-routing.
Connectivity can be modeled as a graph G (V, E) where V is the set of vertices (nodes) and E the set of edges (links). This graph is said to be k-connected if there are at least k disjoint paths between every pair of nodes u, v ∊ V. Connectivity is a measure of fault tolerance or diversity of paths in the network. The need for 1-connectivity of the network graph is a fundamental condition of it being operational. The connectivity of a network can be expressed as follows [15].

$$\mu(R) = \frac{N.\pi.R^2}{A} \tag{1}$$

where R is the radius of transmission, A the area and N the number of nodes in the area A. Kleinrock and Silverster have shown that when connectivity μ(R) reaches 6 nodes, the probability that a node is connected tends to 1, i.e. that the network forms a connected graph [14].

3.2. Fault Tolerance

Wireless sensor networks are commonly deployed in hostile environments and are susceptible to numerous faults in several layers of the system. Figure 1 depicts the source of these failures and demonstrates the potential for propagation to higher layers. The source of failures in this classification is divided in to four layers: node, network, sink and the base station.

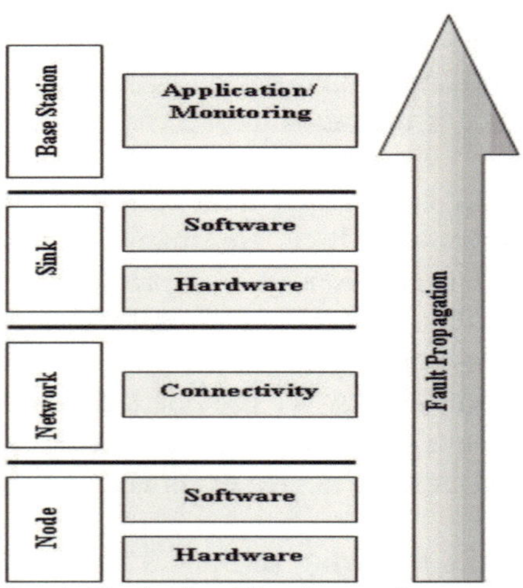

Fig. 1- Fault classification and propagation

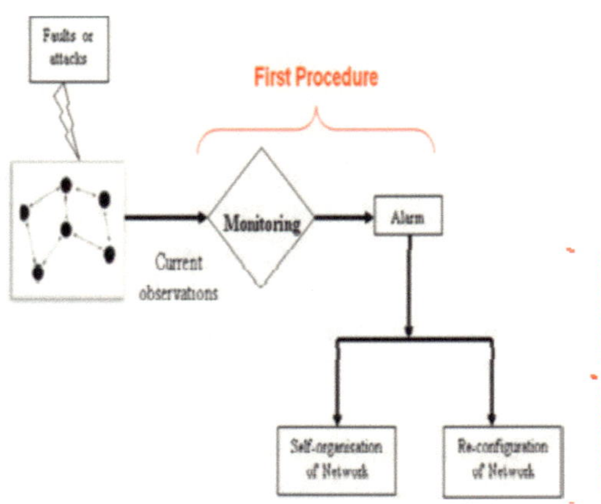

Fig. 2- Monitoring systems

To address these problems it is useful to implement a system that allows monitoring of the network, show figure 2. At any moment such a system must be able to provide the operational status of different devices and to establish mechanisms that provide fault tolerance. By definition fault tolerance [14] is a technique that has been proven to make systems capable of providing a good service, even in the presence of accidental phenomena

such as disturbance of the environment (external faults), failure of hardware components (internal physical faults), or design faults, particularly software faults (bugs). Under the terms of dependability, faults are the causes of errors, mistakes are part of the abnormal state of the system and when errors are propagated to the system interface - i.e. when the service provided by the system is incorrect - this results in a failure. When mistakes are accidental and sufficiently rare, it is possible to tolerate them. This requires detecting errors before they occur, with error handling in case they can't be rectified. We must also make a diagnosis, in other words identify the fault, isolate faulty components, replace or repair and reset the system in case there is no alternative.
In a wireless sensors network, fault tolerance is the ability to ensure the functionality of the network in the face of any interruption due to failures of sensor nodes.

3.3. Modeling the Problem

In most cases a wireless sensors network can be modelled as a unit graph G (V, E) where V is the set of nodes (with each sensor in the network a vertex in the graph) and E the set of all arcs giving opportunities for direct communication between nodes (we assume that the communication is symmetric, meaning that if a node can hear another, it can also be understood by him). The corresponding graph is undirected. If we set d(u, v) as the physical distance between nodes u and v, and Rc the radius of communication, then E is defined as follows [16, 20].

$$E = \{(u, v) \in V2 \mid d(u, v) \leq Rc\} \quad (2)$$

For sensor coverage - i.e. the collection of information by sensors - we need the coverage radius rs, with Rc ≥ 2rs. Figure 3 shows these two ranges (connectivity and coverage).

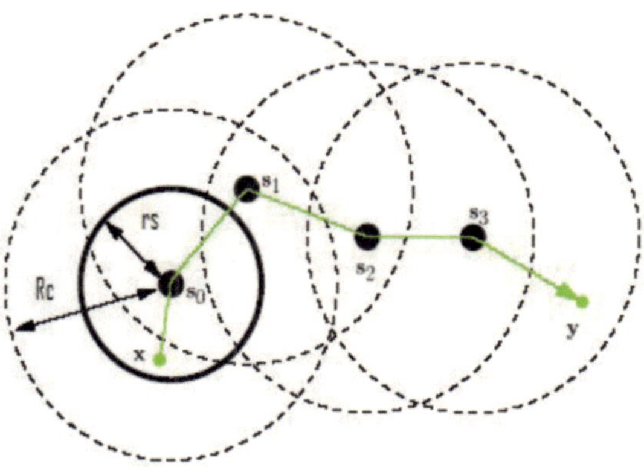

x, y are Events — S_i are sensor nodes
R_c is the radio range — r_s is the sensing range

Fig. 3. Connectivity and coverage in wireless sensor network.

3.4. Connectivity Strategy : Our Approach

In this section we will consider methods used for predicting the partitioning of the network. The prediction algorithm acts as a tool to help provide fault tolerance, aimed at improving the life of the service by detecting critical nodes that might induce a breach of network connectivity should they fail. The mobility of nodes, energy loss, vulnerability to attack and the limited range of their communication implies that the existence of such nodes may result in it becoming impossible to find a route between a source and destination nodes.

The algorithm that we propose for the prediction of partitioning of the network includes the following steps.

- Assess the robustness of the link between nodes.
- If this robustness is below a given threshold, send an alert to self organize the network.

For the assessment of the robustness of communication links, we propose an evaluation based on sets of node-disjoint paths and properties of k-connected graphs.

Theorem (Menger, 1927): In an undirected graph the maximum number of node-disjoint paths from a nonadjacent summit x and summit y is equal to the minimum number of nodes to remove to disconnect x of y [18].

The search for node-disjoint paths between pairs of nodes can be reduced to the search for nodes whose removal disconnects them. Such nodes are called critical points or articulation points and can be detected using a centralized in-depth search algorithm [19]. Figure 4 illustrates this idea.

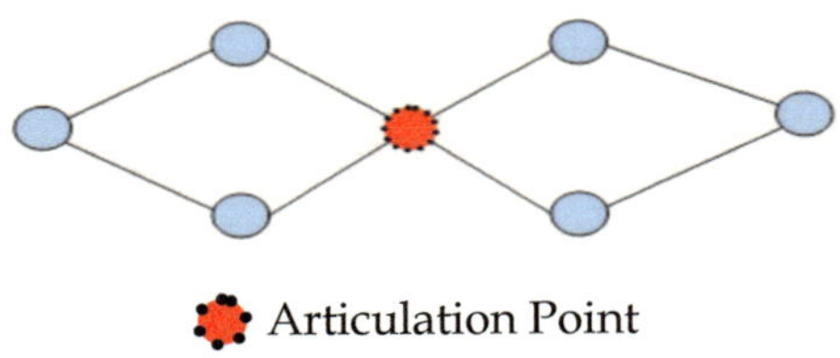

Fig. 4. Topology with an articulation point

Our case is limited to 2-connectivity: we require at least two paths between the source and destination to ensure fault-tolerant connectivity.

Definition 1: A graph is biconnected if for each pair of summits u, v with u ≠ v, there are two summit-disjoint paths that join u and v [20].

Property: A graph is biconnected if and only if it has no articulation point [20].

Algorithm 1: Detection of articulation points in an undirected graph.

Input : G (V, E) Unit Disk Graph

Output : Set of articulation points

- Depth search in graph G and generation of spanning tree T, (in which back edges are shown as dotted lines) to facilitate computing articulation points.
- A vertex x is not an articulation point if it has no successor, or if each of its successor admits a descendant who has a back edge to an ancestor of x in the tree,
- Particular case: the root is an articulation point if it has more than one successor in the tree.

This algorithm has a binomial complexity of the order of O (N + M) for a graph with N vertices and M edges.

3.5 Self Organizing Network

Recent scientific study has considered the behavior of birds, insects and viruses and their capacity to organize themselves. Noting also the pervasive presence and potential benefits of self-organization in natural systems, many researchers have now begun to look at how such models of self-organization can be applied to the design of distributed systems. The mechanisms of self-organization have the potential to provide many solutions in wireless sensor networks. For example, self-organization can be used to change the density of sensor nodes and traffic patterns, or help to reconfigure the network topology in cases where nodes fail or relocate. Inspired by the behavior of ants that organize themselves (moving to form a bridge) and the capabilities of sensors to move or raise their range of connectivity, we propose the following algorithm to allow the self organization of the network, especially around the articulation points discussed above (AP). Our approach is hybrid : content centralized and distributed algorithms . The mechanism for articulation points detection is lunched by the base station, but the self-organisation is lunched by each articulation point. Figure 5 shows this hybrid approach.

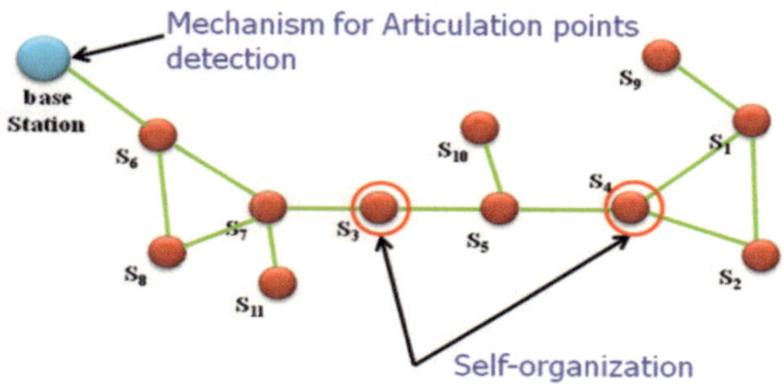

Fig. 5. Hybrid approach for monitoring connectivity in WSN.

Algorithm 2: Self-organization: the principle
Input: G (V, E), with the set of articulation points (AP) previously detected
Output: G (V, E), with a minimum set of articulation points so that G will be at least biconnected.

1. For any articulation point (AP) do

- If there is a neighbour redundant of (AP) then turn on and go to the (AP) following (step 1).
- Else discover the neighbours of (AP) at one hop,
    - If neighbours have redundant nodes, select at least one node with the greatest energy capacity, and move it to the coordinates (x, y) of the (AP) or increase its communication range; go to step 1.
    - Else " no solution at one hop of (AP) "; go to step 1.

End For

This algorithm is demonstrated in the example shown in Figure 6, this algorithm applied to the network to auto-organize and increase connectivity around articulation points.

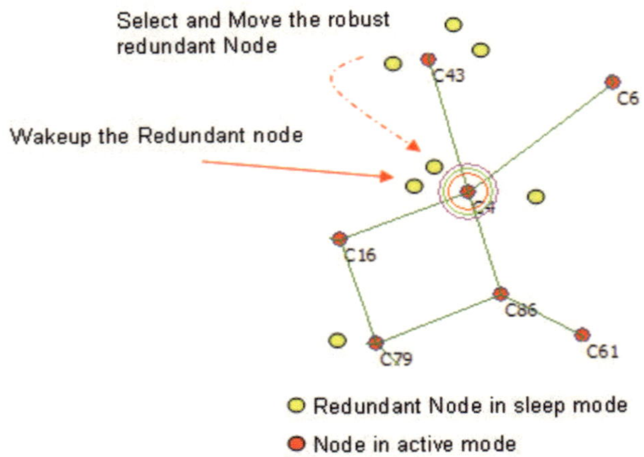

Fig. 6. Self-organization around articulation point

3.6. Simulation and Results

We have tested and validated our algorithm using a simulator implemented in C++, which operates in discrete time. One hundred sensor nodes are distributed randomly on a surface without obstacles. Adjacent nodes at a distance Rc can communicate to form a unit disk graph. The result in figure 7 shows the detection of articulation points (the points surrounded by circles).

A self-organization of the network around the articulation points can increase the degree of network connectivity, the disappearance of the articulation points and finally a fault tolerant network.

We have also simulated the detection of certain targets deployed on the same surface (see figure 10). Consequently any event distant to a sensor with radius r_s will be captured. Figures 11 and 12 give us an idea of the strength of ties between coverage targets. As can be seen in figure 9, every target is covered by at least 2 sensors, ensuring a fault tolerant network. In other words even if some sensors fail there are always other sensors able to provide coverage.

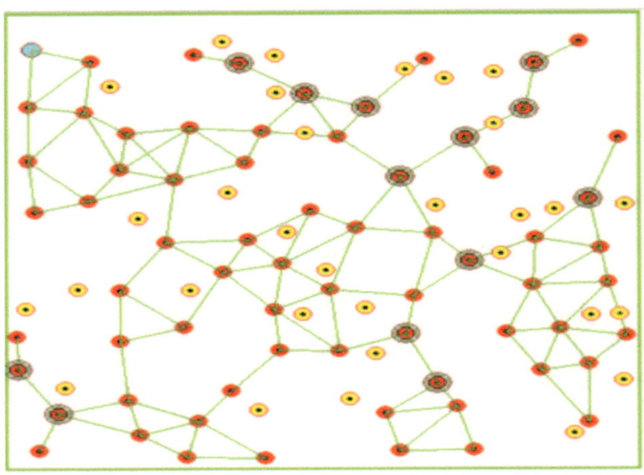

Fig. 7. Articulation point detection

After launch of self-organization algorithm, in first iteration some of articulation points are disparate by wake-up or move redundant nodes near articulation points. Following screenshot illustrate this.,

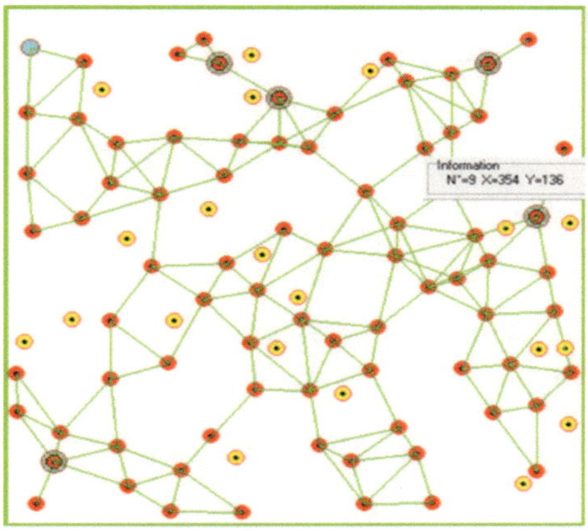

8.a

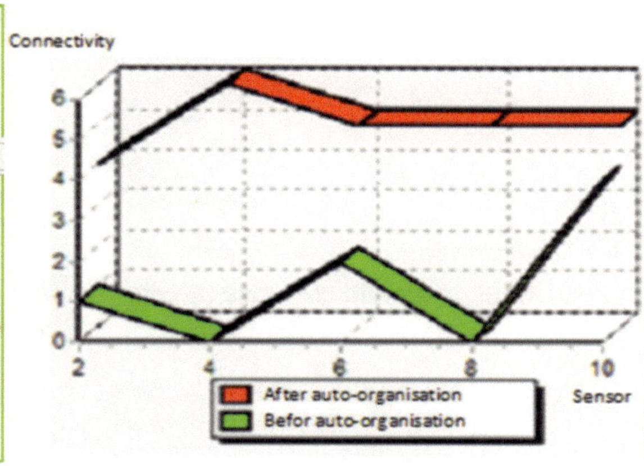

8.b

Fig. 8.a, 8.b. Self-organization after the first iteration.

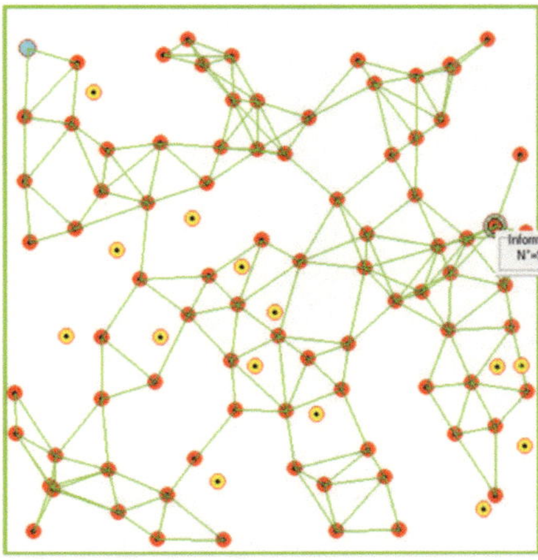

9.a

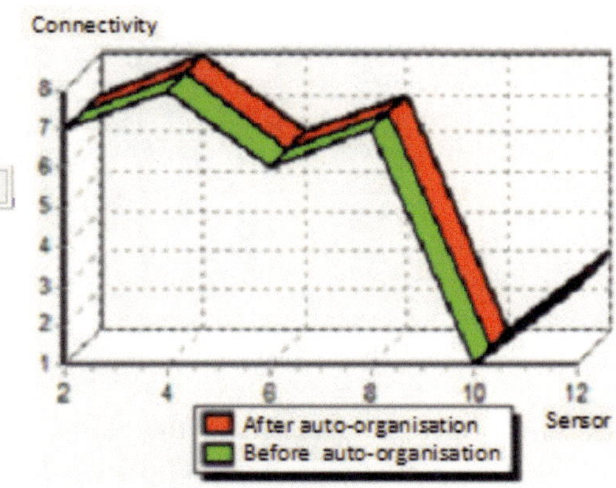

9.b

Fig. 9.a, 9.b Self-organization after the last iteration

Per example: the node number 26 which was the articulation point has become a normal node after self-organization. The degree of connectivity around the point of articulation is increased, as shown in the figure 8; the green graph shows the connectivity before self-organization, the red graph shows the connectivity after self-organization. For next's iterations of self- organization we see the same for nodes number 9, 30, 31, 11 and 72. In the last iteration of self-organization, on notice that it remains one articulation point unresolved. As shown in figure 9, graphs of the degree of connectivity are the same.

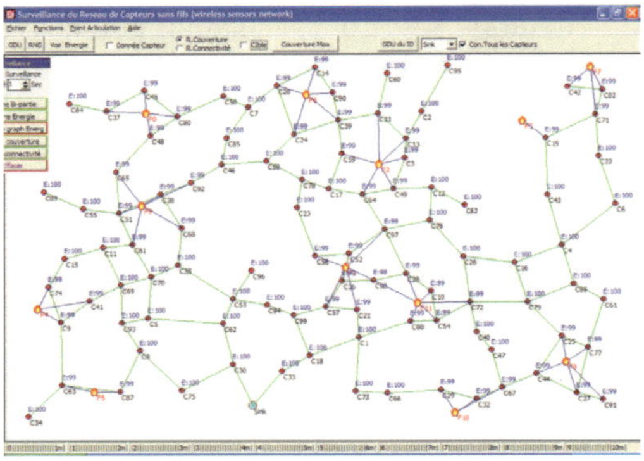

Fig. 10. Deployment and coverage targets

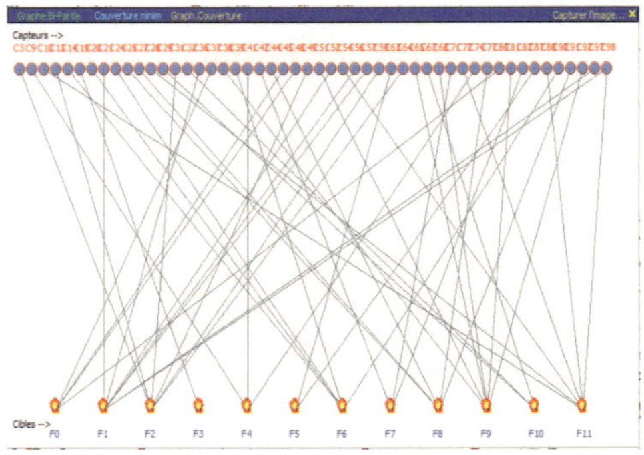

Fig. 11. Bipartite graph showing the maximum coverage of the various targets

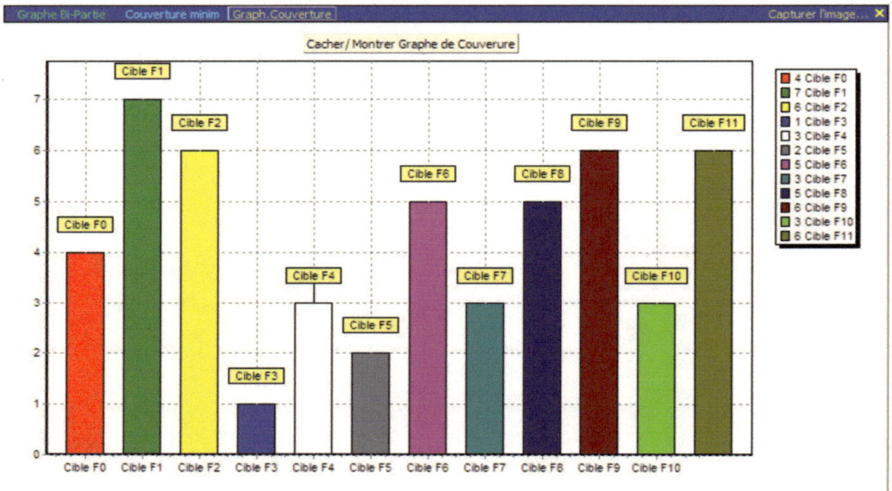

Fig. 12. Statistical state of the cover of each target

Figure 12 shows us results for the statistical state of target coverage. This highlights the heavily covered targets, indicating increased fault tolerant areas. Those that are poorly covered may require a reconfiguration or self-organization of the network.

4. Security in Wireless Sensor Networks Using Distributed Monitoring Mechanisms

4.1 Introduction

Wireless sensors networks are becoming increasingly interesting in recent years. These networks typically consist of hundreds or thousands of small sensors with limited resources (battery, bandwidth, processor, memory), to monitor some phenomena. The characteristics of such networks, such as fault tolerance, self-organization, the detection of high fidelity, low cost and rapid deployment have created many new applications of these networks, such as monitoring of wildlife, disaster response, military surveillance, industrial quality control and buildings intelligent, etc. [1]. However, the open nature of wireless communication, lack of infrastructure deployment in hostile environments where they are highly exposed to physical vandalism and cooperation for the transmission of data, makes them very vulnerable to a wide type of attacks [53,54,55], including attacks against control traffic data as the Wormhole attack, the Rushing attack, the Sybil attack, Sinkhole attack, and the Hello flood attack.

An attack against data traffic includes Blackhole attack and the Selective forwarding attacks. Conventional techniques security, such as antivirus, IDS, encryption mechanisms, can not only prevent these attacks because many of them, such as Wormhole and rushing attacks can be launched without violating of any cryptographic mechanisms. To address these attacks, many researchers have used the concept of centralized monitoring, where a control center is responsible for monitoring all network nodes (such as base station, the central controller or manager, and sink) [56]. Other researchers have used a decentralized approach to monitor network nodes as fault detection through the coordination of neighboring [57,58,59]. The use of watchdog to detect misbehaviour neighbors [60], nodes guards are normal nodes in the network that perform basic operations such as the capture event, in addition to monitoring. Other research using the local monitoring between neighbouring nodes [61, 62]. In local monitoring, nodes monitor some traffic entering and leaving their neighbours for the detection of malicious behaviour.

From this work, nobody thought to use monitoring based on cluster architecture where each cluster member node performs a periodic calculation of certain metric necessary for making local decision at level cluster head. At each change of its state, member nodes sends its report to the cluster Head with a synchronization mechanism between nodes to minimize interference and reduce the number of packets delayed in transmission. Then a mechanism for optimizing the selection of a Cluster Head, who it is responsible for taken a local decision and monitoring the cluster members nodes. Finally, the base station aggregates the results received from different clusters Head and begins a global and centralized monitoring of network status, which can detect abnormalities that require global information network, reducing the flow of communication and the number false alarms. The main challenge of our works is to have a distributed monitoring for security reasons based on a clustered architecture, using a set of rules for diagnosing the status of sensors. The remainder of the sub chapter is organized as follows: Section 4.2 summarizes security in sensor networks, with detailed study of some attacks; the details of our approach are described in Section 4.3. In Section 4.4, we present our simulation results. Finally, we conclude this paper in Section 5.

4.2 Security in Wireless Sensor Networks

A. Attacks on sensor networks

For securing the Wireless Sensor Networks, it is necessary to address the attacks on WSN. This section lists and gives brief discussion about the major attacks against Wireless Sensor Network. Basically attacks are classified as active attacks and passive attacks [63,64,65,66,67,68,69]:

1) Passive Attacks

The monitoring and listening of the communication channel by unauthorized attackers are known as passive attack, such as attacks against privacy.

2) Active Attacks

The unauthorized attackers monitors, listens to and modifies the data stream in the communication channel are known as active attack. The following attacks are active: Figure 1 shows the attacks classification on Wireless Sensor Networks.

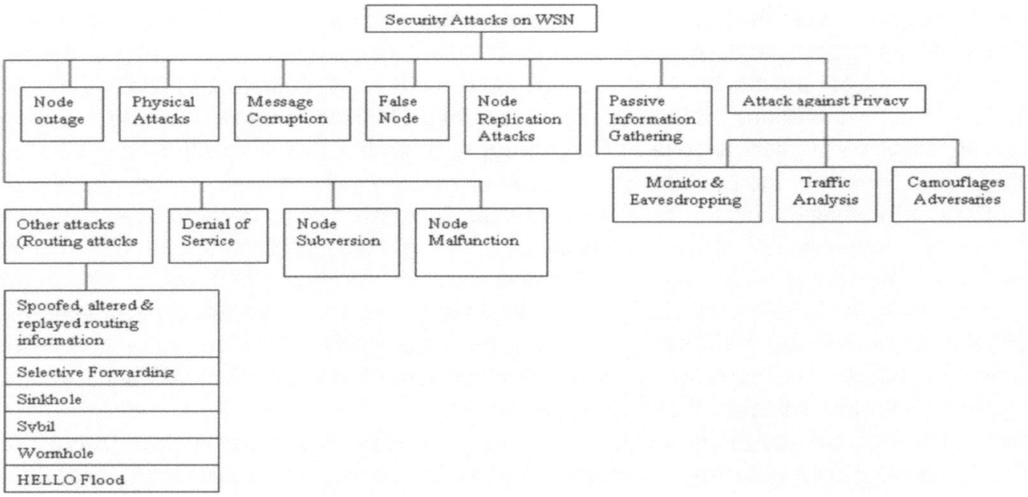

Fig. 1. Attacks classification on WSN.

B. Misbehavior in Wsn

Misbehavior in Wireless sensor network can be classified into two main categories, namely selfish nodes and malicious nodes [70]. These two types can cause real security threats in that they are the main reason for two of the main attacks that can damage MANET, and can be difficult to detect. Selfish node is the one behind the drop packets attack, where as malicious node is the one causing the denial of service (DoS) attack.

C. Anomaly Detection

Traditionally, intrusion detection techniques are classified into two broad categories: misuse detection and anomaly detection. Misuse detection works by searching for the traces or patterns of well-known attacks. Clearly, only known attacks that leave characteristic traces can be detected that way. Anomaly detection, on the other hand, uses a model of normal user or system behavior and flags significant deviations from this model as potentially malicious. This model of normal user or system behavior is commonly known as the user or system profile. A strength of anomaly detection is its ability to detect previously unknown attacks.

4.3. Our Approach

Network monitoring is an interesting approach that allows collecting the required information in order to analyze the behavior of the network. Monitoring in wireless sensor networks can be local with respect to a node or global with respect to the network. In sensor networks, local monitoring is not sufficient to detect some types of errors and security anomalies. For this reason we adopt in this paper a hybrid approach, the global monitoring approach based on a distributed monitoring. In general the existing failure detection approaches in WSNs is classified into two types: centralized and distributed approach. In our case, the observers are the network nodes themselves. They perform a collaborative observation action. At first each node collects its security metrics (local traffic trace, resources consumption) and sends it to the local observer. We assume here that all the nodes have the collector and analyzer program running on their systems.

A. System Architecture

An example of our approach is illustrated in Figure 2. It consists of several coordinating components, namely: a large number of sensing nodes, several monitoring nodes, and base station.

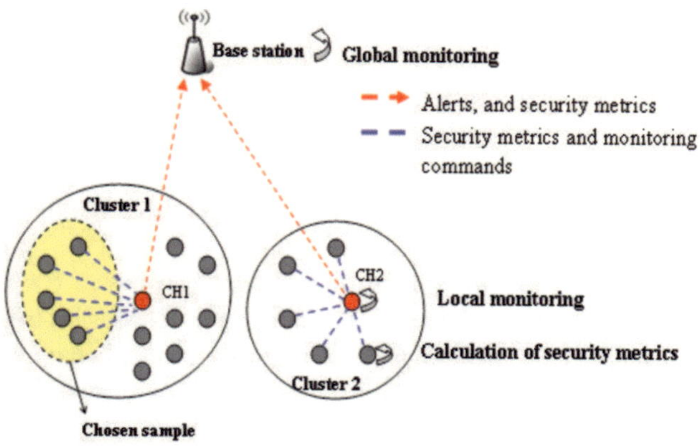

Fig. 2. Distributed Monitoring

Sensing nodes: Sensing nodes are small, resource constrained sensor nodes such as the Mica mote. They organize themselves into a network, sense and relay real-life measurements toward the networks.

Monitoring Nodes: Monitoring nodes have processing and communication capabilities. Each monitoring node covers a portion of the network topology (a cluster), where the sensor network will organize into a cluster formation, with each cluster head at a monitor.

Base Station: The main role of the base station is to make filtering and correlation of alerts and information sent by different monitors (the cluster-heads). Then thereafter it performs a more overall monitoring to detect hidden abnormalities that require an overview of information from the network. All these different entities are indispensable to our distributed Monitoring system. The system complexity and resource requirements increase progressively from sensing nodes, monitoring nodes, to base station.

B. Selection of a sample

The target population to be monitored is usually too large and for reasons of cost, and time, it is practically impossible to analyse all the member nodes in a population of a cluster. In general, we use the formula (3) for compute the number of nodes in the chosen sample [71]:

$$n = \frac{n_0}{1 + \frac{n_0}{N}} \tag{3}$$

Where :

- N is the size of population
- n is the size of chosen sample
- n0 a fixed value, n0 = 385

n = 385 / (1+385/N) to find the size needed (so the margin of error in estimating the proportion is less than 5% and, for a confidence level of 95%). The objective is to construct a sample so that observations can be generalized to the entire population. It is necessary that the sample has the same characteristics as the target population. In other words, it is representative. If this is not the case, the sample is biased.
The attribute state-sc(S_J), indicates the participation of sensor node S_J in the sample or not. For each sensor node $S_J \in$ cluster i, we have:

$$State - sc(S_J) = \begin{cases} 1 \; if \; S_J \; participate \; in \; the \; sample \\ \\ 0 \;\; Otherwise \end{cases} \tag{4}$$

Example: if the number of member node N in the cluster i is 385, in this case the chosen sample n it equal to 192. For each period of monitoring the cluster- head can monitor 192 nodes.

C. Calculation of security metrics

This operation is done at each member node of a chosen sample in the cluster. The node performs after every epoch of time a calculation on its metrics of security, to assess their health status, such a level of energy consumption, level of memory usage, behavior of the nodes, etc. Figure 3 shows the process of metrics computing in member nodes. This node manages functions such as capturing, sending and receiving data messages, in addition to the functions of calculation of a security metrics like: the number of incoming and outgoing packet in a time interval, number of dropped packets, etc. Among the population of member nodes in the cluster, one representative sample of the population is chosen randomly. This sample will be analyzed in the period of ongoing monitoring. Each node in a chosen sample performs a calculation of his status. Once a difference in status between two time intervals is detected a calculated indicators values of security will be sent to the cluster Head for analyses.

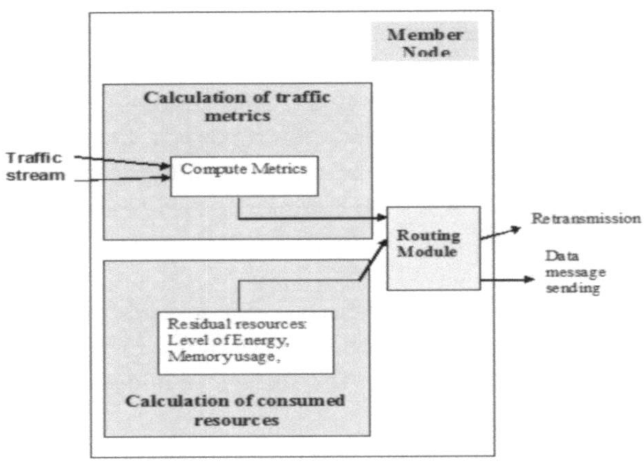

Fig. 3. Calculation of security metrics in each member node of a chosen sample

When sensor data are transmitted to the cluster head, nodes do not transmit sensor data if their data are not changed since last reported. For example, at the current round, sensor member S1 does not transmit its data to the cluster head because its data equal the collected data at the next round.

D. Local Monitoring in Cluster Head

The Cluster Head in figure 4, manages only the functions: self-monitoring of its state, local monitoring of the results obtained from the member nodes of its cluster, the reception and the emission of the messages, but does not manage, the function of capture of event. Cluster head is good at making decision because it has both network-level information and host-based information of all its nodes. The Cluster Head aggregates the results and send them to the base station for more global analysis; this strategy reduces the number of alerts gone up towards the base station.

- Cluster head can monitor its nodes thus to save their resources, or it can collect monitoring report from nodes and do some additional work.
- Cluster head is good at making decision because it has both network-level information and host-based information of all its nodes.

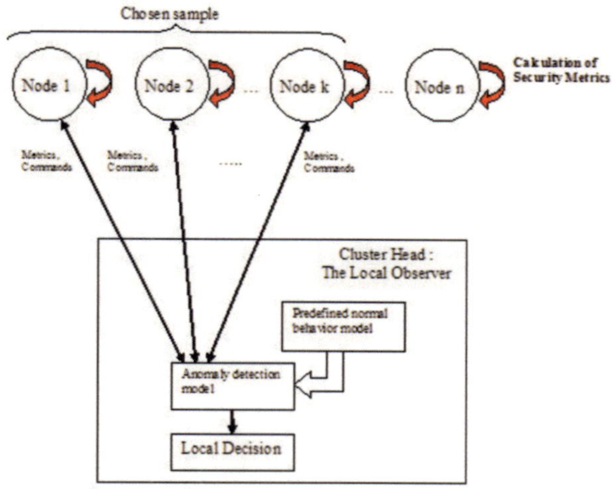

Fig. 4. Local Monitoring

E. Global Monitoring

The global observer receives the local traces collected by the local observers (the clusters-head) in order to analyze them. The first step toward performing this analysis is to correlate the traces and order them chronologically. In the network, all the nodes run with the same clock value allowing thus to perform the trace correlation.

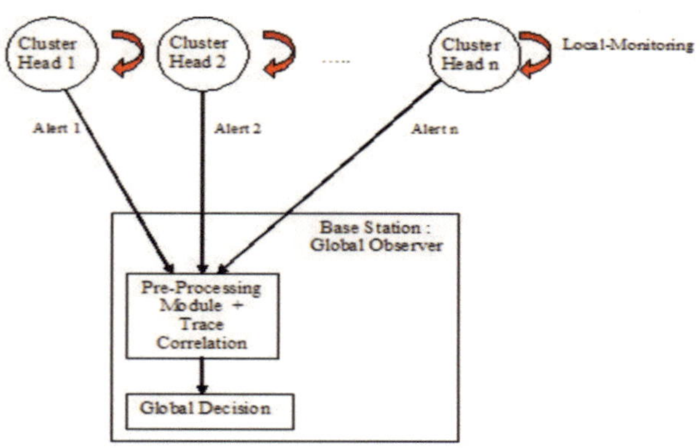

Fig. 5. Global Monitoring in Base Station

In First, the global observer collected alerts, have to be analyzed using a pre-processing module that performs the following tasks:

- Filtering the collected alerts keeping only the relevant information.
- Alert correlation and the construction of a unique global trace file.

F. Distributed Monitoring based clustering architecture

Clustering facilitates the distribution of control over the network. Clustering saves energy and reduces network contention by enabling locality of communication.

In our case, sensor networks are divided into cluster. The reorganization of the cluster will be made for a security reason, where each cluster Head monitors the member nodes of their cluster, which also facilitates the risen of alerts and reduces latency problems. These clusters are generated automatically after an epoch of clusters formation. Every cluster is assigned a cluster head CH, by election with some metrics. We opted for an election of cluster head according a new metrics based on multiple criteria decision approach to decision support for the selection of CHs, the criteria are: the criterion of density (the degree of connectivity of each node), the criterion of energy (the level of residual energy in each node), the distance between nodes in the cluster, the behavior level of each node and the index of mobility. Each node calculates its metrics locally, then evaluates a function of weight according to these metric (each node is limited to the closest neighbors), and diffuses the value of this function to its neighbors. Cluster Head of each cluster is then elected of these results. Three constraints which are the fact, that two CH cannot be coast at coast, and that if a node belongs to two clusters, it must belong with the nearest cluster (by using a parameter of distances), finally if a node is completely isolated it becomes automatically a cluster Head.

1) Clustering algorithm metric

We describe in this section, the metric used in our algorithm for clustering formation, then we present its election protocol and update policy. The updating policy is locally called after mobility or -adding new nodes in the network. To decide how much a node is suited for being a cluster head to offer security services, we take into consideration the following characteristics:

The node behaviour level B(i,t): Nodes with a behaviour level less than a threshold behaviour-Min will not be accepted as candidate for being cluster heads even if they have other interesting characteristics as high energy, high degree of connectivity or low mobility. First of all each nodes are assigned a same static behaviour level B=1. However, this level can be decreased by the anomaly detection algorithm if a nodes are misbehaving B=B – rate. Classification of the behaviour value takes the following values:

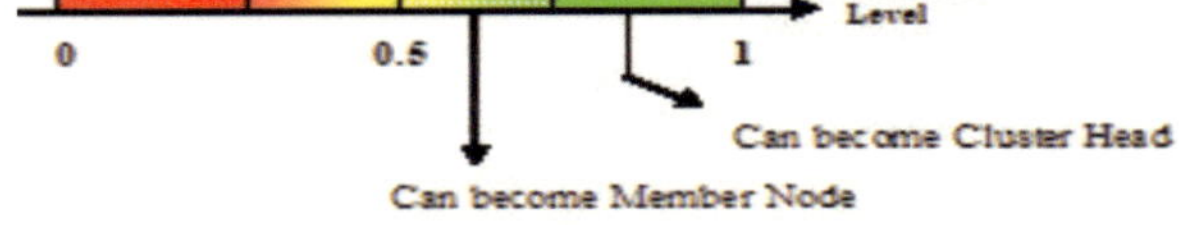

Fig. 6. Behavior Level, $B \in [0,1]$

Classification of the behaviour value takes the following values:

$$\begin{cases} Normal\ Node: 0.8 \le B \le 1 \\ Abnormal\ Node \text{(but not malicious)}: 0.5 \le B < 0.8 \\ Suspect\ Node: 0.3 \le B < 0.5 \\ Malicious\ Node: 0 \le B < 0.3 \end{cases} \tag{5}$$

The node mobility M(i,t): We aim to have stable clusters. So, we should elect nodes with low relative mobility as cluster heads. To characterize the instantaneous nodal mobility, we will use a simple heuristic mechanism [71,72] where each node i estimates its relative mobility index Mi by implementing the following procedure:
Compute the running average of the speed for every node i till current time T. This gives a measure of mobility and is denoted by M_i, as:

$$M_i = \frac{1}{T}\sum_{t=1}^{T}\sqrt{(x_t - x_{t-1})^2 + (y_t - y_{t-1})^2} \tag{6}$$

Where (x_t, y_t) and (x_{t-1}, y_{t-1}) are the coordinates of the node v at time t and (t -1) , respectively.

The distance to neighbors D(i,t): It is better to elect the node with the nearest members as a cluster head [73,74].
For every node i, compute the sum of the distances, D_i, with all its neighbors j , as :

$$D_i = \sum_{j \in N(i)} \{dist(i, j)\} \tag{7}$$

The node remaining energy E(i,t): We should elect nodes with high remaining battery power as cluster heads. The radio spends $E_{Tx\text{-}elec} = E_{Rx\text{-}elec} = E_{elec}$ energy to run receiver and transmitter electronics. Therefore the transmission cost to transfer k-bit message to a distance d is given by the equation (8) [75]:

$$E_{Tx}(k,d) = \{kE_{elec} + kE_{amp}d^2 \tag{8}$$

Where E_{amp} is a required amplifier energy. Similarly, the receiving cost can be given by equation (9) :

$$E_{Rx\,elec}(k) = kE \tag{9}$$

The node connectivity degree C(i,t):
N(i) is the neighbors of node i , defined as [52] :

$$N[i] = \bigcup_{j \in V,\, j \neq i} \{j \mid dist(i, j) < tx_{range}\} \tag{10}$$

Find the neighbors of each node i which defines its degree d_i as :

$$C_i = |N(i)| = \left| \sum_{j \in V, j \neq i} \{dist(i, j) < tx_{range}\} \right| \quad (11)$$

We should elect nodes with very high connectivity as cluster heads.
Each node S_i computes its **weight P_i** according to the method of weighted sum decision model, given by equation (12) :

$$P_i = w_1 * B_i + w_2 * Er_i + w_3 * M_i + w_4 * C_i + w_5 * D_i \quad (12)$$

where w_1, w_2, w_3,w_4,w_5 are the weighing factors for the corresponding system parameters, such that ($w_1+w_2+w_3+w_4+w_5=10$), and since our goal is to monitor sensor we taken a high coefficients for the behavior B_i and the remaining energy Er_i, as follows: $w_1=4$, $w_2=3$, $w_3=1$, $w_4=1$, $w_5=1$.

2) Node Status

A node in wireless sensor network can be in one of the 3 possible states: MEMBER (ME), HEAD (CH), Monitor Node or Guard node (MO). Initially, every node is in ME state. It starts election and may become CH node if it does not have link to any CH node, otherwise it still a member ME.

3) Proposed Methodology

Our goal is to detect malicious activities in the network caused by the attacks and the failure of nodes. We will offer primarily an organization of cluster network, where the cluster- head of each cluster is responsible for monitoring the member nodes of its cluster. Subsequently we propose a system for detecting anomalies based on a distributed approach.

4.4 Simulation and Results

In this section, we present the simulation model and results of our work.

4.4.1 Simulation model

We developed a wireless sensor network simulator to create an environment to evaluate our work. It is a discrete event simulator written in C++. A network generator was built, which generates networks comprised of normal nodes plus malicious node, all located in an square field. Each node has randomized x and y coordinates. No two different nodes share the same coordinates. In our simulation, the sensor nodes are randomly distributed in a 880mx360m square field, the communication range is 150m. The scenario simulation consists of two steps: the first is for the formation of cluster, the second is to monitor the network by different cluster head and the detection of the abnormal behaviour. For the simulation of abnormal behaviour in the network, we generated a number of malicious nodes that their state will move from a normal node with green colour to a abnormal node with yellow colour, to a suspicious node of red colour , and lastly, a malicious node with black colour. All the states of member nodes are detected by their cluster head. Malicious cluster head are detected by the base station.

4.4.2 Results

In the following, we present and discuss the simulation results.

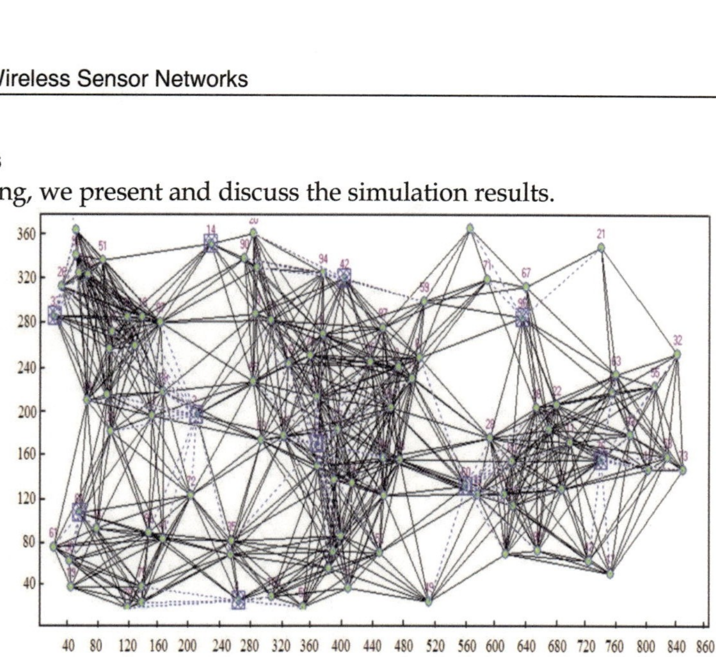

Fig. 7. Random deployment and graph connectivity of 100 nodes in square field.

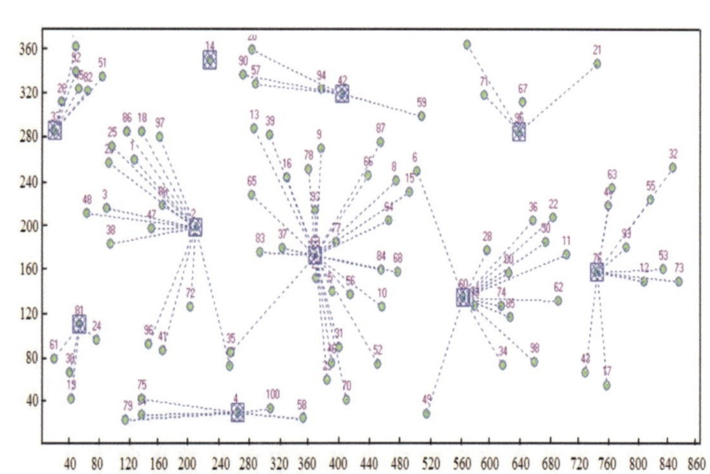

Fig. 8. Network after Clustering Formation

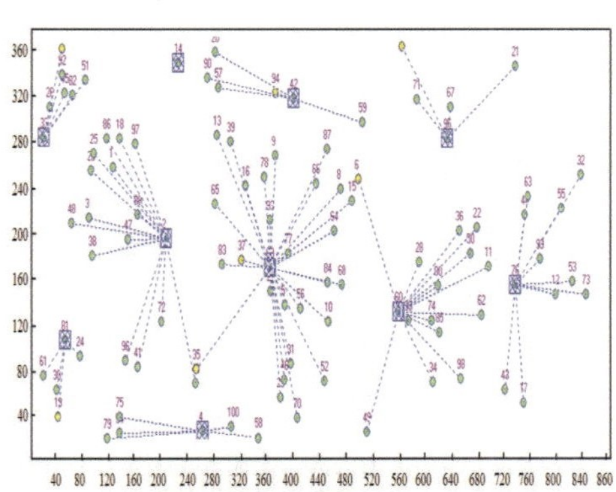

Fig. 9. Sensors with yellow colour are abnormal but not malicious

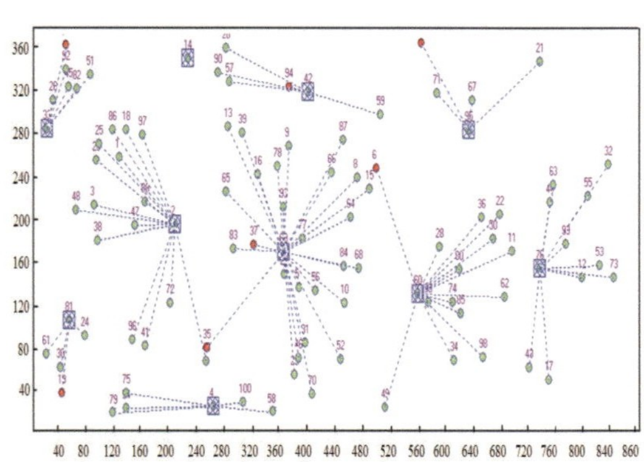

Fig. 10. the red sensors have a suspect behaviour

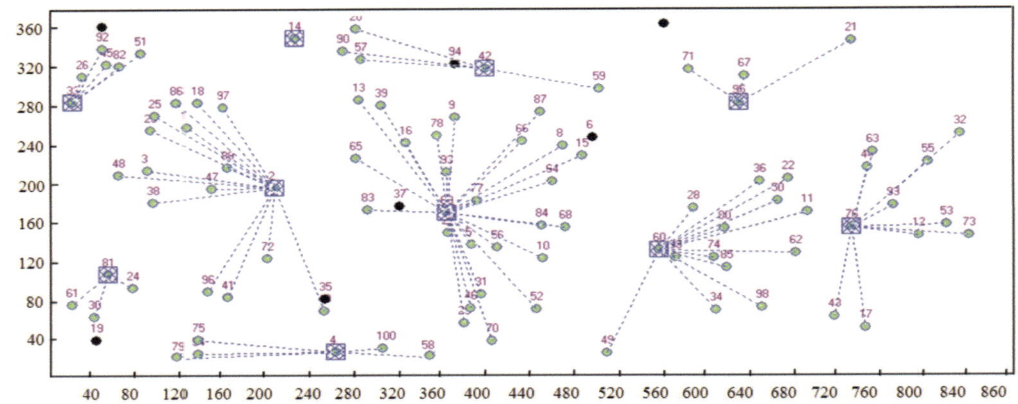

Fig. 11. The sensors with black color are compromised and have an malicious behavior

The black sensors will be placed in a black list and will be disconnected from the network, as shown in Figure 11.

5. Conclusion

In this chapter we started with the presentation of the overview of the mechanisms of monitoring a wireless sensor networks, for the following reasons: topology control (connectivity and the coverage), and the security in wireless sensor networks. Then we have developed a new monitoring mechanism to guarantee strong connectivity in wireless sensors networks, this mechanism is based on the distributed algorithms. The mechanism monitors sensor connectivity and at any time is able to detect the critical nodes that represent articulation points. Such articulation points are liable to cause portions of the network to become disconnected and we have therefore also developed a mechanism for self-organization to increase the degree of connectivity in their vicinity, by increasing fault tolerance. Since connectivity is closely related to the coverage of targets, we have also developed a way to monitor the robustness of the coverage between fixed targets and sensor nodes. The main advantage of our approach is the ability to anticipate disconnections before they occur. We are also able to reduce the number of monitoring node and assume mechanisms for fault tolerance by auto organization of nodes to increase connectivity. Finally, we have demonstrated the effectiveness of our approach and algorithms with satisfactory results obtained through simulation.

After that we have presented our second contribution for the security of a wireless sensor networks based on the distributed monitoring mechanisms. We have presented a decentralized approach to monitor the status and behavior in a wireless sensor network. For this we have developed a completed distributed monitoring mechanism for securing wireless sensor networks. Based on a flexible weight clustering algorithm, a number of parameters of nodes were taken into consideration for assigning weight to a node and election cluster-head. The proposed algorithm chooses the robust cluster-heads who is the responsibility to monitor a chosen sample of nodes in their cluster, and maintains clusters locally. A second algorithm analyzes and detects a specific misbehavior in wireless sensor networks. This algorithm insures the update of a behavior-level metric and isolates the misbehaving node. The advantage of our approach is the minimization of the communication between the monitor's nodes and the normal nodes.

6. References

[1] I. F. Akyildiz, W. Su, Y. Sankarasubramaniam, E. Cyirci, "Wireless Sensor Networks: A Survey.", Computer Networks, vol. 38, no.4, pp. 393-422, 2002.

[2] L. Kleinrock and J. Silvester. "Optimum transmission radio for packet radio networks or why six is a magic number. In National Telecommunications Conference, Birmingham, Alabama, pages 4.3.2-4.3.5, December 1978.

[3] A. Cerpa and D. Estrin, "Ascent: Adaptive self-configuring sensor networks topologies" IEEE Transactions on Mobile Computing, vol. 3, no. 3, pp. 272-285, 2004.

[4] N. Li and J. C. Hou, "Improving connectivity of wireless ad hoc networks", in MOBIQUITOUS '05: Proceedings of the The Second Annual International Conference on Mobile and Ubiquitous Systems: Networking and Services. Washington, DC, USA: IEEE Computer Society, 2005, pp. 314-324.

[5] M. Dunbabin, P. Corke, I. Vasilescu, and D. Rus, "Data muling over underwater wireless sensor networks using an autonomous underwater vehicle.", in IEEE International Conference on Robotics and Automation (ICRA), 2006, May 15- 19 2006, pp. 2091-2098.

[6] K. Benahmed, H. Haffaf , M. Merabti, D. Llewellyn-Jones, "Monitoring Connectivity in Wireless Sensor Networks ", International Journal of Future Generation Communication and Networking, Vol. 2, No. 2, 2009.

[7] G. Yang, L.-J. Chen, T. Sun, B. Zhou, and M. Gerla, "Ad-hoc storage overlay system (asos): A delay-tolerant approach in manets.", in Proceeding of the IEEE MASS, 2006, pp. 296-305.

[8] N. Rao, W. Qishi, S. Iyengar, and A. Manickam, "Connectivity-through-time protocols for dynamic wireless networks to support mobile robot teams.", in IEEE International Conference on Robotics and Automation (ICRA), 2003, vol. 2, Sept 14-19 2003, pp. 1653-1658.

[9] D. Ganesan, R. Govindan, S. Shenker, and D. Estrin Highly-Resilient, "Energy-Efficient Multipath Routing in Wireless Sensor Networks.", Mobile Computing and Communications Review, 1(2), 1997.

[10] D. Spanos and R. Murray, "Motion planning with wireless network constraints.", in Proceedings of the 2005 American Control Conference, 2005, pp. 87-92.

[11] D. Desovski, Y. Liu, and B. Cukic. "Linear randomized voting algorithm for fault tolerant sensor fusion and the corresponding reliability model.", In IEEE International Symposium on Systems Engineering, pages 153-162, October 2005.

[12] A. Boukerche, "Handbook of Algorithms and Protocols for Wireless and Mobile Networks", Chapman CRC/Hall, 2005.

[13] N. Li and J. C. Hou. "FLSS: A Fault-Tolerant Topology Control Algorithm for Wireless Networks.", In Proceedings of the 10th Annual International Conference on Mobile Computing and Networking, pages 275-286, 2004.

[14] J. L. Bredin, E. D. Demaine, M. Hajiaghayi, and D. Rus. "Deploying Sensor Networks with Guaranteed Capacity and Fault Tolerance.", In Proceedings of the 6th ACM international symposium on Mobile ad hoc networking and computing, pages 309-319, 2005.

[15] Bahramgiri, M., Hajiaghayi, M., and Mirrokni, "Fault-tolerant and 3-dimensional distributed topology control algorithms in wireless multi-hop networks.", 2002.

[16] N. Li, J. Hou, and L. Sha. "Design and analysis of an mst-based topology control algorithm." , In Proceedings of the IEEE INFOCOM, 2003.

[17] Xiang-Yang Li, Peng-Jun Wan, Yu Wang, and Chih-Wei Yi. "Fault tolerant deployment and topology control in wireless networks.", In Proceedings of the 4th ACM international symposium on Mobile ad hoc networking & computing (MobiHoc), pages 117.128, 2003.

[18] Michaël Hauspie , "Contributions à l'étude des gestionnaires de services distribués dans les réseaux ad hoc ", Thèse de doctorat, Université des Sciences et Technologies de Lille, 2005.

[19] Bruno Courcelle, " Introduction à la théorie des graphes: Définitions, applications et techniques de preuves ", Université Bordeaux 1, LaBRI (CNRS UMR 5800), 20 Avril, 2004.

[20] R. Tarjan., "Depth First Search and linear graph algorithms.", SIAM Journal of Computing, 1:146_160, 1972.

[21] Wies law Zielonka , "Algorithmique ", LIAFA, Université Denis Diderot, Septembre 2006.

[22] K. Chakrabarty, S. S. Iyengar, H. Qi, E. Cho, "Grid coverage for surveillance and target location in distributed sensor networks," IEEE Transactions on Computers, 51(12):1448-1453, December 2002.

[23]S. Meguerdichian and M. Potkonjak. "Low Power 0/1 Coverage and Scheduling Techniques in Sensor Networks." UCLA Technical Reports 030001. January 2003.

[24]S. Meguerdichian, F. Koushanfar, M. Potkonjak, and M. Srivastava, "Coverage Problems in Wireless Ad-Hoc Sensor Networks." IEEE Infocom 2001, Vol 3, pp. 1380-1387, April 2001.

[25] S. Meguerdichian, F. Koushanfar, G. Qu, and M. Potkonjak, "Exposure in Wireless Ad Hoc Sensor Networks." Procs. of 7th Annual International Conference on Mobile Computing and Networking (MobiCom'01), pp. 139-150, July 2001.

[26] T. Couqueur, V. Phipatanasuphorn, P. Ramanathan and K. K. Saluja, "Sensor Deployment Strategy for Target Detection," Proceeding of The First ACM International Workshop on Wireless Sensor Networks and Applications, Sep. 2002.

[27] D. Tian and N.D. Georganas, "A Coverage-preserved Node Scheduling scheme for Large Wireless Sensor Networks," Proceedings of First International Workshop on Wireless Sensor Networks and Applications (WSNA'02), Atlanta, USA, September 2002.

[28] A. Cerpa and D. Estrin, "ASCENT: Adaptive Self-Configuring Sensor Networks Topologies," International Annual Joint Conference of the IEEE Computer and Communications Societies (INFOCOM 2002), New York, NY, USA, June 23-27 2002.

[29] B. Chen, K. Jamieson, H. Balakrishnan, and R. Morris, "Span: An Energy-Efficient Coordination Algorithm for Topology Maintenance in Ad Hoc Wireless Networks," ACM/IEEE International Conference on Mobile Computing and Networking (MobiCom 2001), Rome, Italy, July 16-21, 2001.
[30] Y. Xu, J. Heidemann, and D. Estrin, "Adaptive Energy-Conserving Routing for Multihop Ad Hoc Networks," Research Report 527, USC/Information Sciences Institute, October 2000.
[31] Y. Xu, J. Heidemann, and D. Estrin, "Geography-informed Energy Conservation for Ad Hoc Routing," ACM/IEEE International Conference on Mobile Computing and Networking (MobiCom 2001), Rome, Italy, July 16-21, 2001.
[32] F. Ye, G. Zhong, S. Lu, and L. Zhang, "PEAS: A Robust Energy Conserving Protocol for Long-lived Sensor Networks". The 23rd International Conference on Distributed Computing Systems (ICDCS'03), May 2003.
[33] A. Perrig, "SPINS: security protocols for sensor networks," In Proc. of ACM MobiCom, 2001.
[34] S. Ganeriwal and M. B. Srivastava, "Reputation-based framework for high integrity sensor networks," In Proc. Of ACM SASN, 2004.
[35] I. Khalil, S. Bagchi, and C. Nina-Rotaru, "DICAS: detection, diagnosis and isolation of control attacks in sensor networks," In Proc. of IEEE SecureComm, 2005.
[36] S.-B. Lee and Y.-H. Choi, "A resilient packet-forwarding scheme against maliciously packet-dropping nodes in sensor networks," In Proc. of ACM SASN, 2006.
[37] I. Khalil, S. Bagchi, and N. Shroff, "LITEWORP: a lightweight countermeasure for the wormhole attack in multihop wireless networks," In Proc. of IEEE/IFIP DSN, 2005.
[38] S. Ganeriwal and M. B. Srivastava, "Reputation-based framework for high integrity sensor networks," In Proc. Of ACM SASN, 2004.
[39] [S. Buchegger and J.-Y. L. Boudec, "Performance analysis of the CONFIDANT protocol: cooperation of nodes fairness in distributed ad-hoc networks," In Proc. of ACM MobiHoc, 2002.
[40] P. Michiardi and R. Molva, "CORE: a collaborativereputation mechanism to enforce node cooperation in mobile ad hoc networks," In Proc. of the IFIP Sixth Joint Working Conference on Communications and Multimedia Security, 2002
[41] K. Ioannis, T. Dimitriou, and F. C. Freiling, "Towards intrusion detection in wireless sensor networks," In Proc. of the 13th European Wireless Conference, 2007.
[42] Y. Huang and W. Lee, "A cooperative intrusion detection system for ad hoc networks," In Proc. of ACM SASN, 2003.
[43] I. Khalil, S. Bagchi, and N. B. Shroff, "SLAM: sleep-wake aware local monitoring in sensor networks," In Proc. Of IEEE/IFIP DSN, 2007.
[44] C. Hsin and M. Liu, "Self-monitoring of wireless sensor networks," Elsevier Computer Communications, vol. 29, pp.462-476, 2006.
[45] T. H. Hail, E.-N. Huh, and M. Jo,"A lightweight intrusion detection framework for wireless sensor networks", Wirel. Commun. Mob. Comput. (2009)
[46] Q. Wang, T. Zhang , "Detecting Anomaly Node Behavior in Wireless Sensor Networks", 21st International Conference on Advanced Information Networking and Applications Workshops, 2007.

[47] K. Ramachandran, E. M. Belding-Royer, and K. C. Almeroth. DAMON: A Distributed Architecture for Monitoring Multi-hop Mobile Networks. In Proceedings of the 1st IEEE International Conference on Sensor and Ad hoc Communications and Networks (SECON), October 2004.
[48] J. Zhao, R. Govindan, and D. Estrin. Residual energy scans for monitoring wireless sensor networks. In IEEE Wireless Communications and Networking Conference (WCNC), 2002.
[49] NIthya Ramanathan, Kevin Chang, Rahul Kapur, Lewis Girod, Eddie Kohler, Deborah Estrin. Sympathy for the Sensor Network Debugger. In 3rd Embedded networked sensor systems. 2005. San Diego, USA: ACM Press.
[50] Y. an Huang and W. Lee, A cooperative intrusion detection system for ad hoc networks, in Proc of the 1st ACM Workshop on Security of Ad hoc and Sensor Networks, 2003, pp. 135–147.
[51] S. Marti, T. J. Giuli, K. Lai, and M. Baker, Mitigating routing misbehavior in mobile ad hoc networks, in Mobile Computing and Networking, 2000, pp. 255–265.
[52] K. Benahmed, H. Haffaf, M. Merabti, D. Llewellyn-Jones, "Monitoring connectivity in Wireless Sensor Networks", IEEE Symposium on Computers and Communications (ISCC'09), Sousse, Tunisia, 5-8 July 2009.
[53] Tanya Roosta, Shiuhpyng Winston Shieh, S. Shankar Sastry. "Taxonomy of Security Attacks in Sensor Networks and Countermeasures ". The First IEEE International Conference on System Integration and Reliability Improvements, December, 2006.
[54] T.Kavitha, D.Sridharan, "Security Vulnerabilities In Wireless Sensor Networks: A Survey", Journal of Information Assurance and Security 5 (2010) 031-044
[55] Song Han, Elizabeth Chang, Li Gao and Tharam Dillon, "Taxonomy of Attacks on Wireless Sensor Networks", Proceedings of the First European Conference on Computer Network Defence School of Computing, University of Glamorgan, Wales, UK, 2005.
[56] M.Yu, H.Mokhtar, M.Merabti,"A Survey on Fault Management in Wireless Sensor Networks", School of Computing & Mathematical Science Liverpool John Moores University. UK, 2007.
[57] Chihfan Hsin, Mingyan Liu. A Distributed Monitoring Mechanism for Wireless Sensor Networks. in 3rd workshopo on Wireless Security. 2002: ACM Press.
[58] Jinran Chen, Shubha Kher, Arun Somani. Distributed Fault Detection of Wireless Sensor Networks. in DIWANS'06. 2006. Los Angeles, USA: ACM Pres.
[59] Anmol Sheth, Carl Hartung, Richard Han. A Decentralized Fault Diagnosis System for Wireless Sensor Networks. in 2nd Mobile Ad Hoc and Sensor Systems. 2005. Washington, USA.
[60] Sergio Marti, T.J.Giuli, Kevin Lai, Mary Baker. Mitigating Routing Misbehavior in Mobile Ad Hoc Networks. in 6th International Conference on Mobile Computing and Networking. 2000. Boston, Massachusetts, USA: ACM.
[61] Y. Huang and W. Lee, "A cooperative intrusion detection system for ad hoc networks," in Proceedings of the 1st ACM workshop on Security of ad hoc and sensor networks, pp. 135-147, 2003.

[62] A. Silva, M. Martins, B. Rocha, A. Loureiro, L. Ruiz, and H. Wong, "Decentralized intrusion detection in wireless sensor networks," in Proceedings of the 1st ACM international workshop on Quality of service & security in wireless and mobile networks, pp. 16-23, 2005.
[63] M. Saraogi, "security in wireless sensor networks" , University of Tennessee, 2005.
[64] J.P. Mäkelä, "Security in Wireless Sensor Networks", Oulu University of Applied Sciences, School of Engineering, Oulu, Finland, 2009.
[65] J. Rehana, "Security of Wireless Sensor Network" Helsinki University of Technology, Helsinki, Technical Report TKK-CSE-B5, 2009.
[66] I. Chatzigiannakis, "A Decentralized Intrusion Detection System for Increasing Security of Wireless Sensor Networks", University of Patras, Greece, 2007.
[67] C. Karlof, D. Wagner, "Secure routing in wireless sensor networks: Attacks and countermeasures". In Proceedings of the 1st IEEE International Workshop on Sensor Network Protocols and Applications (Anchorage, AK, May 11, 2003).
[68] Al-Sakib Khan Pathan, Hyung-Woo Lee, Choong Seon Hong, "Security in Wireless Sensor Networks: Issues and Challenges", Proceedings of 8th IEEE ICACT 2006, Volume II, February 20-22, Phoenix Park, Korea, 2006, pp. 1043-1048.
[69] John Paul Walters, Zhengqiang Liang, Weisong Shi, and Vipin Chaudhary ,"Wireless Sensor Network Security: A Survey", in Distributed, Grid, and Pervasive Computing, Yang Xiao (Eds.), 2006.
[70] E. Z. Ang, "Node Misbehaviour in Mobile Ad Hoc Networks," National University of Singapore, 2004.
[71] A. H. Hussein, A. O. Abu Salem, S. Yousef ,"A Flexible Weighted Clustering Algorithm Based on Battery Power for Mobile Ad Hoc Networks", IEEE, 2008.
[72] C. Li, Y Wang, F. Huang, D. Yang," A Novel Enhanced Weighted Clustering Algorithm for Mobile Networks", IEEE 2009.
[73] B. Kadri, A. M'hamed, M. Feham , "Secured Clustering Algorithm for Mobile Ad Hoc Networks", IJCSNS , VOL.7 No.3, March 2007.
[74] M. Chatterjee, S. K. DAS, D. Turgut, "WCA: A Weighted Clustering Algorithm for Mobile Ad Hoc Networks", Cluster Computing 5, 193–204, 2002.
[75] Z. J.-wu, J. Y.-ying, Z. J.-ji, Y. C.-lei,"A Weighted Clustering Algorithm Based Routing Protocol in Wireless Sensor Networks ", ISECS 2008.

Part 2

Communications and Networking

4

Diversity Techniques for Robustness and Power Awareness in Wireless Sensor Systems for Railroad Transport Applications

Mathias Grudén, Magnus Jobs and Anders Rydberg
Uppsala University
Sweden

1. Introduction

During the last decades business and industry has been constantly optimizing time in production and transportation. This implies that the margins when doing business are decreasing and when margins are decreasing more information is necessary so that the right decisions can be made on time. This is especially important for the transport sector; in all production there is a need to know when the freight with the components is arriving so that the work can be planned. But as the system grows more sensitive to delays it also implies that delays are getting very expensive. The transport of goods on e.g. trains has therefore to be monitored carefully in order to retrieve information on delays. Theses delays can be either due to normal circumstances occurring in transports such as scheduling of time tables or due to mechanical faults. Ball bearings used in the trains are vulnerable to damage which also stands for a large fraction of the mechanical faults that contribute to transport delays by causing costly emergent stops.

Recently the Swedish Transport Administration evaluated a system for monitoring the temperature of the ball bearings (Gruden M., et al, 2009). The evaluation was performed within the Uppsala VINN Excellence Center for Wireless Sensor Networks (WISENET). The evaluation was performed during 2008 by mounting wireless temperature sensors on the ball bearings and with a wireless gateway onboard the train. The positions of the sensors can be seen in Fig. 1.1. This system was monitoring the ball bearing of the wheels and air temperature. The measured temperature of the ball bearing was continuously presented on a webpage. By monitoring the temperature it is possible to see trends of heating and predict if the train wagon needs maintenance or not. This type of monitoring system can greatly increase the reliability of the overall system.

One problem noticed with this system onboard the train was the wireless robustness. Due to the metal parts the wireless connection was partially intermittent. One technique which can be used to improve the robustness of a system is the use of multiple antennas at the receiver or transmitter. As the received signal might suffer severe variation from fading phenomena, techniques must the implemented to mitigate these effects. The choice of techniques can generally be classified into two parts, hardware and software. Software solutions to the fading phenomena usually involve various coding techniques to improve the reliability but this causes slower data rates. Hardware solutions can be found using diversity techniques where two or more antennas are used and then combining the signal using certain schemes can yield significantly increased performance.

Fig 1.1 The position of the wireless sensor.

In this book chapter we will first present the issues of having wireless sensor nodes in train environments. We will also present wave propagation theory to explain why there is a need to introduce diversity techniques to improve the signal quality. In section 2 various well known diversity techniques and implementations will be briefly presented. Due to their intelligence and possibility of decision making, hence high energy consumption and complexity, these types are not suitable for wireless sensor nodes. In section 3 a new diversity combination technique is presented together with some real world measurements that give insight into what kind of performance gain can be expected using the diversity. The new technique presented were developed at Uppsala University, Sweden, as part of the WISENET project on improved wireless communication and wireless sensors in physical and electromagnetic hostile environments. Due to the lower power consumption and simplicity of design this solution is optimized to be use in wireless sensor nodes. First results on this research were presented at EuCAP in 2010 (M. Jobs, et al, 2010). As the need for various wireless devices is increasing exponentially the WISENET group has committed considerable resources to produce new hard- and software technologies to help improve both the robustness and power consumption in wireless devices. Several other, often commercial, forms of wireless devices are gaining ground such as various entertainment systems and sporting gear.

1.1 Wave Propagation Theory

In wave propagation there are many different phenomena that will affect the signal. In this section we describe the models used to characterize the radio channel.

1.1.1 Path Loss

The well-known Friis transmission formula (Balanis, C.A. 2005) shows a dependence on the frequency, distance between transmitter and receiver, and the antenna gains. The wording "Path Loss" might be slightly misguiding as the phenomenon is based around the fact that, assuming an omnidirectional propagation, the energy is spread out over an increasingly larger volume as the distance from the transmitter grows. This causes the received power in a fixed area to decrease exponentially,

$$P_r = \frac{P_t G_t G_r \lambda^2}{(4\pi d)^2} \tag{1}$$

Path loss models are described in the references (Hata, M., 2980) , (EURO-COST 231, 1991), (Kita, N., et al, 2009) and (ITU-R, 2009). The need of expanded models of the Friis transmission equation are motivated by the fact that the basic equation (1) is intended for an ideal environment (with spherical wave propagation and no reflections) which may not be suitable for a real world environments with phenomena such as e.g. losses and various forms of fading. These expanded models are statistical models which determine the attenuation in different environments, mostly in cities and suburban areas. Equation (1) is a special case with losses in an environment without obstacles and multipath propagation. By reformulate equation (1) slightly into

$$P_r = \frac{P_t G_t G_r \lambda^2}{\left(4\pi \frac{d}{d_o}\right)^n}. \tag{2}$$

Where d_0 is a distance where reference signal is measured and n is the path loss exponent. It is then possible to reforumlate equation (2) into an equation with levels in dB

$$L_p = P_t - P_r = 10n\log_{10}\left(\frac{d}{d_0}\right) + K \quad [dB], \tag{3}$$

where n is the path loss exponent, K is an offset value, and d_0 is a reference distance. In Eq. (1) the path loss exponent is equal to 2 but this is only valid for free-space losses The earlier and more well known models (Hata, M., 2980), (EURO-COST 231, 1991) have similar variables determined by experiments. By inspecting the formula it is seen that the equation is linear. The variable K is the offset of the function and is determined by measure the signal level at a reference distance of d_0. The variable n is the path loss exponent and is determined by the slope over distance in the measured sequence. This is simply a coefficient of the losses over the distance. Larger coefficient implies greater losses and vice versa. These two variables are determined later in this chapter, and d_0 is preset to 3 m in this case. The variables are determined at two frequencies, 434 MHz and 2450 MHz. The value of K can not be neglected thus a statistical analyze will be performed which implies that there might be some offset in the linear path loss function.

1.1.2 Multipath Propagation and Fading

Multipath propagation is expected in train environments, because of the large amount of metal surfaces. Measurement determines the path losses and the fading environment. This helps when designing the system of wireless sensor nodes. It gives information about where to place the nodes and if there will be problems with the signal quality due to fading. As the electromagnetic waves transmitted will propagate into virtually all directions this will causes some signals to reach the receivers directly while other impinges on various metal surfaces in the environment. These waves will be reflected by the metal surfaces and hit the receiver slightly delayed in time, causing a fast fading superposition of the waves reaching the receiver. This will create a total received signal that might experience severe distortion in amplitude and phase. This fast fading resulting from the multipath propagation can be modeled by the m-parameter in the Nakagami distribution (A.Goldsmith, 2005)

$$p(r|m,\bar{x}) = \frac{m^m}{\bar{x}^m \Gamma(m)} r^{2m-1} e^{-\frac{m}{\bar{x}}r^2} \qquad (4)$$

The lowest possible value of m is m=0.5. Rayleigh distribution corresponds to m=1, which is a severe multipath environment. A large value of m indicates less fading, which means stronger line of sight. In this paper the measurements are assumed to be Nakagami distributed and m is determined by fitting the measured data to the theoretical distribution.

1.2 Setup

1.2.1 Environment

The measurements are carried out at a railway yard in Borlänge, Sweden. The railway yard is located next to a maintenance hall which is a large brick building with some parts made of metal such as ports and small buildings next to the main building. The ground next to the maintenance hall is asphalt and the rail is built on gravel. East of the railway there is a bank which is a few meters high and mostly covered by small trees and bushes , see Fig. 1.2. The setup of wagons in the 434 MHz and 2450 MHz measurements are different because the measurements were performed at different days and the wagons were moved due to ordinary maintenance work at the site. However, the setup in the two cases was made as similar as possible.

1.2.2. 434 MHz Measurements

In the 434 MHz measurements all wagons near the measurement path except one wagon on a track next to the train are open wagons made for transporting timber. These wagons are located from the mark of "Test Site 1" in Fig. 1.2 and south-west bound. The wagon on the track next to the train is a metal tank and is located next to the marking of "Test Site 1".

1.2.3 2450 MHz Measurements

The 2450 MHz measurements are carried out at "Test Site 2" in Fig. 1.2. There are several different types of wagons at this position. The wagon where the transmitting antenna is positioned is an open wagon made for transporting metal. The wagons next to the transmitting antenna are located northeasterly and are covered wagons, with both soft cover and cover of metal. Fig.1.3 shows a more detailed view of the positions of the transmitting and receiving antennas at this frequency.

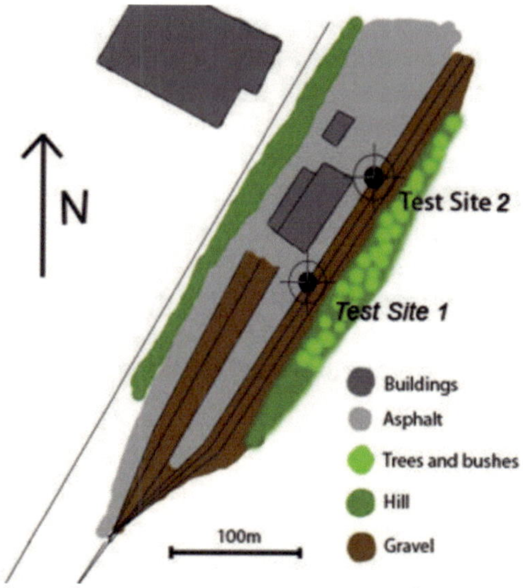

Fig. 1.2 Map of the area where the measurements are carried out.

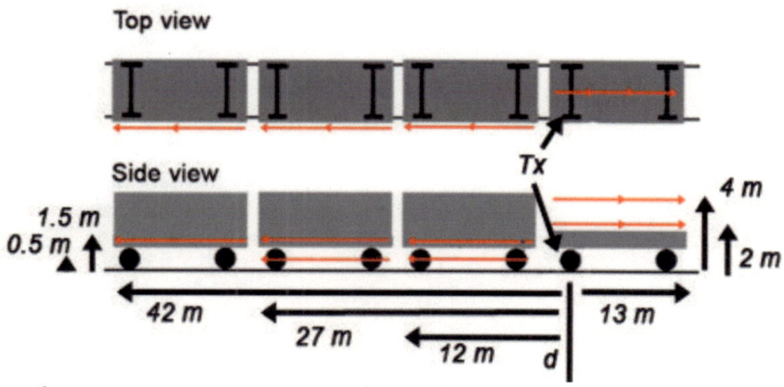

Fig. 1.3. Paths where measurements are performed.

1.2.4 Equipment

In the case of 434 MHz, a signal generator connected to an antenna on the transmit side and an antenna connected to a spectrum analyzer on the receive side, are used. In the case of 2450 MHz, the signal generator is connected to a 30 W amplifier to increase signal strength. The power level of the signal generator is set to 0 dBm for both cases, but as mentioned, amplified at 2450 MHz. The increased power level will not affect the results since the path loss results are relative. The amplifier was only used in order to increase the dynamic range in the measurement. The antennas used are matched dipoles. The equipment is portable to enable easy change of antenna locations.

1.2.5 Measurement Procedure

In all measurements the transmit antenna is fixed and the receiving antenna is moved along a path while recording the signal level. Each measurement consists of a few seconds of stationary measurements in the beginning. After that, a walk of a certain distance and in the end of the walk the receiving antenna is placed in a static position again for a few seconds, hence it is easy to see where the measurement starts and ends. The total length of a measurement is 20 seconds. The starting distance and distance of movement is recorded. The value of d is noted at the start and the end. During the movement of the antenna it is assumed that the velocity is constant. Although measurements are recorded as amplitude versus time, in the post-processing the data is converted to amplitude versus distance, thereby making it possible to determine the path loss as a function of distance. A reference measurement is performed at d_0 (3 m from transmitter in this case), and this value is subtracted from all measured samples.

The data acquired by the above procedure is analyzed using a linear regression on the same form as Eq. (2). From this linear regression the values of n and K are found. The offset value is determined by subtracting the reference value from the value of the linear regression at d_0.

1.3 Measurement Results

Measurements are performed along different paths and at different locations, cf. Fig. 1.3. Both measurement paths are close to the wagon, one of them along the side of the wagon and one on top of the wagon (if it is an open wagon). The results of all measurements are analyzed and compared depending on location, e.g. all measurements beside the wagon are combined, and so forth. Two typical measurement is seen in Fig. 1.4, one at a frequency of 2450 MHz and one at 434 MHz. It clearly seen in the figure that the fading is more severe at higher frequencies.

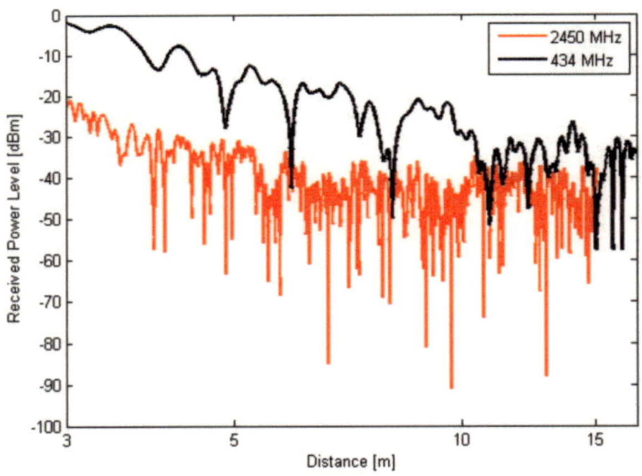

Fig. 1.4. Typical measurement at 434 MHz and 2450 MHz.

The resulting values of n and K in the case of measurements beside the wagon are seen in Table 1, and m is seen in Table 2.

Freq. [MHz]	No. Measure ments	n		K [dB]	
		Mean	Range	Mean	Range
434	39	3.67	1.56 to 4.72	-6	-15 to 0
2450	26	2.22	1.37 to 3.03	-5	-25 to 5

Table 1. Path loss exponent and offset beside the wagon.

Freq. [MHz]	m	
	Mean	Range
434	2.6	1.3 to 7.3
2450	1.3	1.2 to 1.5

Table 2. Fading parameter along the side of the wagon.

Along the second path where measurements are performed on top of an open wagon, the results are slightly different. The results for this path are seen in Table 3 and Table 4.

Freq. [MHz]	n		K [dB]	
	Mean	Range	Mean	Range
434	2.27	1.06 to 3.82	-13	-20 to -7
2450	0.32	-0.33 to 1.85	-2	-10 to 5

Table 3. Path loss exponent and offset on top of the wagon.

Freq. [MHz]	m	
	Mean	Range
434	2.1	1.4 to 3.5
2450	1.5	1.2 to 2.1

Table 4. Fading parameter on top of the wagon.

The smaller path loss exponent at the higher frequency is due to the metal details on the train wagon. They are at a size of a wave length or larger at 2450 MHz but most of the details are smaller compared to the wave length at 434 MHz. The sizes of the details make them to passive radiators at 2450 MHz but not at 434 MHz. This helps the communication link so it is having lower path exponent loss at a higher frequency, see Fig. 1.4.

1.4 Comparison with Simulations

Simulations are performed at 434 MHz using CST Microwave Studio. A properly simulated and verified model provides a powerful tool fast evaluation of proposed systems and antenna concepts. As such it is important to compare measured and simulated data to create reliable model which should be as well validated as possible. The simulation model is a simplified wagon with two bogies with two wheels on each bogie, as seen in Fig. 1.5. Next to the wagon is another truncated wagon that only contains one bogie. The transmitting antenna is placed near this bogie and is vertically oriented. The average power is monitored and data is acquired along the same paths as the measurements. An example of the results is shown in Fig. 1.5.

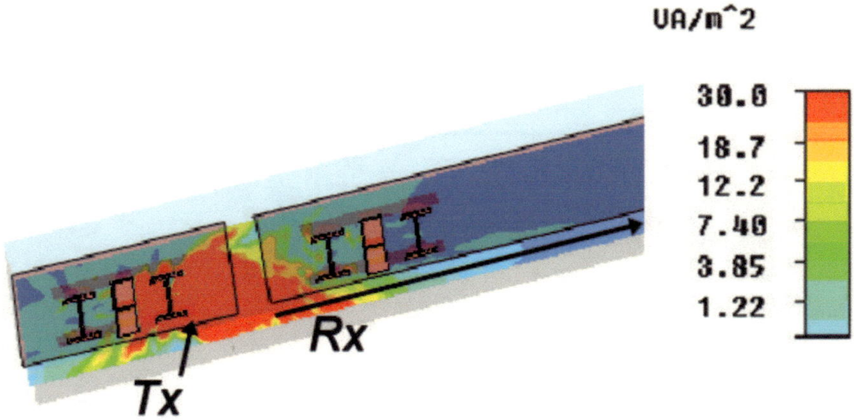

Fig. 1.5. A simulation result showing the field strength. Visualizing plane is approximately at a height of 0.5 m above ground level.

As seen in Fig. 1.6 the simulated level is higher than the measured values, and the simulated values show no fast fading. This is due to the fact that the simulated and presented values are total absolute values of the amplitudes measured on all three polarizations. The simulated path loss exponent is roughly equal to the measured one. One could expect the simulated effects of slow fading, i.e. shadowing or losses in environment to be better correlated to the measured ones. Fast fading on the other hand is highly dependent on the environment, like number and location of reflectors etc., and as such unless a very well defined environment is used for measurements very good correlation will be harder to achieve. The model is as good and detailed as it can bee with the current computer technology.

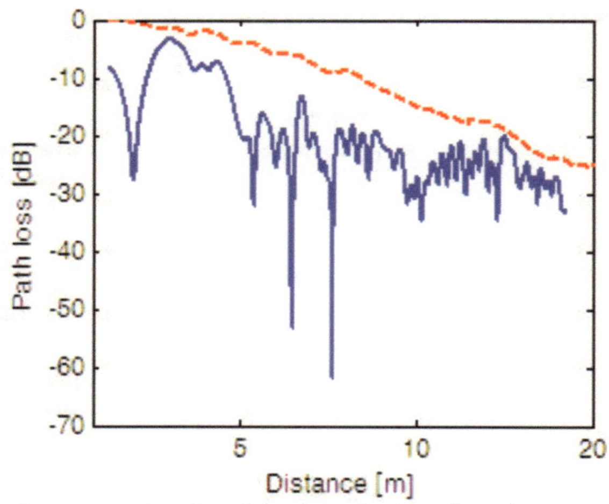

Fig. 1.6. Comparison between simulated (red dashed line) and measurement (blue line) data.

1.5 Motivation to Introduce Diversity

The fast fading seen in Fig 1.4 is one of the most important issues to deal with when improving the wireless communication. By using only one antenna transmitted data can be lost due to severe fading dips. Imagine having two antennas with spatial diversity, and one of the antennas is placed in one of the fading dips. The other antenna will most probably be located outside this fading dip and the signal level can be up to 50-70 dB higher for the antenna outside the fading dip. This prevents packet loss and limits the need to retransmit packages, this lowers the overall power consumption.

2. Common Diversity Techniques

The general explanation of a diversity system is a wireless system that uses several independent channels to communicate in order to increase the reliability of the system. Choosing to use diversity could be considered making a tradeoff by increasing the overall power-consumption in order to get more reliable communication. A diversity system has to be implemented with two parts. One part consists of a diversity antenna, the second part is the combiner which consists of electronic components and includes an intelligent control system. It also exist diversity by using frequency or time coding. But these will not be analysed in this chapter. There are many different types of solutions for both the design of the antennas and for the combiner. How they work individually are described in section 2.1 and 2.2. It will be clear that these techniques are not always suitable for wireless sensor nodes due to the required power to feed the controlling circuitry. The solution later presented in this chapter is only a solution for the combining technique not the antennas. The new technique is less intelligent than the common ones but more suitable for wireless sensor nodes.

2.1 Combining Techniques

One part of the diversity systems has to consist of electronic circuitry. This part has to include some sort of intelligence to enable signal improvement. The general idea about how the combination technique is implemented is seen in figure 2.1. As can be seen that some type of feedback network is used to allow adaptive control of the incoming signals which will increase the overall signal reliability.

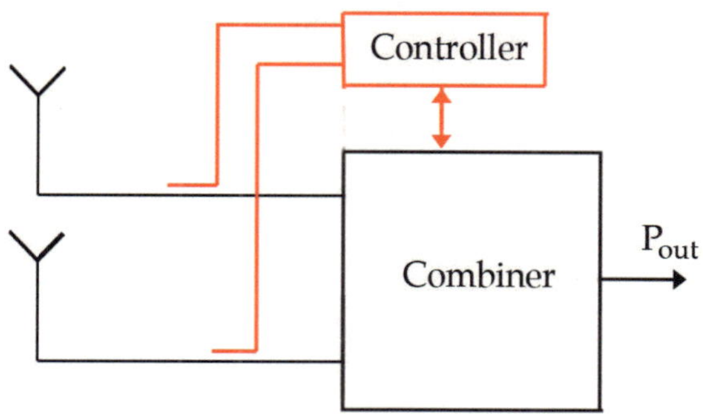

Fig. 2.1. The general idea of combining techniques.

In general the standard form of combining circuits include some network of controllable phase shifters and amplifiers with a combining circuit which will superposition the incoming signals. The difference between the types of combining techniques is mainly dependent on how the controlling algorithms are set-up to handle phase shifting and amplification of incoming signals before they are combined together to create one unique signal. The drawback with these systems is their energy consumption and the complexity.

2.1.1 Selection Combining

The selection combining is the most simple combination technique that can be implemented in a circuit. When having two branches the controller is detecting the received signal level in each branch. The decision is made to choose the branch with the highest signal level at the moment. A sketch of the technique is seen in figure 2.2.

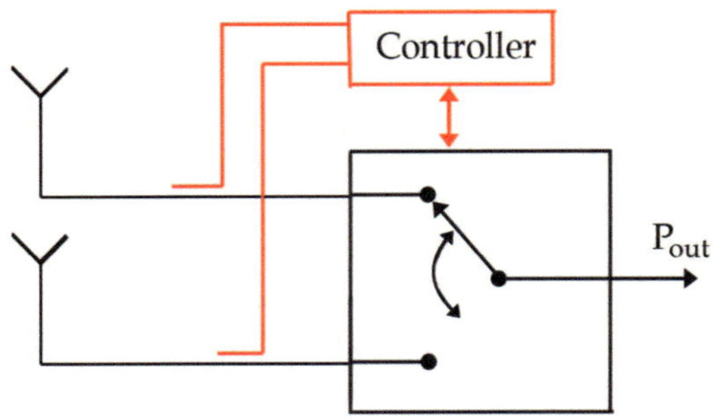

Fig. 2.2 Selection combining.

2.1.2 Equal Gain Combining

Equal gain combining is one of the more advanced techniques. This technique is based on one phase shifter per diversity branch and one combiner/summation. The controller circuit is controlling the relative phase shift of the branches and is shifting the phase so when the signals are combined they are in phase and do not have destructive interference.

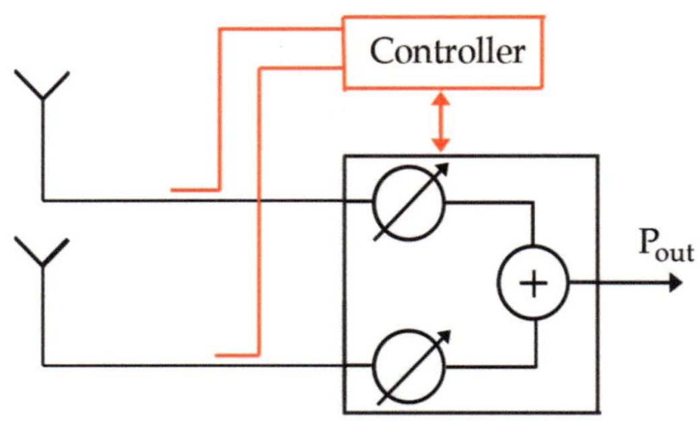

Fig. 2.3. Sketch of equal gain combining.

2.1.3 Maximum Ratio Combining

The maximum ratio combining is probably the most advanced sort of diversity circuits. The controlling circuit is as usual determining the amplitude and phase of the branches. In this stage the circuit is controlling both an amplifier and a phase shifter on each branch. The signals are adaptively amplified and phase shifted before they are constructively combined.

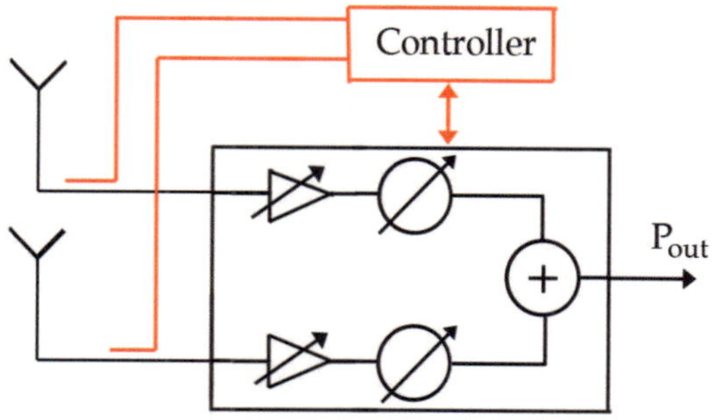

Fig. 2.4 Sketch of maximum ratio combining.

2.2 Antenna Diversity Techniques

As mentioned previously the diversity receiver/transmitter consists of two different parts, the combination techniques and the antenna design. To achieve a good reception and a fully working circuit there is a need for a good antenna design. When considering an antenna design some parameters are more important when the antenna shall be used for a diversity implementation. In this section three of the key parameters for a good diversity antenna are listed, correlation, polarization and spatial diversity. The correlation is probably the most important parameter of these three, it describes the performance of the antennas by comparing how well a signal received at one the antennas couples to the other. Idealy each antenna should be considered a independent channel in which we would have no correlation between them.

2.2.1 Correlation

The level of correlation between antenna elements is the most important parameter when designing a diversity system. This part is, however, not independent of the other two design parameters polarization and spatial properties for the antenna system.
When talking about correlation we simplify the discussion to a system with only two branches since the available space for on the sensor node is very limited. Limited area to use for antennas also implies that the spacing between the antenna elements can not be adjusted which would help minimise the correlation. In the case of two antennas that are correlated the signal level at the output port of one antenna can be determined based on the signal on the other antenna. For the uncorrelated case this is not possible. For two antenna elements on a sensor node falls in between the two cases due to the close distance between the elements. However even though there exists a strong correlation between the antenna elements due to the close spacing in-between the improvement in a multipath environment it is shown that e.g. in the case of polarisation diversity the wireless link budget can be improved by several tens of dB by changing the polarisation in case of fading dip (Buke,A., et al, 1999). Even small improvements in the link budget can be important in creating a

robust communication in the sensor network. The correlations between the output signals from two antennas are described in (Simon, M.K., 2002) the time domain by the eq. (5).

$$\rho = \frac{E[(X_1(t)-\mu_x)(X_2(t)-\mu_Y)]}{\sigma_{X_1}\sigma_{X_2}} \tag{5}$$

The correlation is calculated by using statistics of measured sequences.

2.2.2 Polarization Diversity

When designing the antennas to be used for diversity there are some ways to decrease the correlation. One of the simplest ways is to design the antennas with polarization diversity. If we use two antennas, this means that the two antennas have perpendicular polarization. For the simplest case, with two dipoles, this means that the antennas shall be perpendicular as well, as seen in Figure 2.5.

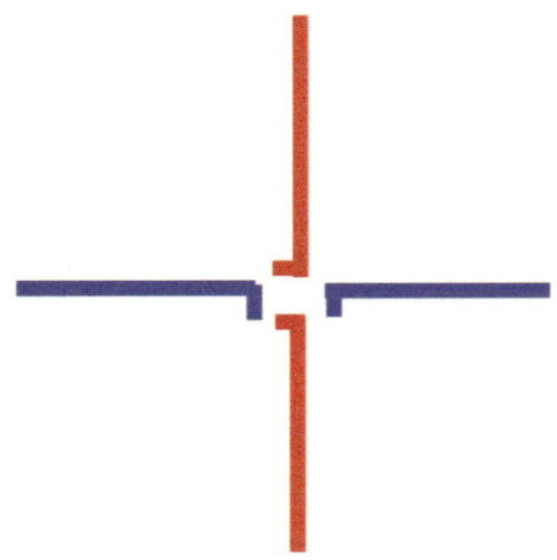

Fig. 2.5. An example of polarization diversity with two dipoles (red and blue).

2.2.3 Spatial Diversity

Spatial diversity is similar to polarization diversity but there is no need to have the antennas in the same position. In this case the antennas have exactly the same radiation pattern, but the distance, *d*, between the antennas will cause the incoming EM-waves to have different amplitude levels at the same moment in time. This can be seen in figure 2.6.

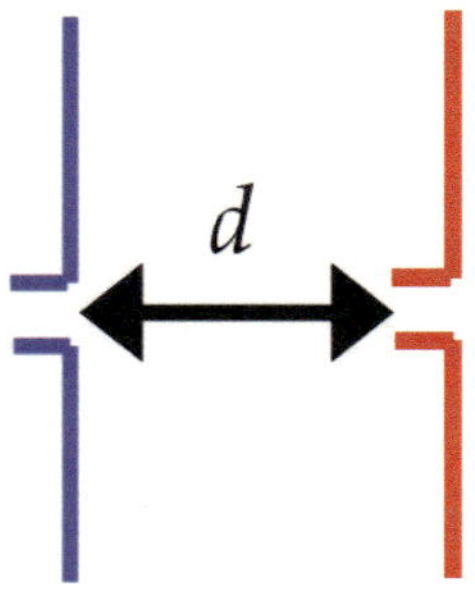

Fig. 2.6. Example of spatial diversity for two dipole antennas.

If the distance *d* is zero the output signal from the antennas are totally correlated. In this case they are totally correlated and no diversity can be achieved. The correlation decreases with distance and is usually low enough at a distance of d=λ/2. Depending of the design of the antenna both larger and smaller distances between the antennas can fulfil the demanded correlation (Valenzuela-Valdes, J.F., et al, 2009).

3. Opportunistic Combining

In previous sections it was determined that diversity had to be designed with some sort of adaptivity to the present environment. It also implies that the diversity system has to be designed along with the receiver itself. The disadvantage of the more well-known types of diversity presented in the previous section is that it demands intelligence to make decisions about how to combine signal or choose the antenna branch with the best signal level. To have an intelligent circuit it demands a micro controller, which drastically increases the energy consumption. Recently a technique has been presented (Jobs, M., et al, 2010) that shows an alternative solution to the standard feedback-based solutions. This new type of combination does not need any sort of intelligence and can be realized with ordinary lumped components, see fig. 3.1. By its simplicity it does not have to be designed to work for one single radio system it can also be added on already existing radio circuits.

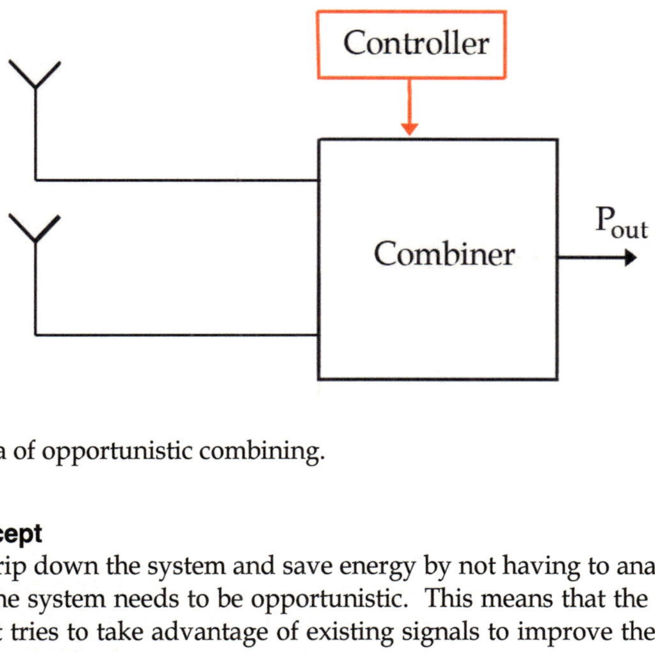

Fig. 3.1 The idea of opportunistic combining.

3.1 Basic Concept

To be able to strip down the system and save energy by not having to analyze data and take any decisions the system needs to be opportunistic. This means that the system behaves in the sense that it tries to take advantage of existing signals to improve the signal quality but do not use adaptive feedback to control the receiving circuit. By using a combiner circuit that cycles through a number of predetermined configurations it is possible to use an averaging or peak detector to create a stronger received signal when considering the time average.

In this case the electronics have a limited number of combinations it will cycle through. As seen in the previous types of combination techniques it is possible to use selection or phase shifting of the signal before adding them together to achieve a good output signal. The proposed technique here is to use a 2 or more uncorrelated received signals from antennas and during each received (transmitted) symbol change the phase shift between the receiver (transmitter) branches so that the total symbol will be a combination of phase shifted signals, as described below:

$$S_{received} = \frac{1}{NrCombinations} \sum_{k=0}^{NrCombinations} \left(\sum_{l=0}^{NrAntennas} A_l e^{jw\varphi_{l,k}} \right) \quad (6)$$

The equation above describes the received signal in an averaging detector connected to an opportunistic diversity switch. During each transmitted symbol all combination of phases are used. This unique sum of signals is created once or more during each received symbol. This could be visualized as sweeping the antenna pattern during each symbol and as such creating angular diversity in the received signal. In a multiscattering environment each direction will receive a unique sum of planar waves which are superimposed and creates a fading signal. By using a directive antenna which changes the position of its main lobe different sums of planar waves will be received and each sum will have a unique signal strength.

In the implementation of such a switch, presented below, 2 antennas were used each with a binary phase shifter. This created a total of 4 different signals during each symbol received. However, using even a limited number of phase shifts significant increase in receive strength can be obtained as long as each phase-shift combination is chosen such as the change in antenna pattern is as large as possible. Depending on the type of antenna array used the phase-shifting circuit is designed for optimal performance of the opportunistic combining.

3.2 Electronics & Performance

The overall goal using an opportunistic combiner is to keep both power consumption and component cost to a minimum. Seen from a pure performance perspective the opportunistic combiner will never be as efficient as, for example, a selection combiner or maximum gain combiner. One of the advantages of the opportunistic combiner, as mentioned previously, is that it keeps power consumption to a minimum. When coupling this to the fact that it can be built using only a small number of components it is possible to implement the advantages of diversity in applications that are both cost sensitive and power constrained. However, the diversity switching system needs to be carefully designed to minimize the insertion loss of the system and improve efficiency.

The phase shifter itself can be implemented using a number of techniques, as the only goal of the phase shifter is to be able to switch between several predefined phase shifts while keeping power consumption low. The two major groups of switching techniques used is normally either PIN-diode (diode consisting of three layers which is P-doped, intrinsic and N-doped creating a current-controlled RF-resistor) based switches or transistor based. Transistor based switches is a good choice due to the fact that each phase shift needs to exist during the time between each shift and a transistor based solutions can provide just that. PIN-diode based switches are very simple to implement but suffer from the fact that they need a continuous current to function. Due to the fact that the resistance is linear as a

function of current the performance is directly linked to power consumption. As such, if a low loss application is desired the power requirements for such a solution will increase.
In the hardware used for the opportunistic combiner proof-of-concept PIN-diodes were used only due to the fact that power consumption was a secondary in this particular implementation. In the circuit seen in Fig. 3.2 two binary phase shifters consisting of two PIN-diodes each coupled to a pi-network provided a phase shift of 90 degrees. This allowed for a total phase shift between the receiving branches of 0°, 90°, 180° and 270°.
In order to complete the system some form of switching circuit needs to be implemented that switches between the different phase shifts. As the system does not need to implement any form of intelligent behaviour, any form of simple multivibrator circuit can be used to keep the phase shifters in continuous rotation. However, care might have to be taken to assure that the frequency of the clock-circuit can be tuned in such a way that it can support all the various baud rates intended in the target application. If the application is such that only a single baud rate is supported this restriction can be alleviated and reduce the complexity of the system even further. The required clock rate from the multivibrator can be described as follows:

$$f_{upper} \geq f_{switch} \geq f_{symbol} \tag{7}$$

$$f_{switch} = n \cdot C \cdot f_{symbol}, \quad \text{when } n = 1,2,3,... \tag{8}$$

The variable C is the number of relative phases, in this case C=4, and n is the number of repetitions of the phases during each symbol. The highest frequency of switching is f_{upper}. If this frequency is to high the system will start deteriorating due to spectrum broadening. However, using only the lower limit of one rotation per received signal should still give good results, also using only one rotation the interference due to switching noise can be minimized. In Fig. 3.3 the received signal during switching is illustrated. In the figure four different phase shifts are cycled through and it can be seen how one of the received signals enters a fading null while the other four phase shift combinations are keeping a relatively constant level.

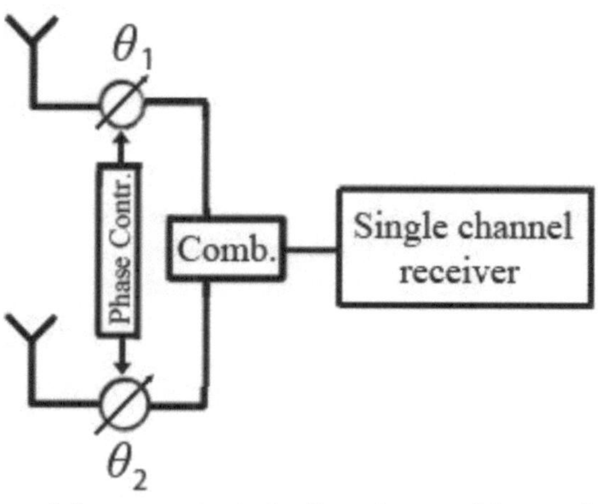

Fig 3.2. Block diagram of the opportunistic diversity combiner, which is connected to the receiver.

Opportunistic combiner should provide a simple mean to implement diversity in a system but it does have some important restrictions. One of the major restrictions is that phase-noise is injected into the system. This causes phase-based modulation techniques to experience severe distortion in the system. As a result it is predicted that phase-modulated systems such as Quadrature Phase Shift Keying (QPSK) and Quadrature Amplitude Modulation (QAM) will be unable to implement opportunistic diversity switch in its simplistic form. Any form of modulation where phase information is discarded such as Amplitude Shift Keying(ASK) and Frequency Shift Keying (FSK) will, however, be good candidates for implementation of opportunistic combining.

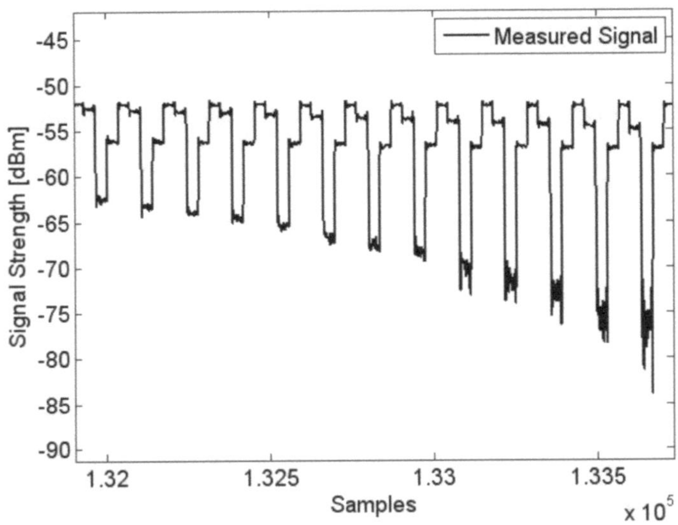

Fig. 3.3. Received signal from diversity combiner in office environment. One of the phase combinations is seen entering a fading dip.

The ultimate diversity gain obtained from a system implementing a opportunistic diversity switch is dependant on the type of detector used in the transceiver system. An opportunistic combiner relies on the detector in the receiver architecture to perform either an averaging or peak detection on the incoming signal. This means that care should be taken when evaluating the performance of a system. If the receiver uses a peak detector the theoretical optimum performance should be obtained as it means that the maximum level of the phase shifter output will be captured and used. If, however, an averaging detector is used the system will not be optimum but should still provide very good protection against deep fading nulls. It is expected that most system considered for implementation of a opportunistic diversity combiner will have some form of averaging combiner.

In Fig. 3.4 seen below all the previously defined parameters has been implemented in a "Proof Of Concept" system. This system was used to evaluate the performance of a opportunistic combiner in a amplitude modulated (ASK) system. The purpose of this rather crude prototype system was to measure the diversity gain obtained in a system based on opportunistic combining. The switching circuit in this case was a small microcontroller rather than a multivibrator in order to give a fully customizable switching signal for the phase shifters for evaluation purposes. The power consumption of the PIN-diodes was controlled externally and during testing it was set to a rather high 10 mA per switch. This is obviously much to high to be considered for a wireless application but with modern Integrated Circuit (IC) RF-switches, pushing power consumption down in the micro-watt ranges, a low power implementation of the system described is not unrealistic.

Fig. 3.4. Photo of the add-on diversity combiner, with SMA connectors for the two diversity antennas and the receiver.

A prototype of the opportunistic combiner was developed and tested in both a fully reflective environment (Reveberation Chamber) and a standard office-type environment. This allowed for evaluation both during theoretically optimum circumstances as well as a "real life" application. Using a pair of uncorrelated diversity antennas based on spatial diversity the signal strength seen in Fig. 3.5 and Fig. 3.6 were obtained. It can be seen that the combiner increases the average signal strength and the mitigates of the deep fading nulls which will otherwise cause packet loss or extend the transmitting range. This would allow the target application to either reduce the number of retransmission necessary or decrease the amount of error-correcting code previously needed to obtain a adequate transmission. The reduced requirements on retransmissions and error-correcting coding can now be used to either provide a more robust communication or allow for an increase in data rate. In each case the implementation of a opportunistic combiner in the system can clearly be seen to have a significant impact of the robustness and overall power consumption of the system.

In general the environment in which a wireless system will be implemented and the requirement of robustness will determine if a diversity based solution is preferred. Rural environments will in general only suffer from slow fading and propagation losses and in such a system diversity switching might not be a good solution as it will require some increase in power consumption and overall cost of the system. However, if the system is to be implemented in a urban and multiscattering environment than the extra cost of adding a diversity switch to the wireless system could prove to both increase the lifespan through energy conservation and reliability of the system by not have to retransmit data.

The results from the evaluation of the diversity combiner prototype showed a significant increase in received signal strength during fading nulls. Using 90 percent signal reliability the diversity gain of such a system was measured in an ideal environment to be 5.5 dB for an averaging detector and 10.3 dB for a peak detector. In the office environment the combiner gave a 1 dB gain using an averaging detector and 5.4 dB using a peak detector. This means the system will experience a significant increase in reliability.

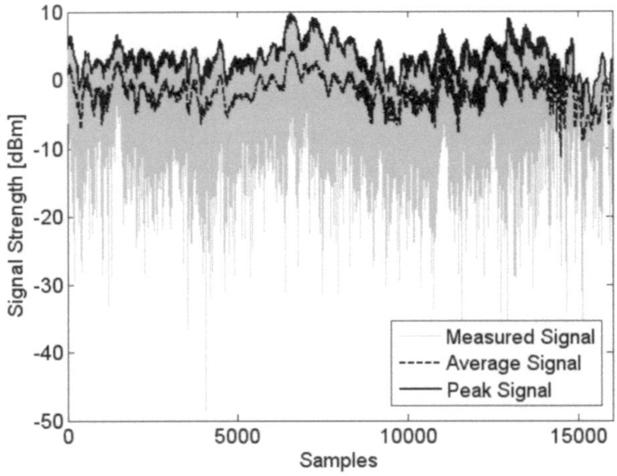

Fig. 3.5. Measured signal in reverberation chamber received from diversity combiner with peak and average signals marked.

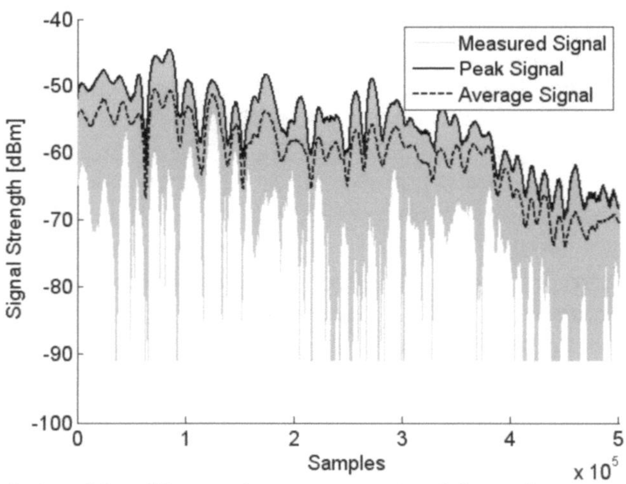

Fig. 3.6. Measured signal in office environment received from diversity combiner with peak and average signals marked.

3.3 Modulation Types

As mentioned in section 3.2 the proposed switching technique is primarily intended to be implemented in an ASK or FSK based sensor system. Both of these modulation types has the information stored in either the amplitude (ASK) or the frequency (FSK) and as such is unaffected by changes in phase of the received and/or transmitted signal. However, as the phase-shifters only shifts through a number of known positions it is should be possible to add a compensation circuit in order to remedy this to some extend. No research has at the time of writing this chapter been presented on this.

4. Implementing diversity in design

The choice to implement some form of diversity should be considered at an early stage of the design of the sensor node. Even though certain diversity techniques such as the opportunistic combiner presented previously is possible to add as an external component this should be avoided if possible. If a wireless node is designed to be low cost, low power and robust this requires that the component count is kept at a minimum as well as board space. Also, if the node houses some form of intelligent processing, i.e. a microcontroller, the need for a multivibrator for the diversity switch can be omitted in favour of direct control by the microprocessor. This further reduces the amount of components.

As mentioned previously in section 3.2 if the design is realized using lumped components instead of an Application Specific Integrated Circuit (ASIC), the phase shifters should be implemented using transistors or an dedicated IC rather than diodes. As most commercial applications are manufactured using lumped components and ICs this would be the most common situation.

5. Conclusion

As we have seen in this chapter there are several important aspects to consider when looking at the reliability and signal robustness of a wireless sensor system in a multiscattering environment. Phenomenon such as propagation losses and fading needs to be modelled correctly in order to give an acceptable representation of the proposed environment. Once a good representation for this has been found various techniques can be implemented that increases the reliability of the system.

By implementing diversity in the system, i.e. using more than one transmitting or receiving antenna, the signal reliability can be increased significantly. However, this comes at the cost of increased complexity and power consumption of system which should be reduced particularly in a sensor node and systems. The use of an opportunistic diversity combiner has been proposed as a way to increase the reliability in system using simpler modulation schemes and fits well where a sensor node is using ASK of FSK based modulation. Such a technology can implement diversity in the system while keeping the amount of additional hardware and power consumption at a minimal level.

Finally some measurements using a prototype system has been presented in which one can clearly see the improved performance when using a opportunistic combiner as opposed to no diversity combining. This combiner was implemented using a minimal amount of external components and as such similar implementations should be considered useful for various embedded wireless devices.

6. Acknowledgements

The authors would like to thank Dr. Paul Hallbjörner at SP Technical Research Institute of Sweden, for all his help and encouragement. The authors would also like to thank Dr. Erik

Björnemo at Uppsala University, department of Signals and Systems, for all good discussions and constructive criticism.

7. References

Allnatt, J.W.; Jones, E.D.J.; Law, H.B. (1957); "Frequency diversity in the reception of selectively fading binary frequency-modulated signals", Proceedings of the IEE - Part B: Radio and Electronic Engineering, Volume: 104 , Issue: 14.

Balanis, C. A.,(2005) *Antenna Theory – Analysis and Design,* 3rd Ed., New Jersey, USA, John Wiley & Sons, Inc., 2005.

Buke, A., Hajian M., Ligthart L.P., Gardener P.,(1999) "Indoor channel measurements using polarisation diversity", Vehicular Technology Conference 1999, pp. 2282 - 2287 vol. 4, Sept. 19-22, 1999

Chin-Lung Y.; Chappell, W.J. (2007); "Angular diversity – prominent for compact devices of wireless sensor network in indoor environments", Antennas and Propagation Society International Symposium, 2007 IEEE

EURO-COST 231, (1991) European Cooperative in the Field of Science and Technical Research "Urban Transmission Loss Models for Mobile Radion in the 900 and 1800 MHz Bands," rev. 2, The Hauge.

Goldsmith, A.(2005), *Wireless Communication*, Cambride, University Press, 2005.

Gruden, M.; Westman, A.; Platbardis, J.; Hallbjorner, P.; Rydberg, A. (2009); "Reliability experiments for wireless sensor networks in train environment", Wireless Technology Conference, 2009. EuWIT 2009, Rome, Italy.

Hata M., (1980)"Empirical Formula for Propagation Loss in Land Mobile Radio Services", IEEE Transactions on Vehicular Technology, Vol. VT-29, No. 3, August 1980, pp. 317-325.

Jobs, M.; Gruden, M.; Hallbjörner, P.; Rydberg, A., (2010); "Antenna Diversity with Opportunistic Combining for ASK Systems with Single Channel Receivers", European Conference on Antennas and Propagation 2010, EuCAP2010, Barcelona, Spain.

Jobs M.; Bestoon J.; Rydberg A.;(2009) "A Wireless Body Area Network (WBAN) Based Tracking and Monitoring Application System" The IET Conference on Body-centric wireless communications, Savoy Place, London, UK.

Kita N., Ito T., Yokoyama S., Tseng M. C., Sagawa Y., Ogasawara M., and Nakatsugawa M., (2009), "Experimental Study of Propagation Characteristics for Wireless Communications in High-Speed Train Cars," 3rd European Conference on Antennas and Propagation (EuCAP2009), pp. 897-901.

Miki, N.; Atarashi, H.; Sawahashi, M. (2004); "Effect of time diversity in hybrid ARQ considering space and path diversity for VSF-OFCDM downlink broadband wireless access", Personal, Indoor and Mobile Radio Communications, PIMRC 2004. 15th IEEE International Symposium on ,Volume: 1.

Recommendation of ITU-R, (2007) "Propagation data and prediction methods for the planning of indoor radiocommunication systems and radio local area networks in the frequency range 900 MHz to 100 GHz,", P.1238-5

Simon, M.K., and Alouini, M.-S.;(2002), "A Unified Performance Analysis of Digital Communication with Dual Selective Combining Diversity over Correlated Rayleigh and Nakagami-m fading channels", Communications, IEEE Transactions on, Volume 47, Issue 1, Pages 33-43.

Valenzuela-Valdes, J.F.; Garcia-Fernandez, M.A.; Martinez-Gonzalez, A.M.; Sanchez-Hernandez, D.A., (2009); "Evaluation of True Polarization Diversity for MIMO Systems", Antennas and Propagation, IEEE Transactions on Volume: 57 , Issue: 9, Page(s): 2746 - 2755.

5

Energy Efficient Transmission Techniques in Continuous-Monitoring and Event-Detection Wireless Sensor Networks

Nizar Bouabdallah, Bruno Sericola, Sofiane Moad
INRIA Rennes-Bretagne Atlantique
France

Mario E. Rivero-Angeles
INRIA Rennes-Bretagne Atlantique / UPIITA-IPN
France / Mexico

1. Introduction

Wireless Sensor Networks (WSNs) can be typically used to achieve Continuous Monitoring (CM) or Event-Detection Driven (EDD) inside the supervised area. For both applications, sensors consume energy for three main reasons: sensing, processing and wireless communicating. The wireless communication refers to data transmission and reception. Among these three operations, it is known that the most power consuming task is data transmission. Approximatively 80% of power consumed in each sensor node is used for data transmission. Hence, unnecessary transmissions and/or unnecessary large data packets reduce the system's lifetime. In this work, we are interested in studying different data transmission schemes that reduce the energy consumption by means of compression, in order to reduce the data packet's length, or by means of avoiding transmission of redundant information.
Continuous-monitoring applications require periodic refreshed data information at the sink nodes. To date, this entails the need of the sensor nodes to transmit continuously in a periodic fashion to the sink nodes, which may lead to excessive energy consumption. In this work, we show that continuous-monitoring does not imply necessarily continuous reporting. Instead, we demonstrate that we can achieve continuous-monitoring using an event-driven reporting approach. For example, consider a continuous-monitoring temperature application, where each sensor node transmits periodically the sensed temperature to the sink node. In such application, it may happen that sensors have very similar reading during long periods of time and it would not be energy-efficient for sensors to continuously send the same value to the sink node. The network lifetime would be greatly increased by programming the sensors to transmit only when they have sensed a change in the temperature compared to the last transmitted information. In doing so, the end user would have a refreshed value of the temperature in the supervised area even if the sensors are not transmitting continuously in a periodic fashion. The final user would have exactly the same information gathered by the WSN as with the classical continuous-monitoring applications, but while the sensors only transmit when there is relevant data.

Building on this, we propose two new mechanisms that enable energy conservation in continuous-monitoring WSNs. The first mechanism can augment any existing protocol, whereas the second is conceived for cluster-based WSNs. With both mechanisms, sensor nodes only transmit information whenever they sense relevant data. Specifically, we refer to these techniques as Continuous-Monitoring based on an Event Driven Reporting (CM-EDR) philosophy. Our proposed CM-EDR mechanisms can be viewed as a particular type of EDD applications, where an event is defined as an important change in the supervised phenomenon compared to the last reading sent to the sink node. However, the main difference with typical EDD applications is that with CM-EDR, the end user would have a continuous reading of the phenomenon of interest, which is not the case with EDD applications.
In Event-Detection Driven applications, on the other hand, once an event occurs, it is reported to the sink node by the sensors within the event area. As such, the reporting nodes are expected to be closer to each other compared to the continuous-monitoring case where all nodes in the system are active simultaneously. Therefore, it is possible to take advantage of the spatial correlation inherit in these conditions. In view of this, we propose a compression technique for clustered-based event driven applications in wireless sensor networks. The main idea behind our proposal is to exploit the spatial correlation of such networks in order to reduce the size of the data packets by means of data compression. Specifically, the proposed scheme is composed of two major operations: Cluster Head (CH) selection and data compression.
Data compression is based on the following reasoning: Since the active nodes are inside the event area, they are usually very close to each other and the data correlation is expected to be high. As such, the data values sensed by the different nodes are most likely very similar. The proposed scheme exploits this correlation since nodes transmit only the difference of their sensed data and a reference value which is transmitted constantly by the node selected as CH. As it is shown, fewer bits are required to encode this difference compared to the case where the complete data value is transmitted. The other important procedure of the proposed scheme is the CH selection. This selection is carried out at the sink node (which is assumed to be outside the system's area and therefore is not energy constrained). The sink node receives a sample value of all active nodes at the beginning of the event and then selects the node that minimizes the aggregated data packet's size. Numerical results show that the proposed scheme achieves significant energy conservation compared to a classical clustering scheme [1].

2. Reference Protocols

As stated before, in this work we focus mainly in cluster-based reference protocols for the introduction of the CM-EDR mechanism. The reason for this is that, as show in section III, clustering sensor nodes provides several advantages compared to the unscheduled case. It allows reducing the energy consumption due to collisions, idle listening and overhearing by coordinating sensor nodes belonging to each cluster with a common schedule. The CH assigns resources by clarifying which sensor nodes should utilize the channel at any time ensuring thus a collision-free access to the shared data channel.
In unscheduled MAC protocol-based WSNs (Kredo et al., 2007), the sensor nodes transmit directly their sensing data to the sink node without any coordination between them.
On the other hand, in cluster-based WSNs (i.e., scheduled MAC protocol-based WSNs) the WSN is divided into clusters. Each sensor communicates information only to the CH, which

[1] This is footnote

communicates the aggregated information to the sink node. In our study, we consider the well known Low Energy Adaptive Clustering Hierarchy (LEACH) (Heinzelman et al., 2002) which is a simple and efficient clustering protocol.

3. Comparison between Cluster-Based and Unscheduled WSNs

In this section, we focus on the analysis of the LEACH protocol as it represents the basic clustering protocol in WSNs.

Results regarding the remaining reference protocols are provided in subsequent sections. Specifically, we explore the main interest of WSN clustering by comparing the LEACH cluster-based model to the basic unscheduled model, where communications are performed directly between the sensor nodes and the sink node.

As a distinguishing future from previous works, we consider in our study the energy consumption due to overhead in the cluster formation phase. We show that the energy consumed in this phase is far from being negligible. Recall that the main philosophy behind clustering is to reduce the energy consumption compared to the unscheduled systems by reducing collisions, idle listening and overhearing at the cost of coordination message overhead during the cluster formation phase.

3.1 Network Model

In our analysis, we consider different variations of the CSMA protocol to arbitrate the access to the medium among the sensor nodes at the cluster formation phase. Specifically, the NP-CSMA, 1P-CSMA and CSMA/CA variations are considered along with different backoff policies are investigated (i.e., GB, UB, BEB and NEB).

According to the CSMA technique, a sensor node listens to the medium before transmission. If the medium is sensed idle, the node starts transmission. Otherwise, in NP-CSMA, the node draws a random waiting time (backoff period) before attempting to transmit again. During this time, the sensor does not care about the state of the medium. In 1P-CSMA, after detecting activity on the medium, the node continues to sense the channel until the end of the ongoing transmission and then immediately transmits. Since in a wireless environment, nodes can not hear collisions, another variant of CSMA called CSMA/CA is used, such as the one used in the Distributed Coordination Function (DCF) of the IEEE 802.11 protocol (IEEE Specification, 1999). Accordingly, the node first senses the medium and if it is idle it does not immediately transmits but rather waits for a certain period of time called Distributed Inter Frame Space (DIFS). If the channel remains idle, the node transmits, otherwise, it continues listening to the channel until it becomes idle for a DIFS period and then enters to the backoff procedure to avoid collisions.

Whenever a collision occurs, sensor nodes must retransmit their packet according to the different backoff policies. For instance, considering the CSMA/CA case, the sending node attempts to send its frame again when the channel is free for a DIFS period augmented by the new backoff value, which is sampled according to the backoff policy. Let W_i (expressed in terms of time slots) be a random variable representing the backoff delay at a node experiencing i consecutive collisions. W_i is distributed as follows according to the different backoff policies:

- UB: W_i is uniformly chosen from the range $[1, w]$.
- BEB: W_i is uniformly chosen from the range $[1, 2^{i-1}w]$, where w is the initial backoff window size. This means that the range of the backoff delay is incremented in a binary exponential manner according to the number of collisions suffered by the packet.

Following each unsuccessful transmission, the backoff window size is doubled until a maximum backoff window size value equal to $2^m w$ is reached, where m is the number of backoff stages.

- GB: W_i is geometrically distributed with parameter q.
- NEB: W_i follows a negative exponential distribution with mean $1/R$.

Based on these random access protocols, a comparison between the LEACH cluster-based WSN and the basic unscheduled WSN is performed using the following assumptions and system parameters:

- The total number of sensor nodes in the system is $N = 100$.
- Sensor nodes are uniformly distributed in an area between $(0,0)$ and $(100,100)$ meters (i.e., square 100×100 area).
- The sink node is situated outside of the supervised area at the coordinate $(50,175)$ as in (Heinzelman et al., 2002).
- All sensor nodes have the same amount of initial energy (2 J).
- Each sensor node senses its area periodically, each $T_{sensing} = 1s$, and transmits the produced data information to the sink node.
- All nodes can transmit with enough power to reach directly the sink node. Additionally, nodes can use power control to vary the amount of transmit power.
- The energy consumed to transmit a packet depends on both the length of the packet l and the distance between the transmitter and receiver nodes d. We use the same model as in (Heinzelman et al., 2002) where:

$$E_{tx}(l,d) = \begin{cases} l \times E_{elec} + l \times \epsilon_{fs} \times d^2, & \text{if } d < d_0 \\ l \times E_{elec} + l \times \epsilon_{mp} \times d^4, & \text{if } d \geq d_0 \end{cases} \quad (1)$$

  where E_{elec} is the electronics energy, $\epsilon_{fs} \times d^2$ or $\epsilon_{mp} \times d^4$ are the amplifier energies that depends on the distance to the receiver, and d_0 is a distance threshold between the transmitter and the receiver over which the multipath fading channel model is used (i.e., d^4 power loss), otherwise the free space model (i.e., d^2 power loss) is considered.
- The energy to receive a packet depends only on the packet size, then, $E_{rx}(l) = l \times E_{elec}$
- Considering LEACH, each CH dissipates energy in reception, transmission and in aggregating the signals received from the CMs. The energy for data aggregation is set as $E_{DA} = 5$ nJ/bit/signal.
- CHs perform ideal data aggregation.
- The expected number N_{CH} of CHs following the cluster formation phase is set equal to 5. In this section, we used the same network topology as in (Heinzelman et al., 2002), where it was demonstrated that LEACH is most efficient when the number of CHs, N_{CH}, is equal to 5 in a 100-node network. Hence, the results shown here for LEACH are obtained by choosing the best parameter value for N_{CH}.
- The rest of the parameters are listed in Table I.

Parameter	Value
ϵ_{fs}	10 pJ/bit/m^2
ϵ_{mp}	0.0013 pJ/bit/m^4
E_{elec}	50 nJ/bit
E_{DA}	5 nJ/bit/signal
Idle power	13.5 mW
Sleep power	15 μW
Initial energy per node	2 J
Transmission bit rate	40 kbs^{-1}
Round time	20 sec.

Table 1. Parameters setting

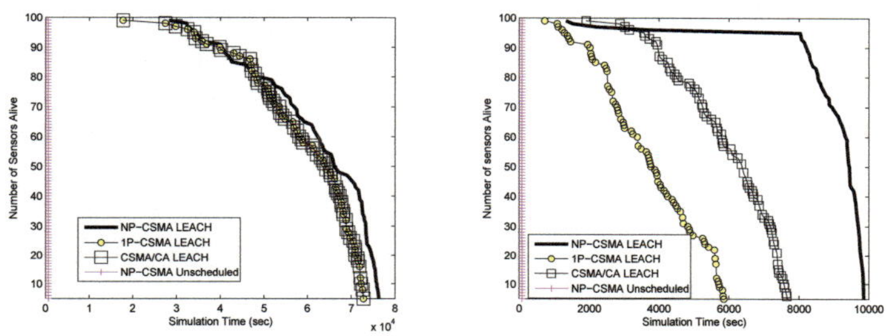

(a) q= 0.01 (b) q = 0.3

Fig. 1. Evolution in time of the number of sensors still alive in the WSN

3.2 Impact of the Random Access Protocol

Figure 1 shows the evolution in time of the number of sensors still alive in the WSN in the LEACH and the unscheduled cases. In the unscheduled case, access is arbitrated using NP-CSMA with GB policy. In the LEACH case, three random access strategies are considered: NP-CSMA, 1P-CSMA and the CSMA/CA, all with the GB policy. We use the same backoff policy (i.e., GB) in order to perceive the impact of the random access strategy on the WSN performance. Typically, we fix the backoff policy and we vary the random access strategy. Note that similar results can be obtained with the other backoff policies.

Let us first focus on the LEACH performance. Figure 1 shows that for low values of q, the different access protocols provide comparable results, whereas for moderate values of q the NP-CSMA is the best (see Fig. 1(b)). Indeed, with low values of the probability q, all the access protocols enable practically collision-free transmission and achieve thus similar energy consumption. It is worth noting that in this range of q, achieving practically collision-free transmission comes at the cost of excessive access delay to the medium. In this context, the energy wasted due to idle listening while waiting to transmit or to receive a packet is dominant compared to the energy wasted due to collisions.

In contrast, for moderate values of q, the energy wasted due to collisions is dominant since collisions are more likely to happen. In this case, NP-CSMA allows the lowest energy consumption. On the other hand, 1P-CSMA presents the highest collision probability leading thus to the highest energy consumption per unit of time when LEACH is enabled as can be

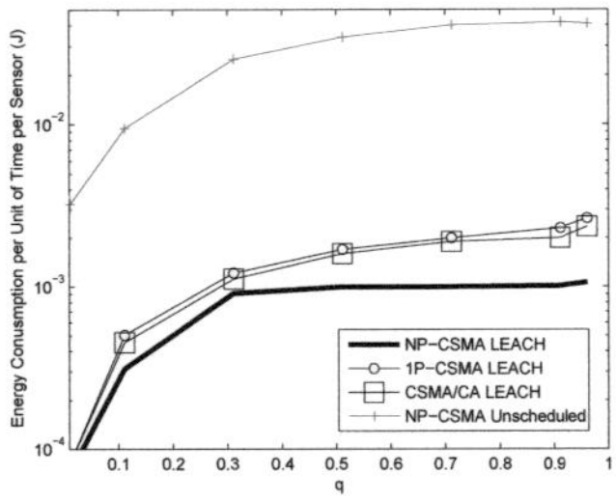

Fig. 2. Average energy consumption per unit of time per sensor node

seen in Fig. 2. In view of this, the WSN experiences the fastest sensor node energy drain with 1P-CSMA (see Fig. 1(b)).

Let us now compare LEACH to the basic unscheduled case from energy consumption perspective. We can see in Figs. 1 and 2 that LEACH achieves always significant gain compared to the basic unscheduled transmission case. This is because LEACH coordinates the sensor nodes' transmissions with a common schedule in the steady phase, which eliminates collisions, idle listening and overhearing. This gain depends on the access protocol choice. For example, Fig. 1(b) shows that using the 1P-CSMA access protocol with LEACH provides the smallest gain. This is because 1P-CSMA causes excessive collisions among the signaling messages at the cluster formation phase. This harmful wastage of energy at the cluster formation phase slows down the gain that achieves LEACH in the steady phase due to its scheduled transmission compared to the unscheduled case.

Let us now focus on the latency performance. Figure 3 depicts the reporting and the cluster formation latencies. The reporting latency is defined as the time between the report generation and its reception by the sink node. The cluster formation latency is the time needed to form the clusters, i.e., to elect the cluster heads and to construct the TDMA frames. Again, NP-CSMA allows the best results when LEACH is enabled. In this case, the reporting latency curve follows the same pace as that of the cluster formation latency curve, which is a convex function of the probability q. The rationale behind this can be explained as follows. For small values of q, the access delay to the medium during the set-up phase is very large, which induces large cluster formation latency. On the other hand, large values of q cause excessive collisions, increasing thus the time needed to transmit correctly a signaling message. Hence, the optimal cluster formation latency is a tradeoff between the above opposite requirements. In our scenario, the minimal cluster formation time is obtained when q ranges between 0.3 and 0.5. It is worth noting that the reporting latency is always lower than the cluster formation latency, since after the set-up phase, packets are transmitted in a contention-free way and sensor nodes only have to wait for their assigned time slots inside the TDMA frame.

Finally, compared to unscheduled case, the NP-CSMA-based LEACH achieves lower latencies thanks to its collision-free transmission during the steady phase.

According to the above results regarding both the energy consumption and the reporting latency, we can draw two important conclusions: i) the cluster-based LEACH architecture performs always better than an unscheduled one and ii) the NP-CSMA behaves better than the 1P-CSMA or CSMA/CA protocols for the different parameters of the backoff policy. Therefore, for the rest of the document, we use the NP-CSMA as access strategy. In the next subsection, different backoff policies are used with the NP-CSMA in order to analyze their performances.

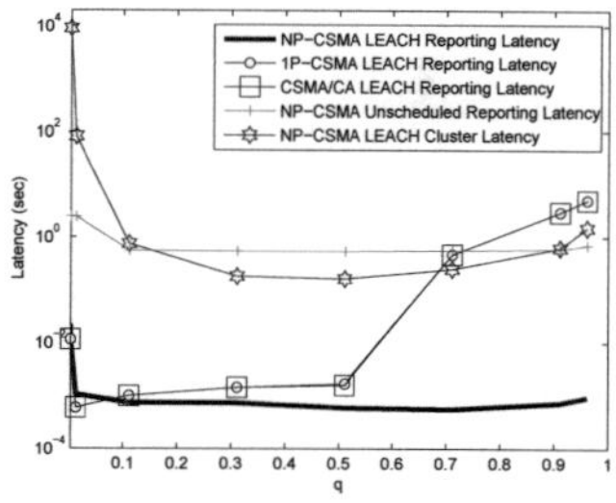

Fig. 3. Average reporting and cluster formation latencies

3.3 Impact of the Backoff Policies

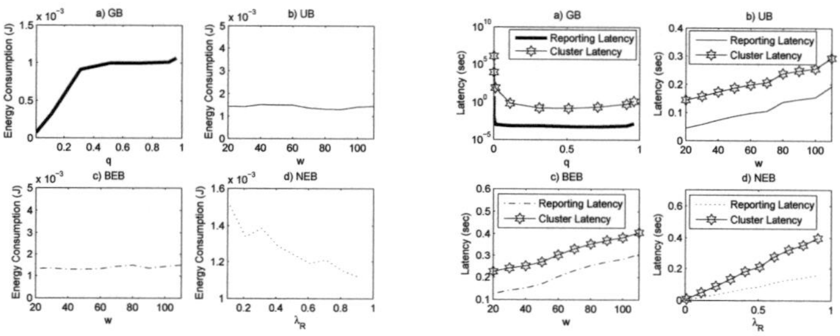

(a) Average energy consumption per unit of time per sensor node when varying the backoff policy

(b) Average reporting and cluster formation latencies when varying the backoff policy

Fig. 4. Impact of the backoff policy on the performance of the system

In this subsection, we analyze the NP-CSMA-based LEACH protocol using different backoff policies. Recall that in the previous subsection, we proved that, using the same access protocol, the cluster-based systems outperform always the unscheduled systems. Moreover, we showed that NP-CSMA stands out as the best access strategy for cluster-based systems. In this subsection, we rather look for the best backoff policy that enables further energy conservation as well as reduced reporting delay.

Figure 4 (a) compare the energy efficiency among the four backoff policies: GB, UB, BEB and NEB. The main observation is that GB provides the lowest energy consumption compared to the remaining policies, which on the other hand exhibit similar results. Specifically, Fig. 4 shows that the energy consumption with the GB policy is always below 1 mJ per unit of time, whereas it is around 1.5 mJ with the other backoff policies.

Figure 4 (b) shows the reporting and the cluster formation latencies for the four backoff policies. Again, using the GB policy the reporting and cluster latencies are convex functions of q, where minimum delays are obtained for q in the range of $[0.3, 0.5]$. Moreover, the GB policy achieves similar results (although sometimes slightly higher) as the remaining backoff policies.

Since the GB policy achieves better results in terms of energy consumption, even at the cost sometimes of slightly higher latencies compared to the other backoff policies, then the NP-CSMA with GB policy will be used as the access strategy for the rest of the manuscript.

4. Mathematical Model for LEACH

In this section, we present a mathematical model for the LEACH-enabled WSNs. Compared to (Heinzelman et al., 2002), we consider the energy consumption and the delay introduced by the cluster formation phase. We present explicit expressions for the average energy consumed per unit of time by a sensor node, the average reporting latency and the average cluster formation time. We consider the LEACH protocol with the NP-CSMA access strategy and the GB policy, where a packet transmission is done with probability q. It is important to note that the results provided by this model will be used as baselines to which the CM-EDR improvements are compared. In the next section, we present the analytical model when the CM-EDR strategy is enabled.

4.1 Energy Consumption Analysis

At the beginning of each new cycle or round, a new set of N_{CH} CHs is elected. The CH role is rotated among all sensor nodes in order to balance the energy consumption inside the WSN. The cluster formation phase can be divided into three steps: CH announcement, CM join and CH schedules. In the first step, each elected CH advertises all the sensor nodes in the WSN. Once the CH announcement step is completed, each sensor node transmits a CM join message to its associated CH. Based on this information, each CH transmits a message indicating the schedule to its associated CMs. In what follows, each step will be analyzed separately.

4.1.1 CH announcement step

At the beginning of the set-up phase, all the elected CHs try to advertise the remaining sensor nodes at the same time, leading thus to a collision occurrence. All the CH nodes undergo hence the backoff procedure. Accordingly, the channel is divided into time slots that can be used by the CHs to transmit their announcement messages. The duration of a time slot t_{sig} is by definition the time that takes a sensor to transmit a control packet.

In order to calculate the energy consumption in the CH announcement step, we consider that at any time slot, the system can be defined according to the number of potential nodes that can initiate transmission, n, and the number of actual transmissions made, m, at the beginning of the time slot. Hence, the system can be described by the duple (n, m). We make use of a transitory Markov chain in order to derive the average number of time slots that the LEACH system remains in the state (n, m) at the cluster formation phase, where n represents the number of CHs with a backlog packet (i.e., CHs that have not yet transmitted correctly their announcement messages) at the beginning of the slot k and $m \in \{0, ..., n\}$ represents the number of nodes that transmit on the slot k.

Let $X(k)$ be the system state at the slot k defined by the tuple (n, m). Then, the event $\{X(k) = (n, 0)\}$ means that no node transmits on the slot k and hence the slot remains free. $\{X(k) = (n, m)\}$ with $m > 1$ means that a collision occurs on the slot k. Finally, $\{X(k) = (n, 1)\}$ means that a successful transmission of a CH announcement message is achieved on the slot k. In this case, the next slot system state will be $X(k+1) = (n-1, m')$ with $m' \in \{0, ..., n-1\}$.

The transmission of each backlog node on a slot is achieved according to a geometric process with a probability q. Hence, the process $\{X(k), k \geq 1\}$ is a discrete time Markov chain with the state space $S = \{(n, m) \mid 0 \leq n \leq N_{CH}, 0 \leq m \leq n\}$ as depicted in Fig. 5. The space state S can be also expressed as follows:

$$S = \bigcup_{n=0}^{N_{CH}} S_n, \text{ with } S_n = \{(n, m) \mid 0 \leq m \leq n\} \tag{2}$$

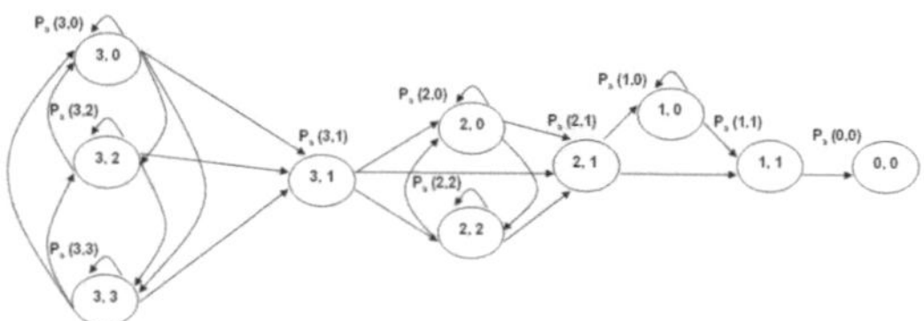

Fig. 5. State transition diagram of the Markov chain X: case $N_{CH} = 3$

To calculate the average energy consumption during the CH announcement step, we need to calculate the average number of visits of each state $s \in S$ before entering the $(0,0)$ absorbing state.

The initial number of backlog CHs is N_{CH}. Hence, the system evolution starts at a state $s \in S_{N_{CH}}$. Specifically, $X(1) = (N_{CH}, m)$, with a probability

$$p_a(N_{CH}, m) = \binom{N_{CH}}{m} q^m (1-q)^{N_{CH}-m}, \; \forall\, m = 0, ..., N_{CH}. \tag{3}$$

Note that $\sum_{m=0}^{N_{CH}} p_a(N_{CH}, m) = 1$.

Any state $s \in S_{N_{CH}}$, i.e., $s \in \{(N_{CH}, m),\ m = 0, ..., N_{CH}\}$, could be visited several times until the system visits the state $(N_{CH}, 1)$, let say at slot k. This signifies that a successful CH transmission occurs at slot k and hence the remaining number of backlog CHs becomes $N_{CH} - 1$. The system evolves thus to the state $X(k+1) \in S_{N_{CH}-1}$ with a probability $p_a(N_{CH} - 1, m)$, $m = 0, ..., N_{CH} - 1$. Again this set of states $S_{N_{CH}-1}$ continues to be visited until the system visits the state $(N_{CH} - 1, 1)$, and so on and so forth.

Building on these observations, we can see that the number of visits to a state $(n, 1), 1 \leq n \leq N_{CH}$, before entering the absorbing state $(0,0)$ is equal to 1. Moreover, calculating the number of visits of the process X to a generic state (n, m), with $1 \leq n \leq N_{CH}$ and $0 \leq m \neq 1 \leq n$, before entering the absorbing state $(0,0)$ turns out at calculating the number of visits of the state (n, m) before entering the state $(n, 1)$, given that the system starts its evolution at the set of states S_n with an initial probability distribution $(p_a(n, 0), \ldots, p_a(n, n))$.

Hence, instead of studying the general process $\{X(k), k \geq 1\}$ to compute the average number of visits of a state (n, m), we can limit our study to the process $Z_n = \{(Z_n(r), r \geq 1\}$. Z_n is a Markov chain on the finite space $S_n = \{(n, 0), \ldots, (n, n)\}$, where $S_n \backslash (n, 1)$ is the set of the transient states and $(n, 1)$ is the absorbing state. This Markov chain can be solved as in (Sericola, 1990), (Bouabdallah, 2009) and the average number of visits of Z_n to the state (n, m) is given by:

$$E\left[N_{\{(n,m)\}}\right] = \frac{p_a(n, m)}{p_a(n, 1)} \tag{4}$$

Accordingly, the total energy consumption in the WSN during the CH announcement step can be calculated as follows:

$$E_{CH\_Announ} = f(N_{CH}, l_{sig}) = N_{CH}E_{tx}(l_{sig}, d_{\max}) + (N - N_{CH})\, E_{rx}(l_{sig})$$
$$+ \sum_{n=1}^{N_{CH}} \sum_{m=1}^{n} E\left[N_{\{(n,m)\}}\right] \left(mE_{tx}(l_{sig}, d_{\max}) + (N - m)\, E_{rx}(l_{sig})\right)$$
$$+ NE_{idle}t_{sig} \sum_{n=1}^{N_{CH}} E\left[N_{\{(n,0)\}}\right] \quad (5)$$

where l_{sig} denotes the size of a control packet, $d_{\max} = \sqrt{2}M$ the diameter of the $M \times M$ square supervised area and E_{idle} the average amount of energy consumed per unit of time by a sensor node in the idle state. We highlight that the first element of (5) corresponds to the energy dissipated in the WSN due to the first collision among all the CHs when attempting to send for the first time all together their announcement messages at the beginning of the set-up phase. The remaining elements of (5) correspond to the energy consumption during the backoff procedure that undergo the N_{CH} CHs.

4.1.2 CM join step

As explained before, once the CH announcement step is completed, each sensor node transmits a CM join message to its associated CH. Similarly to the CH announcement step, the $N - N_{CH}$ sensor nodes try to join their CHs at the same time, leading thus to a collision occurrence. Then, the sensor nodes enter in backoff procedure to transmit their CM join messages. Following the same reasoning as in the CH announcement step (i.e., using (5)), we obtain the average energy dissipated during the CM join step as $E_{CM\_Join} = f(N - N_{CH}, l_{sig})$.

4.1.3 CH schedules step

In this step, each CH transmits a message indicating the schedule to its associated CMs. Using the same reasoning as before, the average energy consumed during the CH schedules step is given by $E_{CH\_Sched} = f(N_{CH}, l_{sig})$.
Finally, the average amount of energy dissipated to form clusters is:

$$E_{Set\text{-}up}(LEACH) = E_{CH\_Announ} + E_{CM\_Join} + E_{CH\_Sched} \quad (6)$$

4.1.4 Energy consumption in the steady phase

Let us now calculate the average amount of energy consumed during the steady phase, where each CH receives periodically a TDMA frame from its CMs. In our study, we assume that the N sensor nodes are uniformly distributed in the supervised area. Hence, there are on average N/N_{CH} nodes, including the CH, in each cluster.
In continuous-monitoring WSNs, each sensor node senses its area periodically, each $T_{sensing}$ period of time, where $T_{sensing} \geq T_{frame}$. We note that $T_{frame} = \frac{N}{N_{CH}} t_{data}$ is the duration of a TDMA frame, where t_{data} is the duration of a time slot needed by a sensor to transmit a data packet of size l_{data}. In the particular case where $T_{sensing} = T_{frame}$, the WSN operates in the saturation regime, i.e., a sensor node always has data to send to the sink node. Since each sensor node wakes up only during its attributed time slot, then the energy consumed by a CM i node during a sensing period $T_{sensing}$ is:

$$E_{CM}(i) = \left(T_{sensing} - t_{data}\right) E_{sleep} + E_{tx}(l_{data}, d_{CM(i)\_CH}) \quad (7)$$

where E_{sleep} is the average amount of energy consumed by a sensor node per unit of time in the sleep state and $d_{CM(i)\_CH}$ is the distance between the CM node i and its associated CH. In (Heinzelman et al., 2002), it was demonstrated that if the density of nodes is uniform throughout the cluster area, then the expected square distance from the CM nodes to the CH is given by $E\left[(d_{CM\_CH})^2\right] = \frac{M^2}{2\pi N_{CH}}$ where M is the side length of the square supervised area. Hence the average amount of energy consumed by a CM node during a sensing period is:

$$E_{CM}(LEACH) = \left(T_{sensing} - t_{data}\right) E_{sleep} + E_{tx}\left(l_{data}, \frac{M}{\sqrt{2\pi N_{CH}}}\right)$$

In turn, each CH consumes energy in receiving and aggregating the data sent by its CMs as well as in the transmission of that aggregated data to the sink node. The energy consumed by a CH node during a TDMA frame is therefore:

$$E_{CH\_frame}(LEACH) = \left(\frac{N}{N_{CH}} - 1\right) E_{rx}(l_{data}) + \frac{N}{N_{CH}} l_{data} E_{DA} + E_{tx}(l_{data}, d_{CH\_SN})$$

where d_{CH_SN} is the average distance from the CH to the sink node. Thus, the energy consumed by a CH node during a sensing period is:

$$E_{CH}(LEACH) = E_{CH\_frame}(LEACH) + \left(T_{sensing} - T_{frame}\right) E_{sleep}$$

The energy consumed in the network during a sensing period is therefore:

$$E_{WSN}(LEACH) = N_{CH}\left(\left(\frac{N}{N_{CH}} - 1\right) E_{CM}(LEACH) + E_{CH}(LEACH)\right)$$

and the total energy consumed in the network during the steady phase is:

$$E_{Steady}(LEACH) = E_{WSN}(LEACH) \times \frac{T_{round} - T_{set-up}(LEACH)}{T_{sensing}}$$

where T_{round} is the round time after which the CH nodes are elected anew and $T_{set-up}(LEACH)$ is the average time spent in the cluster formation phase, which will be derived in the next subsection.

Finally, we obtain the average amount of energy consumed by each sensor node in the WSN per unit of time when the basic LEACH clustering is adopted:

$$E_{sensor}(LEACH) = \frac{E_{Steady}(LEACH) + E_{Set-up}(LEACH)}{NT_{round}} \tag{8}$$

4.2 Latency Analysis

In this subsection we derive both the average cluster formation time and the average reporting latency.

4.2.1 The average cluster formation time

It is the time needed to form the clusters, i.e., to perform the CH announcement, the CM join and the CH schedules steps. Using the same model introduced in the previous section, the CH announcement time is simply the time elapsed from the beginning of the cluster formation procedure to the instant where all the CHs successfully transmit their announcement message. As such, the CH announcement time can be expressed as follows:

$$T_{CH\_Announ} = g(N_{CH}, t_{sig}) = \left(1 + \sum_{n=1}^{N_{CH}} \sum_{m=0}^{n} E\left[N_{\{(n,m)\}}\right]\right) t_{sig} \quad (9)$$

We highlight that (9) is the sum of the time lost due to the first collision among all the CHs when attempting to send for the first time all together their announcement messages (i.e., t_{sig}) and the average duration of the backoff procedure that undergo the N_{CH} CHs.
Following the same reasoning, we obtain the average time spent in the CM join and the CH schedules steps as follows:

$$T_{CM\_Join} = g(N - N_{CH}, t_{sig}) \quad (10)$$

$$T_{CH\_Sched} = g(N_{CH}, t_{sig}) \quad (11)$$

Finally, the average time needed to form clusters is:

$$T_{Set-up}(LEACH) = T_{CH\_Announ} + T_{CM\_Join} + T_{CH\_Sched} \quad (12)$$

4.2.2 The average reporting latency

It is the time needed by a generated report to be received by the sink node. In continuous-monitoring WSNs, the sensor nodes produce data information at the beginning of each sensing period. In the steady phase, the average reporting time is simply the transmission time of a TDMA frame. Considering the extra delay spent in the construction of the clusters, the reporting latency increases slightly as follows:

$$T_{reporting}(LEACH) = T_{frame} + \frac{T_{set-up}(LEACH)T_{sensing}}{T_{round}} \quad (13)$$

5. Energy Efficient Protocols for Continuous-Monitoring Applications

This section introduces our CM-EDR scheme. In the previous section, we presented a mathematical analysis for the classical continuous-monitoring LEACH WSNs. In this section, we analyze the corresponding CM-EDR-aware extension. Comparing the new results, i.e., the average energy consumption, the average reporting latency and the average cluster formation time, to that obtained with the classical approach, we can gauge the benefits introduced by the proposed CM-EDR technique.

5.1 The CM-EDR Scheme

The main idea behind the CM-EDR introduction is avoiding the extra transmission of non relevant data information, typical in classical continuous-monitoring WSNs. With CM-EDR, continuous-monitoring does not imply indeed continuous reporting. By reporting only relevant data, the sink node would gather exactly the same information as with classical continuous-monitoring applications while receiving less reports and thus dissipating less energy.

Enabling the CM-EDR technique, each sensor node continues to produce periodically data information. However, the sensed information is reported to the sink node only if it differs from the last transmitted data information. In doing so, the sensor node dissipates also less energy in communications, achieving thus significant energy conservation. Clearly, the energy consumption will greatly depend on the rate of variation of the phenomenon that the sensors are monitoring.

With CM-EDR, each sensor node needs to storage the last transmitted data (i.e., only a single packet). Evidently, this does not entail the need to increase the memory capacity of sensor nodes. Following to each periodic observation, the sensor node compares the new reading to the stored one. If both readings are similar, the new generated data packet is discarded. Otherwise, the new information is reported to the sink node and the stored information is updated. In this case, we deal with relevant data, referred to us also as an event.

It is worth noting that our approach can be seen as a new alternative to reduce the transmission of redundant information, by profiting from the natural temporal correlation among the sensed data information. Our technique complement the data fusion or aggregation techniques (Intanagonwiwat et al., 2000) – (Larrea er al., 2007) and the spatial-correlation based schemes (Bouabdallah et al., 2009) – (Vuran et al., 2006).

5.2 Analytical Model for the CM-EDR-enabled LEACH WSNs

This subsection extends the analysis done in section IV to the case where the CM-EDR technique is enabled. Since the CM-EDR technique does not affect the set-up phase, the analysis for this phase remains unchanged. Hereafter, we focus on the analysis of the steady phase.

Assume that the variations on the sensed information, for example the temperature around a sensor node, happen following a Poisson process of rate λ. In other words, the time between two variations of the temperature is exponentially distributed. In our case, each sensor node senses its area periodically, each $T_{sensing}$ period of time. $T_{sensing}$ is chosen by the administrator such that the probability that two or more changes on the sensed information occurs during $T_{sensing}$ be negligible, i.e., be below a certain threshold ε as follows:

$$\Pr\{N_{event} \geq 2\} = 1 - e^{-\lambda T_{sensing}} - \lambda T_{sensing} e^{-\lambda T_{sensing}} \leq \varepsilon \tag{14}$$

where N_{event} is the number of changes that occurs on the sensed information during $T_{sensing}$. As such, $T_{sensing}$ must verify:

$$T_{sensing} \leq \sup\{t \mid 1 - e^{-\lambda t} - \lambda t e^{-\lambda t} \leq \varepsilon\} \tag{15}$$

Hence, the probability that the sensed information be relevant, for example the temperature changes between two observations, i.e., during the last $T_{sensing}$ period, is given by:

$$P_{event} \simeq \Pr\{N_{event} = 1\} = \lambda T_{sensing} e^{-\lambda T_{sensing}} \tag{16}$$

Based on this model, during the steady phase each CM-EDR-enabled sensor node transmits on its reserved slot (i.e., uses the current frame) according to a geometric process of probability P_{event}. Assuming that a CM node enters the sleep mode during the sensing period and wakes up only on its associated slot if it has relevant data to transmit, the average amount of energy consumed by a CM node during a sensing period is:

$$E_{CM}(CM\text{–}EDR) \quad = \quad P_{event} E_{CM}(LEACH) + (1 - P_{event})\, T_{sensing} E_{sleep}$$

On the other hand, each CH consumes energy in receiving and aggregating the data sent by its CMs as well as in the transmission of that aggregated data to the sink node. The average amount of energy dissipated by a CH node in the reception of a frame can be given by:

$$E_{CH\_rec} = \sum_{k=0}^{\lceil \frac{N}{N_{CH}} \rceil -1} \binom{\lceil \frac{N}{N_{CH}} \rceil -1}{k} (P_{event})^k (1-P_{event})^{\lceil \frac{N}{N_{CH}} \rceil -1-k} \times \left(kE_{rx}(l_{data}) + t_{data}E_{idle}\left(\left\lceil \frac{N}{N_{CH}} \right\rceil -1-k \right) \right)$$

Assuming perfect data aggregation, the average amount of energy dissipated by a CH node due to aggregation is:

$$E_{CH\_agg} = \sum_{k=0}^{\lceil \frac{N}{N_{CH}} \rceil} \binom{\lceil \frac{N}{N_{CH}} \rceil}{k} (P_{event})^k (1-P_{event})^{\lceil \frac{N}{N_{CH}} \rceil -k} \times (kl_{data}E_{DA})$$

The average amount of energy dissipated by a CH for a possible transmission of the aggregated data to the sink node is:

$$E_{CH\_tr} = \left(1 - (1-P_{event})^{\frac{N}{N_{CH}}} \right) E_{tx}(l_{data}, d_{CH\_SN})$$

Hence, the total energy consumed by a CH node during a TDMA frame when CM-EDR is enabled is:

$$E_{CH\_frame}(CM\text{–}EDR) = E_{CH\_rec} + E_{CH\_agg} + E_{CH\_tr} \tag{17}$$

and the energy consumed by a CH node during a sensing period is:

$$E_{CH}(CM\text{–}EDR) = E_{CH\_frame}(CM\text{–}EDR) + \left(T_{sensing} - T_{frame} \right) E_{sleep}$$

The energy consumed in the network during a sensing period is therefore:

$$E_{WSN}(CM\text{–}EDR) = N_{CH}\left(E_{CH}(CM\text{–}EDR) + \left(\frac{N}{N_{CH}} - 1 \right) E_{CM}(CM\text{–}EDR) \right)$$

and the total energy consumed in the network during the steady phase is:

$$E_{Steady}(CM\text{–}EDR) = E_{WSN}(CM\text{–}EDR) \times \frac{T_{round} - T_{set-up}(LEACH)}{T_{sensing}}$$

Finally, we obtain the average amount of energy consumed by each sensor node in the WSN per unit of time when the CM-EDR option is enabled:

$$E_{sensor}(CM\text{–}EDR) = \left(E_{Steady}(CM\text{–}EDR) + E_{Set-up}(LEACH) \right) \frac{1}{NT_{round}}$$

With regard to the latency performance, it is worth noting that the CM-EDR scheme does not impact the latency compared to the classical LEACH case. Indeed, a relevant data packet is received by the sink node at the same time whether the CM-EDR mechanism is enabled or not. The CM-EDR mechanism avoids only the transmission of non relevant data.

5.3 Optional Mechanism for CM-EDR-enabled Cluster-Based WSNs

Using CM-EDR, a CH node transmits to the sink node only if it senses or receives relevant data from its CMs. As such, the CH may not transmit to the sink during a long period if it does not receive any relevant information. Even though, it dissipates energy due to idle listening. The energy wasted due to idle listening is far from being negligible and can account for a significant portion of the energy a sensor dissipates in some cases (Woo et al., 2001).
To achieve further energy conservation, the CH will be allowed with the optional CM-EDR (OCM-EDR) to enter sleep mode during N_{sleep} sensing periods if it does not receive any relevant data during N_{idle} consecutive frames. The CH assumes indeed that the supervised environment is *"calm"* and it is improbable that an event occurs in the next sensing periods. In this case, the CH advertises its CMs that it will undergo the sleep state during N_{sleep} sensing periods. However, during this period, a CM node may sense a relevant data that needs to be reported immediately (i.e., in the current frame) to the sink node, otherwise continuous-monitoring property is lost. To do so, the sensor node is allowed to transmit directly this information to the sink node during its reserved slot.
Let us now calculate the average energy consumption by a sensor node when this optional mechanism is enabled. Let $Y(k)$ be the CH state at the sensing period k of the steady phase defined by the tuple (i, j), where $i = 0$ if the CH is in the sleep state and $i = 1$ otherwise. Moreover, if $i = 0$, $j = 1, ..., N_{sleep}$ signifies that the CH has been for j sensing periods in the sleep state (including the current sensing period); otherwise (i.e., if $i = 1$) $j = 1, ..., N_{idle}$ indicates the number of consecutive empty (non relevant) frames that has received the CH. The process $Y = \{Y(k), k \geq 1\}$ is a discrete time Markov chain with the state space $S = \{(i, j) \mid 0 \leq i \leq 1, 1 \leq j \leq N_{sleep}1_{\{i=0\}} + N_{idle}1_{\{i=1\}}\}$. For every $s \in S$, we denote by

$$\Pi_s = \lim_{k \to +\infty} \Pr\{Y(k) = s\}$$

where $\Pi = [\Pi_s]$ is the steady state distribution of the Markov chain Y, which satisfies

$$\Pi P = \Pi \text{ and } \sum_{s \in S} \Pi_s = 1, \tag{18}$$

and $P = (P(s, s')), s = (i, j), s' = (i', j') \in S$, is the transition probability matrix of Y given by:

$$P(s,s') = \begin{cases} P_{free} & \text{if } \left(i = i' = 1 \text{ and } j' = j + 1\right); \\ 1 - P_{free} & \text{if } \left(s' = (1,1) \text{ and } s = (1,j) \text{ with } j < N_{idle}\right); \\ 1 & \text{if } \begin{cases} \left(i = i' = 0 \text{ and } j' = j + 1\right) \\ \text{or } \left(s = (1, N_{idle}) \text{ and } s' = (0,1)\right) \\ \text{or } \left(s = (0, N_{sleep}) \text{ and } s' = (1,1)\right); \end{cases} \\ 0 & \text{otherwise.} \end{cases} \tag{19}$$

where P_{free} is the probability that the CH node does not transmit to the sink node since it has not any relevant data to forward. P_{free} is given by:

$$P_{free} = (1 - P_{event})^{\frac{N}{N_{CH}}} \tag{20}$$

Let $K = \left\lceil \frac{T_{round}}{T_{sensing}} \right\rceil$ denote the number of sensing periods during a round. We denote by P_{CH_sleep} the percentage of sensing periods in a round, during which a CH is in the sleep state. P_{CH_sleep} can be expressed as follows:

$$P_{CH\_sleep} = \frac{1}{K} \sum_{j=1}^{N_{sleep}} V_{(0,j)}(K) \tag{21}$$

where $V_{(0,j)}(K)$ is the number of visits to the state $(0, j)$ during a round, i.e., during the K first transitions of process Y. Then, P_{CH_sleep} is given by:

$$\begin{aligned} P_{CH\_sleep} &= \frac{1}{K} \sum_{j=1}^{N_{sleep}} \sum_{k=1}^{K} \Pr\{Y(k) = (0,j)\} \\ &= \frac{1}{K} \sum_{j=1}^{N_{sleep}} \sum_{k=1}^{K} \left(\alpha P^k\right)_{(0,j)} \end{aligned} \tag{22}$$

where α is the initial probability distribution of Y and $\left(\alpha P^k\right)_{(0,j)}$ is the $(0, j)$ element of the vector αP^k. Note that when K goes to the infinity, P_{CH_sleep} denotes the probability that a CH is in the sleep state during a sensing period, i.e.,

$$\lim_{K \to +\infty} P_{CH\_sleep} = \sum_{s \in S} \Pi_s 1_{\{i=0\}} = \sum_{j=1}^{N_{sleep}} \Pi_{(0,j)} \tag{23}$$

Deriving the steady state distribution of the Markov chain Y, we get

$$\begin{aligned} \lim_{K \to +\infty} P_{CH\_sleep} &= \sum_{j=1}^{N_{sleep}} N_{sleep} \left(P_{free}\right)^{N_{idle}-1} \Pi_{(1,1)} \\ &= \frac{N_{sleep}\left(1 - P_{free}\right)\left(P_{free}\right)^{N_{idle}-1}}{1 - \left(P_{free}\right)^{N_{idle}} + N_{sleep}\left(1 - P_{free}\right)\left(P_{free}\right)^{N_{idle}-1}} \end{aligned} \tag{24}$$

Now, we can derive the average amount of energy consumed by a CM node during a sensing period as follows:

$$\begin{aligned} E_{CM}(OCM\text{–}EDR) = &(1 - P_{event})\, T_{sensing} E_{sleep} \\ &+ P_{event}\left(1 - P_{CH\_sleep}\right) E_{CM}(LEACH) \\ &+ P_{event} P_{CH\_sleep}\Big(E_{tx}\left(l_{data}, d_{CM\_SN}\right) \\ &+ \left(T_{sensing} - t_{data}\right) E_{sleep}\Big) \end{aligned} \tag{25}$$

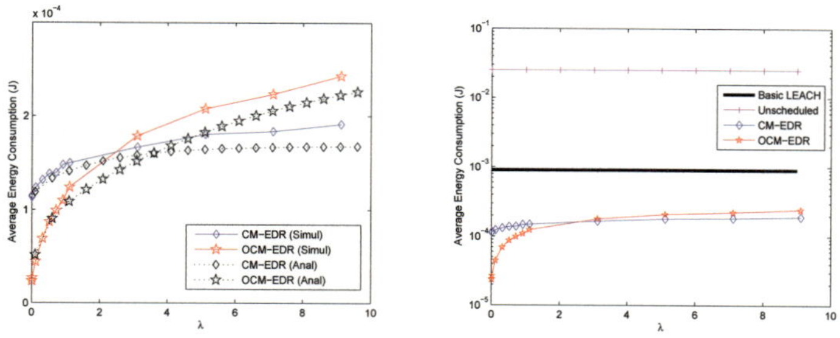

(a) Proposed protocols (b) Comparison with LEACH

Fig. 6. Average energy consumption per unit of time per sensor node

where d_{CM_SN} is the average distance between a CM node and the sink node. On the other hand, the average energy consumed by a CH node during a sensing period with OCM-EDR is:

$$E_{CH}(OCM\text{–}EDR) = \left(1 - P_{CH\_sleep}\right) E_{CH}(CM\text{–}EDR) + P_{CH\_sleep} T_{sensing} E_{sleep}$$

Using the expressions of $E_{CM}(OCM\text{–}EDR)$ and $E_{CH}(OCM\text{–}EDR)$ given by (we derive in the way as in (18) the average energy consumed by a sensor node with OCM-EDR.

5.4 Numerical Results

We now evaluate the efficiency of our proposed mechanisms We first study the gain that they introduced using four baseline examples: the case of unscheduled WSNs and three variants of cluster-based WSNs. Then, we compare between the CM-EDR and OCM-EDR mechanisms.
A simulation model has been developed in order to validate the analytic results. The system of WSNs was implemented as a discrete event simulation. Numerous evaluations were performed in order to confirm the analytic results. In all cases, the results matched very closely. Figure 6 (a) compares the simulation results of the energy consumption with CM-EDR to that given by equation (18) as a function of the rate λ. In this case, $T_{sensing}$ is chosen such that it verifies the constraint given by (15) with $\varepsilon = 10^{-4}$. For the OCM-EDR mechanism, Figure 6 compares the simulation results of the energy consumption as a function of λ. In this case, we consider $N_{idle} = 1$, $N_{sleep} = 10$ and $\varepsilon = 10^{-4}$. Figure 6 (a) shows that there is a good fit between the simulation and analytical results, which exhibits the accuracy of our analysis.
For the remainder of the results, it has been confirmed that there is a good fit between the simulation and analytical results. Therefore, for presentation purposes, all remaining figures show only the simulation results. We assume the same network topology used in the previous sections, i.e., 100 sensor node-network. We assume also that $\varepsilon = 10^{-4}$, i.e., $T_{sensing} = \sup\{t \mid 1 - e^{-\lambda t} - \lambda t e^{-\lambda t} \leq 10^{-4}\}$. Moreover, unless explicitly notified, we consider $q = 0.3$, $N_{idle} = 1$ and $N_{sleep} = 10$. The parameters setting in our experiments are listed in table I.
According to the results presented in Fig. 6 we can draw three main observations:

- Clustering achieves always significant gain in terms of energy Further energy conservation can be achieved when the CM-EDR mechanisms are enabled, which brings us to the second observation.

- The sensor node lifetime is increased considerably when enabling our CM-EDR mechanisms. Clearly, the CM-EDR abilities provide an advantage over the classical WSNs, by preventing the transmission of redundant data. For reference, Fig. 6 (b) shows the relative decrease in the energy consumption by a sensor node per unit of time of the CM-EDR networks compared to the classic networks. The magnitude of the increase regarding the sensor node lifetime decreases as the rate λ grows. In other words, the relative improvement decreases when the supervised area becomes agitated since less non relevant data are transmitted by the classical WSNs.
- The OCM-EDR mechanism outperforms the CM-EDR one, when we deal with calm WSNs, whereas in agitated WSNs, it is better to use the basic CM-EDR mechanism. The rationale behind this can be explained as follows. Allowing the CHs to go to sleep with OCM-EDR results in the occurrence of expensive direct transmissions from the CMs to the sink node. In agitated environment, the energy conservation achieved at the CHs due to their asleep abilities is dominated by the additional energy consumed at the CM nodes due to frequent direct communications to the sink node. These direct communications become rare in calm WSNs.

Clearly, the CM-EDR systems are a major improvement over the classic networks. Figure 6(b) shows the average amount of energy consumed by a sensor node per unit of time as a function of the rate λ. Again, we can observe that the CM-EDR abilities provide significant energy conservation, notably in calm WSNs. This improvement decreases with λ. Moreover, enabling the optional version OCM-EDR is helpful only for small to moderate values of λ; otherwise, the basic version of CM-EDR performs better.

Figure 7 provides more insight into the effectiveness of using the OCM-EDR extension instead of the basic CM-EDR mechanism in the context of cluster-based WSNs. In this case, the two variants of the CM-EDR technique are introduced over a classical LEACH WSN. Note that similar results can be obtained when using the remaining clustering protocols. Figure 7 shows the performance of OCM-EDR as a function of the setting parameters N_{idle} and N_{sleep} for various values of the rate λ. Recall that with the optional OCM-EDR, the CH enters the sleep mode during N_{sleep} sensing periods if it does not receive any relevant data during N_{idle} consecutive frames.

The energy consumption with OCM-EDR is a convex function of N_{idle} (see Fig. 7(a)). For low values of N_{idle}, the CHs enter frequently to the sleep mode. Hence, the sensor nodes are most likely transmitting directly to the sink node instead of passing through the CHs. On the other hand, when N_{idle} gets large values, the CHs almost never enter the sleep mode and can not profit from the calm periods of the supervised environment. Hence, the energy consumption increases. For moderate values of N_{idle}, the CHs enter the sleep mode without really penalizing the sensor nodes. In our scenario, setting $N_{idle} = 25$ enables the minimal energy consumption in the network (see Fig. 7(a)).

In the same way, the energy consumption with OCM-EDR is a convex function of N_{sleep} (see Fig. 7(b)). Decreasing N_{sleep}, the CHs enter into the sleep state for very short periods of time and hence can not really profit from the calm periods of the supervised environment. In our example, the energy consumption is minimal when $N_{sleep} = 36$ (see Fig. 7(b)).

With regard to reporting latency, we can see that OCM-EDR achieves always better results than the basic CM-EDR. This is because the OCM-EDR mechanism replaces some relatively long multi-hop transmissions (i.e.,

To conclude this study, we can state that the CM-EDR philosophy enables significant energy conservation while ensuring continuous-monitoring applications. The decision to use the op-

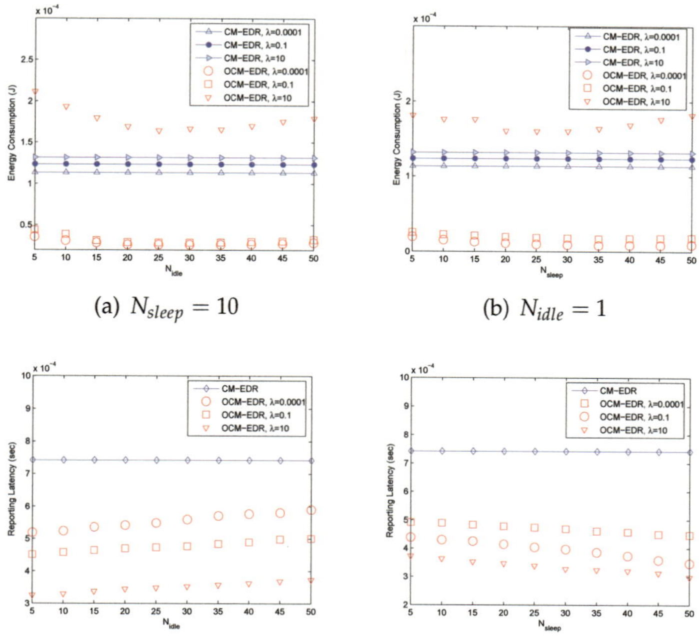

(a) $N_{sleep} = 10$ (b) $N_{idle} = 1$

(c) $N_{sleep} = 10$ (d) $N_{idle} = 1$

Fig. 7. Comparison between OCM-EDR and CM-EDR

tional OCM-EDR instead of the basic CM-EDR mechanism depends on the supervised environment, whether it is calm or agitated. When OCM-EDR is preferred, the optimal parameter values of N_{idle} and N_{sleep} should be used to configure the sensor nodes.

6. Energy Efficient Protocols for Event-Driven Applications

For the event-driven applications, a new compression technique is proposed. Since the benefits of the clustering technique have been studied, the proposed scheme presented in this section is also clustered-based. The main idea behind our proposal is to exploit the spatial correlation of such networks in order to reduce the size of the data packets that will be sent by means of data compression. The proposed clustering scheme is based on selecting a data value as reference while the rest of the active nodes transmit only the difference between their sensed value and this reference value. Hence, one major issue in the proposed mechanism is to appropriate select the reference node that achieves the highest reduction of the packet size among all active nodes. The proposed scheme is evaluated analytically and by simulations. The results show that the proposed scheme may reduce as much as 11 times the energy consumption compared to a classical clustering scheme.

6.1 Network model

We consider an event-driven WSN consisting of N sensors deployed over a vast field We denote the *i-th* sensor node by n_i and corresponding sensor node set $S = \{n_1, n_2, ..., n_N\}$ where $|S| = N$. Some assumptions about the sensor nodes and the underlaying network model are now presented:

- Nodes are uniformly distributed in an $A \times A$ area with (x, y) coordinates. Nodes are homogenous and have the same capabilities. Each node is assigned a unique identifier ID.
- Sensor nodes and the Base Station (BS) are all stationary after deployment. The BS can be reached by sensor nodes under a single high transmission range R_t meters.
- Nodes have two power controls to vary the transmission power which depends on the distance to the receiver. Each node n_i can reach any other node with a transmission range R_c. The BS can be reached with transmission range $R_t > R_c$.
- CHs use the average operation as the aggregation to eliminate the data redundancy. Other aggregation techniques, such as those proposed in (Azim et al., 2010) can also be implemented.

The center of the event is located in a random uniformly distributed point with coordinates (x_{event}, y_{event}) within the network's area. The range of the event area, i.e., the area where sensors can detect the event is *R_event* meters, where $R\_event \in [1, A] meters$. We also suppose that an event has a duration of *T_event* seconds. Additionally, in our model, only one event can be active inside the systemńs area and the data value C at the center of the event is constant, i.e, a stationary model in which the measured data do not change during the *T_event* seconds that the event is active is used.
A clustered based WSN is considered, where only one CH is elected for each event. The clustering process is triggered whenever an event is sensed by the nodes inside the event area.
The spatial correlation of the data from the different active nodes (the nodes that sense the event) can be considered according to the following models:

1. Diffusion propriety (Faruque, et al.).
2. Data is jointly gaussian with the correlation being a function of the distance (Vuran et al., 2006).
3. Data is a function for their joint entropy (Pattem et al., 2004).
4. Correlation is calculated from realistic environmental monitoring and testbeds.

We use the diffusion propriety to model the spatially correlated data (Jindal et al., 2004). The model considered in this paper is the same as in (Faruque, et al.) in which the data reading at a distance d from the center of the event is $D \propto f(1/d)$. Specifically, the data reading in any point at distance d from the center of the event is $D = C/(d+1)^{\alpha}$, where C is a constant representing the value at the center of the event, and α is the diffusion parameter which depends on the particular environment and phenomenon under surveillance, e.g., for light $\alpha = 2$, heat $= \alpha \simeq 1$.
Fig. 8 shows the data reading using the aforementioned model, with different values of α and $C = 250$. On one hand, when $\alpha \geq 1$, we observe a relatively big difference between the value at the event center of the event and the values observed at distance d far away from the center of the event. On the other hand, when $\alpha < 1 (\alpha = 0.1, 0.01$ and $0.001)$, the data readings away from the center of the event are very similar. In our study, we are interested on the type of events where data values are highly correlated.
We use Henizelman's energy consumption model (Heinzelman et al., 2002).

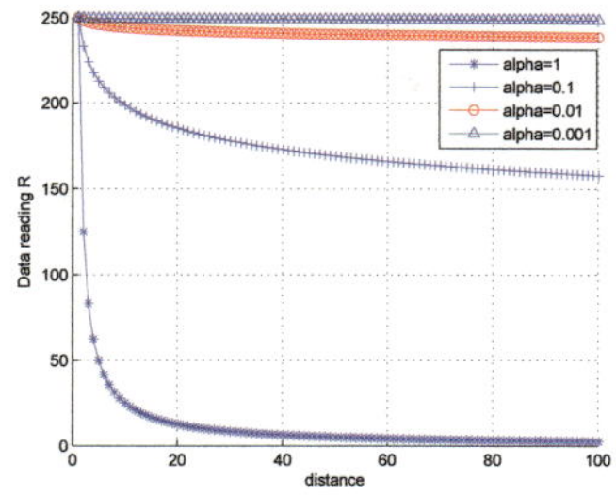

Fig. 8. Variation of data reading with distance d from the center of the event.

6.2 Classical clustering protocol

A classical clustering process is composed by two phases: set up phase and steady state phase. When an event occurs in a random (uniformly distributed) point of the network, nodes inside the event area weak up and start the clustering process. At the beginning of this phase, active nodes compete among each other to become CH. Specifically, active nodes transmit their control packet to the BS according to the specified random medium access protocol. For this protocol the NP-CSMA control protocol is used since it has been proven to be the most energy efficient protocol. The control packet only comprises the node's ID and no data are transmitted at this point. The first node that successfully transmits this packet becomes the CH. All nodes involved in the event reporting immediately send their signaling message to the BS. Therefore, the BS selects the first node that transmitted successfully the signaling message and broadcasts a signaling message over the network for a CH notification. Thus the rest of the nodes inside the event area become CMs. In the steady phase, CMs send their data in a scheduled fashion using a Time Division Multiple Access (TDMA) protocol. The CH aggregates the data values received from its CMs with its own data and sends the resulted data to the BS.

6.3 Proposed clustering protocol

The proposed clustering process is also composed of the same two phases, namely: set up phase and steady state phase. As in the classical protocol, the set up phase is triggered whenever an event occurs in a region of the network. However, in the proposed scheme, the active nodes send their first measured data value to the BS, i.e., they no longer send just their control packet. Instead, active nodes send a data packet. The reason for this is that, this sensed data is used for the CH selection procedure. Indeed, this entails an extra energy consumption at the set up phase compared to the classical protocol. However, this first data transmission allows important energy saving in the steady state phase.

It is important to notice that, our proposed scheme is best suited for environments where the event conditions are fairly stable during the event duration. This is due to the fact that the CH is chosen according to the first sensed data. Hence, if the event conditions suffer a high variation, the originally selected CH may no longer render acceptable energy savings. Example of such applications includes fire surveillance forest, in which when a fire occurs in a region, the temperature remains stationary for the duration of that fire in this region. Another example of application includes target tracking. In this kind of application, the target is the source of the measured data at sensor nodes, such as light or temperature. Here, the measured data remains the same whenever the target stays in the same place and hence the sensor nodes sense the same measured data during the presence of the target. Next, we describe the set up and the steady phase of the proposed algorithm.

- In the set up phase, after reception of the first data packets of all active nodes, the BS calculates the difference between the data of node n_i and the data of node n_j $(i \neq j$, and $i \leq N, j \leq N)$. Next, these differences are summed over. We call this sum of the data value differences S_i. Then, the BS selects as CH the node which minimizes the total difference calculated value S_i between each node n_i and node n_j $(i \neq j$, and $i \leq N, j \leq N)$. Finally, the BS broadcasts a control message to the active nodes to notify the node selected as CH. Therefore, the rest of the nodes consider themselves as CMs. Note that there is no need for the CMs to send any extra packets since the BS already knows the active nodes.
- In the steady state phase, CMs send the difference between their sensed data and the CH's data value, which corresponds to a compressed value, called Δ_i, rather than the complete data packet, $value\_CM_i$. Therefore, $\Delta_i = |value\_CM_i - value\_CH|$ represents the difference between the *i-th* CM's data value $value\_CM_i$, and the corresponding CH data value $value\_CH$. In order to perform this compression, the CH sends its complete sample data value to the CMs at the beginning of each round. Therefore, the CMs send only the Δ_i to the CH. The main advantage of the proposal scheme is that the S_i calculation is centralized at the BS, which is not energy constrained.

6.4 Mathematical Model

In this section, the mathematical model for the clustering protocol is described. For reasons of clarity, the random access protocol is not considered in this analysis. First, because it has been studied in detail in the previous sections. Second, because we are interested on studying the effect of the compression scheme without the extra energy consumption of the collisions and idle listening and third, because the effect of the energy consumption is considered in the simulation results.
The total energy consumed in the network E_{total} for a duration of the event can be calculated as follows:

$$E_{total} = E_{competing} + E_{reporting} \tag{26}$$

where $E_{competing}$ is the energy consumed during the cluster formation phase and $E_{reporting}$ is the energy consumed during the steady state phase. We calculate hereafter $E[E_{total}]$, the average energy consumed through the network for both the classical and the proposed protocol as $E[E_{total}] = E[E_{competing}] + E[E_{reporting}]$.

6.4.1 Classical protocol

We first calculate $E[E_{competing}]$. The energy consumed at the cluster formation phase is due to the signaling packet transmission of the active nodes in the event area directly to the BS plus the reception of the signaling packet from the BS to the active nodes, then:

$$E[E_{competing}] = m \times [E_{tx}(S, R_t) + E_{rx}(S)] \tag{27}$$

where $m = N\pi(R\_event)^2/A^2$, is the average number of active nodes in the dish of radius R_event and N is the total number of nodes in the network. $S = 24bit$ is the size of signaling message, $m \times E_{tx}(S, R_t)$ is the energy consumed to sent by the m compete messages to the BS, and $m \times E_{rx}(S)$ is the energy consumed by the resulting compete message sent from the BS thought the network. On the other hand, the average energy consumption in the steady phase per event can be calculated as:

$$E[E_{reporting}] = Number\_report \times [E_{tx}(S, R_c) + (m-1) \times E_{rx}(S) + (m-1) \times E_{tx}(fixe, R_c) + (m-1) * E_{rx}(fixe) + E_{DA} \times fixe + E_{tx}(fixe, R_t)]$$

where *fixe* is the size of the full data packets of 32bits, $Number\_report = 29$ is the number of packet sent during the steady phase, $E_{tx}(S, R_c)$ is the energy consumed due to the signaling messages sent by the CH to its CMs to begin the event reporting, $(m-1) \times E_{rx}(S)$ is the energy consumed by CMs to receive this message, $(m-1) \times E_{tx}(fixe, R_c)$ is the energy consumed by the CMs to send the data to the CH, $(m-1) \times E_{rx}(fixe)$ is the energy consumed by the CH to receive the data sent by the CMs, $E_{DA} \times fixe$ is the energy consumed by the CH due to the data aggregation, and $E_{tx}(fixe, R_t)$ is the energy consumed by the CH to send the aggregated data to the BS

6.4.2 Proposed protocol

Note that, at the cluster formation phase, the proposed scheme behaves in the same manner as the classical protocol with the important difference that the nodes transmit the data packet instead of the signaling packet, then:

$$E[E_{competing}] = m \times [E_{tx}(fixe, R_t) + E_{rx}(S)] \tag{28}$$

where $m \times E_{tx}(fixe, R_t)$ is the energy consumed to send the m data packets to the BS and $m \times E_{rx}(S)$ is the energy consumed by the transmission of the compete packets from the BS to the active nodes in the network. The energy consumption in the steady state can be found as follows:

$$E[E_{reporting}] = E_{tx}(fixe, R_c) + (m-1)E_{rx}(fixe) + Number\_report[E_{tx}(S, R_c) + (m-1)E_{rx}(S) + (m-1)E_{tx}(S + log_2(E[\Delta_i]), R_c + (m-1)E_{rx}(S + log_2(E[\Delta_i])) + fixeE_{DA} + E_{tx}(fixe, R_t)])$$

where, $E_{tx}(fixe, R_c)$ is the energy consumed by the data packet transmission that is used as the reference value from the CH to the CMs, $(m-1) \times E_{rx}(fixe)$ is the energy consumed by the CMs to receive the aforementioned reference value, $E_{tx}(S, R_c)$ is the energy consumed from a signaling message sent by the CH to its CMs in order to send their data, $(m-1) \times E_{rx}(S)$ is the energy consumed by CMs to receive this signaling message, $(m-1) \times E_{tx}(S + log_2(E[\Delta_i]), R_c)$ is the energy consumed by the CMs to send the compressed data to the CH, $(m-1) \times E_{rx}(S + log_2(E[\Delta_i]))$ is the energy consumed by the CH to receive the compressed data from the CMs, $E_{DA} \times fixe$ is the energy consumed by the CH due to the data aggregation procedure, $E_{tx}(fixe, R_t)$ is the energy consumed by the CH to send the aggregated data to the BS, $E[\Delta_i]$ is the average data packet size which corresponds to the difference between the CMs' data and the CH's data. It is worth noting that considering a uniform node distribution with a large N, the node that minimizes the distance in the *R_event* region will be located in the center of *R_event*. Therefore, to calculate $E[\Delta_i]$ let us first calculate the average distance between active nodes and the CH, $E[d_{toCH}]$.

$$\int_0^{R\_event} r2\pi r dr / \int_0^{R\_event} 2\pi r dr = 2R\_event/3$$

If the density of the nodes is uniform through the *R_event* area. We now calculate $E[\Delta_i]$. The average data difference between the data at the CM and the reference value at the CH C is described by:

$$E[\Delta_i] = C\left|1 - \frac{1}{(1+2R\_event/3)^{\alpha}}\right| \tag{29}$$

6.5 Numerical Results

We first present the some important results derived from the analytical model. According to the previous analysis, Fig. 9(a) and 9(b) shows the average energy consumed in the network when $N = 1000$ for different values of R_t and R_c respectively. The results show that our proposal is suitable when the R_t is lower than $830 meters$ and R_c is higher than 230 meters. Exceeding these thresholds makes the competing process very costly due to the complete data packet sent to the BS during the set up phase. Remember that the classical protocol only transmits a control packet in this phase. Therefore the proposed protocol has a higher energy consumption when the distance from the cluster to the BS is high.

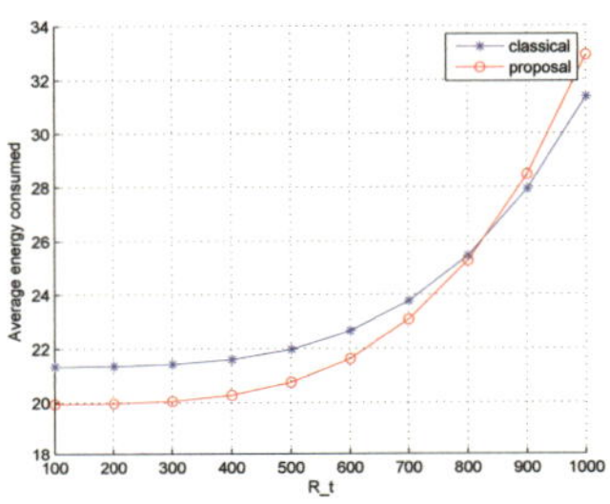

(a) Varying R_t.

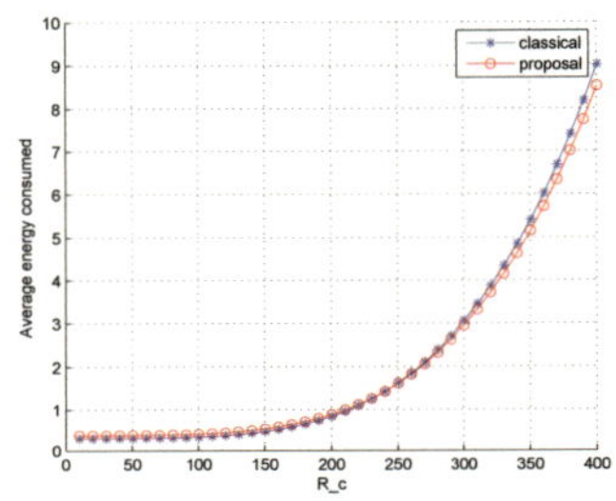

(b) Varying R_c.

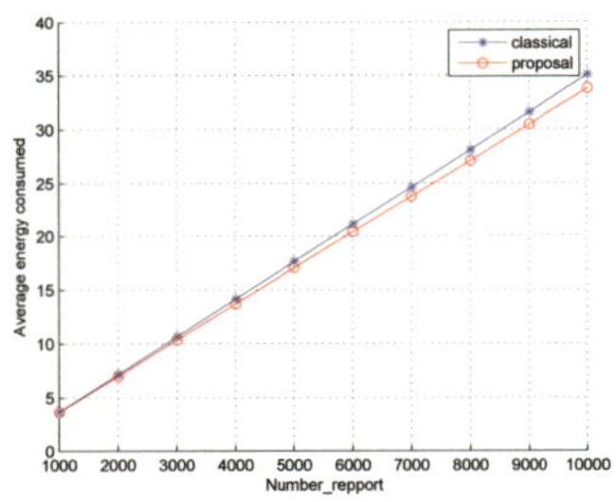

(c) Varying Number_report.

Fig. 9. Analytical results of the energy consumption.

Fig. 9(c) show the average energy consumed in the network varying the *Number_report* parameter. Here R_t and R_c are set to $400m$ and $100m$ respectively. The result shows that significative energy saving can be achieved by increasing the number of reports sent from the CMs to the CH. For the simulation results, we use TinyOS (Levis et al., 2003) as a simulation tool. The parameters used for this set of results are as follows: Signaling packet length (S) = 24 bits, Data value at the center of the event (C) = 250°, Initial energy per node = $10J$, T_event = 200 sec, $R\_event = 60m$, $R_c = 100m$, and $R_t = 400m$.

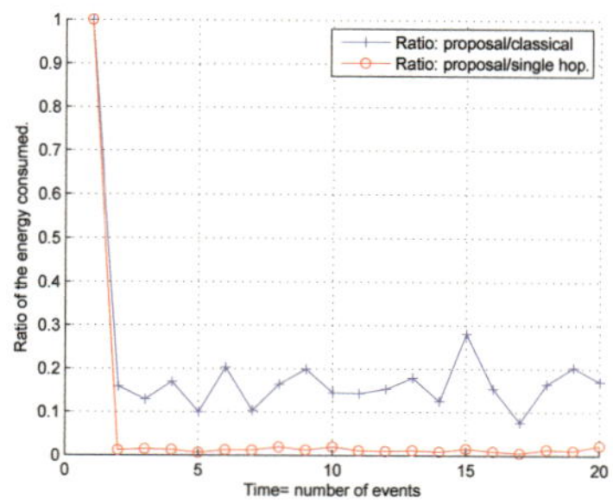

Fig. 10. Average energy consumed in time for direct transmission.

As in the Continuous Monitoring applications, we are interested on investigating the system performance under a cluster based architecture compared to the case where sensor nodes directly transmit to the BS. In order to explore the benefits of the clustering architecture, a scenario where all the nodes transmit directly to the BS is presented with the following modifications to the proposed scheme. All active nodes transmit their initial packet to the BS in order to choose the reference node (note that in this case there is no CH). Then, the BS selects the node that minimizes the data difference as explained in the previous section and then transmits a control packet indicating the ID of the reference node. Following this, in the steady state phase, the active nodes only transmit their difference Δ_i directly to the BS. The results presented in Fig. 10 clearly demonstrate that the proposed scheme conserves more energy compared to the classical scheme. Also, it is clear that the choice of the clustering scheme offers more energy savings than the single hop scheme. The gain ratio may reach up to 11 times more energy conservation than the classical scheme and 119 times more energy conservation than the single hop scheme, which are high benefit ratios.
Fig. 11 (a) shows the average energy consumed in the network per unit of time for different number of nodes. In this case the number of simulated events is 20. The results clearly demonstrates that our proposal outperforms the classical scheme. It can be seen that as the number of nodes in the system increases, also the energy consumption increases. Indeed, when the number of nodes in the network is high, the number of nodes that sense the event is also high. Hence, the number of packet transmissions (both control and data packets) is much higher than for the case where just a few nodes are active per event. The main reason for the better performance of the proposed protocol is that while all active nodes transmit the complete data packet in the steady phase in the classical protocol, for the proposed protocol, only the difference Δ_i is transmitted. Also note that this difference between the classical and proposed protocols increases for higher network densities. The rationale behind this is that for high network densities, the nodes are closer to each other, which in turns entails a higher correlation degree among their sensed data. This in turns renders smaller packet size. Conversely, for the classical scheme, since the packet size is fixed, a higher network density only increases the number of packets transmitted, consuming a lot of energy.
Fig. 11 (b) shows the average energy consumed for different values of *R_event*. When *R_event* is varied, also the number of active nodes per event is modified accordingly. Fig. 11 (c) shows the number of active nodes per event. It can be seen that the average number of active nodes for both the classical and the proposed scheme is approximately the same. Indeed, the proposed mechanism has no impact on the number of active nodes. Note that by increasing the

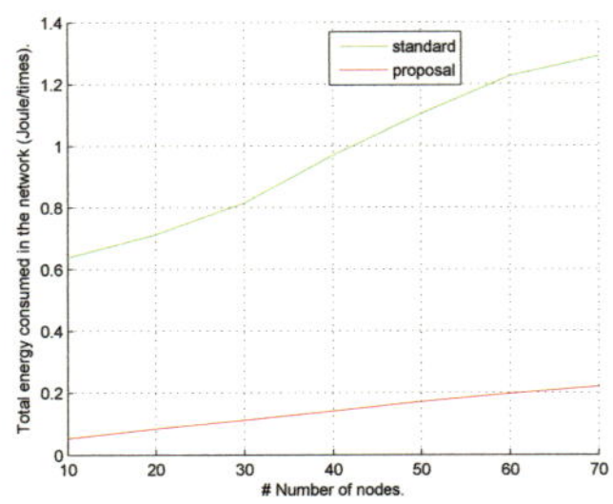

(a) Average energy consumption for unit of time vs number of nodes

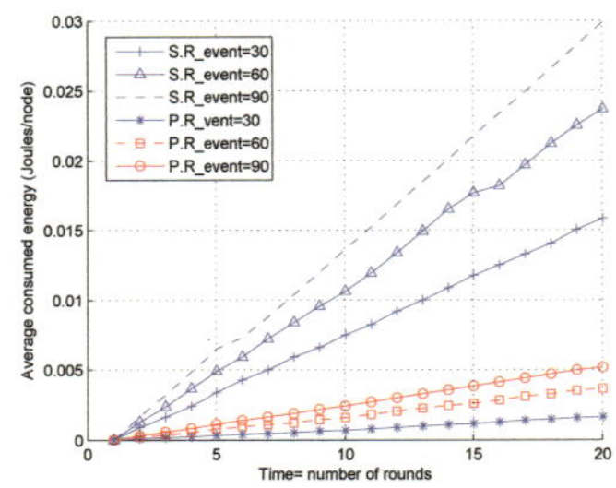

(b) Average energy consumed with varying the *R_event* region

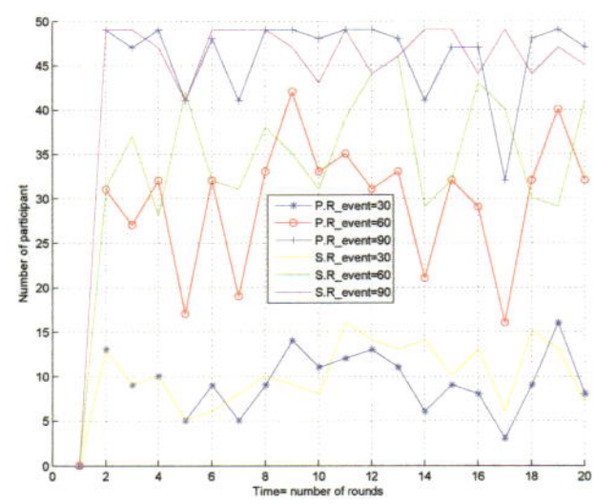

(c) Number of active nodes for each round

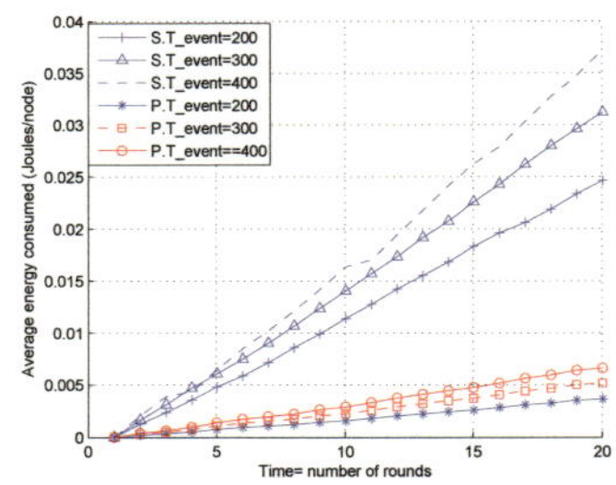

(d) Average energy consumption vs *T_event*

Fig. 11. Average energy consumption for unit of time for different parameters.

number of active nodes the energy consumption also increases. Observe for instance that the energy consumption when $R\_event = 30$ is less than the consumption when $R\_event = 60$ and 90. In each scenario, we observe that enabling our compression scheme reduces the energy consumption over the network and therefore extends the network lifetime.

Fig. 11 (c) shows the average energy consumed for different values of the *T_event* period. Increasing *T_event* also increases the period of the steady state phase and the number of data reported, therefore it can be seen an increase on the energy consumption. That explains why the energy consumed for $T\_event = 200$ sec is less than the energy consumed for $T\_event = 300$ and 400 sec. In each scenario, we observe that enabling our compression scheme reduces the energy consumption over the network and therefore extends the network lifetime. It is important to note that the proposed mechanism is particularly energy efficient for high event duration times. This is due to the fact that as the event duration increases, the CMs in the classical scheme transmit many full length packets while for the proposed mechanism, the CMs also transmit many packets but with a much smaller length. This renders a slight increase of energy consumption for the proposed mechanism while for the classical scheme there is an important augmentation in the energy consumption when the event duration increases.

7. Conclusion

This work has focused on studying the benefits to the energy consumption that can be gained by adding CM-EDR capabilities to systems of classical, unscheduled and cluster-based WSNs. The resulting continuous-monitoring WSN has been modeled, analyzed, simulated and studied.

It has been verified that CM-EDR can allow for an improvement in the network lifetime while ensuring the continuous-monitoring task. More significantly however, it has been shown that for calm supervised environment, it is more convenient to use the optional OCM-EDR, whereas in agitated environment, it is better to use the basic CM-EDR mechanism.

It is worth noting that enabling the CM-EDR and OCM-EDR mechanisms reduces always the energy consumption. On the other hand, the OCM-EDR mechanism has superior performance in terms of energy consumption for low values of λ while for higher values of λ the CM-EDR mechanism provides lower energy consumption.

For both low and moderate values of λ, with low or high values of N_{idle}, OCM-EDR provides good performance. Considering moderate values of N_{idle} the performance is superior since the energy consumption decreases. A similar effect can be seen when N_{sleep} is varied. In this case, OCM-EDR has the best performance when λ is low. On the other hand, when the value of λ is high, OCM-EDR presents relatively bad performance for any values of N_{idle} and N_{sleep} since the energy consumption is higher than the basic CM-EDR mechanism.

For the event detection applications, a new cluster-based compression technique has been proposed. The clustering scheme is based on selecting the node that reduces the packet size among all active nodes in the system. The BS selects the node which minimizes the total amount of data as a CH, therefore it increases the efficiency of the compression technique by sending only the difference, rather than the complete data value to the CH. By varying different parameters of the system, simulation and analytical results conclude that considering the spatial correlation in the communication of WSNs achieves significant energy conservation compared to a classical clustering scheme. The ratio benefit may reach up to 11 times the classical scheme. The proposed scheme extends the network lifetime. In addition, an approximate mathematical model is developed which validate the results.

8. References

K. Kredo II, P. Mohapatra, *Medium access control in wireless sensor networks*, Computer Networks, Volume 51, Issue 4, 14, pp. 961–994, March 2007.

W. B. Heinzelman, A. P. Chandrakasan, H. Balakrishnan, *An application-specific protocol architecture for wireless microsensor networks*, IEEE Transactions on Wireless Communication, vol. 1, no. 4, pp. 660–670, Oct. 2002.

Wireless LAN Medium Access Control (MAC) and Physical Layer (PHY) Specifications, ISO/IEC IEEE 802.11 Standard, 1999.

B. Sericola, *Closed form solution for the distribution of the total time spent in a subset of states of a Markov process during a finite observation period*, Journal of Applied Probability, 27, pp. 713–719, 1990.

N. Bouabdallah, M. Rivero-Angeles, B. Sericola, *Continuous monitoring using event-driven reporting for cluster-based wireless sensor networks*, IEEE Transactions on Vehicular Technology, vol. 58, No. 7, pp. 3360-3479, Sep. 2007.

C. Intanagonwiwat, R. Govindan, and D. Estrin, *Directed Diffusion: A Scalable and Robust Communication Paradigm for Sensor Networks*, (2000), in Proc. Sixth Annual International Conference on Mobile Computing and Networks (MobiCom 2000), Boston, Massachusetts, August 2000.

M. Larrea, C. Martin, J. J. Astrain, *Hierarchical and fault-tolerant data aggregation in wireless sensor networks*, in Proc. IEEE Second International Symposium on Wireless Pervasive Computing (ISWPC '07), pp. 531–536, Feb. 2007.

F. Bouabdallah, N. Bouabdallah, and R. Boutaba, *Towards Reliable and Efficient Reporting in Wireless Sensor Networks*, IEEE Transaction on Mobile Computing, 2009.

M. C. Vuran, and I. F. Akyildiz, *Spatial correlation-based collaborative medium access control in wireless sensor networks*, IEEE/ACM Transactions On Networking, vol 14, Issue 2, pp. 316–329, April 2006.

A. Woo, D.E. Culler, *A transmission control scheme for media access in sensor networks*, in Proc. of the International Conference on Mobile Computing and Networking (MobiCom 2001), pp. 221–235, 2001.

M. A.l Azim, S. Moad and N. Bouabdallah, *SAG: Smart Aggregation Technique for Continuous-Monitoring in Wireless Sensor Networks*, In Proc. of the IEEE International Comunication Conference (ICC), Cap town, May 2010.

J. Faruque and A. Helmy, *Rugged: Routing on fingerprint gradient in sensor network*, in IEEE international confrenece on pervasive Service (ICPS'2004), July 2004.

S. Pattem, B. Krishnmachari, R. Govindan and J. Heidemann, *The impact of spatial correlation on routing with coprerssion in wireless sensor networks*, in Synposium on information Porcessin in Sensor Networks," (IPSN), April 2004.

A. Jindal and K. Psounis,*'Modeling spatially-correlated sensor network data*, In the Proceedings of IEEE International Conference on Sensor and Ad hoc Communications and Networks (SECON'04), Santa Clara, CA, October 2004.

P. Levis, N. Lee, M. Welsh and D. Culler, *TOSSIM: Accurate and Scalable Simulation of Entire TinyOS Applications*, In Proc. of the First ACM Conference on Embedded Networked Sensor Systems (SenSys), 2003.

6

On Clustering in Sensor Networks

Michel Marot, Alexandre Delye and Monique Becker
Télécom SudParis
France

Why to build clusters in sensor networks ? Agregating nodes in clusters allows to reduce the complexity of the routing algorithms, to optimize the medium resource by letting it to be locally managed by a cluster head, to make easy the data fusion, to simplify the network management and particularly the address allocation, to optimize the energy consumption, and at last to make the network more scalable. Using clusters allows also to stabilize the topology if the cluster size is large in comparison to the speed of the nodes. This chapter is dedicated to clustering in sensor networks. First, the state of the art is presented, followed by the detailed presentation of one of the best and most cited cluster formation method with its validation and correction. Then, the next parts of the chapter are dedicated to some considerations on cluster modelling. In the last part, a method to assign addresses to the nodes within a cluster is presented.

1. Overview of the state of the art on cluster construction and cluster head election

The following state of the art has been partially established from the work of Deosarkar Deosarkar et al. (2008), Kumarawadu Kumarawadu et al. (2008) and Abbasi Abbasi & Younis (2007) and their colleagues. This subject has been the matter of a huge number of publications, and we do not pretend to be exhaustive. Nevertheless, it gives a good overview of the main problematics.
Cluster formation algorithms can be classified into:

- implicit (the nodes congregate in groups) or explicit (the nodes congregate around a cluster head);
- active (the clusters are the results of the execution of dedicated protocols) or passive (the clusters are formed spontaneously by deducting the information about the network topology by hearing the MAC messages used to transmit the data traffic);
- hierarchical (clusters of clusters) or non hierarchical;
- centralized or distributed, the distributed algorithms possibly being emerging if they allow to obtain from local behaviors a global result which is predictible in a deterministic or a stochastic fashion.

The role of the cluster head may vary from an architecture to another. It is generally the cluster head which manages the cluster (address assignment, possible assignment of the time slots and resources to the nodes, etc.). It is also the cluster head which restransmits the sensed data to the base station, either by sending directly the data to the base station or in a multi-hop fashion by sending the data to other cluster heads which then relay them in turn, or by retransmitting the data to nodes which may be simple nodes and not necessary cluster heads.

At last, the cluster heads generally fusion the data before to transmit them towards the base station. This shows how the cluster head must bear a heavy processing and retransmission load which make it subject to an energy consumption larger than the other simple nodes. During the cluster formation phase, the choice of the cluster heads impacts then a lot the performance of the network.
From the beginning of the first studies on clustering, it has been quickly understood that the burden of this energy consumption had to be spread on the nodes by rotating the role of cluster heads between the nodes: it is the main contribution of the LEACH algorithm (cf. Heinzelman et al. (2002)). The authors of LEACH propose a self-configurable architecture based on clusters minimizing the energy consumption of the nodes. The cluster heads transmit directly the data to the base station, and thus they spend more energy than the other nodes which send the data to their cluster heads closer than the base station. To spread the energy consumption, they propose an algorithm where any node becomes periodically cluster head with a probability which is increased in function of the duration spent for the last time at which they were cluster heads, and chosen so that the average number of clusters is a parameter of the algorithm. As it gives only guarantees in average of the number of clusters and their locations, a centralized version of the algorithm (LEACH-C) is also proposed. It allows to determine the optimal configuration to minimize the spent energy from the exact location of the nodes by using a simulating annealing (the problem is NP-hard). The choice of the cluster heads is done by LEACH in a random way, which may result in a bad spatial distribution of the cluster heads, clusters with inequal sizes, and a non optimal distribution of the energy consumption since this criterion is not taken into account in the choice of the cluster head.
It is the idea to rotate the burden of cluster head between the nodes which did the success of LEACH, but the network performance can still be improved first by letting a sensor to send its message to its nearest neighbor instead of letting it to send it to the cluster head which may be a little bit farer and second by letting a single node to send all the data to the base station instead of having several cluster heads to send parts of the network data to the base station: it is the goal of PEGASIS Lindsey & Raghavendra (2002) which organizes the sensors in a chain, where the nodes, when they have a packet to send, send it to the next node in the chain, which agregates the received data with the ones it must send, and does the same until the data arrive to a sensor which plays in the chain the role of cluster head. This role is rotated in a deterministic fashion between the sensors of the chain: they must play this role in turns. In this case, it is necessarily a multi-hop algorithm.
PEGASIS and LEACH inspired several variants (cf. 1.2), taking into account the residual energy, the possibility of agregation in trees, or combining both LEACH and PEGASIS: Ye et al. (2005), Jung et al. (2007), Satapathy & Sarma (2006), Huang et al. (2007), Tian et al. (2007) (algorithm ECR), Hao et al. (2008), TCCA in Selvakennedy & Sinnappan (2007) Liang & Yu (2005), Handy et al. (2002), Depedri et al. (2003) (LEACH-B), Gupta & Dave (2008), energy-LEACH in Xiangning & Yulin (2007). Yiming & Jianjun (2007), Jang et al. (2007), Qing et al. (2006), LEACH-ET in Lijun et al. (2006), LEACH-F in Heinzelman (2000), Wang et al. (2007), ERA in Chen et al. (2007), Loscri et al. (2005).
The choice of the cluster heads (cf. 1.1) is crucial for the performance of the network. Research about the choice itself or methods to reduce the overhead due to the signalling necessary to renew the cluster heads have been carried, either by letting the cluster heads directly transmit to their successors the role of cluster head based on the information exchanged the first time to choose the first cluster heads, (e. g. Rajiullah & Shimamoto (2007), Nam & Min (2007)), or by using an information transmitted in the data frames (cf. Gerla et al. (2000)). Other methods

have been proposed like the Türing morphogenesis (cf. Henderson et al. (2004) et Henderson et al. (1998)), the aim of which being to constraint the shape of the clusters (cf. 1.3). At last, building multi-hop clusters (cf. 1.4) combines both the difficulty of the cluster head election and the one of the cluster construction (routing trees inside the clusters) once the cluster heads chosen.

1.1 On the criterion for the choice of the cluster head

Clusters can be built without any constraint on the node which becomes cluster head. Nagpal and Coore propose CLUBS in Nagpal & Coore (1998) which allows to build single hop clusters. After a random time, nodes broadcast a message advertising they are luster heads. The clusters can overlap.
Considering that the cluster formation algorithm must be simple, Xu and Gerla propose in Xu & Gerla (2002) RCC, for multi-hop networks, which allows each node to elect itself as a cluster head and then to broadcast after a random time an advertisement message, the other nodes joining the first node having sent such an advertisement. This algorithm is more stable than another one using the node degree of the highest address because, in these cases, if a sensor having a larger degree or address comes in the network, the cluster head election mechanism must be executed again. RCC is also extended to the multi-hop case where the sensors relay at most K times a cluster head advertisement. The network is then seen as a set of local networks interconnected by the cluster heads and the authors evaluate the performance of proactive routing algorithms within the sensors and on demand routing algorithms between clusters.
The drawback of LEACH is that, since the nodes elect themselves as cluster heads with a certain probability, it is possible that there be not the same number of cluster heads in function of the time, and even that there be no cluster head at all at certain times. Thus, a criterion is used to determine the cluster head and it is generally the address of the node, the remaining energy or the number of neighbors. The easiest way to build clusters, besides the use of the geographical information, is to choose as cluster heads the nodes which have the smallest identifiers. It is exactly what C.R. Lin and M. Gerla propose in Kwon & Gerla (1999), where they present a distributed algorithm allowing to construct two-hop clusters (where each node can reach any other node in the cluster in at most one hop from the cluster head: the cluster is constituted of the cluster head and all its neighbors). The goal of the authors is to allow the spatial reuse of the bandwidth through clustering, to control the bandwidth within each cluster and to have a more stable topology. To build two-hop clusters allows to reuse the power control algorithms developed in the context of cellular networks (cf. Lin & Gerla (1997)).
In Rajiullah & Shimamoto (2007), Rajiullah and Shimamoto propose to decrease the traffic and processing load necessary to renew the choice of the cluster head by letting the first cluster heads to decide which nodes will be cluster heads later. A cluster head, as soon as it has reached a low energy threshold, pass the baton to the next one, which advertises its neighborhood that it becomes the new cluster head. The nodes under its coverage radius choose then the nearest cluster head, on the basis of the strength of the signal they receive. The criterion determining the choice of the cluster head is the smallest node identifier. The same idea is proposed by Nam and Min in Nam & Min (2007) where LEACH is used to build clusters at the beginning, but the cluster heads designate then themselves their successors which are chosen each one in its turn within theirs clusters. The criterion of the choice of the cluster head is the address identifier. Liu, Lee and Wang, in Liu et al. (2007) use also the node identifier. This algorithm is described below.

Tillapart and his colleagues present in Tillapart et al. (2005) LMSSC: a method to partition the nodes in clusters which consists in defining the clusters on the basis of the highest degree nodes, but the cluster heads are chosen in a centralized way by the base station by minimizing the criterion of the ratio of the remaining energy of a node over the sum of the squared distances of this node to all the others plus its squared distance to the base station.
Chang and Kuo propose in Chang & Kuo (2006) an algorithm for the choice of the cluster heads, MECH, where all the nodes send "hello" messages, which are not retransmitted, and the nodes which receive a number of such messages at least equal to a certain threshold elect themselves as cluster heads. It is thus the node degree which is used here as the selectino criterion. The same idea is used in Wen & Sethares (2005) where the nodes declare themselves cluster heads if they do not have one in their neighborhood and after the expiration of a time out randomly generated and decremented at each reception of a "hello" message, these messages being periodically sent by each sensor. Kim et al. (2008) or Chan & Perrig (2004) described below use also the node degree.
The criterion for the choice of the cluster head can also be based on a weigth function of several such criteria (cf. DWEHC in Ding et al. (2005), HEED in Younis & Fahmy (2004), Fan & Zhou (2006), WCA in Chatterjee et al. (2001), Li et al. (2006)). Using weights allows also to take into account the speed (like DMAC in Basagni (1999b), Basagni (1999a), Basagni et al. (2004), Bettstetter (2004), or Chinara & Rath (2008)).
In Ding et al. (2005), Ding, Holliday and Celik propose a cluster formation method, DWEHC, where the sensors elect themselves cluster heads if they have the highest weigth in their one hop neighborhood, which is the product of their remaining energy and the average of their distances to their neighbors. The cluster heads broadcast then an advertisement message and the sensors join the nearest candidate cluster head, in a multi hop way up to the limit of the size of the clusters. In Klaoudatou et al. (2008), the speed is used a criterion (cf. below).
Generally, cluster formation algorithms are made of a cluster construction phase and a maintenance phase, which is particularly important when the nodes are mobiles. Often the first phase assumes that the nodes are almost motionless. To relax this assumption, S. Basagni (cf. Basagni (1999b) and Basagni (1999a), cf. also Basagni et al. (2004) and Bettstetter (2004) for considerations on the performance) proposed an algorithm (DMAC) which associates a weigth to each node and so that the cluster heads are the highest weight nodes and never can be neighbors. The weight may be dependent on the speed or the power level.
These criteria are generally transmitted in a dedicated signaling, but to reduce the overhead due to the control information conveyed to build the clusters, M. Gerla, T.J.Kwon and G. Pei (cf. Gerla et al. (2000)) propose and algorithm using a minimal information transmitted in the data MAC frames. Listening to all the node traffic in their neighborhood allows each node to get this information. This algorithm is suitable in situations where the nodes are very mobiles.

1.2 Variations in LEACH or PEGASIS modes

After LEACH a multitude of variants have been published, like in Hao et al. (2008), where the base station receives the geographical positions of the nodes, and then it partitions the network into different geographical zones which become clusters. LEACH is used in this paper to choose the cluster heads within the clusters. Between the cluster heads, a multi hop mechanism is used to transmit the data. Generally speaking, all the variants are more or less motivated by the necessity to take into account the remaining energy of the nodes in the probability for a node to be elected as a cluster head. In Selvakennedy & Sinnappan (2007) TCCA is proposed, for which the clusters are multi hop ones. Liang & Yu (2005), Handy et al. (2002), Depedri et al. (2003) (LEACH-B), Gupta & Dave (2008), Xiangning & Yulin (2007), Yiming & Jianjun (2007), and Jang et al. (2007), Qing et al. (2006) (which adaptes LEACH to the case where the sensors have initially heterogeneous energy levels) may also be cited. Xiangning & Yulin (2007) propose energy-LEACH which selects as cluster heads the nodes which have a remaining energy level higher than a given threshold and multihop-LEACH whose behavior is the same as LEACH but with the possibility to communicate in multi hop from cluster heads to cluster heads towards the base station. Jang et al. (2007) weights, as the other algorithms, the pobability to become cluster head with the remaining energy, but first this weight intervenes only when the nodes have consummed 50% of their energy otherwise the operation is exactly the same as LEACH, and, second, it also defines a cost function used by any simple node to choose its cluster heads, which is a function of the received signal strength as in LEACH but also of the remaining energy of the cluster head.

In Lijun et al. (2006), the authors propose (LEACH-ET) to trigger the cluster head changes only when a node has reached a given energy threshold or if a cluster head has emptied its battery instead of doing it periodically. In Wang et al. (2007), the authors, noticing that the number of nodes varies in function of the time, propose to dynamically adapt the probability to become cluster head in function of the actual number of sensors in the network. Another LEACH variant aiming at prolonging the network lifetime is proposed in Chen et al. (2007): ERA. The cluster heads are chosen as in LEACH, but the nodes, instead of joining the nearest cluster heads (which has the highest received signal strength), join the cluster head for which the remaining energy on the path (remaining energy of the nodes on the path minus the energy necessary to join the base station) is the highest. At last, Loscri and his colleagues extend LEACH to a two level hierarchy in Loscri et al. (2005). This increases the efficiency of the data fusion and thus the economized energy, if the base station is far from the network.

PEGASIS presents several problems, and particularly to take into account neither the remaining energy of the nodes in the choice of the cluster heads nor the distance to the base station: it is the sensor i mod N which is chosen at the ith round as a cluster head. Moreover, the greedy algorithm used by PEGASIS, by nature, can lead to non optimal chains, the total agregation time may also be very long. That is why the authors of Jung et al. (2007) propose that the base station broadcasts a set of thresholds which corresponds to different signal levels and which define reception zones around the base station. The sensors, in function of these received thresholds, can then be distributed in concentric zones around the base station. Within each zone, the sensors apply PEGASIS: they constitute a chain internal in the zone where they play the role of cluster heads each one in its turn and agregate the data from neighbors to neighbors in the chain. However, the cluster head of the zone i agregates its data with its own data which are then transmitted in a single packet to the cluster head of the zone $i-1$.

An idea similar to PEGASIS is presented in Satapathy & Sarma (2006) where the use of trees replaces the chain of PEGASIS. As soon as the root of a tree dies because of an empty battery, it is changed. The use of a tree necessitates more than a fusion operation, and thus less packet transmissions, and thus an energy gain. At last, the authors of Huang et al. (2007) propose to combine the advantages of the use of clusters in LEACH with the advantages of the trees presented in Satapathy & Sarma (2006) by using several clusters constituting each one a tree. The cluster head is chosen the closest possible to the base station (all the sensors are assumed to know their geographical positions) under the constraint of a minimal remaining energy.

Tian, Wang and Zhang propose in Tian et al. (2007) an algorithm, ECR, which combines the advantages of LEACH and PEGASIS: it allows to have several clusters at the same time to decrease the latency due to the agregation, to use chains within the clusters and between the cluster heads to perform the agregation and at last it uses the remaining energy of the nodes to select the head of the cluster heads (i.e. the cluster head of the chain of the cluster heads, that we call protocaryomme in the following). The sensors are assumed to know their relative positions, a coordinate system of which the "Y" axis is so that the base station if far from the network in this direction and the number N of clusters. The clusters are defined in bands parallel to the X axis knowing N and the sensor broadcast their identifiers, their cluster identifiers and their positions. They can then constitute in a distributed fashion the chain within their clusters. The base station assigns the role of protocaryomme to a sensor which is by the way the cluster head of its cluster, and a greedy algorithm is used to constitute the chain of the cluster heads from the protocaryomme. Besides the data, each node inserts the maximum between its remaining energy level and the one it has received from its neighbor, with the identifier of the node corresponding to the retained energy level. One thing leading up to another, the packet arriving to the base station contains the identifier of the node having the highest remaining energy level which is then elected as the protocaryomme.

1.3 On the methods constraining the shape of the clusters in terms of numbers of nodes, laying out, etc

The drawback of LEACH is that, as the nodes elect themselves cluster heads with a certain probability, it is possible that there be not the same number in function of the time, and even that there be no cluster head at all. To solve this problem, O. Younis and S. Fahmy (cf. Younis & Fahmy (2004)) propose the HEED algorithm which allows to select a cluster head in function of its remaining energy and a cost function defined, depending of the target objectives, either on the number of neighbors or on the average of the minimal power necessary to be reached by the neighbors. Either very dense clusters or clusters with a well distributed load can thus be obtained. In Fan & Zhou (2006), partly inspired from WCA presented in Chatterjee et al. (2001) and which does not take into account the residual energy of the node, the cluster heads are chosen with weights functions of the inverse of the node residual energy, their degree, the sum of the distances to their neighbors and the distance to the base station. This function, when it is minimized, leads to choose sensors having the highest residual energy, having a degree as close as possible to a value which is a parameter of the algorithm and minimizing the distance between the nodes and the base station. A similar intuition leads the authors Li et al. (2006) to propose an algorithm where the cluster heads are chosen by maximizing a cost function of the residual energy, the number of neighbors and the time spent for the last time the node was cluster head. Initially, the base station defines the perimeter of the clusters and chooses the first cluster heads, but, later, the clusters pass the baton by choosing themselves the next cluster heads by taking the nodes which maximize this function in the clusters. Then the new cluster heads send an advertisement message and the nodes join their new cluster heads in function of the signal strength level. The authors of Guo et al. (2007) propose to extend HEED to the case where the routing between cluster heads is in a multi hop fashion to the base station (CMRP algorithm).

Gupta and Younis consider in Gupta & Younis (2003b) an heterogeneous network of which the cluster heads are the nodes having no energy constraint and which can all communicate together. To build the clusters, they discover their neighbors (i.e. the sensors for which they are in visibility), and then they distribute them between them in order to minimize the total transmission cost of the sensors to their cluster heads and to distribute almost evenly the number of sensors. It is an iterative process where a cluster head attributes itself the sensors which are in its coverage progressively increased by the minimum of the distances between the cluster heads and its neighbors to the median of the distances. All the nodes are equiped with a GPS. In the same context, in Gupta & Younis (2003a), the same authors address the issue of the cluster head failure. The network uses a TDMA like transmission mechanism for which some slots are dedicated to the cluster heads to communicate their status. When all the cluster heads have no more information about one of them, they distribute its sensors between them. For this purpose, any cluster head has two lists: a list of sensors of its own cluster and another list of other sensors for which it is the backup cluster head. The first list is obtained according to the method proposed in Gupta & Younis (2003b), the second one is obtained with a simple visibility condition between the cluster head and a sensor.

In Klaoudatou et al. (2008), Klaoudatou et al. consider medical surveillance sensor networks of which the nodes are mobiles. They select the closest node to the base station (in ad-hoc environment during the emergencies on the spot or using access points in the hospital) as a cluster head. Actually, they notice that the mobility allows then to turn this role of cluster head between the different sensors. Chinara & Rath (2008) considers also the case of mobile sensors. They estimate then their speed during the last time period and the least mobile ones are chosen as cluster heads.

Liu, Lee and Wang, in Liu et al. (2007), propose two algorithms. The first on, ACE-C (Algorithm of Cluster head Election by counting), aims at determining the cluster head on the basis of the node identifier: there are N nodes in the network, C cluster heads are required in the network, a node x is then a cluster head all the N/C periods. At the beginning of a new period, a cluster head broadcasts a message to all the nodes advertizing it becomes cluster head and containing its geographical position and its speed vector and the others choose the nearest cluster head. For this purpose, they estimate their relative distance from the position of the cluster heads between the time of the current election and the previous one. The use of the speed vector is not clear in the paper. If the battery of a node is empty when it must become cluster head, all the nodes are informed and the nodes integrate that information in their calulations. This algorithm having the drawback of a possible bad distribution of the cluster heads, a second one is proposed: ACE-L (for "Localization"). Fix anchors are distributed in the network. Any node evaluates its distance to the anchor, which is used to proportionately generate a time out after what the node emits a message advertizing it is a cluster head. The first emitting node is the closest one to the anchor and it becomes then a cluster head.

Another proposal is given by Kim and his colleagues in Kim et al. (2008) to distribute in the middle the cluster heads, that is to avoid that the cluster heads be grouped at the same place. A predifined cluster head number is chosen at the network initialization, possibly misplaced. Each cluster head broadcasts under its coverage an advertisement message. Any node receiving it counts the number of received messages. The cluster heads choose then the cluster head in their coverage which should replace them either by designating a node having received few advertisement messages if the cluster is sparse or, contrary, a node having received a large number of such messages if the cluster is dense. This causes a repulsion effect between cluster heads which tends to give a homogeneous coverage of the network by the clusters. The cluster head selection criterion is then the number of cluster heads in the coverage before the new election.

H. Chan and A. Perrig, in Chan & Perrig (2004), propose a similar algorithm which allows to obtain perfectly homogeneous clusters by minimizing the overlaps, of which the complexity depends only on the sensor density. It then counts the number of loyal followers, that is the number of nodes which would have only it as a cluster head if it became a cluster head. If this number is larger than a certain threshold, it becomes cluster head. By so counting the number of loyal followers, and not only the number of sensors able to belong to several clusters, the chosen candidate cluster head is the one for which the cluster has a minimal cluster overlap. This causes a repulsion effect between clusters, and thus a better distribution of the clusters.

Another proposal aiming at avoiding a non even distribution of the clusters in LEACH is presented in Ye et al. (2005). The candidate cluster heads elect themselves with a fixed probability T, they broadcast an advertisement message, which contains a residual energy level. If such a candidate receives such a message for which the level is greater than its one, it effaces itself, otherwise it proclaims cluster head. An ordinary node joins then the cluster head which minimizes a cost function taking into account its distance to the cluster head and the distance between the cluster head and the base station.

The BCDCP algorithm presented by Muruganathan et al. in Muruganathan et al. (2005) consists in selecting, among the ones having a residual energy greater than the average two nodes which have a maximal distance between them, in distributing between them the nodes of the network in a manner as even as possible and in iterating this process until the desired number of cluster. This allows to ensure there is well the desired number of cluster heads with almost the same number of sensors in each cluster. The nodes have power levels which can vary and they transmit directly their data to their cluster heads which fusion them and send them to the base station from cluster heads to cluster heads. This partitioning and cluster head election algorithm is centralized at the base station.

In ya Zhang et al. (2007), the clusters are obtained by the base station with the algorithm of the k-means for the classification of the nodes in clusters. The choice of the cluster head is done by minimizing the distance between the nodes and the cluster head (this distance is also minimized in the classification) at the beginning, then the clusters remains the same all along the network lifetime, but, periodically, the node having the highest residual energy in the cluster replace the cluster head. The idea is thus to build "natural" clusters corresponding to the node agregates. There is a predifined number of clusters but also a limit threshold for the cluster size which allows to split them into several clusters if they have reached this limit. The idea to build the clusters at the beginning and to leave them after without changing them but to only turn the cluster head role between the nodes of a same cluster is also proposed in an evolution of LEACH-C: LEACH-F (cf. Heinzelman (2000)) which uses at the beginning of the network lifetime the same method as LEACH-C.

Demirbas and his colleagues present FLOC in Demirbas et al. (2004). The nodes can communicate according to two modes: in i-band, a reliable manner, in the limit of a certain unit radius and in o-band beyond this radius but in non reliable mode and still within the limit of another larger radius. The nodes elect themselves candidate cluster heads after a random time and broadcast then an advertisement. If a sensor receives this message and if it is already in the i-band of another cluster head C, the candidate renounce its pretension to become cluster head and it joins C possibly in o-band mode. If a sensor receives this message and if it is in the i-band of the candidate but also in the o-band of a cluster head C, it leaves C to join the candidate. This proposal aims to guarantee clusters having the "solid disk" property: all the nodes at a unit distance of a cluster head are in its cluster or, in other words, there is no overlap of unit radius clusters. This allows to bound the number of clusters, to decrease the signaling (a cluster head has not to listen to all the sensors which are in its coverage but which belongs to other clusters), to obtain a better spatial coverage for the data agregation, etc.

In Zhang & Arora (2003), Zhang and Arora assume the sensor to have a perfect knowledge of the geography and they constitute hexagonal cells. A root node finds its ideal position from the center of its neighbors cells and selects as cluster heads of these cells the closest

node to its ideal position. If there is no such a node (if the coverage radius is to small), the sensors of the cell are distributed among the neighboring cells. The underlying motivation for this perfectly geographical hexagonal partitioning is multiple: numerous sensor network applications give identical results per geographical zones, easy compression by geographical zones, better frequency reuse, etc.
The idea to spread the clusters according to a partition can be extended to a non geographical space. Actually, the notion of cluster is still more important when the agregation (data fusion) is taken into acount. In Vlajic & Xia (2006), the authors propose a cluster grouping based on the similarity of the sensed data: The nodes which sense the same physical characteristics are naturally grouped allowing a maximal compression per data fusion. They propose then in Xia & Vlajic (2006) an algorithm, LNCA, for the multi-hop cluster formation consisting for each sensor in listening to the data transmitted by their neighbors. If the data are the same as their own data, they increment a counter and they insert the neighbor into a list. They broadcast then this counter with a time to live field n to limit the retransmission of the message to n hops and it is the node which has the highest value of this counter which is retain as a cluster head.
An original idea has been proposed by T.C. Henderson and his colleagues in Henderson et al. (2004) and Henderson et al. (1998). It consists in using the Türing's morphogenesis process to give to a very dense network a certain configuration. The idea consists in propagating the result of a certain function from sensors to sensors, this result being used in input of the function on the next sensor. By well choosing the function, a mechanism can be implemented to initialize a variable producing a totally predetermined global configuration. This method is expected for example to radio control robots. If the number of sensors is very large on a given surface, with a certain function, bands can be drawn which can be used as traces to guide robots. This morphogenesis could be used to find more complex configurations.

1.4 The multi-hop case

Apart the cases where the nodes group themselves by affinities (for example on the basis of similar sensed data like in Vlajic & Xia (2006) or LNCA in Xia & Vlajic (2006)) or implicitely like in Kawadia & Kumar (2003), or in a centralized fashion (like the extension of BCDCP, also centralized allowing multi hop communications in clusters thanks to routing trees within the clusters in Huang et al. (2006)), the multi hop cluster formation is doubly complicated: first the question is raised how to choose the cluster heads and, second, how to build the parentage between the ordinary nodes and their cluster heads.
In Kawadia & Kumar (2003), V. Kawadia and P.R. Kumar propose an algorithm integrating routing, power control and implicit clusterization, CLUSTERPOW and tunnelled CLUSTERPOW, for networks of which the node distribution is homogeneous. It is a multi hop routing algorithm where each node has several power levels and where it chooses the smallest possible one to reach its destination. Each power level defines then a cluster: to reach a far destination, the node must send the information by using its largest power level, which is the same to transmit to another cluster when the network is not evenly distributed.
Some approaches not considering the choice of the cluster heads aim only to split the whole network into clusters. Some consist in building spanning trees which are later split into subtrees, the important task being the good distribution of the clusters: Banerjee & Khuller (2001), Fernandess & Communication (2002).
Banerjee and Khuller serach in Banerjee & Khuller (2001) to constitute clusters of which the size is between k and 2k, except a single one allowed to be smaller, and such as the number of clusters a sensor belongs to is bounded. For this purpose, they build spanning trees on

the network and, from the leaves of the tree, they take sub-trees with size between the two bounds. Two versions, centralized and distributed, are proposed. The problem of the cluster head election is not really the main concern of the authors.

In Fernandess & Communication (2002), the authors propose to make a partition into k hop clusters by building a minimum connected dominating set. They obtained next a spanning tree from this set. They add as leaves the nodes of the remaining part of the graph. This tree is later split into sub-trees with a diameter k. Building such a spanning tree gives more balanced clusters than other known techniques.

The same goal is targeted in Youssef et al. (2006) (algorithm MOCA). Youssef and his colleagues (among who there is Younis) put in Youssef et al. (2006) the problem of the necessity to have overlapping clusters, in order to facilitate the routing between clusters (among other reasons), and they define the concept of k-dominating set with overlap: any node is at most at a k hop distance and belongs to at least two clusters. The cluster heads elect themselves with a predetermined probability, and they broadcast an advertisement message, which is retransmitted at most k times. A node receiving this message answers even if it already belongs to a cluster. A sensor can thus belong to more than two clusters at a time. Note that it is always possible that the nodes be isolated and thus belong to only a single cluster; their own. It is the MOCA algorithm.

In Dai & Wu (2005), Dai and Wu propose three algorithms to build a k-connected k-dominating set. It is a set such as first any node is in this set or has at least k neighbors inside, and, second, if $k-1$ nodes are removed, it remains connected. For the first algorithm, each node elects itself as a member of the k-connected k-dominating set with a given probability p. For example, with 200 nodes spread over a 1000×1000 surface and with $k = 2$, $p = 50\%$, this process leads to a 2-dominating set with a probability 98,2%. The second algorithm is deterministic and it consists in removing each node of the k-connected k-dominating set if there exists k disjoint backup paths between every couple (u, v) of its neighbors, via nodes having greater identifiers than v. The third algorithm combines both approaches: it consists for any node to be colored with a certain probability with a color given among k ones, and the deterministic condition is applied but between the nodes of a same color. The cluster heads are arbitrary chosen. This proposal aims to ensure a certain reliability.

The solution of Dai & Wu (2005) rather aims to ensure a certain reliability, but the approach aiming to build independant k-dominating sets is more suited to sensor networks because it leads to a more efficient use of the energy, at the expense of a certain lesser reliability. The work presented in Banerjee & Khuller (2001) and Fernandess & Communication (2002) are methods to partition a graph, but not to elect a cluster head from a given criterion, contrary to the papers McLaughlan & Akkaya (2007) and Nocetti et al. (2003). Nevertheless, in the case of these papers, clusters mades of nodes separated from their cluster heads by paths containing nodes belonging to other clusters can be obtained! To avoid that, Prakash and his colleagues propose in Amis et al. (2000) a heuristic which allows to build k-dominating sets using the address of the nodes as a criterion and made of two phases. The first one is analogous to the classical step of the broadcast of the highest value of the criterion in a d hop neighborhood. The second one consists in broadcasting in a k neighborhood the minimum of these maximums. That allows the cluster heads having not the highest value of the criterion in their k-neighborhood, and thus separated from their members by nodes belonging to other clusters, to gain members in their clusters.

Nevertheless, the choice of the cluster heads impacts the performance and should no be neglected. The simplest method is the one where each node elects itself as a cluster head independantly of its neighbors: for example with a certain probability (cf. Xiangning & Yulin (2007), RCC in Xu & Gerla (2002), Bandyopadhyay & Coyle (2003), EMCA in Qian et al. (2006), Wang et al. (2005), SWEET in Fang et al. (2008), McLaughlan & Akkaya (2007)), and then broadcasts messages which are retransmitted k times at maximum. In McLaughlan & Akkaya (2007), each node diffuses "alive" messages to its k hop neighborhood. The sensors elect cluster heads themselves with a probability which is decreased with the proximity of a cluster board (i.e. the board of the k hop neighborhood of a cluster head) and is increased with the number of neighbors. Then they broadcast to their k hop neighborhood a "dominator" message which, when it reachs a node situated at exactly k hops, triggers this later node to send a "board" message. This message allows the other sensors to determine their proximity to a cluster board.

In Bandyopadhyay & Coyle (2003), the authors propose a multi-hop algorithm where the sensors also elect themselves as cluster heads with a given probability p, then they advertize they are cluster heads. These advertisement messages are retransmitted at most k times. The authors calculate p to optimize the energy consumption in the system. k is fixed with a relationship obtained from the stochastic geometry and which is a function of the probability that the radius of a sphere centered on the cluster head and containing its Voronoï cell be larger than a certain value $r \times k$. An extension of LEACH to the multi-hop case (for the transmission between a sensor and its cluster head) is proposed in Qian et al. (2006): EMCA. The cluster heads are chosen in the same way as in LEACH. Then, they broadcast a message advertizing they are cluster heads. This message is retransmitted a given maximum number of times. A MAC method for the TDMA slots is also proposed.

The authors of Wang et al. (2005) propose a multi-hop cluster formation algorithm oriented towards the attributes. To make easier the data query, the clusters are first geographically defined and second they are defined within a same geographical zone by attributes (temperature, pressure, concentration, age,...). A cluster hierarchy embedded into each others is then defined, each cluster having its own cluster head: the hospital, the floor i of the hospital, the room j of this floor, the pressure sensor k of this room, etc. At the beginning, a node advertizes it is a general cluster head then this information is retransmitted through all the hierarchy by the others after a certain time which is a function of the residual energy. After this random delay, a sensor receiving this information advertizes it is a cluster head if there is still no cluster head in the hierarchy. The cluster heads transmit then the information of the composition of their clusters to the cluster head of higher level, which also gives a routing information used during the query. The idea to announce to be a cluster head after a certain random time inversely proportional to the residual energy is also proposed in Fang et al. (2008) (SWEET). A method proposed to be more efficient constists in comparing between the sensors a certain criterion: node identifier, residual energy, weights, etc. (cf. KHOPCA in Brust et al. (2008), CABCF in Liu et al. (2009), Rasheed et al. (2007), MaxMin in Amis et al. (2000),...).

Variants are proposed but, finally, the same method is always used: either a node elects itself with a given probability and it broadcasts an advertisement until k hops or it broadcasts weights until k hops. In Brust et al. (2008) (KHOPCA), Brust and his colleagues propose a mechanism which consists in decrementing a weight or changing it from MIN to MAX values depending on the values of the neighbors weights. This causes the weights to be spread so that they be separated by a good number of hops. the change from MIN to MAX is done in function of the neighboring weights, and thus not depending on a criterion like the energy of the node degree. In Liu et al. (2009), the authors propose CABCF where each node has a weight function of the residual energy, the degree and the distance to the sink. The nodes are

then grouped into clusters step by step by combining themselves with larger weight sensors. The multi-hop communication is also set up by using this heuristic within the clusters.
It is possible that two nodes have the same criterion value. For this situation, the authors of Nocetti et al. (2003), propose an algorithm which consists in that the sensors having the highest degree and the smallest address elect themselves as cluster heads and broadcast an advertisement untill k hops.
This simple k hop broadcast is omnipresent in the literature, for example in Rasheed et al. (2007). but it is a problem because of the interdependance between the k hop neighborhoods. Actually, when building multi-hop clusters, the question arises sooner or later to know how to let a maximum distance between the cluster heads while ensuring any ordinary sensor to be at most at k hops wide a cluster head, that is how to build an optimal k-dominating independant set. Unfortunately, to find such a set is an NP-hard problem (cf. Amis et al. (2000)), that is why heuristics have been proposed.
It is intuitive that the nodes having the highest criterion value be elected cluster heads. There are two ways to implement that. Either the nodes exchange this criterion information so that each node gets the list of its neighbors and their criterion values, or a node broadcasts the couple of its identifier and its criterion value which is retransmitted by its neighbor if its own value is smaller or after having replaced the received value by its own if it is larger. In this case, all the nodes have finally a single information: the identifier of the node which has the highest value of the criterion in its k-hop neighborhood with this value but no more information on the neighborhood.
The drawback of the first approach is that some nodes become orphans and have no other solution than proclaim themselves cluster heads. Actually, let us consider the weights given on figure 1 and let us assume two-hop clusters. Applying this method leads nodes 5, 4 and 3 to know that the node 5 has the highest criterion in its two-hop neighborhood. Neither 4 nor 3 broadcasts any cluster head advertisement, but 5 does it. 5 is thus a cluster head of the cluster (5,4,3). 4 has not broadcast any cluster head advertisement, the same for 3 and 2, because it noticed it did not have the highest criterion value. The result is that 2 becomes alone. The only solution is to declare 2 cluster head of the cluster containing the only node (2). If there should be a cluster with only one node, it would be more intelligent to choose (5) and (4,3,2). In short, the more appropriate candidate in the neighborhood of 2 does not declare itself as a cluster head because it already belongs to another cluster but any node (e.g. node 2) counts on the node having the highest criterion value in its neighborhood (e.g. node 4) as a cluster head.

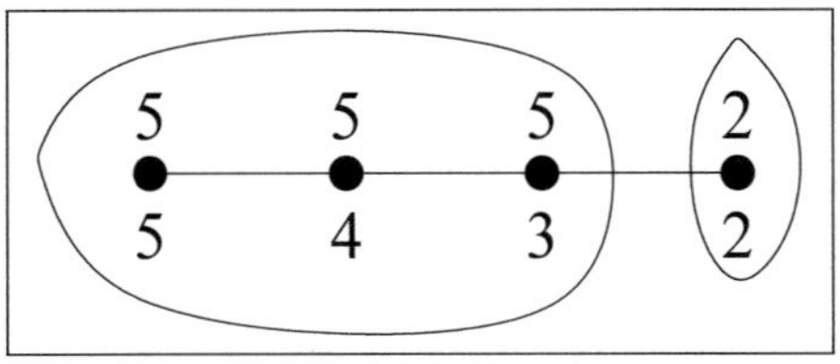

Fig. 1. Case of a bad cluster head selection

In the second case, where a single couple of identifier and criterion is broadcast and possibly overwritten by a node having a higher value, the choice of the cluster head leads to that each node A elects necessarily as a cluster head the node B which has the highest criterion in its k-hop neighborhood. Nevertheless, it is possible that B itself has already elected another node

C in its own k-hop neighborhood but not in A's neighborhood because C has a criterion value higher than the criterion value of B. In this case, a sensor elects a cluster head which does not consider as such a cluster head. On the example of figure 1, 2 would choose 4 as a cluster head which itself would choose 5.

To summarize, either a node does not elect its cluster head but it waits for that another node anounces itself as a cluster head, with the risk that this one is already a member of another cluster that is with the risk to be without cluster head and then to be obliged to be cluster head with a small criterion value, or it decides to elect another node with the risk that this latter is already in another cluster and thus the risk that it is a follower of a node which is not a cluster head. The whole problem comes from the interdenpendance of the k-hop neighborhoods which makes it NP-hard. To give a heuristic is exactly to distribute this problem by relaxing the independence and thus it is exactly to accept either a non optimality or inconsistencies. This fundamental problem has not really been considered in the literature. Scientists have focus their research mainly on finding a good criterion rather than on the method without realizing that an optimal criterion with a bad method could lead to a disastrous performance or to functional problems. It was urgent to consider this problem.

Prakash and his colleagues proposed then in Amis et al. (2000) a heuristic allowing to build d-dominating sets with the criterion of the node identifier and made of two phases. The first one is analogous to the classical broadcast of the highest criterion value in a d-neighborhood with overwritting. The second one consists in doing the same thing than in the first phase but by transmitting in a d-neighborhood the minimum of the exchanged values instead of the maximum. This gives to the cluster heads having not necessarily the highest value, and thus the cluster heads separated from their members by other nodes belonging to other clusters, to gain new members. This allows to solve the problem of the nodes having as cluster heads others which do not consider as such. On the example of figure 1, this algorithm leads to two clusters (5) and (4,3,2).

Of course, it would be naive to think that a NP-hard problem could be solved in a so simple way! this algorithm, by accepting that the minimum of some maximums are chosen, accepts not to be optimal, but since this minimum is chosen among maximums, the performance remains good. Nevertheless two other problems appear. First, as the algorithm has two steps, a phase where the maximums are exchanged until d hops followed by another one where the minimums are exchanged, it is possible to have a cluster head two hops away. Moreover, it is still possible that a node is separated from its cluster head by a father which belongs to another cluster. It is thus necessary to add rules after the phases "Max" and "Min" to avoid that.

The authors of Amis et al. (2000) decide that a node which finally received its own identifier at the end of the algorithm decides it is a cluster head: it is the rule 1. This node has then the highest criterion value in its d-hop neighborhood. They want also that, if a node does not find its identifier, and thus that another node would be a better cluster head, this node be chosen under the condition that it is in its d hop neighborood, and thus that it appears also during the "Max" phase. The node chooses then as a cluster head the node which appeared in both "Min" and "Max" phases, but, for reasons of a better balancing of the number of sensors in the clusters, they impose also that it is the smallest pair which is chosen if several are possible (because the algorithm tends to favor the cluster heads having the highest criterion value): it is the rule 2. At last, if it is in none of both preceding cases, a sensor chooses as a cluster head the node which appeared at the end of the "Max" phase: it is the rule 3.

This solution seems to solve enough problems to give satisfaction. Unfortunately, no validation has been given. In the next sections this heuristic is formally evaluated and it is shown how it still poses a problem. Nevertheless interesting lessons are drawn by this study and solutions are proposed.

2. The Maxi-Min d-cluster formation: election of cluster heads

The deployment of hierarchical sensor networks organized in clusters is of highest importance for applications requiring several hundreds of sensors. This actually allows to set up scalable protocols. Amis et al.'s proposal allows to build multi-hop hierarchical clusters with a bounded depth. The set of the cluster heads constitutes then a d-dominating set on the graph of the network. This notion is formalized in the following paragraphs.
Let $\mathcal{G} = \{V, E\}$ be a graph where E is the set of the edges and V the set of the vertices. In this context, the cluster heads constitute a subset S of V which is *d-dominating* with respect to the graph $\mathcal{G}$. A subset S of V is *d-dominating* when any vertex in E can join a vertex in S via edges in E in less than d hops. Amis et al. have proved that for $\mathcal{G}$, d and an integer k given, it is difficult to know if there exists a set of d-dominating subsets with a size smaller or equal to k. More precisely, the authors have proved that this problem is NP-hard. They propose an algorithm, the "Max-Min d cluster formation", which allows to build a d-dominating set and the tree associated to each cluster head.
To date, this algorithm is one of the very rare ones to propose a wireless network organization as multi-hop clusters and it is very important as already said in the previous section. Moreover, this algorithm is noticeable because the nodes exchange only few informations to build the d-hop dominating set. More precisely, the algorithm is divided into two steps. The first one allows to choose the d-dominating set and to let the simple nodes to know their cluster heads. The second one allows each node to know which node is its father, i.e. to know how to join its cluster head [1]. We first look at the selection of the d-dominating set, that is at the first part of the algorithm proposed by Amis et al. The clusters built with this algorithm depend on the addresses of the nodes. the cluster heads have often[2] the highest address. This means that the clusters formed by the algorithm are not the same for two networks which differ only by their node addresses. Moreover, there is no reason to select cluster heads in function of their addresses and it would certainly be more intelligent to use other criteria. Other criteria could be the node degree, its residual energy, etc. This led us to generalize the first part of this algorithm in order to build clusters of which the cluster heads have often the highest chosen criterion. The criterion becomes thus a parameter of the algorithm, as the maximal depth d. It is this generalized version which is presented here.

2.1 Notations and introduction to the algorithm

This part is extending the results published in CRAS Delye de Clauzade de Mazieux et al. (2006) (Compte Rendu à l'Académie des Sciences).
Let $\mathcal{G} = \{V, E\}$ be a graph with sets of vertices V and edges E . The clusterheads form a subset, S of V which is a $d - dominating$ set over $\mathcal{G}$. Indeed, every vertex not in S is joined to at least one member of S through a path of d edges in E.

[1] In fact, there is a misteake in this second part, as it will be shown in the next sections
[2] This notion will be specify later, see equation 1, p. 18

Let us consider $x \in V$, $\mathcal{N}_i(x)$ is the set of neighbors which are less than i hops from x ; $(\mathcal{N}_i(x))_i$ is an increasing sequence for set inclusion. Let Y be a set on which a total order relation is defined. Let v be an injective function of V in Y. Let X be the image set of V by v ; v is a bijection of V over X. The reverse function is denoted v^{-1}: $\forall x \in V \quad v^{-1}(v(x)) = x$.

The presented algorithm (cf. Delye de Clauzade de Mazieux et al. (2006)) generalizes the one proposed by Amis et al. The algorithm includes $2d$ runs. The d first runs constitute the *Max phase*. The d last runs constitute the *Min phase*. Each node updates two lists *Winner* and *Sender*, of $2d+1$ records. Winner is a list of elements of X. Sender is a list of elements of V. Let us denote $W_k(x)$ and $S_k(x)$ the images in x of the functions W_k and S_k, defined by induction.

The basic idea of the $d-$*dominating* setting is the following: during the first phase, the Max phase, a node determines its dominating node (for the i given criterion) among its d hop neighbors ; a second phase, the Min phase, lets a node know whether it is a dominating node for one of its neighbor nodes. If it is the case, this node belongs to the S set. For a given criterion, the only dominating set is built from this very simple process.

Initial Phase: $k = 0$

$$\forall x \in V, \quad W_0(x) = v(x) \quad S_0(x) = x$$

Max Phase: $k \in [\![1; d]\!]$
Let us assume that the W_{k-1} and S_{k-1} functions have been built.
For $x \in V$, let $y_k(x)$ be the only node of $\mathcal{N}_1(x)$ which is such that:

$$\forall y \in \mathcal{N}_1(x) \setminus \{y_k(x)\}, \quad W_{k-1}(y_k(x)) > W_{k-1}(y)$$

W_k and S_k are derived from:

$$\forall x \in V, \quad W_k(x) = W_{k-1}(y_k(x)) \quad S_k(x) = y_k(x)$$

Min phase: $k \in [\![d+1; 2d]\!]$
Let us assume that the W_{k-1} and S_{k-1} functions have been built.
For $x \in V$, let $y_k(x)$ be the only node of $\mathcal{N}_1(x)$ which is such that:

$$\forall y \in \mathcal{N}_1(x) \setminus \{y_k(x)\}, \quad W_{k-1}(y_k(x)) < W_{k-1}(y)$$

W_k and S_k are derived from:

$$\forall x \in V, \quad W_k(x) = W_{k-1}(y_k(x)) \quad S_k(x) = y_k(x)$$

Definition 2.1. *Let S be the set defined by:*
$S = \{x \in V, W_{2d}(x) = v(x)\}$ [3]

Theorem 2.1. *Each node $x \in V \setminus S$ can determine at least one node of S which is in $\mathcal{N}_d(x)$. It needs only to derive it from its* Winner *list:*

- *if x finds a pair $(v(y))$ in its* Winner *list (that is to say that $v(y)$ appears at least once in each of the two phases), then $y \in S \cap \mathcal{N}_d(x)$. If the node x find several pairs, it chooses the node y with the smallest value $v(y)$ among the pair values that it found.*

[3] This definition is not the same as the one that is given in Amis et al. (2000) but both definitions are equivalent(see Th. 2.5 page 17).

- *if not, let y be the node such that* $v(y) = W_d(x)$. *Then* $y \in S \cap \mathcal{N}_d(x)$.

The preceding theorem, whose proof will be given in the next part, lets us immediately derive the following corollary.

Corollary 1. *S is a d-dominating set for the G graph.*

2.2 Formal validation of the algorithm

It is necessary to check that all the definitions are coherent, i.e. a node chosen as a cluster head by another node is actually a cluster head (with respect to the construction of the set S), and that this node is in the *d*-hop neighborhood of the cluster head.
We shall not prove the three first lemmas which derive directly from the definitions.

Lemma 1. $\forall (x,k) \in V \times [\![1;d]\!]$

- $W_k(x) = Max\{W_{k-1}(y), y \in \mathcal{N}_1(x)\}$
- $S_k(x)$ *is the only element y in* $\mathcal{N}_1(x)$ *such that* $W_{k-1}(y) = W_k(x)$

Lemma 2. $\forall (x,k) \in V \times [\![d+1;2d+1]\!]$

- $W_k(x) = Min\{W_{k-1}(y), y \in \mathcal{N}_1(x)\}$
- $S_k(x)$ *is the only element y in* $\mathcal{N}_1(x)$ *such that* $W_{k-1}(y) = W_k(x)$

Lemma 3. $\forall (x,k) \in V \times [\![0;d]\!]$
$W_k(x) = Max\{v(y),\ y \in \mathcal{N}_k(x)\}$

Definition 2.2. *Let us denote* $M(x)$ *the value* $W_d(x)$.

Theorem 2.2. $\forall x \in V \quad \forall y \in \mathcal{N}_d(x) \quad M(y) \geqslant v(x)$

Proof. Let us assume $x \in V$ and $y \in \mathcal{N}_d(x)$. From Lem. 3, it follows:
$M(y) = W_d(y) = Max\{v(z),\ z \in \mathcal{N}_d(y)\}$. And from $x \in \mathcal{N}_d(y)$, it may be deduced that $Max\{v(z),\ z \in \mathcal{N}_d(y)\} \geqslant v(x)$. □

Lemma 4. $\forall (x,k) \in V \times [\![d+1;2d]\!]$
$W_k(x) = Min\{M(y),\ y \in \mathcal{N}_{k-d}(x)\}$

Proof. The proof is an induction on k, after having chosen x. □

Lemma 5. $\forall (y,k) \in V \times [\![d+1;2d]\!]$
$\exists!x \in \mathcal{N}_{k-d}(y) \quad M(x) = W_k(y)$

Proof. $W_k(y) = Min\{M(z),\ z \in \mathcal{N}_{k-d}(y)\}$. So it exists x in $\mathcal{N}_{k-d}(y)$ such that $M(x) = W_k(y)$. x is unique since the v application is injective. □

Theorem 2.3. *Let us consider* $x \in V$. *Let y be the only node such that* $M(x) = W_d(x) = v(y)$. *Then* $y \in S$.

Proof. >From Def. 2.1 it follows that it has to be proven that $W_{2d}(y) = v(y)$. The node y is among the d hop neighbors of x since $W_d(x){=}v(y)$, so in the other way round, x is among the d hop neighbors of y. Firstly, $Min\,\{M(z),\, z \in \mathcal{N}_d(y)\} \leqslant v(y)$ since $x \in \mathcal{N}_d(y)$ and $M(x) = v(y)$. Secondly it follows from Th. 2.2 that: $\forall\, z \in \mathcal{N}_d(y) \quad M(z) \geqslant v(y)$. So $Min\,\{M(z),\, z \in \mathcal{N}_d(y)\} \geqslant v(y)$.
A conclusion is $Min\,\{M(z),\, z \in \mathcal{N}_d(y)\} = v(y)$ and $y \in S$. □

Corollary 2. *Let us consider $x \in V$. Let y be the only node such that $M(x) = W_d(x) = v(y)$. Then $y \in S \cap \mathcal{N}_d(x)$.*

Proof. Theorem 2.3 proves that $y \in S$ and from the proof it appears that $y \in \mathcal{N}_d(x)$. □

Theorem 2.4. *Let us consider $y \in V$ and $k \in [\![d+1;2d]\!]$. Let $x \in V$ be the only node such that $v(x){=}W_k(y)$. Then $x \in S$.*

Proof. >From Lem. 5 it may be derived that
$\exists! z \in \mathcal{N}_{k-d}(y) \quad M(z) = W_k(y)$. It follows $M(z) = v(x)$. When applying Th.2.3 to z and x, it follows: $x \in S$. □

Corollary 3. *Let us consider $x \in V$. Let us assume that there is an $y \in V$ such that the $v(y)$ value appears again at least once in the Max phase and at least once in the Min phase for the node x. Then $y \in S \cap \mathcal{N}_d(x)$.*

Proof. Theorem 2.4 proves that $y \in S$ because $v(y)$ appears in the Min phase. And since $v(y)$ appears at least once in the Max phase, then $y \in \mathcal{N}_d(x)$. So $y \in S \cap \mathcal{N}_d(x)$. □

Remark 1. *>From the first point of Th. 2.1, it seems reasonable to choose the k-dominating node corresponding to the smallest pair, when there are several ones. This choice leads to sets that are dominated by a smaller criterion value node.*

This definition of S (see Def. 2.1) is different from the definition given in Amis et al. (2000). For them, S' is defined as: $S' = \{x \in V,\, \exists k \in [\![d+1;2d]\!]\ W_k(x) = v(x)\}$.
Clearly, $S \subset S'$. The next theorem proves that the reverse inclusion is also true.

Theorem 2.5. $S = S'$.

Proof. Let us consider $x \in S'$. $W_{2d}(x) \leqslant W_k(x)$ is a consequence of Lem. 2. So $W_{2d}(x) \leqslant v(x)$. Let us assume that $W_{2d}(x){<}v(x)$. Lemma 5 implies:
$\exists\, y \in \mathcal{N}_d(x) \quad M(y) = W_{2d}(x)$. So $y \in \mathcal{N}_d(x)$ and $M(y) < v(x)$. But Th. 2.2 says that it is not true since $\forall y \in \mathcal{N}_d(x)\ M(y) \geqslant v(x)$. So $W_{2d}(x) = v(x)$ and $x \in S$.

Corollaries 2 and 3 prove Th. 2.1. Our definition is equivalent to the definition in Amis et al. (2000). Our definition is more performing since the whole Min phase does not need to be run. □

2.3 Algorithm characterisation

The building of the d-dominating set is distributed, because it is not necessary to know the whole topology, nor the criterion value on each node. The number of computations which have to be completed for each node, is scalable: if the node distribution is Poisson of parameter λ on a plane, if R is the transmission rate and if an edge between two nodes exists only when their distance is less than R, then the number of communications from one node is equal to $2d(1+\lambda\pi R^2)$. The time necessary to build a d-dominating set is $2d$ steps. For this d-dominating set:

$$
\begin{aligned}
& x \in S \Leftrightarrow \\
& \mathcal{N}_d(x) = \varnothing \; or \\
& \qquad \exists y \in \mathcal{N}_d(x)\; v(x) = Max\,\{W_d(z),\, z \in \mathcal{N}_d(y)\}
\end{aligned}
$$

Theorem 2.6. *Let us consider a graph (it may be finite or infinite) and d the maximal depth chosen, let us denote S_d the d-dominating set derived from the algorithm. For the same graph and for $d+1$, let S_{d+1} be the dominating set derived from the algorithm. Then $S_{d+1} \subset S_d$.*

Proof. Let us consider $x \in V \setminus S_d$. $\mathcal{N}_d(x) \neq \varnothing$ so $\mathcal{N}_{d+1}(x) \neq \varnothing$. Let us consider $y \in \mathcal{N}_{d+1}(x)$ and $w \in \mathcal{N}_d(x) \cap \mathcal{N}_1(y)$. $w \in \mathcal{N}_d(x)$ so $\exists z \in \mathcal{N}_d(w)\; v(z) > v(x)$. $z \in \mathcal{N}_d(w)$ and $w \in \mathcal{N}_1(y)$ so $z \in \mathcal{N}_{d+1}(y)$ and $v(z) > v(x)$.
It follows: $\mathcal{N}_{d+1}(x) \neq \varnothing$ and
$\forall y \in \mathcal{N}_{d+1}(x) \quad \exists z \in \mathcal{N}_{d+1}(y) \quad v(z) > v(x)$ so
$x \in V \setminus S_{d+1}$. It may be derived that: $S_{d+1} \subset S_d$. □

2.4 A few criteria that might be useful

The node degree $d(i)$ (i.e. number of neighbors) may be used as a criterion to select the cluster heads: the criterion may be the couple (node degree, node id) and a total order relation may be defined by:

$$
\begin{aligned}
(d(x),x) > (d(y),y) \quad & \Leftrightarrow \\
& (d(x) > d(y)) \quad or \quad (d(x) = d(y)\; and\; x > y)
\end{aligned}
$$

The residual energy of a sensor in a sensor network may also be a good criterion when building the d-dominating set which is the set of the clusterheads.

Simulations of the mechanism have been run for n nodes randomly and uniformly distributed distributed over a $100 \cdot 100$ surface, and a coverage radius R equal to 5.
It can be observed on figure 2 that the number of cluster heads converges towards a constant when the density of nodes increases. Actually, for a given area and a fixed transmission radius, the number of nodes a cluster can be constituted of is not bounded. Consequently, there is a density from which the number of cluster heads stops to increase when the total number of nodes increases. The figure 2 shows that to choose the degree of the nodes (curve "Node degree" on the figure) allows to obtain less cluster heads than the node identifier. The percentage of the number of cluster heads in function of the mean degree of a node is presented on figure 3.

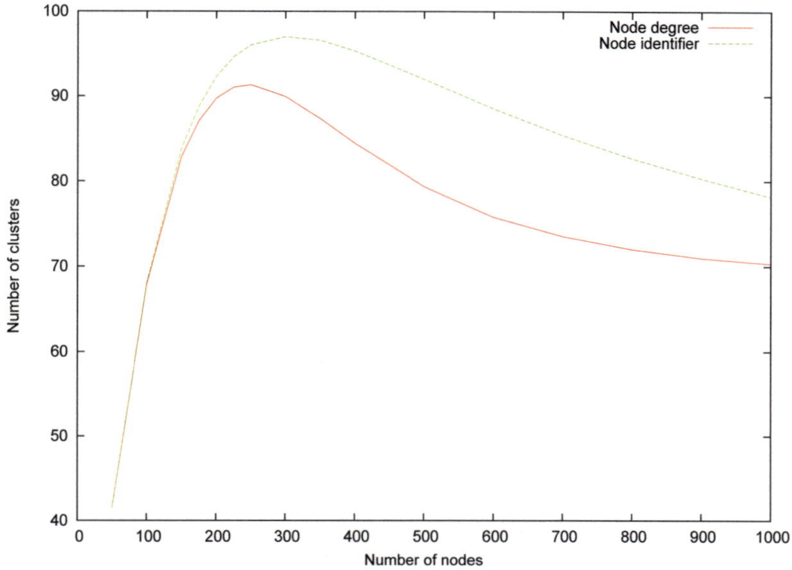

Fig. 2. Number of cluster heads for $S = 100 \cdot 100$, $R = 5$, $d = 3$

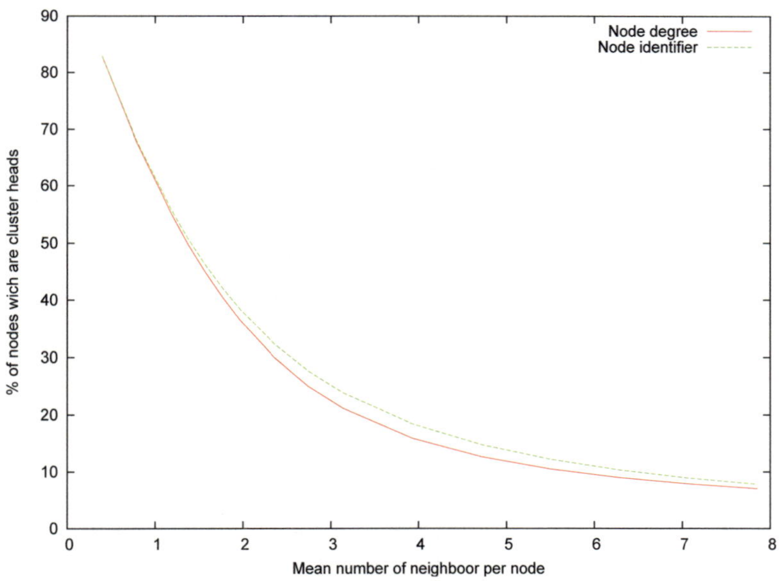

Fig. 3. Percentage of cluster heads for $S = 100 \cdot 100$, $R = 5$, $d = 3$

3. The Maxi-Min d-cluster formation: formation of the clusters

In the previous section, we proved that the nodes can determine a d-dominating set over the graph, for any given criterion. To join a cluster x, with a given $c(x)$ clusterhead, nodes must establish a path to reach $c(x)$ provided all nodes in the path belong to the same cluster x. Therefore, it is necessary to find an algorithm to partition the topology in the connected components, called clusters. In this section the formation of these clusters is studied. In paper Amis et al. (2000), the authors proposed a formation of the above path, at the end of the formation of the d-dominating set. We have proved that there exist some cases for which the formation of the path is not valid.

Max-Min d-cluster formation proposal. The authors of paper Amis et al. (2000) proposed the following algorithm to determine the father of each node. The rules are examined in sequence and the algorithm stops for the node x where x be a node of E, as soon as one of the rules is verified.

- *Rule 1:* if $x \in S$, then x is a cluster of which it is the clusterhead and selects itself as a father;
- *Rule 2:* Else, if x finds a pair $(v(y))$ in its *Winner* list (i.e. if $v(y)$ appears at least once in each of the two phases), then x selects y as a clusterhead [4]. If the node x finds several pairs, it selects the node y whose value $v(y)$ is the smallest, among the found pairs, as a clusterhead. Let $k \in [\![1;d]\!]$ be such as $W_k(x) = v(y)$. x chooses then $S_k(x)$ as a father [5].
- *Rule 3:* Else, let the node y be such as $v(y)=W_d(x)$. Then x selects y as a clusterhead [4]. x selects $S_k(x)$ as a father [5].

Therefore, in some cases it is necessary to use an additional rule to make sure that node $p(x)$ the father of the node x and x are in the same cluster $c(x)$. It may be that following the application of the three preceding rules: $c(p(x)) \neq c(x)$. This rule is named *convergecast* in paper Amis et al. (2000) and it is quoted below:

> "Once a node has identified itself as a gateway node, it then begins to inform (convergecast) its clusterhead by sending a list formed with its node id, all neighboring gateway nodes and their associated clusterhead to its father. A node uses its SENDER table to determine its father. The process continues with the father which adds its own id to the previous list and sends it to its own father. When the clusterhead has heard each of its neighbors, it knows all the links between it and nodes in its cluster. Moreover it knows all the links between its cluster and the other neighboring clusters thanks to the data provided by the gateway nodes."

Consequently, the above rule introduces a new condition. It is necessary that: $\forall x \in E \; p(p(x)) \neq x$. In the contrary case, the rule would lead to an infinite loop.

We now show that cases exist where this condition is not always true because of the fact that loops may appear and we give a necessary and sufficient condition for these loops to occur. This necessary and sufficient condition is due to rule 2. We also show that to remove the loops

[4] We proved in the first part (cf. Th. 2.1 page 15.), that in this case, the node y is well in $S \cap \mathcal{V}_d(x)$. The application of the *Rule 1* thus makes it a cluster.

[5] By definition, $S_k(x) \in \mathcal{N}_1(x)$, cf. page 15.

(by removing this rule 2) is not sufficient to allow the use of this cluster construction heuristic as proposed by their authors. Actually, removing the rule 2 leads to other problems: a node may have as a father a node belonging to another cluster than its own. We deduce than we can (and we must) keep the heuristic to select the cluster heads but the way the clusters are built must be set up differently from what they propose.

3.1 On an example where the algorithm leads to a bug

Let the parameter d be chosen as 5. The 11 nodes are numbered from 1 to 11. An edge is set between the nodes 11 and 1, 1 and 2, 2 and 3, 3 and 5, 5 and 10, 10 and 4, 4 and 6, 6 and 7, 7 and 8, 8 and 9, 6 and 2. Based on number of the node as the criterion and after application of rules 1, 2 and 3, Table 1 depicts the result of the father and clusterhead selection algorithm. In

	1	2	3	4	5	6	7	8	9	10	11
$Max1$	11	6	5	10	10	7	8	9	9	10	11
$Max2$	11	11	10	10	10	10	9	9	9	10	11
$Max3$	11	11	11	10	10	11	10	9	9	10	11
$Max4$	11	11	11	11	11	11	11	10	9	10	11
$Max5$	11	11	11	11	11	11	11	11	10	11	11
$Min1$	11	11	11	11	11	11	11	10	10	11	11
$Min2$	11	11	11	11	11	11	10	10	10	11	11
$Min3$	11	11	11	11	11	10	10	10	10	11	11
$Min4$	11	10	11	10	11	10	10	10	10	11	11
$Min5$	10	10	10	10	11	10	10	10	10	10	11
$Clusterhead$	11	11	10	10	11	10	10	10	10	10	11
$Father$	11	1	5	10	3	4	6	7	8	10	11

Table 1. Max-Min d-cluster formation heuristic applied to the example

this example, at the end of rules 1, 2 and 3, the node 3 has node 5 as a father and node 10 as a clusterhead. However, the node 5 has node 3 as father and node 11 as clusterhead.
Hence, the use of the *convergecast* rule is not possible, as the loop is introduced by the nodes 3 and 5, both of which are gateway nodes also. The next paragraph proves that this phenomenon is due to the use of the *Rule 2*.

3.2 A necessary and sufficient condition for loops to appear

Note that if a node i is such that $v(c(i)) < M(i)$ then the *Rule 2* was used. Now, the necessary conditions for the phenomenon of loops to appear are investigated. Let us assume that there is a loop and let us prove that *Rule 2* was used.
Let us consider node i, $c(i)$ its clusterhead and $p(i)$ its father, selected according to the paper Amis et al. (2000). If i and j are two nodes, let us denote $d(i,j)$ the distance, i.e. the smallest number of hops between i and j. Now, let x, y and z be the three nodes. If the shortest path between x and y is in k_1 hops and between y and z is in k_2 hops, then the shortest path between x and z is in less than $k_1 + k_2$ hops: $d(x,z) \leq d(x,y) + d(y,z)$.
Then, for any node such that $c(i) \neq p(i)$:

$$d(i,c(i)) = d(p(i),c(i)) + 1$$

since $p(i)$ is the node allowing i to know $c(i)$.
Let i and j the be two nodes such as $p(i) = j$ and $p(j) = i$. i and j are thus not clusterhead since they each one have a different father. The preceding equality applies to i and j:
$d(i,c(i)) = d(j,c(i)) + 1$ and $d(j,c(j)) = d(i,c(j)) + 1$

The following deduction proves ab absurdo that $c(i) \neq c(j)$. Assume that $c(i) = c(j) = l$, then $d(i,l) = d(j,l) + 1 \quad d(j,l) = d(i,l) + 1$ which is absurd, so $c(i) \neq c(j)$.

Let us suppose, without any generality restriction, that $v(c(i)) > v(c(j))$. Node i belongs obviously to the d hop neighborhood of $c(i)$. Therefore, according to the equality true for all the nodes, $p(i)$ also is in the d hop neighborhood of $c(i)$, that is to say, here: $j \in \mathcal{V}_d(c(i))$. Thus $c(i) \in \mathcal{V}_d(j)$. So, $M(j) \geq v(c(i))$ and then $M(j) > v(c(j))$. Hence, the *Rule* 2 was used according to what precedes.
In other words, the application of the Rule 2, as proposed by the paper can lead to insolvable problems. Then, let us continue by investigating whether removing the *Rule* 2 could be appropriate and indeed it is proved as follows, that by removing the *Rule* 2 there is no further loop problem. Notice first that the suppression of the *Rule* 2 leads to a new property: if the node i is not a clusterhead, then $v(c(i)) = M(i)$ (*Rule* 3).
Let i be a node which belongs to a loop. Without any generality restriction, let us show that a loop with a length 5 cannot occur. Let j, k, l, m and i be the father of i, j, k, l and m respectively. Since, j is father of i, j belongs to the d hop neighborhood of $c(i)$. So, $M(j) \geq v(c(i))$. But $v(c(i)) = M(i)$ thus $M(j) \geq M(i)$.
So, $M(j) \geq M(i)$, $M(k) \geq M(j)$, $M(l) \geq M(k)$, $M(m) \geq M(l)$ and $M(i) \geq M(m)$.
It may then be deduced that $M(i) = M(j) = M(k) = M(l) = M(m)$ then $c(i) = c(j) = c(k) = c(l) = c(m) = c$. Therefore, it can be written (by applying to each node the general equality $d(i,c(i)) = d(p(i),c(i)) + 1$ since no node among i, j, k, l is clusterhead):

$$\begin{aligned}
d(i,c) &= d(j,c) + 1 \\
d(j,c) &= d(k,c) + 1 \\
d(k,c) &= d(l,c) + 1 \\
d(l,c) &= d(m,c) + 1 \\
d(m,c) &= d(i,c) + 1
\end{aligned}$$

which is absurd. The same kind of demonstration can be applied for any other loop for any given length.
Hence, if the *Rule* 2 is removed, which is necessary, there is no more problem of loops.

3.3 The "convergecast" rule is not sufficient to solve the problems

The following example shows that if the suppression of the *Rule* 2 implies that there are no more loop in the algorithm, this suppression does not remove all the problems. Indeed, the following example shows that we can have $j = p(i)$ and $c(j) \neq c(i)$.

Let us use the parameter $d = 2$. Let us consider 5 nodes, numbered from 1 to 5. The edges are between nodes 1 and 2, 2 and 3, 3 and 5, 2 and 4. The used criterion is the number of the node. The result of the father and clusterhead selection algorithm, after application of rules 1, 2 and 3 is given in Table 2.

	1	2	3	4	5
Max1	2	4	5	4	5
Max2	4	5	5	4	5
Min1	4	4	5	4	5
Min2	4	4	4	4	5
Clusterhead	4	5	5	4	5
Father	2	3	5	4	5

Table 2. Max-Min d-cluster formation heuristic applied to the example

It can be noticed that the node 1 has node 2 as a father and is in the cluster 4 whereas the node 2 is in the cluster 5. It is not possible to go from sons to fathers and to be sure to go through son's clusterhead before the father be attached to another clusterhead. This appears clearly on the above example when going from the node 1. This type of problem thus still exists. The convergecast is thus not a solution to the fact that a node i, such as $c(i) \neq c(p(i))$ can exist.

3.4 Another proposal for the formation of the clusters

Let us start with the clusterheads: if the node i is a clusterhead, after application of the *Rule 1*, then node i informs its neighbors that it is a clusterhead. The neighbors who have not already chosen a clusterhead choose i as a clusterhead. Then, they also transmit a message to their neighbors saying that they are at one hop from the clusterhead i. The neighbors of these nodes which did not already choose a clusterhead then choose i as clusterhead by attaching themselves to one of the neighbors of i and proceed in the same way by informing their neighbors that they are two hops away from i. This process is repeated d times so as not to exceed d hops. This mechanism guarantees that there cannot be a loop and that all the connected components, which are the clusters, are trees and that the roots of these trees are the clusterheads. Because of the second part of the theorem 2.1, each non isolated node which is not cluster head is guaranteed to have a cluster head in its d-neighborhood.

4. On cluster modelling

Having a method to build clusters, it is natural to search to characterize these clusters. It is presented in this section results on cluster modelling. Bounds for the number of clusters are first given. Then, the size of the cluster is investigated in function of parameters of the network like the node density and their coverage radius. At last, the validity of the Voronoï model to model clusters is checked. Actually, in most of the papers dealing with clusters, they are modeled by the Voronoï cell centered in the cluster head. But is it valid? By the way, it is proved that the only quantity of interest when dealing with nodes distributed according to a Poisson process with intensity λ and a coverage radius R is $\lambda \pi R^2$

4.1 Analysis of the number of clusters

We searched to bound the number of cluster heads obtained with the MaxMin algorithm. Actually, to calculate exactly the average number is a very difficul problem related to the percolation theory. It can be shown that (cf. Delye (2007)):

$$\mathbf{E}\left[\text{Number of cluster heads in a surface } \mathcal{S}\right] \geq \lambda \cdot \mathcal{S} \cdot \exp\left(-\lambda \pi R^2\right)$$

In the case where the criterion is uniformly and independently distributed, and in the simplest case where the parameter d is equal to 1,

$$\mathbf{P}\left[O \text{ is cluster head}\right] \leq \left(1+\sum_{i=1}^{\infty}\frac{1}{n}\frac{(\lambda\pi R^2)^n}{n!}\right)\exp\left(-\lambda\pi R^2\right)$$

In the case of a parameter $d > 1$, and still if the criterion is uniformly distributed, for a surface $\mathcal{S}$ and by denoting $E = \lambda\pi R^2$,

$$\begin{aligned}&\mathbf{E}\left[\text{Number of cluster heads in}\mathcal{S}\right]\\&\leq \lambda\cdot\mathcal{S}\cdot\left(1+\sum_{i=1}^{\infty}\frac{1}{n}\frac{E^n}{n!}\right)\exp\left(-E\right)\end{aligned}$$

4.2 An empirical model of the size of the clusters

We have later searched to characterize the number of elements in a cluster. It is a very difficult problem which is to date not already solved. Actually, researchers face the problem to derive a simple law because of the strong dependence of the random variables in the considered process. This problem is related to the percolation theory. In this section, we begin by presenting the known results about coverage, connectivity and percolation. Then, we present our empirical work on the characterization of the size of clusters in a network, which gives at the same time interesting results on percolation.

4.2.1 Coverage, connectivity and percolation

Generally, it is common to consider that the sensors are spread over a plane surface according to a Poisson distribution and that they have a circular coverage, generally with the same radius. In the case of nodes distributed according to a Poisson process with a transmission radius R, we can then find the distribution of the law of the number N of neighbors of a node. It is the same distribution:

$$\mathbf{P}[N = k] = \frac{(\lambda\pi R^2)^k}{k!}\cdot e^{-\lambda\pi R^2}, \mathbf{E}[N] = \lambda\pi R^2 \tag{1}$$

Another newer model, *the Blinking Poisson Model*, was introduced in 2004 by Dousse et al. in the paper Dousse et al. (2004). The idea is to consider a distribution of sensors following a Poisson process with rate λ and a transmission radius R_i for each node. The R_i are independent of the Poisson process and their average is $\mathbf{E}[R_i]$. The sensors switch on (*on* period) and off (*off* period) independently from each other. It is assumed that *on* and *off* periods are independent. The period *on* is distributed according to any distribution with mean t_{on} The *off* period can be distributed according to a deterministic or exponential law.

This short state of the art does not pretend to be exhaustive. The reader can refer to the works of Werner Wendelin, Oded Schramm, Gregory Lawler, François Baccelli, Bartek Blaszczyszyn, Patrick Thiran, etc. for more details.

The problem of coverage

The most interesting paper dealing with this subject is Philips et al. (1989) dating from 1989. The authors study (among other things) the probability that each point of the plane be covered when the used model is a "Poisson blob-model" with parameter λ and R. To do this, they consider a finite surface A and they make tend this surface towards the infinity. The results are the following ones :

Theorem 4.1.

$$\forall \varepsilon > 0 \quad R = \sqrt{\frac{(1-\varepsilon)\ln A}{\pi\lambda}} \Longrightarrow \lim_{A\to\infty} Pr[A \textit{ is covered}] = 0$$

Theorem 4.2.

$$\forall \varepsilon > 0 \quad R = \sqrt{\frac{(1+\varepsilon)\ln A}{\pi\lambda}} \Longrightarrow \lim_{A\to\infty} Pr[A \textit{ is covered}] = 1$$

To demonstrate the first theorem, the authors build a grid of points on the surface A. These are spaced by twice the communication radius. This way, these points are covered, or not, independantly. These points are such that there is $A/4R^2$ points in a surface included in A. So, the probability that there is 0 point not covered is $(1-\lambda\pi R^2)^{A/4R^2}$ which tends towards 0 when A tends towards the infinity.

These are important theorems to dimension a sensor network. Indeed, it is necessary that the area covered by the sensors is good. The last theorem shows that for a given surface A, there must be a given number of neighbor nodes in average to ensure coverage. More precisely, knowing that the average number of neighbors under these conditions is $\lambda\pi R^2$, it is immediate to see it must be a little more than $\ln(A)$ neighbors to be almost sure that the surface is covered.

The problem of connectivity

In the same paper Philips et al. (1989), the authors show also a theorem about the connectivity. Connectivity and coverage must not be confused. The following theorem is proved:

Theorem 4.3.

$$\forall \varepsilon > 0 \quad R = \sqrt{\frac{(1-\varepsilon)\ln A}{\pi\lambda}}$$
$$\Longrightarrow \lim_{A\to\infty} Pr[\textit{the network is connected}] = 0$$

This proves that if the average number of neighbors is given, then it is sure for a large enough surface to have a network disconnected. The authors did not succeed to demonstrate that when the number of neighbors were on average a little larger than $\ln(A)$ connectivity was ensured. The consequence of this theorem is that it can not exist magic number of neighbors. In particular, 6 is not a magic number for the network rate, contrary to what Kleinrock and Sylvester claimed in Kleinrock & Silvester (1978) in 1978.

Using "slotted ALOHA" protocols and by requiring that the transmission power of the nodes be the same for all, Kleinrock and Sylvester Kleinrock & Silvester (1978) suggested that the number six is considered a magic number. Later in 1984, the magic number changed and 8 became the newly elected one Takagi & Kleinrock (1984). In this same paper, Tagaki and Kleinrock also found two other magic numbers (5 and 7) considering other transmission protocols. Considering that the nodes can adapt their transmission radius, the authors of Hou & Li (1986) proposed in 1986 the magic numbers 6 and 8. This is the pantheon of a belief which became false in 1989!

Indeed, none of these analyses dealt with the problem of network connectivity. When Tantawi et al. looked à this problem in 1989 Philips et al. (1989), they proved that no number can be magic. The authors showed that whatever the average number of neighbors is chosen, the network will almost surely be disconnected if this number is constant ...

The authors of Gupta & Kumar (1998) deal with the problem of connectivity on a finite circular surface with unit area in which N nodes are randomly placed. The node density is then $\lambda = N$. The transmission radius of a node n is r. The authors show that if $\pi r^2 = (log(N) + c(N))/N$ then the network is asymptotically connected (ie. when N tends to the infinity) with a probability 1 if and only if $c(N)$ tends also to the infinity. This leads the authors of Shakkottai et al. (2003) to say that the transmission radius must be of the order of $\sqrt{log(N)/N}$ for the network to be connected.

The authors of Shakkottai et al. (2003) study a network of sensors placed on a unit area square surface. When n nodes constitute the network (n is supposed to be a squared number), they are placed on a grid such as the distance between two nodes able to communicate is $\sqrt{1/n}$. When the transmission radius is of the order of $\sqrt{1/n}$, the connectivity is ensured. These authors continue their study by supposing that the nodes are on with probability $p(n)$. They show then that the connectivity is asymptotically ensured when $\sqrt{p(n)r(n)} \sim \sqrt{log(n)/n}$. Moreover, they show that the diameter of such a network where the nodes can crash is of the order of $\sqrt{n/log(n)}$.

To choose the good number of neighbors is important. Indeed, this choice impacts not only the network connectivity but also its capacity: the presence of a large number of links between the nodes is not necessarily advantageous. Indeed, if a link exists between i and j, it is an advantage in terms of energy consumption since i can send a packet to j in a single hop. However, when i sends a message to another neighbor, it causes interferences at j which would not have existed without the link that connects them. Therefore there is a trade-off. More specifically, when the transmission radius increases, the number of retransmissions decreases, but the value of the interferences increases. In the paper Gupta et al. (2000), P. Gupta and P. R. Kumar showed that the number of retransmissions increased as $O(1/r)$ (when r increases) but that the interferences were "only" on the order of $O(r^2)$. Thus, the product of both quantities leads to assert that the "net" effect is about $O(r)$. This means that it is better to choose a radius of little value. However, if the radius is too low, then of course the network is disconnected!

For the moment, the best results are those of Xue Feng and P. R. Kumar presented in 2004 in the paper Xue & Kumar (2002). In a network with n nodes randomly placed (uniformely), the number of neighbors of each node must be of the order of $\Theta(log(n))$ so that the network is connected. More precisely, the network is asymptotically disconnected when this number is less than $0,075 \cdot \log n$ and is asymptotically connected when this number is greater than $5.1774 \cdot \log n$ neighbors.

The problem of percolation
It is dealt here with an issue that seems very simple and raised in 1963 by E. N. Gilbert, the issue of critical density in percolation in a network of clusters. Gilbert is one of the first to propose a modelling of wireless networks. His model is a particular case of modelling

with Boolean networks. He deals with nodes placed in the (infinite) plan according to a two dimensions Poisson process. He demonstrates that in such a plan, there is a critical threshold beyond which the probability to belong to an infinite size cluster is not zero. It is said in this case that there is percolation.

Gilbert introduced in 1961 Gilbert (1961) a modelling of these networks using a graph formalism. The vertices of this gaph are the nodes. All the nodes are supposed to be in a same plan (dimension 2). He assumes that two nodes can communicate if and only if their distance is less than a given value R. An edge exists in the graph of the vertices if and only if the respective nodes of each vertice can directly communicate. Gilbert builds such a network with a Poisson process with intensity λ, on an infinite plan. Each connected component is called cluster.

Let be the quantity $E = \lambda \pi R^2$, expectation of the number of points in a circle of radius R and $P(N)$ the probability that a node belongs to a cluster of size larger than $N-1$. Gilbert shows the following theorem :

Theorem 4.4.

$$\exists E_c \in \mathbb{R} \quad \forall E \in \mathbb{R} \quad \begin{cases} P(\infty) = 0 & \textit{if } E < E_c \\ P(\infty) > 0 & \textit{if } E > E_c \end{cases}$$

He bounds also E_c :

$$1.64 \approx \frac{1}{\frac{1}{3} + \frac{\sqrt{3}}{2\pi}} \leq E_c \leq 8\pi \log_e(2) \approx 17.4$$

In fact, Gilbert wrote 1.75 instead of 1.64, but it is a typo in the paper. He shows also by simulation that $E_c \approx 3.2$. He suggests a beginning for a demonstration which would help to prove that $E_c \leq \frac{26\pi}{3\sqrt{3}} \log_e 2 \approx 10.9$ but he does not succeed to conclude.

Kirkook and Wayne Kirkwood & Wayne (1983) and Hall Hall (1985) showed that $2.186 < E_c < 10.588$.

In 1989, Tantawi et al. Philips et al. (1989) proved that the critical value E_c is in the interval : $2.195 < E_c < 10.526$. The demonstration, non explicitly given in the paper, uses an analogy with the $M/D/1$ queue. The instability of this system corresponds to the existence of an infinite component.

These results are summarized in Fig. 4. The probability to obtain an infinite size cluster is zero for R and λ under the point of the curve at $E_c = 2.195$, and it is non zero for all points above $E_c = 10.588$ To the best of our knowledge, there exists no better bounds than the ones given by Tantawi et al. This very simple problem, dating from 1963, is not already solved.

The authors of Dousse et al. (2004) give the results about percolation for the previously described model. They show that there exists a critical density λ_c, function of R, λ and π_{on} such as the network is constituted almost surely of a unique infinite component for $\lambda > \lambda_c$ and almost surely of an infinity of finite components for $\lambda < \lambda_c$. In addition, if λ^* denotes Gilbert's critical density for R given, then these authors show that $\lambda^* = \pi_{on}\lambda_c$. Consequently, it means that if the sensors are placed according to a two dimension Poisson process with intensity λ and if they switch on and off each independently of one another according to

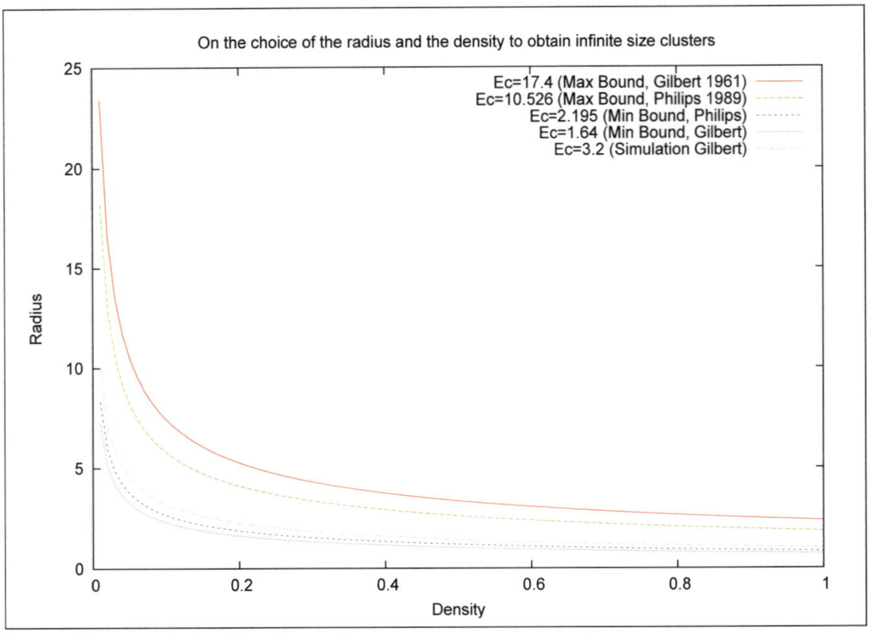

Fig. 4. Synthesis of the results

on and off periods with a ratio π_{on}, then a communication is possible between two sensors almost surely if the density of the Poisson process is strictly greater than Gilbert's constant: $\lambda > \lambda_c = \frac{\lambda^*}{\pi_{on}}$. From the viewpoint of percolation, it is as if there had been a Poisson process with density $\lambda \pi_{on}$.

Moreover, the authors study the transmission delay between two nodes X and Y belonging to the infinite component, where $\lambda > \lambda_c$. Indeed, this delay is theoretically zero when $\pi_{on} = 1$ (classical model) but this is no more true with the Blinking Poisson Model since sensors switch on independently from each other. The result they demonstrate is the following:

$$\exists \eta > 0 \quad (1-\varepsilon)\eta < \frac{T(X,Y)}{|X-Y|} < (1+\varepsilon)\eta$$

where $T(X,Y)$ is the time necessary to transmit and $|X-Y|$ their euclidian distance. The result is true for $|X-Y|$ sufficiently high. This result shows that under this transmission model (which does not take into account, of course, the interferences), the time needed to deliver a message increases linearly with the Euclidean distance. The value of η depends only on the parameters of the model and may therefore be determined by simulation.

4.2.2 Evaluation by simulation of the cluster size: an empirical result

Here, we present an empirical work on the the size of the clusters in terms of number of nodes per cluster. This, by the way, gives a result on percolation in sensor networks: the size of the cluster diverges above a certain density of nodes. placed on the plan through a Poisson process with rate λ and with a transmission radius R. The number of nodes N is a function of λ and R, but, more precisely, a function only of the quantity $E = \lambda \pi R^2$ proposed by Gilbert.

This can be verified on Fig. 6. This is exactly to say that $N = f(\lambda, R) = g(\lambda\pi R^2)$

Note that, the cluster size seems not to diverge at $E_c = 2.3$ as thought Gilbert but rather at $E_c = 4.4$. This is most certainly due to the board effects on small surfaces chosen by him. However, we verified that the divergence is well within the range proposed by Gilbert.

The inverse of the natural logarithm of the number of nodes in function of E can be approximated by a straight line. We determined its coefficients empirically. This allows therefore approaching the number of nodes by the following formula:

$$N = \exp(1/(-0.155E + 0.787))$$

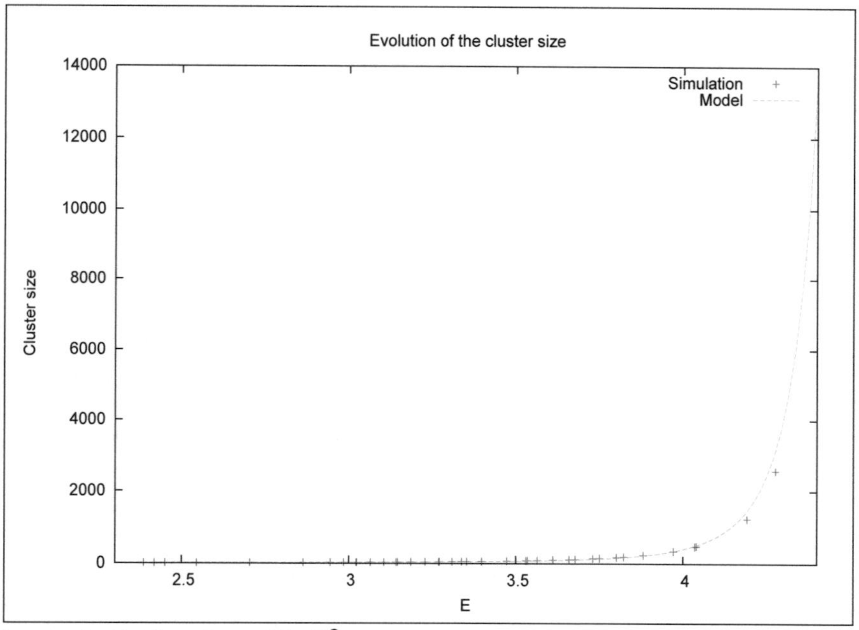

Fig. 5. Cluster size in function of $\lambda\pi R^2$

Comparisons of simulation and heuristic in two dimensions and in three dimensions can be found on the figures 5 and 6.

4.3 Voronoï's modelling

The Voronoï's theory Voronoï (1907) can be used to obtain analytical results in ad-hoc and sensor networks. Actually, the known results of this theory are often applied to model these networks: the comparison between the analytical results given by a Voronoï modelling of the network and the ones obtained by simulation may be interesting. For example, the authors of Bandyopadhyay & Coyle (2003) give an analysis of the performance of their cluster formation algorithm. All their work assumes that the clusters formed with their algorithm can be modellized by Voronoï cells.

It is nevertheless important to check if it is acceptable to consider that a cluster is modelled by a Voronoï cell of which the seed is the cluster head. Actually, a Voronoï tessellation is the

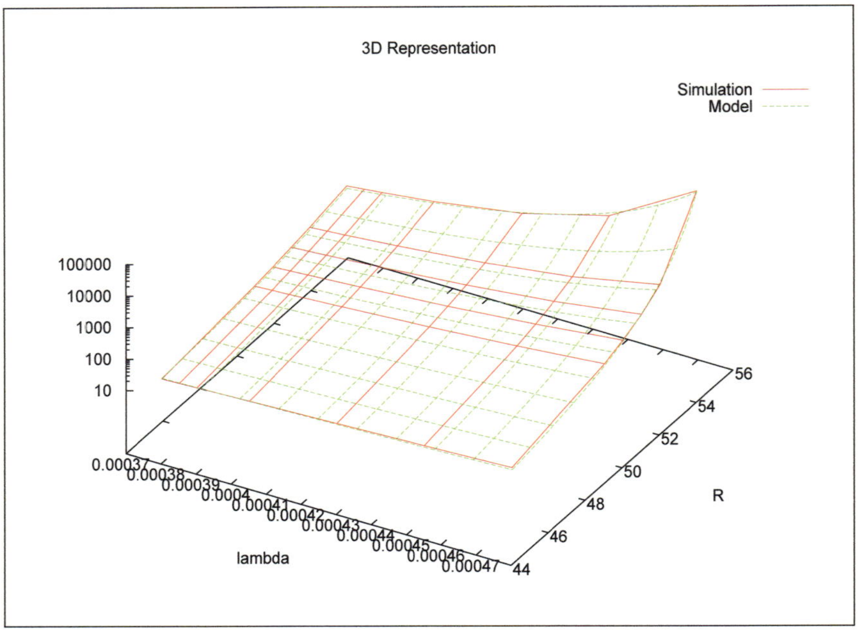

Fig. 6. 3D Comparison

partition in convex polygons generated by seeds: a polygon contains exactly one seed and all the points inside the polygon are closer to this seed than any other seed. Consequently, the simple hop cluster modelling with a Voronoï model is correct for a certain cluster formation policy. This is not at all obvious for all the multi-hop clusters of which nodes can be attached to a cluster head but belong to the Voronoï cell of another cluster head.

This highlights the importance of the choice of the cluster head for the cluster formation mechanism. It seems actually that a Voronoï modelling is more or less false depending on the spatial distribution of the cluster. First of all we assumed that the distribution of the cluster heads is uniform. It is for example the case in the paper Bandyopadhyay & Coyle (2003). Without a generalization of our work, the presented results are thus a priori restricted to this context.

Moreover, it is obvious that the cluster head choice mechanism once the clusters built is also of the highest importance. Let us consider for example cluster heads distributed according to a Poisson process with a density $p \cdot \lambda$ on a surface and a cluster formation policy such that the clusters are single-hop and have a bounded number of children k. It is obvious that some nodes cannot be attached to the nearest cluster head because it has already reached its maximal number of children. Then they can join another cluster head but they do not belong to its Voronoï cell. It is clear that for this policy, the higher the value of λ and the smaller p is, the falser the Voronoï modelling is.

We must then choose the cluster formation policy to partially answer the question. The policy we use is the policy we call canonical: a node which is a neighbor of a cluster head joins the nearest cluster head (the probability that two cluster heads are at the same distance r is $o(\lambda r)$).

This node is then said clustered. A node which has no cluster head among its neighbors joins its nearest clustered neighbor. This policy is debatable but we think it is both more favorable to the use of the Voronoï modelling by choosing the nearest cluster head and more realistic by choosing the nearest clustered neighbor. We also could choose a policy for which the clustered neighbor having the nearest cluster head would have been chosen but in practice it is not possible without the use of a triangulation. Note also that it is not easy to estimate the distance to a neighbor (and thus the nearest one) from the reception power because the wireless medium is by nature quite instable.

The Voronoi model is good for a node x if x belongs to its Voronoï cluster. In this case, it is set $V(x) = 1$. The accuracy of the modelling of the clusters by Voronoï cells can be assessed by observing the percentage of nodes for which the Voronoï model is good. This criterion is *a priori* a function of R, λ and p where R is the transmission radius, λ is the density of the network and p the ratio between the number of cluster heads and the number of nodes, $p\lambda\pi R^2$ being the density of cluster heads. The criterion is denoted by $C = f(\lambda, R, p)$.

We have first evaluated the proportion of ordinary nodes belonging to a cluster. Actually, some nodes can belong to a strongly connected component while there is no cluster head in this component. We have evaluated the percentage of nodes for which the Voronoï modelling is good, that is for which the Voronoï cell in which they are is well centered on their cluster head. These results allow to evaluate in which conditions the Voronoï modelling is acceptable. We have found that it is true for a large interval of densities. A node is said to belong to its Voronoï cluster if and only if its cluster head is the seed of the Voronoï cell which it belongs to.

4.3.1 Probability for a node to belong to a cluster: a simulation study

The cluster formation is done according to the canonical policy above described: first a single hop cluster is formed with the cluster heads and their neighbors. The neighbors of the cluster head can then build a two hop cluster by associating their neighbors, and so on. Two types of nodes exist once the clusters constructed: those actually attached (possibly indirectly via several hops) to a cluster (clustered nodes) and those who are not clustered.

We simulated a Poisson process on a surface area $A = 10000 \cdot 10000$. The transmission radius R and the density λ of the Poisson process are the two main parameters. The percentage of cluster heads is another fundamental parameter, as it will be explained later and is denoted by p. A node is a cluster head with probability p. Then, we know that the cluster heads are distributed according to a Poisson process with intensity $\lambda \cdot p$ while the other nodes are distributed according to a Poisson process with intensity $\lambda \cdot (1 - p)$. In addition, another result states that both processes are independent from the Poisson process with intensity λ. The cluster heads constitute a set denoted S_0 while the set of the other nodes is denoted S'_0.

For the parameters λ,R and p, the probability for a node belonging to S'_0 to be clustered is denoted $\mathbf{P}_{\lambda,R,p}[C(x) = 1]$. This probability has been simulated for

$$\lambda = \{0.001\ 0.0012\ 0.0014\ 0.0016\ 0.0018\ 0.002\},$$
$$R = \{5\ 15\ 25\ 35\ 45\ 55\ 65\ 75\ 85\ 95\}$$

and

$$p = \{0.05\ 0.1\ 0.15\ 0.2\ 0.25\ 0.3\ 0.35\ 0.4\ 0.45\ 0.5\}.$$

As expected, (cf. Sec.4.3.1), $\mathbf{P}_{\lambda,R,p}[C(x) = 1]$ is only a function of $E = \lambda\pi R^2$ and p, probability for a node to be cluster head. The probability $\mathbf{P}_{\lambda,R,p}[C(x) = 1]$ is thus denoted $\mathbf{P}_{\lambda\pi R^2,p}[C(x) = 1]$. The simulations of this probability are given on figure 7. It can be

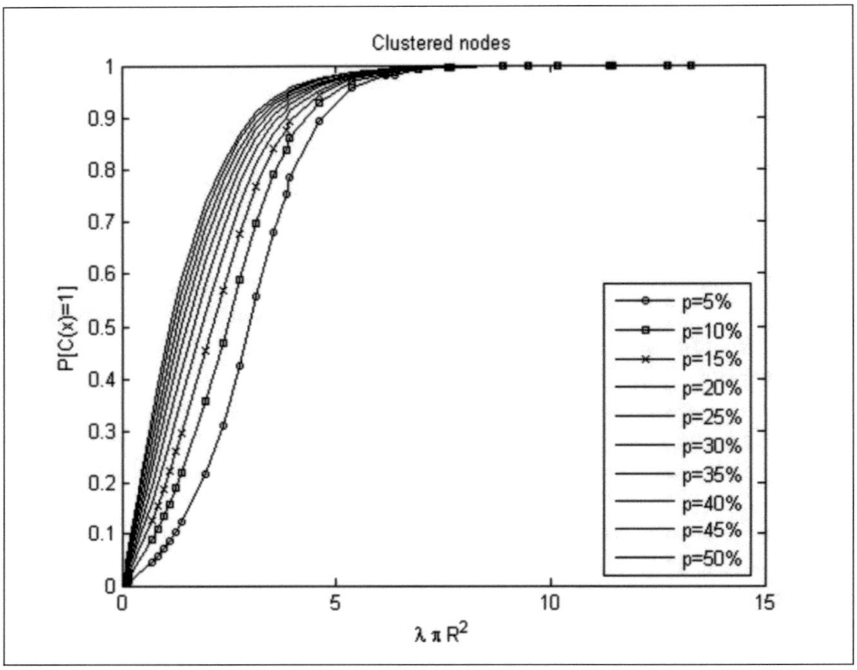

Fig. 7. Probability for a node belonging to S_0' to be clustered

observed that, wathever the value of the parameter p, $\mathbf{P}_{\lambda\pi R^2,p}[C(x) = 1]$ is an increasing function of $\lambda\pi R^2$, which was expected. It can be noticed that, for the "magic number" 6 and $p \geq 0.05$, the probability that a node is clustered is greater than 95%.

Figure 7 gives then limits for the use of the modelling of these clusters by Voronoï cells. Actually, a node must at least be clustered with a good probability to be used in a Voronoï model. The Voronoï modelling must thus not be used for a wireless network distributed according to a Poisson process with intensity λ and a communication radius R such as $\lambda\pi R^2 \leq 5$ which is a lower bound for $\mathbf{P}_{\lambda,R,p}[C(x)=1] \geq 85\%$ when $p \geq 5\%$.

4.3.2 Probability for a node to belong to a cluster: analytical results

In this section, the probability that a node, which is not a cluster head, is clustered is expressed and approximated.

The probabilities ψ and Ψ

We want to evaluate the probability $\psi(X)$ that a node X is clustered knowing that it is not a cluster head. Let $\Psi(S)$ be the probability that a surface S contains at least a clustered node. By observing that a node X is clustered if and only if the open subset of center X and radius R (denoted $B(X,R)$) contains at least a clustered node, the following equality is true: $\psi(X) = \Psi(B(X,R))$.

$\Psi(S) = \psi(X)\nu(S)$ cannot be written since $\psi(X) = \Psi(B(X,R))$. Moreover, a node is clustered if and only if there exists a path from this node to its cluster head. These cluster heads constitute a two dimension Poisson process with density $p\lambda$, that is the number of cluster heads in a surface R follows a Poisson law with parameter $p \cdot \lambda \cdot \nu(S)$.

Another way to calculate the probability that a node is clustered is to consider it is clustered if and only if at least one of its one hop neighbors is a cluster head or at least one of its two hop neighbors is a cluster head, and so on. We will search a lower bound of this probability by calculating the probability that at least one of its neighbors at less than two hops is a cluster head.

Calculation of the supplementary surface $S(n,R)$ brought by n nodes in a disk with radius R to the ring $A(X,R,2R)$

Let be a disk with centre X and radius R. Let's assume that n nodes are uniformly distributed inside. These nodes bring a supplementary surface $S(n)$ to $B(X,R)$ which is a portion of the ring $A(X,R,2R)$ (surface of the disk of radius $2R$ centered in X minus the disk of radius R centered in X). What is the supplementary surface $S(n)$? When $n = 0$, $S(0)$ should be 0. When n tends towards the infinity, all infinitesimal element of $B(X,R)$ contains exactly one node. In this case, $\lim_{n\to\infty} S(n) = 3\pi R^2$, area of the ring. It can be shown that (cf. Delye (2007)):

$$S(n,R) = 3\pi R^2 - 2\pi \int_{r=R}^{2R} \left(1 - \frac{I(r,R)}{\pi R^2}\right)^n r dr \tag{2}$$

with

$$I(r,R) = 2 \cdot R^2 \cdot \arccos\left(\frac{r}{2R}\right) - r \cdot \sqrt{R^2 - \frac{r^2}{4}} \tag{3}$$

When thinking about the quantity $P_n(R) = S(n,R)/(3\pi R^2)$ from a probabilistic viewpoint, it can be conjectured that this quantity does not depend on R. Indeed, $P_n(R)$ represents the percentage of the surface of the ring $A(R,2R)$ brought by the n nodes uniformly placed on the disk $B(0,R)$. Actually, it can be shown (cf. Delye (2007)):

$$P_n(R) = 1 - \frac{2}{3} \int_{u=1}^{2} \left(1 + \frac{u}{\pi} \cdot \sqrt{1 - \frac{u^2}{4}} - \frac{2}{\pi} \cdot \arccos\left(\frac{u}{2}\right)\right)^n udu$$

This last equation thus shows that $P_n(R)$ is independent of R. It can also be shown (cf. Delye (2007)) that the probability that a node has n_2 two hop neighbors is:

$$\mathbf{P}[n_2 = k] = \sum_{i=0}^{\infty} \frac{(3P_i E)^k}{k!} e^{-3P_i E} \frac{E^i}{i!} e^{-E}$$

Its expecation is then:

$$\mathbf{E}[n_2] = \sum_{i=0}^{\infty} P_i \frac{E^i}{i!} e^{-E} 3E$$

Calculation of the probability to be clustered for two-hop clusters

Let be Φ a two hop Poisson process with intensity λ. Let be R the transmission radius and p the probability that a node is cluster head. Let be $x = E = \lambda\pi R^2$ the mean number of neighbors and $\psi(x)$ the probability for a node which is not a cluster head, to be clustered. This probability can be seen as the sum of two other ones :

- the probability to be clustered and that there is at least a cluster head among the neighbors : it is probability $p1$;
- the probability to be clustered and that there is no cluster head among the neighbors : it is probability $p2$.

For the one hop case, the process of the cluster heads is a two dimension Poisson process Φ_C with density $p\lambda$ and which is independent of the first process. Then :

$$p1 = \mathbf{P}^{\circ}\left[\Phi_C(B'(X,R)) = 0\right] = 1 - e^{-p\lambda\pi R^2} = 1 - e^{-px}$$

And for the two hop case, it can be shown (cf. Delye (2007)):

$$p2 = e^{-x}\left[1 - e^{-(1-p)x} - \sum_{i=1}^{\infty}\frac{((1-p)x)^i}{i!}e^{-p\lambda S(i)} + \epsilon\right]$$

with

$$\epsilon = \sum_{i,j=1}^{\infty}\frac{((1-p)x)^i}{i!}\frac{(\lambda S(i)(1-p))^j}{j!}e^{-\lambda S(i)}\psi(i,j)$$

We did not succeed to calculate exactly $\psi(i,j)$. We can only give a lower bound of the probability of clustering of a node :

$$\mathbf{P}\left[C(X) = 1\right] \geq$$
$$1 - e^{-px} + e^{-x}\left[1 - e^{-(1-p)x} - \sum_{i=1}^{\infty}\frac{((1-p)x)^i}{i!}e^{-p\lambda S(i)}\right]$$

4.3.3 Evaluation of the Voronoï model

Here the probability that a node belongs to its Voronoï clusteri is presented. We simulated cluster head distributions and the "canonical" policy for each one of the following triplets :

$$\lambda = \{0.001\ 0.0012\ 0.0014\ 0.0016\ 0.0018\ 0.002\},$$
$$R = \{5\ 15\ 25\ 35\ 45\ 55\ 65\ 75\ 85\ 95\}$$

and

$$p = \{0.05\ 0.1\ 0.15\ 0.2\ 0.25\ 0.3\ 0.35\ 0.4\ 0.45\ 0.5\}.$$

These values lead to $6 \cdot 10 \cdot 10 = 600$ parameters. The average of the criterion is obtained on 1000 simulations for each parameter.

Figure 8 shows the probability that a node belonging to S_0' is in its Voronoï cluster in function of E and p where $E = \lambda\pi R^2$ and p. E is the average of the number of neighbors per node. The

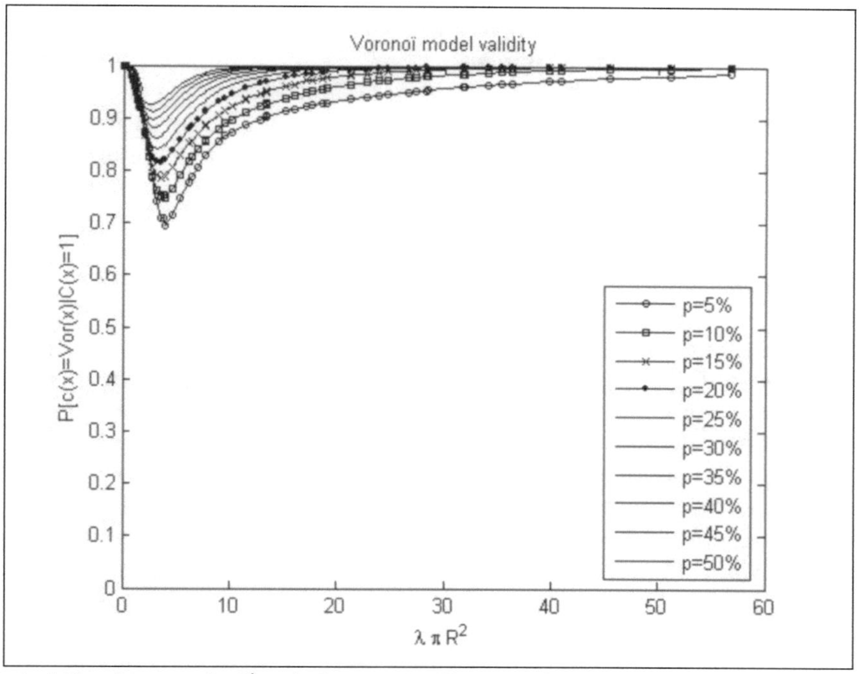

Fig. 8. Probability for a node S'_0 to belong to its Voronoï cluster

validity of the Voronoï model is only function of E and p. $C = f(E, p)$. $V(E, p)$ is equal to 1 when E is small since there is only a single node per cluster, the cluster head itself. When E is large, $V(E, p) = 1$ since every node is connected to its Voronoï cluster because the density is very large. Since the density of the cluster heads is larger than 5% the probability that a node belongs to its Voronoï cluster is larger than 72%.

5. An address assignment mechanism

Addressing nodes is an important step which itself consumes energy and we searched an addressing mechanism allowing to economize energy compared to "naive" protocols like the Cluster Tree Protocol proposed by the Zigbee Alliance.

In Weniger & Zitterbart (2004), the authors define a classification of the different addressing mechanisms. This classification is used in all the papers dealing with this subject. They are separated into two families: the "statefull" protocols and the "stateless" ones. The protocols of the second family do not use allocation tables like the protocols of the first family but they use random addresses or addresses based on a serial number. The protocols of the first type are classified into two subsets: the ones using a centralized allocation table (*Centralized Autoconfiguration CAC*) and the protocols using a distributed allocation table. The protocols MANETconf, Boleng's and Prophet Allocation Zhou et al. (2003) belongs to this last category. At last, a third hybrid family is proposed, in which are the protocols HCQA Yuan Sun et al. (2003) and PACMAN.

We have proposed an address assignment algorithm which is based neither on probabilistic considerations or serial numbers, nor on an address table storage (distributed or not). Moreover, this protocol minimizes the number of exchanges allowing to obtain an address: when a node wants to obtain an address from another one, a single exchange is necessary between these two nodes. An economy of emission, reception and storage is thus gained. This work has been presented in Delye de Clauzade de Mazieux et al. (2009). At the same time, the ZigBee Alliance retained an algorithm very close to this one. This algorithm is based only on a single constaint: the a priori knowledge of the maximum number of children of the vertices in the graph. The idea is the following.

We consider a tree structure (i.e. the cluster has physically a tree structure). We have designed a distributed addressing algorithm on this tree. For a node i of this tree, let $@_i$ denote its address and $e_i(t)$ the number of its children at time t. Let d be a fixed integer. Assume the highest degree of the root is $d-1$ and that the other nodes have degrees less or equal to d. This means that all the nodes have at most $d-1$ children. The root is assigned the address 0. Assume that following an event, node j, without address, queries node i at time t in order to obtain an address. Since $e_i(t) < d-1$, the node i increments the number of its children ($e_i(t+dt) = e_i(t)+1$) and the node i attributes to the node j the address $@_j = d * @_i + e_i(t+dt)$.

This addressing mechanism has interesting properties. First, in terms of efficiency, as already noticed, it is more efficient than the Cluster Tree Protocol. Second, it allows to set up self-routing. Actually, from the only knowledge of the destination address and its own address, every node can determine to which next hop to send the packet to be routed. It is very similar to what allowed once the Banyan networks. The interested reader can refer to Delye de Clauzade de Mazieux et al. (2009) for more details.

6. References

Abbasi, A. A. & Younis, M. (2007). A survey on clustering algorithms for wireless sensor networks, *Comput. Commun.* **30**(14-15): 2826–2841.

Amis, A. D., Prakash, R., Thai, H. P., Dung, V. & Huynh, T. (2000). Max-min d-cluster formation in wireless ad hoc networks, *Proceedings of IEEE INFOCOM 2000*, Dan Panorama Hotel, Tel-Aviv, Israel, pp. 23–41.

Bandyopadhyay, S. & Coyle, E. (2003). An energy efficient hierarchical clustering algorithm for wireless sensor networks, *Proceedings of the Twenty-Second Annual Joint IEEE Computer and Communications Societies. IEEE INFOCOM*, pp. 1713–1723.

Banerjee, S. & Khuller, S. (2001). A clustering scheme for hierarchical control in multi-hop wireless networks, *Proceedings of IEEE Infocom*, Anchorage, Alaska.

Basagni, S. (1999a). Distributed and mobility-adaptive clustering for multimedia support in multi-hop wireless networks, *Proceedings of the IEEE 50th International Vehicular Technology Conference, VTC 1999-Fal*, Amsterdam, The Netherlands, pp. 889–893, vol. 2.

Basagni, S. (1999b). Distributed clustering for ad hoc networks, *Proceedings of the 1999 International Symposium on Parallel Architectures, Algorithms, and Networks (I-SPAN'99)*, Ed. A.Y. Zomaya, D.F. Hsu, O. Ibarra, S. Origuchi, D. Nassimi, M. Palis, IEEE Computer Society, Perth/Fremantle, Australia, pp. 310–315.

Basagni, S., Mastrogiovanni, M. & Petrioli, C. (2004). A performance comparison of protocols for clustering and backbone formation in large scale ad hoc network, *Proceedings of*

The 1st IEEE International Conference on Mobile Ad Hoc and Sensor Systems, MASS 2004, Fort Lauderdale, FL.

Bettstetter, C. (2004). The cluster density of a distributed clustering algorithm in ad hoc networks, *Proceedings of IEEE Intern. Conf. on Communications (ICC)*, Paris, France.

Brust, M., Frey, H. & Rothkugel, S. (2008). Dynamic multi-hop clustering for mobile hybrid wireless networks, *Proceedings of the 2nd international Conference on Ubiquitous information Management and Communication*, ACM, New York, NY.

Chan, H. & Perrig, A. (2004). ACE: An emergent algorithm for highly uniform cluster formation, *Proceedings of the European Workshop on Wireless Sensor Networks (EWSN 2004)*.

Chang, R. & Kuo, C. (2006). An energy efficient routing mechanism for wireless sensor networks, *Proceedings of the 20th international Conference on Advanced information Networking and Applications - Volume 02*.

Chatterjee, M., Das, S. K. & Turgut, D. (2001). Wca: A weighted clustering algorithm for mobile ad hoc networks, *Cluster Computing (Special Issue on Mobile Ad hoc Networks)* **5**: 193–204.

Chen, H., Wu, C.-S., Chu, Y.-S., Cheng, C.-C. & Tsai, L.-K. (2007). Energy residue aware (era) clustering algorithm for leach-based wireless sensor networks, *Proceedings of Second International Conference on Systems and Networks Communications*, pp. 40–40.

Chinara, S. & Rath, S. (2008). Mobility based clustering algorithm and the energy consumption model of dynamic nodes in mobile ad hoc network, *Proceedings of International Conference on Information Technology*, pp. 171–176.

Dai, F. & Wu, J. (2005). On constructing k-connected k-dominating set in wireless networks, *Proceedings of the 19th IEEE international Parallel and Distributed Processing Symposium (Ipdps'05)*, IEEE Computer Society, Washington, DC.

Delye, A. (2007). Etude théorique des clusters multi-sauts dans les réseaux de capteurs sans fil, *Thèse de Doctorat de L'Université Pierre et Marie Curie (Paris VI)*.

Delye de Clauzade de Mazieux, A., Marot, M. & Becker, M. (2006). Construction d'un ensemble d-dominant sur un graphe en utilisant un critère donné, *C.R. Mécanique, Académie des Sciences, Elsevier (10.1016/j.crme.2006.09.001)* **334**.

Delye de Clauzade de Mazieux, A., Marot, M. & Becker, M. (2009). Caac mechanism a cluster address auto-configuration mechanism, *WIRE Special Journal Issues on 'Wireless Networks: Performance Modelling and Evaluation' & 'Wireless Networks: Mobility Management and QoS'. To Appear*.

Demirbas, M., Arora, A. & Mittal, V. (2004). Floc: A fast local clustering service for wireless sensor networks, *Proceedings of Workshop on Dependability Issues in Wireless Ad Hoc Networks and Sensor Networks (DIWANS/DSN)*.

Deosarkar, B., Yadav, N. & Yadav, R. (2008). Clusterhead selection in clustering algorithms for wireless sensor networks: A survey, *Proceedings of the International Conference on Computing, Communication and Networking, 2008. ICCCn 2008.*, pp. 1 –8.

Depedri, A., Zanella, A. & Verdone, R. (2003). An energy efficient protocol for wireless sensor networks, *Proceedings of the Conference on Autonomous Intelligent Networks and Systems (AINS 2003)*, Menlo Park, CA.

Ding, P., Holliday, J. & Celik, A. (2005). Distributed energy efficient hierarchical clustering for wireless sensor networks, *Proceedings of the IEEE International Conference on Distributed Computing in Sensor Systems (DCOSS'05)*, Marina Del Rey, CA.

Dousse, O., Mannersalo, P. & Thiran, P. (2004). Latency of wireless sensor networks with uncoordinated power saving mechanisms, *Proceedings of Mobihoc, 2004*, pp. 109–120.

Fan, Z. & Zhou, H. (2006). A distributed weight-based clustering algorithm for wsns, *Proceedings of the International Conference on Wireless Communications, Networking and Mobile Computing, 2006. WiCOM 2006.*, pp. 1 –5.

Fang, S., Berber, S. & Swain, A. (2008). Performance of a clustering algorithm for high density wireless sensor networks, *Proceedings of the 2008 IEEE Region 10 Conference. TENCON 2008*, pp. 1 –6.

Fernandess, Y. & Communication, D. (2002). K-clustering in wireless ad hoc networks, *Proceedings of the second ACM international workshop on Principles of mobile computing, POMC âĂŹ02*, ACM Press, pp. 31–37.

Gerla, M., Kwon, T. J. & Pei, G. (2000). On-demand routing in large ad hoc wireless networks with passive clustering, *Proceedings of the 2000 IEEE Wireless Communications and Networking Conference, 2000. WCNC.*, Vol. 1, pp. 100 –105 vol.1.

Gilbert, E. N. (1961). Random plane networks, *SIAM J.* **9**: 533–543.

Guo, S.-j., Zheng, J., Qu, Y.-G., Zhao, B.-H. & Pan, Q.-K. (2007). Clustering and multi-hop routing with power control in wireless sensor networks, *The Journal of China Universities of Posts and Telecommunications* **14**(1): 49 – 57.
URL: *http://www.sciencedirect.com/science/article/B8H14-4NHNFT2-B/2/fe068a0f1badd13663dd3677da5a8238*

Gupta, G. & Younis, M. (2003a). Fault-tolerant clustering of wireless sensor networks, *Proceedings of the 2003 IEEE Wireless Communications and Networking, 2003. WCNC 2003.*, Vol. 3, pp. 1579 –1584 vol.3.

Gupta, G. & Younis, M. (2003b). Load-balanced clustering of wireless sensor networks, *Proceedings of the IEEE International Conference on Communications, 2003. ICC '03.*, Vol. 3, pp. 1848 – 1852 vol.3.

Gupta, P. & Kumar, P. R. (1998). Critical power for asymptotic connectivity in wireless networks, *Stochastic Analysis, Control, Optimization and Applications: A Volume in Honor of W.H. Fleming, Ed. W.M.McEneaney , G. Yin, Q Zhang, Birkhauser, Boston*, pp. 547–566.

Gupta, P., Member, S. & Kumar, P. R. (2000). The capacity of wireless networks, *IEEE Transactions on Information Theory* **46**: 388–404.

Gupta, S. & Dave, M. (2008). Real time approach for data placement in wireless sensor networks, *Proceedings of the World Academy of Science, Engineering and Technology*, Vol. 28.

Hall, P. (1985). On continuum percolation, *Ann. Probab.* **9**: 1250–1266.

Handy, M., Haase, M. & Timmermann, D. (2002). Low energy adaptive clustering hierarchy with deterministic cluster-head selection, *Proceedings of the 4th International Workshop on Mobile and Wireless Communications Network, 2002.*, pp. 368 – 372.

Hao, X., Kang, Y. & Wang, Y. (2008). Geographical-based multihop clustering algorithm for distributed wireless sensor network, *Proceedings of the 7th World Congress on Intelligent Control and Automation, 2008. WCICA 2008.*, pp. 3304 –3309.

Heinzelman, W. B. (2000). *Application-specific protocol architectures for wireless networks*, PhD thesis. Supervisor-Chandrakasan, Anantha P. and Supervisor-Balakrishnan, Hari.

Heinzelman, W., Chandrakasan, A. & Balakrishnan, H. (2002). An application-specific protocol architecture for wireless microsensor networks, *Wireless Communications, IEEE Transactions on* **1**(4): 660 – 670.

Henderson, T., Dekhil, M., Morris, S., Chen, Y. & Thompson, W. (1998). Smart sensor snow, *Proceedings of the 1998 IEEE/RSJ International Conference on Intelligent Robots and Systems, 1998.*, Vol. 3, pp. 1377 –1382 vol.3.

Henderson, T., Venkataraman, R. & Choikim, G. (2004). Reaction-diffusion patterns in smart sensor networks, *Proceedings of the 2004 IEEE International Conference on Robotics and Automation, 2004.*, Vol. 1, pp. 654 – 658 Vol.1.

Hou, T.-C. & Li, V. (1986). Transmission range control in multihop packet radio networks, *IEEE Transactions on Communications*, **34**(1): 38 – 44.

Huang, G., Li, X. & He, J. (2006). Dynamic minimal spanning tree routing protocol for large wireless sensor networks, *Proceedings of the 2006 1ST IEEE Conference on Industrial Electronics and Applications*, pp. 1 –5.

Huang, K.-C., Yen, Y.-S. & Chao, H.-C. (2007). Tree-clustered data gathering protocol (tcdgp) for wireless sensor networks, *Proceedings of Future Generation Communication and Networking (FGCN 2007)*, Vol. 2, pp. 31 –36.

Jang, K. Y., Kim, K. T. & Youn, H. Y. (2007). An energy efficient routing scheme for wireless sensor networks, *Proceedings of the International Conference on Computational Science and its Applications, 2007. ICCSA 2007.*, pp. 399 –404.

Jung, S.-M., Han, Y.-J. & Chung, T.-M. (2007). The concentric clustering scheme for efficient energy consumption in the pegasis, *Proceedings of the 9th International Conference on Advanced Communication Technology*, Vol. 1, pp. 260 –265.

Kawadia, V. & Kumar, P. (2003). Power control and clustering in ad hoc networks, *Proceedings of the Twenty-Second Annual Joint Conference of the IEEE Computer and Communications. INFOCOM 2003. IEEE Societies*, Vol. 1, pp. 459 – 469 vol.1.

Kim, N., Heo, J., Kim, H. S. & Kwon, W. H. (2008). Reconfiguration of clusterheads for load balancing in wireless sensor networks, *Comput. Commun.* **31**(1): 153–159.

Kirkwood, J. R. & Wayne, C. E. (1983). *Percolation in continuous systems*.

Klaoudatou, E., Konstantinou, E., Kambourakis, G. & Gritzalis, S. (2008). Clustering oriented architectures in medical sensor environments, *Proceedings of the Third International Conference on Availability, Reliability and Security, 2008. ARES 08.*, pp. 929 –934.

Kleinrock, L. & Silvester, J. (1978). Optimum transmission radii for packet radio networks or why six is a magic number, *Proceedings of the National Telecommunications Conference, Birmingham, Alabama*, pp. 4.3.2–4.3.5.

Kumarawadu, P., Dechene, D., Luccini, M. & Sauer, A. (2008). Algorithms for node clustering in wireless sensor networks: A survey, *Proceedings of the 4th International Conference on Information and Automation for Sustainability, 2008. ICIAFS 2008.*, pp. 295 –300.

Kwon, T. J. & Gerla, M. (1999). Clustering with power control, *Proceedings of the IEEE Military Communications Conference Proceedings, 1999. MILCOM 1999.*, Vol. 2, pp. 1424 –1428 vol.2.

Li, L., Dong, S.-S. & Wen, X.-M. (2006). Adaptive clustering for mobile wireless networks, *The Journal of China Universities of Posts and Telecommunications* **13, issue 3**: 71–75.

Liang, Y. & Yu, H. (2005). Energy adaptive cluster-head selection for wireless sensor networks, *Proceedings of the Sixth International Conference on Parallel and Distributed Computing, Applications and Technologies, 2005. PDCAT 2005.*, pp. 634 – 638.

Lijun, L., Hongtao, W. & Peng, C. (2006). Discuss in round rotation policy of hierarchical route in wireless sensor networks, *Proceedings of the International Conference on Wireless Communications, Networking and Mobile Computing, 2006. WiCOM 2006.*, pp. 1 –5.

Lin, C. & Gerla, M. (1997). Adaptive clustering for mobile wireless networks, *Selected Areas in Communications, IEEE Journal on* **15**(7): 1265 –1275.

Lindsey, S. & Raghavendra, C. (2002). Pegasis: Power-efficient gathering in sensor information systems, *Proceedings of the IEEE Aerospace Conference, 2002.*, Vol. 3, pp. 3–1125 – 3–1130 vol.3.

Liu, C.-M., Lee, C.-H. & Wang, L.-C. (2007). Distributed clustering algorithms for data-gathering in wireless mobile sensor networks, *J. Parallel Distrib. Comput.* **67**(11): 1187–1200.

Liu, Y., Gao, J., Zhu, L. & Zhang, Y. (2009). A clustering algorithm based on communication facility in wsn, *Proceedings of the 2009 WRI International Conference on Communications and Mobile Computing*, IEEE Computer Society, Washington, DC, USA, pp. 76–80.

Loscri, V., Morabito, G. & Marano, S. (2005). A two-levels hierarchy for low-energy adaptive clustering hierarchy (tl-leach), *Proceedings Vehicular Technology Conference, 2005. VTC-2005-Fall. 2005 IEEE 62nd*, Vol. 3, pp. 1809 – 1813.

McLaughlan, B. & Akkaya, K. (2007). Coverage-based clustering of wireless sensor and actor networks, *Proceedings of the IEEE International Conference on Pervasive Services*, pp. 45 –54.

Muruganathan, S., Ma, D., Bhasin, R. & Fapojuwo, A. (2005). A centralized energy-efficient routing protocol for wireless sensor networks, *Communications Magazine, IEEE* **43**(3): S8 – 13.

Nagpal, R. & Coore, D. (1998). An algorithm for group formation in an amorphous computer, *Proceedings of the Tenth International Conference on Parallel and Distributed Systems (PDCS)*.

Nam, D.-H. & Min, H.-K. (2007). An efficient ad-hoc routing using a hybrid clustering method in a wireless sensor network, *Proceedings of the Third IEEE International Conference on Wireless and Mobile Computing, Networking and Communications, 2007. WiMOB 2007.*, pp. 60 –60.

Nocetti, F. G., Gonzalez, J. S. & Stojmenovic, I. (2003). Connectivity based k-hop clustering in wireless networks, *Telecommunication Systems* **22**: 1–4.

Philips, T., Panwar, S. & Tantawi, A. (1989). Connectivity properties of a packet radio network model, *Information Theory, IEEE Transactions on* **35**(5): 1044 –1047.

Qian, Y., Zhou, J., Qian, L. & Chen, K. (2006). Highly scalable multihop clustering algorithm for wireless sensor networks, *Proceedings of the 2006 International Conference on Communications, Circuits and Systems Proceedings*, Vol. 3, pp. 1527 –1531.

Qing, L., Zhu, Q. & Wang, M. (2006). Design of a distributed energy-efficient clustering algorithm for heterogeneous wireless sensor networks, *Comput. Commun.* **29**(12): 2230–2237.

Rajiullah, M. & Shimamoto, S. (2007). An energy-aware periodical data gathering protocol using deterministic clustering in wireless sensor networks (wsn), *Proceedings of the Wireless Communications and Networking Conference, 2007.WCNC 2007. IEEE*, pp. 3014 –3018.

Rasheed, T., Javaid, U., Meddour, D.-E., Reynaud, L. & Al Agha, K. (2007). An efficient stable clustering algorithm for scalable mobile multi-hop networks, *Proceedings of the 4th IEEE Consumer Communications and Networking Conference, 2007. CCNC 2007.*, pp. 89 –94.

Satapathy, S. & Sarma, N. (2006). Treepsi: tree based energy efficient protocol for sensor information, *Proceedings of the 2006 IFIP International Conference on Wireless and Optical Communications Networks*, pp. 4 pp. –4.

Selvakennedy, S. & Sinnappan, S. (2007). An adaptive data dissemination strategy for wireless sensor networks, *Int. J. Distrib. Sen. Netw.* **3**(1): 23–40.

Shakkottai, S., Srikant, R. & Shroff, N. (2003). Unreliable sensor grids: coverage, connectivity and diameter, *Proceedings of the Twenty-Second Annual Joint Conference of the IEEE Computer and Communications. INFOCOM 2003. IEEE Societies*, Vol. 2, pp. 1073 – 1083 vol.2.

Takagi, H. & Kleinrock, L. (1984). Optimal transmission ranges for randomly distributed packet radio terminals, *IEEE Trans. Commun* **32**: 246–257.

Tian, Y., Wang, Y. & Zhang, S.-F. (2007). A novel chain-cluster based routing protocol for wireless sensor networks, *Proceedings of the International Conference on Wireless Communications, Networking and Mobile Computing, 2007. WiCom 2007.*, pp. 2456 –2459.

Tillapart, P., Thammarojsakul, S., Thumthawatworn, T. & Santiprabhob, P. (2005). An approach to hybrid clustering and routing in wireless sensor networks, *Proceedings of the 2005 IEEE Aerospace Conference*, pp. 1 –8.

Vlajic, N. & Xia, D. (2006). Wireless sensor networks: To cluster or not to cluster?, *Proceedings of the 2006 International Symposium on on World of Wireless, Mobile and Multimedia Networks*, IEEE Computer Society, Washington, DC, USA, pp. 258–268.

Voronoï, G. (1907). Nouvelles applications des paramètres continus Ãă la théorie des formes quadratiques, *Journal für die Reine und Angewandte Mathematik* pp. 133:97–178.

Wang, H., Yu, X., Kong, D., Yan, X. & Ma, X. (2007). Route protocol of wireless sensor networks based on dynamic setting cluster, *Proceedings of the International Conference on Information Acquisition, 2007. ICIA '07.*, pp. 112 –117.

Wang, K., Ayyash, S. A., Little, T. D. C. & Basu, P. (2005). Attribute-based clustering for information dissemination in wireless sensor networks, *Proceedings of The 2nd Annual IEEE Communications Society Conf. on Sensor and Ad Hoc Communications and Networks (SECON'05)*.

Wen, C.-Y. & Sethares, W. A. (2005). Automatic decentralized clustering for wireless sensor networks, *EURASIP J. Wirel. Commun. Netw.* **2005**(5): 686–697.

Weniger, K. & Zitterbart, M. (2004). Adress autoconfiguration in mobile ad hoc networks: Current approches and future directions, *IEEE network* **18**(4): 6–11.

Xia, D. & Vlajic, N. (2006). Near-optimal node clustering in wireless sensor networks for environment monitoring, *Proceedings of the Canadian Conference on Electrical and Computer Engineering, 2006. CCECE '06.*, pp. 1825 –1829.

Xiangning, F. & Yulin, S. (2007). Improvement on leach protocol of wireless sensor network, *Proceedings of the International Conference on Sensor Technologies and Applications, 2007. SensorComm 2007.*, pp. 260 –264.

Xu, K. & Gerla, M. (2002). A heterogeneous routing protocol based on a new stable clustering scheme, *Proceedings of the MILCOM 2002 Conference.*, Vol. 2, pp. 838 – 843 vol.2.

Xue, F. & Kumar, P. R. (2002). The number of neighbors needed for connectivity of wireless networks, *Wireless Networks* **10**: 169–181.

ya Zhang, W., ze Liang, Z., guang Hou, Z. & Tan, M. (2007). A power efficient routing protocol for wireless sensor network, *Proceedings of the 2007 IEEE International Conference on Networking, Sensing and Control,*, pp. 20 –25.

Ye, M., Li, C., Chen, G. & Wu, J. (2005). Eecs: an energy efficient clustering scheme in wireless sensor networks, *Proceedings of the 24th IEEE International Performance, Computing, and Communications Conference, 2005. IPCCC 2005.*, pp. 535 – 540.

Yiming, F. & Jianjun, Y. (2007). The communication protocol for wireless sensor network about leach, *Proceedings of the International Conference on Computational Intelligence and Security Workshops, 2007. CISW 2007.*, pp. 550 –553.

Younis, O. & Fahmy, S. (2004). Distributed clustering in ad-hoc sensor networks: a hybrid, energy-efficient approach, *Proceedings of the Twenty-third AnnualJoint Conference of the IEEE Computer and Communications Societies, INFOCOM 2004.*, Vol. 1, p. 640.

Youssef, A., Younis, M., Youssef, M. & Agrawala, A. (2006). Wsn16-5: Distributed formation of overlapping multi-hop clusters in wireless sensor networks, *IEEE Proceedings of the Global Telecommunications Conference, 2006. GLOBECOM '06.*, pp. 1 –6.

Yuan Sun, E., Sun, Y. & Belding-Royer, E. M. (2003). Dynamic address configuration in mobile ad hoc networks, *Technical report*, Computer Science, UCSB, Tech. Rep.

Zhang, H. & Arora, A. (2003). Gs3: scalable self-configuration and self-healing in wireless sensor networks, *Computer Networks* pp. 459–480.

Zhou, H., Ni, L. & Mutka, M. (2003). Prophet address allocation for large scale manets, *Proceedings of the Twenty-Second Annual Joint Conference of the IEEE Computer and Communications. INFOCOM 2003. IEEE Societies*, Vol. 2, pp. 1304 – 1311 vol.2.

7

Cluster-based Routing Protocols for Energy Efficiency in Wireless Sensor Networks

Moufida Maimour, Houda Zeghilet and Francis Lepage
CRAN laboratory, Nancy University, CNRS
France

1. Introduction

Thanks to recent advances in micro-electronics and wireless communications, wireless sensor networks (WSN) are foreseen to become ubiquitous in our daily life and they have already been a hot research area. A WSN is made of large number of low cost sensor nodes with processing and communication capabilities. While sensors are small devices with limited power supply, a WSN should operate autonomously for long periods of time in most applications. In order to better manage energy consumption and increase the whole network lifetime, suitable solutions are required at all layers of the networking protocol stack. In particular, energy-aware routing protocols at the network layer have received a great deal of attention since it is well established that wireless communication is the major source of energy consumption in WSN.

The network layer in WSN is responsible for delivery of packets and implements an addressing scheme to accomplish this. It mainly establishes paths for data transfer through the network. Compared to traditional ad-hoc networks, routing is more challenging in wireless sensor networks due to their limited resources in terms of available energy, processing capability and communication, which are major constraints to all sensor networks applications. These constraints yield frequent topology changes making route maintenance to be a non-easy task. Additionally, the typical mode of communication is many-to-one, from multiple sources to a particular sink rather than from one entity to another. Finally, since data related to one phenomena may be collected by multiple sensors, a significant redundancy is likely to be present and has to be considered. This is why routing protocols proposed for ad-hoc networks in recent years are not suitable for wireless sensor networks. Alternative approaches that take the above limitations into account with energy-awareness are required. Due to that, multiple routing protocols for WSN have been proposed (Akkaya & Younis, 2005; Al-Karaki & Kamal, 2004).

From network organization perspective, routing protocols can coarsely be classified in two main classes : flat network routing and hierarchical network routing. In a flat topology, each node plays the same role and has the same functionality as other sensor nodes in the network. When a node needs to send data, a flat routing protocol attempt to find a route to the sink hop by hop using some form of flooding. The most popular flat-based routing in WSN are data-centric protocols like SPIN (Heinzelman et al., 1999) and Directed Diffusion (DD) (Intanagonwiwat et al., 2003). Data-centric routing protocols were shown to save energy through in-network data aggregation. In order to limit energy consumption due to un-

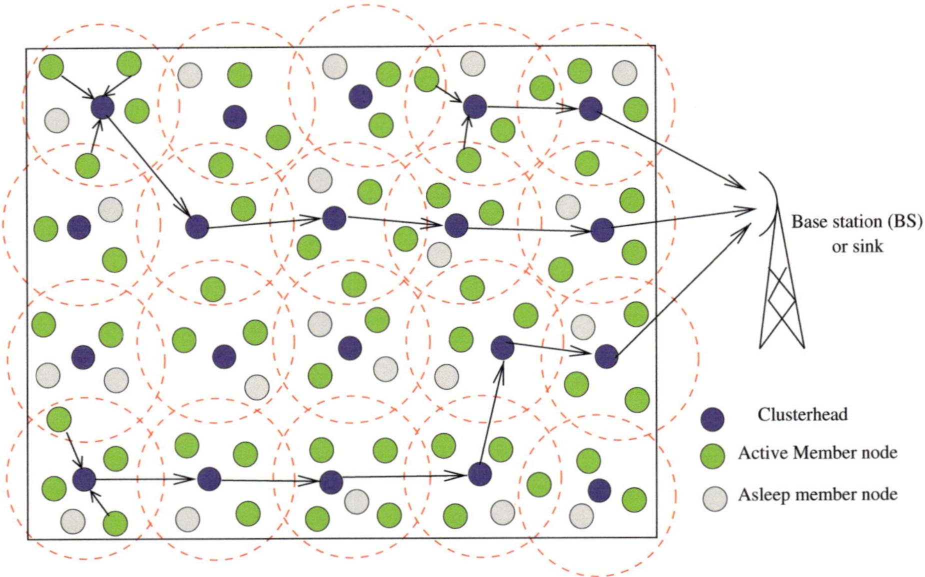

Fig. 1. Cluster-based topology

necessary flooded messages, some routing protocols, mainly geographic ones (Ko & Vaidya, 2000; Lin & Stojmenovic, 2003; Rodoplu & Ming, 1999; Y. Yu & Govindan, 2001) with location awareness, restrict flooding to localized regions. Other protocols that are neither data-centric nor location-based can be qualified as topology-based (Frey et al., 2009). This is the case of routing protocols like those proposed in (He et al., 2003; Sohrabi et al., 2000; Ye et al., 2001).

Flat routing protocols are quite effective in relatively small networks. However, they scale very bad to large and dense networks since, typically, all nodes are alive and generate more processing and bandwidth usage. On the other hand, hierarchical routing protocols have shown to be more scalable and energy-aware in the context of WSN. In hierarchical-based routing, nodes play different roles in the network and typically are organized into clusters. Clustering (Figure 1) is the method by which sensor nodes in a network organize themselves into groups according to specific requirements or metrics. Each group or cluster has a leader referred to as *clusterhead* (CH) and other ordinary member nodes (MNs). The clusterheads can be organized into further hierarchical levels.

As opposed to a flat organization, clustering allows a hierarchical architecture with more scalability, less consumed energy and thus longer lifetime for the whole network. this is due mainly to the fact that most of the sensing, data processing and communication activities can be performed within clusters. Numerous are WSN applications that require simply an aggregate value to be reported to the sink. In such applications, data aggregation at the clusterheads helps to alleviate congestion and save energy. Clustering allows intra-cluster and inter-cluster routing which reduces the number of nodes taking part in a long distance communication, thus allowing significant energy saving in addition to smaller dissemination latency.

In this chapter we consider cluster-based routing protocols to achieve energy efficiency in WSN. Section 2 focuses on clustering from the perspective of data routing and a new classification of cluster-based routing protocols into two classes is proposed. Some representatives of

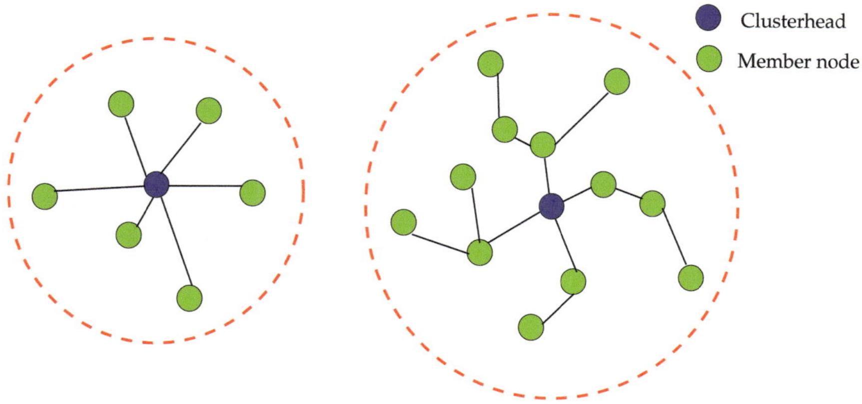

(a) One-hop intra-cluster connectivity (b) multi-hop intra-cluster connectivity

Fig. 2. One-hop toward the sink

both classes are summarized in respectively Sections 3 and 4. Section 5 concludes the chapter with some future research directions.

2. Clustering and Routing in WSN

From a routing perspective, clustering allows to split data transmission into *intra-cluster* (within a cluster) and *inter-cluster* (between clusterheads and every clusterhead and the sink) communication. This separation leads to significant energy saving since the radio unit is the major energy consumer in a sensor node. In fact, member nodes are only allowed to communicate with their respective clusterhead, which is responsible for relaying the data to the sink with possible aggregation and fusion operations. Moreover, this separation allows to reduce routing tables at both member nodes and clusterheads in addition to possible spatial reuse of communication bandwidth.

Intra-cluster communications

Most of the earlier work on clustering assume direct (one-hop) communication between member nodes and their respective clusterheads (*Energy-efficient communication protocol for wireless sensor networks*, 2000; Younis & Fahmy, 2004). All the member nodes are at most two hops away from each other (Figure 2(a)). One-hop clusters makes selection and propagation of clusterheads easy, however, multi-hop intra-cluster connectivity is sometimes required, in particular for limited radio ranges and large networks with limited clusterhead count. Multi-hop routing within a cluster (Figure 2(b)) has already been proposed in wireless ad-hoc networks (Lin & Gerla, 1995). More recent WSN clustering algorithms allow multi-hop intra-cluster routing (Bandyopadhyay & Coyle, 2003; Ding et al., 2005).

Inter-cluster Routing

Earlier cluster-based routing protocols such as LEACH (*Energy-efficient communication protocol for wireless sensor networks*, 2000) assume that the clusterheads have long communication ranges allowing direct connection between every clusterhead and the sink (Figure 3). Although simple, this approach is not only inefficient in terms of energy consumption, it is

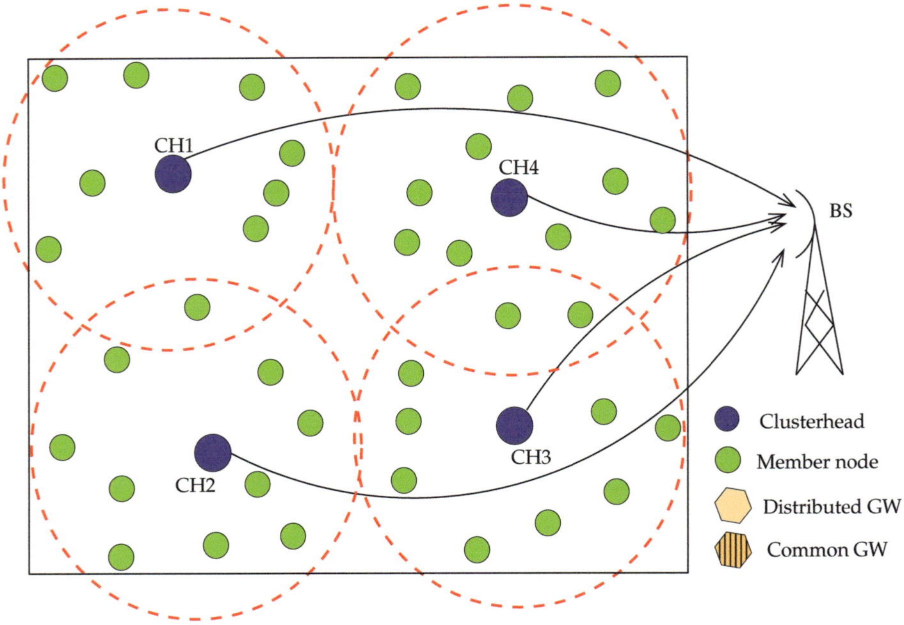

Fig. 3. One-hop toward the sink

based on irrealistic assumption. The sink is usually located far away from the sensing area and is often not directly reachable to all nodes due to signal propagation problems. A more realistic approach is multihop inter-cluster routing that had shown to be more energy efficient (Mhatre & Rosenberg, 2004a). Sensed data are relayed from one clusterhead to another until reaching the sink (Figure 1).

Direct communication between clusterheads is not always possible especially for large clusters (multihop clusters for instance). In this case, ordinary nodes located between two clusterheads could act as *gateways* (GW) allowing the clusterheads to reach each other (Figure 4). A gateway node is either *common* or *distributed*. A common (ordinary) gateway is located within the transmission range of two clusterheads and thus, allows 2-hop communication between these clusterheads. When two clusterheads do not have a common gateway, they can reach each other in at least 3 hops via two distributed gateways located in their respective clusters. A distributed gateway is only reachable by one clusterhead and by another distributed gateway of the second clusterhead cluster.

Inter-cluster communication in several proposals is achieved through organizing the clusterheads in a hierarchy (Figure 5) as done in (Bandyopadhyay & Coyle, 2003) and (Manjeshwar & Agarwal, 2001). Multiple level hierarchy allows better energy distribution and overall energy consumption. However, maintaining the hierarchy could be costly in large and dynamic networks where nodes die as soon as their energy supply is completely discharged.

2.1 Energy Efficiency and Load-balancing

One of the most important objectives of hierarchical organization in sensor networks is energy efficiency that allows longer network lifetime. A clusterhead can perform aggregation and fusion operations on data it receives before relaying it to the base station. In very dense networks, a subset of nodes may be put into the low-power sleep mode provided that these

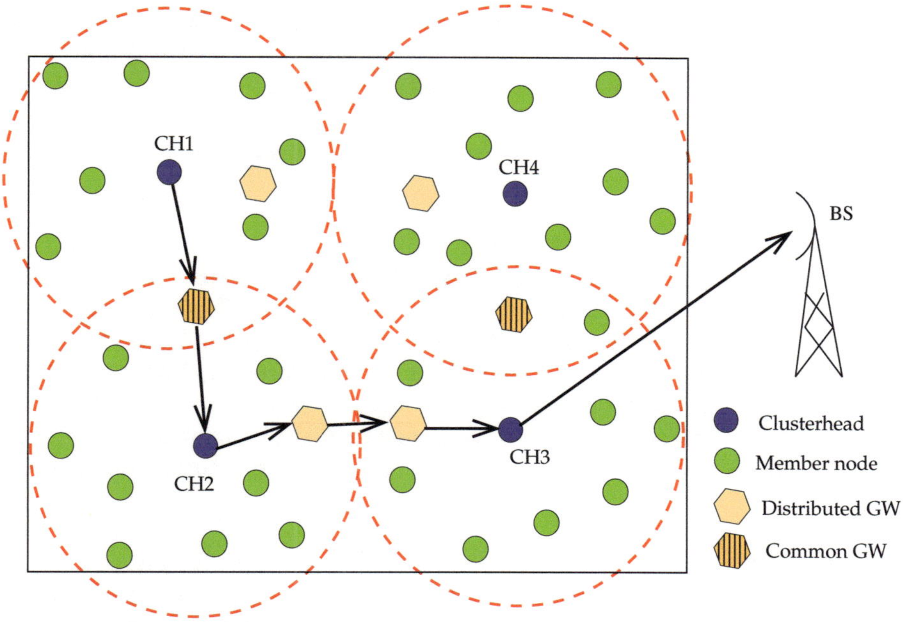

Fig. 4. Multi-hop inter-cluster communication

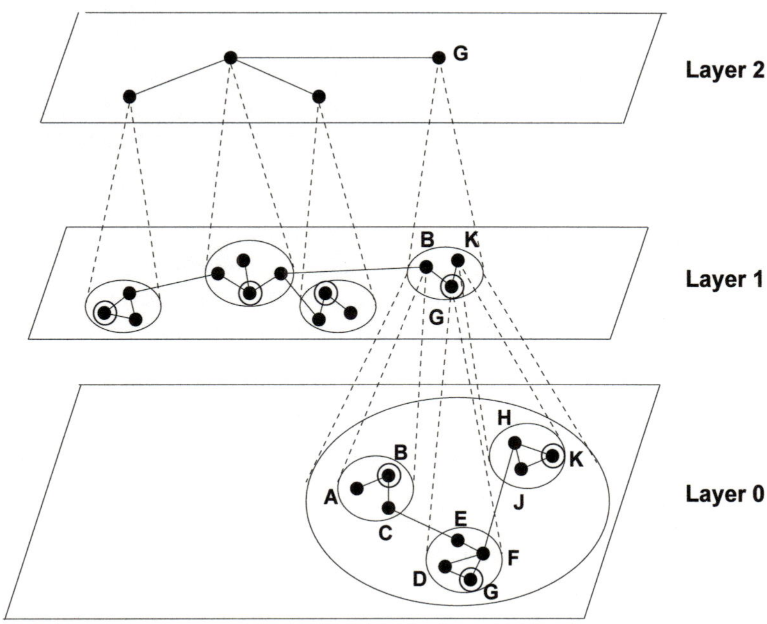

Fig. 5. 3-level hierarchy (redrawn from (Banerjee & Khuller, 2001)

nodes are chosen without affecting the network coverage and connectivity. In this context, a clusterhead can efficiently schedule its member nodes states. Furthermore, medium access collision can be prevented within a cluster if a round-robin strategy is applied among the member nodes. Collisions may require that nodes retransmit their data thus wasting more energy.
Minimizing energy consumption on a per sensor basis is not sufficient to get longer network lifetime, load-balancing is required.

2.1.1 Load-balancing among all nodes

Intra-cluster communications where a member node sends data to its clusterhead for further relaying toward the sink, put a heavy burden on the clusterheads. These Latter have, additionally, the responsibility of in-network data operations such as aggregation and fusion. Even if clusterheads are equipped with more powerful and durable batteries, this heavy burden could result in fast battery depletion at the clusterheads and thus shorter lifetime compared to other sensor nodes. This is one possible load unfairness situation that may occur in cluster-based routing. This issue is usually addressed through clusterhead rotation among nodes in each cluster.

2.1.2 Load-balancing among clusterheads

In order to give each clusterhead equivalent burden in the network, many algorithms focus on balancing the intra-cluster traffic load through the formation of nearly equal size (uniform) clusters. In fact, in clusters of comparable coverage and node density, the intra-cluster traffic volume is more likely to be the same for all clusters.
Regarding inter-cluster communication, balanced intra-cluster traffic results in a highly skewed load distribution on clusterheads. In single-hop communication where clusterheads use direct link to reach the base station, the farther the clusterhead, the more energy it consumes and the earlier will die. Even if multi-hop inter-cluster communication is adopted, the nodes close to the base station are burdened with heavier traffic load leading to the so-called *hot spot problem*. This is due to the many-to-one traffic paradigm that characterizes WSN. Nodes in the hot spot area deplete faster their energy and die much faster than faraway clusterheads. This may lead to serious connectivity (network partition) and coverage problems at the base station vicinity.
As a consequence, both intra-cluster and inter-cluster traffic have to be considered jointly when designing a cluster-based routing algorithm. In other words, one have to consider minimizing energy consumption around the sink instead of minimizing the overall consumed energy in the network in order to achieve longer network lifetime. We will report on some work that dealt with this issue in Section 3.5.

2.2 Clustering Algorithms Taxonomy

In the literature, there have been several different ways to classify Clustering algorithms for WSNs. In (Younis et al., 2006), the classification is performed based on parameter(s) used for electing clusterheads and the execution nature of a clustering algorithm which can be either probabilistic or iterative. In iterative clustering techniques, a node waits for a specific event to occur or certain nodes to decide their role (e.g., become clusterheads) before making a decision. Probabilistic Clustering Techniques enables every node to independently decide on its role in the clustered network while keeping the message overhead low. Considering how the cluster formation is carried out, a clustering algorithm is either executed at a central point or in a distributed fashion at local nodes. Centralized approaches are used by few earlier proposals like LEACH-C (Chandrakasan et al., 2002). They require global knowledge of the network topology and are inefficient in large-scale topologies. A distributed approach, however, is more scalable since a node is able to take the initiative to become a clusterhead or to join an already formed cluster without global topology knowledge.

Authors of (Abbasi & Younis, 2007) classify clustering algorithms according to their convergence rate into two classes : variable and constant convergence time algorithms. The former algorithms have a convergence time that depends on the number of nodes in the network and thus are more suitable to relatively small networks. Constant convergence time algorithms converge in a fixed number of iterations, regardless of the size of the nodes population.

Clustering algorithms can also be classified into homogeneous or heterogeneous (Mhatre & Rosenberg, 2004b) depending on the nature of the deployed sensor network. In heterogeneous environments, the clusterhead roles can be preassigned to nodes with more energy, computation and communication resources. In a homogeneous environment, the clusterheads can be designated in a random way or based on one or more criteria. It is worth mentioning, that even in a homogeneous network, heterogeneity can occur simply in terms of available energy at nodes. As time goes on, some nodes depending on their role and environmental factors, will discharge more quickly their batteries. This is why energy and clusterhead rotation have to be considered in the process of clustering.

Since we report, in this chapter, on clustering techniques and their use to achieve energy efficient routing in WSN, we adopt a different classification. Most proposed cluster-based routing protocols rely on already formed clusters. Afterwards, the inter-cluster communication is generally ensured using traditional flooding among only clusterheads or by recursively executing the clustering algorithm to obtain a hierarchy of clusterheads rooted at the sink. We qualify these protocols as *pre-established* cluster-based routing algorithms. Protocols that build clusters based on packets flowing in the network without a priori construction are qualified as *on-demand* cluster-based algorithms. It is worth mentioning that the second class had always been omitted in surveys like (Younis et al., 2006) (Abbasi & Younis, 2007) and (Mamalis et al., 2009). On-demand clustering by exploiting existing traffic to piggyback cluster-related information, eliminates major control overhead of traditional clustering protocols. Besides, there is no startup latency even if there is a transient period before getting maximum performances.

3. Pre-established Cluster-based Routing Algorithms

In this section, we review most important clustering algorithms. Even if they are limited only to the clusters formation and do not address explicitly inter-cluster routing. It is generally straightforward to apply on top of the clustered topology a routing protocol taking into account only the clusterheads in the route discovery phase.

3.1 Low Energy Adaptive Clustering Hierarchy (LEACH)

Low-Energy Adaptive Clustering Hierarchy (LEACH) (*Energy-efficient communication protocol for wireless sensor networks*, 2000) is one of the most popular hierarchical routing algorithms for sensor networks. LEACH is a cluster-based protocol with distributed cluster formation with random clusterhead election. A sensor node chooses a random number between 0 and 1. If this random number is less than a threshold value, $T(n)$, the node becomes a clusterhead for the current round. This threshold value is calculated using :

$$T(n) = \begin{cases} \frac{P}{1-P(r \bmod \frac{1}{P})} & \text{if } n \in G \\ 0 & \text{otherwise} \end{cases} \quad (1)$$

where P is the desired fraction of nodes to be clusterheads, r is the current round and G is the set of nodes that have not been clusterheads in the last $\frac{1}{P}$ round. The elected clusterheads broadcast an advertisement message to inform other nodes about their states. Based on the received signal strength of the advertisement, a non-clusterhead node decides to which cluster it will belong for this round and sends a membership message to its clusterhead. Based on the number of nodes in the cluster, a clusterhead creates a TDMA schedule and assigns each node a time slot in which it can transmit. This schedule is broadcast to all the cluster nodes. This is the end of the so-called *advertisement* or *setup phase* of LEACH. Then begins the *steady state* where different nodes can transmit their sensed data.

In order to save energy, in the steady phase, the radio of each member node can be turned off until the node's allocated transmission time. Moreover, clusterheads can perform data processing such as fusion and aggregation before relaying to the base station. To evenly distribute energy load among nodes, clusterheads rotation is insured at each round by entering a new advertisement phase and by using equation (1).

LEACH is completely distributed and requires no global knowledge of network. However, it forms one-hop intra and inter cluster topology, which is not applicable to large region networks. Clusterheads are assumed to have a long communication range so they can reach the sink directly. This is not always a realistic assumption since the clusterheads are regular sensors and the sink is often located far away. Furthermore, dynamic clustering brings extra overhead due to the advertisements phase at the beginning of each round, which may diminish the gain in energy. Since the decision to elect a clusterhead is probabilistic without energy considerations, LEACH clusterhead rotation assume a homogeneous network and can not ensure real load-balancing in case of nodes initially with different amount of energy. A node with very low energy becomes a clusterhead for the same number of rounds as other nodes with higher energy and will die prematurely. This could affect network coverage and connectivity.

LEACH-C

LEACH-C (Chandrakasan et al., 2002) is a centralized version of LEACH where only the advertisement phase differs. At this phase, each node sends information about its current location and residual energy level to the sink. Based on nodes location, the sink builds clusters using the simulated annealing algorithm (Murata, 1994) so the amount of energy required by member nodes to transmit their data to their respective clusterhead is minimized. Collected information about nodes energies allows the sink to discard those with energy below the average network energy. Consequently, energy load is evenly distributed among all the nodes.

3.2 Energy Efficient Hierarchical Clustering (EEHC)

Energy Efficient Hierarchical Clustering (EEHC) (Bandyopadhyay & Coyle, 2004) can be seen as an extension of LEACH with multi-hop intra clusters and a hierarchy of clusterheads to route data to the sink. In the single-level clustering of EEHC, each sensor in the network becomes a *Volunteer* clusterhead with probability p. It announces this to the sensors within k hops radio range. Any sensor that receives such advertisements and is not itself a clusterhead joins the closest cluster. If a sensor does not receive a clusterhead advertisement within a certain time duration it can infer that it is not within k hops of any volunteer clusterhead and hence becomes a *forced* clusterhead. Data transmission to the sink can be performed using multi-hop routing through clusterheads organization in a multi-level hierarchy rooted at the sink. To do so, the single-level clustering is repeated recursively at the level of clusterheads. This distributed process allows EEHC to have a time complexity of $O(k_1 + k_2 + ... + k_h)$ where h is the number of levels and k_i is the maximum number of hops between a member node and its clusterhead in the ith level of hierarchy. Since spent energy in the network depends on p and k, the authors provide methods to compute the optimal values of these parameters that ensure minimum consumed energy. Simulation results showed significant energy saving when using the optimal parameter values.

3.3 Hybrid Energy-Efficient Distributed Clustering (HEED)

Both EEHC and LEACH do not consider energy in selecting clusterheads. HEED (Younis & Fahmy, 2004) brings one more step toward energy-efficient cluster-based routing with explicit consideration of energy. Selected clusterheads in HEED have relatively high average residual energy compared to member nodes. Additionally, HEED aims to get a well-distributed clusterheads set over the sensor field. Indeed, in HEED, the probability that two nodes within the transmission range of each other to be clusterheads is small. It is worth mentioning that the main drawback of LEACH is that the random election of clusterheads does not ensure their even distribution in the sensing field. It is quite possible to get multiple clusterheads concentrated in a small area. In this case, this area sensors are likely to exhaust their energy more quickly which may lead to insufficient coverage and network disconnection. Distributing clusterheads evenly in the sensing area is one important goal to be met in order to ensure load balancing and hence longer network lifetime.

HEED periodically selects clusterheads according to a hybrid of their residual energy and intra-cluster communication cost. Initially, to limit the initial clusterhead announcements, HEED sets an initial percentage C_{prob} of clusterheads among all sensors. The probability that a sensor becomes a clusterhead is $CH_{prob} = C_{prob}\, E_{residual} / E_{max}$ where $E_{residual}$ is the current energy in the sensor, and E_{max} is its maximum energy. Afterwards, every sensor goes through several iterations until it finds the clusterhead that it can transmit to with the least transmission power. If it hears from no clusterhead, the sensor elects itself to be a clusterhead and sends an announcement message to its neighbors. Each sensor doubles its CH_{prob} value and goes to the next iteration until its CH_{prob} reaches 1. Therefore, there are two types of status that a sensor could announce to its neighbors:

- **Tentative status:** The sensor becomes a tentative clusterhead if its CH_{prob} is less than 1. It can change its status to a regular node at a later iteration if it finds a lower cost clusterhead.
- **Final status:** The sensor permanently becomes a clusterhead if its CH_{prob} has reached 1.

At the final phase, each sensor makes a final decision on its status. It either picks the least cost clusterhead or pronounces itself as clusterhead. Simulation results showed that HEED outperforms LEACH with respect to the network lifetime and energy consumption distribution. However, HEED suffers from a consequent overhead since it needs several iterations to form clusters. In each iteration, a lot of packets are broadcast.

Clustering Method for Energy Efficient Routing (CMEER)

CMEER (Kang et al., 2007) is another attempt to achieve well distributed Cluster heads. In CMEER, a node declares itself as a candidate to be a clusterhead using equation (1) where P is chosen higher than adopted values in LEACH. Each candidate advertises its intention to be a clusterhead within its radio range. Each node (even candidate to be a clusterhead) decides to join a given clusterhead based on the received signal strength of the advertisement message. In this way, the authors try to avoid redundant creation of clusterheads in a small area. The simulation results showed that CMEER outperforms LEACH in terms of energy consumption and network lifetime.

3.4 Distributed Energy Efficient Hierarchical Clustering (DWEHC)

Distributed Energy Efficient Hierarchical Clustering (DWEHC) (Ding et al., 2005) aims to improve HEED by generating balanced cluster sizes and optimizing the intra-cluster topology thanks to its location awareness. DWEHC creates a multi-level (instead of one-hop in HEED) structure for intra-cluster communication and limits a parent node's number of children.
Each sensor s calculates its weight after locating the neighboring nodes in its area using :

$$W_{weight}(s) = \frac{E_{residual}(s)}{E_{initial}(s)} \times \sum_{u} \frac{R-d}{6R} \tag{2}$$

where $E_{residual}(s)$ and $E_{initial}(s)$ are respectively residual and initial energy at node s, R is the cluster range (a system parameter that corresponds to how far a node inside a cluster can be from the clusterhead) and d is the distance between s and neighboring node u. In a neighborhood, the node with largest weight would be elected as a clusterhead and the remaining nodes become members. At this stage member nodes are considered as 1-level nodes and communicate directly with the clusterhead. If a member node can reach its clusterhead using more than one hop while saving energy, it will become an h-level member where h is the number of hops required to achieve the clusterhead. Required energy to communicate in a cluster can be computed using node's knowledge of the distance to its neighbors. The cluster range R is used to limit the number of levels.
Even if HEED considers energy reserve in clusterhead selection and aims to a well distributed clusterheads, simulation results showed that clusters generated by DWEHC are more well-balanced and that DWEHC achieves significantly lower energy consumption in intra-cluster and inter-cluster communication than HEED. However, location information required by DWEHC are not necessarily and easily available. Many other location-aware clustering techniques have been proposed in the literature :

Geographic Adaptive Fidelity (GAF)

GAF (Xu et al., 2001) is an energy-aware routing algorithm designed primarily for mobile ad hoc networks, but may be applicable to sensor networks as well. GAF is generally cited as a location based routing protocol but may be considered as a hierarchical protocol where the clusters are based on geographic location. The network area is divided into fixed zones (clusters) that form a virtual grid in which nodes collaborate with each other to play different roles. The virtual grid is defined such that for any two adjacent zones A and B, all nodes in A are able to communicate with all nodes in B, and vice versa. By assuming an ideal radio propagation model and choosing appropriate side length of zones according to the radio transmission range, GAF ensures that a connected backbone network can be formed as long as just one node at time need to be active. That node play a role of a CH and each node uses its location to associate itself with a node in the virtual grid. The clusterhead is responsible for monitoring and reporting data to the Base station. The Nodes associated with the same point on the grid are considered equivalent in terms of the cost of packet routing. Such equivalence

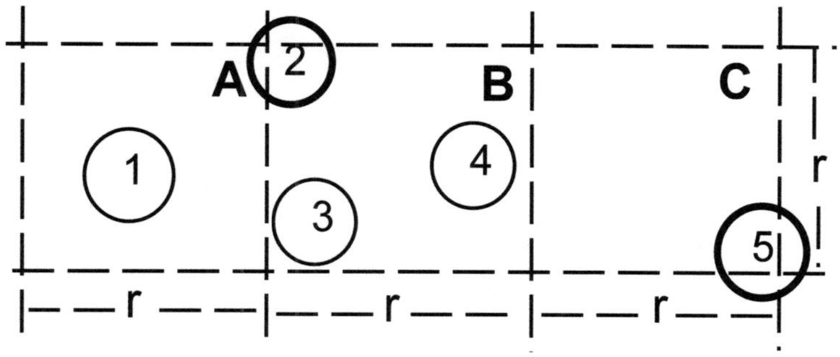

Fig. 6. GAF virtual grids

is exploited in keeping these nodes in sleeping state in order to save energy. Thus, GAF can substantially increase the network lifetime as the number of nodes increases. A sample situation is depicted in Figure 6 redrawn from (Xu et al., 2001). In this figure, node 1 can reach any of 2, 3 and 4 and nodes 2, 3, and 4 can reach 5. Therefore nodes 2, 3 and 4 are equivalent and two of them can sleep.

Nodes change their states from sleeping to active in turn so that the load is balanced in the network. There are three states defined in GAF : (i) *discovery*, for determining the neighbors in the grid,(ii) *active* reflecting participation in routing and (iii) *sleep* when the radio is turned off. The sleeping time is application dependent parameter which is tuned during the routing process. In order to handle the mobility, each node in the grid estimates its leaving time of a grid and sends it to its neighbors. The sleeping neighbors adjust their sleeping time accordingly in order to keep the routing fidelity. Before the leaving time of the active node expires, sleeping nodes wake up and one of them becomes active (a clusterhead). Simulation results showed that GAF performs at least as well as a normal ad hoc routing protocol in terms of latency and packet loss and increases the lifetime of the network by saving energy.

Position-based Aggregator Node Election (PANEL)

PANEL (Buttyan & Schaffer, 2007) is a position-based clustering routing algorithm for WSN. It elects one aggregator node for reliable and persistent data storage applications. PANEL assumes that the sensor nodes are deployed in a bounded area partitioned into geographical clusters. The clustering is determined before the deployment of the network, and each sensor node is pre-loaded with the geographical information of the cluster to which it belongs. At the beginning of each epoch, a reference point is computed in each cluster by the nodes in a completely distributed manner depending on the epoch number. Once the reference point is computed, the nodes in the cluster elect the node that is the closest to the reference point as the aggregator (clusterhead) for the given epoch.

The reference points of the clusters are re-computed and the aggregator election procedure is re-executed in each epoch. This ensures load balancing in the sense that each node of the cluster can become aggregator with nearly equal probability. The communication overhead used in the election procedure is also used to establish the routing tables within the cluster. At the end of the aggregator node election procedure, the nodes also learn the next hop towards the aggregator elected for the current epoch.

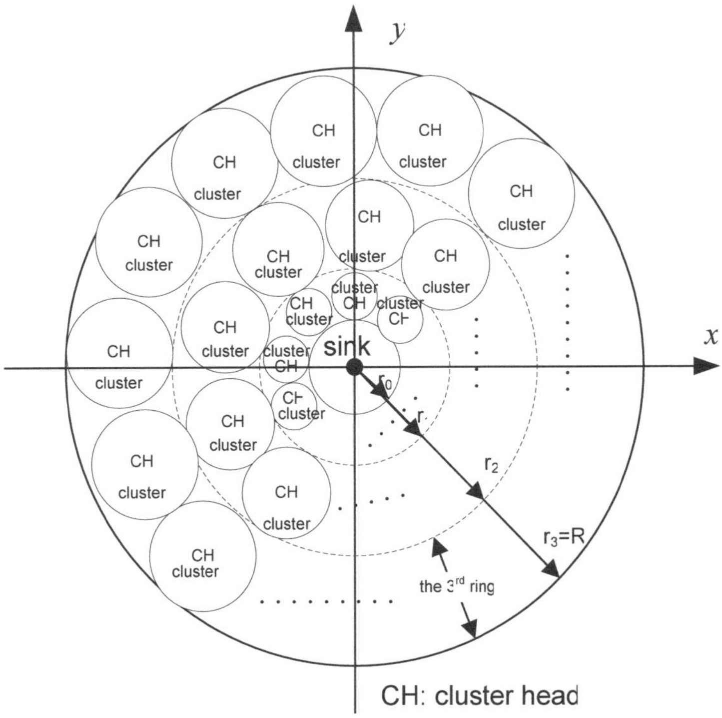

Fig. 7. Unequal size clusters (redrawn from (Shu et al., 2005)

PANEL can be integrated with any position-based routing protocol for inter-cluster communications. The authors proposed to experiment PANEL with the Greedy Perimeter Stateless Routing (GPSR) protocol (Karp & Kung, 2000). Simulation results showed that PANEL outperforms LEACH by about 67% to 83% in terms of network lifetime. This performance gain can be explained by the reduction of the number of transmissions and receptions thanks to data aggregation. However, the main limitation of PANEL is its assumption that the clusters are determined before deployment and thus can not adapt to WSN dynamics.

3.5 Unequal clustering

All the previously cited clustering algorithms form clusters with fixed or variable radius without any consideration of the hot spot problem introduced in Section 2.1.2. One possible solution of this issue is to form unequal clusters depending on how far is a clusterhead from the sink. The rational behind this is that main spent energy by a clusterhead is due to both inter-cluster and intra-cluster communication and hence have to be considered jointly. On the one hand, intra-cluster communication cost is proportional to the number of member nodes in a cluster. On the other hand, in a multihop network, inter-cluster communication cost depends on the experienced forwarding load by a given clusterhead. In the many-to-one communication pattern of WSN, the closer to the sink, the greater forwarding load a clusterhead have to handle. As a consequence, more uniform load distribution among clusterheads in a network can be achieved through smaller clusters near the base station. Figure 7 redrawn from (Shu et al., 2005) illustrates the main idea behind unequal clustering.

(Soro & Heinzelman, 2005) proposed an Unequal Clustering Size (UCS) model for network organization in order to balance energy consumption of clusterheads in multihop sensor networks, thus increasing network lifetime. Clusterheads are deterministically deployed and are assumed to be much more expensive (super nodes) than simple sensor nodes with the ability to move to adjust their locations, managing at the same time the size of their clusters and the expected load from other clusters further away.
In UCS, the sensing field is assumed to be circular and is split into two concentric circles, called layers. Soro et al. showed through both theoretical and experimental analysis, that the size of the cluster in the inner layer should be reduced to get more uniform energy consumption. For both homogeneous and heterogeneous networks, they showed that UCS achieves an improvement of about 10-30% over the Equal Clustering Size (ECS) scheme, depending on the aggregation efficiency of the clusterheads.
(Shu et al., 2005) aimed to design optimal power allocation strategies to achieve power balance among clusterheads that maximize the network lifetime, defined as the time until one clusterhead runs out of battery. The problem of balancing energy consumption among clusterheads is formulated as a signomial optimization problem. Like (Soro & Heinzelman, 2005), Shu et al. split the monitoring area into layers and studied how to achieve load balance by assigning larger cluster sizes to clusterheads that are responsible for less data forwarding as shown by Figure 7. They derived optimal parameters, such as the cluster radius of each layer and the relay probabilities of clusterheads, to prolong the network lifetime. The study demonstrates the significance of simultaneously considering the impacts of intra- and inter-cluster traffic.
Shu et al. stressed the importance of joint design of clustering strategies and routing since the volume of relayed traffic is also affected by the underlying routing scheme. They provided two schemes for balancing power consumption : routing-aware optimal cluster planning and clustering-aware optimal random relay. The former is essentially a clustering approach that is developed in the context of shortest-hop-count inter-clusterhead routing. For this scheme, the optimal cluster size and location are obtained. The latter is essentially a routing strategy for "load-balanced" clustered topologies (i.e., all clusters are of the same size). According to this approach, a clusterhead probabilistically chooses to either relay the traffic to the next-hop clusterhead or to deliver it directly to the sink.
For practical deployment of such schemes, several issues are still open for research, mainly how to optimally select cluster sizes without knowledge of the node locations and without assuming deterministic clusterheads deployment.

3.6 QoS-aware Cluster-based Routing protocols

Numerous routing protocols try to achieve QoS requirements such as end-to-end delay and available bandwidth when building paths in a sensor network. Threshold sensitive Energy Efficient sensor Network protocol (TEEN) (Manjeshwar & Agarwal, 2001) is one of cluster-based routing protocols that aims to responsiveness to sadden changes in time-critical applications. TEEN builds a 2-tier clustering topology as depicted in Figure 8 and relies on broadcasting hard and soft thresholds by each clusterhead to its member nodes. Hard threshold is the absolute value of the attribute beyond which, the node sensing this value must switch on its transmitter and report to its clusterhead. The nodes will next transmit data only when the current value of the sensed data is greater than the hard threshold and differs from the previously sensed value by an amount equal to or greater than the soft threshold. This allows significant decrease of the number of transmissions. Hard and soft threshold values can be adjusted so the data traffic can be controlled.

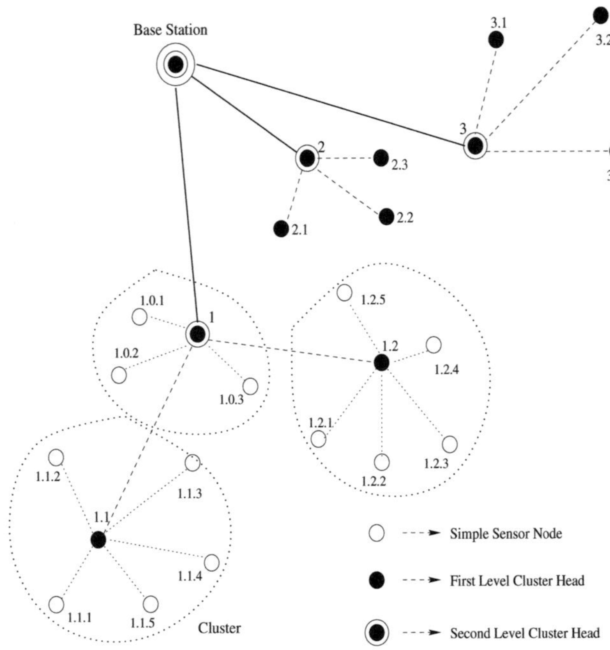

Fig. 8. TEEN hierarchy clustering (redrawn from (Manjeshwar & Agarwal, 2001))

TEEN is quite limited in applications where periodic reports are needed since the user may not get any data at all if the thresholds are not reached. The Adaptive Threshold sensitive Energy Efficient sensor Network protocol (APTEEN) (A. Manjeshwar, 2002) is an extension to TEEN aiming to handle applications with periodic data collections while being sufficiently reactive to time-critical events.

Recent research effort aimed to guarantee WSN specific requirements such as connectivity and coverage in cluster-based routing protocols while being energy efficient. (Soro & Heinzelman, 2009) tackled the problem of clusterhead election with entire area coverage preservation. Based on different coverage-aware cost metrics, nodes more important to the network coverage task are less likely to be selected as clusterheads. The same metrics are used to find the set of active sensor nodes that provide full network coverage, as well as the set of routers that forward the clusterheads' data load to the sink. Soro et al. showed that clustering in sensor networks should be performed with joint consideration of remaining energy and coverage redundancy. Their proposed approach showed to maintain full coverage of the monitored area from 25% to 4.5× with respect to a traditional approach where only residual energy or coverage redundancy are considered separately.

Authors of (Chamam & Pierre, 2009) argue that coverage, connectivity of sensors to clusterheads and routing have to be taken into account within the same global planning process in building a clustering topology. When coverage and connectivity are dealt with separately, the obtained configuration may not be optimal. For example, an optimal covering subset of sensors can fail to guarantee network connectivity because some nodes are switched off or the optimally designated clusterheads may belong to the set of switched-off sensors. Motivated by this fact, Chamam et al. addressed the global problem of maximizing network lifetime under the joint clustering, routing, and coverage constraint. They formulated the problem as an Integer Linear Programming model, proved that it is NP-Complete and implemented a Tabu search heuristic to tackle the exponentially increasing computation time of the exact resolution.

4. On-demand Cluster-based Routing Algorithms

In this class of cluster-based routing algorithms, the clustering topology is built in parallel with the routing discovery phase.

4.1 Passive Clustering (PC)

Passive clustering (PC) (Kwon & Gerla, 2002) is an *on demand* clustering algorithm. It provides scalability and practicality for choosing the minimal number of forwarding nodes in the presence of dynamic topology changes. PC constructs and maintains the cluster architecture based on outgoing data packets piggybacking *cluster related information*. Passive clustering eliminates setup latency and major control overhead of traditional clustering protocols by introducing two innovative mechanisms for the cluster formation: *"first Declaration wins"* rule and *"gateway selection heuristic"*. With the *"first Declaration wins"* rule, a node that first claims to be a clusterhead *rules* the rest of nodes in its clustered area. The *"gateway selection heuristic"* provides a procedure to elect the minimal number of gateways.
The algorithm defines several states in which a node can be. At cold start, all nodes are in the initial state. Nodes can keep internal states such as *clusterhead-ready* or *gateway-ready* to express their readiness to be respectively a clusterhead or gateway. A candidate node finalizes its role as a clusterhead, a gateway (Full-GW or Dist-GW) or an ordinary node. Additional fields suggested by PC in the message header of each packet are :

- *id* : the identity of the originator of this message,
- *state* : this packer sender status in the network,
- $CH1$ and $CH2$: these two fields are only used by a gateway to announce its two clusterhead addresses,

The reactive nature of PC motivated its combination with on demand routing protocols. Originally, PC was applied to reactive routing protocols like AODV (C. Perkins, 1999) and DSR (Johnson et al., 2001). The major overhead in these routing protocols is caused by the flooding of route queries. It was suggested to allow only non-ordinary nodes to rebroadcast query messages.
The PC algorithm presents some shortcomings that have been targeted by several works. In (Rangaswamy & Pung, 2002), the authors proposed to add alive packets to keep the cluster stability as it depends highly on the data packet traffic. Also, a sequence numbering to synchronize packets arriving from a source node is proposed. In fact, if packets containing different states arrive out-of-order at the destination (i.e., the sending node changed its state between transmission of multiple packets) then the destination node will be misled about the true state of the source node. In addition, unnecessary rebroadcasts are eliminated when the final destination of the message is a cluster member.
In WSN, the PC algorithm was proposed in combination with directed diffusion (DD) in (Handziski et al., 2004) to mainly achieve energy efficiency. The main idea of the combination is to save energy in the flooding phases by allowing only clusterheads and gateways to

participate in them. Member nodes are only allowed to send data messages in the data sending phase. Under different network size and load, the combination showed best performances in terms of delivery ratio and average dissipated energy.
Motivated by the results shown in (Handziski et al., 2004) when applying the original PC along with directed diffusion paradigm other works have been proposed in order to achieve better performance of the combination. In (Mamun-or-Rashid et al., 2007), the selection of clusterheads and gateways are done using a heuristic of residual energy and distance. By using residual energy the flooding nodes are chosen in an energy efficient manner. Distances are used to reduce overlapping region and so the number of gateways. The solution proposes to apply a periodic sleep and awake among cluster members. This technique is similar to the one proposed in LEACH and requires a synchronization process between nodes.

4.2 Energy Level-based Passive Clustering (ELPC)

The main idea in combining PC to DD is to reduce energy consumption by minimizing flooding. As this process is known to be very costly, the energy expenditure of the flooding nodes will be much higher than those of ordinary nodes. This will cause a variance in available energy at the nodes in the network and by that a fast partitioning of the network. In PC, topology construction is done according to the lowest ID. The drawback of doing so is its bias towards nodes with smaller IDs leading to their fast battery drainage.
In (Zeghilet et al., 2009), ELPC (Energy Level-based Passive Clustering) is proposed to achieve energy efficiency in terms of network lifetime and not only in terms of energy consumption. This is done through alternating flooding nodes role (clusterheads and gateways) among nodes depending on their energy. The aim of doing so is to have the same amount of energy at all the nodes at a given time which increases substantially the whole network lifetime.
In ELPC, the node's battery is split into levels. One can make a correspondence between different energy levels of a node and virtual sub-batteries it consumed sequentially. The energy level (l) of a node can be computed using :

$$l = \left\lceil L \, \frac{E_r}{E_i} \right\rceil \tag{3}$$

where E_r is the remaining energy, E_i is the initial one and L is the suggested number of levels. The notion of *candidature* to be a clusterhead or a gateway is introduced by defining the *network energy level (nel)* parameter. A node is not allowed to declare itself as a clusterhead (or a gateway) if its energy level is lower than this parameter. A clusterhead (or a gateway) can keep its role as long as its energy level is higher than the *nel*. Otherwise, it gives up its role and passes to the initial or ordinary state according to whether it knows or not a clusterhead in its vicinity.
The network energy level depends on the energy level of the network nodes and can be viewed as the minimum level of energy necessary for a node to be a clusterhead or a gateway. Zeghilet et al. suggested to take an initial value that corresponds to the half of the battery charge. This value is decreased locally each time the condition to be a clusterhead is not satisfied. The local network energy level is then propagated within outgoing packets header. Each time a node receives a smaller *nel* value, its updates its local value accordingly.
ELPC uses the same states as suggested in (Kwon & Gerla, 2002) where a node is initially at the *initial* state. Nodes form and maintain the clustering topology by changing their internal and external states based on outgoing messages. When sending the next message, the node

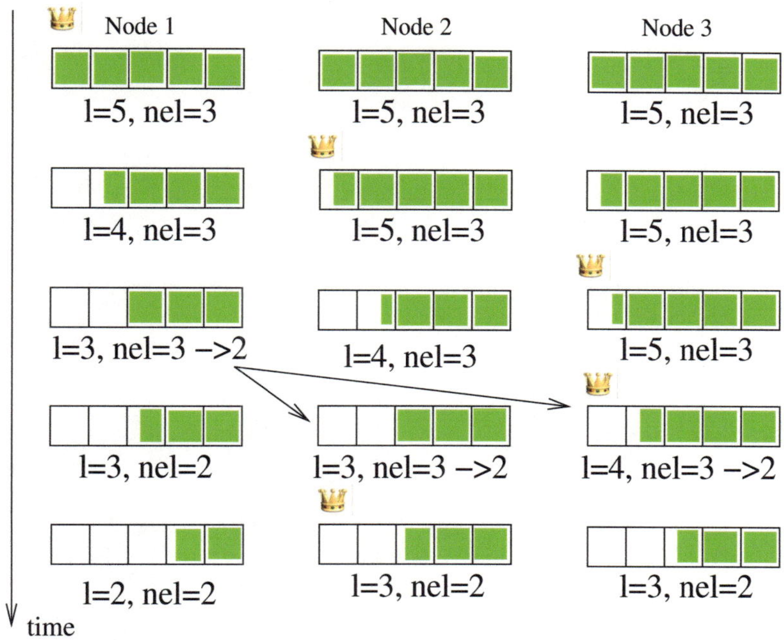

Fig. 9. ELPC and load-balancing feature

announces its external state which becomes visible in the network. ELPC adds the following fields to the packet header :

- *l*, node's energy level
- *nel*, the network energy level
- *give-up*, as in (Handziski et al., 2004) is set when the node is a clusterhead that gives-up its role. It is used to replace the *give-up* message proposed in (Kwon & Gerla, 2002). In ELPC, this field is set when the energy level of a clusterhead drops bellow the (*nel*).

Figure 9 illustrates how clusterhead rotation is achieved in ELPC. Assume that three nodes 1, 2 and 3 (with same initial amount of energy) are contending to be a flooding node (CH in this example). If we use PC algorithm, node 1 will be selected to be a clusterhead since it has the smallest ID. In ELPC, assume that the number of energy level is 5 and that the *nel* is initially set to 3. We can see that the clusterhead role is alternated between the three nodes depending on their energy levels. When two nodes have the same energy level, then the nodes' identities are used to solve conflict in declaring roles. At step 3, we can note that node 1 decreases its *nel* to 2 and propagates this new value to its neighbors so all nodes can have same estimation of the network energy level.

Figure 10 shows the establishment of routing structures of directed diffusion when this latter is used in combination with ELPC. At initialization, all nodes in the network are in the Initial state. Nodes will use the first interest messages to establish the new topology. A possible topology is illustrated in Figure 10(a-b). After establishing the gradient (Figure 10(c)) and path reinforcement (Figure 10(d)), the source begins sending the sensed data. When the energy level falls under the network energy level at node A (Figure 10(d)), it gives-up its role as clusterhead. Thus, a new topology is established (Figure 10(e)). This is done using next

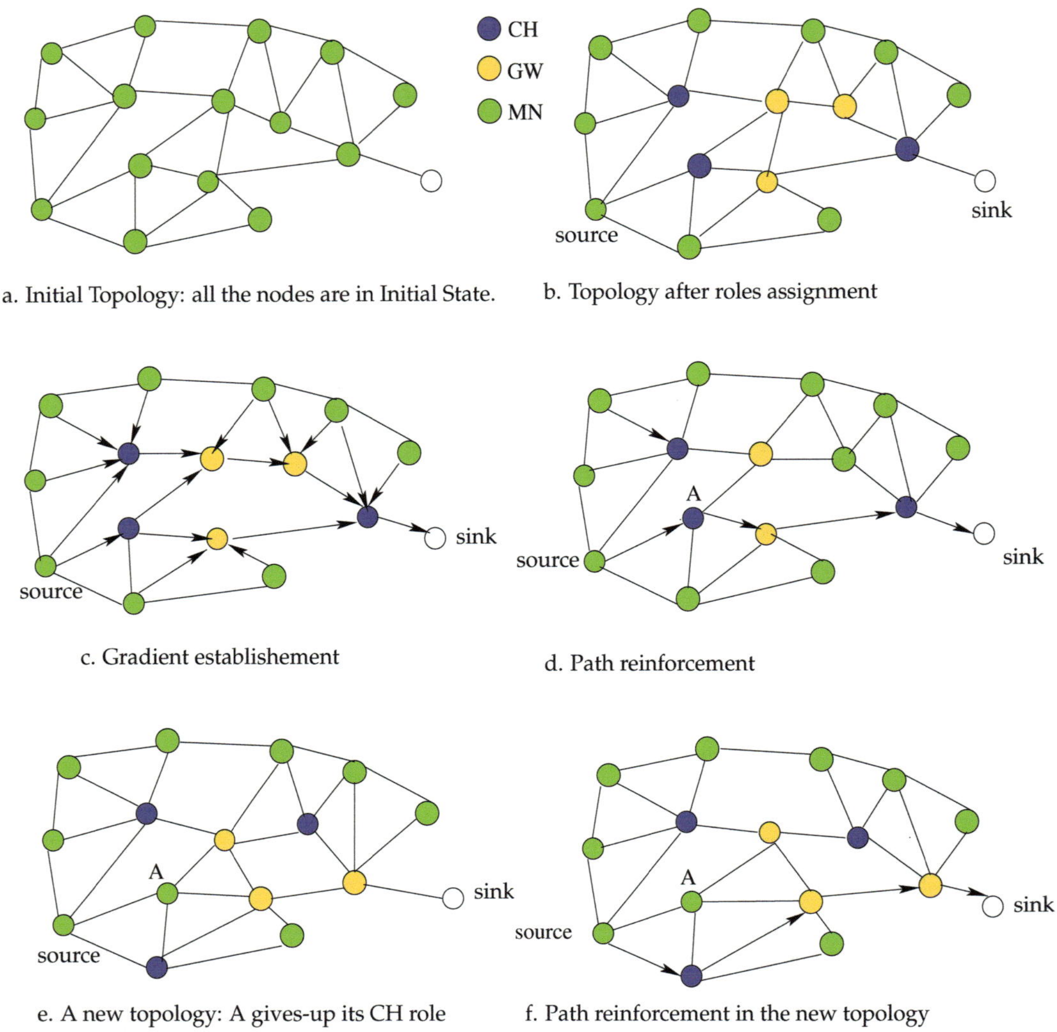

Fig. 10. ELPC illustrated

circulating messages in the network (data messages, interests, explorers data). The resulting PC can be applied to any routing protocol in sensor networks as they mostly rely on flooding and particularly with DD. This not only reduces energy consumption as in (Handziski et al., 2004), but increases the whole network lifetime.
Simulation results showed that ELPC outperforms PC and PCDD (a PC/DD combination without considering energy) in terms of energy consumption and network lifetime thanks to its energy-aware flooding nodes rotation. Figure 11 plots as the network size increases, the network lifetime for the three protocols and shows that ELPC achieves better performances compared to the two others.

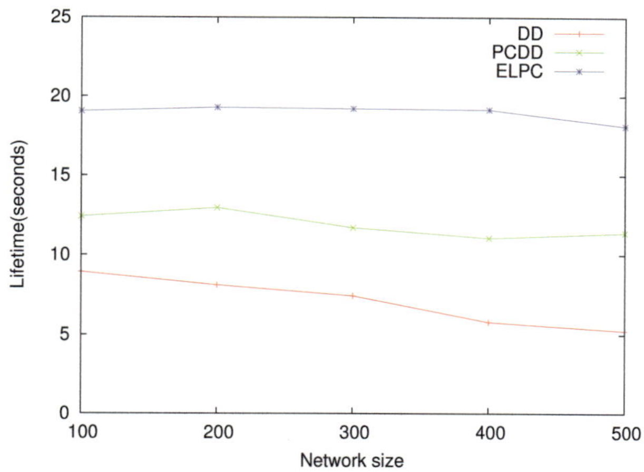

Fig. 11. Network lifetime of ELPC compared to PCDD and DD

4.3 CLIQUE

The work in (Forster & Murphy, 2009) presents CLIQUE, an approach for clusterhead selection based on machine learning (Q-learning). The authors observed that a clusterhead may require less energy than its direct neighbors in a multi-hop intra-cluster topology. They conclude that clusterhead role assignment must take into account not only the current state of the selected clusterheads, but also those of its neighbors and nodes on the paths to the clusterhead. in CLIQUE, clusterhead roles are neither explicitly assigned nor do the nodes need to agree on a clusterhead. Instead, each node decides on a per-packet basis whether to act as clusterhead (aggregating some packets then sending the result to the sinks) or to forward the packet to a better suited neighbor. Authors claimed that this role-free scheme makes the algorithm flexible and robust and eliminates the need for multiple clusterhead selection rounds.

Forster and Murphy focused on the clusterheads selection process and assumed that clusters are predefined (rectangular grids) and that each node knows the identity of the cluster to which it belongs. They targeted a traditional, periodic data reporting application and a multiple sinks network. The sinks flood the network with DATA REQUEST packets announcing their data interest. These packets can carry some routing information that is further used by nodes to estimate the routing cost to the sinks. The routing cost is calculated using a combination of hop counts to reach the sinks and battery status of the nodes on the routes to the sinks. Each sensor node is an independent learning agent, and actions are routing options using different neighbors as the next hop toward the clusterhead. The clusterhead is defined as the cluster node with the best (lowest) routing cost to all sinks.

Even if CLIQUE may incur more energy consumption due to possible coexistence of multiple clusterheads in one cluster, the authors showed through simulations that CLIQUE saves up to 25% of consumed energy thanks to its lower overhead. However, CLIQUE is more suitable for regular data reporting and its performances are to be proved for other types of applications such as event driven ones.

5. Conclusion

Hierarchical (cluster-based) routing protocols hold a great potential toward energy efficiency in WSN. Clustering algorithms have been a hot research area in the last few years. In this chapter, we introduced a new classification of clustering techniques from the perspective of data routing process. We summarized some protocols that we found to be representative of both classes and that give solution (even partial) of a given problem or requirement of energy efficient clustering.

Managing energy consumption individually at each sensor is far from being sufficient to maximize the WSN lifetime. A global management strategy with load balancing feature is required. to do so, clustering techniques have to provide low overhead clusterhead rotation as well as optimal traffic distribution among clusterheads while keeping network connectivity and coverage. Unequal clustering where both intra-cluster and inter-cluster communications are considered is very promising. However, practical techniques need to be developed to build such clusters without knowledge of the global network topology.

Optimal (or even approximate) parameters estimation for successful clustering is very important but is not an easy task since WSN-specific constraints like energy, coverage and connectivity have to be satisfied. These parameters include mainly clusterheads rotation frequency that allows the best load balance with the lowest overhead, in addition to the number of clusters and their size that maximize the network lifetime.

Finally, network dynamics have to be handled appropriately. Network dynamics include possible nodes or sink mobility and topology changes due to death of one or more sensors in the field of interest. Suitable and very reactive solutions have to be provided mainly when a clusterhead dies leaving orphan sensors, possible uncovered area and lack of inter-cluster connectivity.

6. References

A. Manjeshwar, D. A. (2002). Apteen: a hybrid protocol for efficient routing and comprehensive information retrieval in wireless sensor networks, *2nd International Workshop on Parallel and Distributed Computing Issues in Wireless Networks and Mobile computing,* Lauderdale, FL.

Abbasi, A. A. & Younis, M. (2007). A survey on clustering algorithms for wireless sensor networks, *Comput. Commun.* **30**(14-15): 2826–2841.

Akkaya, K. & Younis, M. (2005). A survey of routing protocols in wireless sensor networks, *Ad Hoc Network (Elsevier)* **3**(3): 325–349.

Al-Karaki, J. & Kamal, A. (2004). Routing techniques in wireless sensor networks: a survey, *IEEE Wireless Commun* **6**(11): 6–28.

Bandyopadhyay, S. & Coyle, E. (2003). An energy efficient hierarchical clustering algorithm for wireless sensor networks, *the 22nd Annual Joint Conference of the IEEE Computer and Communications Societies (Infocom),* San Francisco, CA.

Bandyopadhyay, S. & Coyle, E. J. (2004). Minimizing communication costs in hierarchically-clustered networks of wireless sensors, *Comput. Netw.* **44**(1): 1–16.

Banerjee, S. & Khuller, S. (2001). A clustering scheme for hierarchical control in multi-hop wireless networks, *IEEE Infocom 2001,* Anchorage, Alaska.

Buttyan, L. & Schaffer, P. (2007). Panel: Position-based aggregator node election in wireless sensor networks, *Proceedings of the IEEE International Conference on Mobile Ad hoc and Sensor Systems, MASS,* pp. 1–9.

C. Perkins, E. Royer, S. D. (1999). Ad hoc on demand distance vector (aodv) routing. **URL:** *http://www.ietf.org/internet-drafts/draft-ieftmanet-aodv-03.txt*

Chamam, A. & Pierre, S. (2009). On the planning of wireless sensor networks: Energy efficient clustering under the joint routing and coverage constraint, *IEEE Transactions on Mobile Computing* **8**(8): 1077–1086.

Chandrakasan, A. P., Smith, A. C., Heinzelman, W. B. & Heinzelman, W. B. (2002). An application-specific protocol architecture for wireless microsensor networks, *IEEE Transactions on Wireless Communications* **1**: 660–670.

Ding, P., Holliday, J. & Celik, A. (2005). Distributed energy efficient hierarchical clustering for wireless sensor networks, *IEEE International Conference on Distributed Computing in Sensor Systems(DCOSS'05)*, Marina Del Rey, CA.

Energy-efficient communication protocol for wireless sensor networks (2000). Hawaii.

Forster, A. & Murphy, A. L. (2009). Clique: Role-free clustering with q-learning for wireless sensor networks, *ICDCS '09: Proceedings of the 2009 29th IEEE International Conference on Distributed Computing Systems*, IEEE Computer Society, Washington, DC, USA, pp. 441–449.

Frey, H., Ruehrup, S. & Stojmenovic, I. (2009). *Routing in wireless sensor networks*, Springer-Verlag, chapter 4, pp. 81–111.

Handziski, V., KÃűpke, A., Karl, H., Frank, C. & Drytkiewicz, W. (2004). Improving the energy efficiency of directed diffusion using passive clustering., *EWSN*, Vol. 2920 of *Lecture Notes in Computer Science*, Berlin, Germany, pp. 172–187.

He, T., Stankovic, J., Lu, C. & Abdelzaher, T. (2003). Speed: a stateless protocol for real-time communication in sensor networks, *Proceedings of International Conference on Distributed Computing Systems*, Providence, RI.

Heinzelman, W., Kulik, J. & Balakrishnan, H. (1999). Adaptive protocols for information dissemination in wireless sensor networks.

Intanagonwiwat, C., Govindan, R., Estrin, D., Heidemann, J. & Silva, F. (2003). Directed diffusion for wireless sensor networking, *IEEE/ACM Transactions on Networking (TON)* **11**(1): 2–16.

Johnson, D. B., Maltz, D. A. & Broch, J. (2001). *DSR: The Dynamic Source Routing Protocol for Multi-Hop Wireless Ad Hoc Networks*, Addison-Wesley, chapter 5, pp. 139–172.

Kang, T., Yun, J., Lee, H., Lee, I., Kim, H., Lee, B., Lee, B. & Han, K. (2007). A clustering method for energy efficient routing in wireless sensor networks, *EHAC'07: Proceedings of the 6th WSEAS International Conference on Electronics, Hardware, Wireless and Optical Communications*, World Scientific and Engineering Academy and Society (WSEAS), Stevens Point, Wisconsin, USA, pp. 133–138.

Karp, B. & Kung, H. T. (2000). Greedy perimeter stateless routing for wireless networks, *ACM Mobicom*, Boston.

Ko, Y.-B. & Vaidya, N. H. (2000). Location-aided routing (lar) in mobile ad hoc networks, *Wireless Networks* **6**(4): 307–321.

Kwon, T. J. & Gerla, M. (2002). Efficient flooding with passive clustering (pc) in ad hoc networks, *SIGCOMM Comput. Commun. Rev.* **32**(1): 44–56.

Lin, C. R. & Gerla, M. A. (1995). A distributed control scheme in multi-hop packet radio networks for voice/data traffic support, *Proceedings of IEEE GLOBECOM*, pp. 1238–1242.

Lin, X. & Stojmenovic, I. (2003). Location-based localized alternate, disjoint and multi-path routing algorithms for wireless networks, *J. Parallel Distrib. Comput.* **63**(1): 22–32.

Mamalis, B., Gavalas, D., Konstantopoulos, C. & Pantziou, G. (2009). *Clustering in Wireless Sensor Networks*, Y. Zhang, L. T. Yang, J. Chen (Eds.), chapter 12, pp. 324–353.

Mamun-or-Rashid, M., Alam, M. M. & Hong, C. S. (2007). Energy conserving passive clustering for efficient routing in wireless sensor network, *9th International Conference on Advanced Communication Technology*, pp. 982–986.

Manjeshwar, A. & Agarwal, D. P. (2001). Teen: a routing protocol for enhanced efficiency in wireless sensor networks, *1st International Workshop on Parallel and Distributed Computing Issues in Wireless Networks and Mobile Computing*.

Mhatre, V. & Rosenberg, C. (2004a). Design guidelines for wireless sensor networks: communication, clustering and aggregation, *Ad Hoc Networks* **2**(1): 45–63.

Mhatre, V. & Rosenberg, C. (2004b). Homogeneous vs heterogeneous clustered sensor networks: a comparative study, *IEEE International Conference on Communications*, Vol. 6, Paris, France, pp. 3646 – 3651.

Murata, T. Ishibuchi, H. (1994). Performance evaluation of genetic algorithms for flowshop scheduling problems, *Proceedings 1st IEEE Conference Evolutionary Computation*, pp. 812 – 817 vol.2.

Rangaswamy, A. & Pung, H. K. (2002). Enhancement of passive cluster based routing protocol for mobile adhoc networks, *Eleventh International Conference on Computer Communications and Networks*, pp. 376–381.

Rodoplu, V. & Ming, T. (1999). Minimum energy mobile wireless networks, *IEEE Journal of Selected Areas in Communications* **17**(8): 1333–1344.

Shu, T., Krunz, M. & Vrudhula, S. (2005). Power balanced coverage-time optimization for clustered wireless sensor networks, *Proceedings of the 6th ACM international symposium on Mobile ad hoc networking and computing (MobiHoc '05)*, Urbana-Champaign, IL, USA, pp. 111–120.

Sohrabi, K., Gao, J., Ailawadhi, V. & Pottie, G. (2000). Protocols for self-organization of a wireless sensor network, *IEEE Personal Communications* **7**(5).

Soro, S. & Heinzelman, W. (2005). Prolonging the lifetime of wireless sensor networks via unequal clustering, *5th International Workshop on Algorithms for Wireless, Mobile, Ad Hoc and Sensor Networks (IEEE WMAN '05)*, Denver, Colorado.

Soro, S. & Heinzelman, W. (2009). Ch election techniques for coverage preservation in wireless sensor networks, *Ad Hoc Networks* **7**(5): 955–972.

Xu, Y., Heidemann, J. & Estrin, D. (2001). Geography-informed energy conservation for ad hoc routing, *ACM MOBICOM*, pp. 70–84.

Y. Yu, D. E. & Govindan, R. (2001). Geographical and energy-aware routing: a recursive data dissemination protocol for wireless sensor networks, *Technical Report UCLA-CSD TR-01-0023*, UCLA Computer Science Department.

Ye, F., Chen, A., Lu, S., Zhang, L. & Chen, F. Y. A. (2001). A scalable solution to minimum cost forwarding in large sensor networks, *Proceedings. Tenth International Conference on Computer Communications and Networks, 2001*, pp. 304–309.

Younis, O. & Fahmy, S. (2004). Heed: A hybrid, energy-efficient, distributed clustering approach for ad hoc sensor networks, *IEEE Trans. Mob. Comput.* **3**(4): 366–379.

Younis, O., Krunz, M. & Ramasubramanian, S. (2006). Node clustering in wireless sensor networks: Recent developments and deployment challenges, *IEEE Network* **20**(3): 20–25.

Zeghilet, H., Badache, N. & Maimour, M. (2009). Energy efficient cluster-based routing in wireless sensor networks, *IEEE Symposium on Computers and Communications (ISCC'09)*, Sousse, Tunisia, pp. 701–704.

An Energy-aware Clustering Technique for Wireless Sensor Networks

Wibhada Naruephiphat and Chalermpol Charnsripinyo
National Electronics and Computer Technology Center,
National Science and Technology Development Agency,
Thailand

1. Introduction

A wireless sensor network (WSN) is a specialized wireless network that composes of a number of sensor nodes deployed in a specified area for monitoring environment conditions such as temperature, air pressure, humidity, light, motion or vibration, and so on. The sensor nodes are usually programmed to monitor or collect data from surrounding environment and pass the information to the base station for remote user access through various communication technologies. Figure 1 shows general wireless sensor network architecture. Typically, a sensor node is a small device that consists of four basic components as shown in Figure 2: 1) sensing subsystem for data gathering from its environment, 2) processing subsystem for data processing and data storing, 3) wireless communication subsystem for data transmission and 4) energy supply subsystem which is a power source for the sensor node. However, sensor nodes have small memory, slow processing speed, and scarce energy supply. These limitations are typical characteristics of sensor nodes in wireless sensor networks.

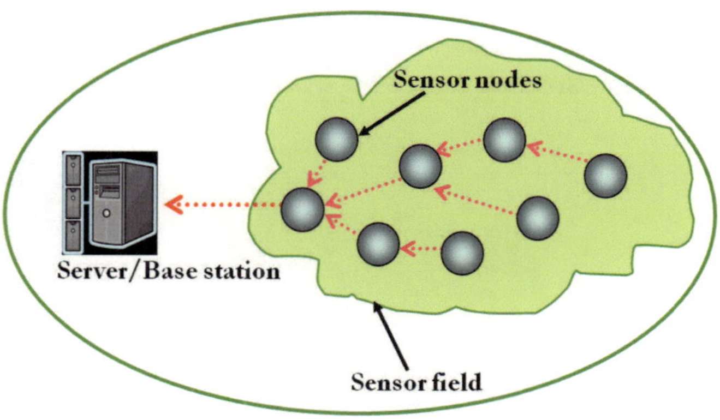

Fig. 1. Wireless Sensor Network

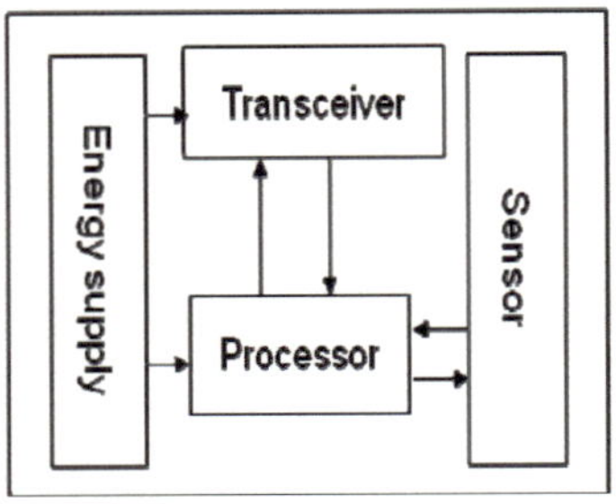

Fig. 2. Overview of sensor node components

A wireless sensor network usually has energy constrained due to each sensor node requires battery with a limited energy supply to operate. In addition, recharging or replacing sensor battery may be inconvenient and impossible in some environments. However, the wireless sensor network should function long enough to accomplish the application requirements. Therefore, energy conservation is a main issue in the design of wireless sensor networks. There are different approaches to preserve energy usage and prolong the network lifetime in WSN. The key approach to improve energy usage in WSNs is the development of energy-aware network protocols.

In this paper we present a review of routing and clustering algorithms for energy conservation in wireless sensor networks. We also present an energy-aware clustering technique for enhancing the network lifetime as well as increasing the number of successfully delivered packets and decreasing the network delay time.

2. Review of Routing and Clustering Algorithms

A routing protocol in wireless sensor networks usually coordinates the activities of sensing nodes in the network for data transmission to the base station. Routing protocols in WSN can be grouped into three models as follows (Ibriq&Margoub, 2004).

1) One-hop model: every node in the network transmits data directly to the base station. This is the simplest model representing direct communication from the sensor node to the base station as shown in Figure 3. However, the direct communication may not be practical for routing in wireless sensor networks because each sensor node has limited transmission range.

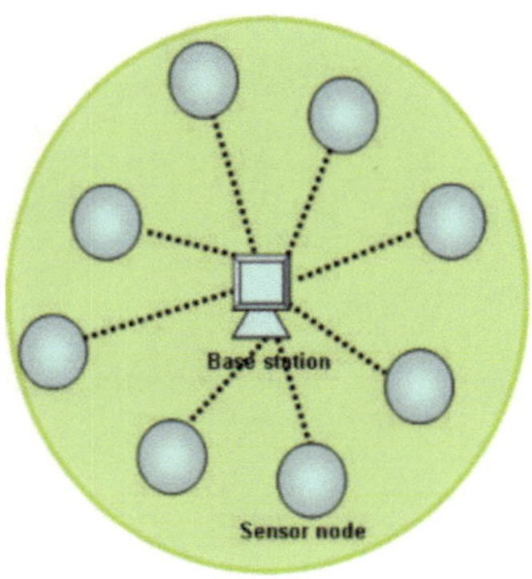

Fig. 3. One-hop model

2) Multi-hop model: a sensor node transmits data to the base station by forwarding its data to one of its neighbors which are closer to the base station. The data packet from the source node is forwarded hop-by-hop from one node to another node until the data packet arrives at the base station as shown in Figure 4.

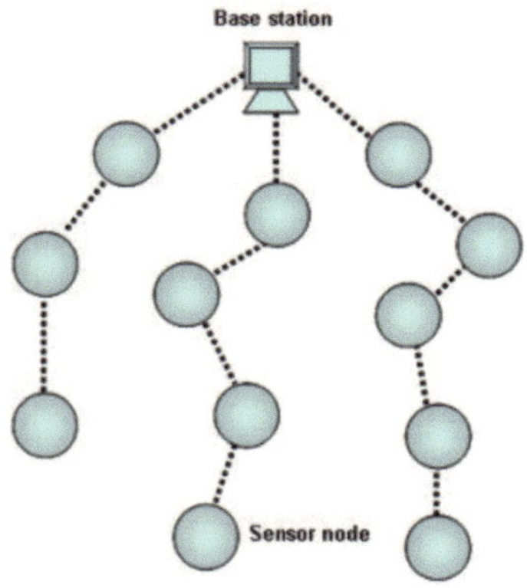

Fig. 4. Multi-hop model

3) Cluster-based Hierarchical Model: each cluster consists of a single cluster head (CH) and multiple member nodes. Nodes are grouped into clusters with a cluster head that has the responsibility of routing data packets from the cluster to another cluster heads toward the base station. A node can be both the cluster head in one cluster, and a member in another cluster which is closer to the base station. The cluster-base hierarchical is shown in Figure 5.

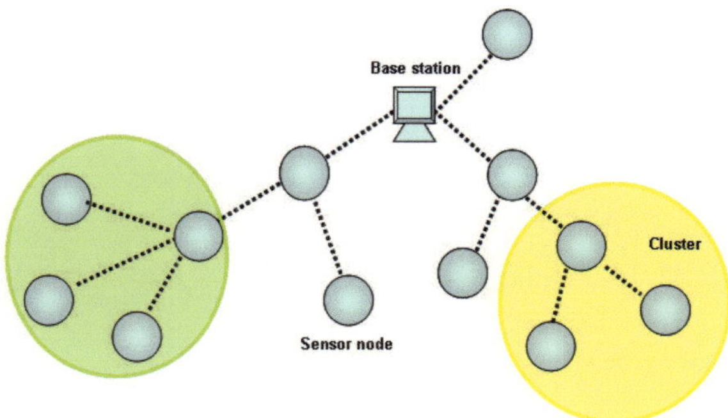

Fig. 5. Cluster-based hierarchical model

Many routing protocols have appeared recently which mainly concentrated on how to find a shorter path between a source and destination node when performing route discovery. The shortest path normally requires minimum number of intermediate forwarding nodes which result in minimum total energy consumption. However it is possible that some particular nodes are unfairly burdened. This hot spot node may consume more energy and stop running earlier than other nodes. (Fedor & Collier, 2007) explored when multi-hop routing is more energy-efficient than direct transmission to the sink and conditions which the two-hop strategy is optimal. The experiments showed that the two-hop communication is more advantageous than the single hop (direct communication) when the relay is equally distant from the source to the sink. (Jia et al., 2007) proposed a novel Hole Avoiding In advance

Routing protocol (HAIR) to decrease both the delay and energy consumption. The proposed protocol has two stages. In the first stage, a node finds barriers and informs its neighbor nodes about holes to avoid the missing path. In the second stage, if few sensor nodes can not find their routes at the first stage, they can find other existing paths in the network. The HAIR protocol can make the packets avoid meeting the "hole" in advance, so it decreases both the routing distance and the energy consumption. (Shen et al., 2009) proposed to improve the Geographical and Energy Aware Routing (GEAR) protocol. The proposed routing mechanism improves the GEAR protocol to reduce the energy consumption and extend the network lifetime. (Hu et al., 2007) proposed to avoid selecting the forwarding node with lower residual battery power than the threshold value. The approach maximizes the lifetime of WSN and equally balances the total energy consumption among all nodes in the network. (Wang et al., 2007) presented a Local Update-based Routing Protocol (LURP) that allows the sink node to move and update its location information. Since the sensor nodes close to the sink deplete their energy quickly by forwarding messages originating from many other nodes, the moving sink node can maintain the energy consumption of sensor nodes close to the sink. (Kai, 2009) proposed an energy-efficient routing called Leaping-Base Routing algorithm. This routing algorithm focuses on the load balancing problems in wireless sensor networks. Its routing table contains the information of neighbor nodes such as nodes' ID, hop length to the base station, and residual energy. A node selects its neighbor by considering the information of routing table.
A routing protocol usually requires updates from path search processes and stores information in the routing table. Therefore, the routing algorithms can affect the processing, memory, and energy consumption. Due to scarce energy supply, less processing power and memory, the routing algorithms should avoid overheads of storing routing table, avoid path search processes to reduce energy usage, and consider energy-efficient approach to preserve energy consumption as shown in Figure 6.

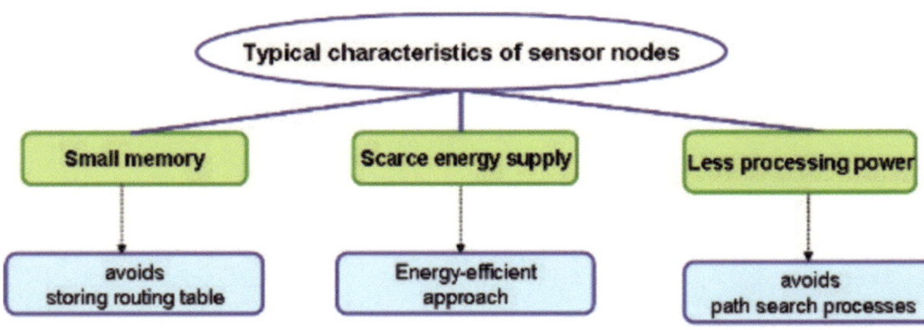

Fig. 6. Energy-aware technique for wireless sensor networks

One of the energy-efficient techniques used in wireless sensor networks is the clustering algorithm. A cluster- based routing protocol can avoids intensive message exchanges of path search update processes and overhead of storing routing table or other information that could be expensive to update.

Typical clustering algorithms divide WSN nodes into two types: member nodes and cluster-heads. The member nodes send data to their cluster-head, then a cluster-head aggregates the data and relays to the base station. Several clustering algorithms have been proposed for wireless sensor networks such as Low-Energy Adaptive Clustering Hierarchy (LEACH)

(Handy et al., 2002) and Hybrid Energy-Efficient Distributed (HEED) Clustering Approach (Younis & Fahmy 2004). (Qiu et al., 2009) presented a tree routing to avoid flooding network with path search and update message in order to conserve energy by using only link information between cluster head and members for packet forwarding. By using the cluster-head and member link information only, it avoids intensive message exchanges of path search update processes and overhead of storing routing table or other information that is expensive to update. An unequal clustering and multi-hop routing scheme was presented by (Gong et al. 2008) to extend the network lifetime of WSNs. The authors presented the cluster head selection approach based on a cost function which considers the distance and energy usage. (Dali & Chan, 2007) proposed an approach to balance and reduce the energy consumption of clustered sensor networks. Since the energy consumption of sensor nodes depend on transmission range, the cluster-heads are normally maintained at the center of cluster. In each cluster, the node located in the center area with the highest residual battery level is selected as the cluster-head. A maximum-Votes and Load-balance Clustering Algorithm (VLCA) was presented by (Zhang et al., 2008) to reduce the number of clusters and prolong network lifetime. To balance the workload among cluster-heads, this algorithm selects the cluster-head by considering the number of member nodes and the residual battery level. (Murthy et al., 2008) proposed a level controlled clustering to reduce the number of messages toward the base station and increase the network lifetime of WSN. This method assumes that the base station is able to transmit at various power levels. The cluster head selection method is also based on the maximum residual battery level.

In previous clustering algorithms discussed above, sensor nodes in the network are assigned to each cluster before the cluster heads are selected. The node with the best parameter value will typically be selected as a cluster head for data gathering and forwarding at each cycle. This could be a heavy burden of the selected cluster head as depicted in Figure 7 where node A and M are selected as the cluster head of other sensor nodes.

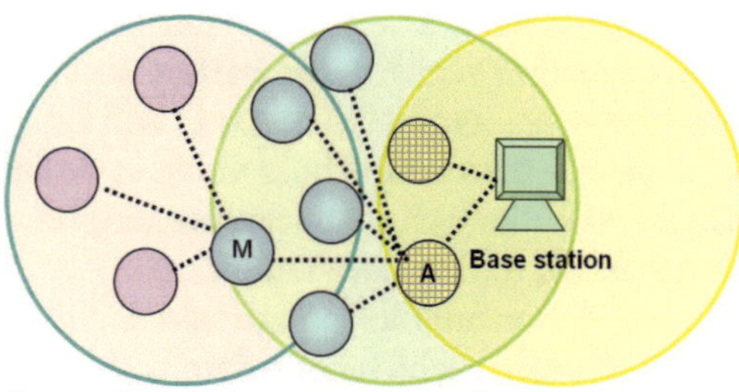

Fig. 7. A cluster-head and member links in typical schemes

3. Efficient Energy-aware Clustering Technique

In the design of energy-aware clustering techniques for wireless sensor networks, a clustering algorithm is used for cluster head selection. A simple clustering algorithm may select a cluster head with minimum distance or maximum residual battery level. A minimum cost function was presented in a previous research work (Chang & Tassiulas, 2004). The minimum cost function combining both energy consumption and battery level for cluster head selection was given as follows.

$$C(i)=(E_{TOT}(i))^{1}(B_{init})^{1}(B(i))^{-1} \tag{1}$$

Where $E_{TOT}(i)$ is the energy consumption at node i, B_{init} is initial battery level of sensor node and B(i) is residual battery at node i.
The minimum cost function algorithm will select a cluster head with minimum cost in order to increase the network lifetime. As a result, the selected cluster head has high residual battery level and low energy consumption.

In this paper, we present a clustering technique called the Limiting member node Clustering (LmC) which considers a maximum number of member nodes for each cluster head. We divide sensor nodes into groups where nodes within the base station's transmission range are defined in "level 1" and nodes far from the base station are defined in a higher level depending on the distance to the base station. Figure 8 shows an example of two-level WSN with limited member nodes for each cluster head.

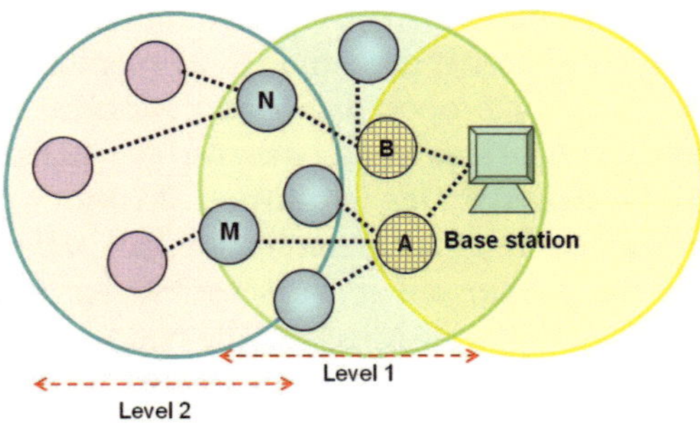

Fig. 8. Cluster-heads and member links in the LmC scheme

In the LmC approach, each sensor node selects a cluster head from the candidate list of cluster heads based on a cost function which takes battery level, energy consumption and distance to the base station into consideration. The LmC will limit the number of member nodes of each cluster head to be less than a threshold value in order to distribute the burden of each cluster head. Consequently, this technique can prolong network lifetime and reduce the time used to forward data packet to the base station.
Each sensor node within "level 1" transmission range selects a cluster head from candidate cluster heads using a new cost function which considers both battery level and distance to the base station. The new cost function is given as

$$C(i)=\frac{B_{init}}{B(i)}+\frac{T_{BS}(i)}{T_{max}} \tag{2}$$

where $T_{BS}(i)$ is the distance between node i and the base station and T_{max} is the maximum transmission range.
Other nodes in a higher level which can not connect to node in "level 1" will select a cluster head from a lower level node which is closer to the base station. The cluster selection will be based on another cost function defined as

$$C(i)=E_{TOT}(i)\frac{B_{init}}{B(i)}+\frac{T_{BS}(i)}{T_{max}} \tag{3}$$

Figure 9 shows the Limiting member node Clustering (LmC) method with cost functions for different layers.

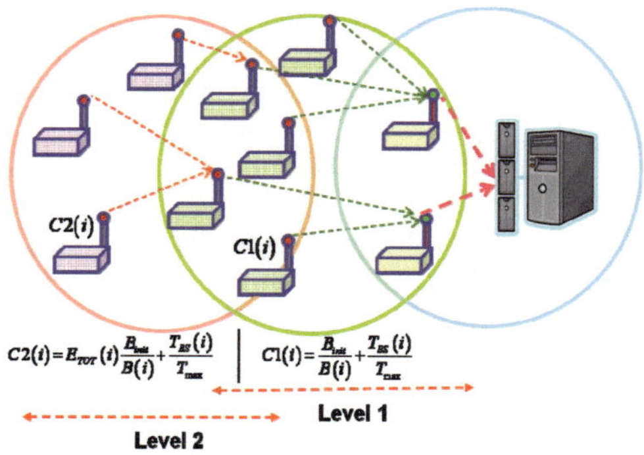

Fig. 9. The Limiting member node Clustering method

Note that the maximum number of member nodes for each cluster head is set to a threshold value. We have investigated different approaches to find the appropriate threshold value. By varying the percentage of the total number of sensor nodes in wireless sensor networks, we found that the appropriate threshold value is around 10 percents.

In LmC algorithm, after a cluster head is selected by nodes in a higher level, the node which has the minimum cost may be disregard if the maximum number of member nodes is attained. The limiting member node clustering algorithm can distribute member nodes to each cluster head. Therefore less data packets will be aggregated in each cluster head. This approach can reduce the time used to send packets to the base station. Since the proposed algorithm selects a cluster head based on the cost function, the selected cluster head can keep high residual battery level and short distance to the base station. The limiting member node clustering is a design approach to enhance network lifetime and also reduce communication delay.

4. Experiments and Performance Evaluation

In this section, we present the experiment results and performance evaluation of our proposed clustering technique. We first describe our experimental design and performance metrics used for evaluating clustering techniques. We then present the experimental results comparing different clustering approaches.

4.1 Experimental design and Performance metrics

We implemented a simulation program using C programming language for evaluating energy-aware clustering techniques. In our experiments, a number of sensor nodes are grouped into clusters where they are within a transmission range.

Nodes select a cluster head and form a cluster according to the self-organized manner. The communication process is described by (Ergen, 2004). Note that the energy usage during the communication process is not considered in our experiments, since we are focusing on the energy usage for sending data packets and the energy used by the communication process is the same amount for all algorithms.

We adopt the "radio model" discussed by (Muruganathan et al., 2005) for the energy consumption of each node in wireless sensor networks. The transmitting and receiving energy required for transmission of a data message of b-bits between two nodes in a transmission range of d meters is given by

$$E_{TOT}(i)=E_{TX}+E_{RX} \tag{4}$$

Where $E_{TOT}(i)$ is the energy consumption at node i, E_{TX} is the energy dissipated in the transmitter of the sending node given by

$$E_{TX}(b,d)=(E_{elec}\times b)+(\varepsilon_{fs}\times b\times d^2) \tag{5}$$

The term E_{RX} is the energy consumption at the receiving node given by

$$E_{RX}(b)=E_{elec}\times b \tag{6}$$

where E_{elec} is the energy expended in the radio electronics which is equal to 50 nJ/bits. ε_{fs} = 10pJ/bit/m^2 is the energy consumed in free space at the output transmitter antenna for a transmitting range of one meter in wireless sensor networks.

We assumed that each node knows the location of its neighbor nodes within the maximum transmission range by using arrival time of "Hello message" during the connection setup process. The information of energy consumption, residual battery level, and distance to the base station (assuming that all nodes know the position of the base station) will be also learnt from the connection setup process.

When each sensor node cooperatively monitors or collects environmental data or conditions (i.e., temperature or humidity), it sends information to a base station via a cluster head selected from a cluster head selection algorithm. We set the length of datagram packets to be 500 Kbits. The data rate for communications is 250 Kbps. The duty cycle is one read per 30 second. The neighbor node information is updated every 60 second. Each sensor node has an initial battery level of 500 J. A node whose battery is depleted will be disconnected from the network and cannot be immediately recharged from any external power supply.
In each experiment, the period of sensing devices to monitor or collect environmental data is 1 day. Each experiment is executed for 10 runs using randomly generated network topologies.

We use the following performance metrics to evaluate and compare among routing/clustering algorithms.
1) Number of successfully delivered packets is the number of times that packets can be successfully delivered to the base station more than 80% of total packets sent by all sensor nodes in the network

2) Network lifetime is the duration from the start up time until the first node is disconnected from the network due to it runs out of battery
3) Delay time is the period of time that the base station takes to receive packets successfully (more than 80% of total packets are delivered)
4) Average number of packets arrived at the base station (Avg_{pkt}) is the average number of packets received at base station (BS). Since each node in the network will send 1 packet at a time, it can be calculated from

$$Avg_{pkt} \frac{\text{Total \# of packet received}}{\text{Total \# of sending times}} \quad (7)$$

4.2 Experimental Results with Different Clustering Techniques

In this section, we present our experimental results with different cluster head selection approaches in order to compare their performances with our proposed Limiting member node Clustering (LmC) technique. We consider three other cluster head selection techniques, namely, Minimum distance Clustering (MdC), Maximum battery Clustering (MbC), and Minimum cost function Clustering (McC).
In the Minimum distance Clustering (MdC) technique, the cluster head selection is based on the distance between sensor nodes and candidate cluster heads. Each sensor node will select a cluster head which has the shortest distance to the sensor node. There is no limit on the number of member nodes for each selected cluster head.
In the Maximum battery Clustering (MbC) technique, the cluster head selection is based on the residual battery level of candidate cluster heads. Each sensor node will select a cluster head which has the maximum residual battery level. There is no limit on the number of member nodes for each selected cluster head.
In the Minimum cost function Clustering (McC) technique, the cluster head selection is based on the minimum cost function previously defined in equation (1) of section 3. Each sensor node will select a cluster head from the candidate cluster heads which has the minimum cost. There is no limit on the number of member nodes for each selected cluster head.

In our experiments, we consider a wireless sensor network with 100 sensor nodes randomly generated and distributed in a square area of 400 meters by 400 meters. The base station is located in the center of the area. Each node has a transmission range of 120 meters. A link is formed between any pair of nodes within this transmission range. The simulation results of the Limiting member node Clustering (LmC) compared with different clustering techniques are illustrated in Figure 10-13 as follows.

A. The network lifetime

The network lifetime of different cluster head selection schemes is shown in Figure 10. The results show that the proposed Limiting member node Clustering (LmC) algorithm has the longest network lifetime and the Minimum distance Clustering (MdC) algorithm has the shortest network lifetime. The reason is because the LmC algorithm considers the distance, energy usage and residual battery level in the cost function for the cluster head selection while other algorithms select the cluster head by considering only energy usage, battery level or distance separately. Since the cluster heads located in the transmission range of the base station will have heavy load from aggregated data packets which are forwarded to the base station, it is suggested that all parameters including energy usage, battery level and distance should be incorporated in the cost function.

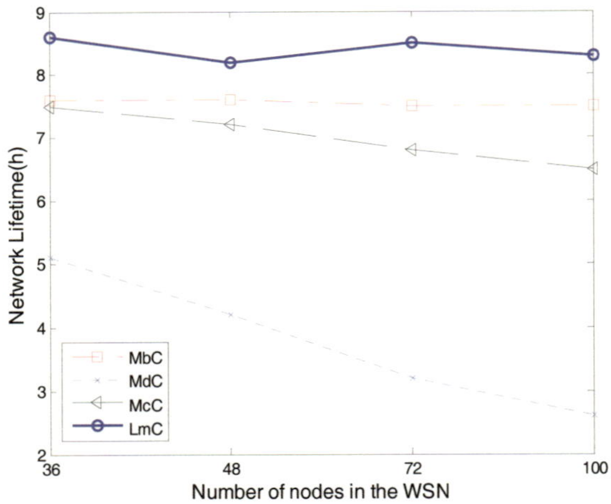

Fig. 10. Comparison of the network lifetime

B. The delay time

Figure 11 shows the delay time of different clustering schemes by varying the number of sensor nodes in the wireless sensor network. It can be seen that the LmC algorithm has the shortest delay time while other algorithms have obviously higher delay time. The reason is because the LmC algorithm can equally balance the number of member nodes for each cluster head. On the other hand, other algorithms select the cluster head based on each parameter constraint which yields a single cluster head in each cycle. Therefore, the single selected cluster head is heavily loaded by aggregated data packets and uses more time to forward those data packets to the base station.

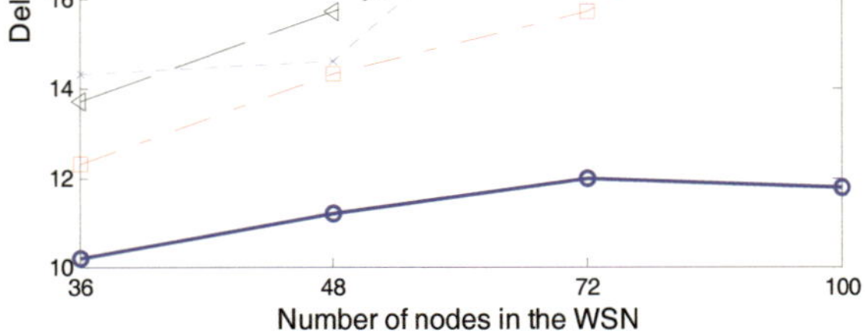

Fig. 11. Comparison of the delay time (s)

C. Number of successfully delivered packets

Figure 12 shows the number of successfully delivered packets for different clustering algorithms. It can be observed that the results of the LmC and MbC algorithms are very close and much higher than the other two methods. Note that the larger number of sensor nodes in the network, a higher number of successfully delivered packets will be attained. The reason is because increasing the number of sensor nodes will also increase the chance to connect with the base station directly and have higher number of candidates for cluster heads.

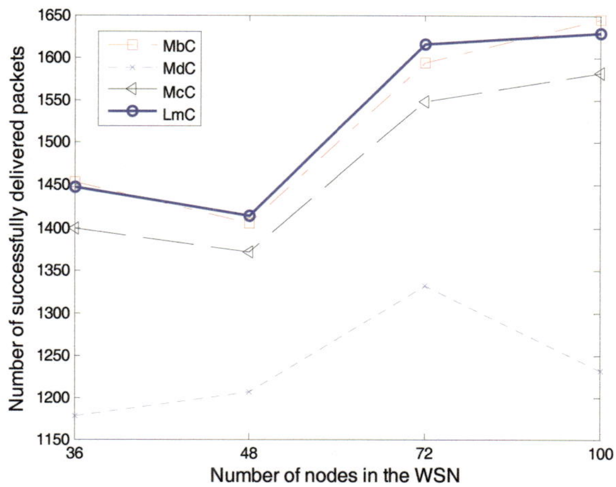

Fig. 12. Comparison of number of successfully delivered packets

D. Average number of packets arrived to the base station per cycle

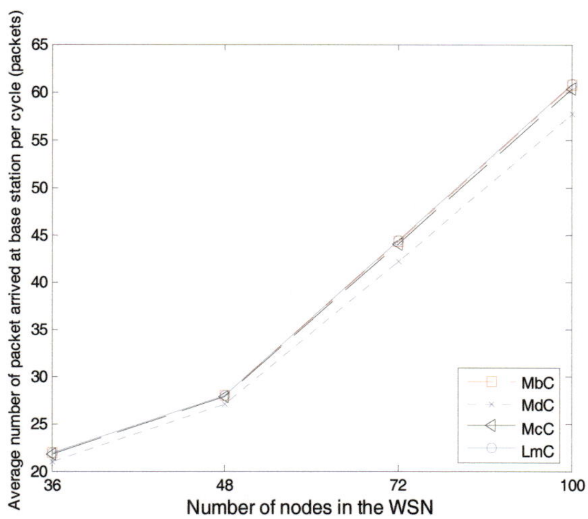

Fig. 13. Comparison of average number of packets arrived to the base station per cycle

Figure 13 compares the average number of packets arrived to the base station per cycle as the number of sensor nodes in the network increases. The results show that the MdC algorithm give the lowest average number of packet arrived to the base station due to this algorithm selects a cluster head based on distance, so the selected cluster head is not changed for static network topology. Therefore, the cluster head with more member nodes will have heavy load and the ability of the cluster head in forwarding packets to the base station is decreased. On the other hand, three remaining algorithms select a cluster head based on a cost function (i.e., energy consumption or battery level) which depends on connection time and energy usage, so cluster head could be changed at each cycle of packets sent. Note that the higher the number of sensor nodes in the network, the higher average number of packets arrived to the base station will be obtained.

4.3 Experimental Results with Transmission Range Extension

In this section, we study the impact of transmission rang control and extension in wireless sensor networks. To evaluate the transmission range control, we consider three scenarios as shown in Figure 14: 1) the base station and each sensor node have a transmission range of 120 meters 2) the base station extends its transmission range to 250 meters while each sensor node has a transmission range of 120 meters 3) the base station and sensor nodes extend their transmission range to 250 meters.

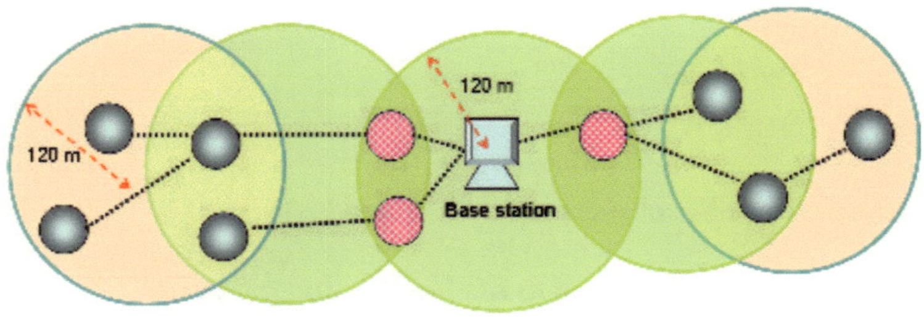

(a) Scenario1: the base station and all sensor nodes have the same transmission range (120m)

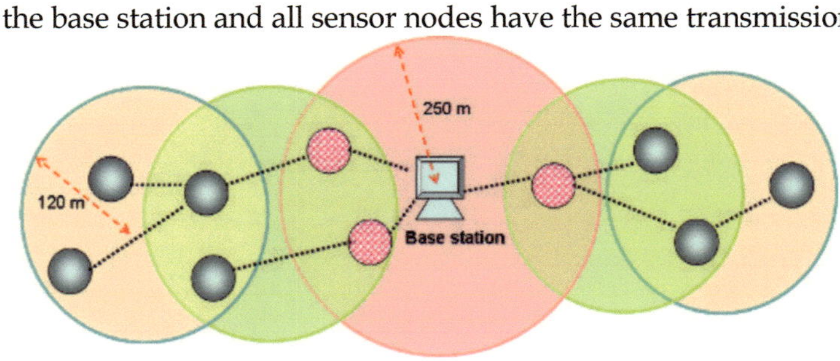

(b) Scenario2: only the transmission range of base station is extended (250m)

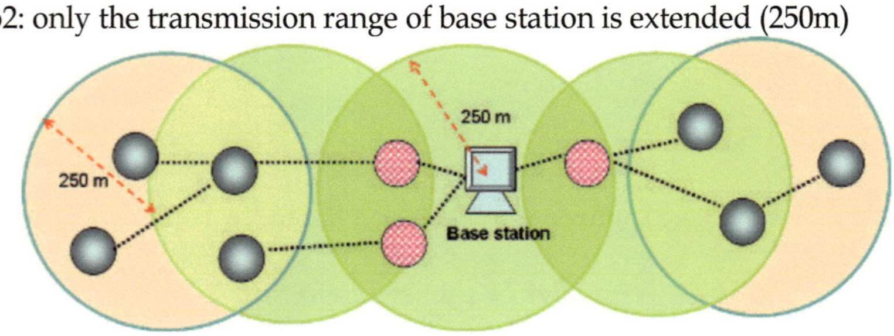

(c) Scenario3: the transmission range of both base station and all sensor nodes are extended (250m)
Fig. 14. WSNs with transmission range control

In our experiments, we compare performances of our proposed Limiting member node Clustering (LmC) with other three clustering techniques, namely, Minimum distance Clustering (MdC), Maximum battery Clustering (MbC), and Minimum cost function Clustering (McC). We conduct experiments in three cases: 1) extending the transmission range of the base station, 2) expanding the network area with the fixed number of sensor nodes, and 3) varying the number of sensor nodes in a fixed area. The simulation results of the three cases are discussed as the following.

4.3.1 Transmission range extension

We consider the impact of extending the transmission range of the base station only by comparing between scenario1 and scenario2.

A. Network lifetime

Figure 15 shows the network lifetime of different clustering techniques using transmission range control for only the base station. It can be observed that all techniques in scenario2 with transmission range extension for the base station have longer network lifetime than scenario1 (without extending the transmission range for the base station). The reason is because extending the transmission range will increase the number of nodes within the base station's transmission range. Therefore, it reduces the amount of aggregated data packets which are forwarded to the base station since nodes can connect with the base station directly.

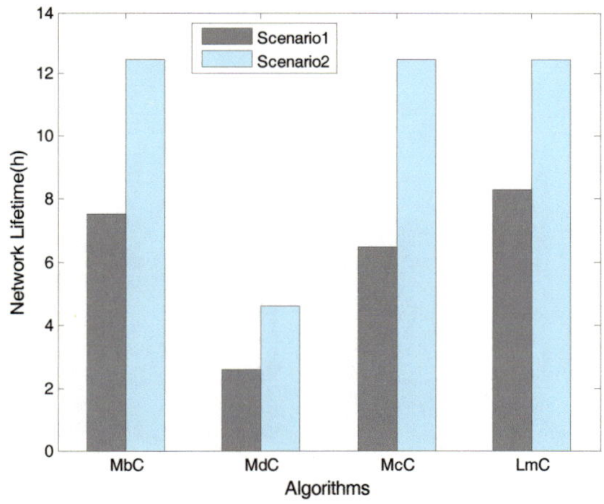

Fig. 15. The network lifetime in scenario1 and 2

B. Delay time

Figure 16 compares the delay time of different techniques. The results show that the extended transmission range of the base station to connect with nodes in "level 1" (scenario2) gives much shorter delay time than the limited transmission range (scenario1). The reason is due to the extension of the transmission range will also increase the number of nodes in "level 1" to connect with the base station directly and reduce the number of member nodes in higher layers.

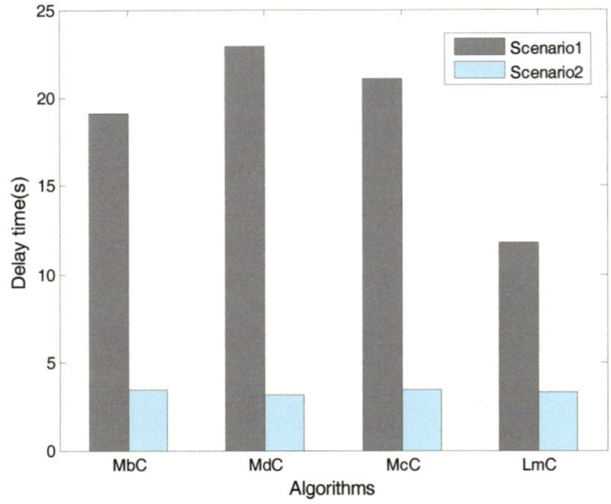

Fig. 16. The delay time in scenario1 and 2

C. Number of successfully delivered packets

Figure 17 compares the number of successfully delivered packets for different algorithms. It can be seen that all algorithms in scenario2 allow more sensor nodes to have direct connectivity with the base station. Therefore, the number of successful packets delivered in the network also increases.

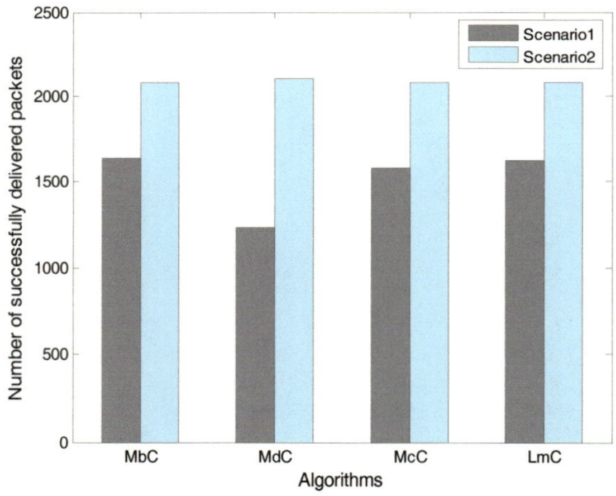

Fig. 17. Number of successfully delivered packets in scenario 1 and 2

4.3.2 Expansion of network area

From the previous case of transmission range control, we found that all clustering techniques perform better when we extend the transmission range of the base station. Therefore, we further extend the transmission range of both the base station and sensor nodes. To study the expansion of network area, the number of sensor nodes is fixed at 100 nodes while the network area is expanded. The simulation results for the scenario2 and scenario3 are compared and discussed as the following.

A. Network lifetime

Figure 18 shows network lifetime of different clustering techniques. It can be observed that, with the area 400x400m (Figure 18a), all techniques in both scenario2 and scenario3 can prolong the network lifetime. However, when we expand the area to 900x900m, the network lifetime is shorter than those in the smaller area. The reason is because in the very large network area, it reduces a chance of sensor nodes to connect with the base station directly. Therefore, each cluster-head has a large number of member nodes and cluster heads near the base station have higher burden to receive and forward data packets. However, when transmission ranges of both the base station and sensors are extended, this can help improving the network lifetime in a large size area (Figure 18b). Note that the Limiting member node Clustering (LmC) technique has the longest network lifetime in a large network area because the proposed technique can balance the number of member node in each cluster head.

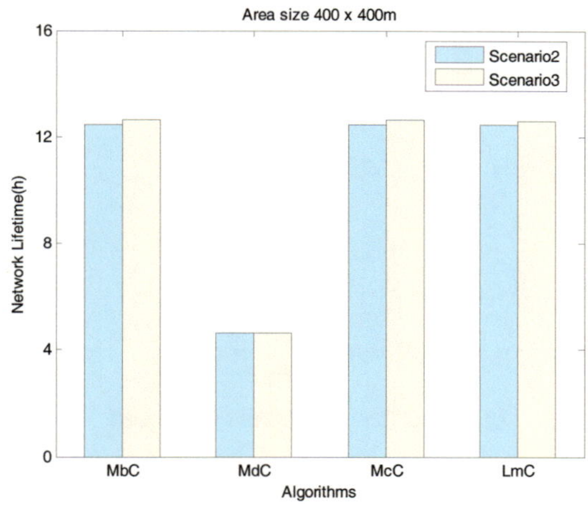

(a) Network area size 400x400m

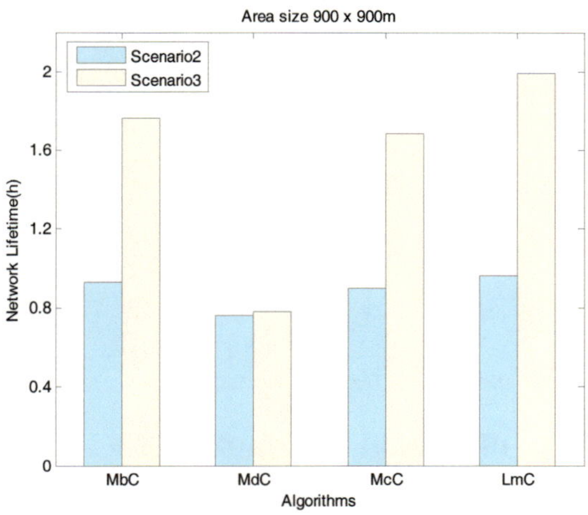

(b) Network area size 900x900m

Fig. 18. The network lifetime with the expansion of network area size

B. Delay time

Figure 19 compares the delay time of different techniques when the network area is expanded. The results show that an extension of the transmission range for both the base station and sensor nodes can reduce the delay time but the expansion of network area increases the delay time. This is because a large number of nodes are in higher levels and there are more packets relayed to the cluster head at each level. Therefore, the cluster heads in "level 1" have higher burden. However, it can be seen that the Limiting member node Clustering (LmC) technique has the shortest delay time while the delay time of other techniques is obviously higher when the size of network area is increased.

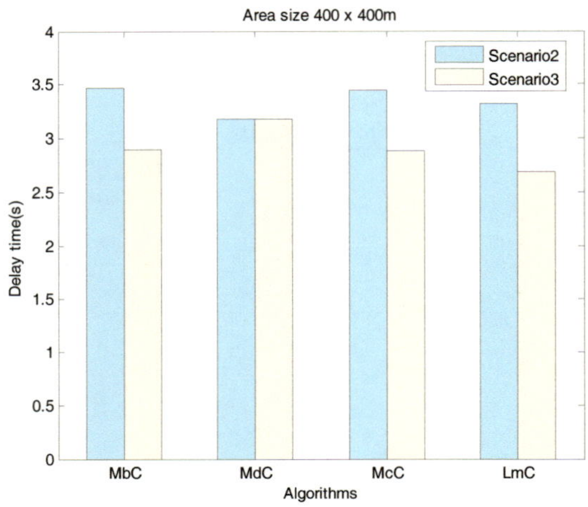

(a) Network area size 400x400m

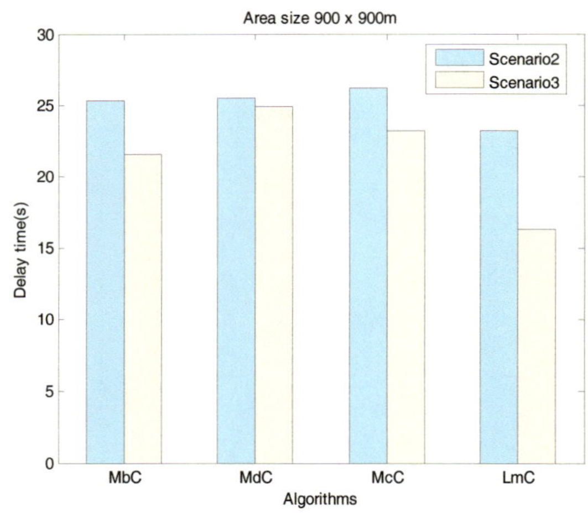

(b) Network area size 900x900m

Fig. 19. The delay time with the expansion of network area size

C. Number of successfully delivered packets

Figure 20 compares the number of successfully delivered packets for different clustering techniques when the network area size is expanded. It can be observed that the number of successfully delivered packets for all clustering techniques is improved due to the transmission range of both the base station and sensor nodes are extended. However, in the larger network area, a lower number of successfully delivered packets will be attained. The reason is because increasing the area size will also reduce the connectivity between sensor nodes in the network. Therefore, it decreases a chance that nodes can connect to the base station directly and have lower number of candidates for cluster heads.

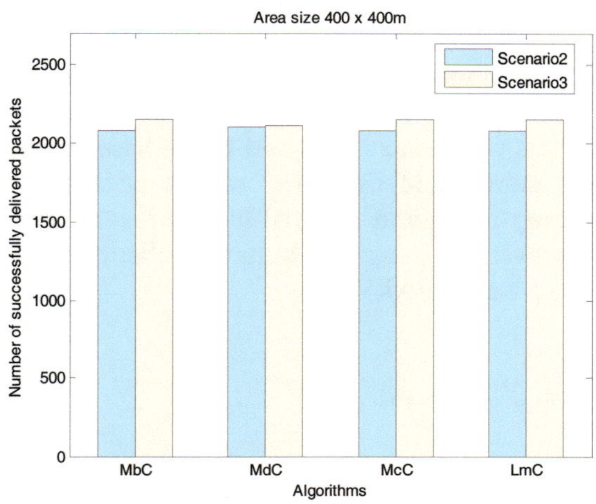

(a) Network area size 400x400m

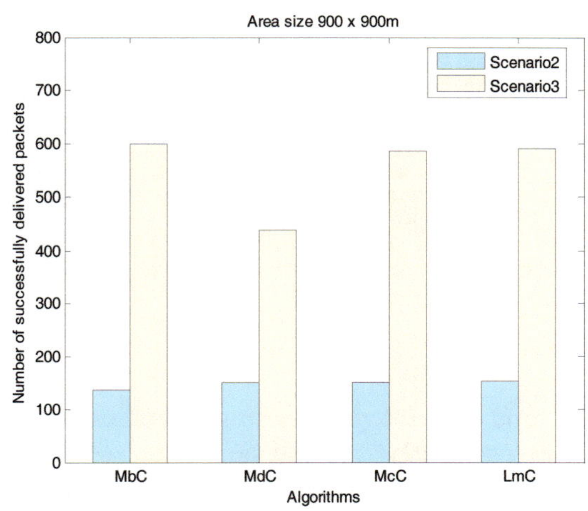

(b) Network area size 900x900m

Fig. 20. The number of successfully delivered packets with the expansion of network area size

4.3.3 Effect of network size

From the simulation results of previous cases discussed above, we found that the performances have been improved in term of the number of successfully delivered packets, the network lifetime and the delay time when we extend the transmission range of both the base station and sensor nodes. To study effect of network size, we vary the number of sensor nodes randomly generated and distributed in a square area of 400 meters by 400 meters. The simulation results of the scenario2 and scenario3 are compared and shown in the following.

A. Network lifetime

Figure 21 compares the network lifetime of clustering techniques for different number of nodes in the network. The results show that Minimum distance Clustering (MdC) has the shortest network lifetime. The reason is because the MdC selects the nearest cluster head so the selected cluster head is often used and the battery level is exhausted quickly. Note that the cluster heads located in the transmission range of the base station will have heavy load from aggregated data packets which are forwarded to the base station. On the other hand, the LmC has the longest network lifetime. The reason is because the LmC technique considers the distance, energy usage and residual battery level in the cost function for the cluster head selection. However, all clustering techniques have improved network lifetime when the transmission range is extended.

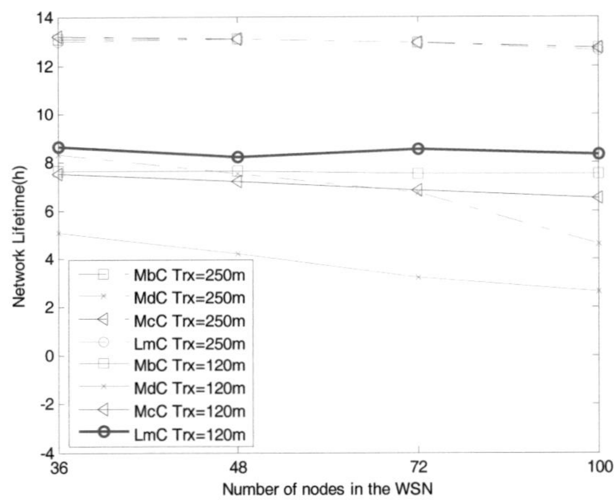

Fig. 21. Network lifetime

B. Delay time

Figure 22 shows the delay time of different clustering techniques by varying the number of sensor nodes in the wireless sensor network. The simulation results show that the LmC has the shortest delay time while other techniques have obviously higher delay time since the transmission range is limited. The reason is because the LmC can equally balance the number of member nodes for each cluster head. On the other hand, other techniques select the cluster head based on each parameter constraint which yields a single cluster head in each cycle. Therefore, the single selected cluster head is heavily loaded by aggregated data packets and uses more time to forward those data packets to the base station.

However, in the case of extending the transmission range to 250m, all techniques have improved delay time to the same level. The reason is because in the small area with the extension of transmission range, most sensor nodes are located within the base station's range so they can connect with the base station directly.

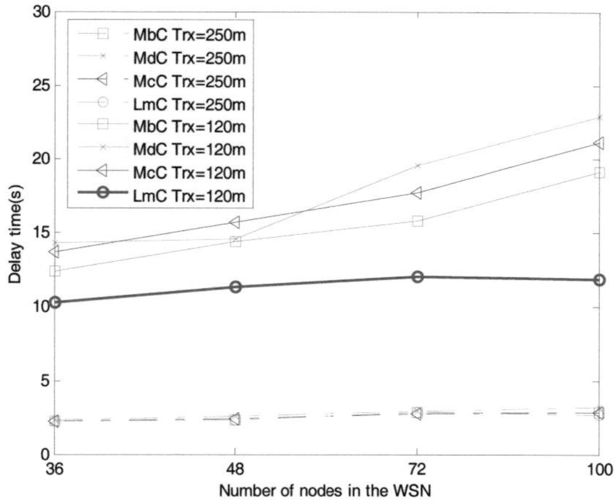

Fig. 22. The delay time

C. Number of successfully delivered packets

Figure 23 shows the number of successfully delivered packets for different clustering techniques. It can be observed that the MdC has less number of successfully delivered packets than the other three techniques. This suggests that the number of successfully delivered packets is related to the network lifetime. Since the MdC cluster head selection based on distance between nodes can not balance the burden of cluster heads, the battery of cluster heads within the base station's range will be exhausted early. Therefore, some packet losses occur at the cluster heads. On the other hand, the LmC can maintain high number of successfully delivered packets.

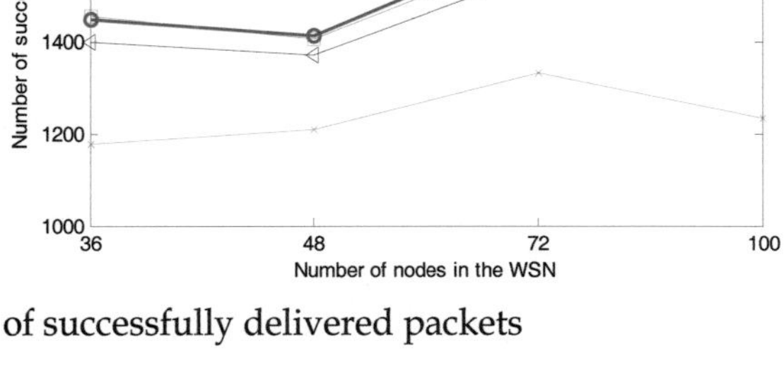

Fig. 23. The number of successfully delivered packets

5. Conclusion

In this chapter, we introduce the background of wireless sensor network and the characteristic of sensor node. A review of routing and clustering algorithms is given. We present a new energy-efficient clustering technique called Limiting member node Clustering (LmC) to balance the burden of each cluster head by limiting the number of member nodes assigned to each cluster head. The proposed LmC technique selects a cluster head based on the cost function which takes residual battery level, energy consumption and distance to the base station into consideration. We also present simulation results to compare the performance of LmC with other three cluster head selection techniques which are Minimum distance Clustering (MdC), Maximum battery Clustering (MbC) and Minimum cost function Clustering (McC). Simulation results show that the proposed limiting member node clustering (LmC) approach can achieve high number of successfully delivered packets as well as the highest network lifetime while give the shortest delay time. Hence, the LmC is an energy-aware clustering technique and capable of providing good performances for cluster head selection in wireless sensor networks.

6. Reference

Yoo S., Kim J., Kim T., Ahn S., Sung J., Kim D.; A2S: Automated Agriculture System based on WSN; Consumer Electronics; 2007. ISCE 2007; Page(s):1 - 5

Galmes S.; Lifetime Issues in Wireless Sensor Networks for Vineyard Monitoring; Mobile Adhoc and Sensor Systems (MASS); 2006 Oct. 2006; Page(s):542 - 545

Guo Y., Corke P., Poulton G., Ark T., Bishop-Hurley G., Swain D.; Animal Behavior Understanding using Wireless Sensor Networks; Local Computer Networks; Proceedings 2006 31st IEEE; Page(s):607 - 614

Xuemei L., Liangzhong J., Jincheng L.; Home healthcare platform based on wireless sensor networks; Technology and Applications in Biomedicine; 2008. ITAB 2008; Page(s):263 - 266

Fariborzi H., Moghavvemi M.; Architecture of a Wireless Sensor Network for Vital Signs Transmission in Hospital Setting; Convergence Information Technology; 2007; Page(s):745 - 749

Arjan D., Mimoza D., Leonard B.; Secure Mobile Communications for Battlefields; Complex Intelligent and Software Intensive Systems, 2008. CISIS 2008; Page(s):205 - 210

Lee S. H., Lee S., Song H., Lee H. S.; Wireless sensor network design for tactical military applications : Remote large-scale environments; Military Communications Conference 2009. MILCOM2009. IEEE; Page(s): 1 - 7

Hongwen X., Qizhi Z.; Wireless Sensors Network Design for Real-time Abrupt Geological Hazards Monitoring; Computer Science and Information Technology; 2008. ICCSIT'08; Page(s):959 - 962

Chayon M., Rahman T., Rabbi M.F., Masum M.; Automated river monitoring system for Bangladesh using wireless sensor network; Computer and Information Technology, 2008. ICCIT 2008.; Page(s):1 - 6

Ibriq J. , Margoub L.; Cluster-Based Routing in Wireless Sensor Networks: Issues and Challenges; SPECTS 2004; Page(s): 759-766

Fedor S., Collier M.; On the Problem of Energy Efficiency of Multi-Hop vs One-Hop Routing in Wireless Sensor Networks; Advanced Information Networking and Applications Workshops, 2007; AINAW '07. 21st International Conference; Page(s): 380 - 385

Jia W., Wang T., Wang G., Guo M.; Hole Avoiding in Advance Routing in Wireless Sensor Networks; Wireless Communications and Networking Conference, 2007.WCNC 2007; Page(s):3519 - 3523

Shen Y., Wu Q., Wang X., Bi H.; Wireless sensor network energy-efficient routing techniques based on improved GEAR; Network Infrastructure and Digital Content, 2009. IC-NIDC 2009. IEEE International; Page(s): 114 - 118

Hu L., Li Y., Chen Q., Liu J. and Long K.; A New Energy-Aware Routing Protocol for Wireless Sensor Networks; Wireless Communications, Networking and Mobile Computing 2007; Page(s): 2444 - 2447.

Wang G., Wang T., Jia W., Guo M., Chen H.-H., Guizani M.; Local Update-Based Routing Protocol in Wireless Sensor Networks with Mobile Sinks; Communications, 2007; Page(s): 3094 - 3099.

Kai L.; A Mine-Environment-Based Energy-Efficient Routing Algorithm for Wireless Sensor Network; Hybrid Intelligent Systems, 2009. HIS '09. Ninth International; Page(s): 215 - 218

Handy M. J., Haase M., Timmermann D.; Low-Energy Adaptive Clustering Hierarchy with Deterministic Cluster-Head Selection; 2002

Younis O., Fahmy S.; HEED:A Hybrid Energy-Efficient Distributed Clustering Approach for Ad-hoc Sensor Networks, Mobile Computing; IEEE Transactions on Volume 3; Issue 4; Oct.-Dec. 2004; Page(s): 366 - 379

Qiu W., Skafidas E., Hao P., Kumar D.; Enhanced tree routing for wireless sensor networks Ad Hoc Networks; Volume 7; Issue 3; May 2009; Page(s): 638-650

Gong B., Li L., Wang S., Zhou X.; Multihop Routing Protocol with Unequal Clustering for Wireless Sensor Networks; Computing, Communication, Control and Management; 2008; ISECS International Colloquium on Volume 2; Page(s): 552 - 556

Dali W., Chan H. A.; Clustering Algorithm to Balance and to Reduce Power Consumptions for Homogeneous Sensor Networks; Wireless Communications; Networking and Mobile Computing, 2007. WiCom 2007; Page(s): 2723 - 2726

Zhang R., Jia Z., Wang L.; A Maximum-Votes and Load-Balance Clustering Algorithm for Wireless Sensor Networks; Wireless Communications, Networking and Mobile Computing, 2008. WiCOM '08. 4th International Conference; Page(s): 1 – 4

Murthy G.R., Iyer V., Radhika B.; Level controlled clustering in wireless sensor networks; Sensing Technology, 2008. ICST 2008. 3rd International Conference 2008; Page(s): 130 -134

Chang J.-H., Tassiulas L.; Maximum lifetime routing in wireless sensor networks; Networking IEEE/ACM Transactions on Volume 12; Issue 4, Aug. 2004; Page(s): 609-619

Muruganathan S.D., Daniel C.F., Bhasin R.I., Fapojuwo A.O.; A centralized energy-efficient routing protocol for wireless sensor networks; Communications Magazine, IEEEVolume 43; Issue 3, March 2005; Page(s): S8 - S13.

Ergen S. C.; ZigBee/IEEE802.15.4 Summary; Sep 2004

EECED: Energy Efficient Clustering Algorithm for Event-Driven Wireless Sensor Networks

Buyanjargal Otgonchimeg[1] and Youngmi Kwon[2]
[1]*Information and Communications Technology and Post Authority (ICTPA)*
[1]*Mongolia*
[2]*Chungnam National University*
[2]*South Korea*

1. Introduction

In recent years, a new wave of networks labelled Wireless Sensor Networks (WSNs) has attracted a lot of attention from researchers in both academic and industrial communities. A WSN consists of a collection of sensor nodes and a base station connected through wireless channels, and can be used for many applications such as military application, building distributed systems, physical environment monitoring, and security surveillance among others. A big advantage of sensor networks is represented by ease of deployment, reducing installation cost, possibility to distribute the tiny sensors over a wide region, and larger fault tolerance (V. Loscri et al., 2005). However, despite the infinite scopes of wireless sensor networks applications, they are limited by the node battery lifetime. Such constraints combined with a typical deployment of large number of sensor nodes have posed many challenges to the design and management of sensor networks and necessitate energy-awareness at all layers of the networking protocol stack (Q. Xue & A. Ganz, 2004). Therefore, energy efficient algorithms have been one of the most challenging issues for WSNs.

Sensor nodes can be in one of four states, namely transmit, receive, idle and sleep. The largest part of a node's energy is consumed while transmitting and receiving. Minimizing the number of communications by eliminating or aggregating redundant sensed data saves much amount of energy (L. B. Ruiz et al., 2003). Among these clustering sensor networks are a very attractive approach because clustering allows for scalability, data aggregation, and energy efficiency. In a clustering network, nodes are grouped into clusters and there are special nodes called cluster head. They are responsible for an efficient way to lower energy consumption within a cluster by performing data aggregation. In a heterogeneous sensor network, two or more different types of nodes with different battery energy and functionality are used. On the other hand, in homogeneous networks all the sensor nodes are identical in terms of battery energy and hardware complexity. As a result, network performance decreases since the cluster head nodes goes down before other nodes do. Thus dynamic, energy efficient and adaptive cluster head selection algorithm is very important.

Sensor networks can be divided in two classes as event-driven and continuous dissemination networks according to the periodicity of communication (L. B. Ruiz et al.,

2004). In continuous dissemination networks, the sink is interested in the conditions of the environment at all times and every node periodically sends data to the sink. In event-driven sensor networks, the sink is only interested in hearing from the network when certain events occur. For example, if the application is temperature monitoring, it could be possible just to report data when the temperature of the area being monitored goes above or below certain thresholds. Configuring the network as event-driven is an attractive option for a large class of applications since it typically sends far fewer messages (C.Intanagonwiwat et al., 2000). This is translated into significant energy saving, since message transmissions are much more energy intensive when compared to sensing and (CPU) processing. Also some existing energy-saving solutions take that into consideration and switch some nodes off, leading the nodes to an inactive state, these are waken up only when interest matches the events "sensed" (J.N.Al-Karaki & A.E.Kamal, 2004). Therefore, event driven protocols are used to conserve the energy of the sensor nodes.

In general, routing in WSNs can be divided into flat-based routing, location-based routing, and hierarchical-based routing depending on the network structure. In flat-based routing, all nodes are typically assigned equal roles or functionality. In location-based routing, sensor nodes' positions are exploited to route data in the network. In hierarchical-based routing, however, nodes will play different roles in the network [8]. Many energy efficient hierarchical or cluster based routing protocols have been proposed in sensor networks for different scenarios and various applications (A. Abbasi & M. Younis, 2007). However, most protocols in the previous literatures have not been considering event driven WSNs and, their focus is on continuous networks. Therefore in this work we focus on energy efficient clustering algorithm for event-driven wireless sensor network. In order to extend the lifetime of the whole sensor network, energy load must be evenly distributed among all sensor nodes so that the energy at a single sensor node or a small set of sensor nodes will not be drained out very soon.

Low Energy Adaptive Clustering Hierarchy (LEACH) is one of the most popular clustering algorithms for WSNs (W. Heinzelman et al., 2000). LEACH guarantees that the energy load is well distributed by dynamically created clusters, using cluster heads elected dynamically according to predetermined optimal probability variable. The rotation is performed by getting each node to choose a random number between 0 and 1. A node becomes a CH for the current rotation round if the number is less than the following threshold:

$$T(n) = \begin{cases} \dfrac{p}{1 - p * \left(r \bmod \frac{1}{p} \right)} & \text{if } n \in G \\ 0 & \text{otherwise} \end{cases} \tag{1}$$

where p is desired percentage of cluster head nodes in the sensor network, r is current round number, and G is the set of nodes that have not been cluster heads in the last 1/p rounds. As long as optimal energy consumption is concerned, it is not desirable to select a cluster head node randomly and construct clusters. However, repeating round can improve total energy dissipation and performance in the sensor network. LEACH has some shortcomings: Firstly, remaining energy of sensor nodes is not considered to construct clusters. The choice of probability for becoming a cluster head is based on the assumption that all nodes start with an equal amount of energy, and that all nodes have data to send during each frame. Accordingly they are hardly applied to the real applications. In real environment, usually non-uniform energy drainage exists due to different distances between sensor and sinks, different quantity of transmission messages and different transmission rate. If nodes have different amounts of energy, then the nodes with more energy should be cluster heads more often than the nodes with less energy, to ensure that all nodes die approximately at the same time. Some researches present a good solution to reduce energy dissipation using cluster head selection algorithm based on sensors' residual energy. But, in many cases, each node sends information about its current location and energy level to the BS. The BS needs to ensure that the energy load is evenly distributed among the all the nodes (Vinh Tran Quang & Takumi Miyoshi, 2008). Another approach is the BS selects cluster head nodes depending on the number of clusters alive in the network (Giljae Lee et al., 2008). Secondly, LEACH does not guarantee the number of cluster head nodes and their distribution because the cluster head nodes are selected stochastically by the value of probability. The different cluster numbers in WSNs will make the node numbers in every cluster different and uneven cluster numbers dissipate uneven energy in each round (Tung-Jung Chan, 2008). In this paper, by applying the optimal cluster numbers to the WSNs, the lifetime of WSNs can be extended very well.

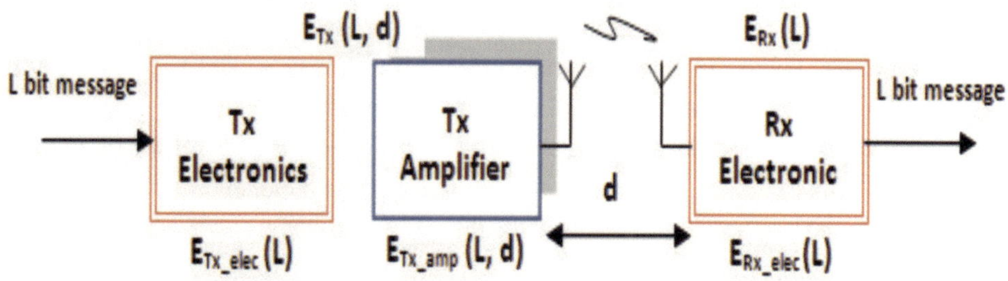

Fig. 1. Radio energy dissipation model

2. Sensor network models

2.1 Network Model

In this work we assume a sensor network model with following properties:

- The sink locates at the centre of sensor nodes and has enough memory and computing capability.
- Sink node is assumed to know all the node locations.
- All sensor nodes are immobile and have a limited energy.
- All nodes are equipped with power control capabilities to vary their transmitting power.
- Also we assume event-driven protocol architecture.

2.2 Radio Model

For the purpose of this study, we use the same condition in LEACH with the simple model for the radio hardware energy dissipation, as a shown Fig.1. L is the number of bits per packet transmission and d is distance between the sender and the receiver. Electronics energy consumption is same for transmitting and receiving the data, is given by,

$$E_{Tx-elec}(L) = E_{Rx-elec}(L) = E_{elec} * L \qquad (2)$$

E_{elec} is the energy dissipated per bit to run the transmitter or the receiver circuit. Transmission cost to transmit L-bit message between any two nodes over distance d is given by the following equation:

$$E_{Tx}(L, d) = E_{Tx-elec}(L) + E_{Tx-amp}(L, d) \tag{3}$$

E_{Tx-amp} (L, d) is the amplifier energy consumption and it can be further expressed in terms of ε_{fs} or ε_{mp}, depending on the transmitter amplifier mode that applied. They are power loss factors for free space (d^2 loss) when $d < d_0$; and multipath fading (d^4 loss) when $d \geq d_0$, respectively. The threshold d_0 can be determined by equating the two expressions, resulting:

$$d_0 = \sqrt{\varepsilon_{fs}/\varepsilon_{mp}} = 87.7\text{m} \tag{4}$$

Thus, to transmit L-bit message within d distance, a node expends:

$$E_{Tx}(L, d) = L*(E_{elec} + \varepsilon_{fs}*d^2) \qquad \text{if } d < d_0 \text{ or}$$

$$E_{Tx}(L, d) = L*(E_{elec} + \varepsilon_{mp}*d^4) \qquad \text{if } d \geq d_0 \tag{5}$$

To receive L-bit message within d distance, a node expends:

$$E_{Rx}(L) = E_{Rx-elec}(L) = E_{elec}*L \tag{6}$$

2.3 Optimal Fixed Number of Cluster

Suppose that there are N sensor nodes randomly deployed into an M x M region. In the k clusters WSN, the squared distance from the nodes to the cluster head is given by (W. Heinzelman et al., 2000):

$$E[d_{toCH}^2] = \frac{M^2}{2\pi k} \tag{7}$$

If assumed that M=100 and the base station locates centre of sensing area, then maximum distance of any nodes from the base station is approximately 70m. Thus, from (4), every time d_{toBS} and d_{toCH} are less than d_o.

Hence, using (5) and (6) the energy consumption for each cluster head, E_{CH}, and energy consumption for non cluster head, E_{nonCH}, can be obtained by:

$$E_{CH} = L\left(\frac{N}{k} - 1\right)E_{elec} + L\frac{N}{k}E_{DA} + LE_{elec} + L\varepsilon_{fs}\, d_{toBS}^2 \tag{8}$$

$$E_{nonCH} = LE_{elec} + L\varepsilon_{fs}\, d_{toCH}^2 \tag{9}$$

respectively, and E_{DA} represents the processing (data aggregation) cost of a bit per signal and L is length of data message. Also we assumed that clusters are equally sized, thus there are average N/k nodes per clusters and (N/k) - 1 non cluster head nodes.

The energy dissipated in a cluster per round, $E_{cluster}$, is expressed by:

$$E_{cluster} = E_{CH} + \left(\frac{N}{k} - 1\right) E_{nonCH} \tag{10}$$

Therefore, the total energy dissipated in the network per round, E_{rnd} , is expressed by:

$$E_{rnd} = k * E_{cluster} = L\left(2NE_{elec} + \varepsilon_{fs}\left(kd_{toBS}^2 + Nd_{toCH}^2\right)\right) \tag{11}$$

By (7) and (11), we can find the optimal cluster number k given by (Tung-Jung Chan, 2008):

$$\frac{\partial E_{rnd}}{\partial k} = 0$$

$$k_{opt} = \sqrt{\frac{N}{2\pi}} \frac{M}{d_{toBS}^2} = \sqrt{\frac{N}{2\pi}} M = \sqrt{N} \tag{12}$$

3. Energy Efficient Clustering for Event-Driven (EECED) Protocol Architecture

Our modified protocol called "Energy Efficient Clustering Algorithm for Event-Driven Wireless Sensor Networks (EECED)" is aimed at prolonging the lifetime of a sensor network by balancing energy usage of the nodes. EECED makes the nodes with more residual energy have more chances to be selected as cluster head. Also, we use elector nodes which take the responsibility of collecting energy information of the nearest sensor nodes and selecting the cluster head. We compared the performance of our EECED algorithm with the LEACH protocol using simulations.

EECED involves three main steps; the initial phase, the clustering phase and the data transmission phase. The initial phase is performed only once at the beginning of network operation. Similar with LEACH, the operation of EECED is divided into round, where each round consists of the clustering phase and the data transmission phase. Each round begins with clustering phase when the clusters are organized, followed by a data transmission phase when data are transferred from the nodes to cluster head and on to the base station (BS). In the following sub-sections we discuss each of these phases in details.

3.1 Initial Phase

The sink selects k_{opt} number of elector nodes using (12), then the sink broadcasts an elector advertisement message (ELEC_ADV) in initial phase, as shown in Fig. 2.

3.2 Clustering Phase

The clustering phase is similar with LEACH protocol, involving cluster head selection part and cluster construction part. During the cluster selection phase, elector node determines CH based on the residual energy of the node. Those with higher residual energy have the advantage during the CH competition. Clusters are created by non-CH nodes choosing to join a CH based on the signal strength of advertisements received from CHs. It has been shown that our protocol reduces energy consumption and improves network lifetime compared to probability schemes.

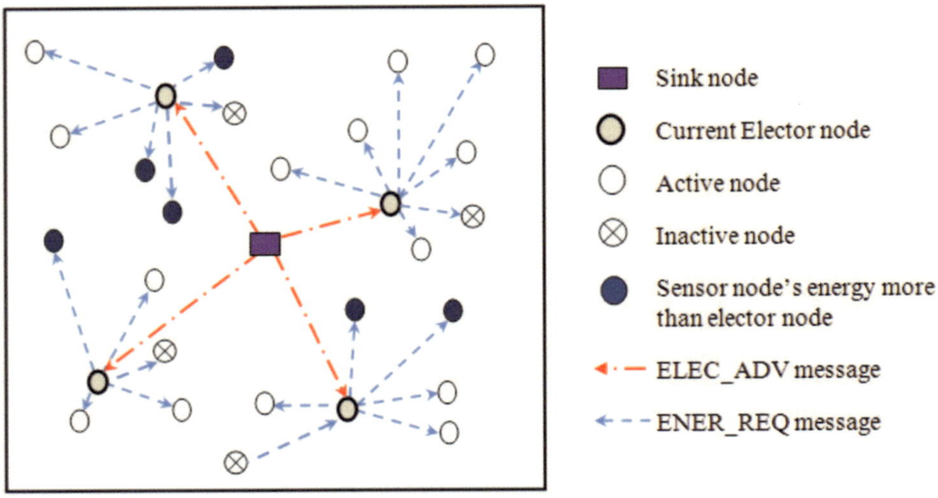

Fig. 2. Initial phase and Energy Request message transmission

3.2.1 Cluster Head Selection

To reduce global communication, a cluster-head may implement one or more optimization functions such as data fusion and transmits to more distant cluster-heads. In a homogeneous network, cluster head uses more energy than non cluster head nodes. As a result, network performance decreases since the cluster head nodes goes down before other nodes do. Clustering schemes have to ensure that energy dissipation across the network should be balanced and the cluster head should be rotated in order to balance the network energy consumption.

Our protocol uses dynamic CH selection algorithm based on higher residual energy. In this part, elector nodes take responsibility for collecting nearest sensors' energy information and selecting cluster head.

- When normal node knows that it has become elector node, then it broadcasts energy request message (ENER_REQ) with its own energy level information to its surrounding nodes, as shown in Fig. 2.
- The normal nodes first compare its own energy level with energy level of most nearest elector node.
- If normal node's energy level is greater than elector node, it sends energy reply (ENER_REP) message, otherwise it waits for cluster head advertisement (CH_ADV) message, as a shown Fig. 3.
- Elector node selects cluster head with maximum residual energy and next elector node with second maximum residual energy.
- Elector node becomes available to become cluster head if its energy is greater than others.

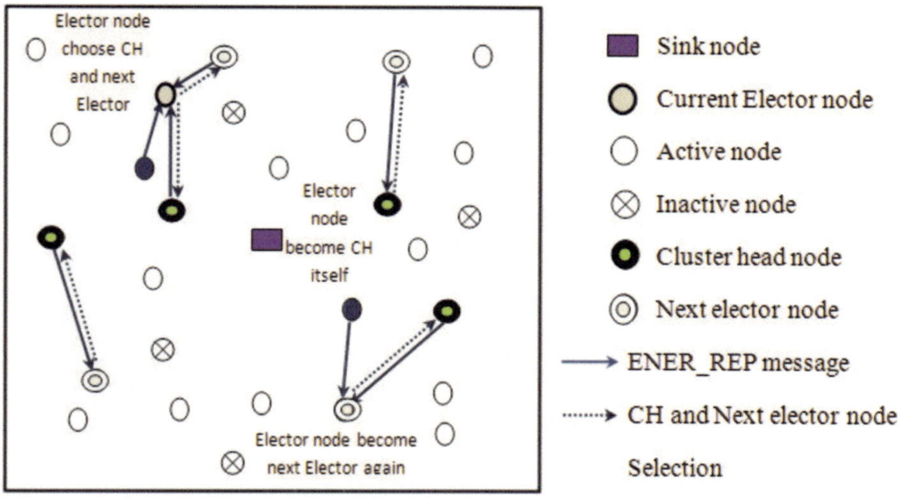

Fig. 3. Cluster head and next elector node selection

3.2.2 Cluster Construction

- After cluster head is selected by elector node, the cluster head node broadcasts a cluster head advertisement message (CH_ADV) containing cluster head ID.
- Non-cluster head sensor nodes then select the most relevant cluster head node according to the signal strength of the advertisement message from the cluster head nodes. Each member node transmits a join request message (JOIN_REQ), as shown in Fig. 4.

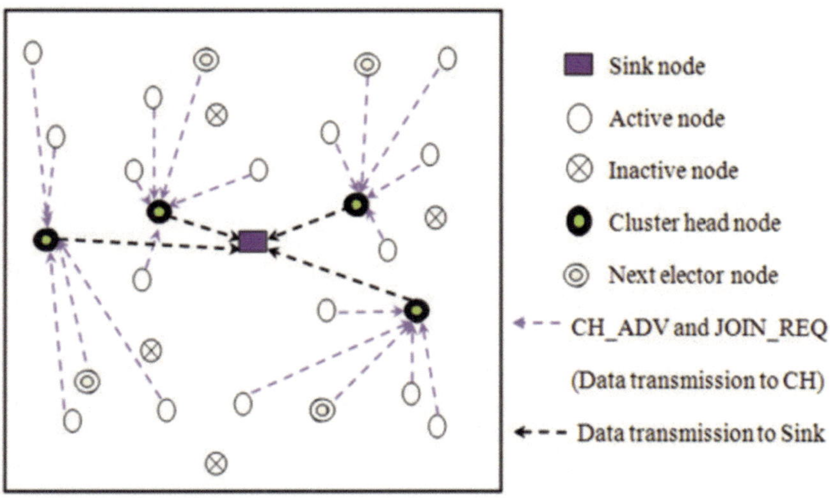

Fig. 4. Cluster construction and Data transmission

Our proposed protocol has some overhead of control message exchanges. To ensure minimizing energy consumption of broadcast messages, we use optimal transmission radius, minimum message length and minimum number of control messages. Elector nodes take responsibility of collecting nearest sensors' energy information, so transmission range is minimized. Also control message size is small containing a header that distinguishes this message type. Some nodes which have energy greater than energy level of elector node ensure minimization of ENER_REP message transmission. Control messages use a non persistent CSMA MAC protocol to avoid collision. Message types and message field size are shown in Table1 and Table2 respectively. Also power consumption can be reduced by assigning the lowest necessary transmission power to the nodes in networks where the nodes exchange control message and are able to adjust their transmission power.

Source ID	Dest ID	Seq ID	Flag	RSSI value	Energy level	Spec info
2 byte	2 byte	2 byte	1 byte	2 byte	2 byte	Variable

Table 1. Control message field size

Flag	Control Message type
1	ENER_REQ (Energy Request)
2	ENER_REP (Energy Reply)
3	CH_ADV (Cluster Advertisement)
4	JOIN_REQ (Join Request)

Table 2. Control and data message type

3.3 Data transmission Phase

In data transmission phase, the cluster head nodes act as local control centers to coordinate data transmission in their cluster, as shown in Fig.4.

Once the selected cluster head node receives the JOIN_REQ message from member nodes, the cluster head set up a TDMA schedule according to their active member nodes. The function of the schedule is to avoid the collision on data transmission and to keep the synchronization among all the nodes within the cluster. Active member nodes wait for receiving the TDMA schedule from the cluster head. Meanwhile, they can turn their radio components off except for their own transmission period. Active sensor nodes exchange their matching data without collision. Inactive nodes go to sleep mode until next round. When cluster heads have aggregated data from their active nodes, they send it to BS directly, because we assume all the nodes are equipped with power control capabilities to vary their transmitted power.

Once the data transmission phase ends, network reforms the cluster head selection procedure in a new round, as shown in Fig.5. For distributed energy dissipation next elector nodes will become elector nodes since the elector nodes dissipate little more energy than normal nodes. Here, we assumed no emergency information occurs in this application.

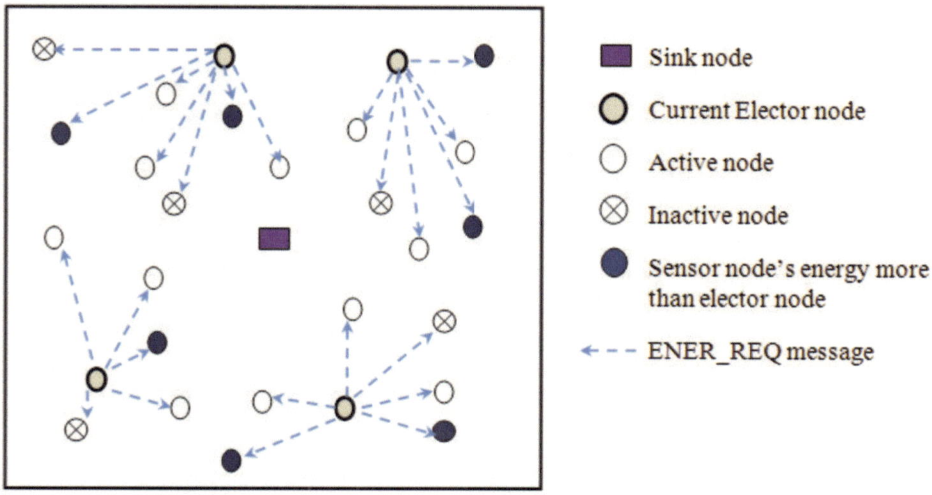

Fig. 5. Condition of next new round

4. Performance evaluation

4.1 Energy Consumption and Simulation Parameters

In this simulation, energy is decreased whenever a node transmits or receives data and whenever it performs data aggregation. We don't decrease energy during carrier-sense operations. For simplicity, we assume that the maximum distance of any node to the cluster head is ≤ d. From (9), we calculated energy dissipation in normal node during a round is given by the following formula:

$$E_{normal} = L_{data} (E_{elec} + \varepsilon_{fs} \, d_{toCH}^2) \quad (13)$$

According to (5) and (8), the energy used in each cluster head node is equal to:

$$E_{CH} = L_{data} (N_{normal} (E_{elec} + E_{DA}) + \varepsilon_{fs} \, d_{toBS}^2) \quad \text{or}$$

$$E_{CH} = L_{data} (N_{normal} (E_{elec} + E_{DA}) + \varepsilon_{mp} \, d_{toBS}^4) \quad (14)$$

where N_{normal} is the number of member nodes in the cluster, E_{DA} is the processing (data aggregation) cost of a bit per signal and L_{data} is length of data message. We do not assume any static energy dissipation, but we calculated energy dissipation of some control messages containing energy level. The energy consumed in each elector node is equal to:

$$E_{elector} = L_{broad} (E_{elec} + \varepsilon_{fs} \, d_{toCH}^2 + E_{elec} \, N_{res}) \quad (15)$$

where L_{broad} is constant and small size of control message and N_{res} is the number of response nodes whose energy level is greater than energy level of elector node. Table 3 shows simulation parameters.

Notation	Description	Value
M x M	Area	100x100
N	Number of the sensors	100 and 300
sinkX, sinkY	Sink node location	50x50
E 0	Initial energy	0.2 J and 0.1J -0.3J
E_{elec}	Electronics energy	50nJ/bit
E_{DA}	Energy of data aggregation	5nJ/bit
d_0	The threshold distance	87m
ε_{fs}	Amplified transmitting energy using free space	10pJ/bit/ m 2
ε_{mp}	Amplified transmitting energy using multipath	0.0013pJ/bit/ m4
L_{data}	Data packet size	500bytes
L_{broad}	Broadcast packet size	25 bytes
p	Probability	0.1

Table 3. Simulation parameter

4.2 Result analysis

We simulated the performance of our EECED algorithm compared to LEACH protocol. We considered both equal initial energy (0.2J) and different initial energy (0.1-0.3J) in each node. Our performance was measured by total residual energy per round, total number of received packet at the BS, network lifetime, total number of nodes alive in the network, and round of first died node. In this simulation, network lifetime is defined as the round interval from the start of operation until death of the last alive node, and the first node died round is defined as the round interval from the start of the network operation until the death of the first sensor node.

Fig. 6 shows the total number of nodes alive per round when total number of sensors in the network varies from 100 to 300, and initial energy is different from each other. From Fig.6, it can be seen that network lifetime and round of the first died node of our algorithm are longer than LEACH. Also EECED shows that the round of the first died node and network lifetime staysare the same regardless of the increase in network size. In case of LEACH protocol, round of the first died node is very fast and linearly decreases until last round. Simulation proves that our algorithm can balance the energy consumption of the entire network compared to LEACH protocol.

Fig. 7 shows the number of nodes alive per round in case of initial energy being same. The Fig. explains that in case of initial energy being the same with all of sensors, EECED performance of round first node is similar with LEACH, but network lifetime is greater than LEACH in case of large network.

Fig. 8 shows number of packets received at the BS per round in the case of total number of nodes being 100 and initial energy being different from each other. From Fig. 8, our targeted algorithm has better performance of data transmission.

The total residual energy per round in case of having total number of nodes as 300 and in case of different initial energy can be seen from the Fig. 9, whereas if initial energy is different, EECED performs better than LEACH.

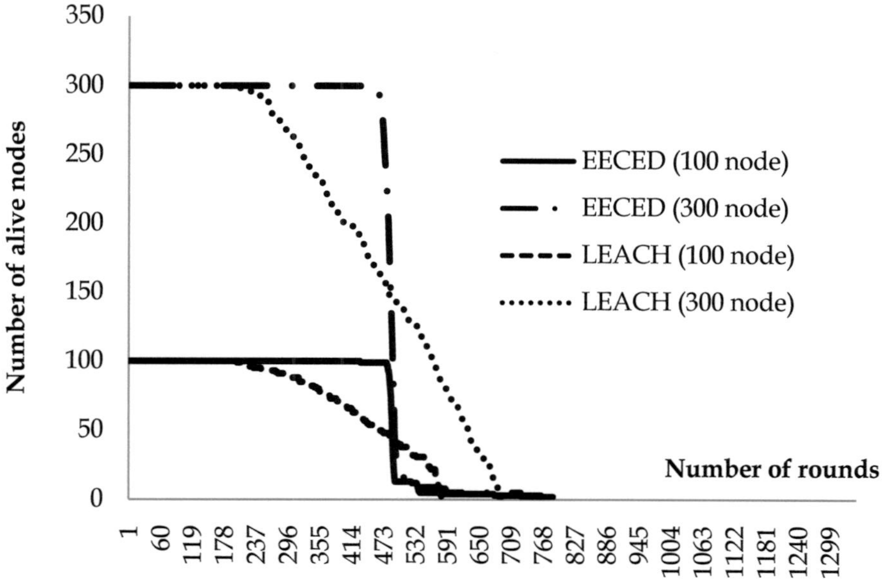

Fig. 6. Number of living nodes in each round with different initial energy is used and total number of nodes 100 and 300

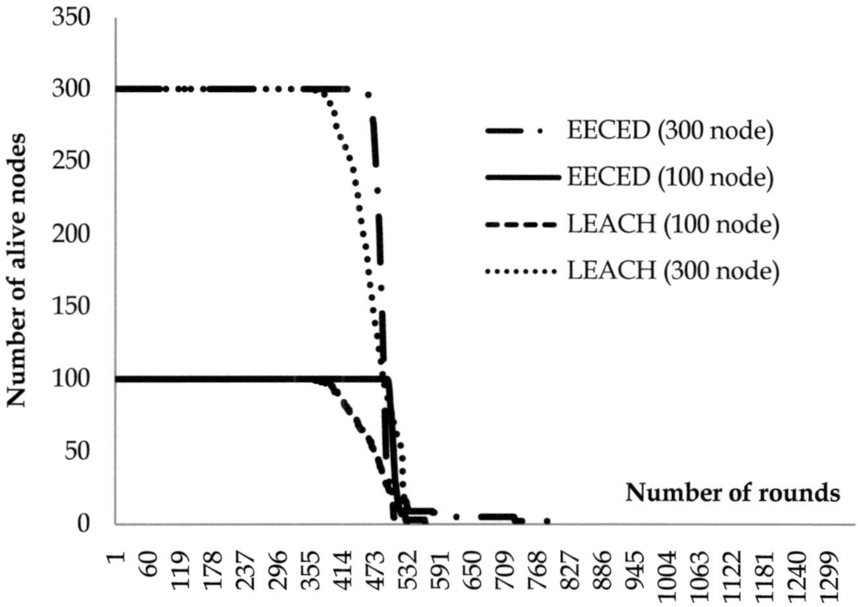

Fig. 7. Number of living nodes in each round with same initial energy is used and total number of nodes 100 and 300

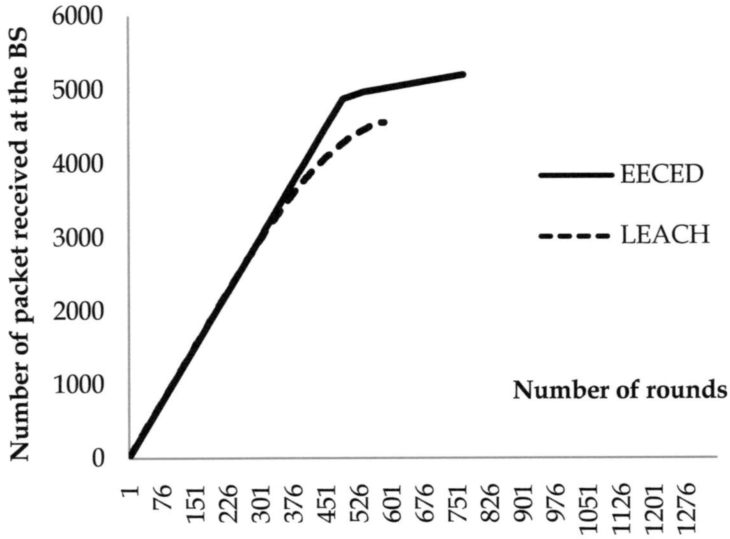

Fig. 8. Number of packet received at the BS (different initial energy and total number of nodes 100)

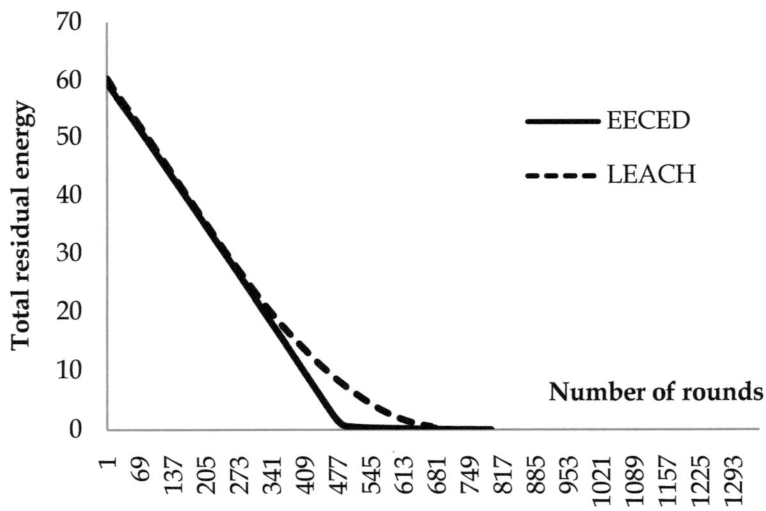

Fig. 9. Total residual energy per round (different initial energy and total number of nodes 300)

Also we simulated the performance changes in large network. The simulation result shows, that the network lifetime decrease rapidly in large area network and the period that the first dead node appears is earlier than those of previous cases. The phenomenon is caused by the fact that the cluster heads waste the considerable amount of energy for transmitting their data to the far away base station. Because in these scheme, all cluster heads transmits aggregated data to the BS directly. In direct transmission, nodes far away from the BS dissipate their energy much faster than those close to the BS, therefore some nodes drained out very soon. We simulated the performance changes of EECED as the size of sensing field increases. The network lifetime when total number of nodes is 100 is shown in Fig. 10. It shows, if size of sensing field increases, the network lifetime of EECED decreases rapidly.

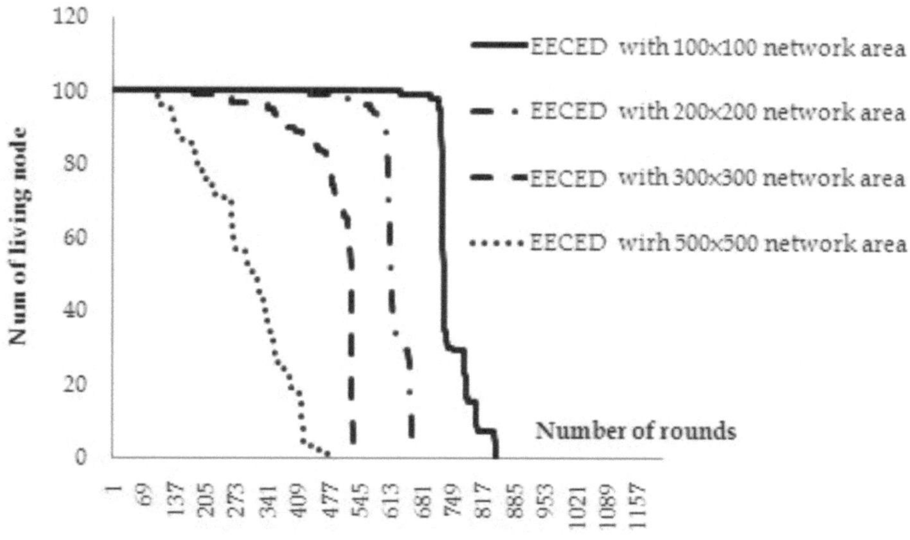

Fig. 10. Number of living nodes in each round with the increase in network

5. Conclusion

Routing in sensor networks is very challenging due to several characteristics that distinguish them from traditional communications and wireless ad-hoc networks since several restrictions, e.g., limited energy supply, computing power, and bandwidth of the wireless links connecting sensor nodes. The major difference between the WSN and the traditional wireless network is that sensors are very sensitive to energy consumption. Introducing clustering into the networks' topology has the goal of reducing the number of message that need to be delivered to the sink in large-scale WSNs.

We proposed an "Energy Efficient Clustering Algorithm for Event-Driven Wireless Sensor Networks (EECED)" to extend the network lifetime of a sensor network by balancing energy usage of the nodes. AEEC improved the energy efficiency of WSNs:

- CHs have higher burdens than member nodes; therefore, rotating the role of the CH shares the burden and thus extending the useful lifetime of those clusters.
- If nodes have different amounts of energy, then the nodes with more energy should be cluster heads more often than the nodes with less energy.

We showed that in many cases our algorithm is more energy efficient than LEACH. The results show that the proposed algorithms can maintain a balanced energy consumption distribution among nodes in a sensor network and thus prolong the network lifetime.

WSNs are increasingly being used for event-driven communications ranging from health care, transportation, manufacturing, and much more. In these kinds of applications, the energy usage is different on all sensors. Cluster-based routing protocols are used in the event driven model, considerable energy can be saved. Our approach is well suitable for the event-driven application in WSNs, because in event-driven sensor network applications, events occur randomly and transiently, and accompanied by the bursts of large numbers of data, therefore, network energy consumption is uneven.

6. References

V. Loscri, S. Marano, G. Morabito (2005). A Two-Levels Hierarchy for Low-Energy Adaptative Clustering Hierarchy (TL-LEACH). *Proceedings of VTC2005,* pp. 1809-1813, Dallas USA, Sep 2005

Q. Xue, A. Ganz. (2004). Maximizing Sensor Network Lifetime: Analysis and Design Guides, *Proceedings of IEEE. MILCOM,* Oct 2004.

K. Intae and R. Poovendran.(2003). Maximizing static network lifetime of wireless broadcast ad hoc networks, *Proceedings IEEE International Conference on Communications,* pp. 2256–2261, May 2003.

L. B. Ruiz, A. A. F.Loureiro and J. M.Nogueira. (2003). Functional and information models for the MANNA architecture. *Proceedings of CFGRS,* pp. 455–470, Feb 2003.

L.B. Ruiz, I.G. Siqueira, L.B. Oliveira, H.C. Wong, J.M.S. Nogueira, A.A.F. Liureiro (2004). Fault management in event-driven wireless sensor networks. *Proceedings of MSWIM'04,* 2004.

C.Intanagonwiwat, R.Govindan, and D.Estrin. (2000). Directed Diffusion: A Scalable and Robust communication Paradigm for Sensor Networks. *Proceedings of 6th ACM/IEEE MOBICOM,* 2000.

J.N.Al-Karaki, and A.E.Kamal. (2004). Routing techniques in wireless sensor networks: A Survey. *IEEE Wireless Communications,* Vol 11, No 6, 2004, pp. 6-28.

A. Abbasi and M. Younis. (2007). A survey on clustering algorithms for wireless sensor networks. *Computer Communications,* Vol. 30, 2007, pp. 2826-2841

W. Heinzelman, A. Chandrakasan and H. Balakrishnan. (2000). Energy-Efficient Communication Protocol for Wireless Microsensor Networks. *Proceedings of HICSS '00,* Jan 2000.

Vinh Tran Quang and Takumi Miyoshi. (2008). Adaptive routing Protocol with Energy efficient and Event Clustering for Wireless Sensor Network. *Transactions on Communications,* Vol. E91-B, No. 9, Sep 2008, pp. 2795-2805

Giljae Lee, Yoonjoo Kwon, Woojin Seok and Minsun Lee. (2008). A Simple dynamic clustering approach to achieve energy efficiency for wireless sensor networks. *International journal of Pervasive Computing and Communications,* Vol 4, No 2, 2008, pp. 185-197.

Tung-Jung Chan, Ching-Mi Chen, Yung-Fa Huang, Jen-Yung Lin and Tair-Rong Chen. (2008). Optimal cluster number selection in Ad-hoc Wireless Sensor networks. *WSEAS Transaction on Communications,* Issue 8, Vol 7, Aug 2008

Guitang Wang, Honglei Zhu, Hui Dai, Liming Wu and Banghong Xiong. (2009). The Clustering Algorithm of Wireless Sensor Networks Based on Multi-Hop between Clusters. *Proceedings of World Congress on Computer Science and Information Engineering,* Vol 03, 2009

10

Topology Control and Routing in Large Scale WSNs

Ines Slama
Telecom SudParis, Institut Telecom
Evry, France

1. Introduction

Achieving maximum lifetime in stationary WSNs by optimally using the energy within sensor nodes has been the subject of significant researches in the last recent years. In this field and as alredy stated, radio transmission and reception operations are being identified as one of the most energy consuming features.

On the other hand, the development of large-scale sensor networks has drawn a lot of attention. Indeed, according to the application, the number of sensor nodes deployed to sense a specific phenomenon may be in the order of hundreds or thousands and can reach a value of millions. Therefore, the large size of wireless sensor networks inevitably introduces significant scalability concerns. One of the main challenges is then to set up new architectures and mechanisms that can efficiently scale up with the growing number of nodes that may be required to ensure adequate coverage of large areas of interest. At the same time, these new architectures and mechanisms should maintain low energy consumption per node so as to get by with energy guaranty acceptable network lifetime. The evaluation of the scalability of an algorithm or protocol is mainly based on a well-known metric for WSNs, which is the network lifetime. The objective is to avoid significant degradations of the network lifetime when the number of nodes composing the WSN increases.

One promising approach to solve the problem of scalabitity is to build hierarchies among the nodes, such that the topology of the network is abstracted (Chen et al., 2006). Indeed, flat topologies are difficult to scale up since communications between thousands or perhaps millions of nodes in a ad hoc fashion lead to degraded performances and hence higher energy consumption. For instance, the routing algorithm proposed in (Slama et al., 2006), requires that the sink have knowledge of the topology of the entire network at the beginning of each round. This requires a lot of signalling and do not scale well with a high number of nodes.

Hierarchical topologies are then recommended in such scenarios. The network protocols or algorithms designed for these architectures are generally highly scalable. Moreover, such architecture enables better resources allocation and improves power control over the network (Rabiner Heinzelman et al., 2000).

Such architectures are consisted of sensor clusters and one or more base stations. Each cluster is managed by a cluster head. Cluster members are generally small sensor nodes that capture relevant information from the area of interest and send it to their cluster head. This latter creates from the received data a comprehensive local view that is then sent to a base station (or sink). By combining all the data received from the different cluster heads, the base stations generate a comprehensive global view for the entire WSN coverage that it finally sent to the application level. In practice, cluster members do not communicate with each others whereas cluster heads can be involved in inter cluster heads relaying. Both cluster members and cluster heads are battery powered and energy constrained. However, cluster heads are considered to have more computing and storage capacities than member nodes. Base stations can be mobile or located in flexible positions and have infinite processing, storage and power resources. In such scenario, if a sensing node (cluster member) is drained out of energy, the cluster head may still be able to provide a comprehensive local-view by data generated by other sensor nodes in the cluster. However, if a cluster head runs out of energy, the whole cluster coverage is broken. More than that, the relaying task performed by this cluster head from/to neighboring cluster heads toward the sink is also lost. This may affect routes and lead in some cases to the disconnection of an entire region of the network, which may be dramatic for the network mission. Therefore, we are more concerned about the energy constraint of cluster heads.

Although cluster heads can have better initial energy provisioning than sensor nodes, cluster heads also consume energy at a considerably higher rate due to the transmission of gathered data over much longer distances (Pan et al., 2003). Moreover, cluster heads relay not only data generated within their own clusters but also data they may receive from their cluster head neighbors. Cluster heads also consume considerable energy for managing their clusters and gathering information from their cluster members (reception operations, data aggregation, signaling, etc). Cluster heads perform then two high energy cost roles. The first is related to forwarding and relaying tasks and the second concerns their own cluster management. To increase the network lifetime, the communication between the cluster heads should be managed in an energy efficient manner. To this end, an energy aware routing scheme should be investigated to fairly balance the energy consumption among the cluster heads according to both their available amount of energy and their role in the network and then route around cluster heads that are dying or need higher energy to manage their clusters.

On the other hand, base stations have infinite processing, storage and power resources. Therefore, an efficient usage of these flexible and mobile base stations to increase the network lifetime should be investigated especially when the sensing nodes and cluster heads are stationary or have very low mobility. The idea behind this is to decrease the distance between each cluster head and the nearest base station. In fact, when a higher number of base stations are distributed within the WSN, the paths length from any cluster head to its nearest base station is decreased leading to lower energy consumption and therefore to higher network lifetime. Moreover, since multi hop communication is used between cluster heads, the ones that are one hop from the base station drain their energy faster than other nodes because they have to relay messages originated from many other nodes (cluster heads) in addition to delivering their own messages. Therefore, using mobile base stations may help distributing the energy consumption over the different relaying cluster heads and then increase the network lifetime. Base stations trajectories can be rather controlled by the application or follow a specific mobility model in which case an estimation of their locations can be computed like in (Chen et al., 2006).

Motivated by these issues, we focus, in this chapter, on first, how to optimally locate the base stations in the network and second, on how to arrange the communication between cluster heads toward the base stations, both in order to guaranty that the gathered information effectively and efficiently reach the application. This is generally referred to as *topology control.* Our goal is to maximize the lifetime of a large-scale and energy constrained WSN.

The remainder of this chapter is organised as follows. A related work is presented in section 2 In section 3, we give the system architecture of a two-tiered WSN, its Cluster Heads energy dissipation model, and the definition of the whole network lifetime. In section 4, we present the approach that optimally locates multiple mobile base stations in the network such that the energy consumption is fairly distributed over the different Cluster Heads. In section 5, by extending the optimal routing appraoch presented in (Slama et al., 2006), we introduce a new optimization scheme that arranges the multi-hop communication between the Cluster Heads while considering their residual amount of energy and the role they play in the network leading to a maximum network lifetime. This new optimal routing scheme takes into account the energy required by each Cluster head to gather data from its cluster as well as to relay the packets coming from its Cluster Head neighbors toward the corresponding base station. We describe then in section 6 the overall dynamic framework. Simulations conducted to evaluate this global approach are described and commented in section 7. Section 8 concludes this chapter and points out the directions for future work.

2. Prior Work

To solve energy constrained problem in wireless sensor networks field, many routing protocols have been proposed; especially, cluster based routing protocols have many advantages such as reducing control messages, maximizing bandwidth reusability, enhanced resource allocation, larger scalability and improved power control.

LEACH (Low-Energy Adaptive Clustering Hierarchy) proposed in (Rabiner Heinzelman et al., 2002) includes a distributed cluster formation technique that enables self-organization of large numbers of nodes and rotating cluster head positions, so that the high energy dissipation in communicating with the base station is spread over all sensor nodes in the sensor network. However, LEACH can suffer from the clustering overhead that may result in supplementary power consumption. Moreover, cluster head rotation requires that all the nodes be capable of performing data aggregation, cluster management and routing decisions. This results in extra hardware complexity in all the nodes.

In contrast, in (Mhatre and al., 2005), the above complexity is embedded in only few nodes (cluster head nodes). In (Mhatre and al., 2005), authors aim to obtain the minimum number of sensing nodes, cluster heads, and battery energy to ensure at least T unit of lifetime. Nodes were divided into two categories: nodes 0 are sensing node and nodes 1 are cluster heads. Analysis results show that the number of cluster heads should be of the order of square root of the number of sensing nodes. However, authors don't give any exact evaluation of the maximum lifetime of the network. Moreover, it was assumed in this work that cluster heads directly communicate (i.e. one hop communication) with the base station, which is difficult to be applied to realistic scenarios. It has also been observed that the nodes close to the cluster heads have high-energy consumption due to packet relaying operations. But, one of the most significant observation is the sharp cutoff effect, to maximize the lifetime does not hold at all time.

In (Pan et al., 2003), a two-tiered wireless sensing network was considered and the authors observed the property that the first tier nodes are important for the lifetime of the entire network. They investigated a topology control approach to maximize the topological lifetime of the network in terms of the base station placement for the two-tiered sensor networks where the nodes are deployed within clusters. However, the optimal base station locations were obtained without considering the inter-cluster heads relaying. Then, once the base stations located, the inter cluster heads relaying may be not applicable in some cases.

In (Luo & Hubaux, 2005), authors have developed an analytical model that describes the communication load distribution in WSNs and proved that base station mobility is a strategy that deserves to be considered when optimizing the network lifetime. They have further shown that the optimum movement strategy for a mobile base station is to follow the periphery when the deployment area is circular.

Network lifetime elongation using mobile base stations has also been investigated in (Wang et al., 2005). The author gave a novel linear programming formulation for the joint problem of determining the movement of the sink and the sojourn time at different points in the network. Simulations have shown that lifetime maximizing solutions are achieved by non-uniform sojourn time distributions among grid points depending on the shape of the deployment area.

In (Gandham et al., 2003), authors propose to divide time into rounds and to dynamically relocate multiple sinks, at different positions along the periphery of the sensed field, at the beginning of each of these rounds. An integer linear program is used to determine the new locations of the different base stations. Results have shown that the energy consumption of individual sensor nodes is better balanced and the overall energy consumption of all sensors is minimized.

In (Kim et al., 2006), authors propose a different approach to find the optimal locations of multiple stationary sink nodes. The proposed scheme allows sensor nodes to communicate with one or multiple sinks through multiple paths in order to improve the network lifetime.

Another approach to solve the problem of multiple mobile base station placements is proposed in (Vincze et al., 2006). An electrostatic model is applied to determine sinks' locations and to coordinates the movements of these sinks considering the network state.

Unfortunately, most of the above base stations locations strategies are proposed and evaluated over small to medium size wireless sensor networks (typically less than 100 nodes). For large-scale wireless sensor networks, where hundreds or thousands of nodes can be deployed, the placement of multiple base stations still requires advanced studies. For instance, and as illustrated on Fig. 1, if we consider the case where the sinks are located along the periphery as stated in (Gandham et al., 2003), the paths between each node and its nearest sink is relatively short when the number of nodes is limited. However, the more the area size increases and/or the number of nodes within it increases, the longer this path is and the shorter the sensor nodes lifetime will be.

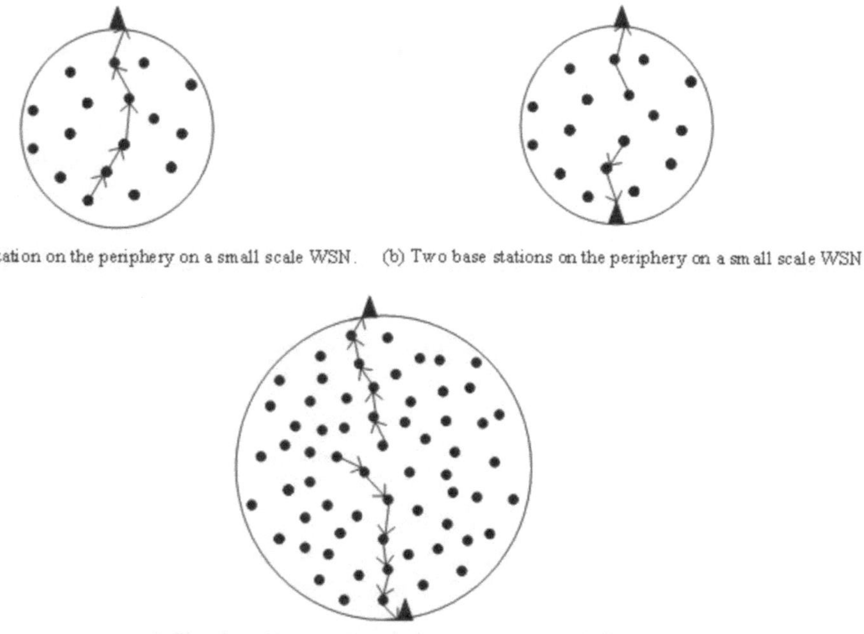

(a) A single base station on the periphery on a small scale WSN. (b) Two base stations on the periphery on a small scale WSN

(c) Two base stations on the periphery on a large scale WSN.

Fig. 1. Multiple Base Stations placement.

3. System Model

3.1 Network architecture

We consider, in this work, a large number of stationary and heterogeneous sensor nodes covering a given area of interest. For scalability management ends, the topology of the network is abstracted and the nodes are organized into a two-tiered WSN as depicted in fig. 2.

It consists of a number of clusters and multiple mobile base stations. Each cluster is composed of a set of Sensing Nodes and one Cluster Head. Sensing Nodes are small, low cost and densely deployed in each cluster. They are responsible of sensing raw data and then forwarding it to their corresponding Cluster Head. We consider that cluster formation is based on neighborhood. Hence, direct transmissions can be used inside each cluster. However, Sensing Nodes do not communicate with other Sensing Nodes in the same or other clusters. Cluster Heads, on the other hand, have much more responsibilities. First, they manage their clusters (send queries, instruct some nodes to be in idle or sleep status…) and gather data from their cluster members. Second, they perform aggregation of this data to eliminate redundancy and minimize the number of transmissions and thus save energy. The aggregated data at each Cluster Head represents a local view of its cluster. Third, they transmit the composite bit-stream towards the nearest base station. Each base station can then generate a comprehensive global view of the entire network coverage by combining the different local view data received from the different Cluster Heads (Pan et al., 2003).

While direct communication is used between Sensing Nodes and Cluster Heads, it may be inappropriate between Cluster Heads and Base Stations. Indeed, distances between them can be large. In such cases, direct transmissions will require high power consumption. In addition, direct transmissions may be simply impossible since Base Stations can be out of the range of some Cluster Heads. Multi-hop communication is then used. Consequently, Cluster Heads can also be involved in inter Cluster Head relaying.

Two main roles/responsibilities can be then defined for Cluster Heads. The first is related to their clusters and consists of managing their cluster members and gathering information from them. The second is related to forwarding activities and consists of relaying their own packets as well as packets they may receive from their Cluster Head neighbors towards the corresponding Base Station.

Although, both Sensing Nodes and Cluster Heads are energy constrained, Cluster Heads are considered to have more computing and storage capacities than member nodes. Base Stations, however, have infinite processing, storage and power resources. They can be mobile or located in flexible positions.

Note that initial energy allocation plays an important role in topology control for WSNs. Attributing different initial energy levels to the nodes according to their position and role in the network can significantly improve the network lifetime (Pan et al., 2003). However, since fixed initial energy sheme is used in practice (among nodes of the same category, Cluster Heads or Sensing Nodes), we adopt this scheme for the rest of this chapter.

In such scenario, Cluster Heads represent the logical bridge between the two tiers of the network architecture: the lower-tier Sensing Nodes and their Cluster Heads and the upper-tier Cluster Heads and Base Stations. They are then vital for the network mission success. Therefore and as stated previously, Cluster heads should be given all our attention in terms of energy efficiency.

Once Cluster Heads and Sensing Nodes are deployed, an immediate challenge is to optimally locate the Base Stations such that the network lifetime is maximized. After Base Stations are located, these latters can compute the inter-Cluster Heads routing scheme. This scheme should instruct Cluster Heads to communicate in an energy efficient manner and fairly distribute the energy consumption among them to achieve a longer network lifetime.

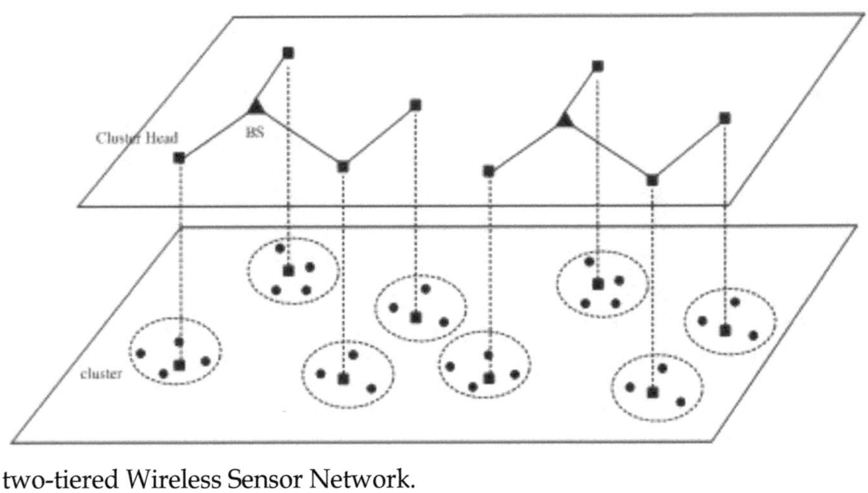

Fig. 2. A two-tiered Wireless Sensor Network.

3.2 Network Graph

We consider that the network is organized into *N* Clusters with one Cluster Head for each Cluster. In the rest of this chapter, we will note a Cluster *i* by C_i and its corresponding Cluster Head by CH_i, *i=1* to *N*. All the Sensing Nodes in cluster C_i can communicate directly with the Cluster Head CH_i.

The network is modelled as an undirected connected graph *G(H,A)* where *H* is the set of Cluster Heads, $H = \{CH_i, i = 1 \; to \; N\}$ and *A* the set of all undirected links (CH_i, CH_j) where CH_i , CH_j are two Cluster Heads of *H*.

Let L_i be the set of Cluster Heads neighbors of Cluster Head CH_i. L_i is composed of all Cluster Heads that can be reached by CH_i using a transmission power equal or less than a maximum predefined threshold. All links are assumed to be bidirectional.

3.3 Energy Model

In this work, we assume the same radio energy model used in (Slama et al., 2006). We remind that E_{elec} and ε_{amp} represent the energy consumed respectively to run the radio electronics and the power amplifier.

Thus, the energy consumed at node i when transmitting information to node j at rate x_{ij} can be written as:

$$E_t(x_{ij}, d_{ij}) = (E_{elec} + \varepsilon_{amp}.d_{ij}^{\;2}).x_{ij} = e_{ij}.x_{ij} \tag{1}$$

And the energy consumed at node j when receiving information from node i at rate x_{ij}:

$$E_r(x_{ij}) = E_{elec}.x_{ij} = e_r.x_{ij} \tag{2}$$

Moreover, since data aggregation (fusion) is performed within the network, the energy consumed at node *j* to aggregate information received from node i at rate x_{ij} can be written as :

$$E_a(x_{ij}) = \gamma.x_{ij} = e_a.x_{ij} \tag{3}$$

Where:

d_{ij} is the Euclid distance between node i and node j.
e_{ij} is the transmission energy required to transmit one data unit from node *i* to node *j*.
e_r is the energy required for the reception of one data unit.
γ is constant and called the aggregation energy consumption coefficient (Chen et al., 2006).
e_a is the energy required to the fusion of one data unit.

3.4 Lifetime Definitions

First, we consider that each cluster in the network dies when no more reliable information can be delivered from the cluster members. A Cluster Head whose cluster is dead, continue performing relaying data from/to neighboring Cluster Heads (i.e., its second role).

Second, we define the lifetime of the whole network as the metric for determining the optimality of the routing algorithm.

In a previous work (Slama et al., 2006), we defined the lifetime of a flat topological small-scale sensor network as the time at which a first node runs out of energy. Analogically and motivated by the same reasons, we define in this work the lifetime of the whole cluster-based sensor network as the period of time that ends when a first Cluster Head runs out of energy. This implies that every Cluster Head is vital for the application and cannot be substituted by others.

Hence, maximizing the network lifetime can be achieved by delaying as much as possible the first Cluster Head death.

4. Base stations placement

We propose in this section to enhance Base Station placement in a two-tiered large-scale WSN.

The idea behind this is to efficiently deploy Multiple Mobile Base Stations within the network. Multiple Base Stations are used to decrease the distance between each Cluster Head and its nearest Base Station. Base Stations are also chosen to be mobile to make the set of Cluster Heads close to the Base Stations and then overloaded, change over the time. This should guaranty a fair distribution of the energy consumption among the different Cluster Heads in the network and hence improve the network lifetime. The Challenge here is then to define the initial locations and the movement trajectories of the different Base Stations.

Since we deal with a large-scale WSN, an intuitively appropriate solution is to decompose the underlying sensor network and then optimize energy usage in each of the sub-networks independently. The objective is to take advantage of the powerful and efficient Base Station placement techniques proposed for small scale WSNs. In order to apply these techniques over large-scale WSNs, we propose to first divide the network into sub-networks according to specific criteria. An adequate Base Station placement technique can then be applied independently within each of the defined sub-networks.

Graph partitioning is a promising approach to split a large sensor network into balanced sub-networks. In practice, different criteria can be considered in order to partition a large-scale two-tiered wireless sensor network. Since multi-hop communication is used between Cluster Heads toward Base Stations, one simple objective is to create balanced sub-networks (in terms of number of Cluster Heads/clusters) that group the Cluster Heads according to their neighborhood. This allows creating smaller two-tiered sub-networks with similar characteristics that can be easily optimized, independently but in the same way.

In graph theory related literature, different approaches and techniques are proposed for balanced graph partitioning.

4.1 Existing Graph Partitioning Techniques

In (Even et al., 1997), a fast approximate graph-partitioning algorithm is proposed. The authors unified the problems of *b-balanced* cuts and *k-multiway* separators using a new approach called minimum capacity *ρ-separators*. They studied the graph partitioning problems on graphs with edge capacities and vertex weights and described a simple approximation algorithm for minimum capacity *ρ-separators* leading to a fast approximation algorithm both for *b-balanced* cuts and *k-multiway* separators. They define a *ρ-separator* as a sub-set of edges whose removal partitions the vertex set into connected components such that the sum of the vertex weights in each component is at most ρ times the weight of the graph.

In (Ito et al., 2006), authors considered three problems to find a *(l, u)-partition* of a given graph. They proposed to partition a graph G into connected components by deleting some edges from G making the total weight of each component equal at least to l and at most to u. The minimum partition problem is to find an *(l, u)-partition* with the minimum number of components, the maximum partition problem is defined in the same way and the *p-partition* problem is to find an *(l, u)-partition* with a fixed number p of components. Authors proved that the three problems are *NP-complete* or *NP-hard.*

In (Chlebikova, 1996), authors studied the approximation of the Maximally Balanced Connected Partition problem (MBCP). They first presented the optimization problem that finds the maximally balanced connected partition for a graph G. It results in a partition (V_1, V_2) of V composed of disjoint sets V_1 and V_2 such that both sub-graphs of G induced by V_1 and V_2 are connected, and maximize an objective function "balance", $B^w(V_1, V_2) = min(w(V_1), w(V_2))$. Authors proved that the problem is *NP-hard.*

In this work, this last approach will be adapted and applied to our Network. Our choice is mainly motivated by the practical approach provided in (Chlebikova, 1996) and based on the use of a polynomial-time algorithm that gives an approximate solution.

In the following the Maximally Balanced Connected Partition (MBCP) technique (Chlebikova, 1996) is adapted and formulated. A corresponding approximate resolution algorithm is then presented.

4.2 Problem formulation

Since, in Base Stations placement scheme, we are considering the communication between the Cluster Heads and the Base Stations (the upper tier of the architecture), only Cluster Heads are concerned by the partitioning scheme. We assume then that if a Cluster Head belongs to a sub-network, then its corresponding Cluster belongs to this sub-network as well.

We consider here the undirected connected graph $G(H,A)$. We remind that H is the set of Cluster Heads, $H = \{CH_i, i = 1 \text{ to } N\}$ and A the set of all undirected links.

The objective is to partition G into connected balanced sub-graphs.

To achieve this objective, let w be a non-negative vertex-weight function representing the balancing criteria. In this case, w will reflect the number of Cluster Heads. Hence $w(H') = |H'|$.

This MBCP problem can then be formulated as follow:

Maximize $B^w(H_1, H_2) = \min(w(H_1), w(H_2))$

Subject to 1. (H_1, H_2) is a partition of H into non-empty disjoints sets H_1 and H_2 such that sub-graphs of G induced by H_1 and H_2 are connected.

$$2.\ w(H') = \sum_{CH_i \in H'} w(CH_i) \quad \forall \quad H' \subseteq H$$

The resolution of this model will result into two balanced sub-networks. Each of them can be partitioned again using the same process.

This partitioning technique should be applied as much as required according to the targeted size for the sub-networks and taking into account the number of available Base Stations to be placed. The final result should be 2^n equivalent smaller connected sub-networks where n is the number of partitioning iterations.

4.3 Problem resolution

To solve this model, we used the polynomial approximation algorithm presented in (Chlebikova, 1996) and which finds an approximate solution for the MBCP problem.

In order to select neighbouring Cluster Heads within the same sub-networks, we adapted this algorithm by sorting the list of candidates for each partition according to their distance (vicinity). Fig. 3 illustrates an example of a WSN partitioned into four sub-networks.

The algorithm can be written as follow:

Input: $G = (H,A)$.
$H= \{CH_1, CH_2, CH_3\ldots CH_N\}$ where $N=|H|$.
0. Initialize $H_1=\{CH_1\}$, $H_2= H\backslash H_1$ such as CH_1 a Cluster Head near the periphery of the network.
1. If $|H_1| >= 1/2\ |H|$ then Step 3 else Step 2.
2 .Let $H_0 = \{CH_i \in H\ /\ (H_1 \cup \{CH_i\}, H_2 \backslash \{CH_i\}, i \in \{1..N\})$ is a connected partition of $G\}$.
Choose CH_i of H_0 such that CH_i the closest element to H_1.
If $|CH_i| < |H| - 2|H_1|$
 then $H_1 := H_1 \cup \{CH_i\}$, $H_2 := H_2 \backslash \{CH_i\}$, Step 1
 else Step 3
3. Return (H_1, H_2).

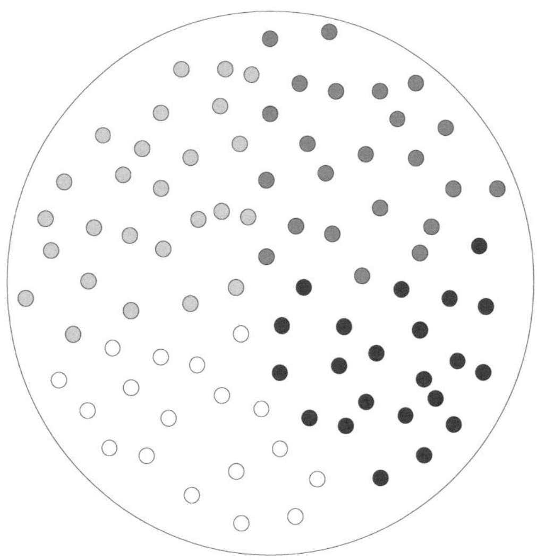

Fig. 3. A Wireless Sensor Network partitioned into 4 sub-networks (the four node patterns represent four different sub-networks).

4.4 Locating Base Stations

Once the network is partitioned into identical smaller two-tiered sub-networks, each of these sub-networks is represented by a disc with the geographic centre of the sub-network as centre and the distance between this centre and the farthest Cluster Head (belonging to this sub-network) from it as radius. Recall that if a Cluster Head belongs to a sub-network, then its Cluster members belong to this sub-network as well.

Base Stations can now be optimally located within each of these sub-networks independently but in the same way.

It has been proven in (Luo & Hubaux, 2005) that the optimum movement for a mobile base station is to follow the periphery when the deployment area is circular.

Motivated by this result, we suggest in this work that one single Base Station be randomly deployed on the periphery of each sub-network. Then each Base Station keeps moving along the periphery of the sub-network in which it was deployed. Note that Cluster Heads can send their data only to the Base Station deployed in the sub-network they belong to.

A mobile base station can move in two different regimes, a fast mobility regime and a slow mobility regime (Luo et al., 2006). In the fast mobility regime, the base station moves in a continuous form with a velocity v along the time without any stop or pause in a particular position. In the slow mobility regime, the base station moves in a discrete form and its trajectory is a sequence of anchor points between which the base station moves with a velocity v and at which it pauses during a period of time (epoch).

The slow mobility regime is considered more realistic and is adopted in many researche studies. Therefore, we assume in this work that each Base Station moves in a slow mobility regime. We propose then to divide time into rounds. Each Base Station moves then at the beginning of each round and remains at its new position until the end of the round.

It is very important to carefully choose the value of the base station velocity. In fact, when the mobile base station velocity is high, the base station will more frequently change its position and visit more the different regions over the area of interest during the network lifetime. Therefore, the energy consumption is efficiently distributed over the cluster heads and the network lifetime is extended. This can be much more efficient in the particular case where the cluster heads buffer the data gathered in their clusters and wait until the base station approaches to deliver it (Chen et al., 2006) which reduces unnecessarily packet forwarding actions since cluster heads are sure of base station arrival before loosing the data (because of buffer size limitation or packet deadline expiration). Besides, the high speed moving base station produces a tolerable data delivery delay especially in the case of fast mobility regime, which can be very important for some specific applications. However, base station high velocity can have negative effects. In fact, it can make the session interval too short to successfully exchange a long data packet and hence the packet loss rate will increase. In slow mobility regime, it is preferred that the epoch (round) be long enough to guaranty long messages exchange.

On the other hand, it seems obvious that the mobility of the base stations will inevitably incur additional overhead in data exchanges since the cluster heads will continuously need to be informed of their corresponding base station location. However, it has been proved in (Luo et al., 2006) that when using a slow mobility regime with an epoch much longer than the base station moving time, the overhead introduced by the mobility of the base station became negligible because amortized across a long epoch. This reinforces our choice in using a slow mobility regime.

After determining the Base Stations placement strategy, we can further prolong network lifetime by instructing Cluster heads to efficiently forward the data to the destination. Hence, at the beginning of each round and after it is located in its new position, each Base Station has to compute the routing scheme that will manage in an energy efficient manner the inter Cluster Heads communication within its corresponding sub-network.

5. Inter-Cluster Head communication

As discussed at the beginning of this chapter, Cluster Heads that are in critical positions run out of energy first. Hence, to further extend the network lifetime, it is necessary to delay as much as possible the first Cluster Heads death.

For small-scale non-clustered WSNs, we proposed in a previous work (Slama et al., 2006) an approach that defines an optimal multi-hop routing. It dynamically distributes flows proportionally to the residual energy available at each node leading to a maximum network lifetime.

The routing scheme is modelled as an optimization algorithm and is computed at the Base Station. Its resolution results in a routing matrix that defines for each node to which of its neighbors it has to send data.

In this section, we propose to extend this approach to two-tiered WSN architectures. In addition to the residual energy at each Cluster Heads, we introduce a new constraint that reflects Cluster Head energy consumption related to its intra-cluster activities (i.e. the first role of Cluster Heads). The idea is to alleviate, from relaying activities (i.e. the second role of Cluster Heads), Cluster Heads requiring higher energy for managing their clusters.

On the other hand, inside each cluster, Sensing Nodes have to provide the information required by the end application. They should be organized such that the QoS is satisfied with minimum cost. Different techniques can be used to achieve this goal. For instance, sensors can be autonomous and self organized (Rabiner, Heizelman et al., 2002, Chatterjee et al., 2002). Another approach is to use a relative central mechanism (e.g. scheduling mechanism) that can take the appropriate decisions on behalf of the Sensing Nodes. For instance, we can consider that within each cluster, one or more Sensing Nodes may be used at any time to provide data to the application, but only certain subsets of available sensors may satisfy channel bandwidth and/or application quality of service constraints (Perillo & Heinzelman, 2003). In this work, we decide to adapt the scheduling mechanism, initially proposed in (Perillo & Heinzelman, 2003) for a flat topological WSNs, to manage communications inside the clusters. This scheduler determines which sensor sets should be used and for how long time so that the lifetime of the cluster is maximized while the necessary quality of service expected from this cluster is always maintained at the application. In addition, Sensing Nodes providing redundant information can be turned off which contributes in energy saving and reduces data flows. Used within each cluster and according to the performance evaluation given in (Perillo & Heinzelman, 2003), this mechanism optimizes individual clusters lifetimes.

In order to achieve a global routing optimization , the inter-Cluster Heads communication approach that we propose should, in addition, take into account these individual clusters lifetimes, as the more a cluster lasts, the more its Cluster Heads requires energy for its management (e.g. reception, data processing and fusion, …).

This inter-Cluster Heads communication approach is modeled within each sub-network as an optimization problem. It is then processed in a centralized manner at the Base Station of each sub-network independently but simultanously. It takes into account the current status and topology of the sub-network and results in a routing matrix that defines the inter-Cluster Heads flows within this sub-network such that the minimum Cluster Head lifetime is optimized.

The inter-Cluster Heads communication approach construction and its details are presented in the following sections.

5.1 Model and Notations

Let's consider N_b Base Stations to be deployed in the network. We note a Base Station k by b_k, $k = 1$ to N_b. The network graph G is then partitioned into N_b equivalent sub-graphs. We consider (H_1, H_2, …, H_{N_b}) the connected partition of G.

Then, each sub-network k corresponding to H_k contains one single mobile Base Station b_k and N_k^{CH} Cluster Heads, $k = 1$ to N_b, $\sum_k N_k^{CH} = N$.

We assume that each sub-network k is modeled as a connected sub-graph $G_k(H_k, A_k)$, $k = 1$ to N_b. H_k is then the set of Cluster Heads belonging to the sub-network k, $H_k = \{CH_{k,i}, i = 1 \text{ to } N_k^{CH}\}$ and A_k the set of the undirected links ($CH_{k,i}$, $CH_{k,j}$) where $CH_{k,i}$ and $CH_{k,j}$ are two Cluster Heads of H_k.

Let $L_{k,i}$ be the set of Cluster Heads neighbors of Cluster Head $CH_{k,i}$ in the sub-network k. $L_{k,i}$ is composed of all Cluster Heads of H_k that can be reached by $CH_{k,i}$. All links are assumed to be bidirectional.

We remind that if a Cluster Head belongs to a sub-network than its corresponding Cluster belongs to this sub-network as well. We will note by $C_{k,i}$ the Cluster of Sensing Nodes corresponding to the Cluster Head $CH_{k,i}$ and then belonging to sub-network k, $i = 1$ to N_k^{CH} and $k = 1$ to N_b.

Each cluster $C_{k,i}$ contains $N_{k,i}^S$ Sensing Nodes. We will refer to the complete set of Sensing Nodes within a cluster $C_{k,i}$ as $S_{k,i} = \{S_{k,il}, l \in \{1...N_{k,i}^S\}\}$.

We remind that all Sensing Nodes in Cluster $C_{k,i}$ can communicate directly with their Cluster Head $CH_{k,i}$ and that all Cluster Heads $CH_{k,i}$ belonging to sub-network k have to forward the gathered data to the Base Station b_k deployed within this same sub-network. Also, Cluster Heads belonging to one sub-network cannot communicate with Cluster Heads belonging to another sub-network.

We finally assume that $E_{k,il}^S$ and $E_{k,i}^{CH}$ are the initial energies of Sensing Node $S_{k,il}$ and Cluster Head $CH_{k,i}$ respectively. In table 1, we list all symbols used in this chapter.

5.2 Flow Conservation

We denote by $r_{k,i}$ the arrival rate of information at $CH_{k,i}$ sensed by the Sensing Nodes within its cluster $C_{k,i}$ and we denote by $v_{k,i}$ the rate of information at $CH_{k,i}$ after aggregation.
Hence, $v_{k,i}$ can be written as, $v_{k,i} = f_a(r_{k,i})$. f_a is a typical linear aggregation function such that $f_a(x) = \beta x$ for some constant β, $0 < \beta < 1$. β is called the data aggregation ratio (Chen et al., 2006).
Let $w_{k,i}$ be the average rate of information that transit through $CH_{k,i}$. It is composed of the generated information rate at $CH_{k,i}$ (sensed by the cluster members and then aggregated at $CH_{k,i}$) plus the information rate received from its Cluster Heads neighbours of $L_{k,i}$.

$w_{k,i}$ is given by:

$$w_{k,i} = v_{k,i} + \sum_{j/CH_{k,j} \in L_{k,i}} p_{k,ji} w_{k,j} \quad \forall (k, i \in \{1 \ldots N_k^{CH}\}) \tag{4}$$

and

$$w_{b_k} = \sum_{i \in \{1 \ldots N_k^{CH}\}} v_{k,i} \tag{5}$$

Where $p_{k,ji} w_{k,j}$ is the proportion of data transmitted by $CH_{k,j}$ to $CH_{k,i}$.

Obviously, $p_{k,ij} \geq 0 \quad \forall (k,i,j)$ and $\sum_{j/CH_{k,j} \in L_{k,i}} p_{k,ij} = 1 \quad \forall (k, i \in \{1 \ldots N_k^{CH}\})$.

We denote by P_k the routing matrix within sub-network k and which can be written as:

$$P_k = \{p_{k,ij}\}$$

Note that Equations (4) and (5) verify the flow conservation condition. The flow conservation condition states that the sum of information generation rate and the total incoming flow must equal the total outgoing flow.

5.3 Lifetime Model

We remind that a cluster dies when no more reliable information can be delivered from the cluster Sensing Nodes. We denote the lifetime of a cluster $C_{k,i}$ by $T_{k,i}^{C}$. Once its cluster dead, each Cluster head continue performing relaying activities until it is over of energy. We then denote by $T_{k,i}^{CH}$, the lifetime of Cluster Head $CH_{k,i}$.
The lifetime of the whole network is defined, as stated in section 4.2.4, as the period of time that ends when a first Cluster Head runs out of energy. We analogically define the lifetime of a sub-network k as the period of time until which the first Cluster Head $CH_{k,i}$ dies and denote it by T_k. Then, T_k can be written as:

$$T_k = \min_{i \in \{1 \ldots N_k^{CH}\}} T_{k,i}^{CH}, \forall k \tag{6}$$

Thus, the network lifetime can be defined as the period of time until which the first sub-network dies.

The network lifetime, denoted by T_{net}, can then be written as follow :

$$T_{net} = \min_{k} T_k = \min_{k,i \in \{1 \ldots N_k^{CH}\}} T_{k,i}^{CH} \tag{7}$$

Hence, maximizing the network lifetime can be achieved by maximizing each sub-network lifetime simultaneously.

5.4 Intra-cluster Communication

As already mentioned, the intra-cluster communication scheme is inspired from (Perillo & Heinzelman, 2003). The communications inside the clusters is managed by an optimized scheduler that determines which sensor sets should be used and for how long time so that the lifetime of the cluster is maximized while the necessary quality of service is respected.

As defined in (Perillo & Heinzelman, 2003), a sensor set is determined to be *feasible* if i) the total bandwidth necessary to support the set is below the capacity of the cluster and the traffic is schedulable and ii) the set provides the necessary reliability to the application. We will refer to the set of feasible sensor sets in a cluster $C_{k,i}$ as $F_{k,i} = \{F_{k,im}, m \in \{1 \ldots N_{k,i}^F\}\}$.

Symbol	Description
N	the number of Cluster Heads/Clusters in the network.
H	the set of N Cluster Heads of the WSN.
A	the set of the undirected links between the Cluster Heads of H
CH_i	a Cluster Head of H.
C_i	the Cluster corresponding to CH_i.
L_i	the set of Cluster Heads neighbours of CH_i.
N_b	the number of base stations deployed in the network.
H_k	a partition of H.
b_k	the base station deployed in sub-graph k.
$CH_{k,i}$	a Cluster Head of H_k.
N_k^{CH}	the number of Cluster Heads in sub-graph k.
$L_{k,i}$	the set of Cluster Heads Neighbors of $CH_{k,i}$ in sub-graph k.
$C_{k,i}$	the cluster in sub-network k corresponding to $CH_{k,i}$.
$N_{k,i}^S$	the number of Sensing nodes in $C_{k,i}$.
$S_{k,i}$	the set of Sensing Nodes in $C_{k,i}$.
$S_{k,il}$	a Sensing Node of $S_{k,i}$.
$E_{k,il}^S$	the initial energy of $S_{k,il}$.

$E^{CH}_{k,i}$	the initial energy of $CH_{k,i}$.
$r_{k,i}$	the arrival rate of sensed data at $CH_{k,i}$.
$v_{k,i}$	the arrival rate of aggregated data at $CH_{k,i}$.
β	the data agregation ratio.
f_a	the aggregation function.
$w_{k,i}$	the average rate of data that transit through $CH_{k,i}$.
w_{b_k}	the average rate of data that transit through b_k.
$p_{k,ij}$	The flow portion transmitted from $CH_{k,i}$ $CH_{k,j}$.
P_k	the routing matrix within sub-network k.
$T^{C}_{k,i}$	the lifetime duration of $C_{k,i}$.
$T^{CH}_{k,i}$	the lifetime duration of $CH_{k,i}$.
T_k	the lifetime duration of sub-network k.
T_{net}	the lifetime duration of the whole network.
$F_{k,i}$	the set of feasible sensor sets in $C_{k,i}$.
$F_{k,im}$	a feasible sensor set of $F_{k,i}$.
$N^{F}_{k,i}$	the number of feasible sensor sets in $C_{k,i}$.
$T^{F}_{k,im}$	the length of time that $F_{k,im}$ is being used in the optimal Schedule of $C_{k,i}$.
$q_{k,il}$	the power consumption at sensor $S_{k,il}$.
E_{elec}	the energy consumed to run the radio electronics.
ε_{amp}	the energy consumed to run the power amplifier.
$e_{k,ij}$	the transmission energy required to transmit one data unit from $CH_{k,i}$ to $CH_{k,j}$.
e_r	the energy required for the reception of one data unit.
e_a	the energy required to the fusion of one data unit.
γ	the aggregation energy consumption coefficient.

Table 1. Notations

The optimal scheduler that maximizes the lifetime of $C_{k,i}$ determines the length of time that each sensor set in $C_{k,i}$ should be used. Let $T^{F}_{k,im}$ represent the length of time that feasible sensor set $F_{k,im}$ is being used in the optimal schedule of $C_{k,i}$. The objective of the problem is to maximize the lifetime of each cluster $C_{k,i}$:

$$T_{k,i}^{C} = \sum_{m} T_{k,im}^{F} \qquad \forall (k, i \in \{1...N_{k}^{CH}\}) \tag{8}$$

We will define $a_{k,ilm}$ as a variable equal to one if sensor $S_{k,il}$ is being used in feasible sensor set $F_{k,im}$ of the cluster $C_{k,i}$ and equal to zero otherwise.
Finally, we define $q_{k,il}$ as a variable that represents the power consumption (sensing and communication) at sensor $S_{k,il}$.
We remind that $E_{k,il}^{S}$ is the initial energy of Sensor Node $S_{k,il}$. This finite energy introduces the following constraint:

$$\sum_{m} a_{k,ilm} T_{k,im}^{F} q_{k,il} \leq E_{k,il}^{S} \qquad \forall (k, i \in \{1...N_{k}^{CH}\}, l \in \{1...N_{k,i}^{S}\}) \tag{9}$$

This scheduling problem has been modeled as a generalized maximum flow graph problem. The same method will be used for each cluster in the network and carried out in a centralized manner by an unconstrained node or at the application level at the beginning of the network deployment and once the clusters are formed (during the set-up phase and before the transmission phase is started). The computation of this optimization scheme defines for each cluster the optimal Schedule that maximizes its lifetime. Each Cluster lifetime value can then be computed and used as an input parameter for the inter-Cluster Heads communication scheme.

To have details about the resolution of this optimization problem the reader is referred to (Perillo & Heinzelman, 2003).

5.5 Maximizing Network Lifetime

According to the scheduling problem described in the last section the lifetime of each cluster $C_{k,i}$ (not including the corresponding $CH_{k,i}$) is $T_{k,i}^{C}$. During this period of time a Cluster Head $CH_{k,i}$ is providing two functionalities: the first concerns internal exchange (receiving and aggregating data coming from its cluster members) and the second concerns external exchange (receiving, transmitting and relaying the data coming from its Cluser Head neighbors).
Once this period achieved, $CH_{k,i}$, if not yet drained out of energy, expend its remaining energy to provide only the second functionality.
During the period of time $T_{k,i}^{C}$, $CH_{k,i}$ expends an amount of energy given by:

$$E_{k,i}^{CH_1} = T_{k,i}^{C} \left(\sum_{j/CH_{k,j} \in L_{k,i}} e_{k,ij} p_{k,ij} w_{k,i} + \sum_{j/CH_{k,j} \in L_{k,i}} e_{r} p_{k,ji} w_{k,j} + (e_a + e_r) r_{k,i} \right) \tag{10}$$

Here, $e_{k,ij}$ is the transmission energy required to transmit one data unit from $CH_{k,i}$ to $CH_{k,j}$ relatively to equation (1).
So, the remaining energy at $CH_{k,i}$ when $T_{k,i}^{C}$ is spent is:

$$E_{k,i}^{CH_2} = E_{k,i}^{CH} - E_{k,i}^{CH_1} \tag{11}$$

Hence, according to the energy model described in section 4.2.3, the lifetime of $CH_{k,i}$ under a given system $P_k = \{p_{k,ij}\}\ \forall(k,i \in \{1..N_k^{CH}\})$ is given by:

$$T_{k,i}^{CH}(P_k) = T_{k,i}^{C} + \frac{E_{k,i}^{CH_2}}{\sum_{j/CH_{k,j}\in L_{k,i}} e_{k,ij} p_{k,ij} w_{k,i} + \sum_{j/CH_{k,j}\in L_{k,i}} e_r p_{k,ji} w_{k,j}}$$

$$= T_{k,i}^{C} + \frac{E_{k,i}^{CH} - T_{k,i}^{C}(\sum_{j/CH_{k,j}\in L_{k,i}} e_{k,ij} p_{k,ij} w_{k,i} + \sum_{j/CH_{k,j}\in L_{k,i}} e_r p_{k,ji} w_{k,j} + (e_a + e_r) r_{k,i})}{\sum_{j/CH_{k,j}\in L_{k,i}} e_{k,ij} p_{k,ij} w_{k,i} + \sum_{j/CH_{k,j}\in L_{k,i}} e_r p_{k,ji} w_{k,j}} \quad (12)$$

Then, T_k, the lifetime of sub-graph k, can be approximated as follow:

$$T_k(P_k) = \min_{i\in\{1 \dots N_k^{CH}\}} T_{k,i}^{CH}(P_k), \forall k \quad (13)$$

Maximizing the lifetime of a sub-network k can be reached by solving the following optimization problem:

$$\begin{array}{lll} \textit{Maximize} & T_k & \\ \textit{Subjectto} & p_{k,ij} \geq 0 \quad \forall i \in \{1...N_k^{CH}\} and \forall j / CH_{k,j} \in L_{k,i} & (14) \\ & \sum_{j/CH_{k,j}\in L_{k,i}} p_{k,ij} = 0 \quad \forall i \in \{1...N_k^{CH}\} & \\ & E_{k,i}^{CH_1} + E_{k,i}^{CH_2} \leq E_{k,i}^{CH} \quad \forall i \in \{1...N_k^{CH}\} & \end{array}$$

The last constraint models energy conservation at each Cluster Head $CH_{k,i}$.

The resolution of this system requires determining the matrix P_k defining, for a fixed position of Base Station b_k, the optimal routing flows that are used by each Cluster Head within sub-network k to forward data to its Neighbors such that the lifetime of this sub-network is maximized. The optimal matrix P_k can then be computed in a centralized fashion at the Base Station b_k.

This optimisation problem is Non Polynomial and can then be solved over Matlab using specific heuristics similar to those used to solve the optimization problem presented in (Slama et al., 2006). Once the different sub-networks lifetimes $T_k, \forall k = 1 to N_b$ are computed, the whole network lifetime can be finally given by:

$$T_{net} = \min_k T_k \quad (15)$$

6. Global Framework

In this section we describe the overall dynamic framework for large two-tiered wireless sensor networks lifetime maximization. The framework is based on the optimisation scheme related to both Base Stations positioning and inter-Cluster Head communication presented previously. A cyclic algorithm is then defined to permit the dynamic adaptation of the optimization process (see Fig. 4).

Once the nodes are deployed in the interested area, the network topology is first abstracted and the overall network is partitioned into equivalent sub-networks that have the same characteristics and where the energy consumption can be optimized independently but in the same way. One mobile base station is then randomly deployed on the periphery of each sub-network. Time is then divided into equal periods of time called rounds or epochs. At the beginning of each round, each base station moves along the periphery of its corresponding sub-network. Once it reached its new position, the base station collects information about the current topology status of its sub-network. These information may include The residual energy at each sensor node, the neighbors list and the positions of each node, sources' throughputs, etc.

In a next step, each base station runs the routing optimization process corresponding to its sub-network as described in the previous section and which results in an updated routing matrix that optimally distributes energy consumption over the different Cluster Heads according to their roles in the sub-network and to the residual amount energy at each of them. Data gathering is then performed by the sensing nodes and the collected data is aggregated and forwarded by the cluster heads toward the corresponding base station using the optimized routing probabilities.

Input: G(H, A).

0.1. *The network is divided into N_b equivalent sub-networks.*

0.2. *One mobile base station is deployed on the periphery of each of these sub-networks.*

0.3. *Initial round duration (epoch) is determined at the application level*

While *(the sensor network is operational for the application)* **do**

{//begin of the round

$\forall k \in \{1 \ldots N_b\}:$

1. *Base station b_k in sub-network k moves to its new position on the periphery*

2. *At base station b_k: Collection of all relevant information from all the cluster heads of H_k concerning the current topology of sub-network k.*

3. *At base station b_k: Run of the optimization process and compute the routing matrix $[P_k]$.*

4. *Base station b_k transmits to each Cluster Head $CH_{k,i}$ the vector $[P_{k,ij}]$*

$(\forall i \in \{1 \ldots N_k^{CH}\}$ *and* $\forall j / CH_{k,j} \in L_{k,i})$.

5. *Each Cluster Head sends the captured/received information to its neighbors toward b_k according to $[P_k]$.*

// end of the round}

Fig. 4. Global Framework.

7. Simulations

This section is dedicated to the evaluation of the performances of first, the Base Stations Placement scheme that optimally locates the different base stations in the network while considering scalability as well as energy efficiency issues and second, the inter-ClusterHead communication approach formulated as an optimization problem that aims to efficiently and fairly distribute the energy among Cluster Heads while taking into account their roles in the network.

7.1 Base Stations placement

The effect of the proposed partitioning technique on the WSN lifetime is investigated using numerical simulations over Matlab environment. A circular large-scale wireless sensor network, with a radius R = 500m is considered. In order to study the performance of the base stations placement scheme, we focused on the upper tier of the network architecture (Base Stations and Cluster Heads) independently of the lower tier (Cluster Heads and Sensing Nodes). 1000 nodes (Cluster Heads) are randomly (uniformly) deployed over a network area. All nodes are similar with a communication range r = 80m and an initial energy of 1000J unit. Base Stations are assumed to have no energy constraints because they have larger batteries or their batteries are rechargeable. We assumed, in this scenario, that the shortest path routing algorithm is used to establish routes from Cluster Heads to base stations. The network lifetime is defined as the moment at which the first node runs out of energy. Time is divided into rounds. Each round is composed of T =100 timeframes. Each sensor node generates one data packet every timeframe.

To evaluate the efficiency of the proposed graph partitioning technique in elongating the network lifetime, three comparative scenarios are considered:

1. Scenario 1:

Case 1: An entire large network (not partitioned) is considered. All the sensors have the same capacity. N base stations are randomly fixed inside the coverage area of interest. Each sensor has to send the data it senses to the nearest base station.
Case 2: The graph-partitioning algorithm (detailed in section 4.3.3) is used to define N smaller sub-networks. One single base station is then randomly fixed in each sub network. Each sensor node sends its data to the base station deployed inside the sub-network the sensor node is belonging to.

2. Scenario 2:

Case 1: The entire network is considered. N mobile base stations are deployed randomly. Then, the base stations start to move inside the area of interest following the random waypoint model (Johnson & Maltz, 1996). At the beginning of each round, each base station moves 60 m.
Case 2: N sub-networks are defined using the graph-partitioning algorithm and one single base station is randomly deployed in each sub network. Then each base station moves 60m each round. The base station cannot go outside the area of the sub-network it belongs to. This area is represented by a disc with the geographic centre of the sub-network as centre and the distance between this centre and the farthest sensor (belonging to this sub-network) from it as radius.

3. Scenario 3:

Case 1: The entire network is considered. N mobile base stations are deployed randomly on the periphery of the network. Then, the base stations start to move along the periphery. In one round each base station moved 60 m.
Case 2: The graph-partitioning algorithm is used to define N smaller sub-networks. One single base station is randomly deployed on the periphery of each sub network. Then each base station moves 60m each round on the periphery.

We consider that the time required by a base station to move to its next position is negligible compared to a round duration.

Several simulations are then run to compare the network lifetime in the two different cases of each of the three different scenarios.

Simulation results are presented in fig. 5, 6 and 7. They respectively compare the performance of the different base stations deployment strategies in the case of partitioned and non-partitioned network (scenario 1, 2and 3).

First, let's notice that the simple use of multiple base stations enhances the network lifetime (with and without partitioning). Indeed, the network lifetime increases proportionally to the number of base stations because the distance between the nodes and their correspondent base stations is shortened. Second, it can be seen that moving the base stations clearly prolong the operation of the network. In fact, figures show that the network lifetime is much longer when the base stations are moving (scenario 2 and 3 with or without partitioning) than when they are fix (scenario1). This result is valid with or without partitioning.
Third, enhancements of the network lifetime can be observed in the case of partitioned large-scale WSNs compared to non-partitioned ones in all the scenarios. But the enhancement is the most significant in the third scenario. This was expected as when one base station is moving along the periphery of each sub-network, the energy consumption is obviously much more distributed over the sensors than when all the base stations are moving along the periphery of the whole network. The nodes that are the closest to the base stations are logically the ones who die first because they not only send their own data but also relay the data of all the nodes in the network. In scenario 3, the nodes who die first in the case of non-partitioned network are the nodes situated all along the periphery whereas in the case of partitioned network, they are the ones situated along the peripheries of the different sub-networks. Then, in this scenario, using the graph partitioning technique to deploy the base stations distributes the load relay and decreases the average distance between the nodes and the base stations. Indeed, the improvement of the network lifetime of the partitioned network is much more important when the number of base stations (or sub-networks) increases.

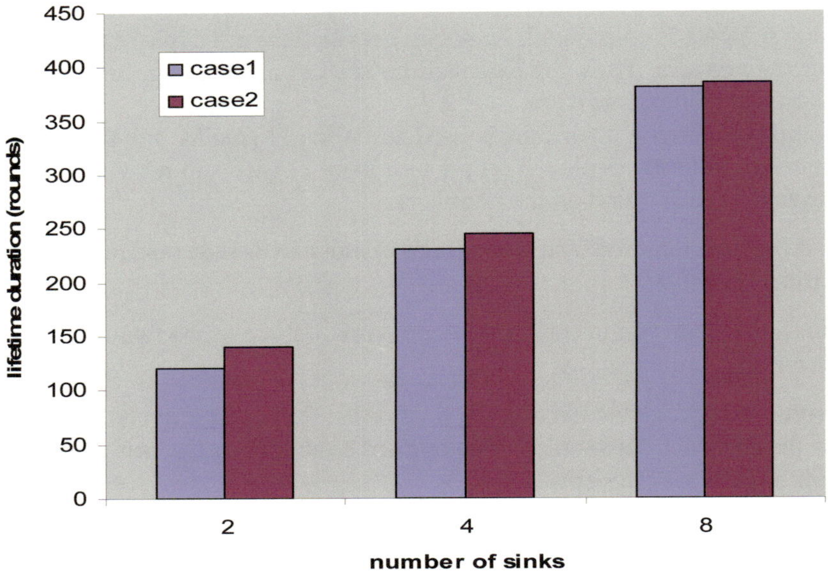

Fig. 5. The network lifetime in the scenario 1.

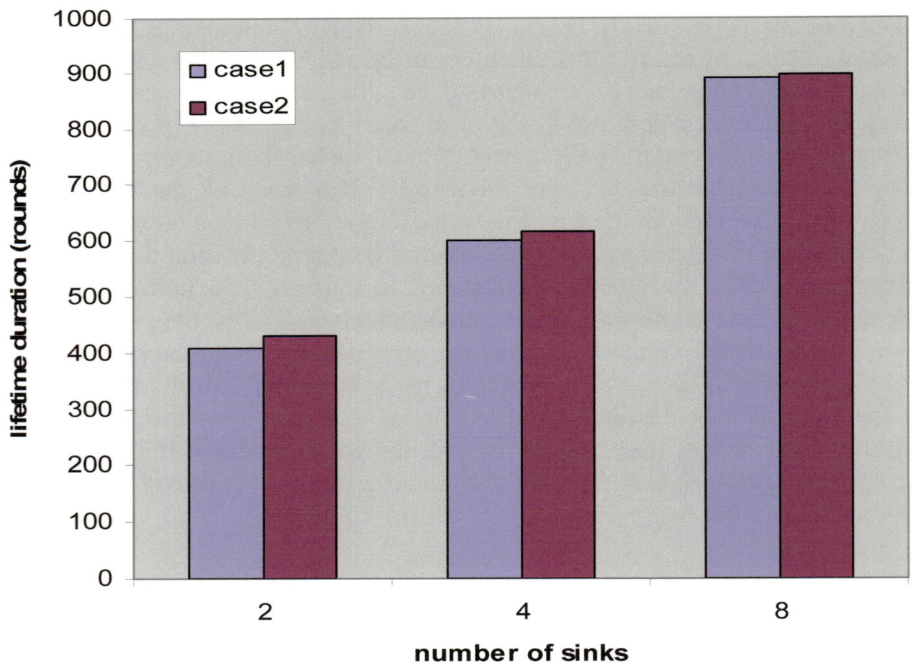

Fig. 6. The network lifetime in the scenario 2.

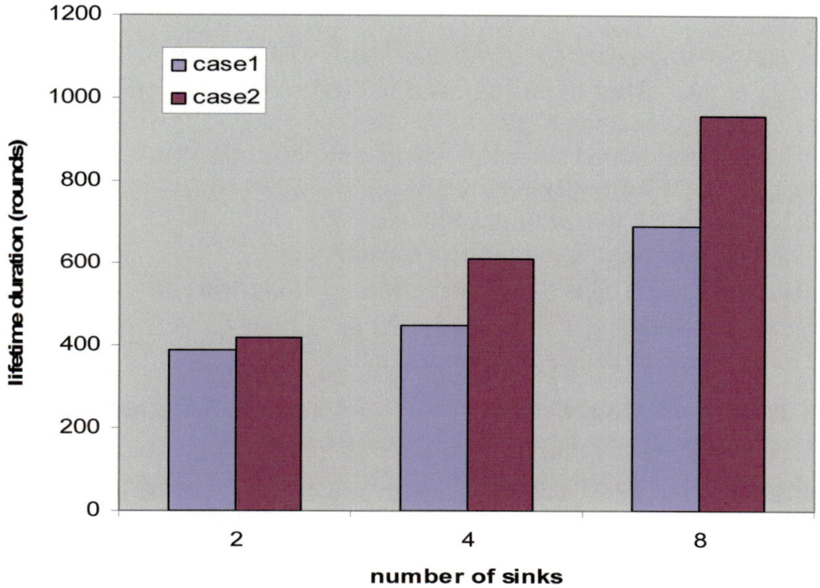

Fig. 7. The network lifetime in the scenario 3.

In the first case of the first scenario, base stations are randomly placed. Hence, they can be in some cases grouped in a small space. As a consequence, the distance between a node and the closest base station may not be really shortened. Whereas, in the second case, where we limited the area in which each base station can be deployed, by partitioning the network into sub networks, this distance is almost always shortened. This can be much more efficient when the base stations move (scenario 2) since the base stations in both cases have the same velocity (60m/round).

However, we notice, from fig. 5 and fig. 6, that the improvement is not so spectacular. This can be explained by the fact that when dividing the network into independent sub-networks, some nodes are bound to send their data to the base station deployed in the sub-network they belong to whereas they are closer to a base station deployed outside (in an other sub-network).

7.2 Inter-Cluster Heads Communication

In this section, we focus on the performance evaluation of the optimization scheme presented in section 4.4 and which manages the communication between Cluster Heads whithin each sub-network to efficiently transmit data toward base stations. The optimization problem is solved using specific heuristics and several simulations were run over Matlab.

Since the same optimal routing process is used in each of the sub-networks, we limit here our simulations to one single sub-network. We consider then a circular sub-network with radius equal to 100m. Cluster Heads and Sensing nodes are assumed to have a maximum communication radius of 80m and 20m respectively. We assume that nodes are, initially, distributed in a random fashion over the sub-area and that the clusterization is based on neighborhood. Feasibles sets are then randomly generated in each cluster of the sub-

network. One base station with no energy constraints is deployed and randomly placed on the periphery of the area.
The same initial energy is assumed for all Cluster Heads and is equal to 1000 J unit. The same initial energy is also assumed for all Sensing Nodes and is equal to 50 J. Power consumption at the Sensing Nodes is 10 μW.
The following values are considered for energy dissipation at Cluster Heads.

E_{elec} =50nJ/bit in the transmit circuitry and

ϵ_{amp} =100pJ/bit/m^2 for the transmit amplifier.

γ = 50nJ/bit for the aggregation energy consumption.

We assume the data aggregation ratio β=25% and a Sensing Node data rate equal to 160bit/s. Figures are obtained by averaging simulation results for a large number of scenarios. For each scenario, a different random node layout is used.

Fig. 8 illustrates the normalized sub-network lifetime. As depicted, the numerical resolution of the proposed model quickly converges to an optimal solution.

To study the effect of the sub-network composition and topology on its lifetime and the interactions between the inter-cluster and intra-cluster communications, we study the scenario where the size of the clusters vary while the number of cluster heads is kept constant. When running the simulations, we randomly generate feasible sets for each cluster. The number of feasible sets in a cluster is randomly chosen. The number of cluster heads is fixed at 20. Initially, we randomly generate the number of sensing nodes in each cluster while keeping the average number equal to 3. Then, we increase the number of sensing nodes similarly in each cluster until it reaches 18 (average size).
The results are presented in fig. 9, which illustrates a sub-network lifetime evolution when increasing the clusters' size and keeping the number of cluster heads constant.
It can be seen that the sub-network lifetime decreases as the clusters size increases. This is expected as when the cluster size increases, the corresponding cluster lifetime increases as well. Hence, each cluster head will spend more time performing both its neighbor's data relay and its own cluster management (its two roles simultaneously). As a result, it expends more quickly its energy which leads to network death in shorter time.

To further explore the performances of the proposed inter-cluster head communication scheme, we propose to study the influence of the clusters lifetime on the choice of the routes to deliver the data from each Cluster Head to the base station. An efficient routing scheme should alleviate from releying tasks cluster heads with long clusters lifetime since these cluster heads will spend longer time and then much more energy to manage their clusters than those with short cluster lifetime. To this end, we voluntarily generate clusters with considerably different lifetimes (through different sizes). This makes the corresponding clusters' lifetime standard deviation be large.

After several simulations, we compute the different cluster head lifetime and we remark that the corresponding standard deviation is considerably small (3.2% of the whole sub-network lifetime). This result proves that the majority of cluster heads die approximately at the same time. This also proves that flows are fairly distributed over the different cluster heads proportionally to the residual energy available at each one of them and also with considering the lifetime of each cluster i.e., proportionally to their role in the sub-network. The objectives of the proposed schemes are obviously attained.

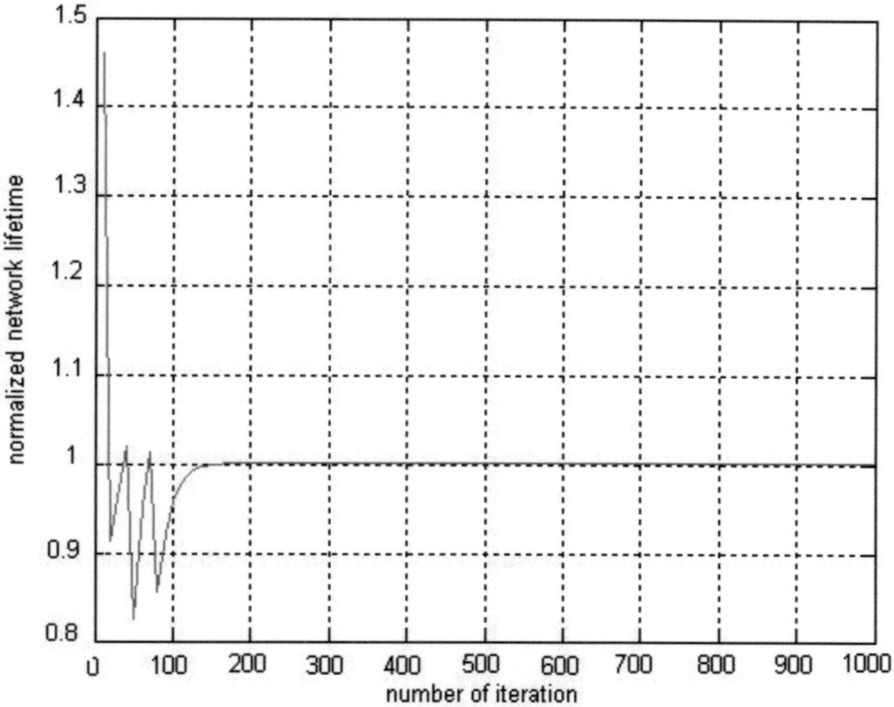

Fig. 8. Lifetime convergence.

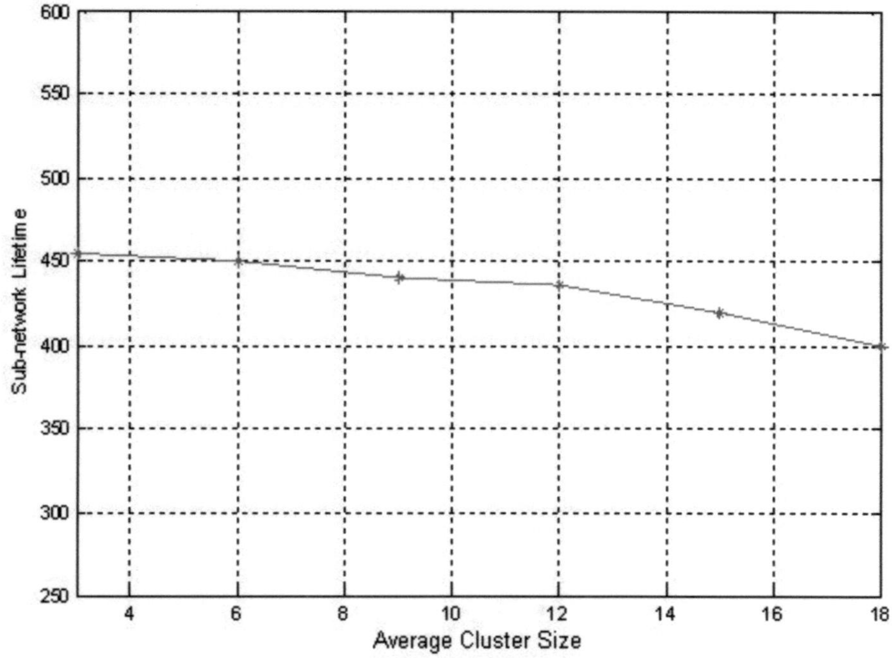

Fig. 9. Sub-network lifetime as a function of the clusters size.

8. Conclusion

The use of multiple mobile base stations in large-scale wireless sensor networks is necessary in order to cover large areas and to minimize energy consumption for data transmission operations. In this chapter, we proposed an energy efficient usage of multiple, mobile base stations to increase the lifetime of a two-tiered large-scale Wireless Sensor Network. Our approach uses a graph-partitioning algorithm to decompose the underlying network into balanced sub-networks. The energy usage is then optimized in each sub-network independently but in the same way using efficient base stations placement techniques that are optimized for small-scale WSNs. Performance results have shown that the proposed technique considerably enhances the network lifetime particularly when the base stations are moving along the periphery.

We have further proposed an optimal multi-hop routing scheme used within each sub-network independently to efficiently manage the communication between the Cluster Heads so that the entire network lifetime is elongated. Different strategies can be used, inside clusters, to manage intra-cluster communications. The proposed scheme simply adapt and fairly distribute the relaying flows according to Cluster Heads residual energy and their corresponding Clusters' lifetime duration, so that Cluster Heads with critical energy situations are alleviated from relaying operations. Simulation results have shown that we can compute a near optimal solution of the routing matrix that defines the optimal flow routing.

The overall dynamic framework that combines the above two schemes has been then described. It is defined as a cyclic algorithm that allows dynamic adaptation of the optimization process according to the current status of the whole network.

Using the graph-partitioning approach to improve energy consumption in large-scale WSNs is promising. We will focus in complementary and future work on more elaborated approaches for optimal multiple mobile base stations placement and WSN partitioning. In addition, efficient tools should be proposed to determine the optimal number of partitions and base stations to be used according to the WSN characteristics, applications' requirements and financial costs.

Moreover, we plan in future work to investigate further the mathematical resolution of the optimization algorithm corresponding to the inter-Cluster Head communication. The effect on energy consumption of the overhead generated by this scheme needs to be more deeply explored.

9. References

Chatterjee, M.; Das, S.K. & Turgut, D. (2002). WCA: A Weighted Clustering Algorithm for Mobile Ad hoc Networks, *Journal of Cluster Computing, special issue on Mobile Ad hoc Networking*, vol. 5, (march 2002), (pp.193-204).

Chen, Y. P.; Liestman, A. L. & Liu, J. (2006). A Hierarchical Energy-Efficient Framework for Data Aggregation in Wireless Sensor Networks, *IEEE Transactions on Vehicular Technology*, vol. 55, no. 3 , (May 2006) (789-796).

Chen, C.; Ma, J. & Yu, K. (2006). Designing Energy-Efficient Wireless Sensor Networks with Mobile Sinks, *Proceeding of ACM Sensys Workshop WSW*, pp. 1-9, USA, Colorado, October 2006, Boulder.

Chlebikova, J. (1996). Approximability of the Maximally balanced connected partition problem in graphs, *Information Processing Letters*, vol. 60, (sept 1996), (pp.225 - 230).

Even, G.; Naor, J.; Rao, S. & Schieber, B. (1997). Fast approximate graph partitioning algorithms, *Proceeding of the 8th Annual ACM-SIAM Symposium on Discrete Algorithms*, pp. 639-648, USA, LA, 1997, New Orleans.

Gandham, S.R.; Dawande, M. ; Prakash, R. & Venkatesan, S. (2003). Energy Efficient Schemes for Wireless Sensor Networks With Multiple Mobile Base Stations, *Proceeding of IEEE GLOBECOM*, pp. 377-381, USA, California, may 2003, San Francisco.

Ito, T.; Zhou, X. & Nishizeki, T. (2006). Partitioning a graph of bounded tree-width to connected subgraphs of almost uniform size, *Journal of discrete algorithms*, Vo. 4, Iss. 1, (March 2006), (pp. 142-154).

Johnson, D. B. & Maltz, D. A. (1996). Dynamic source routing in ad hoc wireless networks, *Mobile Computing*, Vol. 353, (August 1996), (pp. 153-181).

Kim, H.; Seok, Y.; Choi, N.; Choi, Y. & Kwon, T. (2006). "Optimal Multi-sink Positioning and Energy-efficient Routing in Wireless Sensor Networks, *Lecture Notes in Computer Science*, Vol.3391, Note(s):XVII, 936,Document:11, (sept 2006), (pp.264-274).

Luo, J.; Panchard, J.; Piorkowski, M.; Grosglausser, M. & Hubaux, J-P. (2006). Mobiroute: Routing towards a Mobile Sink for Improving Lifetime in Sensor Networks, *Proceeding of the International Conference on Distributed Computing in Sensor Systems*, pp. 480-497, USA, California, June 2006, San Francisco.

Luo, J. & Hubaux, J.-P. (2005). Joint Mobility and Routing for Lifetime Elongation in Wireless Sensor Networks, *Proceeding of IEEE INFOCOM*, pp. 1-10, USA, March 2005, Miami.

Mhatre, V.; Rosenberg, C.; Kofman, D.; Mazumdar, R. & Shroff, N. (2005). A Minimum Cost Heterogeneous Sensor Network with a Lifetime Constraint, *IEEE Transaction on Mobile Computing*, vol. 4, no. 1, (sept 2005), (pp 4-15).

Pan, J.; Hou, Y.; Cai, L.; Shi, Y. & Shen, X. (2003). Topology control for wireless sensor networks, *Proceeding of the 9th ACM Conference on Mobile Computing and Networking*, pp. 286-299, USA, CA, September 2003, San Diego.

Perillo, M. & Heinzelman, W. (2003). Optimal Sensor Management Under Energy and Reliability Constraints, *Proceedings of the IEEE Wireless Communications and Networking Conference, pp.1-6*, USA, Louisiana, March 2003, New Orleans.

Rabiner Heinzelman, W.; Chandrakasan, A. & Balakrishnan, H. (2000). Energy-Efficient Communication Protocol for Wireless Microsensor Networks, *Proceedings of the 33rd International Conference on System Sciences*, pp. 3005-3014, USA, January 2000, *Hawaii.*

Rabiner Heinzelman, W.; Chandrakasan, A. & Balakrishnan, H (2002). An Application-Specific Protocol Architecture for Wireless Microsensor Networks, *IEEE Transaction in Wireless Communications*, vol. 1, no. 4, (Oct. 2002), (pp. 660-670).

Slama, I.; Jouaber, B. & Zeghlache, D. (2006). Routing for wireless sensor networks lifetime maximization under energy constraints, *Preceeding of the 3rd International Conference on Mobile Technology, Applications and Systems*, pp.1-5, Thailand, October 2006, Bangkok.

Vincze, Z. ; Fodor, K. ; Vida, R. & Vidacs, A. (2006). Electrostatic Modelling of Multiple Mobile Sinks in Wireless Sensor Networks, *Proceeding of IFIP Networking Workshop on Performance Control in Wireless Sensor Networks*, pp. 30-37, Portugal, May 2006, Coimbra.

Wang, Z. M.; Basagni, S.; Melachrinoudis, E. & Petrioli, C. (2005). Exploiting Sink Mobility for Maximizing Sensor Networks Lifetime, *Proceedings of the 38th Annual Hawaii International Conference on System Sciences,* pp. 287.1, USA, January 2005, Washington, DC.

11

Dynamic Routing Framework for Wireless Sensor Networks*

Mukundan Venkataraman and Mainak Chatterjee
University of Central Florida
U.S.A.

Kevin Kwiat
Air Force Research Laboratory
U.S.A.

Abstract

Numerous routing protocols have been proposed for wireless sensor networks. Each such protocol carries with it a set of assumptions about the traffic type that it caters to, and hence has limited interoperability. Also, most protocols are validated over workloads which only form a fraction of an actual deployment's requirement. Most real world and commercial deployments, however, would generate multiple traffic types simultaneously throughout the lifetime of the network. For example, most deployments would want all of the following to happen concurrently from the network: periodic reliable sense and disseminate, real time streams, patched updates, network reprogramming, query-response dialogs, mission critical alerts and so on. Naturally, no one routing protocol can completely cater to all of a deployments requirements.

This chapter presents a routing framework that captures the communication intent of an application by using just three bits. The traditional routing layer is replaced with a collection of routing components that can cater to various communication patterns. The framework dynamically switches routing component for every packet in question. Data structure requirements of component protocols are regularized, and core protocol features are distilled to build a highly composable collection of routing modules. This creates a framework for developing, testing, integrating, and validating protocols that are highly portable from one deployment to another. Communication patterns can be easily described to lower layer protocols using this framework. One such real world application scenario is also investigated: that of predictive maintenance (PdM). The requirements of a large scale PdM are used to generate a fairly complete and realistic traffic workload to drive an evaluation of such a framework.

1. Introduction

First generation wireless sensor networks (hereafter 'sensornets') passively transported bits from one end to another. The subjective requirements of a payload are opaque to the network

*Approved for Public Release; Distribution Unlimited: 88ABW-2010-5582 dated 18 October 2010.

protocols, and the role of in-network processing is limited. Various routing protocols for transporting data in sensornets have been proposed: protocols for reliable routing (2; 7; 24; 27; 28), real time communication (10; 11; 14), energy aware communication (1; 12; 29), load balanced communication, aggregation centric approaches (19) and so on to name a few. Each such protocol typically optimizes a certain set of chosen parameters in making routing decisions, and is likewise validated over a workload that only generates that type of network traffic. This means that a deployment that adopts any given protocol has to build its *entire* deployment logic using the traffic type for which the protocol is optimized. To make sensornets a viable solution to real world problems, applications need to be built on top of arbitrary communication patterns, often with conflicting requirements. For example, a meaningful case of habitat monitoring would mostly demand all of the following communication patterns to co-exist: periodic network reports using reliable sense and disseminate, critical real time alerts when anomaly is detected, aggregation to suppress duplicates, network reprogramming to transfer bulk data, patched updates for continuous customization of the sensornet, best effort communication to transfer redundant information, interactivity with the network in the form of request-reply dialogs and so on. Naturally, no *one* routing protocol can cater to such varied application requirements within a given deployment. In other words, a given deployment can be viewed as a collection of various tasks (applications) which have very different, and often conflicting, communication requirements. The deployment goal is met when the goals of its constituent applications are fulfilled.

Secondly, there is little synergy across research efforts. Pressed by scarcity of energy and a need to focus on performance, protocols are developed with little thoughts to modularity and interoperability. Though a new application deployment would have a plethora of routing protocols to choose from, these protocols cannot be readily wired together to form a communicating framework due to compatibility problems. Compatibility problems largely arise because of the assumptions made on interface and data structure requirements. In general, and as Culler *et. at.* (5) note, a framework for testing, integrating and proposing protocols is largely missing.

This chapter presents a routing framework that makes an application's communication requirements visible to the lower layers, and allows activation of application specific processing. The traditional routing layer is replaced by a highly composable collection of routing decisions. Now, the routing logic is dynamically wired as per each packets requirements. The effectiveness of this strategy is demonstrated by gathering requirements and validating a fairly complete deployment scenario: predictive maintenance (PdM) using sensornets (15).

2. Protocol Description

2.1 Routing Framework Overview

Application payload presents a three bit preamble to the framework that describes it communication intent, The framework dynamically switches routing decisions based on the preamble bits. The routing layer is now a composable set of routing components that perform similar functions, but are optimized for different classes of traffic. The routing components carry a similar three bit signature that lets the framework know their applicability for a certain class of traffic. The selection of a routing component is hence a mapping between what the application demands and what the component has to offer. The routing components house core protocol features that cater to a particular application type. This allows components to evolve independently, and owing to their composable nature, allows seamless migration from one deployment experience to another. To unify interface assumptions, the routing components share a universal neighbor lookup table that houses relevant information about various neighbors which saves storage space and makes way for consistent interface assumptions. Component protocols make ranged queries into the neighbor table to derive the best candidate hop for a given application payload. Designing such a framework requires addressing challenges of *composing* routing protocols into core components, and *regularizing* the various interface and data structure requirements. But before that, we begin with the fundamental problem of making visible an applications demands to the stack.

2.2 Specifying Communication Intent

Applications need a way to express their communication requirements in a format that is both completely expressive and minimal. Various approaches have been taken to increase application visibility to the communication framework, and much of the effort has been to prioritize data. For example, the SP architecture (20) argues for a one bit descriptor that describes the urgency of a packet. On the Internet, DiffServ uses a class of service (CoS) field, three bits in length, to specify a priority value between 0 (for best effort traffic) to 7 (real time traffic). ISP's in the present day Internet also use similar tags to differentially route packets from preferred customers and offer them a higher quality of service. However, assigning a priority for any class of traffic is a highly subjective task, and these assignment rules would not be consistent from one deployment to another. Application naming should instead revolve around fundamental communication requirements rather than blatant priorities. This would allow one to construct meaningful and consistent inferences of a packets requirement for virtually any deployment. In effect, the following question is posed: *What is the minimum number of bits to let a deployment specify its fundamental communication requirements?*

To best characterize an application to the communication framework, various possibilities exist: number of recipients (anycast, multicast, broadcast), loss tolerance, delay tolerance, priority, sensitivity to congestion or link losses, soliciting retransmissions, tagging packets for aggregation, tagging packets for load balancing, control information v/s data packets and so on. However, communication patterns can be best described using three fundamental axes: nature of payload, reliability, and time criticality.

Nature of payload: Traffic in the network can be broadly classified as data or control traffic. Data traffic is all of the push based traffic generated by a mote. Control traffic is traffic used for control plane management (like beacons, ACK's etc.), and to a certain extent, user generated traffic. Sensor networks are more than just a collection of data gathering elements that autonomously report values. There is a need to accommodate a human element into the network for a variety of reasons. Users would want interact with the network with queries, and would want to receive responses in short turn around times. More importantly, administrators see the need to continuously customize the network with updates or network reprogramming. Interplay of data and control traffic in the network needs to be closely modeled as per a deployments requirements.

Reliability: The second axis to consider is tolerance to loss. Since a deployment consists of a host of nodes that are primarily involved in data gathering, some level of redundancy is sensed values is inherent. However, dictated by application requirements, some sensed values might be loss intolerant. For example, a deployment might want to construct a time series plot of sensed values from every mote for statistical purposes. This would require every mote to reliably transfer data periodically with minimum losses. Loss intolerance is also required for network re-programming or bulk reliable transfers, where binary updates need to be transmitted to node(s) reliably over the network.

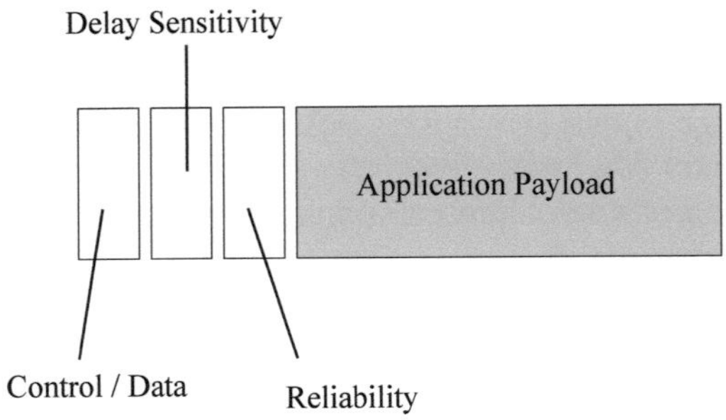

Fig. 1. 3-bit preamble appended to the application payload

Time Criticality: Some sensed value might be of no use if it does not make it to the destination within strict time bounds. Consider a deployment of sensors that report real time co-ordinates of moving objects to a camera which then pans and zooms to that area. If the co-ordinates reach the destination late, the camera might be unable to capture the desired frames of interest. Again, dictated by application requirements, there are time bounds for values of interest to make it to the destination.

2.3 Preamble Bits

Figure 1 shows the three axes of communications reduced to a simple three bit scheme that each packet carries. These bits are set or unset by the application programmer using simple API calls. The three bits, taken in combination, provide a characteristic description of an applications requirements. Figure 2 provides a complete combination of the preamble bits and their inference. For example, a beacon packet would carry a signature of [1,0,0], denoting a control packet that is loss tolerant and insensitive to delay. A data packet demanding reliability over delay would publish [0,0,1]. Similarly, a real time packet which only demands speed of delivery would publish [0,1,0]. An anomalous case is made when a packet demands both reliability *and* speed of delivery (bits [0,1,1] and [1,1,1]). Ensuring reliability inherently adds delay in transit, and such packets are interpreted as "mission critical", which see the need to both make it to the destination and in as short a time as possible. In general, the bits when combined with other information available in the packet headers make way for a powerful expression of precise communication demands. Note that the bits do not convey any notion of relative priority amongst packets, just a set of actual communication requirements.

2.4 Shared Neighbor Table: Unified view for routing protocols

Since the routing layer is now a collection of independent routing components, each component assumes the presence of various state information to be available to perform routing decisions. The neighbor table houses values such as the node-ID, energy available, congestion level, depth, link-quality estimate and a 'last heard' bit. The columns can easily be extended by future protocols. Since routing components share this table, it decouples core protocol features from interface assumptions and regularizes data structure requirements. This leaves the routing layer with a composable set of routing components that can be seamlessly ported across various research efforts.

Data(0)/ Control(1)	Real-Time(1)/ Non-Real-Time(0)	Reliable(1)/ Un-Reliable(0)	Inference
0	0	0	Unreliable, non Real Time, Data Packet
0	0	1	Reliable, non Real Time, Data Packet
0	1	0	Time Critical, Unreliable, Data Packet
0	1	1	Mission Critical Data packet
1	0	0	Unreliable, non Real Time, Control Packet
1	0	1	Reliable non Real Time, Control Packet
1	1	0	Real Time, Unreliable, Control Packet
1	1	1	Mission critical control packet

Fig. 2. Combinations of the preamble bits and their inferences for eight traffic types

2.4.1 Universal Beacon Packet

The wireless medium is unreliable, link qualities show time varying fluctuations, and node failures are not uncommon. Hence, there is a need to continuously monitor the state of the network by exchanging beacons at regular intervals. The various beacons assumed by component protocols are regularized by the use of a universal beacon packet. The creation and maintenance of the neighbor table is performed by the exchange of this universal beacon at periodic intervals.

The beacon packet contains information such as the ID of the node, the advertised depth, a one bit congestion indicator, and a one bit energy available indicator. A node sets the congestion bit if 75% of its buffer capacity is full. Likewise, the node sets the energy bit if less than 25% of its battery life is available.

2.4.2 Neighbor Table Creation and Maintenance

Even though there may be many potentially good neighbors available in the vicinity, there is a limit to the number of neighbors a mote can maintain due to limited resident memory . It is crucial to be able to identify the best possible neighbors and retain them in the neighbor list (27). A statistical approach to identify the goodness of a neighbor is employed.

At the start of the network, a node aggressively enters into its neighbor table every entry that advertises a depth lesser than its own. As time progresses, and owing to the presence of "gray" areas (27; 31), a node continues to hear more often from certain neighbors and less frequently from others. Since the beacon exchange is uniform among all participating nodes, a node maintains a ratio of the number of beacons received to the number that *should* have been received since the entry was made. This ratio (p) gives a good indication of the quality of link to a neighbor by estimating the number of transmissions required as $1/p^2$ (6; 27). Link estimations apart, this ratio is helpful in establishing the true depth of a node. Consider a node which receives a beacon from a neighbor advertising a depth of n with a reception ratio of 0.3, and another neighbor with a depth $n+1$ and a reception ratio of 0.9. The given node correctly infers its true depth to be $n+2$, since it has a stronger and more stable link to the latter neighbor. In general, a node infers its true depth to be one greater than the neighbor with the largest reception ratio and least depth.

The maintenance of the neighbor table is governed by a timer driven "scan and update" (SAU) module. The module is invoked each time the timer fires, whose periodicity is equal to the beacon interval, and offset of 10 seconds. As a node continues to receive beacons from neighbors present in the table, the various fields advertised in the beacon are used to update entries in the table. Associated with every entry is a "last heard" bit which is set to true upon the reception of a beacon. SAU unsets the bit each time it is invoked. SAU also updates the reception ratio of a neighbor by using the last heard bit. An entry is evicted from a neighbor table if one of two things happen: (i) a node discovers that a neighbor's depth is greater than its own; or, (ii) the reception ratio to that neighbor drops below 0.3 for a minimum statistical interval of 100 beacon cycles. Nodes evicted from the table are entered into a pool of blacklisted neighbors, who are not considered as potential neighbors for the next 500 beacon intervals. This prevents stale neighbors re-appearing in the table and gives an opportunity to other potential neighbors in the vicinity.

The notion of a "good" neighbor quickly blurs when there are multiple route selection components, each with their own yard of goodness measure. In general, entries are driven by their depth in the network more than by any other factor. This usually results in a good blend of neighbors with various values of depths, link qualities, congestion levels and energy values, and all of whose depths are lesser than that of the given node. Routing modules make ranged queries into the neighbor table to derive the next hop for a particular packet. However, it may so happen that a routing module fails to find any potential next hop candidate from the table. This could be either because the neighbor table is starved of good neighbors for that routing component, or due to unavailability of neighbors in the vicinity. When this happens, the SAP module is triggered with information about the routing component, and it marks for an insertion of an entry that matches the routing components needs from future beacons. This could possibly lead to an eviction of an entry from the table. The least recently used (LRU) algorithm is used to choose an entry for eviction.

2.5 Decomposing Routing Protocols to Core Components

The routing layer consists of a collection of routing components, with each component optimized for a certain class of traffic. Like the application payload, the component protocols also publish three bits that advertise their suitability for a particular application. It is crucial to decompose component protocols at the right granularity to allow rapid protocol development. As Cheng *et. al.* (3) note, choosing the granularity at which a protocol is to be decomposed is highly challenging: fine grained decomposition will result in un-necessary run time over head, while too large a granularity will fail to leverage code sharing among components and could result in significant re-implementations. They show that a composable set of of protocols that share common code result in smaller memory footprints, and are in general lucrative

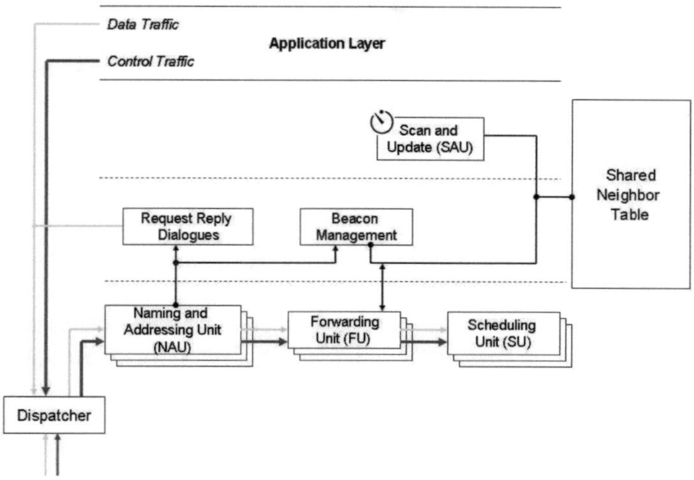

Fig. 3. Internal components of the routing protocols

considering resource scarcity on a mote. While the precise needs of future protocols are highly debatable, we decompose a routing protocol to the following components: the dispatcher, the naming and addressing unit, the forwarding unit, and the scheduling unit. The internal layout of a typical component protocol is shown in Figure 3.

Dispatcher: The first step in accepting a packet is to check the preamble bits to establish the right unit to which the packet is to be forwarded. The dispatcher is agnostic to intricacies of a protocols implementation or its naming and addressing format, and is shared by all the component protocols.

The *naming and addressing unit* (NAU) is the component that understands the addressing format for a protocol. Different addressing units are used by various protocols: while some protocols assume flat ID's, some other expect gradients or location information. Owing to the broadcast nature of the wireless medium, a node may receive a packet that it is not intended to. In effect, the NAU determines if a packet has made it to the destination, or if it needs to be forwarded further.

The *forwarding unit* (FU) is primarily concerned with establishing the next hop for a packet. Packets make it to this unit if the given node is not already the destination. The FU arrives at the next hop by making a raged query into the shared neighbor table. For example, a routing protocol that selects a route based on link qualities, congestion and depth at its neighbors would make a ranged query of the form $f(link\_quality, congestion, depth)$ into the neighbor table to arrive at a list of potential neighbors. The FU then applies its set of optimizations to arrive at the best neighbor(s) to which the packet is to be forwarded.

The *scheduling unit* (SU) is a buffer data structure that can order packets awaiting transmission. It can internally schedule the order in which packets are to be forwarded to the link layer for injection into the wireless medium. Certain protocols might need to buffer packets for potential retransmission, while other might need to hold packets to perform aggregation stalling for similar information to arrive. Certain other protocols might want to reschedule the order of injecting packets into the link layer to transmit mission critical alerts ahead of regular traffic.

Apart from these units, there are certain other units in the routing layer that accept packets for further processing. For example, beacon packets make it to insertion/eviction module, which considers the beacon as a potential neighbor. Likewise, control traffic originating from the base station is a typical query into the node, to which the node responds by performing certain local computations. The shared neighbor table is controlled by the scan and update unit, which periodically scans through the neighbor table to evict stale neighbors and enter them into a blacklisted pool.

2.6 The Dynamic Routing Framework

The effectiveness of using three bits to drive customization and protocol selection within the dynamic routing framework is next investigated. Apart from communication requirements of reliability or delay tolerance, the bits fundamentally divide the traffic as being of control or data type. Control or data traffic do not necessarily demand differential routing in terms of shorter or longer paths. In fact, most of control traffic (like beacons or ACKs) are for one hop use only. Beacons in particular are simply broadcast, and require no route selection or optimization. Communication requirements for control and data packets are best characterized by a need to be *scheduled* differentially inside the framework, and then, routed optimally. Differential treatment is achieved by the use of separate virtual queues for control and data traffic, while routing components are dynamically switched based on demands for reliability and delay. A detailed interaction diagram of packet with the framework is shown in Figure 4. Application presents a blend of control (C0-C3) and data packets (D0-D3). The suffix indicates the status of the reliability and real time bits. For example, packet D3 would have preamble bits set to [0,1,1], with the first bit indicating a data packet and last two bits account for the suffix 3. Selection of a routing component is driven by the suffix number. In other words, both Dx and Cx are offered the same routing component. Data or control traffic, however, are scheduled differently after route selection is established. Two virtual queues, one each for data and control traffic, take the incoming packets and schedule them for transmission to the lower layers of the stack.

2.6.1 Routing Protocols

An implementation of the component routing protocols in the framework is next discussed. Classifying traffic on reliability and delay leads to four combinations of routing components. A possible route selection strategy for four classes of traffic are described as follows:

Reliable, non Real Time Routing: Numerous approaches have been taken for this class of routing in the past (7; 24; 27; 28), to name a few. These routing protocols emphasize reliable delivery over time to deliver. A maximum of three link level retransmissions are used to ensure successful reception. Since a retransmission is costly in terms of energy and bandwidth, the protocol selects the strongest link to a neighbor closer to the destination. High estimations of link quality, or a form of ETX (7), is the most lucrative option here. The packet takes numerous short hops of high quality links to make it to the destination with minimum losses. Hence, speed is compromised by the use of numerous short and reliable hops. Delay tolerance for this class of traffic make it highly conducive to aggregation, since the SU can stall for self-similar data to arrive. Nodes with growing congestion and low energy are avoided as much as possible as the next hop.

Unreliable, Real Time Routing: Certain classes of traffic have a time associated with them, after which the data is virtually of little or no use. Speed of delivery is emphasized, and losses can be tolerated to a certain degree. Various protocols have been proposed for this class of

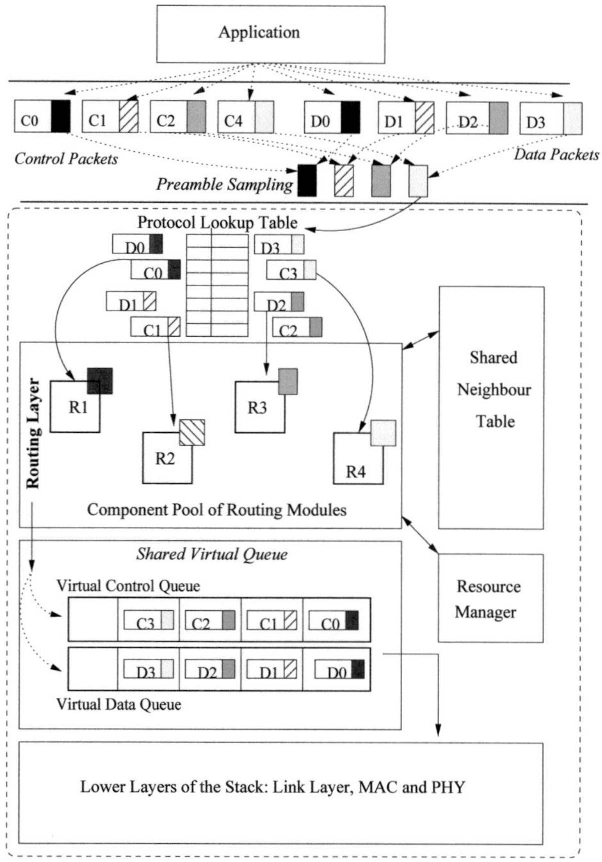

Fig. 4. Overview of the dynamic routing framework

traffic (4; 11; 23). This approach greedily forwards traffic to the base station primarily based on depth, avoiding congestion and low energy nodes, and ignoring link quality estimates or indications of poor packet reception.

Reliable, Real Time Routing: A contradictory requirement where packets need to maximize their chances of reception and yet want the shortest path to a destination. Waiting time for ACKs and retransmissions only add to the delay of the packet in transit. To both meet delivery deadlines and thwart link losses, the routing protocol first chooses nodes with the minimum depth. Of this list, nodes with the highest link quality are chosen, and the protocol injects *multiple* copies of the same packet to minimize link losses. The exact number of duplicates is based on the estimated link quality of a particular node. For example, if a neighbor with a link quality of p is chosen, the protocol transmits $\lceil 1/p \rceil$ packets with the same data. This form of routing is costly in terms of resources, but this type of traffic is reserved for mission critical alerts which is arguably rare in the network.

Unreliable, non Real Time Routing: This form of routing is applied to packets that neither demand reliable transmission, nor demand a speedy delivery. A good fraction of sense and disseminate traffic would fall into this category, where data values exhibit significant redundancy. Every packet adds little value to already existing information in the network, perhaps even repeating known observations. This form of traffic is also predominant in delay tolerant networks. This form of traffic is an excellent candidate for aggregation, where the SU unit can stall for long windows of time to effectively aggregate observed results to reduce the traffic to a handful of packets. This form of routing makes a compromise on both link quality and depth to route packets to the destination. Alternatively, this form of data centric routing could also utilize a minimum Steiner tree to maximize chances of aggregation.

2.6.2 Virtual Queues

Packets that egress the routing modules would need to be passed onto the link layer queue for subsequent transmission. Scheduling packets into the link layer queue is controlled by two virtual queues, one each for data and control traffic. The virtual queues do not maintain any independent buffer of their own: they only maintain the order in which packets shall be forwarded to the link layer. While packets await their transmission, they are physically housed in the scheduling unit of their respective component protocols. The scheduling unit internally marks a packet to be ready for transmission after it performs its set of optimizations (e.g., aggregation, reordering etc.). Fair scheduling is employed amongst the four scheduling units into the virtual queue, and from the virtual queues themselves to the link layer. This ensures that an even mix of control and data traffic are output from the framework, even though one or the other type may have a *different* packet arrival rate. In effect, this schedules both traffic types evenly by policing their resource sharing.

3. Related Research

The idea of classifying network traffic and offering differential services is not new to the networking community. The Internet witnessed the proposal of the DiffServ mechanism. DiffServ assigned integer numbers to derive Class of Service (CoS). The CoS varied between 0 and 7, and three bits were used to specify a value in that range. Each router would then apply a differential treatment to each class of traffic. Likewise, the two bit differentiated service architecture (17) is also in line with our approach. However, DiffServ in general required intelligence embedded in the routers, something which contradicted Internet's philosophy of end system intelligence. It was virtually impractical to maintain behavioral consistency amongst the millions of routers deployed in the Internet. Also, with the advent of optical interconnection lines, link errors were reduced to a bare minimum and the network capacity in terms of bandwidth shot up. The Internet could very well handle most of the demands placed on it in terms of voice, video and data traffic placed on it without the need for DiffServ. It was hence pondered that the DiffServ was a "technical solution to a technical problem that did not exist". The conditions in sensornets, however, are highly fertile for a differentiated service mechanism. Links are error prone, bandwidth is scarce, energy conservation is of utmost importance, and hop-by-hop processing and intelligence are the key to building networking technologies. Service differentiation is beyond a mere luxurious add-on in sensornets, it is more of a basic requirement. Meeting the subjective requirements of various traffic classes is a key to building successful deployments.

On the sensornet side, there has been an excellent series of work that address the problem of promoting synergy and providing sufficient abstraction for rapid protocol development. Work by Culler *et. al.* (5) argued that the abstraction that IP provides in the Internet can be provided at the link layer for sensornets, and this was substantiated by the SP architecture (20). The case for a modular network layer, where core network layer protocols were decomposed into an abstract set of modules, was then proposed by Cheng *et. al.* (3). Dunkels

et. al. (8) proposed an architecture that could adapt to heterogeneous protocols. The idea of building a dynamic routing framework also aligns with the ability to build dynamic networking stacks (18) with a major focus on performance and efficiency, which was proposed to be built on top of the x-kernel (13). However, none of the above architectures raise the issue of conflicting application requirements, and the presence of such a collection of applications in real world deployments. A logical extension of the above efforts is to address ways to specify precise communication patterns for every application, and building a dynamic protocol framework that can cater to such requirements.
The issue of supporting concurrent applications, with each independent application sharing the resources of a mote, has been addressed in Mate (16) and Melete (30). Mate is primarily concerned with code dissemination for network reprogramming. Melete significantly extends the Mate architecture, and addresses the issues of reliable storage and runtime sharing of multiple concurrent applications. Though it effectively substantiates the possibility of running concurrent applications on a mote, it does not address the conflicting communication requirements of various applications. Our work complements their by utilizing that fact that multiple concurrent applications can reside on a mote, and we address every application's communication requirements.
As far as routing protocols themselves are concerned, sensornets are at a stage where sufficient exploratory work has been performed. As each designer explored a potential use of sensornets, new protocols began to emerge. However, the journey has been fraught with severe challenges. As Fonseca *et. al.* (9) note, the trivial case of flooding and tree construction on motes took "three years and five successive implementations". We are not aware of dynamic and composable protocols that can cater to a wide range of application demands in sensornets. However, in our quest to build such a framework, we leverage the vast body of work available in sensornets.

4. Application Background

With the dynamic routing framework in place, the various intricacies of a fairly complete large scale real world deployment of Predictive Maintenance (PdM) is analyzed as an example target application. We begin by gathering the various communication demands of this deployment. The three bit scheme is employed in exposing these requirements to the communication framework. Using these requirements, we chart a comprehensive traffic workload to drive an evaluation of the dynamic routing framework.

4.1 Predictive Maintenance

Productivity is a key weapon for manufacturing companies to stay competitive in a growing global market. Increased productivity comes through increased availability. The need to be continuously available has driven many companies to focus on effective maintenance strategies. PdM is a set of techniques which help determine the condition of in-service equipment in order to predict when maintenance should be performed. PdM uses a variety of vibration analysis, oil analysis, infrared thermography and ultrasonic detection with an objective of reducing catastrophic equipment failures, and the associated repair and replacement costs much before the failure occurs. PdM enables machinery stakeholders to monitor, access, predict and in general understand the working of physical assets.
One alternate way of implementing PdM is by use of online motoring systems of wired sensors that can continuously gather statistics. Wired sensors are constrained by a 1:1 point-to-point connectivity links, and such systems are expensive to procure and install. This trend in

general has limited their penetration into the market. In fact, most small to medium sized companies would not procure them owing to their higher cost and lower return for investment, resulting in their mere 10% market penetration (15). Sensornets break ground by promising to be cheaper, providing higher returns on investment, flexibility owing to a broadcast wireless communication pattern, self-healing abilities when it comes to node failures and adapting to changing network conditions. Hence, there is great potential for sensornets to tap into the market provided they can completely satisfy all of a deployers requirements by allowing them a greater flexibility in communication patterns.

The case for PdM as a potential killer application for sensornets has been investigated by Krishnamurthy *et. al.* (15). They create an excellent groundwork for PdM using sensornets by demonstrating a viable return on investment. However, their focus is primarily on the effect of hardware architectural impact on performance. Their study is limited to PdM vibration analysis, where the deployment generates just one type of traffic: vibration signatures that need to be reliably transported to a base station. We significantly extend PdM's requirements in terms of complete *communication* requirements from a deployers point of view by enumerating various communication patterns possible. Accommodating the various communication patterns in such deployments is an important problem to solve. This is because the cost of deploying and running such an infrastructure would be dominated by *software and service management*, while diminishing hardware costs would make it cheap to network hundreds of motes. In effect, our study builds upon and complements their work, and we show that sensornets can be a powerful platform to perform PdM.

4.2 Requirements

Most deployers want a system of sensors to autonomously report sensed values and raise alerts, while they also want the presence of a human element in the process. Human intervention can be best characterized by an ability to interact with the deployment (query-response), apply patched code updates, re-program the network with newer protocols, manage or change various threshold values and so on. In general, there is a constant need to evolve, interact and update the deployment with changing requirements, which allows for continuous customization of the sensornet behavior.

Based on our survey of PdM requirements, we have found that deployers wanting to adopt sensornets as a means to build their maintenance regime demand the following aspects:

Req. 1. *Periodic reports*: Deployers expect the network to report sensed values on a regular interval. Such statistics make way for monitoring, analysis, trend forecasting and prediction of equipment behavior which can help them chart maintenance schedules. Apart from maintenance predictions, it also helps assess performance of new machinery within a warranty period.

Req. 2. *Streaming Real Time Values*: Apart from periodic updates of sensed values, there are numerous instances when there is a need to report values in *real time*. Examples could include protected zones with cameras which need to capture a video of a personnel when he walks into the facility. This would require sensors to report co-ordinates in real time to camera installations, which would then pan and zoom to capture streaming video.

Req. 3. *Query-Response*: Users or administrators need to query and get a response from the network at will. As with any interactive system, there is a need to maintain short turn-around times within user irritation levels. In fact, there are various levels of accuracy and delay associated with each response. Some responses need to be time critical, while others demand accuracy with more acceptable levels of turn around times. In general, while keeping turn

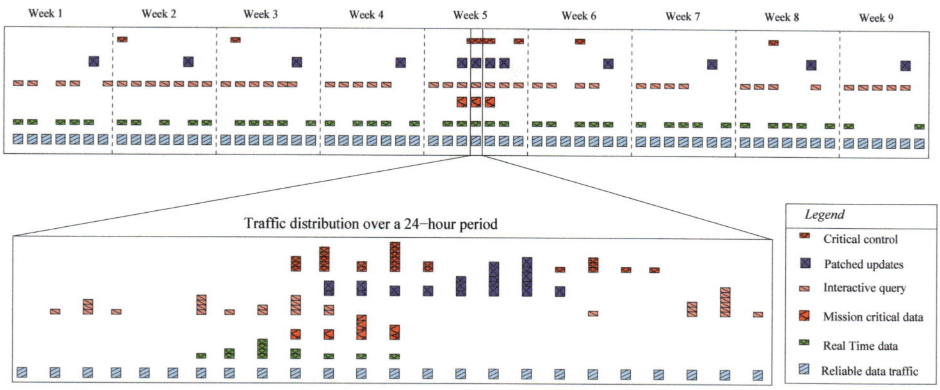

Fig. 5. Traffic generation for a 9-week PdM derived from requirements. The workload generates periodic samples, real time streams and mission critical alerts. The user/administrator interacts with the deployment at will with queries, weekly patched updates, and one major network reprogramming.

around times low, there is a need to model such subtle variations even within these interactive dialogs. Deployers further stress on the need to keep high levels of interactivity at all stages of the deployment, especially when mission critical events occur.

Req. 4. *Continuous Customization*: Requirements are not static, and would continuously evolve with time. This means that administrators see a need to constantly update or change network behavior. Examples could include changing threshold values for alerts, commanding specific motes to raise or lower sensing fidelity and so on. This in general emphasizes robust communication between an administrator and the network, and could even involve minor and incremental changes to network software.

Req. 5. *Network Reprogramming*: Change is inevitable, and must so with network protocols. Deployers want the liberty to completely reprogram an entire sensornet deployment when a more stable or efficient communication suite is available. Such modes of communication emphasizes on the need to reliably bulk transfer large pieces of code to the entire deployment.

Req. 6. *Mission Critical Alerts*: Apart from the above modes of communication, there are certain mission critical events which need to be reported reliably in as short a time as possible. Such events may trigger a cascade of events within the network. In fact, there is a need to accommodate a large number of interactive queries and commands to various motes when such events occur. In short, when serious anomaly occurs in the deployment, there is a need to gracefully alert, shutdown and recover as much as possible before the situation exacerbates.

These set of requirements are not isolated to the case of PdM alone. Virtually any practical sensornet deployment, if put under a microscope, would usually result in an enumeration of similar requirements to take place *concurrently* throughout the lifetime of a network with varying levels of emphasis on one requirement or another.

The requirements reveal significant variations in terms of communication needs. Variations exist in terms of reliability, time criticality, mission critical alerts and levels of interactivity with the deployment. We proceed to show how these set of requirements can be made visible to our framework using just three intent bits.

4.3 Translating Requirements using Three Bits

We show how just three bits can be used to make visible all of PdM's requirements to the communication framework. We proceed by identifying the nature of traffic (data/control) from the requirements, and subsequently analyze requirements of reliability and delay in transit. The first step is to distill control from data traffic. We broadly define data traffic to be traffic that a mote creates using a sensed value, for consumption at its destination. Control traffic is all of those packets that are used for control plane management (e.g., beacons), and traffic from end users. User generated traffic includes queries, responses, commands, and code updates. This is done to ensure that user traffic, along with control plane traffic, is scheduled differently from sensed data. While traffic is normally low during normal operations, critical events in the facility are usually coupled with a surge of traffic in the network. It is precisely at such times that congestion begins to surface (26). And again, at such times, it is highly likely that an end user or an administrator will issue queries, commands or updates into the network. In effect, while the data plane reports values of interest, user interactivity and network management at such times are not compromised. For the rest of the discussion on mapping various requirements using three bits, refer to Figure 2.

Data traffic can be easily grouped as per their requirements of reliability and speed of delivery. PdM requires periodic reports from every sensor to build a statistical record of the facility (Req. 1). This type of traffic publishes [0,0,1]: data traffic demanding high reliability and no urgency of delivery. Real time streams (Req. 2) would publish [0,1,0], stressing on urgency and lesser demands of reliability from the network, since the payload could be rendered useless if reception is delayed. Mission critical traffic (Req. 6) would publish [0,1,1], emphasizing on both urgency and reception.

Beacon packets publish [1,0,0]. Since they are usually broadcast into the medium, there is no notion of reliability or time to deliver. Interaction with the network (Req. 3), however, can be modeled at very fine granularities with these three bits. Using the axes of reliability and urgency, four levels of interaction are made possible. For example, queries of type [1,1,0] would result in very fast turn around times, but with compromised accuracy. Interactions using [1,0,1] preamble would be high on accuracy, but with longer turn around times. Critical real time alerts or updates shall present [1,1,1], which would ensure short turn around times *and* high accuracy. Packets for patched updates, commands or network reprogramming (Reqs 4 and 5) would publish [1,0,1], which stress on reliable delivery. To summarize, we tabulate the various requirements mapped as one or other traffic type in Figure 6.

We conclude our mapping by summarizing the criterions to validate PdM's requirements: (i) Types 1 and 5 should have high delivery ratios compared to other forms of traffic, with an acceptable compromise on transit delay; (ii) Types 2 and 6 should have short transit times, with a compromise on reliability; (ii) Types 3 and 7 traffic should have high delivery ratios *and* short transit time; (iv) interactivity with the deployment should be maintained at all time, even in times of growing congestion and mission critical events. This in general emphasizes differential treatment to data and control traffic.

4.4 A Realistic Traffic Workload

We use PdM's requirements to create a realistic traffic generation scenario of a typical operation to drive our evaluation. We are interested in fairly large scale operations, and seek to identify the various activities that shall typically take place over a large window of time. We reword the requirements in terms of traffic generation, and create a comprehensive workload to drive our evaluation.

All traffic from the motes converge towards the base station. The base station issues queries to random motes, and replies are likewise routed back to the base station. The base station applies patched updates to specific motes, while network reprogramming is applied to every mote. Every mote generates a sensed value at a periodicity of one hour (Type 1), which needs to be reliably transported to the base station. This would help the deployment build a statistical monitoring record of the facility to drive maintenance. Nodes also generate asynchronous real time streams of value, 5 packets at a time (Type 2). Real time streams are generated at a mote with a probability of 0.1 at any instant of time, which needs to be transported to the base station as fast as possible. End users generate queries at will modeled as a Poisson process. Each query has a random flavor to it (Type 4-7), some queries demand high interactivity (small turn around times), while other demand high accuracy (commands, queries or patched updates). The administrator performs a network wide patched update roughly once a week, when the network is fine tuned to slightly update or customize behavior (Type 5). Nodes observe mission critical data with a probability of 0.05, and generate 5 packets at a time (Type 3). Each time this happens, it triggers a surge of real time streams of values at motes close to the phenomena. The situation is responded to by the user, who generates multiple queries to the zone, and with a probability, issues numerous commands and patched updates.
We use this pattern to generate synthetic workloads for a 9-week evaluation period of a large scale plant employing PdM. The generation is better captured in Figure 5. Unless otherwise stated, we consistently use this traffic workload for every experiment.

5. Simulation Results

We evaluate our dynamic routing framework by both simulations and experiment on testbeds of MicaZ motes. We resort to a simulation based study to analyze behavior at scale where we simulate behavior for thousands of motes. Though a simulator tends to hide certain physical characteristics of actual motes, it allows us to demonstrate the effectiveness of our framework for operations at various scales and densities. Nevertheless, we make every effort to model network behavior to as close to reality as possible. Link behavior between any pair of nodes is closely modeled around results by Woo (27) and Zhao (31). Using that model, we also implement the typical broadcast behavior of all packet transmissions and receptions that take place in the network. As our next round of results, we also implement our framework and evaluate it on a 40-node MicaZ testbed.

5.1 Evaluation Criteria

Evaluation is driven by a comprehensive 9-week long workload detailed in Section 4.4. The workload generates a good blend of the various traffic, while carrying out the requirements of PdM. Our first set of results analyze behavior of a dynamic routing framework at scale. We vary the number of nodes from 4 to 1024, and show how PdM's subjective requirements for various traffic types are met at scale. In effect, we show that our framework can adapt to the demands of: (i) reliability for traffic Types 1 and 5; (ii) the real time nature of Types 2 and 6; (iii) the mission critical nature of Types 3 and 7, which demand both reliability *and* short turn-around times; and, (iv) interactivity with the networks using Types 5, 6, and 7[1]. We show that such diverse communication patterns can co-exist, and also meet their subjective

[1] We do not generate Type 0 traffic, since none of PdM's requirements map to this. Also, we do not profile Type 4 since it is made up of beacon and ACK packets in our workload.

Preamble bits	Inference	PdM Req	Traffic Type
[0,0,0]	Unreliable, non real time packet; Exhibits significant redundancy	X	Type 0
[0,0,1]	Reliable data traffic, demands high throughput; agnostic to delay	Req. 1	Type 1
[0,1,0]	Real time data stream, demands short delivery time; agnostic to loss	Req. 2	Type 2
[0,1,1]	Mission critical data, demands reliability *and* speedy delivery	Req. 6	Type 3
[1,0,0]	Unreliable, non real time control packet;	Beacons, ACKs	Type 4
[1,0,1]	Reliable control traffic; agnostic to loss	Req. 3, 4, 5	Type 5
[1,1,0]	Real time control packet; demands speedy delivery	Req. 3	Type 6
[1,1,1]	Mission critical control; demands reliability *and* speedy delivery	Req. 6	Type 7

Fig. 6. PdM's requirements mapped to various traffic types

demands. Since the workload used mimics PdM operations, our results in turn validate the deployments goal for operations at scale.

5.2 Delivery Ratio

We begin by analyzing the average end to end success rate for all the traffic generated by the application (Figure 7).

We see good delivery ratios for reliable data traffic (Type 1), reliable control traffic (Type 5) and mission critical control information (Type 7). High delivery ratio for Type 1 traffic validates PdM's expectation of samples from every mote at periodic intervals. Type 5 traffic denotes interactivity with demands on accuracy, which are likewise met. Reliable traffic in general is driven by a retransmission upon failure, which significantly raises chances of successful packet reception. Similarly, the odds that mission critical traffic (Type 7) makes it to the destination are also high. Mission critical control traffic would be routed through the control queue, and it maintains a high delivery ratio of around 0.9. Delivery ratio is poor for real time data traffic (Type 2), which is routed greedily based on a shortest path algorithm.

While similar routing is applied to both Type 1 and Type 5 traffic, the delivery ratio of reliable data (Type 1) dips a little with increasing number of nodes as compared to Type 5. This is mostly because data traffic is mapped onto a separate virtual queue than control traffic. While Type 1 would face rising congestion with increasing number of nodes, Type 5 traffic hardly witnesses any of this. Likewise, delivery ratio for real time queries (Type 6) are significantly higher than real time data streams (Type 2). Mapping data and control traffic in separate virtual queues has enabled us to provide high levels of interactivity with the network.

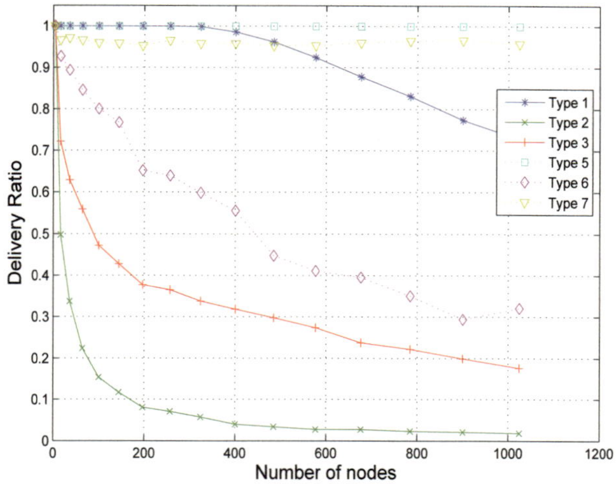

Fig. 7. Delivery ratio for different traffic types

5.3 End to end delay

We profile the average end to end delay experienced for all traffic that makes it to the destination. Delay in the network is primarily a function of the number of hops a packet takes, and the queuing delay at every hop. As a packet increases its hop count to destination, end to end delay intuitively increases.

The average end-to-end delay for all traffic types for increasing number of nodes is shown in Figure 8. All traffic demanding speedy delivery experience short transit time. Real time data experiences least delay (Types 2 and 6), reliable traffic experience maximum delay (Types 1 and 5), while mission critical traffic experiences delay that is only marginally more than real time data.

Small delay profiles for Types 2 and 6 directly validate PdM's requirements of real time streaming communication. Likewise, short delivery times of mission critical traffic (Types 3 and 7), coupled with their high delivery ratios, validates PdM's requirements for mission critical communication. High delay values for Types 1 and 5 only emphasizes on their ability to select numerous short hops of high quality.

The interplay of control and data traffic, which receive differential scheduling due to separate virtual queues, is also captured in the plot. Overall, data traffic experiences more delay than control traffic. For example, though Types 1 and 5 are offered the same routing, the overall delay for Type 1 is almost twice that of Type 5. This highlights the ability of the framework in meeting PdM's overall requirements of maintaining network interactivity. With PdM's overall objectives met, we now analyze the causes of losses in the network.

5.4 Link Losses

Effectiveness of the various routing components in their ability to choose the right path for every traffic type is best characterized by profiling the link loss distribution. Link loss is a typically attributed to the unreliable wireless medium, where packets are corrupt or lost while in transit. Routing components that stress on reliability need to understand the nature

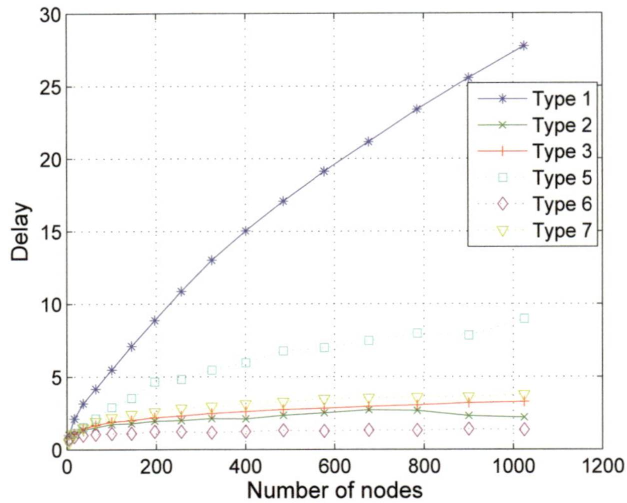

Fig. 8. End to end delay for different traffic types

of links, and use these to derive a suitable next hop while keeping the requirements of the payload consistent.

We profile link losses for various traffic types in Figure 9. As the number of nodes in the network increases, so does the effective number of hops that a packet takes to reach its destination. This in effect increases the probability of a link loss. Real time data streams (Type 2) experience maximum link losses, largely because of the nature of route selection which greedily forwards traffic to nodes closest to the base station. Reliable traffic (Types 1, 5), however, make ranged queries into the neighbor table with high thresholds of link estimates. Likewise, they experience nearly zero link related losses in the network. Because of inter-node spacing in this experiment (10 feet), neighbors closest to a node do not fall over into the gray area. Mission critical alerts (Type 7), likewise experience low values of link losses since they thwart link error by multiple copies per packet transmission.

5.5 Congestion losses

Congestion occurs when nodes inject more packets than the network can handle. While our workload generates traffic that can normally be serviced by the network, congestion does occur for a variety of reason. First, all data traffic is destined to one node (base station). Hence, all of the network's traffic converges towards nodes closer to the base station to be routed via them. Even though we try to avoid congested nodes in route selection, a point comes when all neighboring options for a node are congested. Congestion particularly increases with rising number of nodes in the network, which simply translates to rising traffic levels for nodes near the base station to service. Based on PdM's requirements, we also notice that congestion is likely to occur when serious anomaly is detected. When a mission critical failure is noticed, a surge of events takes place in the network. Nodes report mission critical alerts, and some other nodes in the vicinity would begin to send streams of real time values. The end user or administrator would add on to this by issues commands, queries and triggering actions. In our workload, both these causes are sufficiently represented. We now analyze the

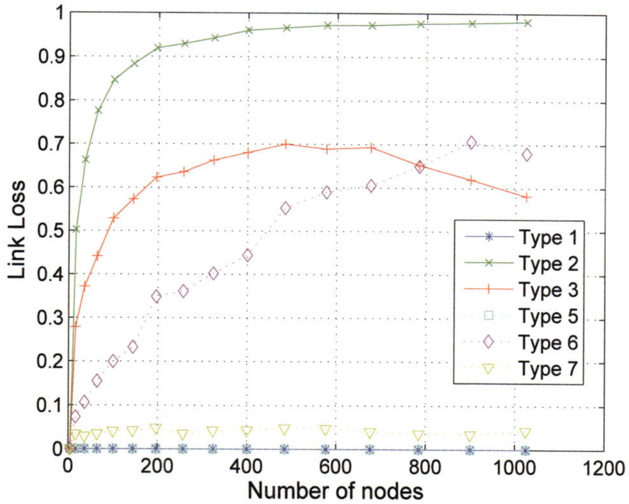

Fig. 9. Fraction of packets loss due to link losses

role congestion plays in the network, and profile the various congestion related losses for the traffic types.

The fraction of packets lost due to congestion are shown in Figure 10. For network scales of a few hundreds of nodes, congestion is not really a pressing problem because of the low duty cycle of nodes. However, congestion starts to surface for networks with more than 300 nodes, primarily because of increased load on nodes closer to base station. We notice that Type 1 traffic witnesses maximum congestion related losses. As packets begin to approach the base station, traffic from other types (real time streams or mission critical alerts) would try to avoid congested nodes nearby and choose low quality links with faster transit times. At this same stage, reliable traffic would take two or three additional hops to ensure high quality links.

It is interesting to see that mission critical data (Type 3) also experiences congestion losses. This has a few implications for congestion control in general. When mission critical anomaly is detected, activity of motes suddenly peaks. Various nodes start to simultaneously inject traffic into the network. Congested links, coupled with multiple copies per packet from Type 3, only makes matters worse for mission critical data. This suggests that dropping *any* packet in a FIFO manner, as most current congestion control schemes do, only undermines performance. In general, utilizing information about nature of payload and dropping packets of relatively lesser importance should be an added metric to future congestion control algorithms.

Lastly, we also observe that control traffic (Types 5, 6, 7) do not experience congestion drops. This means that even in times of congestion, interactivity is kept high because control traffic is offered differential scheduling. This further validates PdM's requirements of maintaining high interactivity with the network even in times of congestion and mission critical events.

5.6 Interactivity with deployment

While the effects of scheduling control and data traffic differentially are brought out, we seek to understand the interplay of various types of interactive control traffic within the virtual 'control' queue. Three levels of interactivity are made possible by the use of preamble bits:

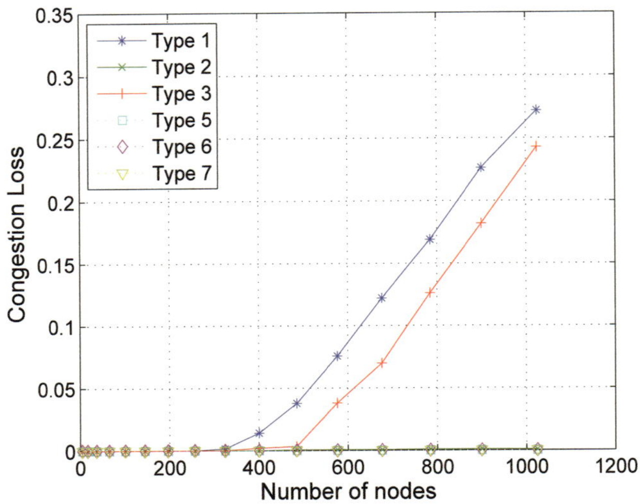

Fig. 10. Packets lost due to congestion for various traffic types. Shown in the figure is the fraction of packets lost due to congestion over all packets lost in transit.

reliability driven queries (Type 5), real time queries (Type 6), and mission critical interaction (Type 7). We analyze the average round trip times (RTT) for various kinds of queries into the network. Our workload generates queries to random motes in the network at various distances. For a 9-week long interaction, we summarize the interactivity times for networks at scale.

The interaction RTTs are plotted in Figure 11. Dynamic routing plays a major role in ensuring that interactivity times are kept low for real time queries (Type 6), acceptable for mission critical queries (Type 7) and relatively higher for reliability driven queries (Type 5). Coupled with high delivery ratios of Types 5 and 7, and short turn around for Type 6, we successfully meet the subtle variations in interactivity demanded by PdM.

5.7 Average Path Distribution

We finally characterize the path distribution statistics for various traffic types in the network (Figure 12). This simulation was run for a collection of 1024 nodes arranged using a 32x32 grid, with a 10 feet inter-node spacing. For every packet received at the base station, we measure the number of hops that it took build a frequency distribution for various hop counts. The curve is representative of route selection since each traffic type generates sufficient number of packets at various distances from the base station.

Requirements of PdM apart, nature of route selection is best captured in this plot. Reliable traffic (Types 1 and 5) take numerous short hops of high quality links, and register large hop counts. Real time traffic (Types 2 and 6), which is routed greedily based on shortest paths, takes the least number of hops. Mission critical data are offered hops that range in between reliable and real time traffic.

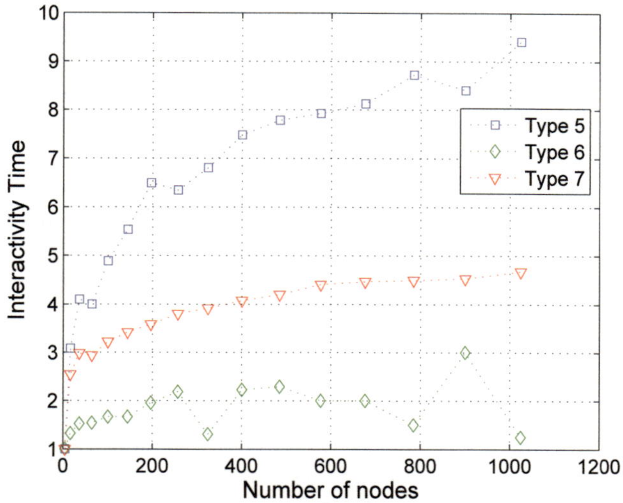

Fig. 11. Average round trip times for interactive queries with the deployment

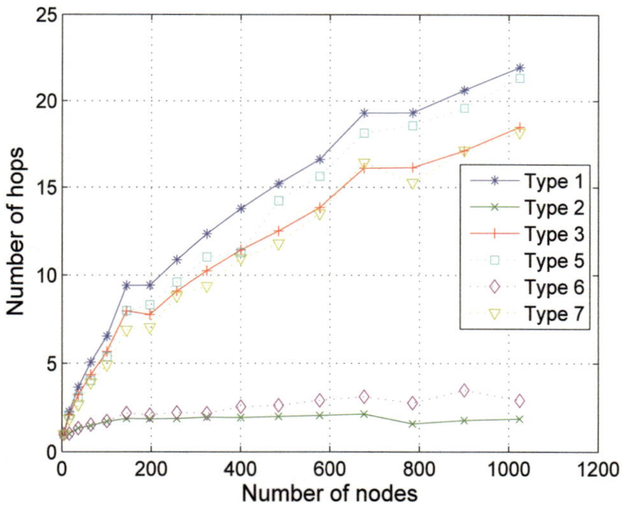

Fig. 12. Path distribution statistics for various traffic types for a deployment of 1000 nodes

6. Discussions

Exposing application requirements creates a plethora of in-networking possibilities. We show the impact of creating a dynamic network architecture with the use of the preamble bits at various levels of the stack: applications, protocol validation, energy efficiency, aggregation, fairness and differentiated services.

Application Programming: With data becoming self identifying, application programming is agnostic to the lower layers of the stack. Since the preambles are not protocol dependent, the scheme is guaranteed to work even when the mapping between the preamble and a particular protocol change over time. The framework in turn understands the nature and requirements of the payload, and accordingly wires a routing module to serve the purpose. We have diverged from priority based approaches, where our three bit scheme provides no notion of relative importance of a packet. We believe this is important, because the subjective notions of a packets relative priority are often debatable, inconsistent and prone to errors. Application programming is virtually error free, since it is not possible to confuse between a packets requirements, whereas it might be really hard to choose between a priority level of 5 or 6 for a range from 0-7 as in the case of DiffServ.
Protocol Validation: Protocols in sensornets are validated over a set of workload at least thought to be representative of the entire application domain. Most protocols are evaluated on a workload for which the protocol is optimized for. For example, a real time routing protocol is evaluated for a workload that emphasizes real time traffic alone. Most practical deployments would generate a workload of which real time communication is only a part of the requirement. Hence, a protocol's behavior in the face of real world deployment traffic is largely unknown. A dynamic routing framework, which can house various types of protocols optimized for various other types of traffic could form the basis of applying real-life workload to evaluate any alternative choice of protocol optimized for a given traffic type.
Energy Efficiency: Energy conservation has been an integral motive of almost every protocol proposed thus far. This trend in general has led to various "energy efficient" protocols with crippled communication abilities. Majority of energy drain happens at a nodes communication interface, and this trend shall continue to hold true well into the future. While computational subunits can be expected to improve in terms of energy per unit computation (e.g. Moore's Law), communication interfaces are governed by static laws of physics. Research by Pottie and Kaiser (21) shows that over 3000 instructions could be executed for the *same* energy cost of transmitting one bit wirelessly by 100 meters. The only foreseeable way to conserve energy is to *compute* more, and communicate *wisely*. With the application's requirements becoming visible, a whole host of in-network processing is now made possible to take the most appropriate action for every packet.
Aggregation: This domain has been widely studied in the sensornet domain, with excellent contributions in literature. However, aggregation cannot be abstracted as a component that generally applies to *any* payload. Aggregation comes with a little cost of delay in terms of processing, and in some cases, stalling for potentially related information to arrive. Delay sensitive data is generally not very amenable to aggregation.
Fairness: Presently, fairness in sensornets is not a well defined notion. Classical notions of fairness, where every player gets an equal share, needs a redefinition in the case of sensor nets. Not all nodes in the sensornet are the same, and neither are all packets equally important. The authors in IFRC (22) raise whether fairness is a reasonable initial design goal in a sensornet. While this may be difficult to answer without extensive deployment experience, what is generally lacking is a *basis* for defining fairness. For example, which packets should be transmitted in what order, or at what power level, or who should be dropped when congestion grows are questions that seek answers.
Differentiated Service: Traditional data networks passively transport bits from one end system to another. To the network, the payload is opaque as far as requirements are concerned, and the role of in-network processing is limited. Protocols and policies ought to act according

to the relative importance of a particular packet in question. Not all packets in a sensornet are of equal importance. For example, during times of congestion, dropping an arbitrary packet makes little sense: a packet carrying a critical alert information is clearly more important than a packet carrying regular sense-and-disseminate data. Similarly, a node with little energy might not receive mundane data, but might be willing to forward critical information when it offers a shorter path. Service differentiation is a strong incentive in sensor networks, largely because typical deployments are governed by higher level logic dictating requirements.
Richer Possibilities: The preamble bits and the dynamic framework provide a basis for adaptive protocols, allowing richer interactions with the deployment. It provides a powerful platform for user driven customization of the infrastructure, allowing new services to be deployed at a faster pace.

7. Conclusions

Typical deployments would consist of multiple concurrent applications, all of whose success leads to the fulfillment of a deployments objective. With every application placing its own subjective communication demand on the framework, there is an urgent need to both expose these requirements to the communication framework, and dynamically customize behavior for every type of application. We have presented a simple scheme of using just three intent bits to completely describe communication patterns the stack, and we use this to drive a dynamic routing framework that customizes its routing behavior for every packet type in the system. We have proved its effectiveness in meeting the demands of a fairly complete deployment of industrial monitoring using PdM, where we analyzed behavior at scale for thousands of nodes, and implemented a prototype of a 40 node wireless testbed.
Diversity in application requirements for sensornets has led to an explosion of network protocols. Protocol developers focus performance for a particular traffic type, and likewise validate protocols for that type of traffic. Our framework allows for rapid protocol development, integration and validation in the face of realistic workloads. With a need to emphasize performance, developers further make assumptions about interfaces and functionalities that further limits synergy across research efforts. In our quest to build a configurable framework, we have regularized interface assumptions to distill core protocol features as individual components. This would ensure that the core components can evolve independently, and research efforts on any component can be seamlessly ported across deployments.
The role of in-network processing is currently limited in sensornets. With the application requirements made visible to the stack, there is great potential to design application specific processing at every node. Our dynamic routing is just one example of using the requirements to switch routing behavior at the network layer. In general, there is excellent potential for designing medium access protocols, scheduling protocols, congestion control algorithms and energy efficiency modules at various layers of the stack using the preamble bits.

8. References

[1] D. Braginsky and D. Estrin. "Rumor routing algorithm for sensor networks", *Proc. First ACM International Workshop on Wireless Sensor Networks and Applications, (WSNA)*, Sept 2002.

[2] Q. Cao, T. Abdelzaher, T. He, and R. Kravets. "Cluster-Based Forwarding for Reliable End-to-End Delivery in Wireless Sensor Networks", *IEEE Infocom*, May 2007.

[3] T. E. Cheng, R. Fonseca, S. Kim, D. Moon, A. Tavakoli, D. Culler, S. Shenker, and I. Stoica. "A modular network layer for sensorsets", *Proc. 7th Symp. on Operating Systems Design and Implementation (OSDI)*, Nov. 2006.

[4] O. Chipara, Z. He, G. Xing, Q. Chen, X. Wang, C. Lu, J. Stankovic, and T. Abdelzaher. " Real-Time power-aware routing in sensor networks", *Proc. IEEE International Workshop on Quality of Service (IWQoS)*, June 2006.

[5] D. Culler, P. Dutta, C. T. Ee, R. Fonseca, J. Hui, P. Levis, J. Polastre, S. Shenker, I. Stoica, G. Tolle, and J. Zhao. "Towards a sensor network architecture: Lowering the waistline", *HotOS X*, June 2005.

[6] D. D. Cuotu, D. Aguayo, B. Chambers, and R. Morris. "Performance of Multihop Wireless Networks: Shortest Path is Not Enough", *First workshop on Hot topics in Networks (HotNets-I)*, Oct. 2002.

[7] D. S. Couto, D. Aguayo, J. Bicket, and R. Morris. "A High-Throughput Path Metric for Multi-Hop Wireless Routing", *ACM Mobicom*, Sept 2003.

[8] A. Dunkels, F. Osterlind, and Z. He. "An adaptive communication architecture for wireless sensor networks", *ACM Sensys*, Nov. 2007.

[9] R. Fonseca, S. Ratnasamy, J. Zhao, T. E. Cheng , D. Culler, S. Shenker, and I. Stoica. "Beacon-Vector Routing: Scalable Point-to-Point Routing in Wireless Sensor Networks", *Proc. Usenix NSDI*, July 2005.

[10] J. L. Gao, "Energy efficient routing for wireless sensor networks", Ph.D. thesis, Electrical and Computer Engineering Department, UCLA, June 2000.

[11] T. He, J.A. Stankovic, C. Lu, and T. Abdelzaher. "SPEED: A Stateless Protocol for Real-Time Communication in Sensor Networks", *Proc. ICDCS'03*, May 2003.

[12] W. R. Heinzelman, A. Chandrakasan, and H. Balakrishnan. "Energy-efficient communication protocol for wireless microsensor networks", *Proc. of 33 Hawaii International Conference on Systems Science (HICSS)*, Hawaii, Jan 2000.

[13] N. C. Hutchison and L. L. Peterson"The X-Kernel: An Architecture for Implementing Network Protocols", *IEEE Trans. on Soft. Engg.*, 17(1), Jan. 1991.

[14] C. Intanagonwiwat, R. Govindan, and D. Estrin. "Directed Diffusion: A Scalable and Robust Communication Paradigm for Sensor Networks", *ACM/IEEE Mobicom'00*, Aug 2000.

[15] L. Krishnamurthy, R. Adler, P. Buonadonna, J. Chhabra, M. Flanigan, N. Kushalnagar, L. Nachman and M. Yarvis. "Design and deployment of industrial sensor networks: experiences from a semiconductor plant and the north sea", *ACM Sensys*, Nov. 2005.

[16] P. Levis and D. Culler, "Mate: A Tiny Virtual Machine for Sensor Networks", *Proc. Intl. Conf. on Architectural Support for Programming Languages and Operating Systems (ASPLOS)*, Oct 2002.

[17] K. Nichols, V. Jacobson, and L. Zhang. "A Two-bit Differentiated Services Architecture for the Internet". Internet Engineering Task Force, RFC 2638, July 1999.

[18] S. W. O'Malley and L. L. Peterson. "A dynamic network architecture", *ACM Transactions on Computer Systems (TOCS)*, 10(2), May 1992.

[19] S. Pattem, B. Krishnamachari, and R. Govindan. "The Impact of Spatial correlation on Routing with Compression in Wireless Sensor Networks", *ACM/IEEE IPSN*, April 2004.

[20] J. Polastre, J. Hui, P. Levis, J. Zhao, D. Culler, S. Shenker, and I. Stoica, "A unifying link abstraction for wireless sensor networks", *ACM Sensys*, Nov 2005.

[21] G. J. Pottie and W. J. Kaiser, "Wireless Integrated Network Sensors", *Communications of the ACM*, Vol. 43(5), May 2000.

[22] S. Rangwala, R. Gummadi, R. Govindan, and K. Psounis. "Interference-Aware Fair Rate Control in Wireless Sensor Networks", *ACM Sigcomm*, Sept 2006.
[23] D. Sharma, V. Zadorozhny, and P. Chrysanthis. "Timely data delivery in sensor networks using whirlpool", *Proc. 2nd international workshop on Data management for Sensor Networks*, Aug. 2005.
[24] F. Stann and J. Heidemann, " RMST: reliable data transport in sensor networks", *First IEEE Intl. Workshop on Sensor Network Protocols and Applications (SNPA)*, May 2003.
[25] M. Venkataraman, K. Muralidharan, and P. Gupta. "Designing New Architectures and Protocols for Wireless Sensor Networks: A Perspective", *IEEE Secon*, Sept 2005.
[26] C.Y. Wan, S.B. Eisenman, and A.T. Campbell. "CODA: Congestion Detection and Avoidance in Sensor Networks", *ACM Sensys*, 2003.
[27] A. Woo, T. Tong, and D. Culler. "Taming the Underlying Challenges of Reliable Multihop Routing in Sensor Networks", *ACM Sensys*, 2003.
[28] C. Y. Wan, A. T. Campbell, and L. Krishnamurthy. "Pump-slowly, fetch-quickly (PSFQ): a reliable transport protocol for sensor networks", *IEEE Journal on Selected Areas in Communication (JSAC)*, 23(4), pp. 862–872, April 2005.
[29] M. A. Youssef, M. F. Younis, and K. Arisha. "A constrained shortest-path energy-aware routing algorithm for wireless sensor networks", *Proc. of IEEE WCNC*, March 2002,
[30] Y. Yu, L. Rittle, J. LeBrun, and V. Bhandari. "MELETE: Supporting Concurrent Applications in Wireless Sensor Networks ", *ACM Sensys*, Nov 2006.
[31] J. Zhao and R. Govindan. "Understanding Packet Delivery Performance In Dense Wireless Sensor Networks", *ACM Sensys*, Nov 2003.
[32] University of California, Berkeley. TinyOS CVS Repository at SourceForge. `http://sf.net/projects/tinyos`. June 2007.
[33] MicaZ motes specification. `www.xbow.com/products/ product_pdf_files/ wireless_pdf/6020-0060-01_a_micaz.pdf`

12

Routing Security Issues in Wireless Sensor Networks: Attacks and Defenses

Jaydip Sen
Innovation Lab, Tata Consultancy Services Ltd.
India

1. Introduction

Wireless Sensor Networks (WSNs) are rapidly emerging as an important new area in wireless and mobile computing research. Applications of WSNs are numerous and growing, and range from indoor deployment scenarios in the home and office to outdoor deployment scenarios in adversary's territory in a tactical battleground (Akyildiz et al., 2002). For military environment, dispersal of WSNs into an adversary's territory enables the detection and tracking of enemy soldiers and vehicles. For home/office environments, indoor sensor networks offer the ability to monitor the health of the elderly and to detect intruders via a wireless home security system. In each of these scenarios, lives and livelihoods may depend on the timeliness and correctness of the sensor data obtained from dispersed sensor nodes. As a result, such WSNs must be secured to prevent an intruder from obstructing the delivery of correct sensor data and from forging sensor data. To address the latter problem, end-to-end data integrity checksums and post-processing of senor data can be used to identify forged sensor data (Estrin et al., 1999; Hu et al., 2003a; Ye et al., 2004).

The design and implementation of secure WSNs must simultaneously address several difficult research challenges. First, wireless communication among the sensor nodes increases the vulnerability of the network to eavesdropping, unauthorized access, spoofing, replay, and *denial-of-service* (DoS) attacks. Second, the sensor nodes themselves are highly resource-constrained in terms of limited memory, CPU, communication bandwidth, and especially battery life. These resource constraints limit the degree of encryption, decryption, and authentication that can be implemented on individual sensor nodes, and call into question the suitability of traditional security mechanisms such as computation-intensive public-key cryptography for such resource-constrained sensor nodes (Carman et al., 2000). Third, WSNs face the added physical security risk of individual sensor nodes falling into wrong hands. Sensor nodes that are physically deployed in the field can be captured by an intruder, and can then be subject to attacks from the potentially well-equipped intruder in order to compromise a single resource-poor node. Following a successful attack, a compromised sensor node could then be used to launch such malicious activities as advertising false routing information, and launching DoS attacks from within the sensor network.

The combined threats introduced by increased physical security risk and severe resource constraints motivate the following design philosophy to achieve secure WSNs: assume that a well-equipped intruder can compromise individual sensor nodes, but secure the overall design of the WSN so that these intrusions can be tolerated and the network as a whole remains functioning despite such localized intrusions. More precisely, the objective is the design of an intrusion-tolerant WSN that has the property that a single compromised node can only disrupt a localized portion of the network, and cannot bring down the entire sensor network. This design objective of intrusion tolerance for secure WSNs must provide protection against two classes of attacks that could bring down an entire sensor network: DoS-type attacks and routing disruption attacks that propagate erroneous control packets containing false routing information throughout the network.

The focus of this chapter is on *routing security* in WSNs. Most of the currently existing routing protocols for WSNs make an optimization on the limited capabilities of the nodes and the application-specific nature of the network, but do not any the security aspects of the protocols. Although these protocols have not been designed with security as a goal, it is extremely important to analyze their security properties. When the defender has the liabilities of insecure wireless communication, limited node capabilities, and possible insider threats, and the adversaries can use powerful laptops with high energy and long range communication to attack the network, designing a secure routing protocol for WSNs is obviously a non-trivial task.

One aspect of sensor networks that complicates the design of a secure routing protocol is *in-network aggregation* (Shrivastava et al., 2004; Madden et al., 2002; Przydatck et al., 2003; Zhu et al., 2004a). In more conventional networks, a secure routing protocol is typically only required to guarantee message availability. Message integrity, authenticity, and confidentiality are handled at a higher layer by an end-to-end security mechanism such as SSH or SSL. End-to-end security is possible in more conventional networks because it is neither necessary nor desirable for intermediate routers to have access to the contents of messages. However, in sensor networks, in-network processing makes end-to-end security mechanism harder to deploy because intermediate nodes need direct access to the contents of the messages. Link layer security mechanisms can help mediate some of the resulting vulnerabilities, but it is not enough: we will now require much more from our protocols, and they must be designed with this in mind.

The organization of this chapter is as follows. In Section 2, we discuss the various resource constraints under which a typical WSN operates. In Section 3, various security requirements of such networks are identified. In section 4, a number of security vulnerabilities of WSNs are presented. Different types of attacks at various layers such as physical, link, network and transport layers are discussed in detail. In particular, various attacks at the network layers are described such as : (i) spoofed routing information (Karlof et al., 2003), (ii) selective packet forwarding (Karlof et al., 2003), (iii) sinkhole (Wood et al., 2002), (iv) Sybil (Newsome et al., 2004), (v) wormhole (Karlof et al., 2003), (vi) hello flood (Karlof et al., 2003), (vii) acknowledgment spoofing etc (Karlof et al., 2003). Section 5 presents a discussion on the defense mechanisms for DoS attacks at the network layer. In particular, schemes such as use of message authentication code (MAC) (Perrig et al., 2002), directional antenna-based defense (Hu et al., 2004a), packet leashes (Hu et al., 2004b), client puzzles (Aura et al., 2001) are discussed. Section 6 discusses secure broadcasting and multicasting techniques based on group key management protocols (Rafaeli et al., 2003) and directed diffusion-based mechanism (Di Pietro et al., 2003) etc. Section 7 presents some of the well-known existing secure routing protocols for WSNs such as *μTESLA* (Liu et al., 2004), INSENS (Deng et al., 2002b), SPINS (Perrig et al., 2002), TRANS (Tanachawiwat et al., 2003), and defense mechanisms against Sybil attack (Newsome et al., 2004; Chan, et al., 2003b; Eschenauer et al., 2002; Du et al., 2003), blackhole and grayhole (Sen et al., 2007b) attacks, a secure and energy-efficient routing protocol (Sen et al., 2010) are also discussed in detail. Finally, in conclusion, some future research directions are discussed.

In summary, the chapter makes the following contributions:

- It proposes threat models and security goals for secure routing in WSNs.
- It identifies various possible attacks on the network layer of a WSN sensor networks
- It demonstrates how attacks against ad-hoc wireless networks and peer-to-peer networks can be adapted into powerful attacks against WSNs.
- It presents a detailed security analysis of all the major routing protocols and energy conserving topology maintenance algorithms for WSNs.
- It presents various defense mechanisms to counter the well-known attacks on the routing protocols of WSNs.

2. Constraints in WSNs

A WSN consists of a large number of sensor nodes which are inherently resource-constrained. These nodes have limited processing capability, very low storage capacity, and constrained communication bandwidth. These limitations are due to limited energy and physical size of the sensor nodes. Due to these constraints, it is difficult to directly employ the conventional security mechanisms in WSNs. In order to optimize the conventional security algorithms for WSNs, it is necessary to be aware about the constraints of sensor nodes (Carman et al., 2000). The major constraints of a WSN are listed below.

(i) *Energy constraints*: Energy is the biggest constraint for a WSN. In general, energy consumption in sensor nodes can be categorized in three parts: (i) energy for the sensor transducer, (ii) energy for communication among sensor nodes, and (iii) energy for microprocessor computation. The study in (Hill et al., 2000) found that each bit transmitted in WSNs consumes about as much power as executing 800 to 1000 instructions. Thus, communication is more costly than computation in WSNs. Any message expansion caused by security mechanisms comes at a significant cost. Further, higher security levels in WSNs usually correspond to more energy consumption for cryptographic functions. Thus, WSNs could be divided into different security levels depending on energy cost (Slijepcevic et al., 2002; Yuan et al., 2002).

(ii) *Memory limitations:* A sensor is a tiny device with only a small amount of memory and storage space. Memory is a sensor node usually includes flash memory and RAM. Flash memory is used for storing downloaded application code and RAM is used for storing application programs, sensor data, and intermediate results of computations. There is usually not enough space to run complicated algorithms after loading the OS and application code. In the SmartDust project, for example, TinyOS consumes about 4K bytes of instructions, leaving only 4500 bytes for security and applications (Hill et al., 2000). A common sensor type- TelosB- has a 16-bit, 8 MHz RISC CPU with only 10K RAM, 48K program memory, and 1024K flash storage. The current security algorithms are therefore, infeasible in these sensors (Perrig et al., 2002).

(iii) *Unreliable communication:* Unreliable communication is another serious threat to sensor security. Normally the packet-based routing of sensor networks is based on connectionless protocols and thus inherently unreliable. Packets may get damaged due to channel errors or may get dropped at highly congested nodes. Furthermore, the unreliable wireless communication channel may also lead to damaged or corrupted packets. Higher error rate also mandates robust error handling schemes to be implemented leading to higher overhead. In certain situation even if the channel is reliable, the communication may not be so. This is due to the broadcast nature of wireless communication, as the packets may collide in transit and may need retransmission (Akyildiz et al., 2002).

(iv) *Higher latency in communication:* In a WSN, multi-hop routing, network congestion and processing in the intermediate nodes may lead to higher latency in packet transmission. This makes synchronization very difficult to achieve. The synchronization issues may sometimes be very critical in security as some security mechanisms may rely on critical event reports and cryptographic key distribution (Stankovic, 2003).

(v) *Unattended operation of networks:* In most cases, the nodes in a WSN are deployed in remote regions and are left unattended. The likelihood that a sensor encounters a physical attack in such an environment is therefore, very high. Remote management of a WSN makes it virtually impossible to detect physical tampering. This makes security in WSNs a particularly difficult task.

3. Security Requirements in WSNs

A WSN is a special type of network. It shares some commonalities with a typical computer network, but also exhibits many characteristics which are unique to it. The security services in a WSN should protect the information communicated over the network and the resources from attacks and misbehavior of nodes. The most important security requirements in WSN are listed below:

(i) *Data confidentiality*: The security mechanism should ensure that no message in the network is understood by anyone except the intended recipient. In a WSN, the issue of confidentiality should address the following requirements (Carman et al., 2000; Perrig et al., 2002): (i) a sensor node should not allow its readings to be accessed by its neighbors unless they are authorized to do so, (ii) key distribution mechanism should be extremely robust, (iii) public information such as sensor identities, and public keys of the nodes should also be encrypted in certain cases to protect against traffic analysis attacks.

(ii) *Data integrity*: The mechanism should ensure that no message can be altered by an entity as it traverses from the sender to the recipient.

(iii) *Availability*: This requirements ensures that the services of a WSN should be available always even in presence of an internal or external attacks such as a *denial of service* (DoS) attack. Different approaches have been proposed by researchers to achieve this goal. While some mechanisms make use of additional communication among nodes, others propose use of a central access control system to ensure successful delivery of every message to its recipient.

(iv) *Data freshness*: It implies that the data is recent and ensures that no adversary can replay old messages. This requirement is especially important when the WSN nodes use shared-keys for message communication, where a potential adversary can launch a replay attack using the old key as the new key is being refreshed and propagated to all the nodes in the WSN. A nonce or time-specific counter may be added to each packet to check the freshness of the packet.

(v) *Self-organization:* Each node in a WSN should be self-organizing and self-healing. This feature of a WSN also poses a great challenge to security. The dynamic nature of a WSN makes it sometimes impossible to deploy any pre-installed shared key mechanism among the nodes and the base station (Eschenauer et al., 2002). A number of key pre-distribution schemes have been proposed in the context of symmetric encryption (Chan et al., 2003b; Eschenauer et al., 2002; Hwang et al., 2004; Liu, et al., 2005a). However, for application of public-key cryptographic techniques an efficient mechanism for key-distribution is very much essential. It is desirable that the nodes in a WSN self-organize among themselves not only for multi-hop routing but also to carry out key management and developing trust relations.

(vi) *Secure localization*: In many situations, it becomes necessary to accurately and automatically locate each sensor node in a WSN. For example, a WSN designed to locate faults would require accurate locations of sensor nodes identifying the faults. A potential adversary can easily manipulate and provide false location information by reporting false signal strength, replaying messages etc., if the location information is not secured properly. The authors in (Capkun et al., 2006) have described a technique called *verifiable multi-lateration* (VM). In multi-lateration, the position of a device is accurately computed from a series of known reference points. The authors have used authenticated ranging and distance bounding to ensure accurate location of a node. Because of the use of distance bounding, an attacking node can only increase its claimed distance from a reference point. However, to ensure location consistency, the attacker would also have to prove that its distance from another reference point is shorter. As it is not possible for the attacker to prove this, it is possible to detect the attacker. In (Lazos et al., 2005), the authors have described a scheme called *secure range-independent localization* (SeRLoC). The scheme is a decentralized range-independent localization scheme. It is assumed that the locators are trusted and cannot be compromised by any attacker. A sensor computes its location by listening to the beacon information sent by each locator which includes the locator's location information. The beacon messages are encrypted using a shared global symmetric key that is pre-distributed in the sensor nodes. Using the information from all the beacons that a sensor node receives, it computes its approximate location based on the coordinates of the locators. The sensor node then computes an overlapping antenna region using a majority vote scheme. The final location of the sensor node is determined by computing the center of gravity of the overlapping antenna region.

(vii) *Time synchronization*: Most of the applications in sensor networks require time synchronization. Any security mechanism for WSN should also be time-synchronized. A collaborative WSN may require synchronization among a group of sensors. In (Ganeriwal et al., 2005), the authors have proposed a set of secure synchronization protocols for multi-hop sender-receiver and group synchronization.

(viii) *Authentication*: It ensures that the communicating node is the one that it claims to be. An adversary can not only modify data packets but also can change a packet stream by injecting fabricated packets. It is, therefore, essential for a receiver to have a mechanism to verify that the received packets have indeed come from the actual sender node. In case of communication between two nodes, data authentication can be achieved through a *message authentication code* (MAC) computed from the shared secret key among the nodes. A number of authentication schemes for WSNs have been proposed by researchers. Most of these schemes are for secure routing and reliable packet. Some of these schemes will be discussed in Section 5.

4. Security Vulnerabilities in WSNs

Wireless Sensor Networks are vulnerable to various types of attacks. These attacks are mainly of three types (Shi et al., 2004):
(i) *Attacks on network availability*: attacks on availability of WSN are often referred to as DoS attacks.
(ii) *Attacks on secrecy and authentication*: standard cryptographic techniques can protect the secrecy and authenticity of communication channels from outsider attacks such as eavesdropping, packet replay attacks, and modification or spoofing of packets.
(iii) *Stealthy attack against service integrity*: in a stealthy attack, the goal of the attacker is to make the network accept a false data value. For example, an attacker compromises a sensor node and injects a false data value through that sensor node.
In these attacks, keeping the sensor network available for its intended use is essential. DoS attacks against WSNs may permit real-world damage to the health and safety of people (Wood et al., 2002). The DoS attack usually refers to an adversary's attempt to disrupt, subvert, or destroy a network. However, a DoS attack can be any event that diminishes or eliminates a network's capacity to perform its expected functions (Wood et al., 2002).

4.1 Denial of Service Attacks

Wood and Stankovic have defined a DoS attack as an event that diminishes or attempts to reduce a network's capacity to perform its expected function (Wood et al., 2002). There are several standard techniques existing in the literature to cope with some of the more common denial of service attacks, although in a broader sense, development of a generic defense mechanism against DoS attacks is still an open problem. Moreover, most of the defense mechanisms require high computational overhead and hence not suitable for resource-constrained WSNs. Since DoS attacks in WSNs can sometimes prove very costly, researchers have spent a great deal of effort in identifying various types of such attacks, and devising strategies to defend against them. Some of the important types of DoS attacks at different layers of WSNs are discussed below:
(a) Physical layer attacks: The physical layer is responsible for frequency selection, carrier frequency generation, signal detection, modulation, and data encryption (Akyildiz et al. 2002). As with any radio-based medium, the possibility of jamming is there. The nodes in WSNs may be deployed in hostile or insecure environments, where an attacker has the physical access. Two types of attacks in physical layer are (i) jamming and (ii) tampering.
(i) *Jamming:* it is a type of attack which interferes with the radio frequencies that the nodes use in a WSN for communication (Wood et al., 2002; Shi et al., 2004). A jamming source may be powerful enough to disrupt the entire network. Even with less powerful jamming sources, an adversary can potentially disrupt communication in the entire network by strategically distributing the jamming sources. Even an intermittent jamming may prove detrimental as the message communication in a WSN may be extremely time-sensitive (Wood et al., 2002).

(ii) *Tampering*: sensor networks typically operate in outdoor environments. Due to unattended and distributed nature, the nodes in a WSN are highly susceptible to physical attacks (Wang et al., 2004a). The physical attacks may cause irreversible damage to the nodes. The adversary can extract cryptographic keys from the captured node, tamper with its circuitry, modify the program codes, or even replace it with a malicious sensor (Wang et al., 2005). It has been shown that sensor nodes such as MICA2 motes can be compromised in less than one minute time (Hartung, et al., 2004).

(b) Link layer attacks: The link layer is responsible for multiplexing of data-streams, data frame detection, medium access control, and error control (Akyildiz et al., 2002). Attacks at this layer include purposefully created collisions, resource exhaustion, and unfairness in allocation.

A collision occurs when two nodes attempt to transmit on the same frequency simultaneously (Wood et al., 2002). When packets collide, they are discarded and need to re-transmitted. An adversary may strategically cause collisions in specific packets such as ACK control messages. A possible result of such collisions is the costly exponential back-off. The adversary may simply violate the communication protocol, and continuously transmit messages in an attempt to generate collisions. Repeated collisions can also be used by an attacker to cause resource exhaustion (Wood et al., 2002). For example, a naïve link layer implementation may continuously attempt to retransmit the corrupted packets. Unless these retransmissions are detected early, the energy levels of the nodes would be exhausted quickly. Unfairness is a weak form of DoS attack (Wood et al., 2002). An attacker may cause unfairness by intermittently using the above link layer attacks. In this case, the adversary causes degradation of real-time applications running on other nodes by intermittently disrupting their frame transmissions.

(c) Network layer attacks: The network layer of WSNs is vulnerable to the different types of attacks such as: spoofed routing information, selective packet forwarding, sinkhole, Sybil, wormhole, blackhole, hello flood, Byzantine attack, information disclosure, resource depletion attack, acknowledgment spoofing, routing table overflow, route poisoning, rushing attack etc. These attacks are described briefly in the following:

(i) *Spoofed routing information*: the most direct attack against a routing protocol is to target the routing information in the network. An attacker may spoof, alter, or replay routing information to disrupt traffic in the network (Karlof et al., 2003). These disruptions include creation of routing loops, attracting or repelling network traffic from selected nodes, extending or shortening source routes, generating fake error messages, causing network partitioning, and increasing end-to-end latency.

(ii) *Selective forwarding*: in a multi-hop network like a WSN, for message communication all the nodes need to forward messages accurately. An attacker may compromise a node in such a way that it selectively forwards some messages and drops others (Karlof et al., 2003).

(iii) *Sinkhole*: In a sinkhole attack, an attacker makes a compromised node look more attractive to its neighbors by forging the routing information (Karlof et al., 2003; Wood et al., 2002; Newsome et al., 2004). The result is that the neighbor nodes choose the compromised node as the next-hop node to route their data through. This type of attack makes selective forwarding very simple as all traffic from a large area in the network would flow through the compromised node.

(iv) *Sybil attack*: it is an attack where one node presents more that one identity in a network. It was originally described as an attack intended to defeat the objective of redundancy mechanisms in distributed data storage systems in peer-to-peer networks (Douceur, 2002). Newsome et al. describe this attack from the perspective of a WSN (Newsome et al., 2004). In addition to defeating distributed data storage systems, the Sybil attack is also effective against routing algorithms, data aggregation, voting, fair resource allocation, and foiling misbehavior detection. Regardless of the target (voting, routing, aggregation), the Sybil algorithm functions similarly. All of the techniques involve utilizing multiple identities. For instance, in a sensor network voting scheme, the Sybil attack might utilize multiple identities to generate additional "votes". Similarly, to attack the routing protocol, the Sybil attack would rely on a malicious node taking on the identity of multiple nodes, and thus routing multiple paths through a single malicious node.

(v) *Wormhole*: a wormhole is low latency link between two portions of a network over which an attacker replays network messages (Karlof et al., 2003). The attacker receives packets at one location in the network, and tunnels them to another location in the network, where the packets are resent into the network. The tunnel between the two colluding attackers is known as the *wormhole*. This link may be established either by a single node forwarding messages between two adjacent but otherwise non-neighboring nodes or by a pair of nodes in different parts of the network communicating with each other. The latter case is closely related to sinkhole attack as an attacking node near the base station can provide a one-hop link to that base station via the other attacking node in a distant part of the network. Due to the broadcast nature of the radio channel, the attacker can create a wormhole link even for packets which are not addressed to it. If proper security mechanisms are not deployed to defend against such attacks, routing in WSN may be impossible.

(vi) *Blackhole* and *Grayhole*: in this attack, a malicious node falsely advertises good paths (e.g. the shortest path or the most stable path) to the destination node during the path-finding process (in reactive routing protocols), or in the route updates messages (in proactive routing protocols). The intention of the malicious node could be to hinder the path-finding process or to intercept all data packets being sent to the destination node concerned. A more delicate form of this attack is known as the grayhole attack, where the malicious node intermittently drops the data packets thereby making its detection even more difficult.

(vii) *Hello flood*: most of the protocols that use Hello packets make the naïve assumption that receiving such a packet implies that the sender is within the radio range of the receiver. An attacker may use a high-powered transmitter to fool a large number of nodes and make them believe that they are within its neighborhood (Karlof et al., 2003). Subsequently, the attacker node falsely broadcasts a shorter route to the base station, and all the nodes which received the Hello packets, attempt to transmit to the attacker node. However, these nodes are out of the radio range of the attacker.

(viii)*Byzantine attack*: in this attack, a compromised node or a set of compromised nodes works in collusion and carries out attacks such as creating routing loops, forwarding packets in non-optimal routes, and selectively dropping packets (Awerbuch et al., 2002). Byzantine attacks are very difficult to detect, since under such attacks the networks usually do not exhibit any abnormal behavior.

(ix) *Information disclosure*: a compromised node may leak confidential or important information to unauthorized nodes in the network. Such information may include information regarding the network topology, geographic location of nodes, or optimal routes to authorized nodes in the network.

(x) *Resource depletion attack*: in this type of attack, a malicious node tries to deplete resources of other nodes in the network. The typical resources that are targeted are: battery power, bandwidth, and computational power. The attacks could be in the form of unnecessary requests for routes, very frequent generation of beacon packets, or forwarding of stale packets to other nodes.
Acknowledgment spoofing: some routing algorithms for WSNs require transmission of acknowledgment packets. An attacking node may overhear packet transmissions from its neighboring nodes and spoof the acknowledgments thereby providing false information to the nodes (Karlof et al., 2003). In this way, the attacker is able to disseminate wrong information about the status of the nodes.
(xi) *Attacks on routing protocols*: most of the routing protocols for WSNs are vulnerable to various types of attacks. Some of these attacks are listed below.

- *Routing table overflow*: in this type of attack, an adversary node advertises routes to non-existent nodes, to the authorized node present in the network. The main objective of such an attack is to cause an overflow of the routing tables, which would in turn prevent the creation of entries corresponding to new routes to authorized nodes. Proactive routing protocols are more vulnerable to this attack compared to reactive routing protocols.
- *Routing table poisoning*: in this case, the compromised nodes in the network send fictitious routing updates or modify genuine route update packets sent to other honest nodes. Routing table poisoning may result in sub-optimal routing, congestion in some portions of the network, or even make some parts of the network inaccessible.
- *Packet replication*: in this attack, an adversary node replicates stale packets. This consumes additional bandwidth and battery power and other resources available to the nodes and also causes unnecessary confusion in the routing process.
- *Route cache poisoning*: in reactive (i.e. on-demand) routing protocols such as ad hoc on-demand distance vector (AODV) (Perkins, et al., 1999), each node maintains a route cache which holds information regarding routes that have become known to the node in the recent past. Similar to routing table poisoning, an adversary can also poison the route cache to achieve similar objectives.
- *Rushing attack*: on-demand routing protocols that use *duplicate suppression* during the route discovery process are vulnerable to this attack (Hu et al., 2003b). An adversary node which receives a *routerequest* packet from the source node floods the packet quickly throughout the network before other nodes which also receive the same *routerequest* packet can react. Nodes that receive the legitimate *routerequest* packets assume those packets to be duplicates of the packet already received through the adversary node and hence discard those packets. Any route discovered by the source node would contain the adversary node as one of the intermediate nodes. Hence, the source node would not be able to find secure routes, that is, routes that do not include the adversary node. It is extremely difficult to detect such attacks in WSNs.

(d) Transport layer attacks: The attacks that can be launched on the transport layer in a WSN are flooding attack and de-synchronization attack.
(i) *Flooding*: Whenever a protocol is required to maintain state at either end of a connection, it becomes vulnerable to memory exhaustion through flooding (Wood et al., 2002). An attacker may repeatedly make new connection request until the resources required by each

connection are exhausted or reach a maximum limit. In either case, further legitimate requests will be ignored.

(ii) *De-synchronization*: De-synchronization refers to the disruption of an existing connection (Wood et al., 2002). An attacker may, for example, repeatedly spoof messages to an end host causing the host to request the retransmission of missed frames. If timed correctly, an attacker may degrade or even prevent the ability of the end hosts to successfully exchange data causing them instead to waste energy attempting to recover from errors which never really exist. The possible DoS attacks and the corresponding countermeasures are listed in Table 1.

Layer	Attacks	Defense
Physical	Jamming	Spread-spectrum, priority messages, lower duty cycle, region mapping, mode change
Link	Collision	Error-correction code
	Exhaustion	Rate limitation
	Unfairness	Small frames
Network	Spoofed routing information & selective forwarding	Egress filtering, authentication, monitoring
	Sinkhole	Redundancy checking
	Sybil	Authentication, monitoring, redundancy
	Wormhole	Authentication, probing
	Hello Flood	Authentication, packet leashes by using geographic and temporal info
	Ack. flooding	Authentication, bi-directional link authentication verification
Transport	Flooding	Client puzzles
	De-synchronization	Authentication

Table 1. Various attacks on WSNs and their countermeasures (Wang et al., 2006)

4.2 Attacks on Secrecy and Authentication

There are different types of attacks under this category as discussed below.

(i) Node replication attack: In a node replication attack, an attacker attempts to add a node to an existing WSN by replicating (i.e. copying) the node identifier of an already existing node in the network (Parno et al., 2005). A node replicated and joined in the network in this manner can potentially cause severe disruption in message communication in the WSN by corrupting and forwarding the packets in wrong routes. This may also lead to network partitioning, communication of false sensor readings etc. In addition, if the attacker gains physical access to the entire network, it is possible for him to copy the cryptographic keys and use these keys for message communication from the replicated node. The attacker can also place the replicated node in strategic locations in the network so that he could easily manipulate a specific segment of the network, possibly causing a network partitioning.

(ii) Attacks on privacy: Since WSNs are capable of automatic data collection through efficient and strategic deployment of sensors, these networks are also vulnerable to potential abuse

of these vast data sources. Privacy preservation of sensitive data in a WSN is particularly difficult challenge (Gruteser et al., 2003). Moreover, an adversary may gather seemingly innocuous data to derive sensitive information if he knows how to aggregate data collected from multiple sensor nodes. This is analogous to the *panda hunter* problem, where the hunter can accurately estimate the location of the panda by monitoring the traffic (Ozturk et al., 2004).

The privacy preservation in WSNs is even more challenging since these networks make large volumes of information easily available through remote access mechanisms. Since the adversary need not be physically present to carryout the surveillance, the information gathering process can be done anonymously with a very low risk. In addition, remote access allows a single adversary to monitor multiple sites simultaneously (Chan et al., 2003a). Following are some of the common attacks on sensor data privacy (Gruteser et al., 2003, Chan et al., 2003a):

(iii) *Eavesdropping and passive monitoring*: This is the most common and the easiest form of attack on data privacy. If the messages are not protected by cryptographic mechanisms, the adversary could easily understand the contents. Packets containing control information in a WSN convey more information than accessible through the location server, Eavesdropping on these messages prove more effective for an adversary.

(iv) *Traffic analysis*: In order to make an effective attack on privacy, eavesdropping should be combined with a traffic analysis. Through an effective analysis of traffic, an adversary can identify some sensor nodes with special roles and activities in a WSN. For example, a sudden increase in message communication between certain nodes signifies that those nodes have some specific activities and events to monitor. Deng et al. have demonstrated two types of attacks that can identify the base station in a WSN without even underrating the contents of the packets being analyzed in traffic analysis (Deng et al., 2004).

(v) *Camouflage*: An adversary may compromise a sensor node in a WSN and later on use that node to masquerade a normal node in the network. This camouflaged node then may advertise false routing information and attract packets from other nodes for further forwarding. After the packets start arriving at the compromised node, it starts forwarding them to strategic nodes where privacy analysis on the packets may be carried out systematically.

It may be noted from the above discussion that WSNs are vulnerable to a number of attacks at all layers of the TCP/IP protocol stack. However, as pointed out by authors in (Perrig et al., 2004), there may be other types of attacks possible which are not yet identified. Securing a WSN against all these attacks may be a quite challenging task.

5. Network Layer Defense on DoS Attacks

A countermeasure against spoofing and alteration is to append a *message authentication code* (MAC) after the message. By adding a MAC to the message, the receivers can verify whether the messages have been spoofed or altered. To defend against replayed information, counters or time-stamps may be introduced in the messages (Perrig et al., 2002). A possible defense against selective forwarding attack is using multiple paths to send data (Karlof et al., 2003). A second defense is to detect the malicious node or assume it has failed and seek an alternative route.

Hu et al. have proposed a novel and generic mechanism called *packet leashes* for detecting and defending against wormhole attacks (Hu et al., 2004b). As mentioned in Section 4.1, in a wormhole attack, a malicious node eavesdrops on a series of packets, then tunnels them through a path in the network, and replays them. This is done in order to make a false representation of the distance between the two colluding nodes. It is also used, more generally, to disrupt the routing protocol by misleading the neighbor discovery process (Karlof et al., 2003). Hu et al. have presented a mechanism that employs directional antenna to combat wormhole attack (Hu et al., 2004a). Wang and Bhargava have used a visualization approach to detect wormholes in a WSN (Wang et al., 2004b). In the mechanism proposed by the authors, a distance estimation is made between all the sensor nodes in a neighborhood. Using multi-dimensional scaling, a virtual layout of the network is then computed, and a surface smoothing strategy is used to adjust the round-off errors. Finally, the shape of the resulting virtual network is analyzed. If any wormhole exists, the shape of the network will bend and curve towards the wormhole, otherwise the network will appear flat.

To defend against flooding DoS attack at the transport layer, Aura et al. have proposed a mechanism using client puzzles (Aura et al., 2001). The main idea is that each connecting client should demonstrate its commitment to the connection by solving a puzzle. As an attacker in most likelihood, does not have infinite resource, it will be impossible for him to create new connections fast enough to cause resource starvation on the serving node.

A possible defense against de-synchronization attack on the transport layer is to enforce a mandatory requirement of authentication of all packets communicated between nodes (Wood et al., 2002). If the authentication mechanism is secure, an attacker will be unable to send any spoofed messages to any destination node.

Some mechanisms for secure multicasting and broadcasting in WSNs are discussed in the following sub-section.

6. Secure Broadcasting and Multicasting Protocols for WSNs

Multicasting and broadcasting techniques are used primarily to reduce the communication and management overhead of sending a single message to multiple receivers. In order to ensure that only legitimate group members receive the multicast and broadcast communication, appropriate authentication and encryption mechanisms must be in place. To handle this problem, several key management schemes have been devised: centralized group key management protocols, decentralized key management protocols, and distributed key management protocols (Rafaeli et al., 2003). First, we will discuss some generic security mechanisms for multicast and broadcast communication in wireless networks. Then we will present some of the well-known propositions specific to WSNs.

In the case of the centralized group key management protocols, a central authority is used to maintain the group. Decentralized management protocols, however, divide the task of group management amongst multiple nodes. In distributed key management protocols, the key management activity is distributed among a set of nodes rather than on a single node. In some cases, the entire group of nodes is responsible for key management (Rafaeli et al., 2003).

An efficient way to distribute keys in a network is to use a logical key tree. Such techniques essentially fall under the category of centralized key management protocols. Some schemes have been developed for WSNs based on logical key tree technique (Di Pietro et al., 2003; Lazos et al., 2002; Lazos et al., 2003). While centralized solutions are not always the most efficient ones, these mechanisms may sometimes be very effective for WSNs, as relatively heavier computations can be usually carried out in powerful base stations.

Di Pietro et al. have proposed a directed diffusion-based multicast mechanism for WSNs that utilizes a logical key hierarchy (Di Pietro et al., 2003). In the logical hierarchy, a central key distributor is at the root of a tree, and the nodes in the network are the leaf level. The internal nodes of tree contain keys that are used in the re-keying process. The directed diffusion is an energy-efficient data dissemination technique for WSNs (Intanagonwiwat et al., 2000). In directed diffusion, a query is transformed into an interest and then diffused throughout the network. The source node then starts collecting data from the network based on the propagated interest. The dissemination technique also sets up certain gradients designed to draw events toward the interest. The collected data is then sent back to the source along the reverse path of the interest propagation. The directed diffusion-based logical key hierarchy scheme as proposed by Di Pietro et al. allows nodes to join and leave groups. The key hierarchy is used to effectively re-establish keys for the nodes below the node that has left the group. When a node declares its intension to join a group, a key set is generated for the new node based on the keys within the existing key hierarchy.

Kaya et al. discuss the problem of multicast group management in (Kaya et al., 2003). In their proposition, the nodes in a network are grouped based on their locality and a security tree is constructed on the groups.

Lazos and Poovendran have presented a tree-based key distribution scheme that is similar to the directed diffusion-based logical key hierarchy proposed by Di Pietro et al. (Lazos et al., 2003). In their proposed scheme, a routing-aware tree is constructed in which the leaf nodes are assigned keys based on all relay nodes above them. As the scheme takes advantage of routing information for construction the key hierarchy, it is more energy-efficient than routing schemes that arbitrarily arrange nodes into a routing tree. The authors have also proposed a greedy routing-aware key distribution algorithm.

In (Lazos et al., 2003), the authors have proposed a mechanism that uses geographic location information (e.g. GPS data) for construction of a logical key hierarchy for secure multicast communication. The nodes, based on the geographical location information, are grouped into different clusters. The nodes within a cluster are able to reach each other with a single hop communication. Using the cluster information, a key hierarchy is constructed in a manner similar to that proposed in (Lazos et al., 2002).

7. Secure Routing Protocols for WSNs

Many routing protocols have been proposed for WSNs. These protocols can be divided into three broad categories according to the network structure: (i) flat-based routing, (ii) hierarchical-based routing, and (iii) location-based routing (Al-Karaki et al., 2004). In flat-based routing, all nodes are typically assigned equal roles or functionality. In hierarchical-based routing, nodes play different roles in the network. In location-based routing, sensor node positions are used to route data in the network. One common location-based routing protocol is GPSR (Karp et al., 2000). It allows nodes to send packets to a region rather than to a particular node. All these routing protocols are vulnerable to various types of attacks such as selective forwarding, sinkhole attack etc as mentioned in Section 4. An elaborate

discussion on various types of attacks on the routing protocols in WSNs is given in (Karlof et al., 2003).
The goal of a secure routing protocol for a WSN is to ensure the integrity, authentication, and availability of messages. Most of the existing secure routing algorithms for WSNs are all based on symmetric key cryptography except the work in (Du et al., 2005), which is based on public key cryptography. In the following sub-sections, some of the existing secure routing protocols for WSNs are discussed in detail.

7.1 Micro TESLA Protocol

The "micro" version of the *Timed, Efficient, Streaming, Loss-tolerant Authentication* ($\mu TESLA$) protocol (Perrig et al., 2002) and its extensions (Liu et al., 2003; Liu et al. 2004) have been proposed to provide broadcast authentication for sensor networks. $\mu TESLA$ is broadcast authentication mechanism which was proposed by Perrig et al. for the SPINS protocol (Perrig et al., 2002). $\mu TESLA$ introduces asymmetry through a delayed disclosure of symmetric keys resulting in an efficient broadcast authentication scheme. For its operation, it requires the base station and the sensor nodes to be loosely synchronized. In addition, each node must know an upper bound on the maximum synchronization error.
To send an authenticated packet, the base station simply computes a MAC on the packet with a key that is secret at that point of time. When a node gets a packet, it can verify that the corresponding MAC key was not yet disclosed by the base station. Because a receiving node is assured that the MAC key is known only to the base station, the receiving node is assured that no adversary could have altered the packet in transit. The node stores the packet in a buffer. At the time of key disclosure, the base station broadcasts the verification key to all its receivers. When a node receives the disclosed key, it can easily verify the correctness of the key. If the key is correct, the node can now use it to authenticate the packet stored in its buffer.
Each MAC is a key from the key chain, generated by a public one-way function F. To generate the one-way key chain, the sender chooses the last key K_n from the chain, and repeatedly applies F to compute all other keys: $K_i = F(K_{i+1})$.

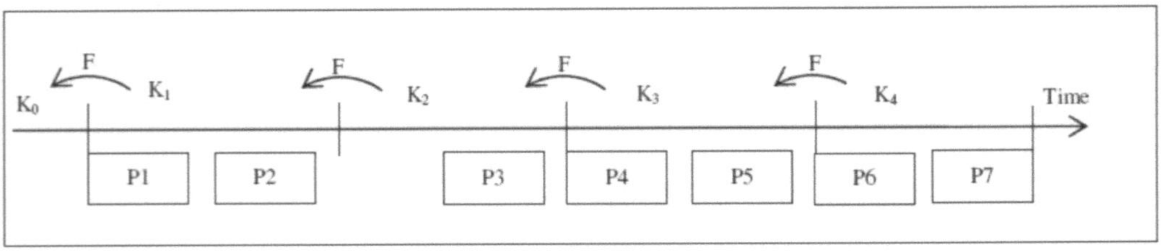

Fig. 1. Time-released key chain for source authentication (Wang et al. 2006)

Fig. 1 shows an example of $\mu TESLA$. The receiver node is loosely time synchronized and knows K_0 in an authenticated way. Packets P_1 and P_2 sent in interval 1 contain a MAC with a key K_1. Packet P_3 has a MAC using key K_2. If P_4, P_5, and P_6 are all lost, as well as the packet that disclosed the key K_1, the receiver cannot authenticate P_1, P_2, and P_3. In interval 4, the base station broadcasts the key K_2, which the nodes authenticate by verifying $K_0 = F(F(K_2))$, and hence know also $K_1 = F(K_2)$, so they can authenticate packets P_1, P_2 with K_1, and P_3 with K_2. SPINS limits the broadcasting capability to only the base station. If a node wants to broadcast authenticated data, the node has to broadcast the data through the base station. The data is first sent to the base station in an authenticated way. It is then broadcasted by the base station.

To bootstrap a new receiver, $\mu TESLA$ depends on a point-to-point authentication mechanism in which a receiver sends a request message to the base station and the base station replies with a message containing all the necessary parameters. It may be noted that $\mu TESLA$ requires the base station to unicast initial parameters to individual sensor nodes, and thus incurs a long delay to boot up a large-scale sensor network. Liu and Ning have proposed a multi-level key chain scheme for broadcast authentication to overcome this deficiency (Liu et al., 2003; Liu et al. 2004).

The basic idea in (Liu et al., 2003; Liu et al., 2004) is to predetermine and broadcast the initial parameters required by $\mu TESLA$ instead of using unicast-based message transmission. The simplest way is to pre-distribute the $\mu TESLA$ parameters with a master key during the initialization of the sensor nodes. As a result, all sensor nodes have the key chain commitments and other necessary parameters once they are initialized, and are ready to use $\mu TESLA$ as long as the staring time has passed. Furthermore, the authors have introduced a multi-level key chain scheme, in which the higher key chains are used to authenticate the commitments of the lower-level ones. However, the multi-level key chain suffers from possible DoS attacks during commitment distribution stage. Further, none of the $\mu TESLA$ or multi-level key chain schemes is scalable in terms of the number of senders. In (Liu et al., 2005b), a practical broadcast authentication protocol has been proposed to support a potentially large number of broadcast senders using $\mu TESLA$ as a building block.

$\mu TESLA$ provides broadcast authentication for base stations, but is not suitable for local broadcast authentication. This is because $\mu TESLA$ does not provide immediate authentication. For every received packet, a node has to wait for one $\mu TESLA$ interval to receive the MAC key used in computing the MAC for the packet. As a result, if $\mu TESLA$ is used for local broadcast authentication, a message traversing l hops will take at least l $\mu TESLA$ intervals to arrive at the destination. In addition, a sensor node has to buffer all unverified packets. Both the latency and the storage requirements limit the scheme for authenticating infrequent messages broadcast by the base station. Zhu et al. have proposed a one-way key chain scheme for one-hop broadcast authentication (Zhu et al., 2004b). The mechanism is known as LEAP. In this scheme, every node generates a one-way key chain of certain length and then transmits the commitment (i.e., first key) of the key chain to each neighbor, encrypted with their pair-wise shared key. Whenever a node has a message to send, it attaches to the message to the next authenticated key in the key chain. The authenticated keys are disclosed in reverse order to their generation. A receiving neighbor can verify the message based on the commitment or an authenticated key it received from the sending node more recently.

7.2 Intrusion Tolerant Routing Protocol in WSNs

Deng et al. have proposed an *intrusion tolerant routing protocol in wireless sensor networks* (INENS) that adopts a routing-based approach to security in WSNs (Deng et al., 2002b). It constructs routing tables in each node, bypassing malicious nodes in the network. The protocol can not totally rule out attack on nodes, but it minimizes the damage caused to the network. The computation, communication, storage, and bandwidth requirements at the nodes are reduced, but at the cost of greater computation and communication at the base station. To prevent DoS attacks, individual nodes are not allowed to broadcast to the entire network. Only the base station is allowed to broadcast, and the base station is authenticated using one-way hash function so as to prevent a possible masquerading by a malicious node. Control information pertaining to routing is authenticated by the base station in order to prevent injection of false routing data. The base station computes and disseminates routing tables, since it does not have computational and energy constraints. Even if an intruder takes over a node and does not forward packets, INSENS uses redundant multi-path routing, so that the destination can still reach without passing through the malicious node.

INSENS has two phases: *route discovery* and *data forwarding*. During the route discovery phase, the base station sends a request message to all nodes in the network by multi-hop forwarding. Any node receiving a request message records the identity of the sender and sends the message to all its immediate neighbors if it has not already done so. Subsequent request messages are used to identify the senders as neighbors, but repeated flooding is not performed. The nodes respond with their local topology by sending feedback messages. The integrity of the messages is protected using encryption by a shared key mechanism. A malicious node can inflict damage only by not forwarding packets, but the messages are sent through different neighbors, so it is likely that it reaches a node by at least one path. Hence, the effect of malicious nodes is not totally eliminated, but it is restricted to only a few downstream nodes in the worst case. Malicious nodes may also send spurious messages and cause battery drain for a few downstream nodes. Finally, the base station calculates forwarding tables for all nodes, with two independent paths for each node, and sends them to the nodes. The second phase of data forwarding takes place based on the forwarding tables computed by the base station.

7.3 Security Protocols for Sensor Networks

SPINS is a suite of security protocols optimized for sensor networks (Perrig et al., 2002). SPINS includes two building blocks: (i) *secure network encryption protocol* (SNEP) and (ii) μTESLA protocol. SNEP provides data confidentiality, two-party data authentication, and data freshness for peer-to-peer communication (node to base station). *μTESLA* provides authenticated broadcast as discussed already.

SPINS assumes that each node is pre-distributed with a master key *K* which is shared with the base station at its time of creation. All the other keys, including a key K_{encr} for encryption, a key K_{mac} for MAC generation, and a key K_{rand} for random number generation are derived from the master key using a string one-way function. SPINS uses RC5 protocol for confidentiality. If *A* wants to send a message to base station *B*, the complete message *A* sends to *B* is:

$$A \rightarrow B : D_{<KencrC>}, \text{MAC}\,(K_{mac}, C \mid D)_{<KencrC>}$$

In the above expression, *D* is the transmitted data and *C* is a shared counter between the sender and the receiver for the block cipher in counter mode. The counter *C* is incremented after each message is sent and received in both the sender and the receiver side. SNEP also provides a counter exchange protocol to synchronize the counter value in both sides.

SNEP provides the flowing properties:

(i) *Semantic security*: the counter value is incremented after each message and thus the same message is encrypted differently each time.
(ii) *Data authentication*: a receiver can be assured that the message originated from the claimed sender if the MAC verification produces positive results.
(iii) *Replay protection*: the counter value in the MAC prevents replaying old messages by an adversary.
(iv) *Weak freshness*: SPINS identifies two types of freshness. Weak freshness provides partial message ordering and carries no delay information. Strong freshness provides a total order on a request-response pair and allows delay estimation. IN SNEP, the counter maintains a message ordering in the receiver side and yields weak freshness. SNEP guarantees weak freshness only, since there is no guarantee to node *A* that a message was created by node *B* in response to an event in node *A*.
(v) *Low communication overhead*: the counter state is kept at each endpoint and need not be sent in each message.

7.4 A Secure Protocol for Defending Cooperative Grayhole Attack

As mentioned in Section 4.1, blackhole and grayhole are two attacks that can severely disrupt routing in WSNs. A blackhole attack typically has two phases. In the first phase, the malicious node exploits the ad hoc routing protocol such as AODV (Perkins et al., 1999) to advertise itself as having a valid route to a destination node, with the intention of intercepting packets, even though the route is spurious. In the second phase, the attacker node drops the intercepted packets without forwarding them.

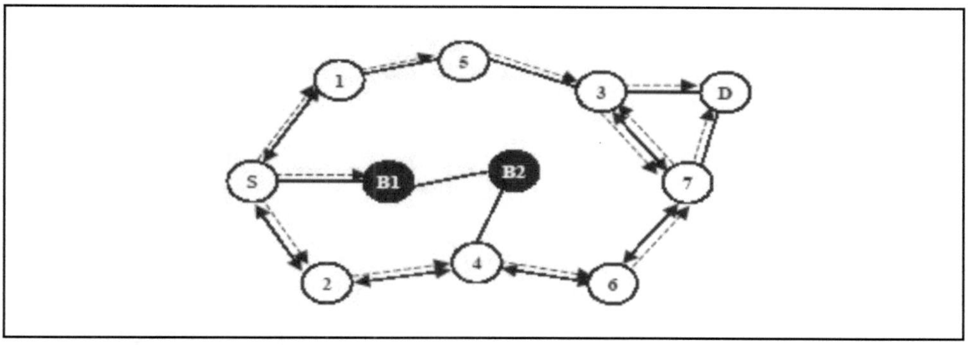

Fig. 2. Network flooding by RREQ and propagation of RREP (Deng et al., 2002a)

In the standard AODV protocol, when the source node *S* (Fig. 2) wants to communicate with the destination node *D*, the source node *S* broadcasts the *route request* (RREQ) packet. Each neighboring active node updates its routing table with an entry for the source node *S*, and checks if it is the destination node or whether it has the current route to the destination node. If an intermediate node does not have the current route to the destination node, it updates the RREQ packet by increasing the hop count and floods the network with the RREQ to the destination node *D* until it reaches node *D* or any other intermediate node that has the current route to *D*. The destination node *D* or any intermediate node that has the current route to *D*, initiates a *route reply* (RREP) in the reverse direction. Node *S* starts sending data packets to the neighboring node that responded first, and discards the other responses. This works fine when the network has no malicious nodes.

In (Deng et al., 2002a), authors have proposed a solution to identify and isolate a single blackhole node. However, the security threat arising out of the situation where multiple blackhole nodes act in coordination has not been addressed. For example, in Fig. 2, when more than one blackhole nodes are acting in coordination with each other, the first black hole node B_1 refers to one of its partners B_2 as the next hop. In the mechanism proposed in (Deng et al., 2002a), the source node S sends *further request* (FRq) to B_2 through a different route ($S \rightarrow 2 \rightarrow 4 \rightarrow B_2$) other than via B_1. Node S asks B_2 if it has a route to node B_1 and a route to destination node D. Since B_2 is cooperating with B_1, its *further reply* (FRp) will be '*yes*' to both the queries. Node S starts sending the data packets assuming that the route $S \rightarrow B_1 \rightarrow B_2$ is secure. However, in reality, the packets are intercepted and then dropped by the node B_1 and the security of the network is compromised.

Sen et al. have proposed a security mechanism that can detect cooperative grayhole attacks in a wireless ad hoc and sensor network (Sen et al., 2007b). As mentioned in Section 4.1, detection of grayholes is more difficult than detection of blackholes since these nodes drop packets intermittently and change their behavior frequently so as to avoid detection. In the proposed mechanism, each node in the network collects the data forwarding information in its neighborhood and stores it in a table known as the *data routing information* (DRI) table.

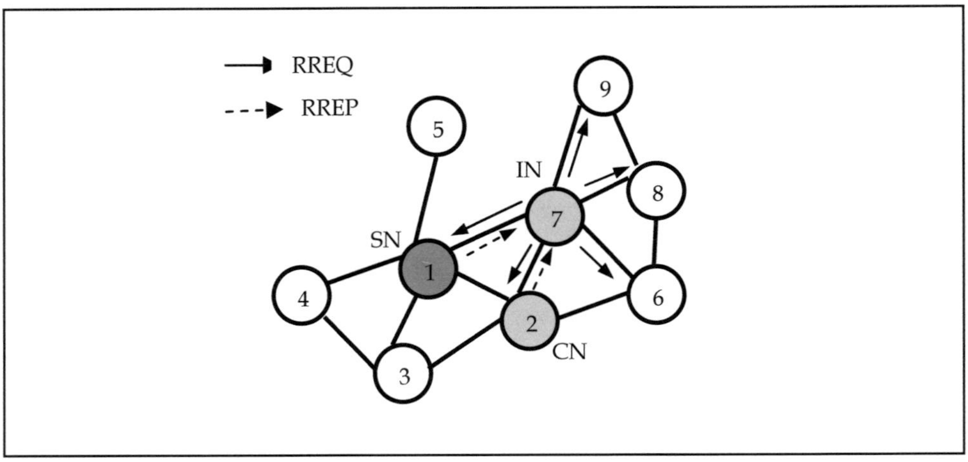

Fig. 3. The topology of a wireless ad hoc and sensor network (Sen et al., 2007b)

The DRI table of node 7 in Fig. 3 is shown in Table 2. In its DRI table node 7 maintains packet routing information of its neighbor nodes 1, 2, 6, 8, and 9. An entry '1' for a node under the column '*From*' implies that node 7 has forwarded data packet coming from that node and an entry '1' for a node under the column '*Through*' implies that node 7 has forwarded data packets to that node. Thus, as per Table 2, node 7 has neither forwarded any data packet from node 1 nor it has forwarded any data packet to node 1. However, node 7 has forwarded data packets to node 2 and also has forwarded data packets that have come from node 2. In this way, each node constructs its DRI table and maintains it. After a certain threshold time interval, each node identifies its neighbors with which it has not interacted, and invokes subsequent detection procedures to probe them further. This identification is done on the basis of the nodes that have '0' entries both in the '*From*' and '*Through*' columns in the DRI table. For example, as shown in Table 2, node 7 has not communicated to node 1. Therefore, the node 7 invokes the local anomaly detection procedure for node 1. The 'RTS/CTS' column in the DRI table gives the ratio of the number of *request to send* (RTS) messages to the number of *clear to send* (CTS) messages for the corresponding node. This gives a rough idea about the number of requests arriving at the node for data communication and number packet transmission that the node is actually doing. The significance of the column 'CheckBit' in the DRI table will be discussed in later in this section.

Node	From	Through	RTS/CTS	CheckBit
1	0	0	15	0
2	1	1	5	1
6	0	1	3	0
8	1	0	6	1
9	0	1	4	0

Table 2. The DRI table of node 7 as depicted in Fig. 3 (Sen et al., 2007b)

The node that initiates the anomaly detection procedure is called the *initiator node* (IN). The IN first chooses a *cooperative node* (CN) in its neighborhood based on its DRI records and broadcasts a RREQ message to its 1-hop neighbors requesting for a route to the CN. In reply to this RREQ message the IN will receive a number of RREP messages from its neighboring nodes. It will certainly receive a RREP message from the *suspected node* (SN) if the latter is really a grayhole (since the grayholes always send RREP messages but drop data packets probabilistically). After receiving the RREP from the SN, the IN sends a probe packet to the CN through the SN. After the *time to live* (TTL) value of the probe packet is over, the IN enquires the CN whether it has received the probe packet. If the reply to this query is affirmative, (i.e., the probe packet is really received by the CN) then the IN updates its DRI table by making an entry '1' under the column '*CheckBit*' against the node ID of the SN. However, if the probe packet is found to have not reached the CN, the IN increases its level of suspicion about the SN and activates the cooperative anomaly detection procedure, as discussed later in this section.

In Fig. 3, node 7 acts as the IN and initiates the local anomaly detection procedure for the SN (node 1) and chooses node 2 as the CN. Node 2 is the most reliable node for node 7 as both the entries under columns 'From' and 'Through' for node 2 are '1'. Node 7 broadcasts a RREQ message to all its neighbor nodes 1, 2, 6, 8 and 9 requesting them for a route to the CN, i.e., node 2 in the example. After receiving a RREP from the SN (node 1), node 7 sends a probe packet to node 2 via node 1. Node 7 then enquires node 2 whether it has received the probe packet. If node 2 has received the probe packet, node 7 makes an entry '1' under the column '*CheckBit*' in its DRI table corresponding to the row of node 1. If node 2 has not received the probe packet, then node 7 invokes the cooperative anomaly detection procedure. The objective of the cooperative anomaly detection is to increase the detection reliability by reducing the probability of false detection.

The cooperative detection procedure is activated when an IN observes that the probe packet it had sent to the CN through the SN did not reach the CN. The IN invokes the cooperative detection procedure and sends a cooperative detection request message to all the neighbors of the SN. When the neighbors of the SN receive the cooperative detection request message, each of them sends a RREQ message to the SN requesting for a route to the IN. After the SN responds with a RREP message, each of the requesting nodes sends a *'further probe packet'* to the IN along that route. This route will obviously include SN, as SN is a neighbor of each requesting node and the IN as well. Each neighbor of the SN (except the IN) now notifies the IN that a *'further probe packet'* has already been sent to it. This notification message from each neighbor is sent to the IN through routes which do not include the SN. This is necessary to ensure that the SN is not aware about the on-going cross checking process. The IN will receive numerous *'further probe packets'* and notification messages. The IN now constructs a *ProbeCheck* table. The ProbeCheck table has two fields: *NodeID* and *ProbeStatus*. Under the NodeID field, the IN enters the identifiers of the nodes which have sent notification messages to it. An entry of '1' is made under the column 'ProbeStatus' corresponding to the nodes from which the IN has received the 'further probe packet'.

NodeID	ProbeStatus
2	0
6	1
8	1
9	1

Table 3. The ProbeCheck table for node 7 (Sen et al., 2007b)

An example ProbeCheck table for node 7 of the network in Fig. 3 is presented in Table 3. It may be observed that node 7 has received the *'further probe packet'* from all the neighbors of the SN (node 1) except node 2. There may be a possibility that the probe packet might have not been maliciously dropped by the SN, rather it has been lost because of collision or buffer overflow. A mathematical estimation can be made for the probability of collision or buffer overflow at the SN (Sen et al., 2007a). However, to avoid complex mathematical computation, we propose a simple mechanism where each node sends three 'further probe packets' interspaced with a small time interval. If none of these three packets from a neighbor are received by the IN, the SN is believed to be behaving like a grayhole for that node during that time. This grayhole behavior may be exhibited for a single node (as for node 2 in Table 3) or may be for a group of nodes.

7.5 A Secure and Energy-Efficient Routing Protocol for WSNs

To address the problem of security and efficiency in routing in WSNs, a scheme has been proposed by Sen et al. that reliably identifies compromised (or faulty) nodes and utilizes a routing path that avoids these nodes (Sen et al., 2010). The protocol utilizes a single-path routing concept and thereby saves energy-consumption. The proposed protocol is a modification of the routing scheme proposed in (Lee et al., 2006). However, it is more energy- efficient and less delay-inducing.

The protocol is based on a robust *neighborhood monitoring system* (NMS). NMS works on promiscuous monitoring of the neighborhood by a node and detection of any possible malicious packet dropping attack by a cooperative algorithm using *neighbor list checking* (Sen et al., 2010). The scheme ensures reliable hop-by-hop delivery of packets in a WSN even in presence of malicious nodes that may launch packet-dropping attack in the routing path. To defend against packet-dropping attack, most of the existing algorithms exploit the concept of multi-path routing, where a single packet is routed through multiple paths from the source to the sink. While this approach ensures reliable packet delivery, it consumes an appreciable amount of energy for delivering each packet. To avoid this problem, the protocol uses a single-path routing mechanism. If a malicious node is encountered, the node is avoided and the packet is routed around it in an efficient manner, still in a single-path mode to the base station. The selection of the new path is based on some broadcast signaling in the neighborhood of the malicious node. The salient features of the protocol are briefly described below:

(i) *Neighbor list checking*: during the neighbor discovery phase, each node exchanges *hello* messages with its neighbor nodes to know its 1-hop and 2-hop neighbors (i.e., neighbors of each of its neighboring nodes). The neighborhood information is subsequently verified by exchange of *neighbor list checking* messages (Sen et al., 2010).

(ii) *One-hop packet forwarding*: when a node u sends a packet to its neighbor, it first keeps a copy of the packet in its buffer, and then forwards it to its next-hop node v before encrypting it with the cluster key of the node u. Since the cluster key is shared between the node and all its neighbors, the packet encrypted and sent by node u to node v can be overheard by all the neighbors of node u.

(iii) *Monitoring nodes selection*: as the packet is forwarded from node u to node v, the neighbors of node u that are also neighbors of node v receive the packet and store it in their buffers. These nodes are designated as the secondary monitoring nodes. For example, in Fig. 4, nodes w and y are the secondary monitoring nodes for node v. The node u is the primary monitoring node. The nodes that are not neighbors of node v but have received the packet because they are neighbors of node u, discard the packet. The primary node knows the secondary monitoring nodes, since every node knows its 1-hop and 2-hop neighbors.

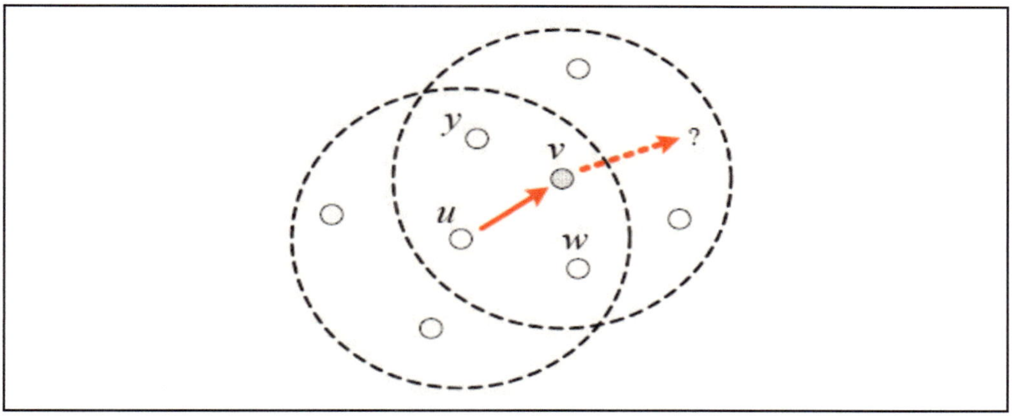

Fig. 4. Neighbor monitor system (secondary nodes w, y ; primary node u) (Sen et al., 2010)

(iv) *Role of secondary monitoring nodes*: the secondary monitoring nodes w and y monitor the traffic from node v and compare the outbound packets from node v with the packets stored in their buffer. The next-hop address of each packet is also verified to check whether the packet's intended next-hop is a really a neighbor of node v, by cross-checking the neighbor list of node v. If both these checks yield positive results, the secondary monitoring nodes remove the packet from their buffer and their role of monitoring is complete for that packet. If any packet is found to remain in the buffer of a secondary monitoring node for more than a threshold period of time, it first sends a broadcast signal in its neighborhood to inform all its neighbors that it is going to forward the packet to its designated next-hop so that other neighbors do not forward the same packet. The secondary monitoring node now forwards the packet to its designated next-hop after encrypting the packet with the cluster key. The role of the secondary node now becomes that of the primary node and its neighbors become the secondary node. This is in contrast to the scheme proposed in (Lee et al., 2006), where all the secondary nodes forward the packet in a multi-path mode.

(v) *Role of primary monitoring node*: the role of a primary monitoring node (node *u*) is identical to that of secondary monitoring nodes (nodes *w* and *y*); the only difference is that it listens not only on the traffic from node *v*, but also on the traffic from the nodes *w* and *y*. If the packet is correctly forwarded by any one of the nodes *v*, *w*, *y*, the node *u* removes the packet from its buffer. The role of node *u* as the primary monitoring node is now complete. If time out occurs for a packet, the primary monitoring node *u* forwards the packet (encrypted with its cluster key) to its next-hop other than node *v*.

As the packet is routed along a path towards the sink, the above steps of NMS algorithm except the *neighbor list checking* are executed at each hop so that reliable packet delivery can happen through a single path. This is in contrast to the previous schemes proposed in (Ye et al., 2005; Morcos et al., 2005; Yang et al., 2005). In these schemes, a node broadcasts a packet without specifying a designated next-hop, and all neighboring nodes with smaller costs (the cost at a node is the minimum energy required to forward a packet from the node to the base station) or within a specific geographic region continue forwarding the packet to the base station. If nodes *v*, *w*, and *y* have smaller costs than node *u* in Fig. 4, then each of them will forward packets received from node *u* following the existing approaches. However, in the proposed scheme, nodes *w* and *y* only observe the packet forwarding activities of node *v*, instead of actively forwarding the packets. In the event of no packet drop, the routing to the base station happens in a single-path, thereby making the process highly energy-efficient. Even in the event of a packet drop, the proposed algorithm works in a single-path mode. This makes it more efficient than the one proposed in (Lee et al., 2006). If the node *v* in Fig. 4 does not forward the packet it has received from node *u*, then one of the secondary monitoring nodes *w* and *y* would forward the packet to its next-hop nodes. The node (either *w* or *y*) that forwards the packet to its next-hop neighbors will first send a broadcast message in its neighborhood so that its other neighbors would not forward the same packet.

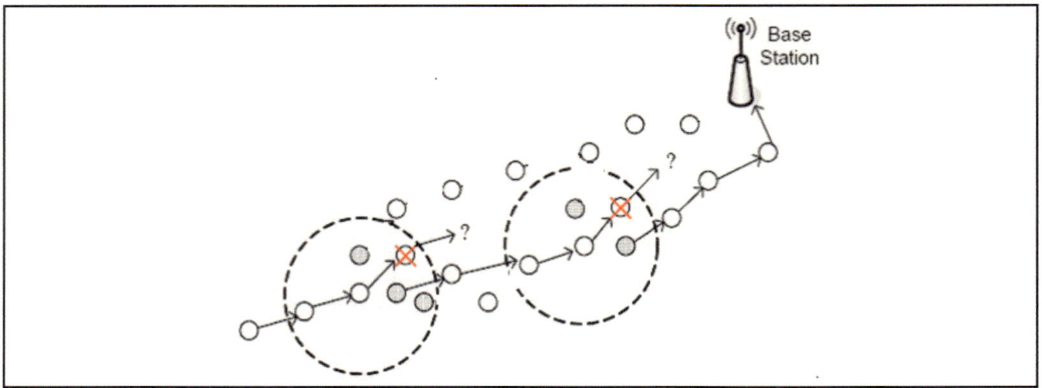

Fig. 5. Two malicious nodes identified by secondary monitoring nodes (Sen et al., 2010)

Fig. 5 shows an example of the application of the scheme, where two malicious (or faulty) nodes are bypassed as the packet is routed to the base station in a single-path.

For the scheme to work, each packet should be encrypted with a cluster key of the forwarding node so that all the neighbors of the forwarding node can decrypt and overhear it. If a link-level encryption was applied between each pair of nodes in the routing path, the scheme would have been more robust, since a compromised node could decrypt only the packets which were destined to it. However, it would have made the scheme less resilient to packet dropping attack. Since encryption with a cluster key provides a reasonable level of robustness to a node compromise, and also supports local broadcast (i.e. resiliency against packet-dropping) it makes the algorithm optimum in its performance (Karlof et al., 2004).
To make the scheme robust to routing disruption attack, where a node intentionally forwards the packets to a spurious address of the next-hop so that the packet is lost in routing, it is necessary that each node should prove that it really has the claimed neighbors. Apparently, a node has the knowledge of its direct neighbors by neighbor discovery and pair-wise key establishment phases discussed earlier. However, in the case of two-hop neighbors, a malicious node v can inform its neighbor u that it also has neighbor node x (any possible id in the network) which in fact is not a neighbor of node v (Fig. 4). Apparently, there is no way node u can detect these false claim of v since x is not in the neighborhood of u. To handle this problem, a scheme has been proposed by the authors using which a node can verify the neighbors of each of its neighboring nodes (Sen et al., 2010).

7.6 Some More Protocols for Secure Routing in WSNs

Inspired by the work on public key cryptography (Gura et al., 2004; Gaubatz et al., 2004; Watro et al., 2004; Wander et al., 2005), Du et al. have investigated the public key authentication problem (Du et al., 2005). The use of public key cryptography eases many problems in secure routing, for example, authentication and integrity. However, before a node A uses the public key from another node B, A must verify that the public key is actually B's, i.e., A must authenticate B's public key; otherwise, man-in-the-middle attacks are possible. In general networks, public key authentication involves a signature verification on a certificate signed by a trusted third party *Certificate Authority* (CA). However, the signature verification operations are very expensive operations for sensor nodes. Du et al. have proposed an efficient alternative that uses only one-way hash function for the public key authentication. The proposed scheme can be divided into two stages. In the pre-distribution stage, A Merkle tree R is constructed with each leaf L_i corresponding to a sensor node. Let pk_i represent node i's public key, V be an internal tree node, and V_{left} and V_{right} be V's two children. The value of an internal tree node is denoted by Φ. The Merkel tree can then be constructed as follows:

$$\Phi\ (L_i) = h\ (id_i,\ pk_i) \text{ for } i = 1,\ldots.N$$

$$\Phi(V) = h\ (\Phi(V_{left})\ ||\ \Phi(V_{right}))$$

In the above expressions, "||" represents the concatenation of two strings and h is a one-way hash function such as MD5 or SHA-1. Let R be the root of the tree. Each sensor node v needs to store the root value $\Phi(R)$ and the sibling node values $\lambda_1,\ldots\ldots. \lambda_H$ along the path from v to R. If node A wants to authenticate B's public key, B sends its public key pk along with the value of $\lambda_1,\ldots\ldots. \lambda_H$ to node A. Then, A can use the same procedure to reconstruct the Merkle tree R` and calculate the root value $\Phi(R`)$. A will trust B to be authentic if $\Phi(R`) =$

$\Phi(R)$.A sensor node only needs $H + 1$ storage units for the extra hash values. Based on this scheme, Du et al. further extended the idea to reduce the height of the Merkle tree to improve the communication overhead of the scheme. The proposed scheme is more efficient than signature verification on certificates. However, the scheme requires that some hash values be distributed in a pre-distribution stage. This results in some scalability issues when new sensors are added to an existing WSN.

Tanachaiwiwat et al. have presented a novel secure routing protocol- *trust routing for location aware sensor networks* (TRANS) (Tanachawiwat et al., 2003). It is primarily meant for use in data centric networks. It makes use of a loose-time synchronization asymmetric cryptographic scheme to ensure message confidentiality. The authors have used *μTESLA* to ensure message authentication and confidentiality. Using *μTESLA*, TRANS is able to ensure that a message is sent along a path of trusted nodes utilizing location aware routing. The base station broadcasts an encrypted message to all its neighbors. Only the trusted neighbors will possess the shared key necessary to decrypt the message. The trusted neighbors then add their locations (for the return trip), encrypt the new message with their shared key, and forward the message to their neighbors closest to the destination. Once the message reaches the destination, the recipient is able to authenticate the source (base station) using the MAC corresponding to the base station. To acknowledge or reply to the message, the destination node can simply forward a return message along the same trusted path from the message was received (Tanachawiwat et al., 2003).

Papadimitratos and Hass have proposed a secure route discovery protocol that guarantees correct topology discovery in an ad hoc sensor network (Papadimitratos et al., 2002). The security relies on the MAC (message authentication code) and an accumulation of the node identities along the route traversed by a message. In this way, a source node discovers the sensor network topology as each node along the route from source to destination appends its identity to the message. In order to ensure that the message has not been tampered with, a MAC is verified at the source and the destination.

A family of configurable secure routing protocols called *secure implicit geographic forwarding* (SIGF) has been proposed in (Wood et al., 2006). SIGF is based on a nondeterministic hybrid routing protocol - IGF (Blum et al., 2003) that is completely stateless. This allows SIGF to handle network dynamics effortlessly, and intrinsically limits the effects of a compromised node to a local area. There are no routing tables to corrupt, since forwarding decisions are made as late as possible - when a packet is ready to transmit over the air. However, the protocol is susceptible to a CTS rushing attack (Hu et al., 2003b).

To defend against route poisoning attack in a multi-hop WSN, a trust-aware routing framework has been proposed in (Zhan et al., 2010). The protocol integrates trustworthiness and energy-efficiency in routing decisions. Each node maintains a neighborhood table with trust level values and energy cost values for certain known neighbors. Once a node is able to decide its next-hop for routing a packet to the base station, it broadcasts its energy-report message that contains the information regarding the energy cost to deliver a packet from the node to the base station. The trustworthiness of a node is computed from its packet forwarding statistics. In this way, a secure and energy-efficient routing is achieved.

Table 4 presents a comparative analysis of some secure routing protocols for WSNs.

Protocols / Attacks	SPINS (Perrig, '02)	LKHW (Di Pietro, '03)	SecDEACH (Han, '10)	KeyChain (Liu, '03)	LEAP (Zhou, '07b)	SecRoute (Sen, '10)	SIGF (Wood, '06)	TARF (Zhan, '10)	RLEACH (Zhang, '08)	TamperSec (Pecho,'09)
Eavesdropping	Yes	Yes	Yes	No	Yes	Yes	Yes	Yes	Yes	Yes
Route poisoning	Yes	Yes	Yes	Yes	No	Yes	Yes	No	Yes	Yes
Sinkhole/ Blackhole	Yes	No	Yes	No	Yes	Yes	Yes	Yes	Yes	Yes
Grayhole	Yes	No	Yes	No	Yes	Yes	Yes	Yes	Yes	Yes
Wormhole	Yes	No	Yes	No	No	No	No	Yes	Yes	Yes
Sybill	Yes	No	No	No	Yes	Yes	Yes	Yes	Yes	Yes
Replay	Yes	Yes	Yes	Yes	Yes	Yes	Yes	No	Yes	Yes
Hello Flood	Yes	No	Yes	Yes	Yes	Yes	Yes	Yes	Yes	Yes
Node impersonation	Yes	Yes	No	No	Yes	Yes	No	Yes	No	Yes
Node replication	Yes	Yes	Yes	No	No	Yes	No	No	No	Yes

Table 4. Comparison of secure routing protocols for WSNs

8. Conclusion

Although research efforts have been made on cryptography, key management, secure routing, secure data aggregation, and intrusion detection in WSNs, there are still some challenges to be addressed. First, the selection of the appropriate cryptographic methods depends on the processing capability of the sensor nodes, indicating that there is no unified solution for all sensor networks. Instead, the security mechanisms are highly application-specific. Second, sensors are characterized by the constraints on energy, computation capability, memory, and communication bandwidth. The design of security services in WSNs must satisfy these constraints. Third, most of the current protocols assume that the sensor nodes and the base stations are stationary. However, there may be situations, such as battlefield environments, where the base station and possibly the sensors need to be mobile. The mobility of the sensor nodes has a great influence on sensor network topology and thus raises many issues in secure routing protocols. Some future trends in WSN security research are identified as follows:

Exploit the availability of private key operations on sensor nodes: recent studies on public key cryptography have shown that public key operations are still very expensive to realize in sensor nodes. A public key cryptography can greatly ease the design of security in WSNs, improving the efficiency of private key operations on sensor nodes is highly desirable.

Secure routing protocols for mobile sensor networks: mobility of sensor nodes has a great influence on sensor network topology and thus on the routing protocols. Mobility can be at the base station, sensor nodes, or both. Current protocols assume the sensor network is stationary. New secure routing protocols for mobile sensor networks need to be developed.

Time synchronization issues: current broadcast authentication schemes such as μTESLA and its extensions require the sensor network to be loosely time synchronized. This requirement is often hard to meet and new techniques that do not have such requirement are in demand.

Scalability and efficiency in broadcast authentication protocols: new schemes with higher scalability and efficiency need to be developed for authenticated broadcast protocols. The recent progress on public key cryptography may facilitate the design of authenticated broadcast protocols.

QoS and security: performance is generally degraded with the addition of security services in WSNs. Current studies on security in WSNs focus on individual topics such as key management, secure routing, secure data aggregation, and intrusion detection. QoS and security need to be evaluated together in WSNs.

9. References

Akyildiz, I.F. ; Su, W. ; Sankarasubramaniam, Y. & Cayirci, E. (2002). A survey on sensor networks. *IEEE Communications Magazine*, Vol. 40, No. 8, pp. 102-114.

Al-Karaki, J.N. & Kamal, A.E. (2004). Routing techniques in wireless sensor networks : a survey. *IEEE Wireless Communications*, Vol. 11. No. 6, pp. 6 – 28.

Aura, T. ; Nikander, P. & Leiwo, J. (2001). DoS-resistant authentication with client puzzles. *Proceedings of the 8th International Workshop on Security Protocols*, pp. 170-177, Springer-Verlag, Germany.

Awerbuch, B. ; Holmer, D. ; Nita-Rotaru, C. & Rubens, H. (2002). An on-demand secure routing protocol resilient to Byzantine failures. *Proceedings of the ACM Workshop on Wireless Security*, pp. 21 – 30.

Blum, B. ; He, T. ; Son, S. & Stankovic, J. (2003). IGF : a state-free robust communication protocol for wireless sensor networks. *Technical Report : CS-2003-11,* University of Virginia, Charlottesville, VA, USA.

Capkun, S. & Hubaux, J.-P. (2006). Secure positioning in wireless networks. *IEEE Journal on Selcted Areas in Communications,* Vol. 24, No. 2, pp. 221-232.

Carman, D.W. ; Krus, P.S. & Matt, B.J. (2000). Constraints and approaches for distributed sensor network security. *Technical Resport No : 00-010,* NAI Labs, Network Associates Inc., Glenwood, MD, USA.

Chan, H. & Perrig, A. (2003a). Security and privacy in sensor networks. *IEEE Computer Magazine,* pp. 103 - 105.

Chan, H. ; Perrig, A. & Song, D. (2003b). Random key pre-distribution schemes for sensor networks. *Proceedings of the IEEE Symposium on Security and Privacy,* p. 197, IEEE Computer Society Press.

Deng, H. ; Li, H. & Agrawal, D. (2002a). Routing security in wireless ad hoc networks. *IEEE Communications Magazine,* Vol. 40, No. 10.

Deng, J. ; Han, R. & Mishra, S. (2002b). INSENS : intrusion-tolerant routing in wireless sensor networks. *Technical Report CU-CS-939-02,* Department of Computer Science, University of Colorado at Boulder.

Deng, J. ; Han, R. & Mishra, S. (2004). Countermeasures against traffic analysis in wireless sensor networks. *Technical Report : CU-CS-987-04,* University of Colorado at Boulder.

Di Pietro, R. ; Mancini, L.V. ; Law, Y.W. ; Etalle, S. & Havinga, P. (2003). LKHW : a direced diffusion-based secure multi-cast scheme for wireless sensor networks. *Proceedings of the 32nd International Conference on Parallel Processing Workshops (ICPPW'03),* pp. 397-406, IEEE Computer Society Press.

Douceur, J. (2002). The Sybil attack. *Proceedings of the 1st International Workshop on Peer-to-Peer Systems* (IPTPS'02).

Du, W. ; Deng, J. ; Han, Y.S. & Varshney, P.K. (2003). A pair-wise key pre-distribution scheme for wireless sensor networks. *Proceedings of the 10th ACM Conference on Computer and Communications Security,* pp. 42-51, New York, USA, ACM Press.

Du, W. ; Wang, R. & Ning, P. (2005). An efficient scheme for authenticating public keys in sensor networks. *Proceedings of the 6th ACM International Symposium on Mobile Ad Hoc Networking and Computing,* pp. 58 - 67, New York, USA, ACM Press.

Eschenauer, L. & Gligor, V.D. (2002). A key-management scheme for distributed sensor networks. *Proceedings of the 9th ACM Conference on Computer and Networking,* pp. 41-47.

Estrin, D. ; Govindan, R. ; Heidemann, J.S. & Kumar. S. (1999). Next century challenges : scalable coordination in sensor networks. *Mobile Computing and Networking,* pp. 263-270.

Ganeriwal, S. ; Capkun, S. ; Han, C.-C. & Srivastava, M.B. (2005). Secure time synchronization service for sensor networks. *Proceedings of the 4th ACM Workshop on Wireless Security,* pp. 97 - 106, New York, USA, ACM Press.

Gaubatz, G. ; Kaps, J.P. & Sunar, B. (2004). Public key cryptography in sensor networks-revisited. *Proceedings of the 1st European Workshop on Security in Ad- Hoc and Sensor networks (ESAS'04).*

Gruteser, M. ; Schelle, G. ; Jain, A. ; Han, R. & Grunwald, D. (2003). Privacy-aware location sensor networks. *Proceedings of the 9th USENIX Workshop on Hot Topics in Operating Systems (HotOS IX).*

Gura, N. ; Patel, A. ; Wander, A. ; Eberle, H. & Shantz, S. (2004). Comparing elliptic curve cryptography and RSA on 8-bit CPUs. *Proceedings of Workshop on Cryptographic Hardware and Embedded Systems (CHES'04).*

Han, Y-J. ; Park, M-W. & Chung, T-M. (2010). SecDEACH : secure and resilient dynamic clustering protocol preserving data privacy in WSNs. *Proceedings of the International Conference on Computational Science and its Applications (ICCSA'10),* pp. 142 - 157, Fukuaka, Japan.

Hartung, C. ; Balasalle, J. & Han, R. (2004). Node compromise in sensor networks : the need for secure systems. *Technical Report : CU-CS-988-04,* Department of Computer Science, University of Colorado at Boulder.

Hill, J. ; Szewczyk, R. ; Woo, A. ; Hollar, S. ; Culler, D.E. & Pister, K. (2000). System architecture directions for networked sensors. *Proceedings of the 9th International Conference on Architectural Support for Programming Languages and Operating Systems,* pp. 93-104, ACM Press.

Hu, L. & Evans, D. (2003a). Secure aggregation for wireless sensor networks. *Proceedings of the Symposium on Applications and the Internet Workshops,* p. 384, IEEE Comp. Soc. Press.

Hu, L. & Evans, D. (2004a). Using directional antennas to prevent wormhole attacks. *Proceedings of the 11th Annual Network and Distributed System Security Symposium.*

Hu, Y. ; Perrig, A. & Johnson, D.B. (2003b). Rushing attacks and defense in wireless ad hoc network routing protocols. *Proceedings of the ACM Workshop on Wireless Security,* pp. 30 – 40.

Hu, Y. ; Perrig, A. & Johnson, D.B. (2004b). Packet leashes : a defense against worm-hole attacks. *Proceedings of the 11th Annual Network and Distributed System Security Symposium.*

Hwang, J. & Kim, Y. (2004). Revisiting random key pre-distribution schemes for wireless sensor networks. *Proceedings of the 2nd ACM Workshop on Security of Ad Hoc and Sensor Networks (SASN'04),* pp. 43-52, New York, USA, ACM Press.

Intanagonwiwat, C. ; Govindan, R. & Estrin, D. (2000). Directed diffusion : a scalable and robust communication paradigm for sensor networks. *Mobile Computing and Networking,* pp. 56 – 67.

Karlof, C. & Wagner, D. (2003). Secure routing in wireless sensor networks : attacks and countermeasures. *Proceedings of the 1st IEEE International Workshop on Sensor Network Protocols and Applications,* pp. 113-127.

Karlof, C. ; Sastry, N. & Wagner, D. (2004). TinySec : a link layer security architecture for wireless sensor networks. *Proceedings of ACM SensSys,* pp. 162 – 175.

Karp, B. & Kung, H.T. (2000). GPSR : greedy perimeter stateless routing for wireless networks. *Proceedings of the 6th Annual International Conference on Mobile Computing and Networking,* pp. 243 – 254, ACM Press.

Kaya, T. ; Lin, G. ; Noubir, G. & Yilmaz, A. (2003). Secure multicast gropus on ad hoc networks. *Proceedings of the 1st ACM Workshop on Security of Ad Hoc and Sensor Systems (SASN'03),* pp. 94 - 102, ACM Press.

Lazos, L. & Poovendran, R. (2002). Secure broadcast in energy-aware wireless sensor networks. *Proceedings of the IEEE International Symposium on Advances in Wireless Communications (ISWC'02).*

Lazos, L. & Poovendran, R. (2005). SERLOC : robust localization for wireless sensor networks. *ACM Transactions on Sensor Networks*, Vol. 1, No. 1, pp. 73 -100.

Lazos, L. & Poovendran, R. (2003). Energy-aware secure multi-cast communication in ad-hoc networks using geographic location information. *Proceedings of the IEEE International Conference on Acoustics Speech and Signal Processing.*

Lee, S-B. & Choi, Y-H. (2006). A resilient packet-forwarding scheme against maliciously packet-dropping nodes in sensor networks. *Proceedings of the 4th ACM Workshop on Security of Ad Hoc and Sensor Networks*, pp. 59-70.

Liu, D. & Ning, P. (2003). Efficient distribution of key chain commitments for broadcast authentication in distributed sensor networks. *Proceedings of the 10th Annual Network and Distributed System Security Symposium*, pp. 263 – 273, San Diego, CA, USA.

Liu, D. & Ning, P. (2004). Multilevel µTESLA : broadcast authentication for distributed sensor networks. *ACM Transactions on Embedded Computing Systems (ECS)*, Vol. 3, No. 4, pp. 800-836.

Liu, D. ; Ning, P. & Li, R. (2005a). Establishing pair-wise keys in distributed sensor networks. *ACM Transactions on Information Systems Security*, Vol. 8, No. 1, pp. 41-77.

Liu, D. ; Ning, P. ; Zhu, S. & Jajodia, S. (2005b). Practical broadcast authentication in sensor networks. *Proceedings of the 2nd Annual International Conference on Mobile and Ubiquitous Systems : Networking and Services*, pp. 118 – 129.

Madden, S. ; Franklin, M.J. ; Hellerstein, J.M. & Hong, W. (2002). TAG : a tiny aggregation service for ad-hoc sensor networks. *SIGOPS Operating Systems Review, Special Issue*, pp. 131-146.

Morcos, H. ; Matta, I. & Bestavros, A. (2005). M2RC : multiplicative-increase /additive-decrease multipath routing control for wireless sensor networks. *ACM SIGBED Reviw*, Vol. 2.

Newsome, J. ; Shi, E. ; Song, D. & Perrig, A. (2004). The Sybil attack in sensor networks : analysis and defenses. *Proceedings of the 3rd International Symposium on Information Processing in Sensor Networks*, pp. 259-268, ACM Press.

Ozturk, C. ; Zhang, Y. & Trappe, W. (2004). Source-location privacy in energy-constrained sensor network routing. *Proceedings of the 2nd ACM Workshop on Security of Ad Hoc and Sensor Networks.*

Papadimitratos, P. & Haas, Z.J. (2002). Secure routing for mobile ad hoc networks. *Proceedings of the SCS Communication Networks and Distributed System Modeling and Simulation Conference (CNDS'02).*

Parno, B. ; Perrig, A. & Gligor, V. (2005). Distributed detection of node replication attacks in sensor networks. *Proceedings of IEEE Symposium on Security and Privacy.*

Pecho, P. ; Nagy, J. ; Hanacke, P. & Drahansky, M. (2009). Secure collection tree protocol for tamper-resistant wireless sensors. *Communications in Computer and Information Science*, Vol. 58, pp. 217 – 224, Springer-Verlag, Heidelberg, Germany.

Perkins, C.E. & Royer, E.M. (1999). Ad hoc on-demand distance vector routing. *Proceedings of IEEE Workshop on Mobile Computing Systems and Applications*, pp. 90 – 100.

Perrig, A. ; Stankovic, J. & Wagner, D. (2004). Security in wireless sensor networks. *Communications of the ACM*, Vol. 47, No. 6, pp. 53 – 57.

Perrig, A. ; Szewczyk, R. ; Wen, V. ; Culler, D.E. & Tygar, J.D. (2002). SPINS : security protocols for sensor networks. *Wireless Networks*, Vol. 8, No. 5, pp. 521-534.

Przydatck, B. ; Song, D. & Perrig, A. (2003). SIA : secure information aggregation in sensor networks. *Proceedings of the 1st International Conference on Embedded Networked Systems (SenSys '08)*, pp. 255-265, ACM Press.

Rafaeli, S. & Hutchison, D. (2003). A survey of key management for secure group communication. *ACM Computing Survey*, Vol. 35, No. 3, pp. 309-329.

Sen, J ; Chandra, M.G. ; Harihara, S.G. ; Reddy, H. & Balamuralidhar, P. (2007b). A mechanism for detection of grayhole attack in mobile ad hoc networks. *Proceedings of the 6th International Conference on Information, Communication, and Signal Processing (ICICS'07)*, pp. 1 - 5, Singapore.

Sen, J. & Ukil, A. (2010). A secure routing protocol for wireless sensor networks. *Proceedings of the International Conference on Computational Sciences and its Applications (ICCSA'10)*, pp. 277 - 290, Fukuaka, Japan.

Sen, J. ; Chandra, M.G. ; Balamuralidhar, P. ; Harihara, S.G. & Reddy, H. (2007a). A distributed protocol for detection of packet dropping attack in mobile ad hoc networks. *Proceedings of the IEEE International Conference on Telecommunications (ICT'07)*, Penang, Malaysia.

Shi, E. & Perrig, A. (2004). Designing secure sensor networks. *Wireless Communication Magazine*, Vol. 11, No. 6, pp. 38 - 43.

Shrivastava, N. ; Buragohain, C. ; Agrawal, D. & Suri, S. (2004). Medians and beyond : new aggregation techniques for sensor networks. *Proceedings of the 2nd International Conference on Embedded Networked Sensor Systems*, pp. 239-249, ACM Press.

Slijepcevic, S. ; Potkonjak, M. ; Tsiatsis, V. ; Zimbeck, S. & Srivastava, M.B. (2002). On communication security in wireless ad-hoc sensor networks. *Proceedings of the 11th IEEE International Workshop on Enabling Technologies : Infrastructure for Collaborative Enterprises (WETICE'02)*, pp. 139-144.

Stankovic J.A. (2003). Real-time communication and coordination in embedded sensor networks. *Proceedings of the IEEE*, Vol. 91, No. 7, pp. 1002-1022.

Tanachawiwat, S. ; Dave, P. ; Bhindwale, R. & Helmy, A. (2003). Routing on trust and isolating compromised sensors in location-aware sensor systems. *Proceedings of the 1st International Conference on Embedded Networked Sensor Systems*, pp. 324-325, ACM Press.

Wander, A.S. ; Gura, N. ; Eberle, H. ; Gupta, V. & Shantz, S.C. (2005). Energy analysis of public-key cryptography for wireless sensor networks. *Proceedings of the 3rd IEEE International Conference on Pervasive Computing and Communication.*

Wang, W. & Bhargava, B. (2004b). Visualization of wormholes in sensor networks. *Proceedings of the 2004 ACM Workshop on Wireless Security*, pp. 51 - 60, New York, USA, ACM Press.

Wang, X. ; Gu, W. ; Chellappan, S. ; Xuan, D. & Laii, T.H. (2005). Search-based physical attacks in sensor networks : modeling and defense. *Technical Report*, Department of Computer Science and Engineering, Ohio State University.

Wang, X. ; Gu, W. ; Schosek, K. ; Chellappan, S. & Xuan, D. (2004a). Sensor network configuration under physical attacks. *Technical Report : OSU-CISRC-7/04-TR45*, Department of Computer Science and Engineering, Ohio State University.

Wang, Y. ; Attebury, G. & Ramamurthy, B. (2006). A survey of security issues in wireless sensor networks. *IEEE Communications Surveys and Tutorials*, Vol. 8, No. 2, pp. 2- 23.

Watro, R. ; Kong, D. ; Cuti, S. ; Gardiner, C. ; Lynn, C. & Kruus, P. (2004). TinyPK : securing sensor networks with public key technology. *Proceedings of the 2nd ACM Workshop on Security of Ad Hoc and Sensor Networks (SASN'04)*, pp. 59 - 64, New York, USA, ACM Press.

Wood, A.D. & Stankvic, J.A. (2002). Denial of service in sensor networks. *IEEE Computer*, Vol. 35, No. 10, pp. 54-62.

Wood, A.D. ; Fang, L. ; Stankovic, J.A. & He, T. (2006). SIGF : a family of configurable, secure routing protocols for wireless sensor networks. *Proceedings of the 4th ACM Workshop on Security of Ad Hoc and Sensor Networks*, pp. 35 - 48, Alexandria, VA, USA.

Yang, H. ; Ye, F. ; Yuan, Y. ; Lu, S. & Arbough, W. (2005). Towards resilient security in wireless sensor networks. *Procedings of ACM MobiHoc*, pp. 34 - 45.

Ye, F. ; Luo, L.H. & Lu, S. (2004). Statistical en-route detection and filtering of injected false data in sensor networks. *Proceddings of IEEE INFOCOM'04.*

Ye, F. ; Zhong, G. ; Lu, S. & Zhang, L. (2005). GRAdient Broadcast : a robust data delivery protocol for large scale sensor networks. *ACM Journal of Wireless Networks (WINET).*

Yuan, L. & Qu, G. (2002). Design space expolration for energy-efficient secure sensor networks. *Proceedings of IEEE International Conference on Application-Specific Systems, Architectures, and Processors*, pp. 88-100.

Zhang, K. ; Wang, C. & Wang, C. (2008). A secure routing protocol for cluster-based wireless sensor networks using group key management. *Proceedings of the 4th International Conference on Wireless Communications, Networking and Mobile Computing (WiCOM'08)*, pp. 1-5, Dalian.

Zhan, G. ; Shi, W. & Deng, J. (2010). TARF : a trust-aware routing framework for wireless sensor networks. *Proceedings of the 7th European Conference on Wireless Sensor Networks (EWSN'10)*, pp. 65 - 80, Coimbra, Portugal.

Zhu, H. ; Bao, F. ; Deng, R.H. & Kim, K. (2004a). Computing of trust in wireless networks. *Proceedings of 60th IEEE Vehicular Technology Conference*, California, USA.

Zhu, S. ; Setia, S. & Jajodia, S. (2004b). LEAP : efficient security mechanism for large-scale distributed sensor networks. *Proceedings of the 10th ACM Conference on Computer and Communications Security*, pp. 62 - 72, New York, USA, ACM Press.

Part 3

Optimization for WSN Applications

13

Optimization Approaches in Wireless Sensor Networks

Arslan Munir and Ann Gordon-Ross
Department of Electrical and Computer Engineering
University of Florida, Gainesville, Florida, USA

1. Introduction

Advancements in silicon technology, micro-electro-mechanical systems (MEMS), wireless communications, and digital electronics have led to the proliferation of wireless sensor networks (WSNs) in a wide variety of application domains including military, health, ecology, environment, industrial automation, civil engineering, and medical. This wide application diversity combined with complex sensor node architectures, functionality requirements, and highly constrained and harsh operating environments makes WSN design very challenging.

One critical WSN design challenge involves meeting *application requirements* such as lifetime, reliability, throughput, delay (responsiveness), etc. for myriad of application domains. Furthermore, WSN applications tend to have competing requirements, which exacerbates design challenges. For example, a high priority security/defense system may have both high responsiveness and long lifetime requirements. The mechanisms needed for high responsiveness typically drain battery life quickly, thus making long lifetime difficult to achieve given limited energy reserves.

Commercial off-the-shelf (COTS) sensor nodes have difficulty meeting application requirements due to the generic design traits necessary for wide application applicability. COTS sensor nodes are mass-produced to optimize cost and are not specialized for any particular application. Fortunately, COTS sensor nodes contain *tunable parameters* (e.g., processor voltage and frequency, sensing frequency, etc.) whose values can be specialized to meet application requirements. However, optimizing these tunable parameters is left to the application designer.

Optimization techniques at different design levels (e.g., sensor node hardware and software, data link layer, routing, operating system (OS), etc.) assist designers in meeting application requirements. WSN optimization techniques can be generally categorized as *static* or *dynamic*. Static optimizations optimize a WSN at deployment time and remain fixed for the WSN's lifetime. Whereas static optimizations are suitable for stable/predictable applications, static optimizations are inflexible and do not adapt to changing application requirements and environmental stimuli. Dynamic optimizations provide more flexibility by continuously optimizing a WSN/sensor node during runtime, providing better adaptation to changing application requirements and actual environmental stimuli.

This chapter introduces WSNs from an optimization perspective and explores optimization strategies employed in WSNs at different design levels to meet application requirements

Design-level	Optimizations
Architecture-level	bridging, sensorweb, tunneling
Component-level	parameter-tuning (e.g., processor voltage and frequency, sensing frequency), MDP-based dynamic optimization
Data Link-level	load balancing and throughput, power/energy
Network-level	query dissemination, data aggregation, real-time, network topology, resource adaptive, dynamic network reprogramming
Operating System-level	event-driven, dynamic power management, fault-tolerance

Table 1. Optimizations (discussed in this chapter) at different design-levels.

as summarized in Table 1. We present a typical WSN architecture and architectural-level optimizations in Section 2. We describe sensor node component-level optimizations and tunable parameters in Section 3. Next, we discuss data link-level Medium Access Control (MAC) optimizations and network-level routing optimizations in Section 4 and Section 5, respectively, and operating system-level optimizations in Section 6. After presenting these optimization techniques, we focus on dynamic optimizations for WSNs. There exists much previous work on dynamic optimizations e.g., (Brooks & Martonosi, 2000); (Hamed et al., 2006); (Hazelwood & Smith, 2006); (Hu et al., 2006), but most previous work targets the processor or cache subsystem in computing systems. WSN dynamic optimizations present additional challenges due to a unique design space, stringent design constraints, and varying operating environments. We discuss the current state-of-the-art in dynamic optimization techniques in Section 7 and propose a Markov Decision Process (MDP)-based dynamic optimization methodology for WSNs to meet application requirements in the presence of changing environmental stimuli in Section 8. Numerical results validate the optimality of our MDP-based methodology and reveal that our methodology more closely meets application requirements as compared to other feasible policies.

2. Architecture-level Optimizations

Fig. 1 shows an integrated WSN architecture (i.e., a WSN integrated with external networks) capturing architecture-level optimizations. Sensor nodes are distributed in a *sensor field* to observe a phenomenon of interest (i.e., environment, vehicle, object, etc.). Sensor nodes in the sensor field form an ad hoc wireless network and transmit the sensed information (data or statistics) gathered via attached sensors about the observed phenomenon to a base station or *sink node*. The sink node relays the collected data to the remote requester (user) via an arbitrary computer communication network such as a gateway and associated communication network. Since different applications require different communication network infrastructures to efficiently transfer sensed data, WSN designers can optimize the communication architecture by determining the appropriate topology (number and distribution of sensors within the WSN) and communication infrastructure (e.g., gateway nodes) to meet the application's requirements.

An infrastructure-level optimization called *bridging* facilitates the transfer of sensed data to remote requesters residing at different locations by connecting the WSN to external networks such as Internet, cellular, and satellite networks. Bridging can be accomplished by overlaying a sensor network with portions of the IP network where gateway nodes encapsulate sensor

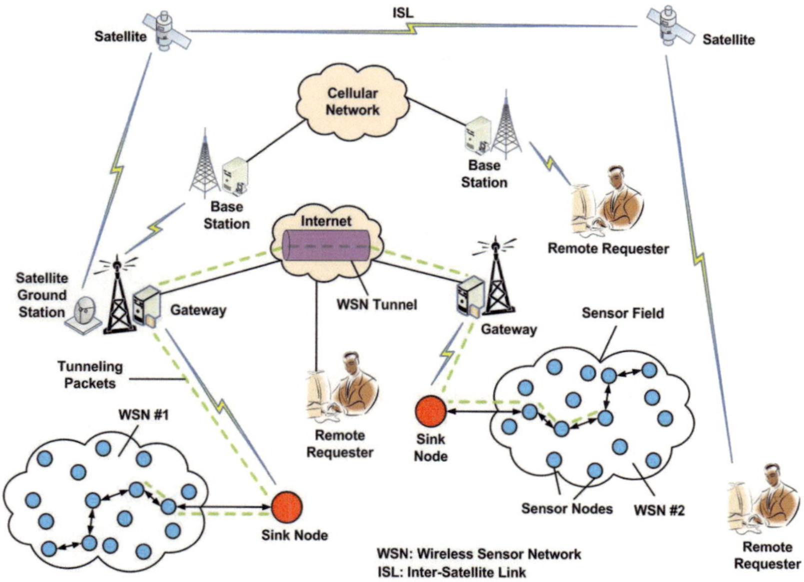

Fig. 1. Wireless sensor network architecture.

node packets with transmission control protocol or user datagram protocol/internet protocol (TCP/IP or UDP/IP).

Since sensor nodes can be integrated with the Internet via bridging, this WSN-Internet integration can be exploited to form a *sensor web*. In a sensor web, sensor nodes form a web view where data repositories, sensors, and image devices are discoverable, accessible, and controllable via the World Wide Web (WWW). The sensor web can use service-oriented architectures (SoAs) or sensor web enablement (SWE) standards (Mahalik, 2007). SoAs leverage extensible markup language (XML) and simple object access protocol (SOAP) standards to describe, discover, and invoke services from heterogeneous platforms. SWE is defined by the OpenGIS Consortium (OGC) and consists of specifications describing sensor data collection and web notification services. An example application for a sensor web may consist of a client using WSN information via sensor web queries. The client receives responses either from real-time sensors registered in the sensor web or from existing data in the sensor data base repository. In this application, clients can use WSN services without knowledge of the actual sensor nodes' locations.

Another WSN architectural optimization is *tunneling*. Tunneling connects two WSNs by passing internetwork communication through a gateway node that acts as a WSN extension and connects to an intermediate IP network. Tunneling enables construction of large virtual WSNs using smaller WSNs (Karl & Willig, 2005).

3. Sensor Node Component-level Optimizations

COTS sensor nodes provide optimization opportunities at the component-level via tunable parameters (e.g., processor voltage and frequency, sensing frequency, duty cycle, etc.), whose values can be specialized to meet varying application requirements. Fig. 2 depicts a sensor node's main components such as a power unit, storage unit, sensing unit, processing unit,

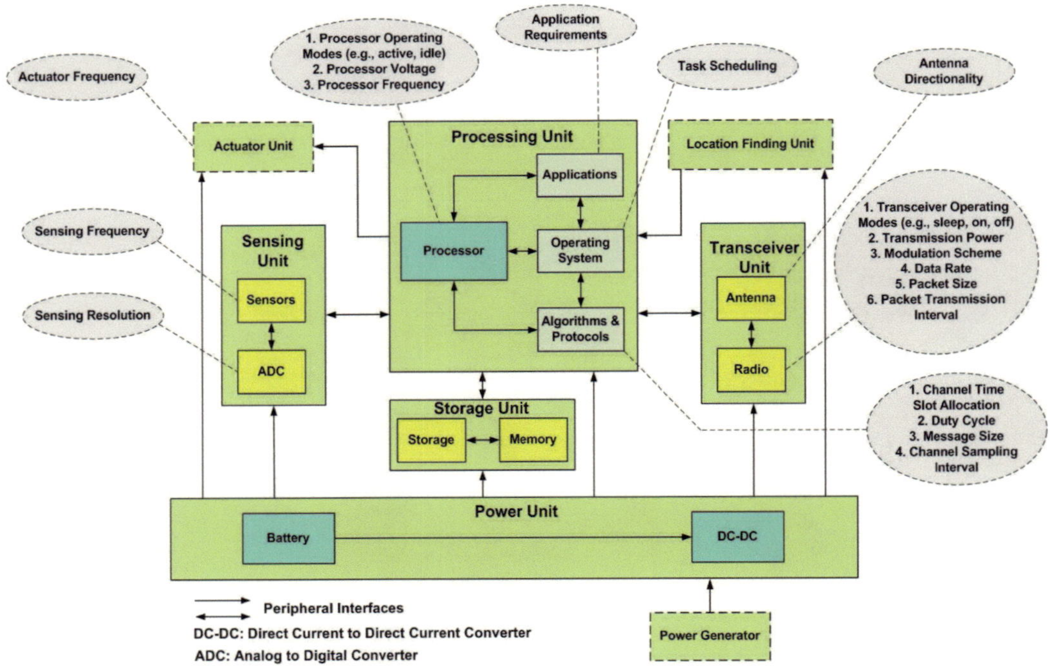

Fig. 2. Sensor node architecture with tunable parameters.

and transceiver unit along with potential tunable parameters associated with each component (Karl & Willig, 2005). In this section, we discuss these components and associated tunable parameters.

3.1 Sensing Unit

The sensing unit senses the phenomenon of interest using sensors and an analog to digital converter (ADC). The sensing unit's tunable parameters can control power consumption by changing the sensing frequency and the speed-resolution product of the ADC. Sensing frequency can be tuned to provide constant sensing, periodic sensing, and/or sporadic sensing. In constant sensing, sensors sense continuously and sensing frequency is limited only by the sensor hardware's design capabilities. Periodic sensing consumes less power than constant sensing because periodic sensing is duty-cycle based where the sensor node takes readings after every T seconds. Sporadic sensing consumes less power than periodic sensing because sporadic sensing is typically event-triggered by either external (e.g., environment) or internal (e.g., OS- or hardware-based) interrupts. The speed-resolution product of the ADC can be tuned to provide high speed-resolution with higher power consumption (e.g., seismic sensors use 24-bit converters with a conversion rate on the order of thousands of samples per second) or low speed-resolution with lower power consumption.

3.2 Processing Unit

The processing unit consists of a processor (e.g., Intel's Strong ARM (StrongARM, 2010), Atmel's AVR (ATMEL, 2009)) whose main tasks include controlling sensors, gathering and processing sensed data, executing WSN applications, and managing communication protocols and algorithms in conjunction with the operating system. The processor's tunable parameters include processor voltage and frequency, which can be specialized to meet power budget and throughput requirements. The processor can also switch between different operating modes (e.g., active, idle, sleep) to conserve energy. For example, the Intel's StrongARM consumes 75 mW in idle mode, 0.16 mW in sleep mode, and 240 mW and 400 mW in active mode while operating at 133 MHz and 206 MHz, respectively.

3.3 Transceiver Unit

The transceiver unit consists of a radio (transceiver) and an antenna, and is responsible for communicating with neighboring sensor nodes. The transceiver unit's tunable parameters include modulation scheme, data rate, transmit power, and duty cycle. The radio contains different operating modes (e.g., transmit, receive, idle, and sleep) for power management purposes. The sleep state provides the lowest power consumption, but switching from the sleep state to the transmit state consumes a large amount of power. The power saving modes (e.g., idle, sleep) are characterized by their power consumption and latency overhead (time to switch to transmit or receive modes). Power consumption in the transceiver unit also depends on the distance to the neighboring sensor nodes and transmission interferences (e.g., solar flare, radiation, channel noise).

3.4 Storage Unit

Sensor nodes contain a storage unit for temporary data storage since immediate data transmission is not always possible due to hardware failures, environmental conditions, physical layer jamming, and energy reserves. A sensor node's storage unit typically consists of Flash and static random access memory (SRAM). Flash is used for persistent storage of application code and text segments whereas SRAM is for run-time data storage. One potential optimization uses an extremely low-frequency (ELF) Flash file system, which is specifically adapted for sensor node data logging and operating environmental conditions. Storage unit optimization challenges include power conservation and memory resources (limited data and program memory, e.g., the Mica2 sensor node contains only 4 KB of data memory (SRAM) and 128 KB of program memory (Flash)).

3.5 Actuator Unit

The actuator unit consists of actuators (e.g., mobilizer, camera pan tilt), which enhance the sensing task. Actuators open/close a switch/relay to control functions such as camera or antenna orientation and repositioning sensors. Actuators, in contrast to sensors which only sense a phenomenon, typically affect the operating environment by opening a valve, emitting sound, or physically moving the sensor node. The actuator unit's tunable parameter is actuator frequency, which can be adjusted according to application requirements.

3.6 Location Finding Unit

The location finding unit determines a sensor node's location. Depending on the application requirements and available resources, the location finding unit can either be global positioning system (GPS)-based or ad hoc positioning system (APS)-based. The GPS-based location finding unit is highly accurate, but has high monetary cost and requires direct line of sight between the sensor node and satellites. The APS-based location finding unit determines a sensor node's position with respect to *landmarks*. Landmarks are typically GPS-based position-aware sensor nodes and landmark information is propagated in a multi-hop fashion. A sensor node in direct communication with a landmark estimates its distance from a landmark based on the received signal strength. A sensor node two hops away from a landmark estimates its distance based on the distance estimate of a sensor node one hop away from a landmark via message propagation. When a sensor node has distance estimates to three or more landmarks, the sensor node computes its own position as a centroid of the landmarks.

3.7 Power Unit

The power unit supplies power to a sensor node and determines a sensor node's lifetime. The power unit consists of a battery and a DC-DC converter. The electrode material and the diffusion rate of the electrolyte's active material affect the battery capacity. The DC-DC converter provides a constant supply voltage to the sensor node.

4. Data Link-level Medium Access Control Optimizations

Data link-level medium access control (MAC) manages the shared wireless channel and establishes data communication links between sensor nodes. Traditional MAC schemes emphasize high quality of service (QoS) (Rappaport, 1996) or bandwidth efficiency (Abramson, 1985); (IEEE Standards, 1999), however, WSN platforms have different priorities (Sohraby et al., 2007) thus inhibiting the straight forward adoption of existing MAC protocols (Chandrakasan et al., 1999). For example, since WSN lifetime is typically an important application requirement and batteries are not easily interchangeable/rechargeable, energy consumption is a primary design constraint for WSNs. Similarly, since the network infrastructure is subject to changes due to dying nodes, self-organization and failure recovery is important. To meet application requirements, WSN designers tune MAC layer protocol parameters (e.g., channel access schedule, message size, duty cycle, and receiver power-off, etc.). This section discusses MAC protocols for WSNs with reference to their tunable parameters and optimization objectives.

4.1 Load Balancing and Throughput Optimizations

MAC layer protocols can adjust wireless channel slot allocation to optimize throughput while maintaining the traffic load balance between sensor nodes. A *fairness* index measures load balancing or the uniformity of packets delivered to the sink node from all the senders. For the perfectly uniform case (ideal load balance), the fairness index is 1. MAC layer protocols that adjust channel slot allocation for load balancing and throughput optimizations include Traffic Adaptive Medium Access Protocol (TRAMA) (Rajendran et al., 2003), Berkeley Media Access Control (B-MAC) (Polastre et al., 2004), and Zebra MAC (Z-MAC) (Rhee et al., 2005).

TRAMA is a MAC protocol that adjusts channel time slot allocation to achieve load balancing while focusing on providing collision free medium access. TRAMA divides the channel access into random and scheduled access periods and aims to increase the utilization of the scheduled access period using time division multiple access (TDMA). TRAMA calculates a Message-Digest algorithm 5 (MD5) hash for every one-hop and two-hop neighboring sensor nodes to determine a node's priority. Experiments comparing TRAMA with both contention-based protocols (IEEE 802.11 and Sensor-MAC (S-MAC) (Ye et al., 2002)) as well as a scheduled-based protocol (Node-Activation Multiple Access (NAMA) (Bao & Garcia-Luna-Aceves, 2001)) revealed that TRAMA achieved higher throughput than contention-based protocols and comparable throughput with NAMA (Raghavendra et al., 2004).

B-MAC is a carrier sense MAC protocol for WSNs. B-MAC adjusts the duty cycle and time slot allocation for throughput optimization and high channel utilization. B-MAC supports on-the-fly reconfiguration of the MAC backoff strategy for performance (e.g., throughput, latency, power conservation) optimization. Results from B-MAC and S-MAC implementation on TinyOS using Mica2 motes indicated that B-MAC outperformed S-MAC by 3.5x on average (Polastre et al., 2004). No sensor node was allocated more than 15% additional bandwidth as compared with other nodes, thus ensuring fairness (load balancing).

Z-MAC is a hybrid MAC protocol that combines the strengths of TDMA and carrier sense multiple access (CSMA) and offsets their weaknesses. Z-MAC allocates time slots at sensor node deployment time by using an efficient channel scheduling algorithm to optimize throughput, but this mechanism requires high initial overhead. A time slot's *owner* is the sensor node allocated to that time slot and all other nodes are called *non-owners* of that time slot. Multiple owners are possible for a given time slot because Z-MAC allows any two sensor nodes beyond their two-hop neighborhoods to own the same time slot. Unlike TDMA, a sensor node may transmit during any time slot but slot owners have a higher priority. Experimental results from Z-MAC implementation on both ns-2 and TinyOS/Mica2 indicated that Z-MAC performed better than B-MAC under medium to high contention but exhibited worse performance than B-MAC under low contention (inherits from TDMA-based channel access). The fairness index of Z-MAC was between 0.7 and 1, whereas that of B-MAC was between 0.2 to 0.3 for a large number of senders (Rhee et al., 2005).

4.2 Power/Energy Optimizations

MAC layer protocols can adapt their transceiver operating modes (e.g., sleep, on and off) and duty cycle for reduced power and/or energy consumption. MAC layer protocols that adjust duty cycle for power/energy optimization include Power Aware Multi-Access with Signaling (PAMAS) (Stojmenović, 2005); (Karl & Willig, 2005), S-MAC (Ye et al., 2002), Timeout-MAC (T-MAC) (Van Dam & Langendoen, 2003), and B-MAC.

PAMAS is a MAC layer protocol for WSNs that adjusts the duty cycle to minimize radio on time and optimize power consumption. PAMAS uses separate data and control channels (the control channel manages the request/clear to send (RTS/CTS) signals or the receiver busy tone). If a sensor node is receiving a message on the data channel and receives an RTS message on the signaling channel, then the sensor node responds with a busy tone on the signaling channel. This mechanism avoids collisions and results in energy savings. The PAMAS protocol powers off the receiver if either the transmit message queue is empty and the node's neighbor is transmitting or the transmit message queue is not empty but at least one neighbor is transmitting and one neighbor is receiving. WSN simulations with 10 to 20 sensor nodes with 512-byte data packets, 32-byte RTS/CTS packets, and 64-byte busy tone signal packets revealed power savings between 10% and 70% (Singh & Raghavendra, 1998). PAMAS optimization challenges include implementation complexity and associated area cost because the separate control channel requires a second transceiver and duplexer.

The S-MAC protocol tunes the duty cycle and message size for energy conservation. S-MAC minimizes wasted energy due to *frame* (packet) collisions (since collided frames must be retransmitted with additional energy cost), overhearing (a sensor node receiving/listening to a frame destined for another node), control frame overhead, and idle listening (channel monitoring to identify possible incoming messages destined for that node). S-MAC uses a periodic sleep and listen (sleep-sense) strategy defined by the duty cycle. S-MAC avoids frame collisions by using virtual sense (network allocation vector (NAV)-based) and physical carrier sense (receiver listening to the channel) similar to IEEE 802.11. S-MAC avoids overhearing by instructing interfering sensor nodes to switch to sleep mode after hearing an RTS or CTS packet (Stojmenović, 2005). Experiments conducted on Rene Motes (Culler et al., 2002) for a traffic load comprising of sent messages every 1-10 seconds revealed that a IEEE 802.11-based MAC consumed 2x to 6x more energy than S-MAC (Ye et al., 2004).

T-MAC adjusts the duty cycle dynamically for power efficient operation. T-MAC allows a variable sleep-sense duty cycle as opposed to the fixed duty cycle used in S-MAC (e.g., 10% sense and 90% sleep). The dynamic duty cycle further reduces the idle listening period. The sensor node switches to sleep mode when there is no activation event (e.g., data reception, timer expiration, communication activity sensing, or impending data reception knowledge through neighbors' RTS/CTS) for a predetermined period of time. Experimental results obtained from T-MAC protocol implementation on OMNeT++ (Varga, 2001) to model EYES sensor nodes (EYES, 2010) revealed that under homogeneous load (sensor nodes sent packets with 20- to 100-byte payloads to their neighbors at random), both T-MAC and S-MAC yielded 98% energy savings as compared to CSMA whereas T-MAC outperformed S-MAC by 5x under variable load (Raghavendra et al., 2004).

B-MAC adjusts the duty cycle for power conservation using channel assessment information. B-MAC duty cycles the radio through a periodic channel sampling mechanism known as low power listening (LPL). Each time a sensor node wakes up, the sensor node turns on the radio and checks for channel activity. If the sensor node detects activity, the sensor node powers up and stays awake for the time required to receive an incoming packet. If no packet is received, indicating inaccurate activity detection, a time out forces the sensor node to sleep mode. B-MAC requires an accurate clear channel assessment to achieve low power operation. Experimental results obtained from B-MAC and S-MAC implementation on TinyOS using Mica2 motes revealed that B-MAC power consumption was within 25% of S-MAC for low throughputs (below 45 bits per second) whereas B-MAC outperformed S-MAC by 60% for higher throughputs. Results indicated that B-MAC performed better than S-MAC for latencies under 6 seconds whereas S-MAC yielded lower power consumption as latency approached 10 seconds (Polastre et al., 2004).

5. Network-level Data Dissemination and Routing Protocol Optimizations

One commonality across diverse WSN application domains is the sensor node's task to sense and collect data about a phenomenon and transmit the data to the sink node. To meet application requirements, this data dissemination requires energy-efficient routing protocols to establish communication paths between the sensor nodes and the sink. Typically harsh operating environments coupled with stringent resource and energy constraints make data dissemination and routing challenging for WSNs. Ideally, data dissemination and routing protocols should target energy efficiency, robustness, and scalability. To achieve these optimization objectives, routing protocols adjust transmission power, routing strategies, and leverage either single-hop or multi-hop routing. In this section, we discuss protocols, which optimize data dissemination and routing in WSNs.

5.1 Query Dissemination Optimizations

Query dissemination (transmission of a sensed data query/request from a sink node to a sensor node) and data forwarding (transmission of sensed data from a sensor node to a sink node) requires routing layer optimizations. Protocols that optimize query dissemination and data forwarding include Declarative Routing Protocol (DRP) (Coffin et al., 2000), directed diffusion (Intanagonwiwat et al., 2003), GRAdient Routing (GRAd) (Poor, 2010), GRAdient

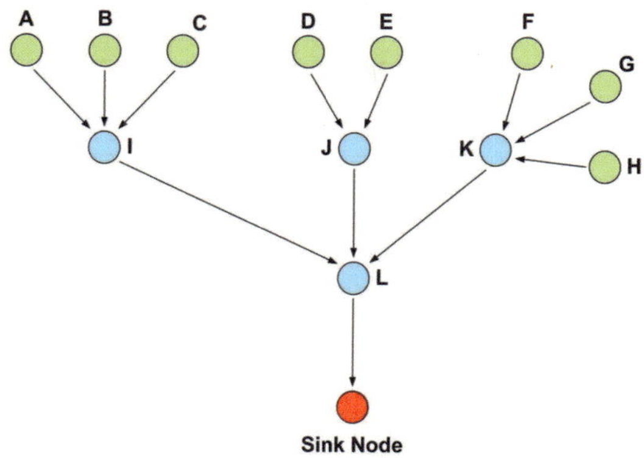

Fig. 3. Data aggregation.

Broadcast (GRAB) (Ye et al., 2005), and Energy Aware Routing (EAR) (Raghavendra et al., 2004); (Shah & Rabaey, 2002).

DRP targets energy efficiency by exploiting in-network aggregation (multiple data items are aggregated as they are forwarded by sensor nodes). Fig. 3 shows in-network data aggregation where sensor node I aggregates sensed data from source nodes A, B, and C, sensor node J aggregates sensed data from source nodes D and E, and sensor node K aggregates sensed data from source nodes F, G, and H. The sensor node L aggregates the sensed data from sensor nodes I, J, and K, and transmits the aggregated data to the sink node. DRP uses reverse path forwarding where data reports (packets containing sensed data in response to query) flow in the reverse direction of the query propagation to reach the sink.

Directed diffusion targets energy efficiency, scalability, and robustness under network dynamics using reverse path forwarding. Directed diffusion builds a shared mesh to deliver data from multiple sources to multiple sinks. The sink node disseminates the query, a process referred to as *interest propagation* (Fig. 4(a)). When a sensor node receives a query from a neighboring node, the sensor node sets up a vector called the *gradient* from itself to the neighboring node and directs future data flows on this gradient (Fig. 4(b)). The sink node receives an initial batch of data reports along multiple paths and uses a mechanism called *reinforcement* to select a path with the best forwarding quality (Fig. 4(c)). To handle network dynamics such as sensor node failures, each data source floods data reports periodically at lower rates to maintain alternate paths. Directed diffusion challenges include formation of initial gradients and wasted energy due to redundant data flows to maintain alternate paths.

GRAd optimizes data forwarding and uses cost-field based forwarding where the cost metric is based on the hop count (i.e., sensor nodes closer to the sink node have smaller costs and those farther away have higher costs). The sink node floods a REQUEST message and the data source broadcasts the data report containing the requested sensed information. The neighbors with smaller costs forward the report to the sink node. GRAd drawbacks include wasted energy due to redundant data report copies reaching the sink node.

GRAB optimizes data forwarding and uses cost-field based forwarding where the cost metric denotes the total energy required to send a packet to the sink node. GRAB was designed for harsh environments with high channel error rate and frequent sensor node failures. GRAB controls redundancy by controlling the width (number of routes from the source sensor node

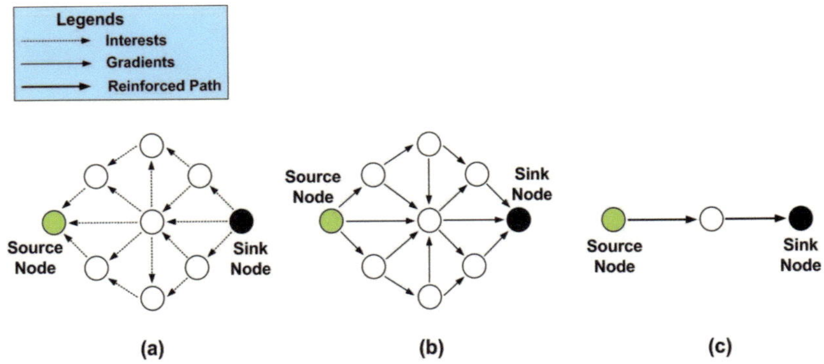

Fig. 4. Directed diffusion: (a) Interest propagation; (b) Initial gradient setup; (c) Data delivery along the reinforced path.

to the sink node) of the forwarding mesh but requires that sensor nodes make assumptions about the energy required to transmit a data report to a neighboring node.

EAR optimizes data forwarding and uses cost-field based forwarding where the cost metric denotes energy per neighbor. EAR optimization objectives are load balancing and energy conservation. EAR makes forwarding decisions probabilistically where the assigned probability is inversely proportional to the neighbor energy cost so that paths consuming more energy are used less frequently (Raghavendra et al., 2004).

5.2 Real-Time Constrained Optimizations

Critical WSN applications may have real-time requirements for sensed data delivery (e.g., a security/defense system monitoring enemy troops or a forest fire detection application). Failure to meet the real-time deadlines for these applications can have catastrophic consequences. Routing protocols that consider the timing constraints for real-time requirements include Real-time Architecture and Protocol (RAP) (Lu et al., 2002) and a stateless protocol for real-time communication in sensor networks (SPEED) (He et al., 2003).

RAP provides real-time data delivery by considering the data report expiration time (time after which the data is of little or no use) and the remaining distance the data report needs to travel to reach the sink node. RAP calculates the desired velocity $v = d/t$ where d and t denote the destination distance and packet lifetime, respectively. The desired velocity is updated at each hop to reflect the data report's urgency. A sensor node uses multiple first-in-first-out (FIFO) queues where each queue accepts reports of velocities within a certain range and then schedules transmissions according to a report's degree of urgency (Raghavendra et al., 2004).

SPEED provides real-time data delivery and uses an exponentially weighted moving average for delay calculation. Given a data report with velocity v, SPEED calculates the speed v_i of the report if the neighbor N_i is selected as the next hop and then selects a neighbor with $v_i > v$ to forward the report to (Raghavendra et al., 2004).

5.3 Network Topology Optimizations

Routing protocols can adjust radio transmission power to control network topology (based on routing paths). Low-Energy Adaptive Clustering Hierarchy (LEACH) (Heinzelman et al., 2000) optimizes the network topology for reduced energy consumption by adjusting the radio's transmission power. LEACH uses a hybrid single-hop and multi-hop communication paradigm. The sensor nodes use multi-hop communication to transmit data reports to a cluster head (LEACH determines the cluster head using a randomized distributed algorithm). The cluster head forwards data to the sink node using long-range radio transmission.

5.4 Resource Adaptive Optimizations

Routing protocols can adapt routing activities in accordance with available resources. Sensor Protocols for Information via Negotiation (SPIN) (Kulik et al., 2002) optimizes performance efficiency by using data negotiation and resource adaptation. In data negotiation, sensor nodes associate metadata with nodes and exchange this metadata before actual data transmission begins. The sensor nodes interested in the data content, based on metadata, request the actual data. This data negotiation ensures that data is sent only to interested nodes. SPIN allows sensor nodes to adjust routing activities according to available energy resources. At low energy levels, sensor nodes reduce or eliminate certain activities (e.g., forwarding of metadata and data packets) (Sohraby et al., 2007).

6. Operating System-level Optimizations

A sensor node's operating system (OS) presents optimization challenges because sensor node operation falls between single-application devices that typically do not need an OS and general-purpose devices with resources to run traditional embedded OSs. A sensor node's OS manages processor, radio, I/O buses, and Flash memory, and provides hardware abstraction to application software, task coordination, power management, and networking services. In this section, we discuss several optimizations provided by existing OSs for sensor nodes (Sohraby et al., 2007).

6.1 Event-Driven Optimizations

Sensor nodes respond to events by controlling sensing and actuation activity. Since sensor nodes are event-driven, it is important to optimize the OS for event handling. WSN OSs optimized for event handling include TinyOS (TinyOS, 2010) and PicOS (Akhmetshina et al., 2002).

TinyOS operates using an event-driven model (tasks are executed based on events). TinyOS is written in the nesC programming language and allows application software to access hardware directly. TinyOS's advantages include simple OS code, energy efficiency, and a small memory foot print. TinyOS challenges include introduced complexity in application development and porting of existing C code to TinyOS.

PicOS is an event-driven OS written in C and designed for limited memory microcontrollers. PicOS tasks are structured as a finite state machine (FSM) and state transitions are triggered by events. PicOS is effective for reactive applications whose primary role is to react to events. PicOS supports multitasking and has small memory requirements but is not suitable for real-time applications.

6.2 Dynamic Power Management

A sensor node's OS can control hardware components to optimize power consumption. Examples include Operating System-directed Power Management (OSPM) (Sinha & Chandrakasan, 2001) and MagnetOS (Barr & et al., 2002), each of which provide mechanisms for dynamic power management. OSPM offers greedy-based dynamic power management, which switches the sensor node to a sleep state when idle. Sleep states provide energy conservation, however, transition to sleep state has the overhead of storing the processor state and requires a finite amount of wakeup time. OSPM greedy-based adaptive sleep mechanism disadvantages include wake up delay and potentially missing events during sleep time. MagnetOS provides two online power-aware algorithms and an adaptive mechanism for applications to effectively utilize the sensor node's resources.

6.3 Fault-Tolerance

Since maintenance and repair of sensor nodes is typically not feasible after deployment, sensor nodes require fault-tolerant mechanisms for reliable operation. MANTIS (Abrach & et al., 2003) is a multithreaded OS that provides fault-tolerant isolation between applications by not allowing a blocking task to prevent the execution of other tasks. In the absence of fault-tolerant isolation, if one task executes a conditional loop whose logical condition is never satisfied, then that task will execute in an infinite loop blocking all other tasks. MANTIS facilitates simple application development and allows dynamic reprogramming to update the sensor node's binary code. MANTIS offers a multimodal prototyping environment for testing WSN applications by providing a remote shell and command server to enable inspection of the sensor node's memory and status remotely. MANTIS challenges include context switch time, stack memory overhead (since each thread requires one stack), and high energy consumption.

7. Dynamic Optimizations

Dynamic optimizations enable in-situ parameter tuning and empowers the sensor node to adapt to changing application requirements and environmental stimuli throughout the sensor node's lifetime. Dynamic optimizations are important because application requirements change over time and environmental stimuli/conditions may not be accurately predicted at design time. Although some OS, MAC layer, and routing optimizations discussed in prior sections of this chapter are dynamic in nature, in this section we present additional dynamic optimization techniques for WSNs.

7.1 Dynamic Voltage and Frequency Scaling

Dynamic voltage and frequency scaling (DVFS) adjusts a sensor node's processor voltage and frequency to optimize energy consumption. DVFS trades off performance for reduced energy consumption by considering that the peak computation (instruction execution) rate is much higher than the application's average throughput requirement and that sensor nodes are based on CMOS logic, which has a voltage dependent maximum operating frequency. Min et al. (Min et al., 2000) demonstrated that a DVFS system containing a voltage scheduler running in tandem with the operating system's task scheduler resulted in a 60% reduction in energy consumption. Yuan et al. (Yuan & Qu, 2002) studied a DVFS system for sensor nodes that required the sensor nodes to insert additional information (e.g., packet length, expected processing time, and deadline) into the data packet's header. The receiving sensor node utilized this information to select an appropriate processor voltage and frequency to minimize the overall energy consumption.

7.2 Software-based Dynamic Optimizations

Software can provide dynamic optimizations using techniques such as duty cycling, batching, hierarchy, and redundancy reduction. Software can control the *duty cycle* so that sensor nodes are powered in a cyclic manner to reduce the average power draw. In *batching*, multiple operations are buffered and then executed in a burst to reduce startup overhead cost. Software can arrange operations in a *hierarchy* based on energy consumption and then invoke low energy operations before high energy operations. Software can reduce *redundancy* by compression, data aggregation, and/or message suppression. Kogekar et al. (Kogekar et al., 2004) proposed an approach for software reconfiguration in WSNs. The authors modeled the WSN operation space (defined by the WSN software components' models and application requirements) and defined reconfiguration as the process of switching from one point in the operation space to another.

7.3 Dynamic Network Reprogramming

Dynamic network reprogramming reprograms sensor nodes to change/modify tasks by disseminating code in accordance with changing environmental stimuli. Since recollection and reprogramming is not a feasible option for most sensor nodes, dynamic network reprogramming enables the sensor nodes to perform different tasks. For example, a WSN initially deployed for measuring relative humidity can measure temperature statistics after dynamic reprogramming. The MANTIS OS provides this dynamic reprogramming ability (Section 6.3).

8. MDP-based Dynamic Optimizations

In this section, we extend our discussion of dynamic optimizations using an MDP-based dynamic optimization (Munir & Gordon-Ross, 2009) as a specific example. MDP is suitable for WSN dynamic optimizations because of MDP's inherent ability to perform dynamic decision making. We propose MDP as a method to perform parameter tuning-based dynamic optimizations. Traditional microprocessor-based systems use DVFS for energy optimizations. DVFS only provides a partial tuning for sensor nodes because sensor nodes are distinct from traditional systems in that they have embedded sensors coupled with an embedded processor. For example, the sensing frequency dictates the amount of processed and communicated data. We propose dynamic voltage, frequency, and sensing frequency scaling (DVFS2) to provide enhanced optimization potential as compared to DVFS for WSNs. Our MDP-based optimization focuses on DVFS2 but is equally applicable for extensive design spaces with more tunable parameters (e.g., transmission power, packet transmission interval, etc.).

8.1 Dynamic Optimization Methodology

Fig. 5 depicts the process diagram for our dynamic optimization, which consists of three logical domains: the application characterization domain, the communication domain, and the sensor node tuning domain.

The *application characterization domain* refers to the WSN application's characterization/specification where the application manager/designer (one who manages/designs a WSN) defines various *application metrics* (e.g., lifetime, throughput, reliability, etc.) based on application requirements. The application manager/designer also assigns *weight factors* to application metrics which signify the weightage or relative importance of each application metric with respect to other metrics. The *objective function* or *reward function* signifies the overall reward (revenue) for given application requirements. The application metrics along with associated weight factors represent the *objective/reward function parameters*.

The *communication domain* (depicted by the sink node in Fig. 5) encompasses the communication network between the application manager and the sensor nodes. The application manager transmits the objective or reward function parameters to the sink node via the communication domain which in turn relays these parameters to the sensor nodes.

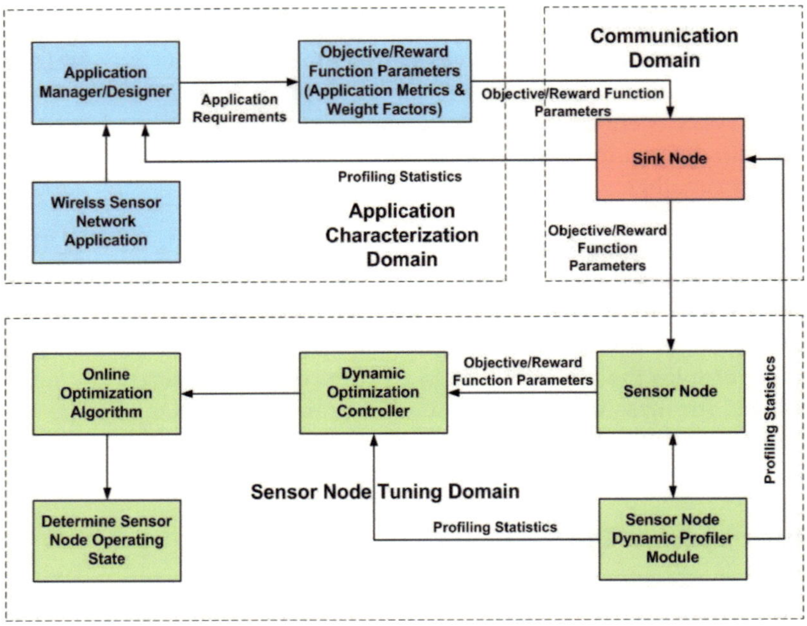

Fig. 5. Process diagram for parameter tuning-based dynamic optimizations for WSNs.

The *sensor node tuning domain* consists of sensor nodes and performs sensor node parameter tuning. Each sensor node contains a *dynamic optimization controller*, which orchestrates the dynamic optimization process. The dynamic optimization controller module receives the reward function parameters and invokes an online optimization algorithm to determine an optimal or near-optimal sensor node *state* (tunable parameter value settings).

Our proposed methodology reacts to environmental stimuli via a *dynamic profiler module*, which monitors environmental changes over time and captures unanticipated environmental situations not predictable at design time. The dynamic profiler module profiles the *profiling statistics* (e.g., wireless channel condition, number of packets dropped, battery energy, etc.). The dynamic profiler module informs the dynamic optimization controller as well as the application manager of the profiled statistics. The dynamic optimization controller processes the profiling statistics to determine if the current operating state meets the application requirements. If the current operating state does not meet the application requirements, the dynamic optimization controller reinvokes the online optimization algorithm (e.g., MDP-based or any other) to determine the new operating state. This feedback process continues to ensure the selection of a good operating state to better meet application requirements in the presence of changing environmental stimuli.

8.2 Dynamic Optimization Formulation

In this subsection, we formulate the constructs of our MDP-based dynamic optimization (Munir & Gordon-Ross, 2009). Although we describe dynamic optimization constructs with reference to MDP, our formulation provides insight into any other dynamic optimization algorithm.

8.2.1 State Space

The state space S for our MDP-based dynamic optimization methodology given N tunable parameters is defined as:

$$S = S_1 \times S_2 \times \cdots \times S_N \;:\; |S| = I \tag{1}$$

where S_i denotes the state space for tunable parameter i, $\forall\ i \in \{1, 2, \ldots, N\}$ and $\times$ denotes the Cartesian product. The state space S consists of a total of I states as given by the state space cardinality $|S| = I$. Each tunable parameter's state space S_i consists of n tunable values:

$$S_i = \{s_{i_1}, s_{i_2}, s_{i_3}, \ldots, s_{i_n}\} \;:\; |S_i| = n \tag{2}$$

where $|S_i|$ denotes the number of tunable values in S_i. S is a set of N-tuples formed by taking one tunable parameter value from each tunable parameter. A single N-tuple $s \in S$ is given as:

$$\begin{aligned} s \;=\; & (s_{1_y}, s_{2_y}, \ldots, s_{N_y}) : s_{i_y} \in S_i, \\ & \forall\, i \in \{1, 2, \ldots, N\},\, y \in \{1, 2, \ldots, n\} \end{aligned} \tag{3}$$

Each N-tuple represents a sensor note state. We point out that some N-tuples in S may not be feasible (such as invalid combinations of processor voltage and frequency) and can be regarded as *do not care* tuples.

For example, given three tunable parameters, S can be written as:

$$S = V_p \times F_p \times F_s \tag{4}$$

where V_p, F_p, and F_s denote the state space for a sensor node's processor voltage, processor frequency, and sensing (sampling) frequency, respectively.

8.2.2 Decision Epochs and Actions

The *decision epochs* refer to the points of time during a sensor node's lifetime at which the sensor node makes a decision regarding its operating state (i.e., whether to continue operating in the current state or transition to another state). We consider a discrete time process where time is divided into *periods* and a decision epoch corresponds to the beginning of a period. The sequence of decision epochs is represented as:

$$T = \{1, 2, 3, ..., N\}, \quad N \leq \infty \tag{5}$$

where the random variable N corresponds to the sensor node's lifetime (each individual time period in T can be denoted as time t).

At each decision epoch, a sensor node's *action* determines the next state to transition to given the current state. The sensor node action in state $i \in S$ is defined as:

$$A_i = \{a_{i,j}\} \in \{0, 1\} \tag{6}$$

where $a_{i,j}$ denotes the action taken at time t that causes the sensor node to transition to state j at time $t + 1$ from the current state i. If $a_{i,j} = 1$, the action is taken and if $a_{i,j} = 0$, the action is not taken.

8.2.3 Policy and Performance Criterion

For each given state $s \in S$, a *policy* π determines whether an action $a \in A_s$ is taken or not at a decision epoch. A *performance criterion* compares the performance of different policies. The sensor node selects an action prescribed by a policy based on the sensor node's current state. The sensor node receives a reward $r(X_t, Y_t)$ as a result of selecting an action Y_t at decision epoch t where the random variable X_t denotes the state at decision epoch t. The *expected total reward* $v_N^{\pi}(s)$ denotes the expected total reward over the decision making horizon N given a specific policy π (Puterman, 2005); (Stevens-Navarro et al., 2008):

$$v_N^{\pi}(s) = \lim_{N\to\infty} E_s^{\pi}\left[E_N\left\{\sum_{t=1}^{N} r(X_t, Y_t)\right\}\right] \tag{7}$$

where E_s^{π} represents the expected reward with respect to policy π and the initial state s (the system state at the time of the expected reward calculation) and E_N denotes the expected reward with respect to the probability distribution of the random variable N. We can write (7) as (Puterman, 2005):

$$v_N^{\lambda}(s) = E_s^{\pi}\left\{\sum_{t=1}^{\infty} \lambda^{t-1} r(X_t, Y_t)\right\} \tag{8}$$

which gives the *expected total discounted reward*. We assume that the random variable N is geometrically distributed with parameter λ and hence the distribution *mean* is $1/(1-\lambda)$ (Stevens-Navarro et al., 2008). The parameter λ can be interpreted as a *discount factor*, which measures the present value of one unit of reward received one period in the future. Thus, $v_N^{\lambda}(s)$ represents the expected total present value of the reward (income) stream obtained using policy π (Puterman, 2005). Our objective is to find a policy that maximizes the expected total discounted reward i.e., a policy π^* is *optimal* if:

$$v^{\pi^*}(s) \geq v^{\pi}(s) \quad \forall\ \pi \in \Pi \tag{9}$$

where Π denotes the set of admissible policies.

8.2.4 State Dynamics

The state dynamics of the system (sensor node) can be delineated by the state transition probabilities of the embedded Markov chain. We formulate our sensor node policy as a deterministic dynamic program (DDP) because the choice of an action determines the subsequent state with certainty. Our sensor node DDP policy formulation uses a *transfer function* to specify the next state. A transfer function defines a mapping $\tau_t(s, a)$ from $S \times A_s \to S$, which specifies the system state at time $t+1$ when the sensor node selects action $a \in A_s$ in state s at time t. To formulate our DDP as an MDP, we define the *transition probability function* as:

$$p_t(j|s,a) = \begin{cases} 1 & \text{if } \tau_t(s,a) = j \\ 0 & \text{if } \tau_t(s,a) \neq j \end{cases} \tag{10}$$

8.2.5 Reward Function

The reward function captures application metrics and sensor node characteristics. Our reward function characterization considers the power consumption (which affects the sensor node

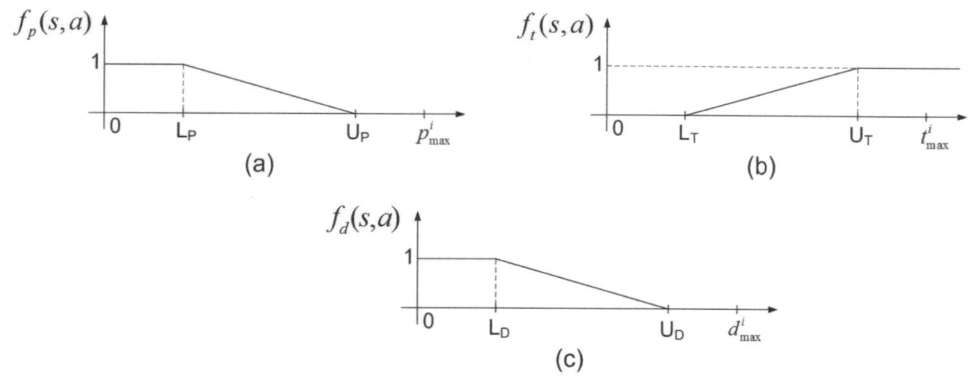

Fig. 6. Reward functions: (a) Power reward function $f_p(s,a)$; (b) Throughput reward function $f_t(s,a)$; (c) Delay reward function $f_d(s,a)$.

lifetime), throughput, and delay application metrics. We define the reward function $f(s,a)$ given the current sensor node state s and the sensor node's selected action a as:

$$\begin{aligned} f(s,a) &= \sum_{k=1}^{m} \omega_k f_k(s,a) \\ \text{s.t.} \quad & s \in S \\ & \omega_k \geq 0, \quad k = 1,2,\ldots,m \\ & \omega_k \leq 1, \quad k = 1,2,\ldots,m \\ & \sum_{k=1}^{m} \omega_k = 1, \end{aligned} \tag{11}$$

where $f_k(s,a)$ and ω_k denote the reward function and weight factor for the kth application metric, respectively, given that there are m application metrics. Our objective function characterization considers power, throughput, and delay (i.e., $m = 3$) (additional application metrics can be included) and is given as:

$$f(s,a) = \omega_p f_p(s,a) + \omega_t f_t(s,a) + \omega_d f_d(s,a) \tag{12}$$

where $f_p(s,a)$ denotes the power reward function, $f_t(s,a)$ denotes the throughput reward function, and $f_d(s,a)$ denotes the delay reward function (Fig. 6); ω_p, ω_t, and ω_d represent the *weight factors* for power, throughput, and delay, respectively.

We define linear reward functions for application metrics because an application metric reward (objective function) typically varies linearly, or piecewise linearly, between the minimum and maximum allowed values of the metric (Stevens-Navarro et al., 2008). However, a non-linear characterization of reward functions is also possible and depends upon the particular application. Our methodology works for any characterization of reward function. We define the power reward function (Fig. 6(a)) in (11) as:

$$f_p(s,a) = \begin{cases} 1, & 0 < p_a \leq L_P \\ (U_P - p_a)/(U_P - L_P), & L_P < p_a < U_P \\ 0, & p_a \geq U_P \end{cases} \tag{13}$$

where p_a denotes the power consumption of the current state given action a taken at time t and the constant parameters L_P and U_P denote the minimum and maximum allowed/tolerated sensor node power consumption, respectively. Similar equations can be written for $f_t(s,a)$ and $f_d(s,a)$.

State transitioning incurs a cost associated with switching parameter values from the current state to the next state (typically in the form of power and/or execution (time) overhead). We define the transition cost function $h(s,a)$ as:

$$h(s,a) = \begin{cases} H_{i,a} & \text{if } i \neq a \\ 0 & \text{if } i = a \end{cases} \tag{14}$$

where $H_{i,a}$ denotes the transition cost to switch from the current state i to the next state as determined by action a. Note that a sensor node incurs no transition cost if action a prescribes that the next state is the same as the current state.

Hence, the overall reward function $r(s,a)$ given state s and action a at time t is:

$$r(s,a) = f(s,a) - h(s,a) \tag{15}$$

which accounts for the power, throughput, and delay application metrics as well as state transition cost.

8.2.6 Optimality Equation

The optimality equation, also known as Bellman's equation, for expected total discounted reward criterion is given as (Puterman, 2005):

$$v(s) = \max_{a \in A_s} \left\{ r(s,a) + \sum_{j \in S} \lambda p(j|s,a) v(j) \right\} \tag{16}$$

where $v(s)$ denotes the maximum expected total discounted reward. The salient properties of the optimality equation are: the optimality equation has a unique solution; an optimal policy exists given conditions on states, actions, rewards, and transition probabilities; the value of the discounted MDP satisfies the optimality equation; and the optimality equation characterizes stationary policies.

The solution of (16) gives the maximum expected total discounted reward $v(s)$ and the MDP-based optimal policy π^* (or π^{MDP}), which gives the maximum $v(s)$. π^{MDP} prescribes the action a from action set A_s given the current state s for all $s \in S$. There are several methods to solve the optimality equation (16) such as value iteration, policy iteration, and linear programming, however in this work we use the policy iteration algorithm. The details of the policy iteration algorithm can be found in (Puterman, 2005).

8.3 Numerical Results

In this section, we compare the performance (based on expected total discounted reward criterion) of our proposed MDP-based DVFS2 optimal policy π^* (π^{MDP}) with several fixed heuristic policies using a representative WSN platform. We use the MATLAB MDP tool box (Chadès et al., 2005) implementation of the policy iteration algorithm (Puterman, 2005) to determine the MDP-based optimal policy. Given the reward function, sensor node state parameters, and transition probabilities, (8) gives the expected total discounted reward. Our reference WSN platform consists of eXtreme Scale Motes (XSM) sensor nodes (Dutta et al.,

Parameter	$i_1 = [2.7, 2, 2]$	$i_2 = [3, 4, 4]$	$i_3 = [4, 6, 6]$	$i_4 = [5.5, 8, 8]$
p_i	10 units	15 units	30 units	55 units
t_i	4 units	8 units	12 units	16 units
d_i	26 units	14 units	8 units	6 units

Table 2. Power consumption p_i, throughput t_i, and delay d_i parameters for wireless sensor node state $i = [V_p, F_p, F_s]$ (V_p is specified in volts, F_p in MHz, and F_s in KHz). Parameters are specified as a multiple of a base unit where one power unit is equal to 1 mW, one throughput unit is equal to 0.5 MIPS, and one delay unit is equal to 50 ms. Parameter values are based on the XSM mote.

2005); (Dutta & Culler, 2005). The XSM motes have an average lifetime of 1,000 hours of continuous operation with two AA alkaline batteries, which can deliver 6 Whr or an average of 6 mW (Dutta et al., 2005). The XSM platform integrates an Atmel ATmega128L microcontroller (ATMEL, 2009), a Chipcon CC1000 radio operating at 433 MHz, and a 4 Mbit serial flash memory. The XSM motes contain infra red, magnetic, acoustic, photo, and temperature sensors. To represent sensor node operation, we analyze a sample application domain that represents a typical security system or defense application (henceforth referred to as a *security/defense system*).

8.3.1 Fixed Heuristic Policies for Performance Comparisons

We consider the following four fixed heuristic policies for comparison with our MDP policy:

- A fixed heuristic policy π^{POW} that always selects the state with the lowest power consumption.
- A fixed heuristic policy π^{THP} that always selects the state with the highest throughput.
- A fixed heuristic policy π^{EQU} that spends an equal amount of time in each of the available states.
- A fixed heuristic policy π^{PRF} that spends an unequal amount of time in each of the available states based on a specified preference for each state. For example, given a system with four possible states, the π^{PRF} policy may spend 40% of the time in the first state, 20% of the time in the second state, 10% of the time in the third state, and 30% of the time in the fourth state.

8.3.2 MDP Specifications

We compare different policies using the *expected total discounted reward* performance criterion. The state transition probability for each sensor node state is given by (10). The sensor node's lifetime and the time between decision epochs are subjective and may be assigned by an application manager according to application requirements. A sensor node's *mean lifetime* is given by $1/(1 - \lambda)$ *time units*, which is the time between successive decision epochs (which we assume to be 1 hour). For instance for $\lambda = 0.999$, the sensor node's mean lifetime is $1/(1 - 0.999) = 1000$ hours ≈ 42 days.

For our numerical results, we consider a sensor node capable of operating in four different states (i.e., $I = 4$ in (1)). Each state has a set of allowed actions prescribing transitions to available states. For each allowed action a in a state, there is a $\{r_a, p_a\}$ pair where r_a specifies the immediate reward obtained by taking action a and p_a denotes the probability of taking action a.

λ	**Sensor Lifetime**	π^{MDP}	π^{POW}	π^{THP}	π^{EQU}	π^{PRF}
0.94	16.67 hours	10.0006	7.5111	9.0778	7.2692	7.5586
0.95	20 hours	12.0302	9.0111	10.9111	8.723	9.0687
0.96	25 hours	15.0747	11.2611	13.6611	10.9038	11.3339
0.97	33.33 hours	20.1489	15.0111	18.2445	14.5383	15.1091
0.98	50 hours	30.2972	22.5111	27.4111	21.8075	22.6596
0.99	100 hours	60.7422	45.0111	54.9111	43.6150	45.3111
0.999	1000 hours	608.7522	450.0111	549.9111	436.15	453.0381
0.9999	10,000 hours	6088.9	4500	5499.9	4361.5	4530.3
0.99999	100,000 hours	60890	45000	55000	43615	45303

Table 4. The effects of different discount factors λ for a security/defense system. $H_{i,j} = 0.1$ if $i \neq j, \omega_p = 0.45, \omega_t = 0.2, \omega_d = 0.35$.

Table 2 summarizes state parameter values for each of the four states i_1, i_2, i_3, and i_4. We define each state using a $[V_p, F_p, F_s]$ tuple where V_p is specified in volts, F_p in MHz, and F_s in KHz. For instance, state one i_1 is defined as $[2.7, 2, 2]$, which corresponds to a processor voltage of 2.7 volts, a processor frequency of 2 MHz, and a sensing frequency of 2 KHz (2000 samples per second). We assume, without loss of generality, that the transition cost for switching from one state to another is $H_{i,a} = 0.1$ if $i \neq a$.

Our selection of the state parameter values in Table 2 corresponds to XSM mote specifications. The XSM mote's Atmel ATmega128L microprocessor has an operating voltage range of 2.7 to 5.5 V and a processor frequency range of 0 to 8 MHz. The ATmega128L throughput varies with processor frequency at 1 MIPS per MHz, thus allowing a WSN designer to optimize power consumption versus processing speed (ATMEL, 2009). Our chosen sensing frequency also corresponds with standard sensor node specifications. The Honeywell HMC1002 magnetometer sensor (Honeywell, 2009) consumes on average 15 mW of power and can be sampled in 0.1 ms on the Atmel ATmega128L microprocessor, which results in a maximum sampling frequency of approximately 10 KHz (10,000 samples per second). The acoustic sensor embedded in the XSM mote has a maximum sensing frequency of approximately 8.192 KHz (Dutta et al., 2005).

Table **??** summarizes the minimum L and maximum U reward function parameter values for application metrics (power, throughput, and delay) and associated weight factors for a security/defense system. We selected reward function parameter values according to typical application requirements for a security/defense system (Akyildiz et al., 2002). For instance, a data sensitive and time critical security/defense system with stringent minimum and maximum tolerable delay might require a comparatively large minimum throughput in order to obtain a sufficient number of sensed data samples for meaningful analysis.

For brevity, we select a single sample WSN platform configuration and application, but we point out that our proposed MDP model and methodology works equally well for any other WSN platform and application.

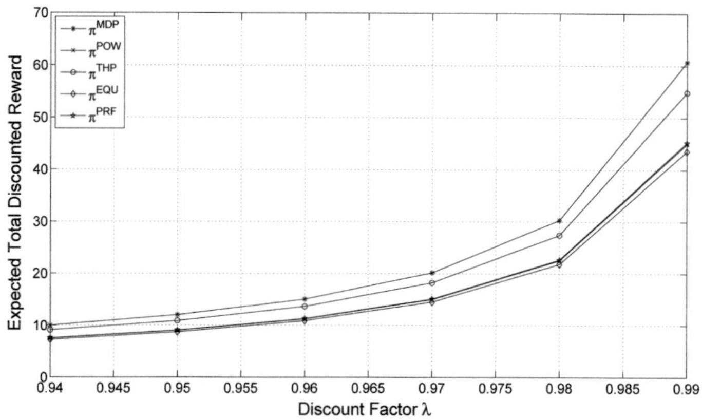

Fig. 7. The effects of different discount factors on the expected total discounted reward for a security/defense system. $H_{i,j} = 0.1$ if $i \neq j$, $\omega_p = 0.45, \omega_t = 0.2, \omega_d = 0.35$.

8.3.3 Results

In this subsection, we present the results for a security/defense system for our MDP-based optimal policy and the fixed heuristic policies (Section 8.3.1). We evaluate the effects of different discount factors, different state transition costs, and different application metric weight factors on the expected total discounted reward. The magnitude of difference in the total expected discounted reward for different policies is important as it provides relative comparisons between the different policies.

Table 4 and Figure 7 depict the effects of different discount factors λ on the heuristic policies and π^{MDP} for a security/defense system when the state transition cost $H_{i,j}$ is held constant at 0.1 for $i \neq j$. Since we assume the time between successive decision epochs to be 1 hour, the range of λ from 0.94 to 0.99999 corresponds to a range of average sensor node lifetime from 16.67 to 100,000 hours $\approx$ 4167 days $\approx$ 11.4 years. Table 4 and Figure 7 show that π^{MDP} results in the highest expected total discounted reward for all values of λ and corresponding average sensor node lifetimes. For instance, when the average sensor node lifetime is 1,000 hours ($\lambda = 0.999$), π^{MDP} results in a 26.08%, 9.67%, 28.35%, and 25.58% increase in expected total discounted reward as compared to π^{POW}, π^{THP}, π^{EQU}, and π^{PRF}, respectively. On average over all discount factors λ, π^{MDP} results in a 25.57%, 9.48%, 27.91%, and 25.1% increase in expected total discounted reward as compared to π^{POW}, π^{THP}, π^{EQU}, and π^{PRF}, respectively.

Figure 8 depicts the effects of different state transition costs on the expected total discounted reward for a security/defense system with a fixed average sensor node lifetime of 1000 hours ($\lambda = 0.999$). Figure 8 shows that π^{MDP} results in the highest expected total discounted reward for all transition cost values. Figure 8 also shows that the expected total discounted reward for π^{MDP} is relatively unaffected by state transition cost. This relatively constant behavior can be explained by the fact that our MDP optimal policy does not perform many state transitions. π^{MDP} performs state transitions primarily at sensor node deployment or whenever a new MDP-based optimal policy is determined as the result of changes in application requirements.

Figure 9 shows the effects of different reward function weight factors on the expected total discounted reward for a security/defense system when the average sensor node lifetime is 1,000 hours ($\lambda = 0.999$) and the state transition cost $H_{i,j}$ is held constant at 0.1 for $i \neq j$. We explore various weight factors that are appropriate for different security/defense system

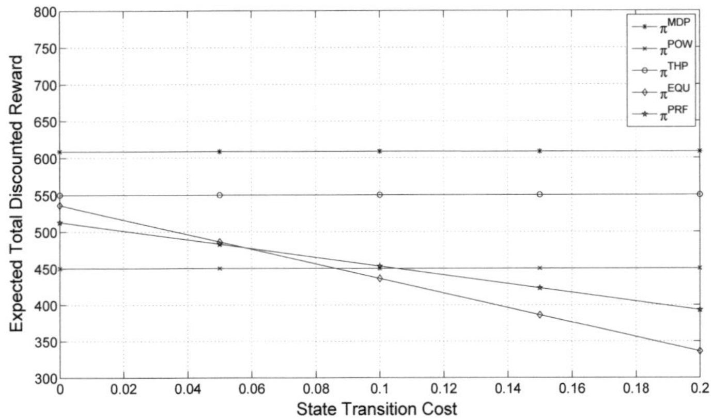

Fig. 8. The effects of different state transition costs on the expected total discounted reward for a security/defense system. $\lambda = 0.999, \omega_p = 0.45, \omega_t = 0.2, \omega_d = 0.35$.

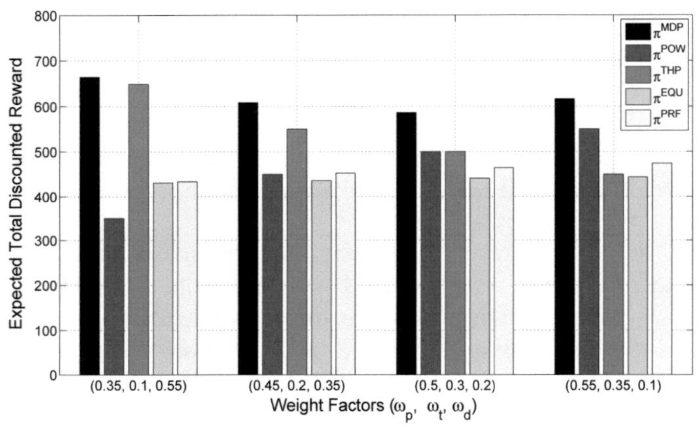

Fig. 9. The effects of different reward function weight factors on the expected total discounted reward for a security/defense system. $\lambda = 0.999$, $H_{i,j} = 0.1$ if $i \neq j$

specifics (i.e., $(\omega_p, \omega_t, \omega_d)$ = {(0.35, 0.1, 0.55), (0.45, 0.2, 0.35), (0.5, 0.3, 0.2), (0.55, 0.35, 0.1)}). Figure 9 reveals that π^{MDP} results in the highest expected total discounted reward for all weight factor variations.

9. Conclusions

WSNs have been employed in diverse application domains each with different and competing application requirements. Given this diversity, meeting application requirements is a challenging design task. Optimization techniques at different design levels help meet these application requirements. In this chapter, we discussed WSNs from an optimization perspective. We presented a typical WSN architecture along with several possible integration scenarios with external IP networks for ubiquitous availability of WSN offered services (e.g., sensed temperature and humidity statistics). We discussed COTS sensor node components

and associated tunable parameters that can be specialized to provide component-level optimizations. We presented data link-level and network-level optimization strategies focusing on MAC and routing protocols, respectively. Our presented MAC protocols targeted load balancing, throughput, and energy optimizations and routing protocols addressed query dissemination, real-time data delivery, and network topology. Different OS optimizations include event-driven execution, dynamic power management, and fault-tolerance.
Even though many of the optimizations offered by MAC, routing, and the OS are dynamic in nature, we focused on dynamic optimizations separately due to their increasing research significance. Traditional DVFS-based optimizations only tune processor voltage and frequency, however, sensor nodes possess other tunable parameters (e.g., sensing frequency, transmission power) whose tuning can increase the potential for meeting application requirements. In this chapter, we proposed an MDP-based dynamic optimization methodology to optimally tune sensor node parameters. Our proposed methodology is adaptive and dynamically determines the new MDP-based optimal policy (sensor node operating state) whenever application requirements change (which may be in accordance with changing environmental stimuli). We compared our MDP-based methodology with four fixed heuristic policies. Numerical results revealed that our MDP-based policy outperformed other heuristic policies for all sensor node lifetimes, state transition costs, and application metric weight factors. Future research trends in WSN dynamic optimizations include the investigation of lightweight online algorithms suitable for sensor nodes with constrained resources and incorporation of profiling statistics to provide feedback to the optimization algorithms.

Acknowledgments

This work was supported by the National Science Foundation (CNS-0834080) and the Natural Sciences and Engineering Research Council of Canada (NSERC). Any opinions, findings, and conclusions or recommendations expressed in this material are those of the author(s) and do not necessarily reflect the views of the NSF or NSERC.

10. References

Abrach, H. & et al. (2003). MANTIS: System Support for Multimodal Networks of In-Situ Sensors, *Proc. of Workshop on Wireless Sensor Networks and Applications (WSNA)'03*, San Diego, California, pp. 50–59.

Abramson, N. (1985). Development of the ALOHANET, *IEEE Trans. on Information Theory* **31**(2): 119–123.

Akhmetshina, E., Gburzynski, P. & Vizeacoumar, F. (2002). PicOS: A Tiny Operating System for Extremely Small Embedded Platforms, *Proc. of Conference on Embedded Systems and Applications (ESA)'02*, Las Vegas, Nevada, pp. 116–122.

Akyildiz, I., Su, W., Sankarasubramaniam, Y. & Cayirci, E. (2002). Wireless Sensor Networks: A Survey, *Elsevier Computer Networks* **38**(4): 393–422.

ATMEL (2009). ATMEL ATmega128L 8-bit Microcontroller Datasheet, *ATMEL Corporation*, San Jose, California.
URL: *http://www.atmel.com/dyn/resources/prod_documents/doc2467.pdf*

Bao, L. & Garcia-Luna-Aceves, J. (2001). A New Approach to Channel Access Scheduling for Ad Hoc Networks, *Proc. of MobiCom'01*, ACM, Rome, Italy.

Barr, R. & et al. (2002). On the Need for System-Level Support for Ad Hoc and Sensor Networks, *ACM SIGOPS Operating Systems Review* **36**(2): 1–5.

Brooks, D. & Martonosi, M. (2000). Value-based Clock Gating and Operation Packing: Dynamic Strategies for Improving Processor Power and Performance, *ACM Trans. on Computer Systems* **18**(2): 89–126.

Chadès, I., Cros, M., Garcia, F. & Sabbadin, R. (2005). Markov Decision Process (MDP) Toolbox v2.0 for MATLAB, *INRA Toulouse*, INRA, France.
URL: *http://www.inra.fr/internet/Departements/MIA/T/MDPtoolbox/*

Chandrakasan, A., Amirtharajah, R., Cho, S., Konduri, J., Kulik, J., Rabiner, W. & Wang, A. (1999). Design Considerations for Distributed Microsensor Systems, *Proc. of IEEE Custom Integrated Circuits Conference (CICC)*, San Diego, California.

Coffin, D., Hook, D., McGarry, S. & Kolek, S. (2000). Declarative Ad-hoc Sensor Networking, *In SPIE Integrated Command Environments.*

Culler, D., Hill, J., Horton, M., Pister, K., Szewczyk, R. & Woo, A. (2002). MICA: The Commercialization of Microsensor Motes, *Sensor Magazine.*
URL: *http://www.sensorsmag.com/articles/0402/40/*

Dutta, P. & Culler, D. (2005). System Software Techniques for Low-Power Operation in Wireless Sensor Networks, *Proc. of IEEE/ACM ICCAD*, San Jose, California.

Dutta, P., Grimmer, M., Arora, A., Bibyk, S. & Culler, D. (2005). Design of a Wireless Sensor Network Platform for Detecting Rare, Random, and Ephemeral Events, *Proc. of ACM IPSN*, Los Angeles, California.

EYES (2010). Energy Efficient Sensor Networks.
URL: *http://www.eyes.eu.org/sensnet.htm*

Hamed, H., El-Atawy, A. & Ehab, A.-S. (2006). On Dynamic Optimization of Packet Matching in High-Speed Firewalls, *IEEE Journal on Selected Areas in Communications* **24**(10): 1817–1830.

Hazelwood, K. & Smith, M. D. (2006). Managing Bounded Code Caches in Dynamic Binary Optimization Systems, *ACM Trans. on Architecture and Code Optimization* **3**(3): 263–294.

He, T., Stankovic, J., Lu, C. & Abdelzaher, T. (2003). SPEED: A Stateless Protocol for Real-time Communication in Sensor Networks, *Proc. of International Conference on Distributed Computing Systems (ICDCS)'03*, IEEE, Providence, Rhode Island.

Heinzelman, W., Chandrakasan, A. & Balakrishnan, H. (2000). Energy-Efficient Communication Protocols for Wireless Microsensor Networks, *Hawaiian International Conference on System Sciences.*

Honeywell (2009). Honeywell 1- and 2- Axis Magenetic Sensors HMC1001/1002, and HMC1021/1022 Datasheet, *Honeywell International Inc.*, Morristown, New Jersey.
URL: *http://www.ssec.honeywell.com/magnetic/datasheets/hmc1001-2_1021-2.pdf*

Hu, S., Valluri, M. & John, L. K. (2006). Effective Management of Multiple Configurable Units using Dynamic Optimization, *ACM Trans. on Architecture and Code Optimization* **3**(4): 477–501.

IEEE Standards (1999). Wireless LAN Medium Access Control (MAC) and Physical Layer (PHY) Specification, IEEE Std 802.11-1999 edition: LAN MAN Standards Committee of the IEEE Computer Society.

Intanagonwiwat, C., Govindan, R., Estrin, D., Heidemann, J. & Silva, F. (2003). Directed Diffusion for Wireless Sensor Networking, *IEEE/ACM Trans. on Networking* **11**(1): 2–16.

Karl, H. & Willig, A. (2005). *Protocols and Architectures for Wireless Sensor Networks*, John Wiley and Sons, Inc.

Kogekar, S., Neema, S., Eames, B., Koutsoukos, X., Ledeczi, A. & Maroti, M. (2004). Constraint-Guided Dynamic Reconfiguration in Sensor Networks, *Proc. of ACM IPSN*, Berkeley, California.

Kulik, J., Heinzelman, W. & Balakrishnan, H. (2002). Negotiation-Based Protocols for Disseminating Information in Wireless Sensor Networks, *ACM Wireless Networks (WINET)* **8**(2/3): 169–185.

Lu, C., Blum, B., Abdelzaher, T., Stankovic, J. & He, T. (2002). RAP: A Real-Time Communication Architecture for Large-Scale Wireless Sensor Networks, *Real-Time and Embedded Technology and Applications Symposium (RTAS)'02*, San Jose, California.

Mahalik, N. (2007). *Sensor Networks and Configuration: Fundamentals, Standards, Platforms, and Applications*, Springer.

Min, R., Furrer, T. & Chandrakasan, A. (2000). Dynamic Voltage Scaling Techniques for Distributed Microsensor Networks, *Proc. of IEEE WVLSI*, Orlando, Florida.

Munir, A. & Gordon-Ross, A. (2009). An MDP-based Application Oriented Optimal Policy for Wireless Sensor Networks, *Proc. of International Conference on Hardware/Software Codesign and System Synthesis (CODES+ISSS'09)*, ACM, Grenoble, France, pp. 183–192.

Polastre, J., Hill, J. & Culler, D. (2004). Versatile Low Power Media Access for Wireless Sensor Networks, *Proc. of International Conference on Embedded Networked Sensor Systems (SenSys)'04*, ACM, Baltimore, Maryland.

Poor, R. (2010). Gradient Routing in Ad Hoc Networks.
URL: *http://www.media.mit.edu/pia/Research/ESP/texts/poorieeepaper.pdf*

Puterman, M. (2005). *Markov Decision Processes: Discrete Stochastic Dynamic Programming*, John Wiley and Sons, Inc.

Raghavendra, C., Sivalingam, K. & Znati, T. (2004). *Wireless Sensor Networks*, Kluwer Academic Publishers.

Rajendran, V., Obraczka, K. & Garcia-Luna-Aceves, J. (2003). Energy-efficient Collision-free Medium Access Control for Wireless Sensor Networks, *Proc. of International Conference on Embedded Networked Sensor Systems (SenSys)'03*, ACM, Los Angeles, California.

Rappaport, T. S. (1996). *Wireless Communications, Principles and Practice*, Prentice Hall.

Rhee, I., Warrier, A., Aia, M. & Min, J. (2005). Z-MAC: A Hybrid MAC for Wireless Sensor Networks, *Proc. of International Conference on Embedded Networked Sensor Systems (SenSys)'05*, ACM, San Diego, California.

Shah, R. & Rabaey, J. (2002). Energy Aware Routing for Low Energy Ad Hoc Sensor Networks, *Proc. of Wireless Communications and Networking Conference (WCNC)*, IEEE, Orlando, Florida.

Singh, S. & Raghavendra, C. S. (1998). PAMAS - Power Aware Multi-Access protocol with Signaling for ad hoc networks, *ACM Sigcomm Computer Communication Review* **28**(3): 5–26.

Sinha, A. & Chandrakasan, A. (2001). Operating System and Algorithmic Techniques for Energy Scalable Wireless Sensor Networks, *Proc. of International Conference on Mobile Data Management*, Hong Kong, pp. 199–209.

Sohraby, K., Minoli, D. & Znati, T. (2007). *Wireless Sensor Networks: Technology, Protocols, and Applications*, John Wiley and Sons, Inc.

Stevens-Navarro, E., Lin, Y. & Wong, V. (2008). An MDP-based Vertical Handoff Decision Algorithm for Heterogeneous Wireless Networks, *IEEE Trans. on Vehicular Technology* **57**(2): 1243–1254.

Stojmenović, I. (2005). *Handbook of Sensor Networks: Algorithms and Architectures*, John Wiley and Sons, Inc.

StrongARM (2010). Intel StrongARM SA-1110 Microprocessor.
URL: *http://bwrc.eecs.berkeley.edu/research/pico_radio/test_bed/hardware/documentation/arm/sa1110briefdatasheet.pdf*

TinyOS (2010). .
URL: *http://www.tinyos.net/*

Van Dam, T. & Langendoen, K. (2003). An Adaptive Energy-Efficient MAC Protocol for Wireless Sensor Networks, *Proc. of International Conference on Embedded Networked Sensor Systems (SenSys)'03*, ACM, Los Angeles, California.

Varga, A. (2001). The OMNeT++ discrete event simulation system, *Proc. of European Simulation Multiconference (ESM)'01*, Prague, Czech Republic.

Ye, F., Zhong, G., Lu, S. & Zhang, L. (2005). GRAdient Broadcast: A Robust Data Delivery Protocol for Large Scale Sensor Networks, *ACM Wireless Networks (WINET)* **11**(2).

Ye, W., Heidemann, J. & Estrin, D. (2002). An Energy-Efficient MAC protocol for Wireless Sensor Networks, *Proc. of INFOCOM'02*, IEEE, New York, New York.

Ye, W., Heidemann, J. & Estrin, D. (2004). Medium Access Control with Coordinated Adaptive Sleeping for Wireless Sensor Networks, *IEEE/ACM Trans. on Networking* **12**(3): 493–506.

Yuan, L. & Qu, G. (2002). Design Space Exploration for Energy-Efficient Secure Sensor Network, *Proc. of IEEE ASAP*, San Jose, California.

14

A k-covered Mobile Target Tracking in Voronoi-based Wireless Sensor Networks

Jiehui Chen[1,2], Mariam B.Salim[2] and Mitsuji Matsumoto[2]
[1]*Global COE Program International Research and Education Center for Ambient SoC sponsored by MEXT,Japan*
[2]*Graduate School of Global Information and Telecommunication Studies Waseda University, Tokyo, Japan*

1. Introduction

Recent advances in micro electro mechanical systems (MEMS) and wireless communication technologies are responsible for the emergence of Wireless Sensor Network (WSN) that deploys thousands of low-cost sensors integrating sensing, processing and communication capabilities. It motivated the use of mobile sensor node in WSNs for many surveillance applications including mission-critical target tracking. One of the most attractive areas is exploited to be the mobile target tracking. Typical examples include establishing survivable military surveillance systems, environmental and industrial monitoring, personnel and wildlife monitoring systems requiring tracking schemes, capable of deducing kinematic characteristics such as position, velocity, and acceleration of single or multiple targets(J. Janssen, et al, 2008) of interest (T.He, et al, 2006). For the above observations, the possible existence of targets can be inductively described as the Fig.1 shows for simplicity.

Despite the fact that sensor deployment sensitive target tracking could ~~both~~ be managed by taking full advantage of Voronoi diagram, less efforts were made so far. Generally, different sensor applications may pose different requirements for how good a network's coverage should be. Previous research (M.Cardei, et al, 2004) has studied sensor coverage problems and categorized them into three types: *area coverage, point coverage,* and *barrier coverage.* The objective of the first, area coverage is to maximize the coverage for a region of interest. The objective of point coverage is similar, but it is to cover a set of points. The latter, barrier coverage, aims to minimize the probability of undetected penetration through a sensor network. The choice of using a particular coverage measurement depends on the purpose of a sensor network. For instance, if the purpose is to monitor moving objects in a field, barrier coverage is more suitable. To measure barrier coverage, we consulted the work (S.Meguerdichian, et al, 2005) in which the worst- and best-case coverages are defined. The detailed design will be given in the next section.

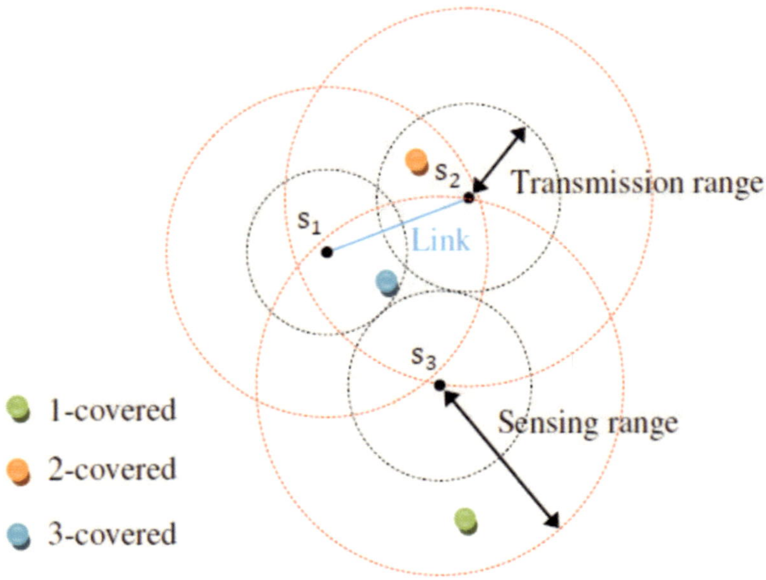

Fig. 1. Higher-order coverage

Sensor communication usually requires the data to be aggregated before being transmitted, which motivates the network to have an efficient clustering in a priority. In literature, Linked Cluster Algorithm (LCA)(D.J.Baker, et al, 1981), a sensor becomes a CH if it has the highest identity among all the one-hop sensors or one-hop sensors of its one-hop neighbors. The Max-Min d-Cluster Algorithm (A.D.Amis, et al, 2000) generates d-hop clusters with a run-time of O(d) round, and achieves better load balancing among the CHs, generates fewer clusters than (A.Ephremides, et al, 1987). (W.R.Heinzelman, et al, 2000) proposed a distributed algorithm for micro-WSNs where sensors elected themselves CHs with some probabilities and broadcast their decisions. However, this algorithm only allows one-hop clusters to be formed, which might lead to a large number of clusters.

In this paper, we proposed a novel clustering algorithm to generate a multi-hop Voronoi diagram-based WSNs, one of the most attractive areas of sensor network called mobile target tracking is exploited to be performed on that base. Obviously, in Figure 1, the situation is getting more and more complicated as the density of network increases. Our motivation is to efficiently monitor the moving multi-covered mobile target in Voronoi-based sensor networks by measuring the moved hop distance before being detected. ~~We~~ By taking full advantage of Voronoi diagram structure, we tactfully utilized trajectory estimation technologies to predict the potential trajectory of the moving target.

Moreover, we designed a optimized barrier coverage and an energy-efficient clustering algorithm for clearing the Vonoroi architecture and better energy conservation. The proposed mobile target tracking scheme (CTT&MAV) was designed to take full advantage of Voronoi-diagram boundary to improve the detectability. we enhanced PRAM algorithm (H. Meyerhenke, et al, 2005) and Final simulation results verified that our proposal outperformances random walk(T.Camp, et al, 2002), random waypoint(B.Liang, et al, 1999), random direction(L.Lima, et al, 2007)and Gauss-Markov(C.Bettstetter, et al, 2003) in terms of reducing average hop distance that the mobile target moved before being detected and lower sensor death rate as well. Finally, we demonstrated that our results are robust to realistic sensing models and also validate the correctness through extensive simulations.

The remainder of this paper is organized as follows: the next section presents the optimized barrier coverage design; Section 3 shows Mathematical modeling of Voronoi-based WSN based on energy consumption in detail; Section 4 illustrates the proposed intelligent mobile target tracking scheme called CTT&MAV; Section 5 conducts experiments in Matlab simulator under multi-covered Voronoi-based clustered sensor network. Finally, section 6 concludes the paper with future perspective.

2. Optimized barrier coverage design

Although maintaining full sensing coverage guarantees immediate response to intrude targets, sometimes it is not favorable due to its high energy consumption. We investigate a new and more efficient approach for deploying sensors in a large scale two dimensional monitoring area.

2.1 New approach for sensor deployment

To monitor an area, WSN should achieve a certain level of detection performance. Due to the highly considerable cost in a given monitoring area, better *detection capacity* and *communication coverage* is critical to sequential deployment of sensors. In this paper, we explored a new approach for sensor deployment (see Figure 2) to improve barrier coverage.

Theorem 1. Let A denotes the area and *f*(A) denotes barrier coverage, namely the fraction of the area that is in the sensing area of one or more sensors where sensors can provide a valid sensing measurement and Γ is the cartographic representation of area .Then,

$$\Gamma_{f(\beta)} \gg \Gamma_{f(\alpha)} \text{ in } G = (V, E) \text{ where } E \neq \emptyset \quad (1)$$

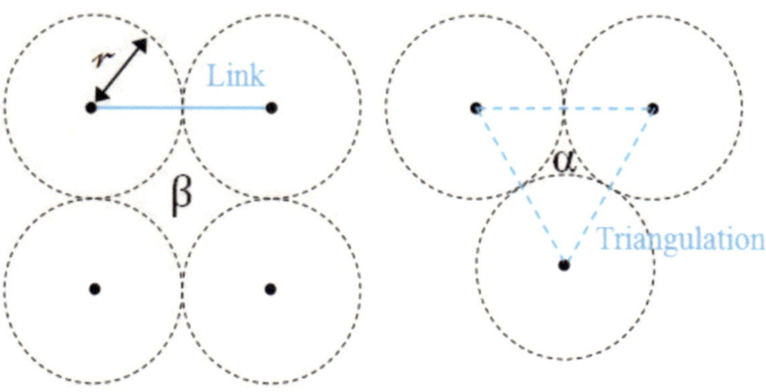

Fig. 2. Detection capacity-based sensor deployment

Proof: In literature, the majority of researches prefer grid-based (see Figure 2(a)) sequential sensor deployment. Instinctively, we get $\Gamma_{f(\beta)}$ is more efficient than $\Gamma_{f(\alpha)}$.The computational evidences are as follows:

$$\Gamma_{f(\beta)}=(2r)^2-4(\frac{\pi r^2}{4}) = (4-\pi)r^2 \approx 0.86r^2 \quad (2)$$

$$\Gamma_{f(\alpha)}= (\sqrt{3}-\frac{\pi}{2})\, r^2 \approx 0.1512r^2 \quad (3)$$

We skipped the considerably simple computation procedure and directly transformed to the result. The unit difference is obviously given by approximately $0.71\mathbf{r}^2$. Although the difference is indistinctive when the value of r is small enough, for monitoring applications, accuracy is vital consideration. The smaller the value of $\mathbf{\Gamma_f}$ is, the higher possibility that a moving object will not be detected, therefore Figure 2 (b) has better detection capacity than Figure 2(a).

Theorem 2. let H_v be a hop distance and p_v^{up}, $p_v^{same}$and p_v^{lower} denotes the possible existence of CHs at the upper, same and lower layer respectively. The Triangle-based is more suitable for our monitoring network in term of higher communication coverage.

Proof: Figure 3 clearly shows that Triangle-based has more relay one hop neighbors (€(v)) to relay than Grid-based at a rate of 6:4. For multi-hops transmission, when receiving a message, a sensor (N_v) should relay it to another sensor at a price of energy consumption. The sensor to relay should be one at the higher layer compared to N_v.

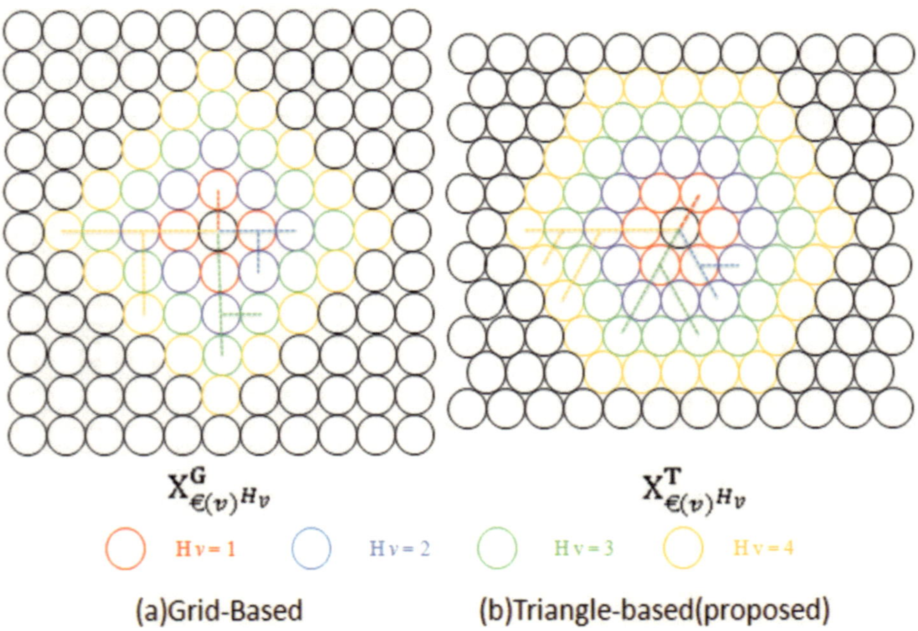

Fig. 3. Communication coverage-based sensor deployment.

Denote H_v^{up}, H_v^{same} and H_v^{lower} represent the number of hops on the shortest routing path from N_v to a sensor at the upper, same and lower layer respectively. On the other hand, within a certain hop distance, the higher possibility of existing sensors to relay, the better. Therefore, the focus is to find out which one has more $€(v)^{H_v}$ between Figure 3 (a) and Figure 3 (b), where $€(v)^{H_v}$: a set of H_v hop distance neighborhood sensors.

Let $X^T_{€(v)^{H_v}}$ and $X^G_{€(v)^{H_v}}$ denote the total number of detectable $€(v)^{H_v}$ of N_v for Triangle-based and Grid-based respectively. According to Fig. 3, we easily get:

$$X^T_{€(v)^{H_v}} = 3(1 + H_v)H_v \quad (4)$$

$$X^G_{€(v)^{H_v}} = 2(1 + H_v)H_v \quad (5)$$

Where $H_v \geq 1$ and get $X^T_{€(v)^{H_v}} \gg X^G_{€(v)^{H_v}}$ that prove Triangle-based is more suitable for G = (V, E) where $E \neq \emptyset$, in terms of higher *communication coverage*.

The above observations show evidences for proving the efficiency of the proposed optimized barrier coverage design.

3. Mathematical modeling of Voronoi-based WSN based on energy consumption

In this section, we present the mathematical modeling of Voronoi diagram for sensor node distribution. The proposed approaches are developed with the following assumptions:

- Static Sensor Nodes are of the same capacity and functionalities. The communication is contention and error free.
- Mobile Sensor Nodes are equipped with binary sensors characterized by a sensing radius R_{s_i} for a sensor node s_i. ($i \leq n$)
- The corresponding sensing range of s_i is a perfect disc denoted by $\Gamma(s_i, R_{s_i})$, and the mobile targets will be detected by s_i if they are in its sensing range(see Figure 1).

A multi-hop WSN was modeled by an undirected graph G = (V, E) where V, |V|=n, is the set of wireless sensor nodes and there exists an edge $\{s_u, s_v\} \in E$, if and only if s_u and s_v can mutually receive each other's transmission. Namely, two sensor nodes are considered neighbors if the Euclidean distance is smaller or equal to the transmission rang r. The set of *k*-hop neighbors of s_v is denoted by $€(s_v)^k$.

Let $\mathcal{M}$ be a metric space, and $\mathcal{M} \times \mathcal{M} \rightarrow \zeta$ denoting the Euclidean distance on $\mathcal{M}$. A set of sensor nodes having their coordinates in $\mathcal{M}$ is denoted by $\chi = \{\phi_i, 1 \leq i \leq n\} \subseteq \mathcal{M}$. The Voronoi diagram associated to χ is the unique subset called Voronoi diagram related to $\{\phi_i, 1 \leq i \leq n\}$. In literature, many algorithms have been proposed to determine the Voronoi diagram in a 2D space. In this section, we define the *k*-Voronoi diagram construction model based on the cooperation of χ elements. Our strategy is based on the PRAM algorithm (H. Meyerhenke, et al, 2005). The major merit is that the algorithm is performed in a recursive manner where the (*k-1*)-Voronoi diagram is used to collaboratively compute the *k*-Voronoi diagram. Let's define the subsets of χ includes the nearest elements to be ψ_i^k which can help finding the elements closer to the most distant neighbouring.

---

1. **Input**: a set of χ of sensor nodes and Voronoi of order (*k*-1)
2. Divide each region by $\psi_i^k \subset \chi$ into subregions
3. Merge equivalent new sub-regions who are tightly relevant to the neighboring ψ_i^{k-1}
4. Update the current edges and vertices.
5. **Output**: *k*-Voronoi diagram

To generate a single level energy-efficient clustering algorithm, suppose that a single event is densely happened in a square area. The number of sensors is a Poisson random variable with E[n] = λA. Since the probability of becoming a CH is p, the CHs and non-CHs are distributed as per independent homogeneous spatial Poisson processes with intensity $\lambda_1 = p\lambda$ and $\lambda_0 = (1-p)\lambda$.. To generate stochastic geometry for the proposed clustering algorithm and minimize energy cost in the network without loss of generality, we present the mathematical model of Voronoi diagram for sensor distribution.

n	The No. of sensors
n_c	The No. of sensors in a single cluster
D_{all}	The total length of segments, all sensors →the sink
$D_{c\to s}$	The total length of segments, all CHs→the sink
$\delta_{c\to s}$	The total energy cost, all CHs→the sink
δ	Total energy cost of data communication between sensors and the sink through a network hierarchy

Table 1. Simulation parameters Setup

Suppose a sensor located at (x_i, y_i),*i*=1,2,…,n. Then get

$$E[D_{all}|N = n] = 12\sum_{i=1}^{R} i^2 = 2R(R+1)(2R+1) \qquad (6)$$

Where, R is the radius of the network area.

Since there are on an average npCHs with their locations independent, therefore, $D_{c\to s} = pD_{all} = 2R(R+1)(2R+1)p$. By arguments similar to (S.G.Foss, et al, 1996), if N_v is a random variable denoting the number of PP0 process points in each Voronoi diagram (e.g. Figure 4) and L_v is the total length of segments that connect the PP0 process points to the nucleus in a Voronoi diagram.

$$E\,[N_v|N=n] \approx E[N_v] = \frac{\lambda_0}{\lambda_1} \qquad (7)$$

$$E\,[L_v|N=n] \approx E[L_v] = \frac{\lambda_0}{2\lambda_1^{3/2}} \qquad (8)$$

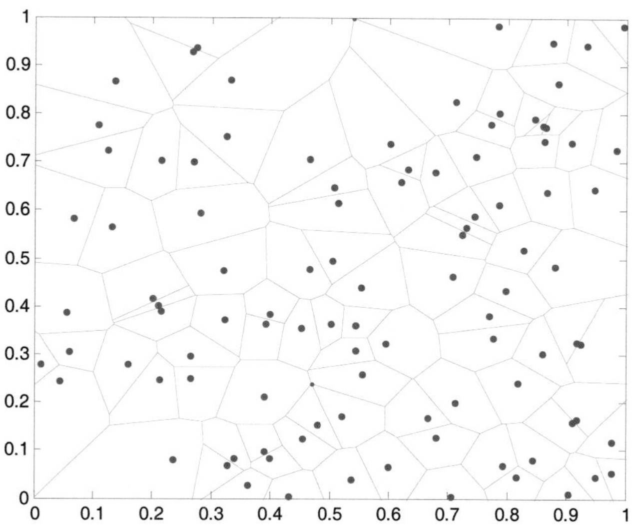

Fig. 4. Voronoi diagram based WSN

Define δ_1 to be the total energy spent by all the sensors communicating 1 unit of data to their CHs, since there are on average $(2R)^2$ CHs, namely, $p(2R)^2$ Voronoi diagrams. Let assume that there exists very small amount of isolated sensors so that ignore them without any bad influence to the accuracy of the algorithm. Therefore, the expected value of δ_1 conditioning on N, is given by

$$E[\boldsymbol{\delta}_1|N=n]=np\frac{E[L_v|N=\alpha^2]}{r}=\frac{2(1-p)R^2}{r\sqrt{\lambda p}} \tag{9}$$

Conditioning on N, total energy spent by all the CHs communicating 1 unit of data to the sink is given by

$$E[\boldsymbol{\delta}_{c\rightarrow s}|N=n]=\frac{E[D_{c\rightarrow s}|N=\alpha^2]}{r}=\frac{2pR(R+1)(2R+1)}{r} \tag{10}$$

Then

$$E[\boldsymbol{\delta}|N=n]=E[\boldsymbol{\delta}_1|N=n]+E[\boldsymbol{\delta}_{c\rightarrow s}|N=n]=[\frac{2(1-p)R^2}{r\sqrt{\lambda p}}+\frac{2pR(R+1)(2R+1)}{r}] \tag{11}$$

E[δ] is minimized by a value of p that is a solution of equation that gives partial derivative to (10) as follow:

$$\frac{2R(R+1)(2R+1)}{r}-\frac{R^2}{r\sqrt{\lambda p}}-\frac{R^2}{rp^{3/2}\sqrt{\lambda}}=0 \tag{12}$$

Then, get

$$-\mu p^{3/2} + p + 1 = 0 \quad (13)$$

Where

$$\mu = \frac{2(R+1)(2R+1)}{R}\sqrt{\lambda} \quad (14)$$

The equation (13) has three roots, two of them are imaginary. The second derivative of the above function is positive only for the real root that is given by

Real Root:

$$\frac{1}{3\mu^2} - \frac{2^{\frac{1}{3}}(-1-6\mu^2)}{3\mu^2(2+18\mu^2+27\mu^4+3\sqrt{3}\,\mu^3\sqrt{27\mu^2+4})^{\frac{1}{3}}} + \frac{(2+18\mu^2+27\mu^4+3\sqrt{3}\,\mu^3\sqrt{27\mu^2+4})^{\frac{1}{3}}}{2^{\frac{1}{3}}(3\mu^2)} \quad (15)$$

Hence, if and only if the value of p is equal to the real root, the algorithm does really minimize the energy cost.

3.1 Simulations on energy efficiency of clustering for generating Voronoi-based WSN

In this section, we simulated the proposed algorithm with totally n distributed sensors in a square of 1000 sq. units. Energy dissipation follows Low Energy Adaptive Clustering Hierarchy (LEACH) protocol. The experiments were conducted with the communication range r was assigned to be 1 unit and total number of sensors n is assigned to be 400, 1600, 2500 with R=10, 20, 25 respectively. Moreover, the processing center is assumed to be at the center of the network area. Don't consider the unexpected errors and influences from outside circumstance.

For the simulation experiments, considered a range of possible value of the probability (*p*) less than 0.1 for most of potentials. For each of possible value of *p*, compute the density of Poisson process λ for generating the network under different network conditions. The results are provided in Figure 5. In figure 5, the proposed algorithm was used to detect the boundary of the network with R=10, R=20 and R=25 respectively. Then vary the value of the density of Poisson process (λ) to get the willing values of *p* for computation on minimized energy cost (**δ** δ). However, it shows that the value of *p* decreases as the value of λ increases stably at an interval {0.03, 0.1}. To achieve *p* with a value of smaller than 0.03, we have to manage the rapidity of changing λ at a high value since clustering algorithm are well working in a densely deployed large scale WSNs, while to achieve *p* in excess of 0.1, we don't need to concern too much because there are few sensors randomly distributed in such a large scale area with λ pretty small that indicates sensors are difficult to get communicate with each other, they are of great potential to be geographically separated. In this case, the algorithm produce huge amount of *isolated sensors* that is object to the assumption and beyond our consideration.

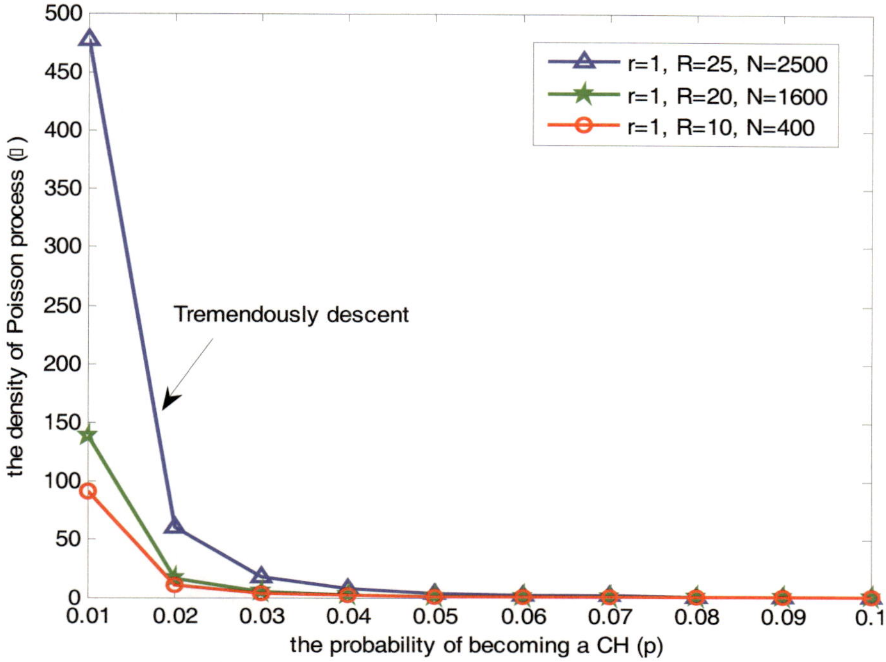

Fig. 5. The computation of parameters {p, λ}

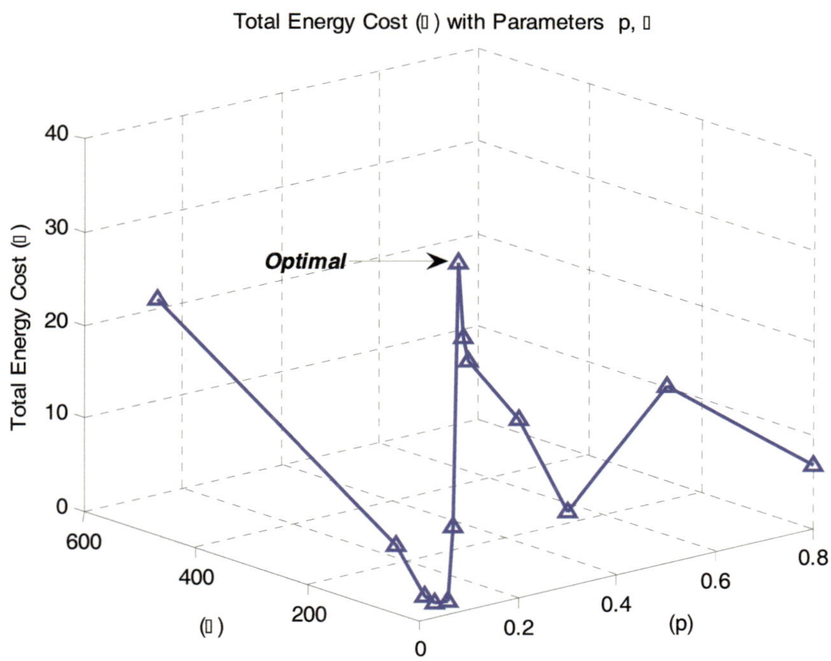

Fig. 6. Optimal value p for minimizing total energy cost $(\delta\delta)$

Each data point in Figure 6 corresponds to the average energy cost over 100 experiments. It is verified that the energy spent in the network is indeed minimized at the theoretically optimal value of p at "0.08" under a network condition of {r=1, R=10, N=400} in a randomly distributed large scale Voronoi cell based WSNs. The optimal value of p here will be of more considerable for the future research. Now, let's do comparative study between popular Max-Min D-Cluster algorithm and the proposed clustering algorithm in terms of minimizing energy cost.

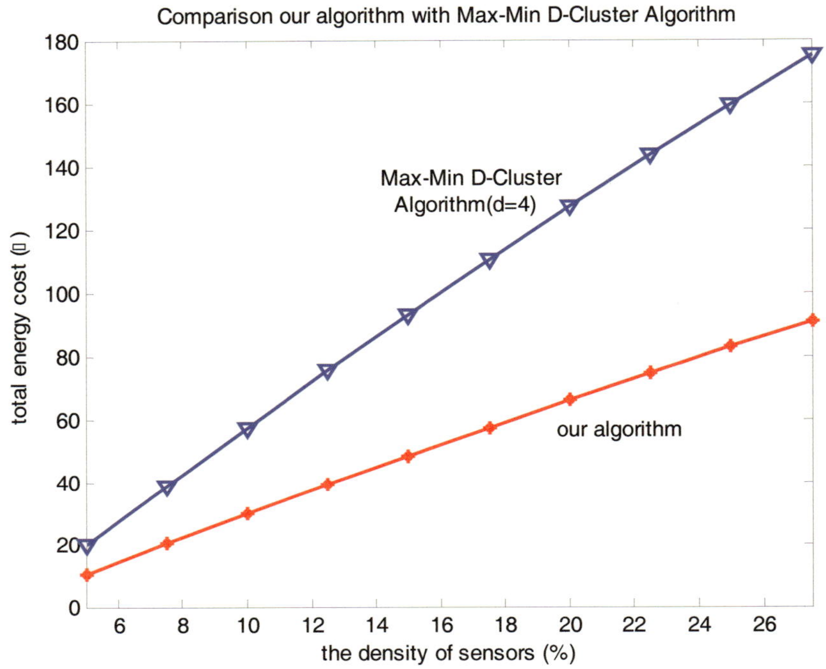

Fig. 7. Comparison with Max-Min D-Cluster algorithm

In Figure 7, the pre-obtained optimal values of all the critical parameters of the proposed algorithm in simulation model are used to evaluate the performance of the algorithm. At same time, we evaluated the Max-Min D-Cluster Algorithm with d=4. The result (e.g. Figure 7) clearly verifies that the algorithm performances better in terms of energy cost in the network under this network specification.

4. Mobility model for k-covered mobile target tracking

In this section, we proposed a mobility model for k-covered target tracking applications based on Voronoi diagram. The following gives a condition for a Voronoi diagram partly uncovered. Let $\mathcal{P}$ be a set of sensor node physical positions. If there exists s_u, such that $d(s_v, s_u) > 2R_{s_v}$, then s_u is not fully covered in $\Gamma(s_v, R_{s_v})$.

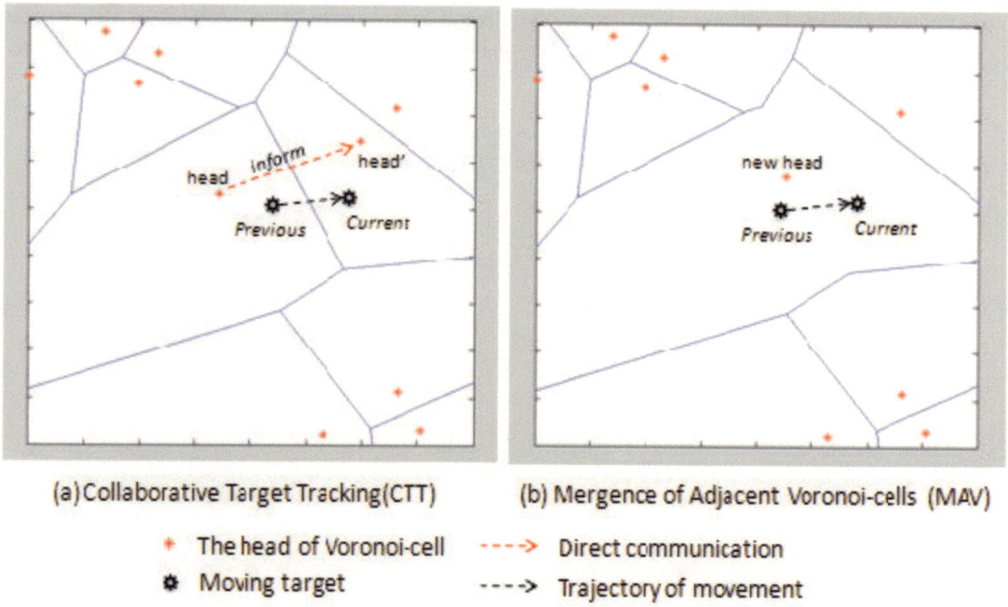

Fig. 8. Mobile target tracking strategy in Voronoi-based WSN

For the situations described in Figure 8, a mobile target moved from one Voronoi diagram to another during a time interval τ. As a result, head cannot detect it any more. To avoid such a sudden undetectability, two intelligent tracking strategies were proposed as follows:

1) **Collaborative Target Tracking** (CTT):
The network topology keeps the same. The major merit is that we adopt a target-closed boundary monitoring that enables the *head* to have a quick knowledge of the boundary line to which the target is current most closed. By using it, the potential mobile target trajectory can be easily predicted by current *head*. Once the mobile target disappeared suddenly from the monitoring area, the current head will immediately inform the head' to be responsible for tracking the entered target. (see Figure 8(a)).

2) **Mergence of Adjacent Voronoi-diagrams** (MAV):
We keep using mobile target-closed boundary monitoring to get knowledge of the potential trajectory of the mobile target. The difference from CTT is that once the mobile target went cross the boundary line, two Voronoi diagrams divided by this boundary line will merge into one larger Voronoi diagram (see Figure 8(b)). Additionally, we do not need to perform the global re-clustering, instead just re-clustering the ~~influenced~~ involved sensor nodes in this case.

5. Simulations of k-covered mobile target tracking in Voronoi-based wireless sensor network

The simulations described in this section have been performed using the Matlab environment. We made a comparison with random walk (T.Camp, et al, 2002), random waypoint (B.Liang, et al, 1999), random direction(L.Lima, et al, 2007)and Gauss-Markov (C.Bettstetter, et al, 2003). The mobile target enters the network one by one continuously by programming.

Parameter	*Value*
Network Area	$(100m)^2$
The sink	(50,50)
No. of sensors	100
Transmission range	20m
Time slots	100 (seconds)
Initial Energy/sensor	2J/battery
Message size	100 Bytes
Mobile target velocity	0~10 m/sec
$\mathbf{E_{elec}}$	50 nJ/bit
$\mathbf{E_{fs}}$	10 pJ/bit/m2
$\varepsilon_{\mathbf{amp}}$	0.0013 pJ/bit/m4
$\mathbf{E_{DA}}$	5 nJ/bit/signal

Table 2. Simulation parameters

For monitoring sensor network, energy conservation plays a dominated role in monitoring efficiency and accuracy. Figure 9 captured the energy levels of 100 sensors. *Note*: that the results represent the average performance of our proposed network over 100 times simulation trials. Obviously, it differs every time, but makes no distinction.

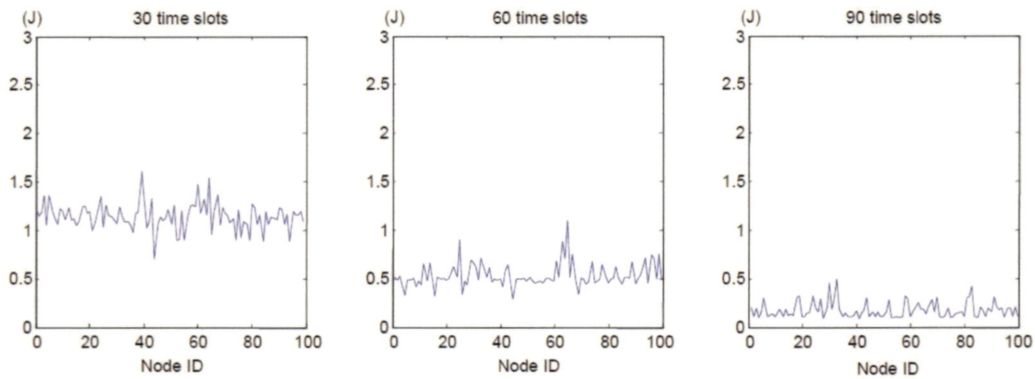

Fig. 9. Energy level of sensors at different timing

Sensor death rate is essential for heterogeneous sensor network. With the number of alive nodes decreasing, the network cannot make more contributions. Thus, the network lifetime should be defined as the time when enough nodes are still alive to keep the network operational. In Figure 10, it is no doubt that our proposed CTT&MAV outperformance *random walk, random waypoint, random direction and Gauss-Markov* mobility models in term of lower sensor death rate. Intuitively, CTT&MAV keeps more sensors alive at any timing. For the 1st half, sensors die very slowly, while for the 2nd half, since few alive nodes cannot fully take advantage of CTT&MAV, they die almost at the same speed as that of other evaluated models.

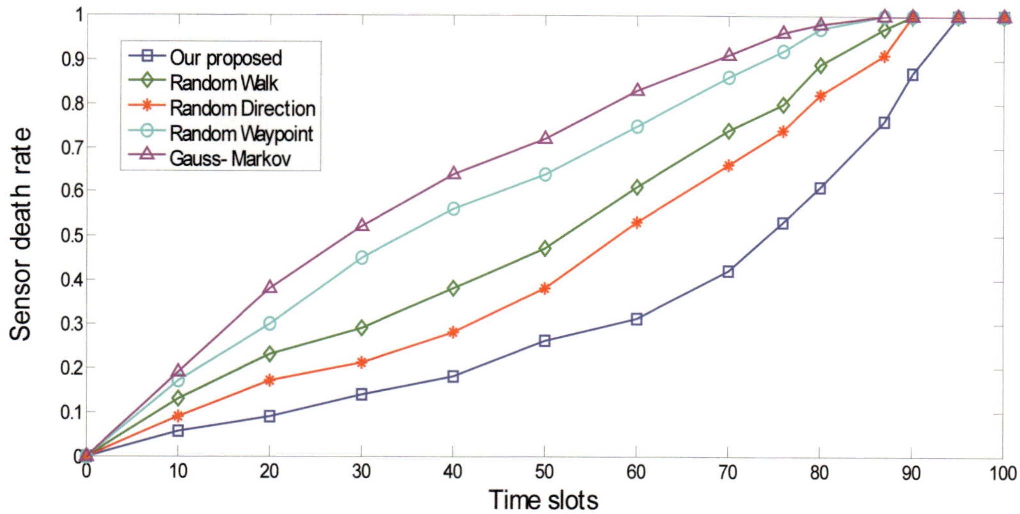

Fig. 10. Sensor death rate based on different time slots (k=2)

In this subsection we present the results of the simulations that have been conducted to assess the efficiency of the proposed CTT&MAV. It is based on estimating the average hop distance that mobile target can make before being detected. Figures 11 show that our proposed has the better performance among the tested models. Apparently, CTT& MAV perform significantly better with the help of Section 2 and Section 3.

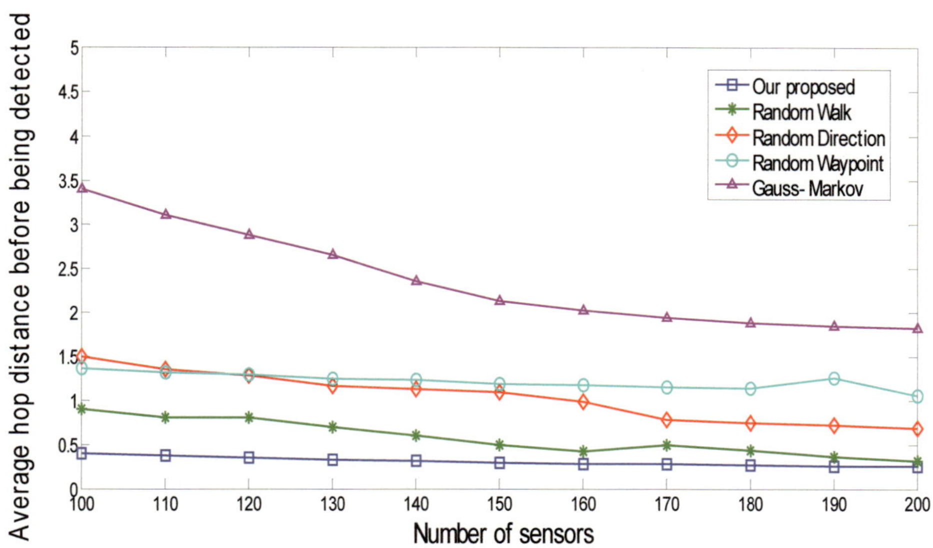

Fig. 11. Average hop distance before being detected (k=2)

6. Conclusion

In this chapter, we proposed two intelligent tracking strategies to monitor the moving multi-mobile target in a *k*-covered Voronoi-based WSNs. The current simulations based on the simplified 2-covered network region with one mobile target show that CCT and MAV performed better than *random direction* in term of average distance that the target moved before being detected. However it is currently insufficient, we will simulate more based on the uncertain *k* and the number of mobile targets to prove our hypothesis. In a word, mobile target tracking using Voronoi diagram is a meaty theme. Our future work will include verification of precision of mobile target trajectory and invention of a new protocol that consider the fast mobility of each sensor as well as destructive sensors or sudden failures in the network connectivity during communication.

7. References

T. He, P. Vicaire, T. Yan, et al, (2006). Achieving real-time target tracking using wireless sensor networks, Proceedings of IEEE Real-Time and Embedded Technology and Applications Symposium, California , USA.

H. Meyerhenke, et al, (2005). Constructing Higher-Order Voronoi Diagrams in Parallel, in Proc. of 21st European Workshop on Computational Geometry (EWCG) 2005, Eindhoven, Netherlands.

L. Lima, J. Barros, (2007). Random Walks on Sensor Networks," the 5th International Syposium on Modeling and Optimization in Mobile, Ad hoc, and Wireless Networks (WiOpt 2007), Limassol, Cyprus.

J. Janssen, M.Ditzel, C. Lageweg, Arne Theil, (2008). Multi-target Data Aggregation and Tracking in Wireless Sensor Networks, Proceeding of journal of networks, VOL. 3, NO. 1, Miami, FL,USA.

Cardei, M.&Wu, J.(2004).Coverage in wireless sensor networks. Handbook of sensor networks: Compact Wireless and Wired Sensing Systems. CRC Press LLC.

Meguerdichian, S.; Koushanfar, F.; Potkonjak, M., & Srivastava, M. (2005). Worst and best-case coverage in sensor networks, Proceeding of IEEE Transactions on Mobile Computing, 4(1), 84–92. doi: 10.1109/TMC.

D. J. Baker & A. Ephremides.(1981). The Architectural Organization of a Mobile Radio Network via a Distributed Algorithm. IEEE Transactions on Communications, Vol. 29, No. 11, pp. 1694-1701.

A.D. Amis; R. Prakash; T.H.P. Vuong & D. T. Huynh(2000). Max-Min D-Cluster Formation in Wireless Ad Hoc Networks, Proceedings of IEEE INFOCOM2000,pp.32-41,March,2000, Tel-Aviv, Israel.

A. Ephremides; J.E. W. & D.J.B.(1987).A Design concept for Reliable Mobile Radio Networks with Frequency Hopping Signaling, Proceeding of IEEE, Vol. 75, pp. 56-73, ISSN:0018-9219, Jan.1987.

W.R. Heinzelman; A.C. & H. Balakrishnan(2000) "Energy-Efficient Communication Protocol for Wireless Microsensor Networks", in Proceedings of IEEE HICSS,pp:3005-3014, print ISBN:0-7695-0493-0, Jan. 2000. Hawaii,USA.

T. Camp; J. Boleng & V. Davies (2002). A Survey of Mobility Models for Ad Hoc Network Research, in Wireless Communication and Mobile Computing (WCMC): Special issue on Mobile Ad Hoc Networking: Research, Trends and Applications, vol. 2, no. 5, pp. 483-502

B. Liang & Z. J. Haas. Predictive Distance-Based Mobility Management for PCS Networks, in Proceedings of IEEE information Communications Conference (INFOCOM 1999),pp:1377-1384, print ISBN:0-7803-5417-6,Apr. 1999,New York,NY,USA.

L. Lima & J. Barros(2007).Random Walks on Sensor Networks, Proceedings of the 5th International Symposium on Modeling and optimization in Mobile, Ad hoc, and Wireless Networks (WiOpt 2007), pp:1-5, print ISSN:978-1-4244-0960-0, April, 2007, Limassol, Cyprus.

C. Bettstetter; G. Resta & P. Santi(2003). The Node Distribution of the Random Waypoint Mobility Model for Wireless Ad Hoc Networks", IEEE Transactions on Mobile Computing, Vol. 2, No. 3, pp. 257-269.

S.G.Foss & S.A.Zuyev(1996).On a Voronoi Aggregative Process Related to a Bivariate Poisson Process, proceeding of Advances in Applied Probability(SGSA), Vol. 28, no. 4, pp. 965-981.

Power Efficient Target Coverage in Wireless Sensor Networks

Dimitrios Zorbas and Christos Douligeris
Department of Informatics, University of Piraeus
Greece

1. Introduction

A Wireless Sensor Network (WSN) can be used in a variety of applications, such as in environmental monitoring and in battlefield surveillance in military applications (Akyildiz et al., 2002). A WSN consists of hundreds or thousands of sensors and, depending on the application, the node deployment and placement can be realised either in a deterministic way or randomly. In hostile environments, for example, sensors may be dropped from an aeroplane, resulting in a random placement, where the likely node density requirements cannot be guaranteed; some areas may contain more sensors than others.

Each sensor can collect data by monitoring a usually small area that it is in its sensing range. We say that the sensor provides *coverage* to this area. A sensor collects data periodically or continuously depending on the nature of the application and forwards the data to a node called the Base Station (BS) or sink which provides the necessary connections to infrastructure networking. A sensor node is equipped with a radio device that supports *connectivity* between two nodes or between a node and the BS.

One of the fundamental problems in WSNs is the coverage of the targets in conjunction with energy efficiency constraints. The problem of coverage in wireless sensor networks has been studied from many different aspects. In (Li et al., 2003; Megerian et al., 2005), the coverage problem is described as a quality of service problem, where the objective is to find how well, in terms of the quality of monitoring data, the field is monitored by the sensors. In (Berman et al., 2004; Cardei & Du, 2005; Cardei, Thai, Li & Wu, 2005; Slijepcevic & Potkonjak, 2001; Zhang & Hou, 2005; Zorbas et al., 2007), the problem is formulated as the maximisation of the network lifetime under the *area* or *target* coverage constraint. In the former formulation (see Figure 1left), the whole area (e.g. a big square region) must be monitored by the sensors, while in the latter the sensors must cover a set of points (targets) lying in the field (see Figure 1right). (Berman et al., 2004; Slijepcevic & Potkonjak, 2001; Wang & Kulkarni, 2008; Zhang & Hou, 2005; Zhong et al., 2002) deal with the area coverage problem. This chapter focuses on the target coverage problem, but it often refers to important works about other types of coverage that can help in the solution of the target coverage problem.

The most important challenge in a WSN is to efficiently manage the battery consumption of the sensors, since WSNs are characterized by limited energy resources and low computational capabilities. Managing the energy consumption in an efficient way can lead to an extension of the total network lifetime. In the case of the deterministic node placement this is translated as an optimal deployment of a set of sensors, where all the targets are covered. When the sensors

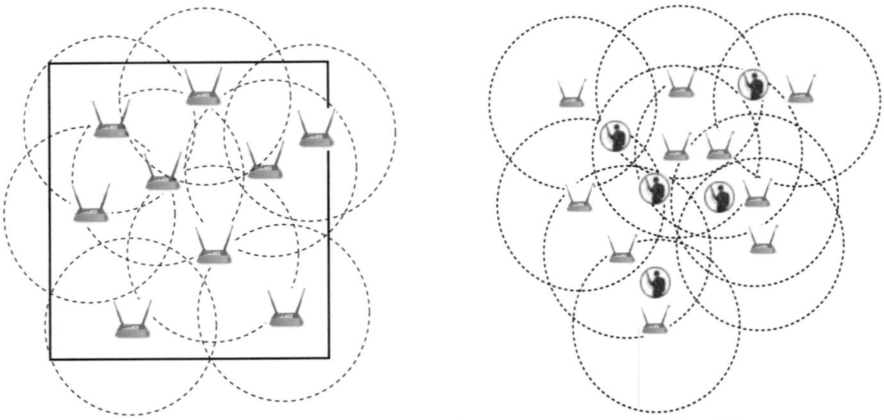

Fig. 1. Two major types of coverage: area and target coverage (The big square on the left denotes the covered area, the soldiers on the right denote the covered targets)

are randomly deployed, the energy management takes advantage of the ability of a sensor to put certain parts of the device into "sleep mode" and, thus, to consume less energy whenever it is not needed to perform monitoring or, more often, to participate in relaying tasks. This is achieved by dividing the sensors into sets, called *cover sets* or *sensor covers*, whereas each cover set can monitor all the available targets. Thus, only one set must be active at any time, while the rest of the sensors can be in sleep mode. Figure 2 illustrates two cover sets that provide full coverage.

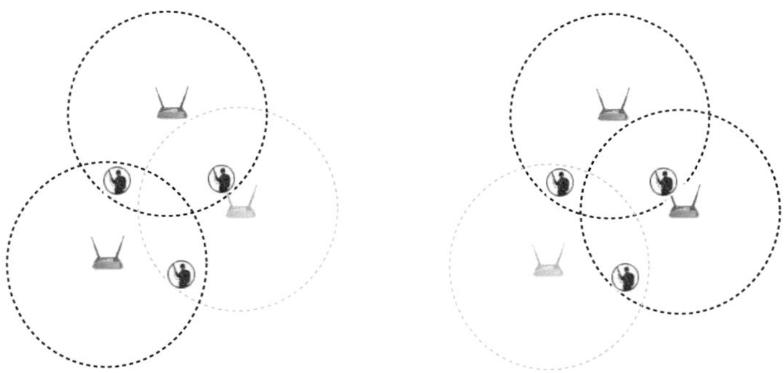

Fig. 2. Two generated cover sets (light grey colour denotes a node in sleep mode)

Next, we present the main works and solutions presented in the literature the past years, paying more attention to the random target coverage problem where a solid piece of work has been done. Furthermore, we classify the proposed solutions according to their objectives and present several variations of the target coverage problem.

2. Random target coverage

Most of the works in target coverage deal with the problem of dividing the sensors into cover sets and scheduling these sets consecutively such as only sensors belonging in one set are active at any time, while the rest are inactive. Assuming a random sensor deployment and the fact that each sensor consumes the same amount of energy in each cover set the coverage

problem is transformed to a problem of finding the optimal number of cover sets. Finding this optimal number is proven to be an NP-Complete problem (Cardei & Du, 2005), hence suboptimal solutions have been proposed in the literature such as algorithms based on linear programming, integer programming, greedy heuristics and branch and bound algorithms.
The proposed solutions can be separated into *centralised* and *distributed*. In a centralised coverage algorithm the monitoring schedule is first calculated on the base station and it is then sent to the sensor nodes for execution. The advantage of this scheduling approach is that it requires very low processing power from the sensor nodes, that usually have limited processing capabilities. A major disadvantage is the fact that the location of the sensors must be known in advance, which means that the sensors must me equipped with a global positioning system. Moreover centralised algorithms are not tolerant to the existence of corrupted nodes that can lead to a loss of data. In distributed algorithms the nodes usually use broadcasting in order to ensure connectivity with their neighbours and to detect failures.

2.1 Centralised algorithms

Below we analyse the basic characteristics of the existing centralised coverage scheduling algorithms that can be used in random sensor deployment scenarios with homogeneous device characteristics in terms of communication and sensing ranges. Many of the existing algorithms that deal with the maximisation of the number of cover sets incorporate a special strategy about the sensors that cover the most poorly covered targets. These targets are called *critical* and set an upper bound on the number of cover sets and, thus, on the achievable network lifetime. As described in (Zorbas et al., 2010) the number of cover sets is reduced as there are double-covered critical targets in a cover set. Moreover, regardless of the algorithmic approach (centralised or distributed) the cover sets can be assumed disjoint or non-disjoint. In disjoint cover sets a sensor can participate in only one cover set, while the opposite holds true in the non-disjoint case. In some cases the non-disjoint approach increases the overall network lifetime, but it incurs a higher complexity.

2.1.1 Disjoint approaches

Slijepcevic and Potkonjak (Slijepcevic & Potkonjak, 2001) propose a centralised algorithm for the area coverage problem. They introduce the idea of the *field* as a set of targets. Two targets belong to the same field if and only if they are covered by the same set of sensors. In particular, the fields are small areas which are produced by the intersection of the coverage limits of sensors and/or the physical limits of the monitoring terrain. As it shown in Figure 3, replacing each field (number) by a unique point (target), the area coverage problem is equivalent to the target coverage problem and, thus, the area coverage algorithms can be used to solve the target coverage problem as well.
Every sensor may cover one or more fields and one field is covered by at least one sensor. Their algorithm initially covers the *critical* fields (targets) and then it excludes all the other nodes that cover the same field. Thus, it is assured (during the construction of a cover set) that only one node covering a particular critical field shall be selected. This is a deterministic strategy in order to avoid the double-covering of the critical targets. The complexity of the algorithm is $O(n^2)$, where n is the total number of sensors.
Cardei *et al.* (Cardei et al., 2002) propose an algorithm to solve the same problem using graphs. They construct an undirected graph $G = (V, E)$, where V is the set of sensors and E the set of edges, such that the edge $(u, v) \in E$ if and only if u and v are within each other's sensing range. The goal is to find the maximum number of *dominating sets*. To achieve this a graph

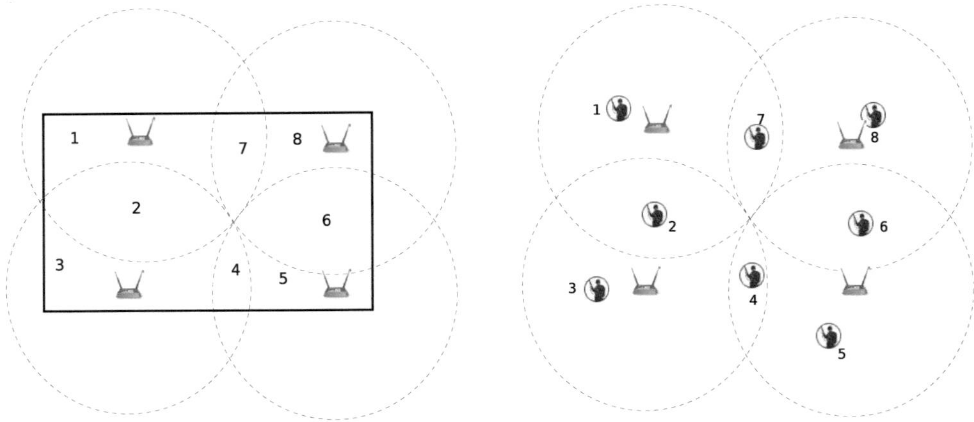

Fig. 3. Relation between area and target coverage

colouring technique is used. As depicted in (Thai et al., 2008), despite the production of more sets than the ones achieved in the proposal of (Slijepcevic & Potkonjak, 2001), the dominating sets do not guarantee the coverage of the whole area. The complexity of the heuristic which computes the disjoint sets from the coloured graph is $O(n^3)$.
Cardei and Du (Cardei & Du, 2005) propose a heuristic algorithm in order to solve the random target coverage problem. This problem is successively formulated as an area coverage problem, which is proved to be an NP-Complete problem. Cardei and Du define the disjoint-set coverage problem, that was first introduced by Slijepcevic and Potkonjak (Slijepcevic & Potkonjak, 2001), as a generalisation of the 3-SAT problem (Garey & Johnson, 1979). They propose a heuristic to compute the disjoint sets. In order to compute the maximum number of covers, they transform the problem into a maximum-flow problem. Then, the result of the maximum-flow problem is solved using *Mixed Integer Programming*, which heuristically produces the final number of cover sets. The results in (Cardei & Du, 2005) show a slight improvement in the number of produced sets in comparison to (Slijepcevic & Potkonjak, 2001) but there is a substantial increase of the execution time. The complexity of this algorithm depends on the complexity of the mixed integer programming technique used.
Finally, the work of (Liu et al., 2005) addresses the problem where given a set of sensors and targets, a sensor can watch only one target at a time. The objective is to schedule sensors to monitor targets, such that the lifetime of the surveillance system is maximized, where lifetime is defined as the duration of time when all targets are covered. This problem does not belong to the NP class, as it can be solved in polynomial time.

2.1.2 Non-disjoint approaches

Cardei *et al.* (Cardei, Thai, Li & Wu, 2005) propose a *Linear Programming* (LP) solution to the target coverage problem for non-disjoint cover sets as well. Although the LP algorithm presents a high complexity $O(m^3n^3)$, where m is the number of cover sets and n the number of sensors, the authors also propose a greedy algorithm with a lower complexity $O(dk^2n)$, where d is the number of sensors that cover the most poorly covered targets and k is the number of targets. The greedy algorithm is called Greedy-MSC and it uses a similar strategy to (Slijepcevic & Potkonjak, 2001) in order to avoid double-covering the critical targets.

In (Kim et al., 2009) the authors solve the same problem using a branch and bound algorithm. Their algorithm incorporates rules that decrease the probability of selecting two sensors that cover one or more similar targets. However, the authors do not give the complexity of their approach. The simulation results show a slight improvement of the network lifetime over the Greedy-MSC algorithm.
An LP technique is proposed by Berman *et al.* (Berman et al., 2004). In this approach first a series of cover sets is computed and then the optimal lifetime for each cover set is deduced. This approach is based on the $(1+\epsilon)$-approximation of the Garg and Könemann algorithm (Garg & Könemann, 1998), with an approximation factor of $(1+\epsilon)(1+2\ln n)$ for any $\epsilon > 0$.
In (Zorbas et al., 2010), the authors present a detailed methodology of how a greedy target coverage algorithm works and how it is possible to maximize the number of cover sets by efficiently managing the coverage status of the sensors and their association with the poorly covered targets. During the construction of a cover set, the sensor nodes are evaluated mainly based on their coverage status. Depending on the number of sensors that have been already covered in the examined cover set, the authors distinguish four kinds of sensor candidates, as shown in Figure 4. Candidates of the top two classes are more preferable as their selection is trouble free concerning the double-coverings, but they are fewer in number during the generation process.

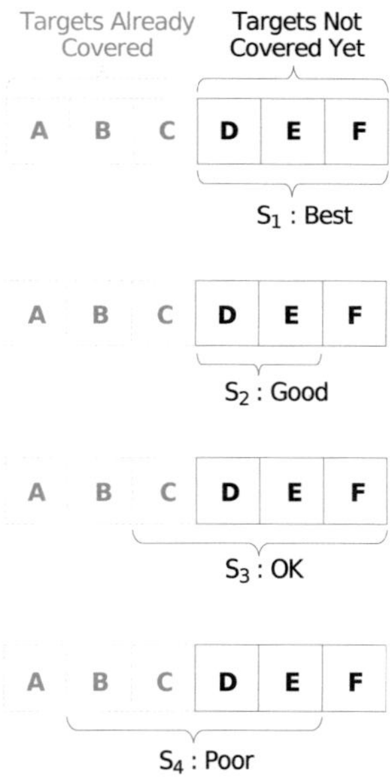

Fig. 4. The four classes of sensor candidates

Moreover, it is considered that all the available targets (i.e T_0) incorporate a degree of criticality based on the number of the sensors that they are covered by. The degree of criticality is formulated using the *badness* attribute. The badness, given by Formula (1), is calculated once for every sensor s_j at the beginning of the algorithm. N_i contains the sensors that cover the target t_i, P_j contains the targets that the sensor s_j covers, while μ is equal to $\max(|N_1|, \ldots, |N_k|), k = |T_0|$.

$$B_j = \sum_{i=1}^{|P_j|} (\mu - |N_i| + 1)^3 \,. \tag{1}$$

The badness attribute of a sensor describes the accumulative criticality level of all the targets covered by this sensor. This attribute can be used to prioritise the selection of sensor nodes that exhibit a low badness value (i.e., they are not as heavily associated with the targets of high degree of criticality). The advantage of the badness attribute compared to the deterministic strategy used in (Slijepcevic & Potkonjak, 2001) and (Cardei, Thai, Li & Wu, 2005) is that it is computed only once, thus achieving lower execution times.

The sensor candidates are evaluated using a complex cost function that takes into account three different characteristics: (a) the coverage status of the nodes, (b) their association to the poorly covered targets (badness) and (c) their remaining energy. Two heuristics are proposed, one based on the badness attribute and one based on the deterministic approach. The complexity of the approaches is $O(wn^2k)$, where w is a value that represents how many times a node can participate in the produced cover sets. The proposed algorithms are evaluated using 2-dimensional and 3-dimensional sensor and target deployments. The results show that the two heuristics present almost the same performance, which is very close or equal to the optimum, thus achieving constant execution times per generated cover set. The algorithm of (Slijepcevic & Potkonjak, 2001) exhibits an exponential increase in execution time, while the greedy heuristic of (Cardei, Thai, Li & Wu, 2005) requires the participation of a node in many cover sets in order to reach a satisfactory result, which comes at a cost in total execution time.

The following table presents a comparison between the algorithm that incorporates the badness attribute and the greedy heuristics of (Cardei, Thai, Li & Wu, 2005; Slijepcevic & Potkonjak, 2001).

	(Zorbas et al., 2010)	(Slijepcevic & Potkonjak, 2001)	(Cardei, Thai, Li & Wu, 2005)
Type of sets produced	Disjoint and non-disjoint	Disjoint	Disjoint and non-disjoint
Critical node handling	Prioritise nodes that cover targets with low criticality	Starts cover set with a critical node. Other nodes covering the same critical targets are ignored.	Starts cover set with a critical node. It does not implement any critical node avoidance strategy.
Candidate node selection criteria	a) # of uncovered targets vs. # of already covered b) # of available targets c) association with poorly monitoring targets d) remaining battery life	# of already covered targets the candidate covers	a) # of uncovered targets the candidate covers b) remaining battery life
Complexity	$O(wn^2k)$	$O(n^2)$	$O(dk^2n)$
Dense node deployment incurs significant penalty in execution time	No	Yes	Yes, due to increased number of participations required for optimal solution

Table 1. A comparison between centralised greedy algorithms of (Cardei, Thai, Li & Wu, 2005; Slijepcevic & Potkonjak, 2001; Zorbas et al., 2010)

2.2 Distributed and localised algorithms

In distributed and localised algorithms the decision whether a node will be in sleep mode or not is taken by the sensors. The nodes perform the required calculations cooperatively by communicating with their neighbours. These schemes may require some processing and a certain communication cost by the sensors involved, but they scale better to accommodate larger networks as well as networks with many sensor failures.

An important issue of the distributed and localised algorithms is the coverage synchronisation. In most cases the process is divided in rounds. During the synchronisation phase that

takes part before each round, the sensors decide whether they will be active or in sleep mode during the next coverage round. In target coverage the sensors exchange messages with their neighbours informing them about their coverage status and their *id*. Usually, the sensors that receive these messages evaluate them according to a cost function and the top scored node of the neighbourhood remains active during the next round. The problem rises when two or more nodes have a similar coverage status or cost function result. The nodes consider their selves as active (or sleep) in the same neighbourhood, leading to double covered targets (or uncovered targets). This issue is addressed in (Cardei & Cardei, 2008) by assigning back-off times to the sensors. The rationale behind this assignment is to give higher priority to sensors that have higher residual energy and cover a larger number of uncovered targets. When this time expires, a node declares itself as a sensing node during the next round. Additionally, it broadcasts this decision to all its 2-hop neighbours.
In (Tian & Georganas, 2002), the authors present a distributed and localised algorithm to solve the area coverage problem. They provide their solution as an extension to the well-known *LEACH* clustering protocol (Heinzelman et al., 2000). The process is divided into rounds and in every round two phases are distinguished: the self-scheduling and the sensing phase. During the self-scheduling phase the nodes investigate the off-duty eligibility rule. The eligibility rule determines whether a node's sensing area is included in its neighbours' sensing areas. Eligible nodes turn off their communication and sensing units to save energy. Non-eligible nodes perform sensing tasks during the sensing phase. The sensing phase is much longer than the self-scheduling phase. The authors incorporate a scheme in order to avoid the appearance of blind points (uncovered areas). They do not deal with the connectivity requirement, leaving this task to the data gathering protocol. Making this assumption this algorithm can be used to solve the target coverage problem as well.
In (Ye et al., 2002), the authors use a probing based scheme in order to determine which sensor will be active. In this scheme a sleeping node wakes up after sleeping for an exponentially distributed period of time specified by a wakeup rate. After a sleeping node wakes up, it broadcasts a probing message within a range r. When hearing a probing message, any working node within this range will locally broadcast a reply message. If the wakeup node hears a reply message, it knows that there is a working node within distance r and the node goes back to the sleeping mode. If the wakeup node does not hear a reply message within a prespecified time interval, it assumes that no working node is within its probing range and it starts working continually. It must be noted that this algorithm controls the active node density and may not provide full coverage. Moreover, since this algorithm is developed for the area coverage problem, a scheme using relay nodes that ensures connectivity must be provided in order to be able to use this algorithm in the target coverage problem.

2.3 Incorporating connectivity

More recent works in the literature take into account the connectivity requirement that appears in multi-hop networks. Considering this requirement it is not possible to use the area coverage algorithms to solve the target coverage problem as well. The problem derives from the fact that in area coverage if the communication range is at least twice the length of the sensing range coverage implies connectivity (Zhang & Hou, 2005). In target coverage the issue that must be addressed concerns the finding of the maximum number of cover sets, while every node in each cover set in multi-hop networks remains connected with the BS using relay nodes (see Figure 5. This problem often translates to the computation of paths with the minimum possible cost since the consumed energy rises with the distance between two nodes.

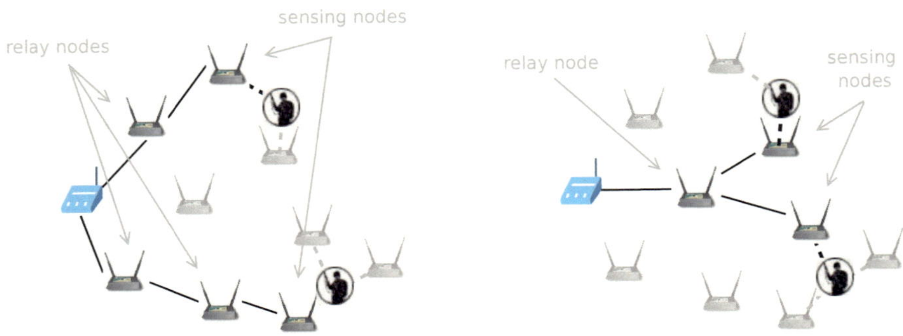

Fig. 5. Two cover sets that are connected with a base station

In (Cardei & Cardei, 2008), centralised and distributed algorithms are proposed for the computation of the connected cover sets. A breadth first search algorithm is used to discover the node-path to the BS through a centralized algorithm, while a minimum spanning tree algorithm is used in the distributed version of the algorithm.

In (Jaggi & Abouzeid, 2006), it is proposed another connected cover set generation algorithm in order to extend the lifetime of the network. They consider that all the cover sets are disjoint and they try to maximize their number, while they compute a shortest path tree to select the relay nodes that manage to retain connectivity in the network.

These two works use a simplified energy consumption model. The energy consumed for communication is predefined for all sensors and it does not depend on the distance between the nodes, which is far from true in a real network environment. It is, also, assumed that each sensor consumes the same amount of energy, regardless of the number of targets it covers. In real-time WSNs the consumed energy increases with the distance between the nodes, while the amount of the transmitted data depends on the size of the packets and the degree of the data aggregation that may used.

In (Zhao & Gurusamy, 2008b), the authors model the connected target coverage problem as a maximum cover tree problem. A theoretical analysis of the problem shows that it is also NP-Complete. An approximation algorithm as well as a greedy one with a lower computation cost are proposed. Connectivity, coverage and a practical energy consumption model that is based on distance are taken into account. The network is modelled as a graph, where the vertices correspond to the nodes and the edges to the links between two sensors. The greedy algorithm applies weights on the edges of the graph of nodes in order to select nodes with high remaining energy and low communication cost. However, it requires a re-computation of all the weights of the graph, each time a new cover set is generated and no policy is applied about the critical targets.

3. Other types of target coverage

3.1 Deterministic target coverage

The objective in deterministic target coverage is to deploy a minimum number of sensors in order to cover a set of targets and apply connectivity to the network. This problem is addressed by (Kar & Banerjee, 2003), where the authors consider a 2-dimensional region with randomly deployed targets. They propose a polynomial time algorithm with a performance ratio of 7.256 when the communication range is equal to the sensing range.

In (Dasgupta et al., 2003), the authors address the problem of placing a given number of sensors in order to cover a set of targets and at the same time to maximise the lifetime of the network. They consider a realistic energy consumption model where the consumed energy increases with the distance. The process is divided in rounds and in every round they try to minimise the energy consumption per node by balancing the traffic among the appropriate relay nodes. The simulation results show an over 40% improvement over the random node deployment case.
Finally, authors in (Wang et al., 2006) formulate the problem considering a sensing model where the sensing signal weakens when the target is in a long distance away from the sensor. Moreover, each target has a degree of importance called the *utility*. Based on these parameters the authors develop a greedy heuristic algorithm to find the optimal positions of the sensors.

3.2 Adjustable sensing ranges

The works of (Cardei, Wu, Lu & Pervaiz, 2005; Dhawan et al., 2006; Lu et al., 2009) deal with the target coverage problem where the sensors can adjust their sensing range in order to conserve energy. The sensor of (Migatron, n.d.) has an adjustable two to six inch sensing range with background suppression, that means that any object within the desired range is detected, while objects out of the desired range are ignored. A network with three sensors with adjustable sensing ranges and four targets is illustrated in Figure 6.

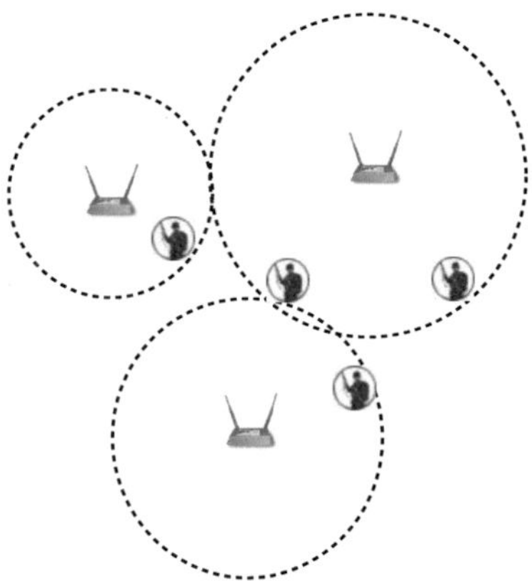

Fig. 6. Three sensors with adjustable sensing ranges cover four targets

In (Cardei, Wu, Lu & Pervaiz, 2005), the authors divide the sensors in cover sets, where the sensors of each cover set adjust their sensing range in order to avoid covering the same targets two or more times. The authors examine the case where the nodes' sensing range has p steps, while they target to maximise the number of cover sets. They present an LP-based solution, a centralised greedy one and a distributed and localised algorithm. They assume a linear and an exponential energy consumption model for the sensing operation. The simulation results show that the consumed energy can be reduced in half compared to the approach where the sensors have the longest possible sensing range.
Unlike (Cardei, Wu, Lu & Pervaiz, 2005), in (Dhawan et al., 2006), it is assumed that each sensor has an infinite number of options concerning its sensing range. The authors propose an approximate algorithm to solve this problem based on Garg and Könemann algorithm (Garg & Könemann, 1998). Their simulation results show an significant improvement over the distributed algorithm of (Cardei, Wu, Lu & Pervaiz, 2005).

The previous two works do not take into account the connectivity requirement. The work of (Lu et al., 2009) presents an extension of the work of (Cardei, Wu, Lu & Pervaiz, 2005), where target coverage and connectivity are maintained. The authors present a distributed algorithm that builds a virtual backbone first to satisfy network connectivity, and they ensure coverage based on that backbone. In providing such a virtual backbone, the authors first construct a connected dominating set and prune redundant sensors by applying the Rule-k algorithm (Dai & Wu, 2003).

3.3 Partial target coverage

In applications where the full coverage is not a critical requirement, algorithms that provide partial coverage can be used. Such applications are those where the data provided by a subset of the available targets are satisfactory for the required measurements. Partial coverage has been first studied for the area coverage problem. Partial area coverage deals with the maximization of the α-lifetime with a minimum amount of energy, where α denotes the coverage percentage of the total monitoring area (i.e. $0 < \alpha \leq 1$). In target coverage this means the percentage of the total number of covered targets.

Abrams et al (Abrams et al., 2004) propose centralised and distributed algorithms to solve the coverage problem, as well as a randomised algorithm with performance guarantee. Each generated cover set does not provide complete coverage and the cover sets must be scheduled successively in order to achieve at least the 72% of the monitored areas.

In (Wang & Kulkarni, 2008), the authors describe a localized protocol for the area coverage problem, called *pCover*. The protocol tries to maintain a high degree of coverage (over 90%), but it also produces an increased surveillance time compared to the full coverage approach. The simulation results provided show an improvement of 2 to 7 times compared to the total coverage duration. However, the connectivity of the network is not guaranteed in this solution.

In (Yan et al., 2003) an adaptive algorithm that adjusts the degree of coverage depending on the problem requirements is proposed. The degree of coverage can be either lower or higher than one, depending on the node deployment density. This approach can be used in applications where the coverage requirements are changing during the monitoring process. The connectivity is still an open problem whenever the degree of coverage is below one.

In (Liu & Liang, 2005), the objective is to not only to find a subset of sensors for partial coverage with a given coverage guarantee, but also to ensure that the communication graph induced by the chosen sensors is connected. To achieve this the authors propose the use of a shortest path tree and a cost function based on the total area that a sensor candidate covers. According to the simulation results the network lifetime can be prolonged over three times compared to the full coverage approach.

Even though the previous works have been developed for the area coverage problem, they can be also used for the target coverage problem without ensuring connectivity. The work of (Zorbas et al., 2009) introduces the partial target coverage problem, where two neighbouring targets may not be covered in the same cover set, as it is considered that they may provide similar data. However, this decision is taken by a parameter that uses the Euclidean distance between two or more targets and not by a scheme that is based on geographical or statistical information from the past data collections. The overall number of covered targets per cover set is controlled by a user-given value. The proposed centralised solution incorporates the connectivity constraint and compared to the full coverage approach of (Zhao & Gurusamy, 2008b) can double the overall network lifetime covering the 90% of the targets.

3.4 Target coverage under QoS constraint

In applications where the quality of the monitoring data is critical in terms of robustness and accuracy, all or some of the targets may be covered by more than one sensor per time. This problem is defined in the literature as the k-coverage problem or the QoS-aware coverage problem. This technique is, also, useful in environments with many node failures as it provides a shield against possible loss of data.

In (Zhou et al., 2004), the connected k-coverage problem is addressed. The authors propose centralised and distributed algorithms that select a minimum number of sensors that provide connectivity and cover each point in a given query region with at least k distinct sensors. The idea is to keep only those sets of sensors active to provide the necessary coverage and connectivity, resulting in a fault-tolerant energy conservation technique.

The works of (Hefeeda & Bagheri, 2006; Simon et al., 2007) and (Vu et al., 2006) deal with the k-coverage problem where the connectivity is ensured if the communication range is at least twice the sensing range. The authors propose efficient centralised and distributed algorithms, but a connectivity scheme is required to use them in the target coverage problem.

In (Zhao & Gurusamy, 2008a), the authors deal with the k-target coverage problem, while they take into account the network connectivity. They develop an optimal solution based on an LP formulation and an efficient approximation algorithm to solve it. They, also, present a low cost greedy heuristic algorithm that is useful for practical implementations. The connectivity in the greedy algorithm is achieved using a shortest path tree. The simulation results of the greedy algorithm are very close to the optimum (LP algorithm).

The problem of coverage where each sensor has different coverage requirements (Q-coverage problem) is examined in (Gu et al., 2009). The authors design a general optimization architecture using linear programming techniques that contains a lifetime upper bound and a column generation based approach. However, the network connectivity is not included in their method.

In addition, the k-coverage problem has been described from many other different aspects. In (Shen & Wu, 2010) a Minimum Movement-assisted k-Coverage deployment problem is formulated, where a minimum set of sensors are selected and relocated to appropriate positions such that each point in the entire region is covered by at least k sensors. In (Liu et al., 2008) the problem of Directional k-Coverage (DKC) in camera-equipped sensor networks is addressed. The DKC problem is different from the one addressed in conventional sensor networks due to the directionality of the sensing model and the effective sensing. In (Kim et al., 2007), a distributed k-coverage algorithm is presented that leaves a small number of areas uncovered. The paper of (Ammari & Giudici, 2009) focuses on the problem of connected k-coverage in heterogeneous wireless sensor networks, while in (Ammari & Giudici, 2009) the k-coverage problem is examined in the presence of sensor mobility. Finally, in (Ammari & Das, 2010), the problem of connectivity and k-coverage in 3D WSNs is addressed.

3.5 Target coverage under bandwidth constraint

In applications where a large amount of data (e.g. video data) is required to be delivered and the time division protocol of the sink has a limited number of available time slots, the sensors of each cover set can transmit a limited amount of bytes. If this number of nodes is large and the interval between consecutive reports of the same target is critical for the applications, a *coverage breach* may occur (Cheng et al., 2007). In a case of a breach several targets may remain uncovered for a period of time.

In (Cheng et al., 2005), the target coverage problem under a bandwidth constraint is formulated as the minimum breach problem where the objective is to divide the sensors in disjoint cover sets while in each cover set the maximum possible number of targets is covered. Each cover set contains a maximum number of sensors equal to the available time slots (i.e. W). The authors prove that it is an NP-Complete problem and they transform it to an integer programming problem.
The work of (Cheng et al., 2007) is an extension to the previous work. The authors analyse three instances of the problem; the minimum breach, the minimum individual breach time and the minimum maximal breach. The objective of the first instance is to find a user given number of cover sets when the cardinality of each cover set must be smaller than $W+1$ and the total breach is minimised. The other two instances consider a maximum allowed breach time and a maximum number of cover sets that must be computed. Two algorithms are proposed to solve the above problems; a greedy one and an LP-based.
In (Wang et al., 2007) two equivalent instances of the coverage breach problem with those of (Cheng et al., 2007) are presented. The objective in the first instance is to achieve a maximum amount of network lifetime by minimizing the total breach time, while in the second one a maximum value of the breach rate is allowed, while the lifetime must be maximised. The authors allow a sensor node to be a member of multiple cover sets. In order to solve the above instances of the problem they propose an LP-based algorithm and a greedy heuristic.
While in the previous approaches the network connectivity is not taken into account, the work of (Zorbas & Douligeris, 2009) presents a greedy heuristic that produces connected cover sets under the bandwidth constraint. The authors compare their solution to the previous approaches in 1-hop environments. The results show that their approach present a slightly lower number of cover sets, but each cover set is capable of monitoring more targets than the other approaches. Concerning the multi-hop networks, the simulation results show that the connectivity constraint and the increased needs of data limit the lifetime of the network due to the energy exhaustion of the nodes that must transmit the data to the sink.

4. Conclusions

This chapter analysed the target coverage problem in wireless sensor networks under the constraint of the power efficiency. Recent works found in the literature have been described and organised according to the objectives of the coverage approach. Depending on the requirements of the particular application many different design approaches have been presented. The classic target coverage problem can be used for military purposes and the sensor deployment may be random or deterministic.
Since the most likely way to achieve energy efficiency is to divide sensors in groups (cover sets), where only one group is active at any time instant, this chapter focused on the works that address the problem of finding the maximum number of disjoint or non-disjoint cover sets. WSNs usually operate in a multi-hop manner, hence the network connectivity is always a critical requirement for the target coverage surveillance.
Moreover, the power efficient target coverage problem can be observed under further practical constraints. These constraints can either reduce the energy consumption, (adjustable sensing ranges and partial coverage), or increase the availability and the reliability of the monitoring data (QoS constraint). Finally, in applications with a high flow of data, as the camera-equipped sensor networks, efficient algorithms that minimise the number of uncovered targets have also been presented in this chapter. Table 2 summarises the works described in this chapter. Since

Reference	Objectives & characteristics	Algorithmic approach	Network connectivity
Slijepcevic & Potkonjak (2001)	Maximise the number of cover sets (1)	Centralised	no
Cardei et al. (2002)	(1)	Centralised	no
Berman et al. (2004)	(1)	Centralised & distributed	no
Cardei & Du (2005)	(1)	Centralised	no
Cardei, Thai, Li & Wu (2005)	(1)	Centralised	no
Zorbas et al. (2007)	(1)	Centralised	no
Kim et al. (2009)	(1)	Centralised	no
Zorbas et al. (2010)	(1)	Centralised	yes, but only between sensing nodes
Liu et al. (2005)	(1), a sensor covers only one target	Centralised	no
Tian & Georganas (2002)	Minimise the number of active nodes (2)	Distributed	no for target coverage
Ye et al. (2002)	(2)	Distributed	no for target coverage
Cardei & Cardei (2008)	(1)	Centralised & distributed	yes
Jaggi & Abouzeid (2006)	(1)	Centralised	yes
Zhao & Gurusamy (2008b)	Maximise the lifetime of each cover set & the network lifetime (3)	Centralised	yes
Kar & Banerjee (2003)	Minimise the number of deployed nodes (4)	–	yes
Dasgupta et al. (2003)	Deploy a number of sensors & maximise the netw. lifetime	Distributed	yes
Wang et al. (2006)	(4)	–	no for target coverage
Cardei, Wu, Lu & Pervaiz (2005)	(1), adjustable sensing ranges	Centralised & distributed	no
Cardei, Wu, Lu & Pervaiz (2005)	(1), adjustable sensing ranges	Centralised	no
Lu et al. (2009)	(1), adjustable sensing ranges	Centralised & distributed	yes
Abrams et al. (2004)	(1), partial coverage	Centralised & distributed	no for target coverage
Wang & Kulkarni (2008)	(2), partial coverage	Distributed	no
Yan et al. (2003)	(2), partial or over-coverage	Distributed	no
Liu & Liang (2005)	(1), partial coverage	Centralised	yes
Zorbas et al. (2009)	(3), Partial coverage	Centralised	yes
Zhou et al. (2004)	(2), *k*-coverage	Centralised & distributed	yes
Simon et al. (2007)	(2), *k*-coverage	Centralised & distributed	no for target coverage
Hefeeda & Bagheri (2006)	(1), *k*-coverage	Centralised & distributed	no for target coverage
Vu et al. (2006)	(1), *k*-coverage	Distributed	no for target coverage
Zhao & Gurusamy (2008a)	(2), *k*-coverage	Centralised	yes
Gu et al. (2009)	(1), *Q*-coverage	Centralised	no
Cheng et al. (2005)	(1) under bandwidth constraint (5)	Centralised	no
Cheng et al. (2007)	(5)	Centralised	no
Wang & Kulkarni (2008)	(5)	Centralised	no
Zorbas & Douligeris (2009)	(5)	Centralised	yes

Table 2. A summary of works related to the target coverage problem

the network lifetime is strongly connected with the coverage and connectivity, these works are considered as an important factor in designing energy efficient sensor networks.
The most recent efforts in the area of coverage involve the usage of mobile sensor nodes. With the growth of robotics the sensors will be able to move across the field and collect data from inaccessible places. This ability will make sensor networks more desirable in dynamic battlefields, where the targets are mobile and multiple unpredictable events occur. Undoubtlessly, the coverage and communication in these environments require the development of robust, reliable and long lived networks that operate in a distributed way. Moreover, the recent advances in microelectronics, chemistry and solar systems will lead to the development of self-powered small devices that will operate for almost infinite period of time increasing the availability of the networks, specially in the case of hostile environments.

5. References

Abrams, Z., Goel, A. & Plotkin, S. (2004). Set k-cover algorithms for energy efficient monitoring in wireless sensor networks, *Proc. of Third International Symposium on Information Processing in Sensor Networks*, ACM, pp. 424–432.

Akyildiz, I., Su, W., Sankarasubramaniam, Y. & Cayirci, E. (2002). Wireless sensor networks: a survey, *Computer Networks* **38**(4): 393–422.

Ammari, H. & Das, S. (2010). A study of k-coverage and measures of connectivity in 3d wireless sensor networks, *Computers, IEEE Transactions on* **59**(2): 243 –257.

Ammari, H. & Giudici, J. (2009). On the connected k-coverage problem in heterogeneous sensor nets: The curse of randomness and heterogeneity, *Distributed Computing Systems, 2009. ICDCS '09. 29th IEEE International Conference on*, pp. 265 –272.

Berman, P., Calinescu, G., Shah, C. & Zelikovsky, A. (2004). Power efficient monitoring management in sensor networks, *Proc. of Wireless Communications and Networking Conference*, Vol. 4, IEEE, pp. 2329–2334.

Cardei, I. & Cardei, M. (2008). Energy efficient connected coverage in wireless sensor networks, *Int. J. Sen. Netw.* **3**(3): 201–210.

Cardei, M. & Du, D.-Z. (2005). Improving wireless sensor network lifetime through power aware organization, *ACM Wireless Networks* **11**(3): 333–340.

Cardei, M., MacCallum, D., Cheng, M. X., Min, M., Jia, X., Li, D. & Du, D.-Z. (2002). Wireless sensor networks with energy efficient organization, *Journal of Interconnection Networks* **3**(3-4): 213–229.

Cardei, M., Thai, M., Li, Y. & Wu, W. (2005). Energy-efficient target coverage in wireless sensor, *Proc. of INFOCOM 05*, Vol. 3, IEEE, pp. 1976–1984.

Cardei, M., Wu, J., Lu, M. & Pervaiz, M. (2005). Maximum network lifetime in wireless sensor networks with adjustable sensing ranges, *Wireless And Mobile Computing, Networking And Communications, 2005. (WiMob'2005), IEEE International Conference on*, Vol. 3, pp. 438 – 445 Vol. 3.

Cheng, M., Ruan, L. & Wu, W. (2007). Coverage breach problems in bandwidth-constrained sensor networks, *ACM Trans. Sen. Netw.* **3**(2): 12.

Cheng, M. X., Ruan, L. & Wu, W. (2005). Achieving minimum coverage breach under bandwidth constraints in wireless sensor networks, *Proc. of INFOCOM 05*, pp. 2638–2645.

Dai, F. & Wu, J. (2003). Distributed dominant pruning in ad hoc networks, *Communications, 2003. ICC '03. IEEE International Conference on*, Vol. 1, pp. 353 – 357 vol.1.

Dasgupta, K., Kukreja, M. & Kalpakis, K. (2003). Topology-aware placement and role assignment for energy-efficient information gathering in sensor networks, *ISCC '03: Pro-*

ceedings of the Eighth IEEE International Symposium on Computers and Communications, IEEE Computer Society, Washington, DC, USA, p. 341.

Dhawan, A., Vu, C., Zelikovsky, A., Li, Y. & Prasad, S. (2006). Maximum lifetime of sensor networks with adjustable sensing range, *Software Engineering, Artificial Intelligence, Networking, and Parallel/Distributed Computing, 2006. SNPD 2006. Seventh ACIS International Conference on*, pp. 285 –289.

Garey, M. R. & Johnson, D. S. (1979). *Computers and Intractability. A Guide to the Theory of NP-Completeness*, Freeman, New York.

Garg, N. & Könemann, J. (1998). Faster and simpler algorithms for multicommodity flow and other fractional packing problems, *Proc. of 39th Annual IEEE Symposium on Foundations of Computer Science*, IEEE, pp. 300–309.

Gu, Y., Ji, Y., Li, J. & Zhao, B. (2009). Qos-aware target coverage in wireless sensor networks, *Wirel. Commun. Mob. Comput.* **9**(12): 1645–1659.

Hefeeda, M. & Bagheri, M. (2006). Efficient k-coverage algorithms for wireless sensor networks, *Technical report*, School of Computing Science, Simon Fraser University.

Heinzelman, W. R., Chandrakasan, A. & Balakrishnan, H. (2000). Energy-efficient communication protocol for wireless microsensor networks, *HICSS '00: Proceedings of the 33rd Hawaii International Conference on System Sciences-Volume 8*, IEEE Computer Society, Washington, DC, USA, p. 8020.

Jaggi, N. & Abouzeid, A. A. (2006). Energy-efficient connected coverage in wireless sensor networks, *4th Asian International Mobile Computing Conference (AMOC)*, pp. 77–86.

Kar, K. & Banerjee, S. (2003). Node placement for connected coverage in sensor networks, *Proc. of WiOpt 2003: Modeling Optimization Mobile, ad hoc Wireless Networks*.

Kim, H., Kim, E. J. & Yum, K. H. (2007). Roal: A randomly ordered activation and layering protocol for ensuring k-coverage in wireless sensor networks, *Wireless and Mobile Communications, 2007. ICWMC '07. Third International Conference on*, pp. 4 –4.

Kim, Y.-H., Lee, H.-J., Han, Y.-H. & Jeong, Y.-S. (2009). Branch and bound algorithm for extending the lifetime of wireless sensor networks, *VTC Fall*.

Li, X., Wan, P. & Frieder, O. (2003). Coverage in wireless ad hoc sensor networks, *IEEE Transactions on Computers* **52**.

Liu, H., Wan, P., Yi, C.-W., Jia, X., Makki, S. & Pissinou, N. (2005). Maximal lifetime scheduling in sensor surveillance networks, *INFOCOM 2005. 24th Annual Joint Conference of the IEEE Computer and Communications Societies. Proceedings IEEE*, Vol. 4, pp. 2482 – 2491 vol. 4.

Liu, L., Ma, H. & Zhang, X. (2008). On directional k-coverage analysis of randomly deployed camera sensor networks, *Communications, 2008. ICC '08. IEEE International Conference on*, pp. 2707 –2711.

Liu, Y. & Liang, W. (2005). Approximate coverage in wireless sensor networks, *Local Computer Networks, Annual IEEE Conference on* **0**: 68–75.

Lu, M., Wu, J., Cardei, M. & Li, M. (2009). Energy-efficient connected coverage of discrete targets in wireless sensor networks, *Int. J. Ad Hoc Ubiquitous Comput.* **4**(3/4): 137–147.

Megerian, S., Koushanfar, F., Potkonjak, M. & Srivastava, M. B. (2005). Worst and best-case coverage in sensor networks, *IEEE Transactions on Mobile Computing* **4**: 84–92.

Migatron (n.d.).
URL: *http://www.migatron.com/products/rps-400-6/rps-400-6.htm*

Shen, W. & Wu, Q. (2010). Minimum sensor relocation for k-coverage in wireless sensor networks, *Communications and Mobile Computing (CMC), 2010 International Conference on*, Vol. 3, pp. 274 –278.

Simon, G., Molnar, M., Gonczy, L. & Cousin, B. (2007). Dependable k-coverage algorithms for sensor networks, *Instrumentation and Measurement Technology Conference Proceedings, 2007. IMTC 2007. IEEE*, pp. 1 –6.

Slijepcevic, S. & Potkonjak, M. (2001). Power efficient organization of wireless sensor networks, *Proc. of International Conference on Communications (ICC'01)*, IEEE, pp. 472–476.

Thai, M. T., Wang, F., Du, H. & Jia, X. (2008). Coverage problems in wireless sensor networks: Designs and analysis, *International Journal of Sensor Networks, Special issue on coverage problems* **3**: 191–200.

Tian, D. & Georganas, N. D. (2002). A coverage-preserving node scheduling scheme for large wireless sensor networks, *Proc. of 1st ACM International Workshop on Wireless Sensor Networks and Applications*, ACM Press, pp. 32–41.

Vu, C., Gao, S., Deshmukh, W. & Li, Y. (2006). Distributed energy-efficient scheduling approach for k-coverage in wireless sensor networks, *Military Communications Conference, 2006. MILCOM 2006. IEEE*, pp. 1 –7.

Wang, C., Thai, M., Li, Y., Wang, F. & Wu, W. (2007). Minimum coverage breach and maximum network lifetime in wireless sensor networks, *Proc. of GLOBECOM 07*, pp. 1118–1123.

Wang, L. & Kulkarni, S. S. (2008). Sacrificing a little coverage can substantially increase network lifetime, *Ad Hoc Netw.* **6**(8): 1281–1300.

Wang, Q., Xu, K., Takahara, G. & Hassanein, H. (2006). Deployment for information oriented sensing coverage in wireless sensor networks, *Proceedings of the 49th IEEE Global Telecommunication Conference*.

Yan, T., He, T. & Stankovic, J. A. (2003). Differentiated surveillance for sensor networks, *SenSys '03: Proceedings of the 1st international conference on Embedded networked sensor systems*, ACM, New York, NY, USA, pp. 51–62.

Ye, F., Zhong, G., Lu, S. & Zhang, L. (2002). Energy efficient robust sensing coverage in large sensor networks, *Technical report*, UCLA.

Zhang, H. & Hou, J. (2005). Maintaining sensing coverage and connectivity in large sensor networks, *Ad Hoc & Sensor Wireless Networks* **1**(1-2).

Zhao, Q. & Gurusamy, M. (2008a). Connected k-target coverage problem in wireless sensor networks with different observation scenarios, *Comput. Netw.* **52**(11): 2205–2220.

Zhao, Q. & Gurusamy, M. (2008b). Lifetime maximization for connected target coverage in wireless sensor networks, *IEEE/ACM Trans. Netw.* **16**(6): 1378–1391.

Zhong, F. Y., Lu, G. S. & Zhang, L. (2002). Peas: a robust energy conserving protocol for long-lived sensor networks, *Proc. of 10th IEEE International Conference on Network Protocols*, pp. 200–201.

Zhou, Z., Das, S. & Gupta, H. (2004). Connected k-coverage problem in sensor networks, *In Proceedings of the International Conference on Computer Communications and Networks (IC3N*, pp. 373–378.

Zorbas, D. & Douligeris, C. (2009). Satisfying coverage and connectivity in bandwidth constrained sensor networks, *Communications and Information Technology, 2009. ISCIT 2009. 9th International Symposium on*, pp. 390 –395.

Zorbas, D., Glynos, D. & Douligeris, C. (2009). Connected partial target coverage and network lifetime in wireless sensor networks, *Wireless Days (WD), 2009 2nd IFIP*, pp. 1 –5.

Zorbas, D., Glynos, D., Kotzanikolaou, P. & Douligeris, C. (2007). B{GOP}: An adaptive coverage algorithm for wireless sensor networks, *Proc. of the 13th European Wireless Conference*, Paris.

Zorbas, D., Glynos, D., Kotzanikolaou, P. & Douligeris, C. (2010). Solving coverage problems in wireless sensor networks using cover sets, *Ad Hoc Networks* **8**: 400–415.

Node Deployment and Mobile Sinks for Wireless Sensor Networks Lifetime Improvement

George Zaki[1], Nora Ali[1], Ramez Daoud[2], Hany ElSayed[1], Sami Botros[1], Hassanein Amer[3] and Magdi El-Soudani[1]
[1] *Cairo University,* [2] *KAMA Trading,* [3] *American University in Cairo*
Egypt

1. Introduction

In the last two decades, and owing to advances in MEMS technologies, wireless communications and low-power electronics, the development of low-cost micro sensor nodes was possible. This enabled the deployment of Wireless Sensor Networks (WSN) comprising large numbers of nodes to monitor various physical phenomena in real-time. This can be of prime importance in several industrial, environmental, health, and military applications (Akyildiz et al., 2002; Tavares et al., 2008).

A WSN may have up to hundreds or even thousands of sensor nodes densely deployed either inside or close to a monitored area. Nodes process data prior to transmission, to ensure acquisition of accurate and detailed information. Processed information is then passed on to a sink node, which transmits necessary data to some base station. Nodes may also be divided into clusters, with nodes in each cluster sending data to a particular sink node. Sensor nodes typically operate in an unattended environment, and are equipped with small, often irreplaceable batteries with limited power capacity. Thus a major consideration in WSN research is to ensure reliable transmission of data while prolonging network lifetime by making maximum use of the available energy in the nodes (Heinzelman et al., 2002).

In this chapter, recent work by the authors in the area of WSN is presented with particular emphasis on maximizing the lifetime of the network. In Section 2, algorithms are described that build upon two well known WSN routing techniques, namely LEACH (Heinzelman et al., 2000) and LEACH-C (Heinzelman et al., 2002) to further optimize network lifetime through carefully planned selection of the sink nodes. Simulation results that illustrate the resulting improvement in network lifetime are presented. The position of sensor nodes need not be predetermined, which allows random deployment in inaccessible terrains. However, in some applications, the deployment of nodes at pre-specified positions is feasible. Taking advantage of this feature is thus considered to achieve further enhancement in network lifetime by considering the effect of various geometrical distributions of nodes and relative sink locations.

Further reductions of the transmission energy requirements can be attained by making use of uncontrolled mobile sinks in addition to the distant fixed sinks. It is not possible to depend solely on mobile sinks as their presence is not guaranteed in any time interval, so a hybrid approach is necessary. A hybrid method for message relaying is presentd in section 3, satisfying efficiency and load balancing requirements. A node either uses a single hop transmission if a nearby mobile sink is present, or a multi-hop transmission to a far fixed node depending on the predicted sink mobility pattern. Analysis is used to adjust system parameters such that all sensor nodes dissipate the same amount of energy. This prevents the problem of losing connectivity as a result of rapid power drainage of the nearest node to the fixed sink. Numerical results indicate the improvements in lifetime compared to other traditional methods.

2. Lifetime optimization

This section focuses on routing protocols that prolong Wireless Sensor Network (WSN) lifetime (Akkaya & Younis, 2005; Mahfoudh & Minet, 2008; Narasimha & Gopinath, 2006). Two of the most famous hierarchical protocols are LEACH and LEACH-C (Heinzelman et al., 2000; Heinzelman et al., 2002). In both protocols, sensor nodes are clustered. A cluster head receives data from all other nodes in the cluster, aggregates it and sends it to a fixed sink. The work presented next describes two algorithms that produce longer system lifetimes when compared to LEACH or LEACH-C. Both algorithms assume that sensors are randomly-distributed in the area under study. Geometric distributions of sensors are studied next as well as different fixed sink locations. More details about these issues can be found in (Botros et al., 2009; Nouh et al., 2010).

2.1 System description

The WSN under study is composed of homogeneous sensor nodes that are deployed in the area of interest. Sensors are randomly distributed in the deployment area. This is the most common case. It may be required that hundreds or thousands of sensors be deployed in a remote, unreachable or dangerous environment. In such cases, sensors may be thrown from an aircraft flying over that dangerous area to extract information in ways that would not have been possible otherwise. The sink node is fixed and located far away from the sensors field.

System lifetime is usually defined as one of the following (Mahfoudh & Minet, 2008): 1) The time to the first node failure due to battery outage. 2) The time to the first network partitioning. 3) The time to the unavailability of application functionality. 4) The time to the failure of certain percentage of the nodes. In this work, the first definition is considered since it does not depend on the type of application and it is suitable for any network architecture, either divided into groups, clusters or not. This definition is also preferred since it guarantees that during the whole lifetime of the network, it is fully covered with active nodes which collect data from all positions in the network. This may be a primary requirement in some application classes, such as security monitoring and node tracking scenarios. The proposed algorithms deal with all sensors as one network. The cluster head approach is used to manage the communications within the network. One of the sensors in the whole area is selected as a master. It is called a Network Master node to differentiate it from the cluster head used in LEACH or other algorithms. The Network Master (NM) for the network receives data from the other sensors in the network. The NM then performs data aggregation and compression to remove redundancy and send the useful information to the sink or base station. This is similar to the idea of LEACH and LEACH-C (Heinzelman et al., 2000; Heinzelman et al., 2002) and some others, but here it is applied to the whole network.

2.2 System assumptions and network parameters

In the model under study, several basic assumptions are considered. These are:

- All sensors in the network are homogeneous and energy constrained.
- All sensors are sensing the environment at a fixed rate, and thus always have data to send.
- All sensors can transmit with enough power to reach the fixed sink if needed.
- Sensors can use power control to vary the amount of transmit power.
- Each sensor has the computational power to perform signal processing functions.
- Sensors have a method to be aware of their position after deployment.

The parameters used are as shown in Table 1. Some values given in the table are based on the electronics of most commonly used sensor nodes while the others are used for the sake of comparison with other algorithms.

Parameter	Symbol	Value
Network Size	M X M	100 × 100 m
Number of Sensors	N	100 Sensors
Transmitter / Receiver Electronics	E_{elec}	50 nJ/bit
Transmitter Amplifier for short distance Transmitter Amplifier for long distance	$E_{amp\text{-}short}$ $E_{amp\text{-}long}$	10 pJ/bit/m^2 0.0013 pJ/bit/ m^4
Pass Loss Factor for short distance Pass Loss Factor for long distance		2 4
Aggregation Energy	E_{agg}	5 nJ/bit/Signal
Data Packet Size		500 B
Overhead Packet Size		125 B

Table 1. Network parameters

Two algorithms are proposed next and it is shown that they produce longer lifetimes than the algorithms presented in (Heinzelman et al., 2000; Heinzelman et al., 2002).

2.3 Algorithm I

Assume that all the sensors are aware of their positions, as assumed in (Heinzelman et al., 2002), and that the sink knows these positions. The algorithm consists of rounds. Each starts with the NM selection by the sink. This node remains as NM for a fixed number of cycles "C", after which a new round starts. Before the selection of the NM, the energies of sensors are compared with two thresholds "En_{Th}" and "En_{ThNM}". The first threshold, "En_{Th}", is the energy required by each sensor to transmit its data to the farthest possible NM node for one complete round. This is calculated for each sensor assuming the worst case that the NM is the farthest node from this sensor. A sensor that has energy below this threshold is not active anymore and cannot perform any useful function. The second threshold, "En_{ThNM}", is the energy required by the sensor to act as an NM, gathering data from sensors, aggregating it and sending the resulting packet to the far away sink. Again, this is energy needed for one

complete round. It is calculated for each sensor according to its distance from the sink. A sensor that has energy below this threshold, cannot act as an NM for the network. Sensors are classified according to these thresholds before NM selection into one of three categories: 1) Active nodes that can act as NMs. 2) Active nodes but cannot act as NMs and 3) Inactive nodes or dead nodes.
Once a node is classified as a dead node, the network is considered dead, according to the definition of lifetime used in this study. The sink has knowledge about the whole network and is responsible for selecting the NM and informs all other sensors about the current NM. It selects a sensor as an NM for the current round according to the following criteria. 1) The node belongs to the first category. 2) The node has energy greater than the average energy of all active nodes and 3) The sum of its distances to the active nodes is least. In this algorithm, it is assumed that a node can be selected as an NM for many rounds throughout network lifetime. A simulation model is built using MATLAB (MatLab) with the same network parameters used in (Heinzelman et al., 2002) and described above. The system is run for different values of the number of cycles "C" per round, and the corresponding network lifetime is as shown in Fig. 1. The figure shows that there is an optimum number of cycles for which each sensor remains acting as NM, before another round starts over and a new NM is selected. For the parameters considered, the longest lifetime is achieved for "C=3", resulting in a lifetime equivalent to "3702" cycles.

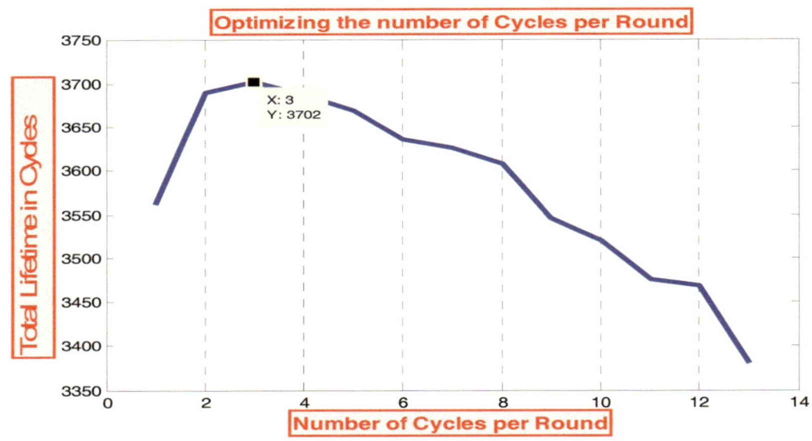

Fig. 1. Network lifetime vs number of cycles per round

2.4 Algorithm II

The previous algorithm selected a fixed optimum number of cycles "C" per round in order to achieve a longer lifetime. It is observed that with this relatively small number of cycles, a sensor is chosen as an NM for many rounds. It is observed also that not all sensors act as NMs for the same number of rounds. So, if these could be gathered together such that each sensor is selected as an NM only once, but without exhausting sensors which require more energy to act as an NM, a longer lifetime for the network will be achieved. Another observation in previous techniques is that after the death of the first node, there is still some residual energy for some sensors. This residual energy is not used efficiently. One reason is that it is distributed to all the sensors, and hence, the share of each sensor is not large enough to work as NM. Another reason is that the full coverage of the network, which may be a primary concern in many applications, is lost. Both observations lead to an algorithm which requires that each sensor be selected as an NM only once, and acts as an NM for a certain number of cycles "C_i", which need not be the same for all sensors. The algorithm also requires the most usage of the available energies for each sensor.

The algorithm is simply run once at the sink based on its knowledge of the locations of the different sensors. The sink can calculate the energy "$E_{txi \text{ to } NMj}$" required by each sensor "i" to transmit its data to any of the other nodes "j" acting as an NM, as well as the energy "E_{NMi}" needed by the node "i" to act as an NM itself. Assuming that each sensor acts as an NM for a certain number of cycles "C_i", before and after which it acts as an ordinary node, the energy consumed by any sensor "i" through the network lifetime can be calculated as:

$$E_{sensor\ i} = C_i \times E_{NMi} + \sum_{\substack{j=1 \\ j \neq i}}^{j=N} C_j \times E_{txi \text{ to } NMj} \tag{1}$$

for $i = 1,2,\cdots,N$

Since each sensor will act as a NM only once for "C_i" cycles, then the total lifetime, in number of cycles, is the summation of the different "C_i"s.

$$T = \sum_i C_i \tag{2}$$

If each sensor node "i" has an initial energy "$E_{o\,i}$", it must be that the energy consumed by any sensor is less than or equal its initial energy. That is:

$$E_{sensor\ i} \le E_{0i} \tag{3}$$

In order to make the best use of the available energies for the sensor, the following set of "N" equations in "N" unknowns, { C_1 , C_2 , C_3 , ….., C_N }, is solved.

$$E_{sensor\ i} = E_{0i} \tag{4}$$

for $i = 1,2,\cdots,N$

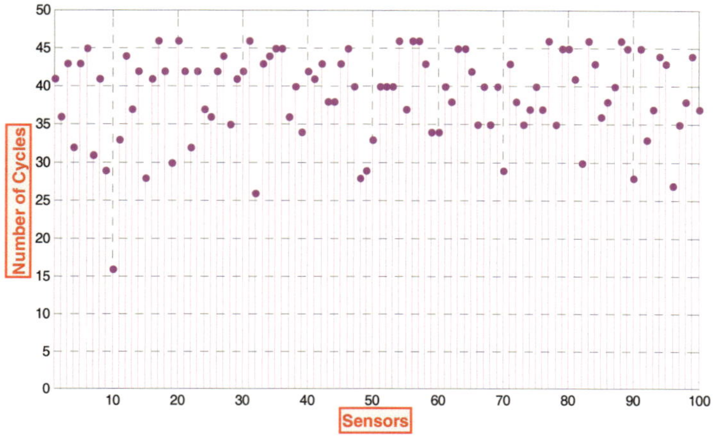

Fig. 2. Number of cycles "C_i" assigned to each sensor to act as a Network Master

The solution set S = {C_i} indicates that the network will have maximum lifetime. Any other set, S′ = {C_i'}, will not be a solution for the set of equations. It should be noted that the solution of such equations does not guarantee integer values for the "C_i"s; therefore, the fractional part of the solution set must be truncated. The simulation environment used before is used for the new scheme. The solution of the set of equations in (4) resulted in the set of "C_i"s shown in Fig. 2 after truncation. It can be observed that the different values of "C_i" range between 16 and 46 cycles per round. The summation of these "C_i"s causes the expected lifetime of the network to be almost 3900 cycles which is higher than the lifetime obtained from the first algorithm.

2.5 Geometric distributions

Random distributions, which were used in (Botros et al., 2009), are more suitable for certain applications where the network locations are inaccessible (Tavares et al., 2008), such as military applications. However, as mentioned before, in some applications (such as urban applications), the deployment of nodes at pre-specified positions is feasible (Onur et al., 2007). Hence, this subsection focuses on geometric distributions instead of random distribution and their effect on maximizing the network's lifetime.

2.5.1 Star topology

The Star topology is one of the most common geometric distributions used in networks (Cheng & Liu, 2004; Bose & Helal, 2008). Therefore star topologies are chosen for testing as geometric distributions. By using the same previous parameters (Botros et al., 2009), it is found that the star with 3 branches and 33 sensors per branch (3×33 star) produces 5% increase in network lifetime. Furthermore, several stars with different numbers of branches are generated for simulation. The main characteristics for the used star distributions in this study are as follows:

- Sensors are distributed in circles from the centre to the borders of the area and each circle has an equal number of sensors.
- Equal angles between branches and equal distances between sensors in the same branch.

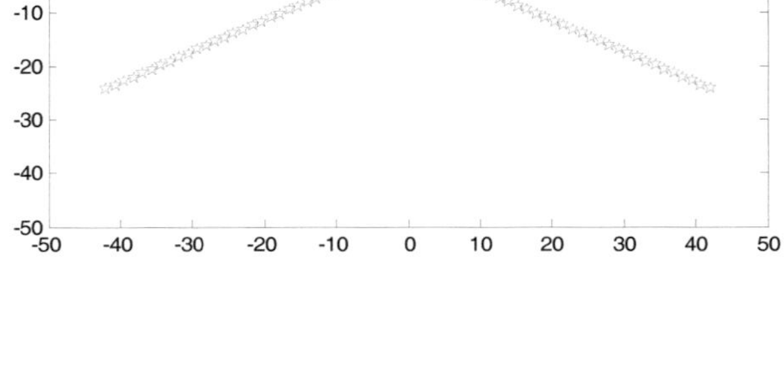

Fig. 3. 3x33 Star

The number of branches that were tested ranges between 3 and 20 with a suitable number of sensors in each circle to constitute the used number of sensors which is N=100 sensors used by (Botros et al., 2009; Minet & Mahfoudh, 2009). The 3×33 star (shown in Fig. 3) has 3 branches, 33 sensors per branch and the 100th sensor is located in the center of the star. The network parameters used in this study are as follows:

- Number of Sensors (N): 100 Sensors
- Initial Energy: 2 J
- Transmitter/ Receiver Electronics: 50 nJ/bit
- Transmitter Amplifier : 100 pJ/bit/m^2
- Path Loss factor: 2
- Aggregation Energy: 5 nJ/bit/Signal
- Data packet size (K): 2000 bits
- Sink location: (0; –125)

2.5.2 Proposed algorithm

A simulation model is built using MATLAB considering the above network parameters. The lifetime in case of geometric distributions is computed by using the algorithm described in section 2.4.

2.5.3 Simulations and results

By simulating the proposed algorithm with different star distributions, it was found that the 3x33 star achieves the maximum lifetime compared to the other star distributions as shown in Table 2. It was found that the 3x33 star extends the lifetime of the network by 35.6% compared to the random distribution used in (Botros et al., 2009). The numbers of sensors that can act as NMs in 3x33 star were 70 out of 100 sensors and the number of cycles allocated for each NM are as shown in Fig. 4. All the simulations results are specific to the orientation of the used topology.

Star Distribution	Lifetime (Cycles)
3x33	4612
4x25	4510
5x20	4278
6x16	4346
7x14	4437
8x12	4399
9x11	4510
10x10	4466
12x8	4314
14x7	4388
20x5	4412

Table 2. Lifetimes of different star distributions

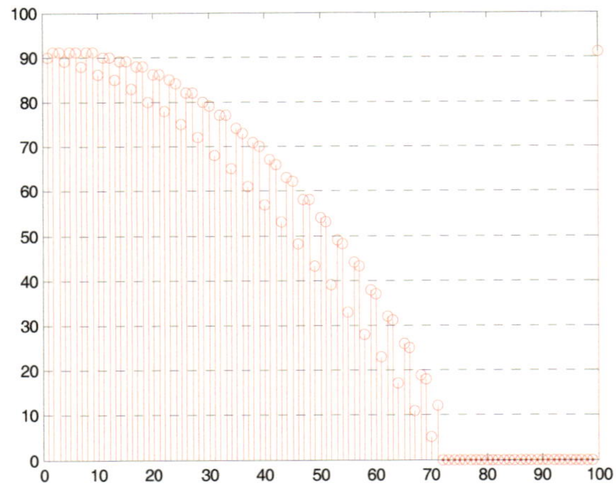

Fig. 4. Number of cycles for each NM in a 3x33star

2.6 Sink locations

The different star distributions used in the previous section were tested to achieve the best distribution with respect to the lifetime using the sink location at (0; –125) which was used by (Botros et al., 2009). The results showed that 3×33 star produces the highest lifetime. This result was taken a step further by applying other sink locations in order to explore the effect of the other sink locations on network lifetime. The sink locations used in this study are (0; 125), (125; 0), (–125; 0), (–125; –125), (125; –125), (125; 125), (–125; 125) and (0; 0).

Simulating the different sink locations on the best star (3×33 star) results in better and worse lifetime with respect to the (0; –125) sink location. But the objective is to increase network lifetime, so sink locations that achieve higher lifetime are of great concern. The (0; 0) sink location increased the network's lifetime of the 3×33 star from 4612 cycles, in the case of the (0; –125), to 5205 cycles, which is an improvement of approximately 13%.

In order to find the reason why changing the sink location to (0; 0) increases the lifetime, some calculations were computed to measure the total distance traveled by data. As mentioned before, each sensor acted as a NM for a certain number of cycles for only one round. This NM collects data from all other sensors, aggregates it then sends the aggregated data to the sink. Therefore, two communication distances must be measured for each sensor as follows:

- $d_{sensor-NM}$;

which is the communication distance between every sensor and the selected NM.

- $d_{NM-Sink}$

which is the communication distance between the selected NM and the sink. By adding all the distances between the sensors and every NM and the distance between every NM and the sink, a new metric is derived as follows:

$$d_{data} = \sum_{j=1}^{M} \sum_{\substack{i=1 \\ i \neq j}}^{N} d_{sensor_i - NM_j} + \sum_{j=1}^{M} d_{NM_j - Sink} \quad (5)$$

where N is the number of sensors and M is the number of NMs. Comparing the distance travelled by data for each sink location, it was found that at sink (0;0), d_{data} was the lowest.

2.7 Uniform distributions

Using the star topologies was successful in prolonging the lifetime of the network. But the star distributions are not suitable for all WSN applications. Some WSN applications such as chemical, environmental and nuclear sensing systems require uniformly distributed sensors (Bestavros et al., 2004). Therefore, some distributions with uniform densities were investigated in this study. The distributions were tested at the different sink locations and it was found that the maximum lifetime was obtained at the (0; 0) sink location. First, the hexagonal distribution was tested due to its wide and comprehensive coverage (Prabh et al., 2009; Gui & He, 2009). The second distribution is the Homogeneous Density Distribution in which a sensor was placed every meter square over the entire area (see Fig. 5). Finally, a circular distribution is tested with uniform density in which the number of sensors per circle increased as they move towards the border of the area. The homogeneous density distribution resulted the highest lifetime compared to the other uniform distributions. It produced 3301 cycle, while the hexagonal and the circular distributions produced only 3293 and 2876 cycles respectively.

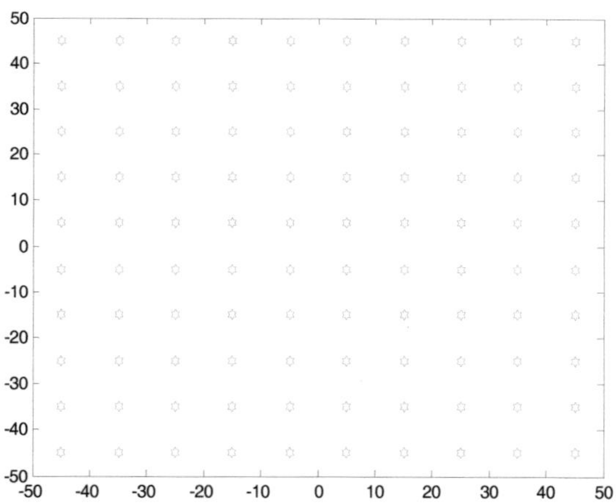

Fig. 5. Homogeneous Density Distribution

3. Relaying data collection

The fact that a sensor drains much of its power in trying to send its data to a fixed sink makes it necessary to use a mobile sink in addition to the fixed one. This is called a hybrid system. This section considers the problem of maximizing system life time (i.e., reducing the energy consumption) by properly choosing the destination; either the fixed sink or the mobile one (which is not controlled). More details about this work can be found in

(Zaki et al., 2008; Zaki et al. 2009). Using a hybrid model for message relaying, an energy balancing scheme is proposed in a linear low mobility wireless sensor network. The system uses either a single hop transmission to a nearby mobile sink or a multi-hop transmission to a far-away fixed sink depending on the predicted sink mobility pattern. Taking a mathematical approach, the system parameters are adjusted so that all the sensor nodes dissipate the same amount of energy. Simulation results showed that the proposed system outperforms classical methods of message gathering in terms of system lifetime. On the single node level, the average total energy consumed by the hybrid system is equalized over all sensors and the problem of losing connectivity due to the fast power drainage of the closest node to the fixed sink, is resolved.

3.1 System description

Fixed wireless sensor networks are described in the form of two tiers: the sensor and the fixed sink (observer). Another approach is the introduction of a third tier which is the mobile sink. Sensors send their data to the mobile sink as the second relay point instead of sending to the fixed sink. There are many benefits of using this approach where the most important is the reduction of power consumption during the transmission phase. The sensor is not required anymore to send its messages to faraway points as the mobile sink approaches the sensor to get the data. This system has many other advantages including robustness against the failure of nodes, higher network connectivity and reduction of the control messages overhead required to set up paths to the observer (Al-Karaki & Kamal, 2004).

The Data Mules (Shah et al., 2003), approach aims at addressing the operation of using existing mobile sinks, termed MULEs (Mobile Ubiquitous LAN Extensions) to collect sensed data in the environment. In a vehicular traffic monitoring application, the vehicles can serve as mobile agents, whereas in a wildlife tracking application, the animals can be used as mobile agents. The MULEs are fitted with transceivers that are capable of short-range wireless communication. They can exchange data with sensors and access points when they move into their vicinity. The main disadvantage of the basic implementation of the Data Mules scheme is its high latency. Each sensor node needs to wait for a MULE to come within its transmission radius before it can transfer its readings. Another disadvantage is that the system assumes the existence of mobile agents in the target environment, which may not always be true. The sensor nodes need to keep their radio receivers on continuously to be able to communicate with MULEs. In this section, a hybrid message transmission system that takes advantages of the data MULEs concept as well as the basic protocols of data routing, is developed. The system solves the inherit disadvantages of the basic MULEs architecture and increases network lifetime by reducing the single node power consumption and by balancing the overall system energy.

A typical three layers architecture for environmental monitoring system in urban areas consists of (Jain et al., 2006):

- The lowest layer consists of different types of sensor nodes.
- The second layer consists of the mobile agent that can be a moving car, a personal digital assistant or any moving device.
- The higher layer consists of the fixed sink. It represents the collection point of the sensed data before its transmission through a WAN to a monitoring point.

Considering this architecture for a city, a large number of fixed sensor nodes are deployed on both sides of the street to monitor different phenomena. Sensors work on their limited energy reservoir. Fixed sinks are the collection points that receive the sensed data directly from the sensor modules or from mobile sinks. They have higher capability than the sensor modules in terms of computational power and connectivity. The number of fixed sinks is usually smaller than the number of sensors; that is why it is not a costly operation to connect them to permanent power supplies or large energy scavenger and different communications facilities. When the sensed data is received by the fixed sinks, it can be forwarded to central databases through the wired or wireless infrastructure network for further processing. The mobile sinks periodically broadcast a beacon to notify nearby sensors of their existence. Upon reception of the beacon message, the sensor module can transmit its data to the nearby mobile node as the next overlay, thus saving its energy. The mobile agent can then send the sensed data to the fixed sink or to the remote database using other communication means.

3.2 Underlying system models

The models used in the system under study are explained next.

3.2.1 Routing, MAC and mobility models

The fixed part of the network operates the routing protocol suggested in (Younis et al., 2002). The basic assumptions are:

1. Appling a MAC protocol that allows the sensor to listen to the channel in a specified time slot as TDMA based protocol that minimizes the idle listening power when routing to fixed points.
2. The gateway which can be seen as the fixed sink has high computational power. All system algorithms are run on the gateway and the system parameter values are then broadcasted to the sensor nodes.
3. The sensor can determine transmission distance to its next hop and adjust its power amplifier correspondingly.
4. The radio transceiver can be turned on and off.

In mobile sink WSN, various basic approaches for mobility are involved: random, controlled and predictable. Random objects such as humans and animals can be used to relay the sensed data when they are in the coverage range. As the main issue in the described system is the moving cars in a street, therefore only one-dimensional uncontrolled mobility is considered. Different techniques are used to model vehicular traffic flows (Hoogendorn & Bovy, 2001). One well known example of mesoscopic model is the headway distribution model where it expresses the vehicular time headway as a probability distribution (Al-Ghamdi, 2001). Typical distributions are negative exponential and gamma distributions. The inter-arrival time T between two successive cars is modeled as a negative exponential distribution with an average β.

$$F(T,\beta)=\frac{1}{\beta}e^{-\frac{T}{\beta}} \tag{6}$$

During a 24-hour period, the traffic flow rate varies between heavy traffic during rush hours and low traffic at the end of day. Therefore, the one day cycle can be divided into several time intervals in which the value of β is considered constant.

3.2.2 Energy model

There are three basic operations in which sensors consume their energy (Shebli et al., 2007). First the sensor node has to convert the sensed phenomena to a digital signal. This is called aquisition. Second, the digital signal may be processed before transmission. Finally the sensor has to wirelessly communicate the data it aquire or receives. In this work, the focus is on the communication operation which is the basic source of power consumption.
The wireless node transceiver may be in one of four states:
1. sending a message,
2. receiving a message,
3. idle listening for a message,
4. in the low power sleep mode.

The linear transceiver model is used where:
1. The energy consumed to send a frame of size m over a distance of d meters consists of two main parts: the first one represents the energy dissipated in the transmitter and the second represents the energy dissipated in the power amplifier.

$$E_{TX}(m,d) = m\left(e_{elec} + e_{amp}d^k\right) \tag{7}$$

   where m is the message length in bits, e_{elec} is the amount of energy consumed by the transmitter circuits to modulate one bit and $e_{anp}d^K$ is the amount of energy dissipated in the power amplifier in order to reach acceptable signal to noise ratio at the receiver that is located d meters away. k is an integer constant that varies between two to four depending on the surrounding medium. e_{anp} takes into account the antenna gain at the transmitter and the receiver:
2. To receive an m bits long message, the receiver then consumes:

$$E_{RX}(m) = m \cdot e_{rx} \tag{8}$$

   where e_{rx} represents the reception energy per bit and m the message length. In order to send a message to a nearby mobile sink, the sensor node has to ensure the presence of the sink. The mobile node continuously sends out a detection message (beacon) to detect a nearby sensor. This requires a sensor to listen for discovery messages.
3. The idle listening energy is dissipated in two cases: when the sensor node communicates to fixed nodes, the suggested MAC protocols require that the nodes wake up in the same time to exchange messages. The second source of idle listening energy consumption is when communicating with a mobile sink. The sensor node stays in the idle listening state until it detects a mobile agent beacon. The low power idle listening protocol proposed in (Polastre et al., 2004) is used where the receiver samples the channel with a duty cycle. Each time the node wakes up, it turns on the radio and checks for activity. If activity is detected, the node powers up and stays awake for the

time required to receive the incoming packet. If no packet is received (a false positive), the node is forced back to sleep. In this model, the sensor has to be in the low power idle listening state for a given amount of time denoted by T. The power dissipated during this period is denoted by P_{idle}. Thus the idle listening energy is given by:

$$E_{idle} = P_{idle} \cdot T \tag{9}$$

4. Finally the low power sleeping state is when the sensor shuts down all its circuitry and becomes unable to neither send nor receive any message. The microcontroller is responsible for waking up the transceiver when the sensor node wants to communicate. This energy is neglected when comparing between any two systems as it does not differ for both systems.
In this hybrid model, the mobile sink only notifies its presence to one hop away nodes only (Zaki et al., 2008). The sensor node decides either to route its message to the next fixed node or to the mobile sink depending on the parameter T_o. After the sensor collects the required data, it goes to the idle listening state for a maximum waiting period of T_o. During T_o, if the sensor receives a beacon, the next relay point will be the mobile sink; otherwise the sensor transmits to the fixed sink after spending T_o seconds in the idle listening state. After sending its message, the sensor node goes to the low power sleeping state. A cycle is defined as the state of the sensor from when it is required to send a message to the next relay point until it sends the message. The sensor energy states versus time graphs are shown in Figs. 6 and 7.

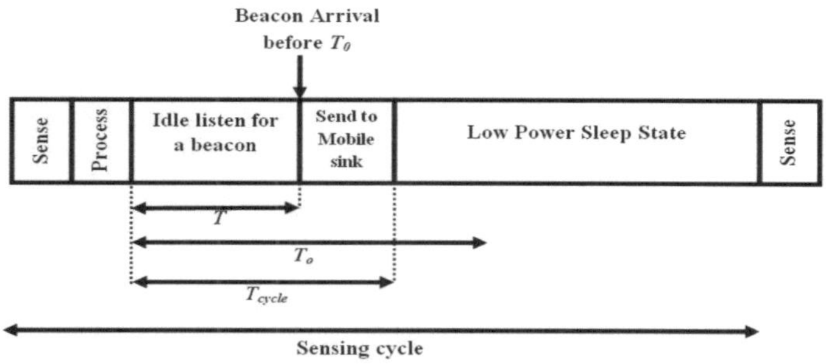

Fig. 6. Sensor states vs time in case of a mobile sink

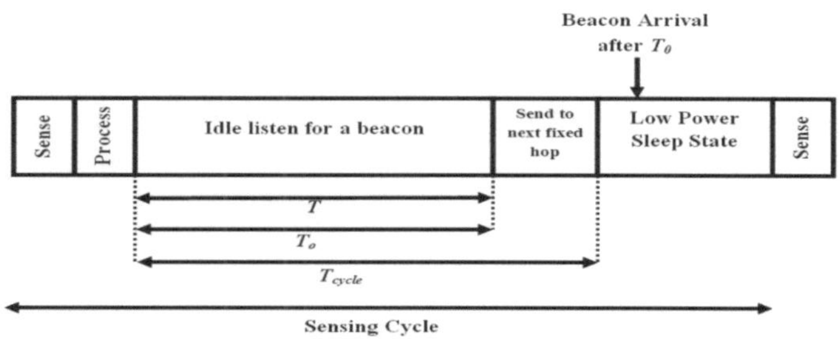

Fig. 7. Sensor states vs time in case of a fixed sink (hop)

Assuming that the beacon message arrives to the sensor after T seconds from the beginning of the listening state, then the energy consumed by the sensor during a cycle W_{cylce} equals:

$$W_{cycle} = \begin{cases} P_{idle} \cdot T + E_s & \text{if } T < T_o \\ P_{idle} \cdot T_o + E_l & \text{if } T > T_o \end{cases} \quad (10)$$

where:

$$E_s = m\left(e_{elec} + e_{amp} D_s^K\right) \quad (11)$$

$$\text{and } E_l = m\left(e_{elec} + e_{amp} D_l^K\right) \quad (12)$$

D_s and D_l are the distances between the sensor and the mobile sink and the fixed relay point respectively. Note that $D_l > D_s$ as D_l is proportional to the street length. D_s is the required distance to communicate with the mobile sink which is proportional to the street width. By investigating the effect of T_o on the system when transmitting a message during W cycles, the energy dissipated in the circuits $m.e_{elec}$ is constant for both interval definition of W_{cycle} and can be neglected. Also the energy required to receive the beacon is neglected as the discovery message is small compared to the sensor message.
There are many advantages of using such methodoly. Some of them are spacial reuse of the bandwith by allowing short range communication, simple scalability of the system, extendability of the system and guaranteed delivery of the sensed message as the there is always an alternative fixed path to route the data.

3.3 Single node simulation

From the sensor point of view, the system can be modeled as shown in Fig. 8.

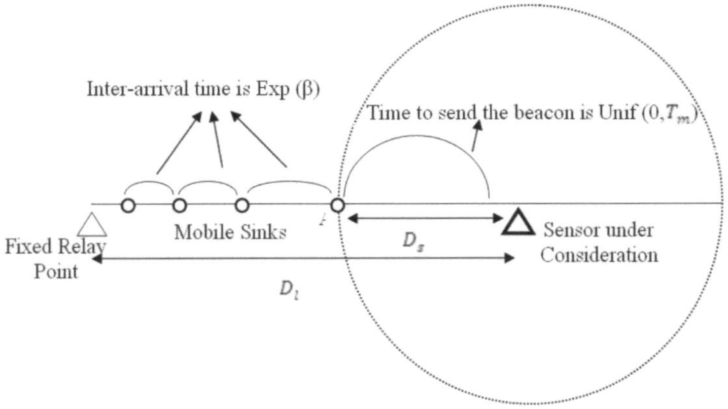

Fig. 8. Beacons transmission time

Point A is taken as the observation point. Given the mobility model described above, the inter-arrival time between the mobile sinks to point A is exponentially distributed with a mean β. In this section, the system is studied for a time interval when β can be considered constant. The mobile sinks periodically send a beacon to the nearby sensor every T_m. It is important to note that very low values of T_m is not a practical solution as the mobile sink will use the channel all the time preventing other communications to take place. The time taken by a mobile sink to send its first beacon after arriving to the sensor coverage area varies uniformly between Zero and T_m. The uniform distribution is assumed as the cars have started their message broadcasting at some points in time that are completely independent. The sensor can receive the beacon if it has been sent from a distance D_s or fewer meters away from it. The cars are assumed to be moving with a velocity V during their journey in the sensor range. MATLAB (MatLab) simulations of the described system is used to model the system kinematics and obtain guidelines on system behavior.

3.3.1 Simulation setup

The energy required to send a message is calculated using the transceiver properties of the Mica2 Motes produced by Chipcon CC1000 data sheet (Chipcon, 2008) and the values mentioned in (Polastre et al., 2004). The transmitter power needed to achieve a dedicated signal to noise ratio at the receiver is highly dependent on the system deployment. $e_{elec} + e_{amp}D_l^K$ and $e_{elec}+ e_{amp}D_s^K$ are taken as the maximum and minimum powers that can be generated from the transceiver respectively. The simulation parameters are as shown in Table 3.

Parameter	Description	Default value
B	Cars inter-arrival mean time	8 to 30 seconds
P_{idle}	Idle listening power	173 μJoules
$R_{bit}\,(e_{elec} + e_{amp}D_l^K)$	Maximum output power per bit	26.7 mA * 3 V
$R_{bit}\,(e_{elec} + e_{amp}D_s^K)$	Minimum output power per bit	6.9 mA * 3 V
M	Number of bits per message	120*8
D_s	Lower sensor transmission radius	22.5 m
T_m	Beacon sending period	3 seconds
V	Moving sink velocity	15 m/s
$Sensing_{cycle}$	Sensor sensing cycle	60 seconds
R_{bit}	Transmission bit rate	19.2 kbps

Table 3. Default simulation parameters

The average energy consumed per cycle during 6500 cycles with respect to the value of T_o is simulated and given in Fig. 9 for exponential distributions with different values of β.

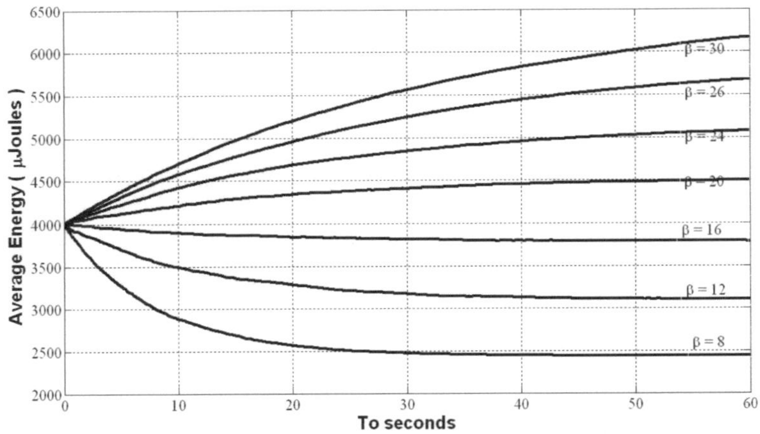

Fig. 9. Average energy for different traffic flow

3.3.2 Single node analysis

It can be seen from Fig. 9 that the optimum values for T_o are infinity for β equals 8, 12, 16; and zero for β equals 20, 24, 26, 30. The Low Traffic state will be applied when the optimum value of T_o equals zero. In this case, the sensor is synchronized by the cluster head (the fixed sink) to previously determined time instants in which it can send its message to the next faraway fixed relay point in the route path. In other words, the sensor will not wait for the mobile sink beacon. In this case the amount of energy dissipated by the sensor equals E_l, where D_l is the inter sensor node distance.

The second case, the High Traffic state, is when the optimum value of T_o equals infinity, i.e., the sensor goes to the idle listening state until it detects a beacon from a nearby mobile sink. Upon reception of the beacon, the sensor sends its message to the mobile sink and goes to the low power sleeping state. It is important to note that T_o equals infinity does not mean that the sensor will wait for an infinite time to receive a beacon, but the sensor is allowed to wait an unconstrained time until it receives the beacon. In Fig. 9, the three curves are for β equals 8, 12 and 16 seconds; the average energy consumed can be considered constant when T_o > 40 seconds. The value of T_o can be constrained by another system performance metric such as latency. When the optimum value is infinity, the average amount of energy dissipated equals:

$$E_{\inf} = E_{idle} \cdot \tau_b + E_s \tag{13}$$

where E_s is the energy required to send its message to the mobile sink. τ_b is the average time during which the sensor will be in the idle state during the W cycles.

From Fig. 9, the threshold value of τ_b that determines the system state can be calculated by getting the minimum of E_l and $E_{\inf}$ where:

$$\tau_{b\ \text{threshold}} = \frac{m \times e_{amp}\left(d_l^K - d_s^K\right)}{P_{idle}} \tag{14}$$

The sensor will be in the Low Traffic state (LTS) when $\tau_b > \tau_b$ threshold and it will be in the High Traffic state (HTS) when $\tau_b < \tau_{b\ \mathrm{threshold}}$.

3.4 Energy balanced linear network with mobile sinks

In the previous section, the energy improvement of a single sensor node using the suggested hybrid system was proven. In this section, the work is extended to investigate the impact on overall network performance. The main goal of environmental monitoring WSN is maximizing the network lifetime while keeping its connectivity. This can be done by several ways on different network layers starting from the physical to the application layer.

3.4.1 Basic problem

In all the possible wireless sensor network topologies, two basic approaches can be used to deliver messages to the sink node: direct transmission and hop-by-hop transmission (Mhatre & Rosenberg, 2004). As shown in Fig. 10, in direct transmission where packets are directly transmitted to the fixed sink without any relay, the nodes located farther away from the sink have higher energy consumption due to long range communication, and these nodes die out first. On the other hand, in multi-hop linear networks, the total energy consumed in the nodes participating in the message relaying is less than the energy consumed in direct transmission; however, it suffers from the fast energy drainage in the nearest node to sink. Both cases inherit the energy unbalance problem of wireless sensor networks due to the many to one communication paradigm. Although all the previously mentioned protocols consider energy efficiency but they do not explicitly take care of the phenomena of unbalanced energy consumption. In such networks, some nodes die out early, thus resulting in the network collapse although there is still significant amount of energy in other sensors.
Next, a new solution using the hybrid message transmission method mentioned previously, is presented.

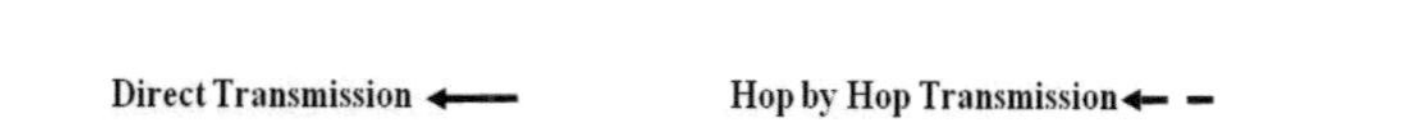

Fig. 10. Direct and Hop by Hop transmission for linear network

3.4.2 Using hybrid message transmission schemes

The problem of unbalanced load distribution in case of multi-hop networks can be manipulated by using a hybrid message transmission system. The basic idea lies in mixing single-hop with multi-hop message transmission. A simple way to implement the hybrid

scheme would be to make the sensor node spend a period of its lifetime using one of the modes while spending the other period using the second mode.
In (Efthymiou et al., 2004; Mhatre & Rosenberg, 2004; Zhang et al., 2007), the authors calculate the optimized ratio of the time by which the sensor decides either to send directly to the fixed sink or to overload its neighbors using hop-by-hop transmission as in Fig. 10. The basic idea is simple: find an alternative -and usually higher energy- way for faraway nodes to send their message to the sink in order to reduce the load on closer nodes. The proposed solutions are efficient for small networks; but for large networks practical limitations can prevent a far-away node from sending a message using high transmission power.
Another approach for message transmission energy reduction is the usage of mobile sinks. As stated previously uncontrolled mobile-sink WSN suffer from energy overhead required to detect the presence of mobile agents. In the previous subsection, the sink detection *controlled* overhead was modeled as the maximum period that the sensor nodes stay in the idle listening state.
In this subsection and based on the results obtained previously, energy balanced linear sensor network with one fixed sink and multiple uncontrolled mobile sinks, is achieved. Based on the system current status and using a hybrid message transmission algorithm, the sensor nodes can decide either to send to the next fixed relay node or to wait for the mobile sink a maximum period of time T_o. Energy balancing is performed for different mobile sinks behaviors. In the low mobility state, every node is assigned a maximum waiting time for the mobile sink before it sends to the fixed relay node. A mathematical formulation is shown to obtain the best waiting time values that balance the energy among all nodes. The system is solved for different parameters' values using a generic numerical algorithm.

3.4.3 Model under study

The environmental monitoring system studied here consists of a linear sensor network with one fixed sink and multiple uncontrolled mobile sinks. The sensor nodes are equidistantly distributed with a distance D_l. The fixed and mobile sinks are assumed to have a continuous power supply while the sensors are energy constrained. Sensors are assumed to be able to adjust their transmit power amplifiers to exactly meet the required signal strength at receivers with different distances. The sensor nodes can receive or send a message to the mobile sink if it is located at a distance that is less than D_s meter away from it. The network model is shown in Fig. 11.

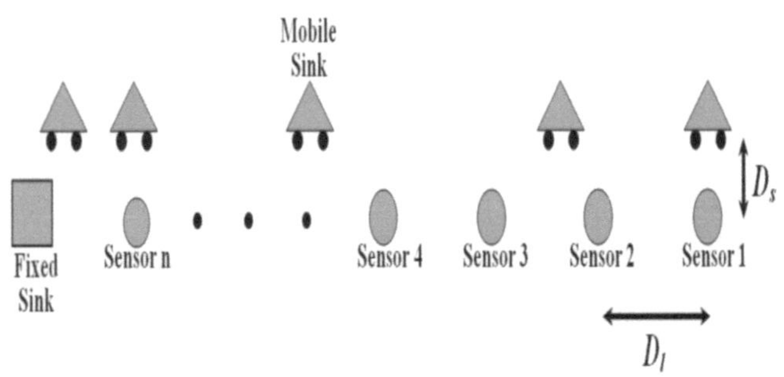

Fig 11. Linear sensor network model with mobile sinks.

3.4.4 Basic notations

Let ε_X denote the expected value of energy consumed for $T_o = X$. For every sensor that has a maximum waiting time of $T_o(i)$; $\varepsilon_{T_o(i)}$ can be obtained by multiplying equation 10 with the PDF of the waiting time T and integrating on the range of T. The resultant points for different valued of T_o are given in Fig. 9 using the equation:

$$\varepsilon_{To} = P_{idle}\beta + E_s + \left(E_l - P_{idle}\beta - E_s\right)e^{\frac{-T_o}{\beta}} \tag{15}$$

When T_o equals infinity, the average energy consumed per cycle can be calculated as:

$$\varepsilon_{\inf} = P_{idle} \times \beta + E_s \tag{16}$$

Finally for T_o = zero,

$$\varepsilon_o = E_s \tag{17}$$

Let $\varsigma(i)$ denote the total energy dissipated by the sensor node i during a sensing cycle. $\varsigma(i)$ takes into consideration two loads: The energy required to send the message generated by the node itself and the energy required to relay possible messages from nearby nodes during a sensing cycle.

Let $E[*]$ represents the expected value of any quantity *. For the mentioned network to be energy balanced, the total expected energy consumed by any sensor node i, $E[\varsigma(i)]$, during the system lifetime must be the same for all the nodes.

From the result shown in Fig. 9, in the HTS the optimum average energy consumed by any sensor node to send its self generated message $E[\omega_{cycle}(i)]$ equals $\varepsilon_{\inf}$. In this case all the sensor nodes always send their message to one of the mobile sinks. Consequently, sensor nodes do not relay messages generated by other sensor nodes. Every sensor dissipates the same average amount of energy: $E[\varsigma(i)] = \varepsilon_{\inf}$;therefore, energy balancing is achieved.

In the LTS the best solution from the sensor point of view is that it directly forwards all incoming packets to the next fixed node. In this case, the total energy consumed by a node i during a sensing cycle equals:

$$\zeta(i) = E_l + (i-1) \times (E_l + E_{rx}) \tag{18}$$

since the node has to send the data message generated by itself and relay $(i-1)$ messages from the other nodes in the queue. In the LTS, $\varepsilon(i) = E_l\,\varepsilon(i)$ obtained by substituting T_o with zero in equation 15. It is assumed that the sensor will wake up in pre-determined time instants to send its message to the next relay point in the routing path. It can be shown that every node dissipates different amount of energy depending on its position where sensor n is the highest loaded node. Energy balancing is required in the LTS.

3.4.5 Balancing the low traffic state

Energy balancing can be done by increasing the energy required by the relatively far-away nodes from the fixed sink for sending a data message, to reduce the number of messages that a relatively nearby node has to relay. This can be done by finding an alternative path to send the message. In the system under study, the alternative is a longer waiting time in the idle listening state for an approaching mobile sink.

For the LTS in the hybrid message transmission system described above, waiting any amount of time for hearing a beacon from a mobile sink increases the average energy required to send the message. It also decreases the probability that a node sends its messages to the next fixed node to relay it (Zaki et al., 2009).

3.4.6 Problem statement

Given a linear wireless sensor network that consists of n sensor nodes, a sensor node i may transmit a data message to the next fixed point or to one of the mobile sinks depending on the maximum waiting time $T_o(i)$. The mobile sinks have an exponentially distributed waiting time with mean $\beta > \beta_{threshold}$. What are the values of $T_o(i)$ for $i = 1,2,\ldots\ldots,n$ that equalize and minimize the total average energy consumed by every sensor causing the maximization of the network life time?

$$E[\zeta(i)] = E[\zeta(j)] \text{ for } i,j = 1,2,\ldots,n \tag{19}$$

3.4.7 Mathematical formulation

Let P_i denote the probability that a node i sends to the mobile sink. Using the exponential distribution as the Probability Density Function (PDF) of the waiting time and the definition of $T_o(i)$, then:

$$P_i = \int_0^{T_o(i)} \exp(t,\beta)dt = 1 - e^{-T_o(i)/\beta} \tag{20}$$

Let $N_r(i)$ denote the number of relayed messages by sensor i. The total energy consumed by sensor i during a sensing cycle is given by:

$$\varsigma(i) = \omega_{cycle}(i) + N_r(i) \times \left(E_{rx} + \omega_{cycle}(i)\right) \tag{21}$$

as $N_r(i)$ depends on the amount of messages relayed from successor nodes for nodes 1 to node $i-1$, and $\omega_{cycle}(i)$ depends on $T_o(i)$. Therefore, $N_r(i)$ and $\omega_{cycle}(i)$ are both independent variables. The expected total energy consumed by node i equals:

$$E[(\varsigma(i))] = E\left[\omega_{cycle}(i)\right] + E[N_r(i)] \times \left(E_{rx} + E\left[\omega_{cycle}(i)\right]\right) \tag{22}$$

For every sensor that has a maximum waiting time of $T_o(i)$, $E[\omega_{cycle}(i)] = \varepsilon_{T_o(i)}$ can be obtained from equation 15.
The average number of the total messages that node i receives from all previous nodes $E[N_r(i)]$ is given by:

$$E[N_r(i)] = \sum_{k=1}^{i-1} \prod_{j=k}^{i-1} (1 - P_j) \tag{23}$$

From equations 22 and 23, the total average energy consumed by the sensor node i equals:

$$E[\varsigma(i)] = \varepsilon_{T_o(i)} + \left[\sum_{k=1}^{i-1} \prod_{j=k}^{i-1} (1 - P_j) \right] \times \left(E_{rx} + \varepsilon_{T_o(i)} \right) \tag{24}$$

where i varies from 1 to n
The energy balancing problem can be solved by equating the above equations 24. Thus, there are *(n-1)* equations. The last equation can be deduced from node n average transmission energy. Knowing that the last sensor will not overload any other subsequent node, the optimum average energy consumption for node n is when $T_o(n) = zero$ or:

$$\omega_{cycle}(n) = \varepsilon_{zero} = E_l \tag{25}$$

3.4.8 Solving the system states

The algorithm shown in Fig. 12 solves N simultaneous equations resulting from equating the equations in 24 and using 25. It can be implemented on a processing unit for any PDF of the arriving beacon time T other than the exponential distribution. It is important to note that using the LTS graph and knowing the value of ε_{To} is sufficient to calculate T_o. Rewriting the general equation of $E[\varsigma(i)]$ in order to get $\varepsilon_{T_o(i)}$:

$$\varepsilon_{T_o(i)} == \frac{E[\varsigma(i)] - \left[\sum_{k=1}^{i-1} \prod_{j=k}^{i-1} (1 - P_j) \right] \times E_{rx}}{1 + \sum_{k=1}^{i-1} \prod_{j=k}^{i-1} (1 - P_j)} \tag{26}$$

The numerical algorithm works a follows:

1. Calculate the LTS graph for T_o varying form zero to Max_T_o with an appropriate resolution.
2. N = Number of sensor nodes.
3. Calculate $\varepsilon_{zero} = E_l$ and ε_{inf} using equations 16 and 17.
4. Assume $\varepsilon_{To(1)} = \frac{\varepsilon_{zero} + \varepsilon_{\inf}}{2} = E[\zeta(i)], \forall i = 1toN$.
5. $\Delta\varepsilon = \frac{\varepsilon_{\inf} - \varepsilon_{zero}}{1000}$.
6. For I=1 to 1000.
   node = 1.
   While ($\varepsilon_{To(node)} > \varepsilon_{zero}$) and (node < N)
      Get $T_o(node)$ and P_{node}.
      Use (26) to calculate $\varepsilon_{To(node+1)}$
      node = node + 1.
   If ($\varepsilon_{To(node)} > \varepsilon_{zero}$)
      $$\varepsilon_{To(1)} = \varepsilon_{To(1)} - \Delta\varepsilon$$
   Else
      $$\varepsilon_{To(1)} = \varepsilon_{To(1)} + \Delta\varepsilon$$
7. $$Error = \left| \frac{\varepsilon_{To(node)} - \varepsilon_{zero}}{\varepsilon_{zero}} \right| \times 100.$$
8. If (Error > 5)
   Repeat from step 6 for $N = N\text{-}1$
9. If node < N
   Assume that all the nodes from 1 to (N-node) work at T_o = infinity and they dissipate ε_{inf}

Fig. 12. Solving the system equations.

From equations 24, it is clear that $\varepsilon_{To(i)} > \varepsilon_{To(i+1)}$ for all values of i. The algorithm starts by assigning node *1* an average energy $\varepsilon_{To(1)}$ in the middle of the LTS curve. All next nodes are solved correspondently. The algorithm iteratively tries to assign the last node (e.g. the closest to the fixed sink) an average energy consumption of E_l.

3.4.9 Simulation results

Using MATLAB, the algorithm was run using the values presented in Table 3 and for a network of 10 sensor nodes. The system is simulated for two cases. Case 1 is when $\beta = 100$ seconds. In this case, the algorithm succeeded to get the values of $T_o(i)$ for all 10 nodes. The percentage of error between $\varepsilon_{To(10)}$ and E_l equals –0.238%. The results are shown in Fig. 13.

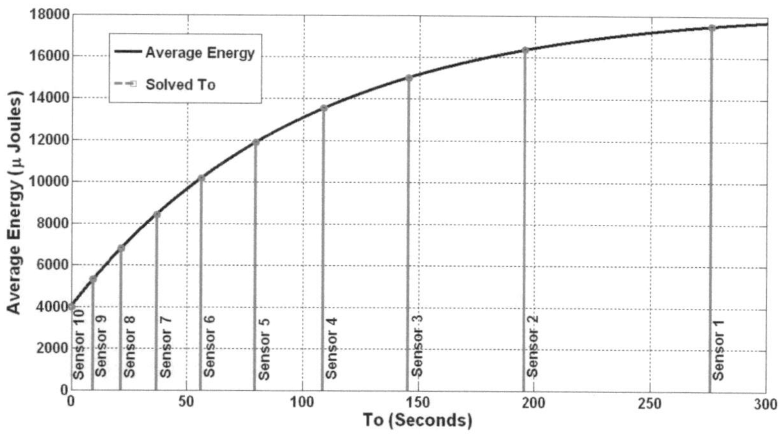

Fig. 13. Waiting time $T_o(i)$ for all the 10 sensor nodes versus $\varepsilon[T_o(i)]$ for $\beta = 100$ seconds.

Case 2 is simulated for $\beta = 40$ seconds (see Fig. 14). The values of $T_o(i)$ for 8 nodes starting from node 3 to node 10 is obtained and the solutions for nodes 1 and 2 are approximated to T_o = infinity (or a relatively high value as mentioned in the graph). In this case, all the sensor nodes dissipate $E[\zeta(i)] \approx \varepsilon_{inf}$, i = 1, 2,….,N and the hybrid message relaying method has the same performance as always relaying to the mobile sink. However, using the hybrid system the upper bound on the delay can be calculated and message delivery is guaranteed.

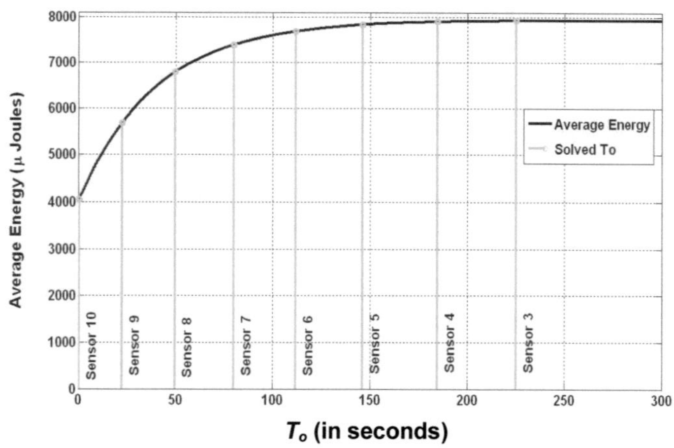

Fig. 14. Waiting time $T_o(i)$ for all the 8 solved nodes versus $\varepsilon[T_o(i)]$ for $\beta = 40$ seconds.

The system life time is calculated as follows: assume that all the nodes are given initially the same amount of energy X. The total number of sensing cycles ψ can be calculated by dividing the total amount of energy by the average total energy consumed in a cycle. Taking the conservative approach mentioned in (Mhatre & Rosenberg, 2004), the system is said to be dead when the first sensor node dies, i.e., the node that consumes the most energy.

The system lifetime is compared to the two classical cases: All-Mobile system when all the nodes always send to the mobile sink, i.e., T_o = infinity, and All-Fixed system when all the nodes always send to the fixed relay node, i.e., T_o = zero. In the All-Fixed system, node n

will die first. In this case, ψ_{fixed} is calculated for the average energy consumed per node n $E[\zeta(n)]_{fixed}$ which is obtained by substituting i with n in equation 18. In the All-Mobile system, all the nodes always send to the mobile sinks; they dissipate the same amount of energy. ψ_{mobile} is calculated by substituting $E[\zeta(i)]_{mobile}$ by $\varepsilon_{inf.}$. Similarly, in the hybrid model described here, all the nodes dissipated the same amount of energy. Ψ_{hybrid} is calculated by substituting $E[\zeta(i)]_{hybrid}$ by $E[\zeta(1)]$.

$$\Psi = \frac{X}{E[\varsigma(\text{most energy dissipating node})]} \tag{27}$$

Fig. 15 shows the average total average energy consumed in the three mentioned systems for different ascending values of β starting from $\beta_{threshold}$. It is clear that the hybrid system moves from the All-Mobile performance to the All-Fixed performance for low and high values of β respectively.

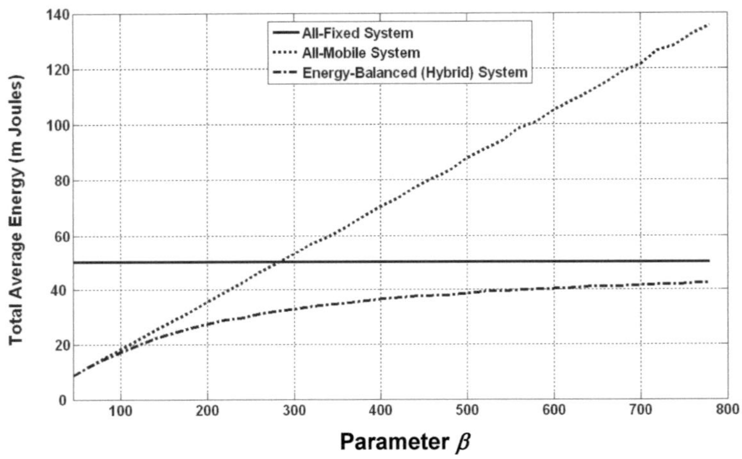

Fig. 15. Comparison between the total average energy consumed in the All-Fixed, All-Mobile and Energy-Balanced systems.

4. Conclusion

The advantages of WSNs using low-cost and low-power sensors in several application areas justify the research interest in network lifetime optimization techniques. In this chapter, results pertaining to this research problem were presented. First, a modification of LEACH-C method was described that obtains the optimum number of cycles for each sensor to act as a network master such that the network lifetime is maximized. Then, it was shown that use can be made of geometric node distributions and sink locations to prolong the network lifetime compared to the case of random node distributions. Then, an energy efficient relaying data collection system is considered. It can be used for different applications such as environmental monitoring in urban areas. Using moving cars as uncontrolled mobile sinks, a hybrid model that proposes a maximum sensor waiting time for the mobile agent before sending to the fixed node was investigated both in high and low traffic states. Also suggested was a hybrid message transmission scheme that decreases the load on nodes nearby to the fixed sink while maximizing the network lifetime. Simulation results that indicate the benefits of each of the proposed techniques were given throughout the chapter.

5. References

Akyildiz, I.F.; Su, W.; Sankarasubramaniam, Y. & Cayirci, E. (2002). Wireless sensor networks: a survey. *Computer Networks Magazine*, Vol. 38, Issue 4, March 2002, pp. 393-422.

Akkaya, K. & Younis, M. (2005). A survey on routing protocols for wireless sensor networks. *Journal of Ad Hoc Networks*, Vol. 3, No. 3, May 2005, pp. 325-349.

Al-Ghamdi, A. S. (2001), Analysis of Time Headways on Urban Roads: Case Study from Riyadh, *Journal of Transportation Engineering*, Volume 127, No. 4, July/August 2001, pp. 289 - 294.

Al-Karaki, J.N. & Kamal, A.E. (2004), Routing techniques in wireless sensor networks: a survey, *IEEE Wireless Communications Magazine*, Volume 11, No. 6, December 2004, pp. 6–28.

Bestavros, A.; Matta, I. & Riga, N. (2004). "DIP: density inference protocol for wireless sensor networks and its application to density-unbiased statistics. *Proceedings of the Second International Workshop on Sensor and Actor Network Protocols and Applications (SANPA)*, Boston, Massachusetts, USA, August 2004.

Botros, S.; ElSayed, H.; Amer, H. & El-Soudani, M. (2009). Lifetime optimization in hierarchical wireless sensor networks, *Proceedings of the IEEE International Conference on Emerging Technologies and Factory Automation ETFA*, Mallorca, Spain, September 2009.

Chipcon Corporation (2008), CC1000 low power FSK transceiver. Data Sheet, March 2008, Available:http://focus.ti.com/lit/ds/symlink/cc1000.pdf

Efthymiou, C.; Nikoletseas, S. & Rolim, J. (2004), Energy Balanced Data Propagation in Wireless Sensor Networks, *Proceedings of the 18th International Parallel and Distributed Processing Symposium (IPDPS'04)*, Santa Fe, New Mexico, USA, April 2004.

Gui, X & He, X. (2009). The localized area coverage algorithm based on delayed start scheme for WSN. *Journal of Software*, Vol.4, No. 3, May 2009, pp. 183-190..

Heinzelman, W.; Chandrakasan, A. & Balakrishnan, H. (2000). Energy-efficient routing protocols for wireless microsensor networks, *Proceedings of the 33rd Hawaii International Conference on System Sciences (HICSS)*, Maui, HI, USA, January 2000.

Heinzelman, W.; Chandrakasan, A. & Balakrishnan, H. (2002). An application-specific protocol architecture for wireless microsensor networks. *IEEE Transactions On Wireless Communications*, Vol. 1, No. 4, October 2002, pp. 660-670.

Hoogendoorn, S.P. & Bovy, P. H. (2001), State-of-the-art of vehicular traffic flow modeling, *Proceedings of the Institution of Mechanical Engineers Part I Journal of Systems & Control Engineering*, Volume 215, Issue 4, August 2001, pp. 283 - 303.

Jain, S.; Shah, R. C.; Brunette, W.; Borriello, G. & Roy, S. (2006), Exploiting Mobility for Energy Efficient Data Collection in Wireless Sensor Networks, *Mobile Networks and Application*, Vol. 11, Issue 3, June 2006, pp. 327-339.

MatLab, Official Site of MatLab: www.mathworks.com

Mahfoudh, S. & Minet, P. (2008). Survey of energy efficient strategies in wireless Ad Hoc and sensor networks, *Proceedings of the Seventh International Conference on Networking (icn 2008)*, Cancun, Mexico, April 2008.

Mhatre, V. & Rosenberg, C. (2004), Design guidelines for wireless sensor networks: communication, clustering and aggregation, *Ad Hoc Networks*, Volume 2, Issue 1, January 2004, pp. 45-63.

Minet; P. & Mahfoudh, S. (2009). Survey of energy efficient strategies in wireless adhoc and sensor networks. *Proceedings of the 2009 International Conference on Wireless Communications and Mobile Computing: Connecting the World Wirelessly*, Leipzig, Germany, June 2009.

Narasimha, N. & Gopinath, K. (2006). A survey of routing algorithms for wireless sensor networks. *Indian Institute of Science*, Vol. 43, No. 4.

Nouh, S.; Abbas, R.; AbouElSeoud, D.; Ali, N.; Daoud, R.; Amer, H. & ElSayed, H. (2010). Effect of node distributions on lifetime of wireless sensor networks, *Proceedings of the IEEE International Symposium on Industrial Electronics ISIE*, Bari, Italy, July 2010.

Onur, E.; Ersoy, C.; Delic, H. & Akarun, L. (2007). Surveillance wireless sensor networks: deployment quality analysis. *IEEE Network*, Vol. 21, No. 5, September 2007, pp. 48-53.

Polastre, J.; Hill, J. & Culler, D. (2004), Versatile Low Power Media Access for Wireless Sensor Networks, *Proceedings of the 2nd International Conference on Embedded Networked Sensor Systems*, pp. 95–107, Baltimore, MD, USA , November, 2004.

Prabh, Sh.; Deshmukh, Ch. & Sachan, Sh. (2009). A distributed algorithm for hexagon topology formation in wireless sensor networks. *Proceedings of the 14th International Conference on Emerging Technologies and Factory Automation*, ETFA, Mallorca, Spain, September 2009.

Shah, R. C.; Roy, S.; Jain, S. & Brunette, W. (2003), Data mules: Modeling a three-tier architecture for sparse sensor networks, *Proceedings of the first IEEE International Workshop on Sensor Network Protocols and Applications (SNPA)*, pp. 30 - 41, May 2003, IEEE, Piscataway

Shebli, F.; Dayoub, I.; Rouvaen, J.M. & Zaouche, A. (2007), A New Optimization Approach for Energy Consumption within Wireless Sensor Networks, *Proceedings of The Third Advanced International Conference on Telecommunications*, Mauritius, May 2007.

Tavares, J.; Felez, F.J. & Ferro, J.M. (2008). Application of wireless sensor networks. *Measurement Science Review*, Vol. 8, Sec. 3, No. 3, 2008.

Younis, M.; Youssef, M. & Arisha, K. (2002), Energy-Aware Routing in Cluster-Based Sensor Networks, *Proceedings of the 10th IEEE/ACM International Symposium on Modeling, Analysis and Simulation of Computer and Telecommunication Systems (MASCOTS2002)*, Fort Worth, TX, USA, October 2002.

Zaki, G.; Elsayed, H.; Amer, H. & El-Soudani, M. (2008) Energy-Efficient Hybrid Wireless Sensor Networks in Urban Areas, In: *International Journal of Factory Automation, Robotics and Soft Computing*, Vol., No., April 2008, pp. 68-74.

Zaki, G.; Elsayed, H.; Amer, H. & El-Soudani, M. (2009) Energy Balanced Model for Data Gathering in Wireless Sensor Networks with Fixed and Mobile Sinks, *Proceeding of the ICCCN workshop on sensor Networks*, San Francisco, CA, USA, August 2009.

Zhang, H.; Shen, H. & Tan, Y. (2007), Optimal Energy Balanced Data Gathering in Wireless Sensor Networks, *Parallel and Distributed Processing Symposium*, California, March 2007.

A Sink Node Allocation Scheme in Wireless Sensor Networks Using Suppression Particle Swarm Optimization

Hidehiro Nakano, Masaki Yoshimura, Akihide Utani,
Arata Miyauchi and Hisao Yamamoto
Tokyo City University
Japan

1. Introduction

A wireless sensor network, which is a key network to facilitate ubiquitous information environments, has attracted a significant amount of interest from many researchers (Akyildiz et al., 2002). A wireless sensor network has a wide range of applications, such as natural environmental monitoring, environmental control in residential spaces or plants, object tracking, and precision agriculture. In a general wireless sensor network, hundreds or thousands of micro sensor nodes, which are generally compact and inexpensive, are placed in a large scale observation area and sensing data of each node is gathered to a sink node by inter-node wireless multi-hop communication. Each sensor node consists of a sensing function to measure the status (temperature, humidity, motion, etc.) of an observation point or object, a limited function on information processing, and a simplified wireless communication function, and generally operates on a resource of a limited power-supply capacity such as a battery. Therefore, a data gathering scheme and/or a routing protocol capable of meeting the following requirements has been mainly studied to prolong the lifetime of a wireless sensor network.

1. Efficiency of data gathering
2. Balance on communication load among sensor nodes

As the scheme that satisfy the above two requirements, gradient-based routing protocol has attracted attention (Xia et al., 2004). However, this does not positively ease the communication load concentration to sensor nodes around a sink node that is the source of problems on the long-term operation of a wireless sensor network. In a large scale and dense wireless sensor network, the communication load is generally concentrated on sensor nodes around a sink node during the operation process. In case sensor nodes are not placed evenly in a large scale observation area, the communication load is concentrated on sensor nodes placed in an area of low node-density. Intensive data transmission to specific nodes, such as sensor nodes around a sink node and sensor nodes placed in an area of low node-density, brings on concentrated energy consumption of them and causes them to break away from the network early. This makes the long-term observation by a wireless sensor network difficult. To solve this communication load concentration problem, a data gathering scheme for a wireless sensor network with multiple sinks has been proposed (Dubois-Ferriere et al., 2004; Oyman & Ersoy,

2004). Each sensor node, in this scheme, sends sensing data to the nearest sink node. In comparison with the case of a one-sink wireless sensor network, the communication load of sensor nodes around a sink node is reduced. In the existing studies, however, the effective locations for sink nodes, which are an important design problem for the long-term operation of a wireless sensor network, have not been discussed.
This chapter discusses a method of suppressing the communication load on sensor nodes by effectively placing a limited number of sink nodes in an observation area. As a technique of solving effective locations for sink nodes, this chapter presents a new search algorithm named the suppression particle swarm optimization algorithm (Yoshimura et al., 2009). This algorithm is based on the particle swarm optimization algorithm (Kennedy & Eberhart, 1995) that is one of the swarm intelligence algorithms. The suppression particle swarm optimization algorithm can provide plural effective allocation sets for sink nodes so that total hops in all sensor nodes are minimized. As their allocation sets are switched dynamically, the above two requirements can be satisfied.
This chapter consists of five sections. In Section 2, the basic particle swarm optimization algorithm is introduced. In Section 3, the suppression particle swarm optimization algorithm is explained. In Section 4, simulation results for two types of wireless sensor networks are presented. Through numerical simulations, effectiveness by using the suppression particle swarm optimization algorithm is confirmed. In Section 5, the overall conclusions of this work are given and future problems are discussed.

2. The Particle Swarm Optimization Algorithm

In this section, the original particle swarm optimization algorithm is outlined. The particle swarm optimization algorithm belongs to the category of swarm intelligence algorithms. It was developed and first introduced as a stochastic optimization algorithm (Kennedy & Eberhart, 1995). Currently, the particle swarm optimization algorithm is intensively researched because it is superior to the other algorithms on many difficult optimization problems. The ideas that underlie the particle swarm optimization algorithm are inspired not by the evolutionary mechanisms encountered in natural selection, but rather by the social behavior of flocking organisms, such as swarms of birds and fish schools. The particle swarm optimization algorithm is a population-based algorithm that exploits a population of individuals to probe promising regions of the search space. In this context, the population is called a *swarm* and the individuals are called *particles*. In the particle swarm optimization algorithm, a multi-dimensional solution space by sharing information between a swarm of particles is efficiently searched. The algorithm is simple and allows unconditional application to various optimization problems.
Assume a D-dimensional search space and a swarm consisting of N particles. Each particle (The i th particle) has a position vector

$$\boldsymbol{x}_i = (x_{i1}, x_{i2}, \cdots, x_{iD})^{\mathrm{T}}, \tag{1}$$

and the velocity vector

$$\boldsymbol{v}_i = (v_{i1}, v_{i2}, \cdots, v_{iD})^{\mathrm{T}}, \tag{2}$$

where the subscript i ($i = 1, \cdots, N$) represents the particle's index. In addition, each particle retains the best position vector $\boldsymbol{pbest}_i$ found by the particle in the search process and the best position vector $\boldsymbol{gbest}$ among all particles as information shared in the swarm in the search

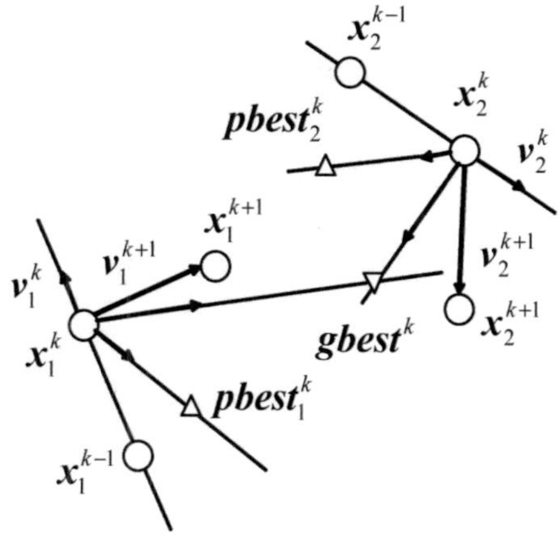

Fig. 1. The movement of particles.

process. In the particle swarm optimization algorithm, each particle produces a new velocity vector $\boldsymbol{v}_i^{k+1}$ by linearly coupling $\boldsymbol{pbest}_i^k$ found by the particle in the past, $\boldsymbol{gbest}^k$ shared in the swarm, and the previous velocity vector $\boldsymbol{v}_i^k$ and moves to the next position $\boldsymbol{x}_i^{k+1}$, where the superscript k indicates the number of search iterations. At the $k+1$ th iteration, the velocity vector $\boldsymbol{v}_i^{k+1}$ and the position vector $\boldsymbol{x}_i^{k+1}$ of the i th particle is updated by the following equations:

$$\boldsymbol{v}_i^{k+1} = w \cdot \boldsymbol{v}_i^k + c_1 \cdot r_1 \cdot (\boldsymbol{pbest}_i^k - \boldsymbol{x}_i^k) + c_2 \cdot r_2 \cdot (\boldsymbol{gbest}^k - \boldsymbol{x}_i^k) \tag{3}$$

$$\boldsymbol{x}_i^{k+1} = \boldsymbol{x}_i^k + \boldsymbol{v}_i^{k+1} \tag{4}$$

where r_1 and r_2 represent random numbers, uniformly distributed within the interval [0,1]. w is a parameter called the inertia weight. c_1 and c_2 are positive constants, referred to as cognitive and social parameters, respectively. The settings of w, c_1, and c_2 affect the performance of the particle swarm optimization algorithm. In Fig. 1, an example on the movement of particles is shown. By iterating the search based on Equations (3) and (4) until the end condition is satisfied, a solution to an objective function $f(\boldsymbol{x})$ can be obtained. The particle swarm optimization algorithm to search the minimization of an objective function $f(\boldsymbol{x})$ is as follows (see Fig. 2):

Step 0 : Preparation

Set the total number of particles N, the particle parameters (w, c_1, c_2), and the maximum number of iterations K_{max}.

Step 1 : Initialization

Set the search iteration counter to $k = 1$. Generate the initial velocity vector $\boldsymbol{v}_i^1$ and the initial position vector $\boldsymbol{x}_i^1$ of each particle from random numbers and determine the initial $\boldsymbol{pbest}_i^1$ and $\boldsymbol{gbest}^1$.

$$\boldsymbol{pbest}_i^1 = \boldsymbol{x}_i^1, \quad i = 1, \cdots, N \tag{5}$$

$$i_g = \arg\min_i f(\boldsymbol{pbest}_i^1) \tag{6}$$

$$\boldsymbol{gbest}^1 = \boldsymbol{pbest}_{i_g}^1 \tag{7}$$

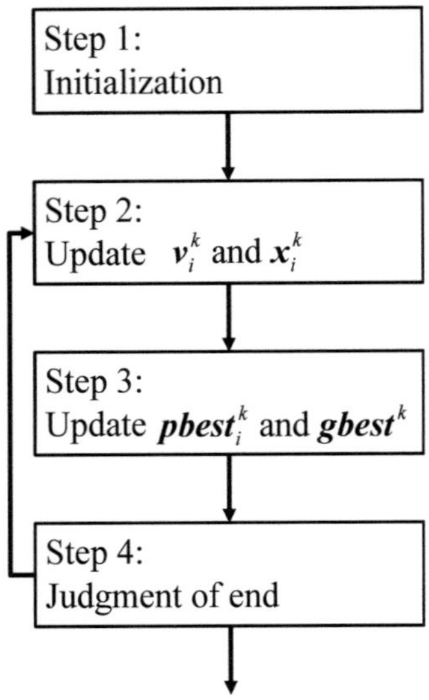

Fig. 2. Flowchart of the particle swarm optimization algorithm.

Step 2 : Update of velocity vector and position vector

Update the velocity vector and the position vector of each particle by Equations (3) and (4).

Step 3 : Update of $\boldsymbol{pbest}$ and $\boldsymbol{gbest}$

Update $\boldsymbol{pbest}_i^{k+1}$ and $\boldsymbol{gbest}^{k+1}$ as follows:

$$I_1 = \left\{ i \;\middle|\; f(\boldsymbol{x}_i^{k+1}) < f(\boldsymbol{pbest}_i^k), 1 \le i \le N \right\} \tag{8}$$

$$\boldsymbol{pbest}_i^{k+1} = \begin{cases} \boldsymbol{x}_i^{k+1}, & i \in I_1 \\ \boldsymbol{pbest}_i^k, & i \notin I_1 \end{cases} \tag{9}$$

$$i_g = \arg\min_i f(\boldsymbol{pbest}_i^{k+1}) \tag{10}$$

$$\boldsymbol{gbest}^{k+1} = \boldsymbol{pbest}_{i_g}^{k+1} \tag{11}$$

Step 4 : Judgment of end

Finish the search when $k = K_{max}$. Otherwise, return to Step 2 by assuming $k = k + 1$.

The particle swarm optimization algorithm can fast solve various optimization problems in nonlinear continuous functions, although the algorithm uses only simple and fundamental arithmetic operations. However, the basic particle swarm optimization algorithm can find only a single solution for a single trial.

3. The Suppression Particle Swarm Optimization Algorithm

In this section, the suppression particle swarm optimization algorithm having a simple self control mechanism is explained (Yoshimura et al., 2009). The overall processing flow of the suppression particle swarm optimization algorithm is shown in Fig. 3. As shown in the figure,

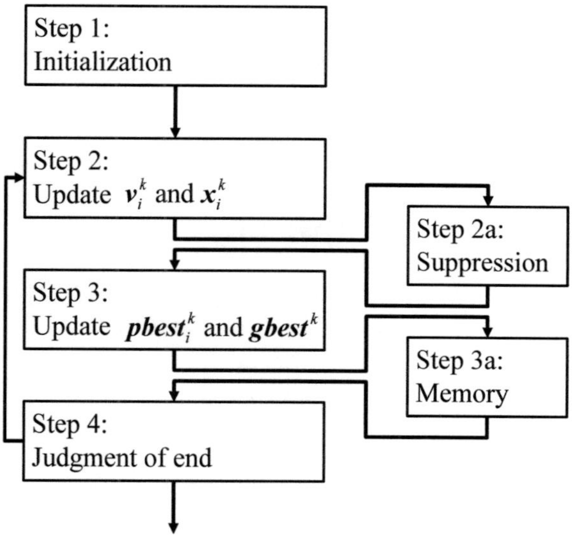

Fig. 3. Flowchart of the suppression particle swarm optimization algorithm.

"suppression" and "memory" are added to the flow of the original particle swarm optimization algorithm. The suppression scheme controls excessive conversion of particles as referring to density of particles. The memory scheme stores copies of position vectors having better evaluation values, which are distant from each other. These schemes can realize to provide various acceptable solutions.

In the suppression particle swarm optimization algorithm, distance between the i th and the j th particles is calculated by

$$distance_{ij} = ||\boldsymbol{x}_i - \boldsymbol{x}_j|| \tag{12}$$

Also, density of the i th particle is calculated by

$$density_i = \frac{1}{N} \sum_{j=1, j \neq i}^{N} \alpha(distance_{ij}; T_d) \tag{13}$$

where N is the number of particles, T_d is a distance threshold parameter, and $\alpha(z;T)$ is the following function.

$$\alpha(z;T) = \begin{cases} 1, & z \leq T \\ 0, & \text{otherwise} \end{cases} \tag{14}$$

That is, the number of particles having shorter distances than the threshold T_d is proportional to the density. Let $\tilde{\boldsymbol{x}}_j^k$ be the j th position vector preserved in the memory scheme at the k th iteration. Set the number of the preserved position vectors to $L = 0$ in Step 1. Additional schemes in the suppression particle swarm optimization algorithm are as follows:

Step 2a: Suppression

Consider the following subset:

$$I_2 = \{i \mid density_i > T_s, 1 \leq i \leq N\} \tag{15}$$

where T_s is a density threshold parameter. Reset the velocity vector $\boldsymbol{v}_i^k$ and the position vector $\boldsymbol{x}_i^k$ to random values if $i \in I_2$ is satisfied.

Step 3a: Memory

Set the position vector $\boldsymbol{x}_i^k$ as a candidate preserved in the memory scheme if the following condition is satisfied:

$$f(\boldsymbol{x}_i^k) < T_{mf} \tag{16}$$

where T_{mf} is a fitness threshold parameter. Store the candidate position vector $\boldsymbol{x}_i^k$ in the memory scheme and let $L = L + 1$ if the following condition is satisfied:

$$\bigwedge_{j=1}^{L} \left(||\boldsymbol{x}_i^k - \tilde{\boldsymbol{x}}_j^k|| > T_d \right) \tag{17}$$

where T_d is the distance threshold parameter explained before. Otherwise, consider the following subset:

$$I_3 = \left\{ j \;\middle|\; ||\boldsymbol{x}_i^k - \tilde{\boldsymbol{x}}_j^k|| \le T_d \right\} \tag{18}$$

Replace the preserved position vectors $\tilde{\boldsymbol{x}}_j^k$ $(j \in I_3)$ with the candidate position vector $\boldsymbol{x}_i^k$ and let $L = L - |I_3|$ if the following condition is satisfied:

$$\bigwedge_{j \in I_3} \left(f(\boldsymbol{x}_i^k) < f(\tilde{\boldsymbol{x}}_j^k) \right) \tag{19}$$

The suppression particle swarm optimization algorithm is based on the artificial immune system which is one of optimization algorithms (de Castro & Timmis, 2002). The living body has a mechanism to reconstruct own genes and generate antibodies which eliminate antigens from outside. The antibodies affect not only antigens but also antibodies themselves. Repeating in such a process between antibodies and antigens, effective antibodies are generated. The artificial immune system mimics such a process. This algorithm can keep a diversity of solutions by a production mechanism of antibodies and a self-control mechanism in an immunity system, and can search plural acceptable solutions. However, the artificial immune system requires large computation costs. The suppression particle swarm optimization algorithm can be regarded as a fusion algorithm which has simple and fast search functions in the particle swarm optimization algorithm, and plural solution search functions in the artificial immune system.

Purpose of this study is to suppress the communication load on sensor nodes by effectively placing a limited number of sink nodes in an observation area. However, the communication load is concentrated on sensor nodes around a sink node during the operation process of wireless sensor networks and causes them to break away from the network early. Therefore, it is needed to find plural allocation sets for sink nodes so that total hops in all sensor nodes are minimized, and to switch their allocation sets dynamically considered energy consumption of each sensor node. The suppression particle swarm optimization algorithm can provide plural effective allocation sets for sink nodes, such that the communication load of each sensor node can be reduced.

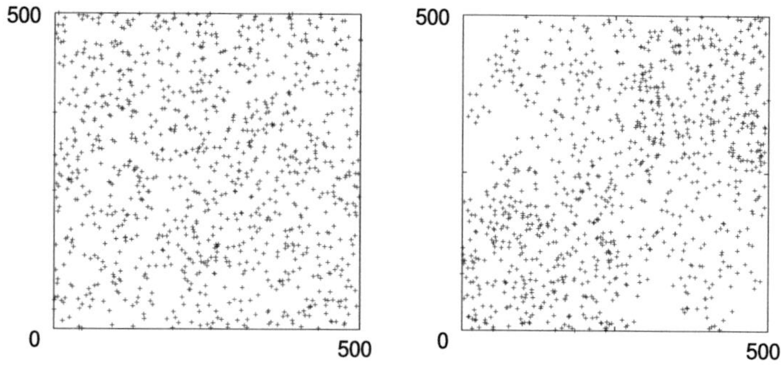

(a) (b)

Fig. 4. Sensor node allocations. (a) Uniform node-density. (b) Nonuniform node-density.

sink nodes	S_{i1}		S_{i2}		S_{i3}		S_{i4}		S_{i5}	
particle	X_{i1}	Y_{i1}	X_{i2}	Y_{i2}	X_{i3}	Y_{i3}	X_{i4}	Y_{i4}	X_{i5}	Y_{i5}

Fig. 5. Coding method to each particle ($M = 5$).

4. Simulation Experiments

In this section, three methods, the suppression particle swarm optimization algorithm, the particle swarm optimization algorithm and the artificial immune system, are applied to a sink node allocation problem, and the solving performances are compared.

4.1 Sink Node Allocation Problem

The problem to allocate M sink nodes in a two dimensional observation area is considered. In the observation area, sensor nodes are allocated randomly as the followings:

1. Uniform node-density as shown in Fig. 4(a); sensor nodes are allocated evenly in whole of the area.
2. Nonuniform node-density as shown in Fig. 4(b); many sensor nodes are allocated in the lower left and upper right area, and few sensor nodes are allocated in the other area.

Sink nodes can be allocated at the arbitrary locations in the area.
For the locations of M sink nodes in the two dimensional area, the expressions of each particle are $2M$ design variables as shown in Fig. 5. In the figure, S_{im} denotes a two dimensional location of the m th sink node which the i th particle has. In order to apply each method to this problem, $distance_{ij}$ in Equation (12) is defined as the minimum value in all Euclidean distances between sink node locations which each particle has (see Fig. 6):

$$distance_{ij} = \min_{m,n} |S_{im} - S_{jn}|, \quad 1 \leq m \leq M, \ 1 \leq n \leq M \qquad (20)$$

where M is the number of sink nodes. Then, the density basically increases when at least two sink nodes which each particle has are contiguous.
The evaluation value (fitness) of each particle is given by total hop counts from all sensor nodes to each nearest sink node. This fitness is used for all the methods, the suppression particle swarm optimization algorithm, the particle swarm optimization algorithm and the artificial immune system.

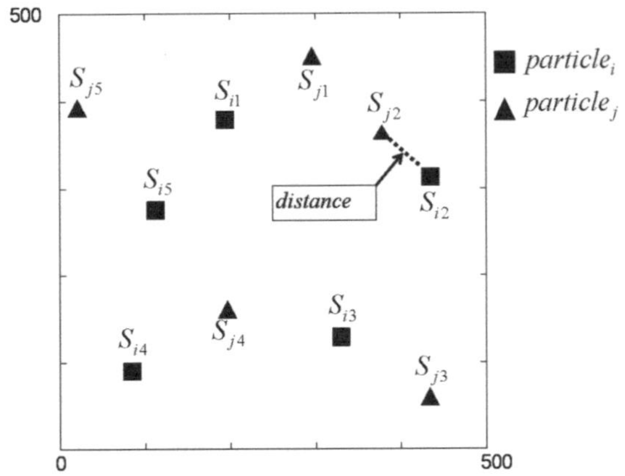

Fig. 6. Definition of distance between each particle ($M = 5$).

Parameter	Value
Area Size $[m^2]$	500×500
Number of sink nodes M	5
Number of sensor nodes	1000
Radio range $[m]$	25

Table 1. Conditions in wireless sensor networks.

The conditions in wireless sensor networks are shown in Table 1, and the parameters in each method are shown in Table 2, which are decided by preliminary experiments.

4.2 Average Delivery Ratio

In sink node allocation sets provided with each method, lifetime of wireless sensor networks is evaluated. Each sensor node periodically transmits sensor information to the nearest sink node. Then, the sensor node and relative relay sensor nodes consume energy (Heinzelman et al., 2000):

$$E_{Rx}(b) = E_{elec} \times b \tag{21}$$

$$E_{Tx}(b,d) = E_{elec} \times b + \varepsilon_{amp} \times b \times d^2 \tag{22}$$

Parameter	Value
Inertia coefficient w	0.9
Weight coefficient c_1	1.0
Weight coefficient c_2	1.0
Threshold of distance T_d	30
Threshold of density T_s	0.9
Threshold of fitness T_{mf}	6250
Number of particles N	30
Total number of iterations K_{max}	100

Table 2. Parameters in each method.

Parameter	Value
Battery capacity [J]	0.5
Processing coefficient E_{elec} [nJ/bit]	50
Transmission coefficient $\varepsilon_{amp}[pJ/bit/m^2]$	100
Data size b [$Byte$]	12
Transmission output d [m]	25
Number of transmissions	900
Number of trials	100

Table 3. Conditions in calculating average delivery ratio.

Algorithm	SPSO	AIS	PSO
Best fitness	5274	5429	5149
Average fitness	5501	5701	5382
Number of solutions	3.73	3.44	1

Table 4. Fitness and the number of solutions for a uniform node-density wireless sensor network. SPSO: the suppression particle swarm optimization. AIS: the artifical immune system. PSO: the particle swarm optimization.

where E_{Rx} and E_{Tx} denote energy consumption in reception and transmission, respectively. b is data size of sensing data. d is transmission output. E_{elec} and ε_{amp} are processing and transmission coefficients, respectively. All sensor nodes have the same battery capacity at first, and simultaneously transmit sensing data with the same data size to each nearest sink node via some relay sensor nodes. The relay sensor nodes are selected so that each sensing data is transmitted in minimum hop counts to each nearest sink node. If plural candidates of the relay sensor nodes exist, one of them is selected randomly. If buttery shutoff occurs in a relay sensor node, the sensor node can not relay sensing data. In such a situation, average delivery ratio for wireless sensor networks is calculated. Table 3 shows conditions in calculating the average delivery ratio.

4.3 Results for a Wireless Sensor Network with uniform node-density

First, simulation results for the wireless sensor network with uniform node-density shown in Fig. 4(a) are presented. Fig. 7 shows transitions of the best fitness (total hop counts) in each method. In the figure, each value corresponds to the best fitness in all particles in each iteration. Table 4 shows the best fitness, the average fitness, and the average number of the solutions preserved in the memory scheme. These are the average values for 100 trials. In the suppression particle swarm optimization algorithm and the artificial immune system, it is possible to search widely in the solution space by the self-control mechanism and each fitness does not converge monotonously. On the other hand, in the particle swarm optimization algorithm, fitness converges to a single solution and it is not possible to search other solutions. As comparing quality of solutions, the particle swarm optimization algorithm is the best in all the methods. However, it should be noted that the suppression particle swarm optimization algorithm and the artificial immune system can search plural acceptable solutions while the particle swarm optimization algorithm can not. Fig. 8 shows three allocation sets for five sink nodes finally obtained by the suppression particle swarm optimization algorithm. As shown in the figure, all the sink nodes are allocated without overlapping. This is very important in

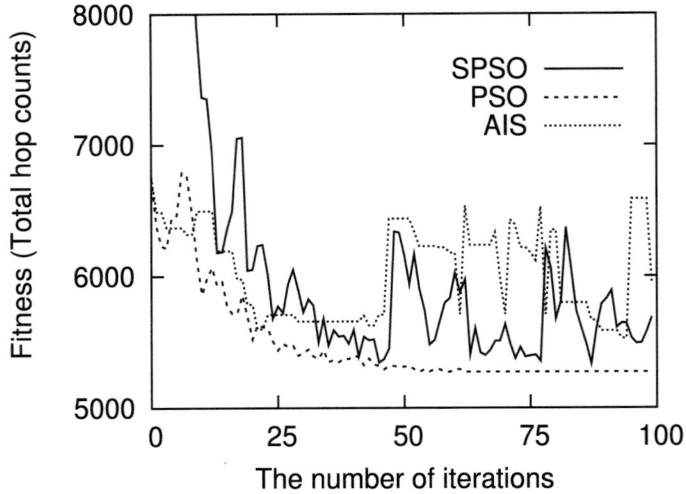

Fig. 7. Fitness in each method for a wireless sensor network with uniform node-density. SPSO: the suppression particle swarm optimization. AIS: the artifical immune system. PSO: the particle swarm optimization.

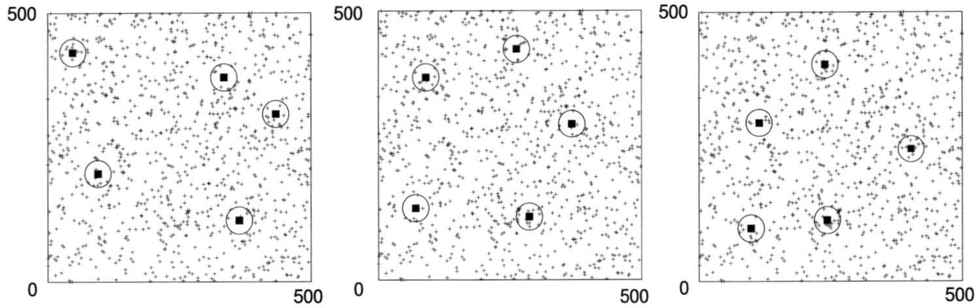

Fig. 8. Three allocation sets for five sink nodes in a uniform node-density wireless sensor network obtained by the suppression particle swarm optimization algorithm.

the viewpoints of suppressing communication load in each sensor node. Fig. 9 shows average delivery ratio for the following three methods:

SPSO: Three sink node allocation sets obtained by the suppression particle swarm optimization algorithm are switched in every 300 transmission.

PSO: The best sink node allocation set obtained by the particle swarm optimization algorithm continues to be used during 900 transmissions.

Regular: The regular sink node allocation set in the area continues to be used during 900 transmissions.

Sink node allocation sets obtained by all the methods are shown in Fig. 10.
It is found that average delivery ratio in the suppression particle swarm optimization method is higher than those in the particle swarm optimization method and the regular allocation method. Because, communication load in each sensor node is distributed by dynamically switching sink node allocation sets. That is, energy consumption of sensor nodes is balanced.

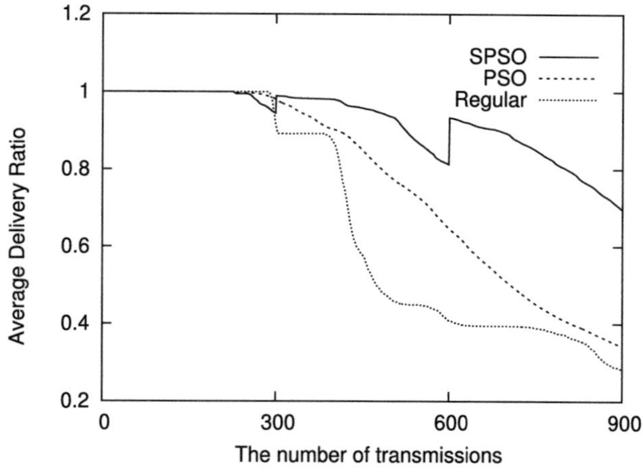

Fig. 9. Average delivery ratio for a uniform node-density wireless sensor network. SPSO: the suppression particle swarm optimization method. PSO: the particle swarm optimization method. Regular: the regular allocation method.

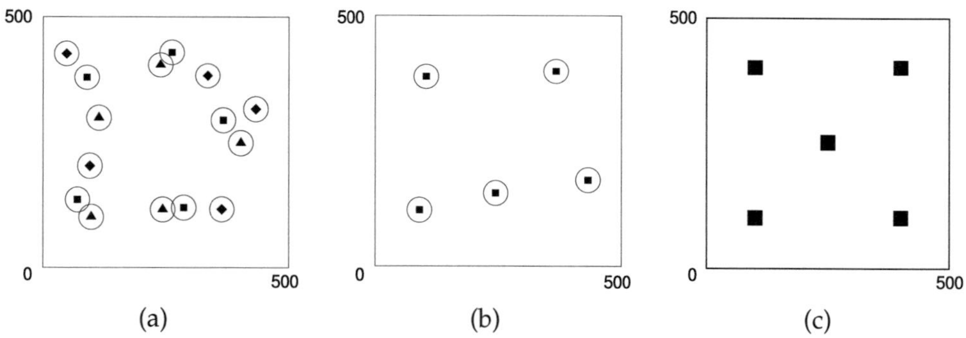

Fig. 10. Sink node allocation sets obtained by each method. (a) SPSO: the suppression particle swarm optimization method. (b) PSO: the particle swarm optimization method. (c) Regular: the regular allocation method.

But, fitness of the suppression particle swarm optimization algorithm is worse than that of the particle swarm optimization algorithm. This means that in order to prolong wireless sensor network lifetime, it is necessary to search for plural distant solutions rather than to search for a single high accuracy solution. Therefore, it is shown that the suppression particle swarm optimization method is effective for the long-term operation of wireless sensor networks.

4.4 Results for a wireless sensor network with nonuniform node-density

Next, simulation results for the wireless sensor network with nonuniform node-density shown in Fig. 4(b) are presented. Fig. 11 shows transitions of the best fitness (total hop count) in each method. In the figure, each value corresponds to the best fitness in all particles in each iteration. Table 5 shows the best fitness, the average fitness, and the average number of the solutions preserved in the memory scheme. These are the average values for 100 trials. As same as the previous experiment, in the suppression particle swarm optimization algorithm and the artificial immune system, it is possible to search widely in the solution space by the

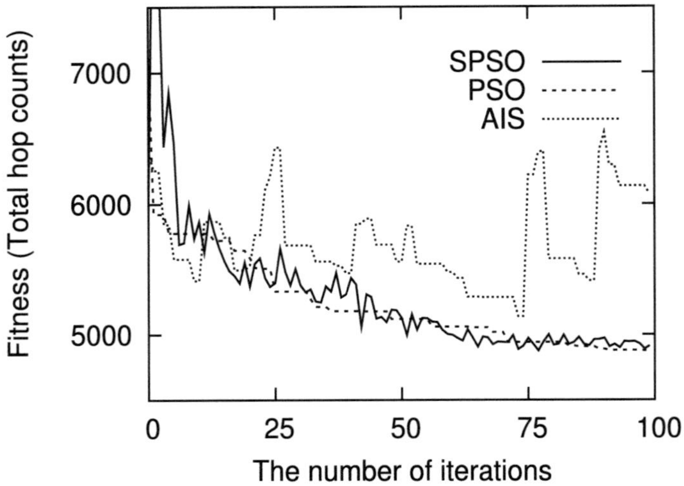

Fig. 11. Fitness in each method for a nonuniform node-density wireless sensor network. SPSO: the suppression particle swarm optimization. AIS: the artifical immune system. PSO: the particle swarm optimization.

Algorithm	SPSO	AIS	PSO
Best fitness	4800	5115	4800
Average fitness	4979	5429	4971
Number of solutions	3.51	6.17	1

Table 5. Fitness and the number of solutions for a nonuniform node-density wireless sensor network. SPSO: the suppression particle swarm optimization. AIS: the artifical immune system. PSO: the particle swarm optimization.

self-control mechanism and fitness does not converge monotonously. On the other hand, in the particle swarm optimization algorithm, fitness converges to a single solution and it is not possible to search other solutions. The number of obtained solutions in the artificial immune system is the most, but fitness is the worst. The fitness in the suppression particle swarm optimization algorithm is almost the same as that in the particle swarm optimization algorithm. Fig. 12 shows three allocation sets for five sink nodes finally obtained by the suppression particle swarm optimization algorithm. Fig. 13 shows average delivery ratio for three methods. Sink node allocation sets obtained by all the methods are shown in Fig. 14.

As same as the previous experiment, the suppression particle swarm optimization algorithm can keep higher average delivery ratio than the other methods. This means that for the nonuniform node-density wireless sensor network, the suppression particle swarm optimization algorithm can also search effective sink node allocation sets. Because, it is possible to widely search on solution space. That is, the suppression particle swarm optimization method is applicable to various wireless sensor networks, and can realize long-term operation of the wireless sensor networks.

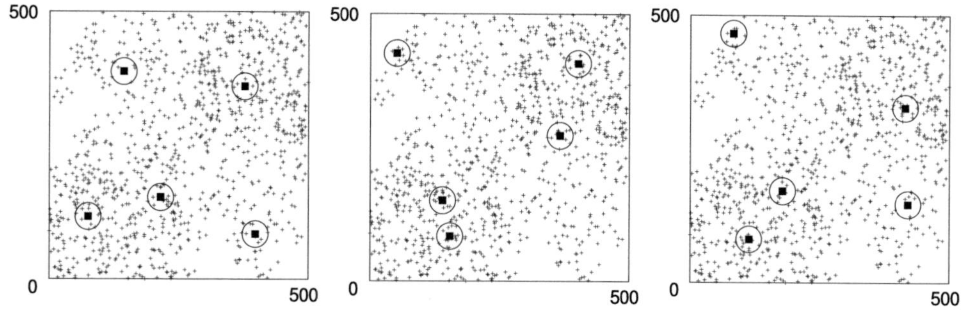

Fig. 12. Three allocation sets for five sink nodes in a nonuniform node-density wireless sensor network obtained by the suppression particle swarm optimization algorithm.

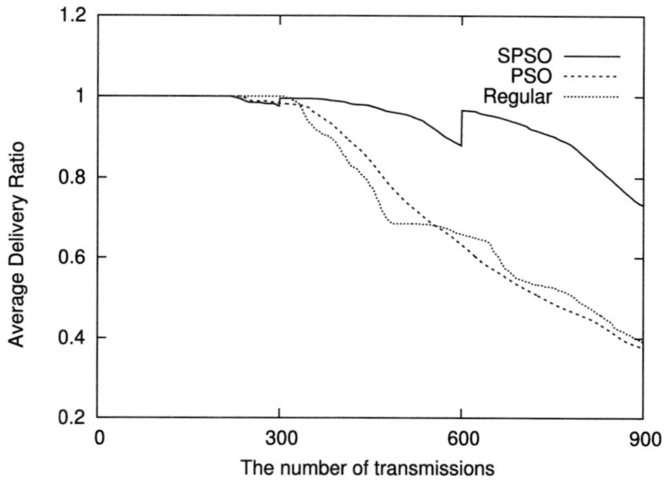

Fig. 13. Average delivery ratio for a nonuniform node-density wireless sensor network. SPSO: the suppression particle swarm optimization method. PSO: the particle swarm optimization method. Regular: the regular allocation method.

5. Conclusions

This chapter has discussed a method of placing sink nodes effectively in an observation area to use wireless sensor networks for a long time. For the effective search of sink node locations, this chapter has presented the suppression particle swarm optimization method, which is a new method based on the particle swarm optimization algorithm, to search several acceptable solutions. In the actual environment of wireless sensor networks, natural conditions or other factors may disturb the placement of a sink node at a selected location or the location effect may be lost due to the appearance of a blocking object. Therefore, it is important to provide several means (candidate locations) for sink nodes by using a method capable of searching several acceptable solutions. In the simulation experiment, the effectiveness of the method has been verified by comparison for the particle swarm optimization algorithm and the artificial immune system. Without increasing the number of search iterations, several solutions (candidate locations) of approximately the same level as that by the existing particle swarm optimization could be obtained. Future problems include evaluation for solving ability of the

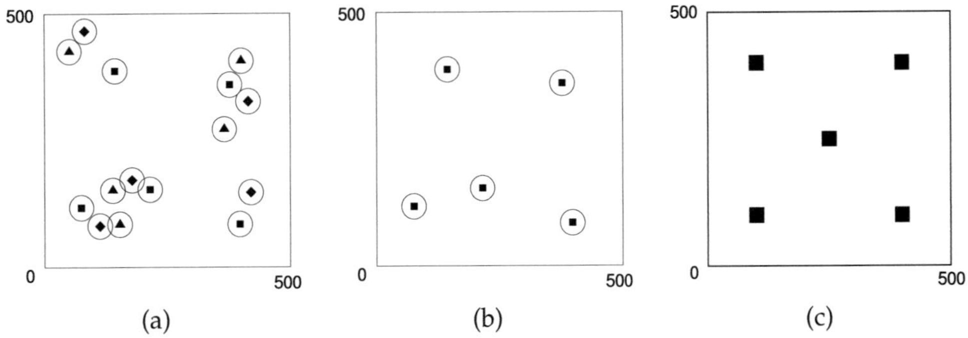

Fig. 14. Sink node allocation sets obtained by each method. (a) SPSO: the suppression particle swarm optimization method. (b) PSO: the particle swarm optimization method. (c) Regular: the regular allocation method.

method in more detail, and fusion with the existing communication algorithms dedicated to wireless sensor networks.

6. References

Akyildiz, I.; Su, W.; Sankarasubramaniam, Y. & Cayirci, E. (2002). Wireless sensor networks: A survey, *Computer Networks Journal*, Vol. 38, No. 4, 393-422

de Castro, L.; Timmis, J. (2002). *Artificial immune systems: A new computational approach*, Springer, London.

Dubois-Ferriere, H.; Estrin, D. & Stathopoulos, T. (2004). Efficient and practical query scoping in sensor networks, *Proceedings of the IEEE International Conference on Mobile Ad-Hoc and Sensor Systems*, 564-566

Heinzelman, W.R.; Chandrakasan, A. & Balakrishnan, H. (2000). Energy-efficient communication protocol for wireless microsensor networks, *Proceedings of Hawaii International Conference on System Sciences*, 3005–3014

Kennedy, J. & Eberhart, R.C. (1995). Particle swarm optimization, *Proceedings of the IEEE International Conference on Neural Networks*, 1942-1948

Oyman, E.I. & Ersoy, C. (2004). Multiple sink network design problem in large scale wireless sensor networks, *Proceedings of the International Conference on Communications*, Vol. 6, 3663-3667

Xia, L.; Chen, X. & Guan, X. (2004). A new gradient-based routing protocol in wireless sensor networks, *Lecture Notes in Computer Science*, Vol. 3605, 318-325

Yoshimura, M.; Nakano, H.; Utani, A.; Miyauchi A. & Yamamoto, H. (2009). An Effective Allocation Scheme for Sink Nodes in Wireless Sensor Networks Using Suppression PSO, *ICIC Express Letters*, Vol. 3, No. 3(A), 519–524

Hybrid Approach for Energy-Aware Synchronization

Robert Akl, Yanos Saravanos and Mohamad Haidar
University of North Texas
Denton, Texas, USA

1. Introduction

Several sensor applications have been developed over the last few years to monitor environmental properties such as temperature and humidity. One of the most important requirements for these monitoring applications is being unobtrusive, which creates a need for wireless ad-hoc networks using very small sensing nodes. These special networks are called wireless sensor networks (WSN). WSNs are built from many wireless sensors in a high-density configuration to provide redundancy and to monitor a large physical area.

WSNs can be used to detect traffic patterns within a city by tracking the number of vehicles using a designated street (Winjie et al., 2005), (Tubaishat et al., 2008). If an emergency arises, the network can relay the information to the city hall and notify police, fire, and ambulance drivers of congested streets. An application could even be designed that suggests the fastest route to the emergency area. When compared to computer terminals in Local Area Networks (LANs), wireless sensors must operate on very low capacity batteries to minimize their size to about that of a quarter. The nodes use slow processing units to conserve battery power. A typical sensor node such as Crossbow's Mica2DOT operates at 4 MHz with 4 KB of memory and has a radio transceiver operating at up to 15 Kbps (MICA2DOT, 2005). Radio transmissions consume by far the majority of the battery's energy, so even with this low-power hardware, a sensor can easily be depleted within a few hours if it is continuously transmitting.

One of the most common uses for wireless sensor networks is for localization and tracking(Patwari et al., 2005), (Langendoen & Reijers , 2003). Tracking of a single object is relatively simple since data can be handed-off from sensor to sensor as the object moves through the network.

Another important aspect is time synchronization in a networked system. The majority of research in this field has concentrated on traditional high-speed computer networks with few power restraints, leading to the Global Positioning System (GPS) and the Network Time Protocol (NTP), (NTP, 2009). Although GPS is an accurate and commonly used synchronization protocol, there are a few requirements that GPS fails to meet. Some of which are that the receiver is 4.5 inches in diameter, more than 4 times the size of a typical sensor node, and also requires an external power source. These two traits counteract the goal of using small and mobile nodes to create a WSN, not to forget the line-of-sight requirement that cripples GPS's use for sensor networks dispersed within a building or in a heavily forested area. On the other hand, NTP is one of the first synchronization protocols used for computer systems, first developed in 1985 (NTP, 2009). This protocol uses a relatively large amount of memory to store data for synchronization sources, authentication codes, monitoring options, and access options. As mentioned earlier, typical wireless sensor nodes have limited onboard memory. A large sensor network will require large files for synchronization sources and codes. If these configuration files can be programmed into each node, it would leave very little memory to hold the data monitored by the sensor, limiting NTP's use for WSNs. Furthermore, NTP's synchronization accuracy is within 10 ms over the Internet, and up to 200 μs in a LAN (NTP, 2009); these specifications are inadequate for most sensor network applications. Therefore, new synchronization methods have been developed specifically for sensor networks, such as the reference broadcast synchronization method (RBS) (Elson et al., 2002) and the timing-sync protocol for sensor networks (TPSN) (Ganeriwal, November 2003), (Ganeriwal, 2003).

RBS and TPSN achieve accurate clock synchronization within a few microseconds of uncertainty nonetheless both are designed for networks with a small number of sensors and are not specifically geared towards energy conservation. Although these algorithms tend to work for larger networks, their energy consumption becomes inefficient and network connectivity is broken once nodes begin lacking power. Simulations show that synchronizing a large sensor network requires a large number of transmissions, which will quickly deplete sensors and reduce the network's coverage area.

A time synchronization scheme for wireless sensor networks that aims to save sensor battery power while maintaining network connectivity for as long as possible is presented based on a hybrid algortihm that combines both TPSN and RBS.

This algorithm is an extension of our previous work presented in (Akl & Saravanos, 2007). It focuses on the following aspects of WSNs:

1. Design a hybrid method between RBS and TPSN to reduce the number of transmissions required to synchronize an entire network.
2. Extend single-hop synchronization methods to operate in large multi-hop networks.
3. Verify that the hybrid method operates as desired by simulating against RBS and TPSN.
4. Maintain network connectivity and coverage.

2. Time Synchronization Algorithms in WSNs

Traditional synchronization methods, that are effective for computer networks, are ineffective in sensor networks. New synchronization algorithms specifically designed for wireless sensor networks have been developed and can be used for several applications (Sivirkaya & Yener, 2004). The authors in (Palchaudhuri et al., 2004) present a probabilistic method for clock synchronization based on RBS. In (Sun et al., 2006), the authors present a level-based and a diffusion-based clock synchronization that is resilient to some source nodes. The authors in (He & Kuo, 2006) propose creating spanning trees with multiple subtrees in which two subtree synchronization algorithms can be performed. Four methods are described in (Qun & Rus, 2006) to achieve global synchronization: a node-based, a hierarchal cluster-based, a diffusion-based, and a fault-tolerant based approach. An Efficient

RBS (E-RBS) algorithm is proposed in (Lee et al., 2006) to decrease the number of messages to be processed and save energy consumption within a given accuracy range.

2.1 The Reference Broadcast Synchronization Method (RBS)

Since GPS and NTP are not very effective in wireless sensor applications, the first major research attempts to create a time synchronization algorithm specifically tailored for sensor networks led to the development of reference broadcast synchronization (RBS) in 2002 (Elson et al., 2002). The algorithm defines a critical path, which is represented by the portion of the network where a significant amount of clock uncertainty exists. A long critical path results in high uncertainty and low accuracy in the synchronization. There are four main sources of delays that must be accounted for to have accurate time synchronization:

- *Send time:* this is the time to create the message packet.
- *Access time:* this is a delay when the transmission medium is busy, forcing the message to wait.
- *Propagation time:* this is the delay required for the message to traverse the transmission medium from sender to receiver.
- *Receive time:* similar to the send time, this is the amount of time required for the message to be processed once it is received.

The RBS algorithm can be split into three major events:

1. *Flooding:* a transmitter broadcasts a synchronization request packet.
2. *Recording:* the receivers record their local clock time when they initially pick up the *sync* signal from the transmitter.
3. *Exchange:* the receivers exchange their observations with each other.

RBS synchronizes each set of receivers with each other as opposed to traditional algorithms that synchronize receivers with senders. These latter algorithms have a long critical path, starting from the initial send time until the receive time. For this reason, NTP's accuracy is severely limited, as discussed previously. RBS uses a relative time reference between nodes, eliminating the send and access time uncertainties. The propagation delay of signals is extremely fast from point-to-point, so this delay can be ignored when dealing in the microsecond scale. Lastly, the receive time is reduced since RBS uses a relative difference in times between receivers. Nonetheless, the time of reception is taken when the packet is first received in the MAC layer, eliminating uncertainties introduced by the sensor's processing unit.

There are two unique implementations of RBS. The simplest method is designed for very high accuracy for sparse networks, where transmitters have at most two receivers. The transmitter can broadcast a synchronization request to the two receivers, which will record the times at which they receive the request, just as the algorithm describes. However, the receivers will exchange their observations with each other multiple times, using a linear regression to lower the clock offset. The other version of the RBS algorithm involves the following steps: the transmitter sends a reference packet to two receivers; each receiver checks the time when it receives the reference packet; the receivers exchange their recorded times. The main problems with this scheme are the nondeterministic behavior of the receiver, as well as clock skew. The receiver's nondeterministic behavior can be resolved by simply sending more reference packets. The clock skew is resolved by using the slope of a least-squares linear regression line to match the timing of the crystal oscillators.

RBS can be adapted to work in multi-hop environments as well. Assuming a network has grouped clusters with some overlapping receivers, linear regression can be used to synchronize between receivers that are not immediate neighbors. However, it is more complicated than the single-hop scenario since there will be timestamp conversions as the packet is relayed through nodes. This extra complication is manifested in larger synchronization errors. Fig. 1 shows how a sensor network is synchronized by using RBS.

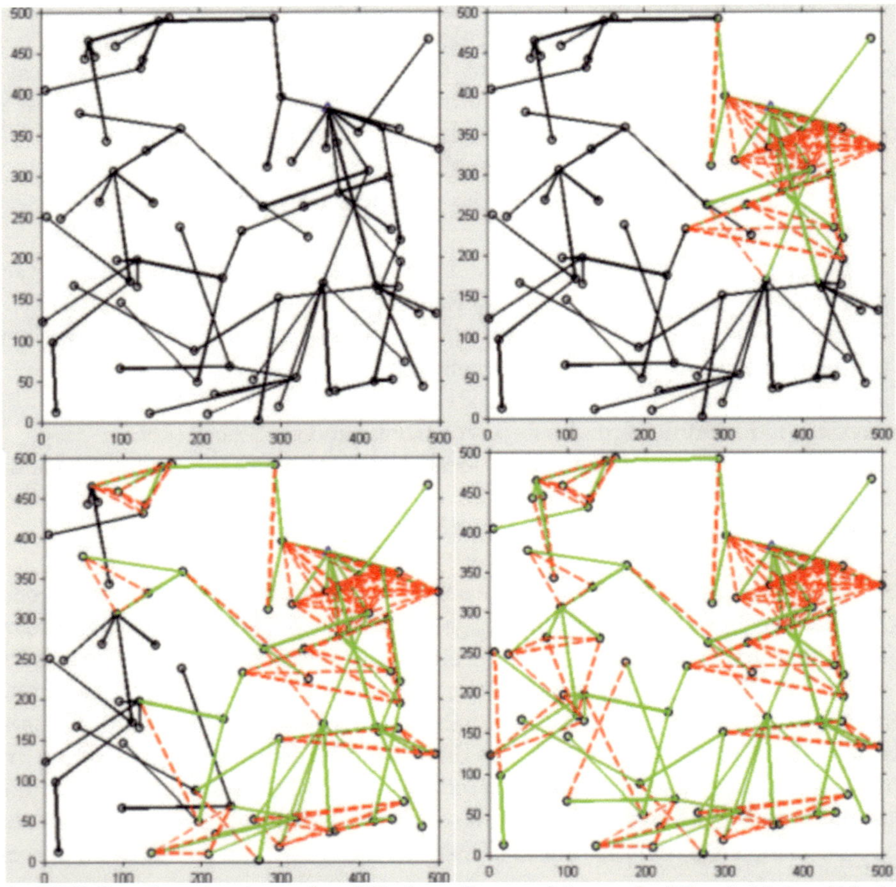

Fig. 1. RBS Synchronization of a Wireless Sensor Network (The initial solid dark lines represent the network's topology after flooding; the solid light lines represent transmitter-to-receivers communication; the dashed lines represent receiver-to-receiver transmissions).

There are some issues with the RBS synchronization algorithm that must be addressed in an energy-aware sensor network. First, the receiver-to-receiver synchronization method is effective at reducing the critical path to increase the accuracy, but RBS scales poorly with dense networks where there are many receivers for each transmitter. Given n receivers for a single transmitter, the number of transmissions increases linearly with n, but the number of receptions increases as $O(n^2)$. The following numbers of transmissions and receptions exist in RBS:

$$TX_{RBS} = n , \tag{1}$$

$$RX_{RBS} = n + \sum_{i=1}^{n-1} i = n + \frac{n(n-1)}{2} = \frac{n^2+n}{2} \quad (2)$$

For a large number of receivers per transmitter, this method becomes infeasible due to energy constraints.

Lastly, RBS does not account for lost network coverage when nodes begin losing power. Should a transmitting node be depleted, all of its receivers will be dropped from the network, so measures should be taken to re-establish connectivity when the coverage decreases beyond some threshold value.

2.2 TheTiming-Sync Protocol

The timing-sync protocol for sensor networks (TPSN) was developed in 2003 in an attempt to further refine time synchronization beyond RBS's capabilities (Ganeriwal, November 2003), (Ganeriwal, 2003). TPSN uses the same sources of uncertainty as RBS does (send, access, propagation, and receive), with the addition of two more:

- *Transmission time:* the time for the packet to be processed and sent through the RF transceiver during transmission.
- *Access time:* the time for each bit to be processed from the RF transceiver during signal reception.

The TPSN works in two phases:

1. *Level Discovery Phase:* this is a very similar approach to the flooding phase in RBS, where a hierarchical tree is created beginning from a root node.
2. *Synchronization Phase:* in this phase, pair-wise synchronization is performed between each transmitter and receiver.

In the level discovery phase, each sensor node is assigned a level according to the hierarchical tree. A pre-determined root node is assigned as level 0 and broadcasts a *level_discovery* packet. Sensors that receive this packet are assigned as children to the transmitter and are set as level 1 (they will ignore subsequent *level_discovery* packets). Each of these nodes broadcasts a *level_discovery* packet, and the pattern continues with the level 2 nodes.

In the synchronization phase, pair-wise synchronization is performed between the transmitter and receiver nodes using a 2-way handshake.

Although RBS removes the uncertainty at the sender by exchanging times amongst receivers, TPSN reduces the remaining uncertainties by a factor of 2 due to the handshake process that averages the clock drift and propagation delay. However, TPSN's uncertainty at the sender can be reduced to an insignificant delay by time-stamping at the MAC layer just before the bits are sent through the transceiver.

The number of transmitters and receivers in TPSN are as follows:

$$TX_{TPSN} = n+1, \quad (3)$$

$$RX_{TPSN} = 2n. \quad (4)$$

Fig. 2 shows how a sensor network is synchronized by using TPSN.

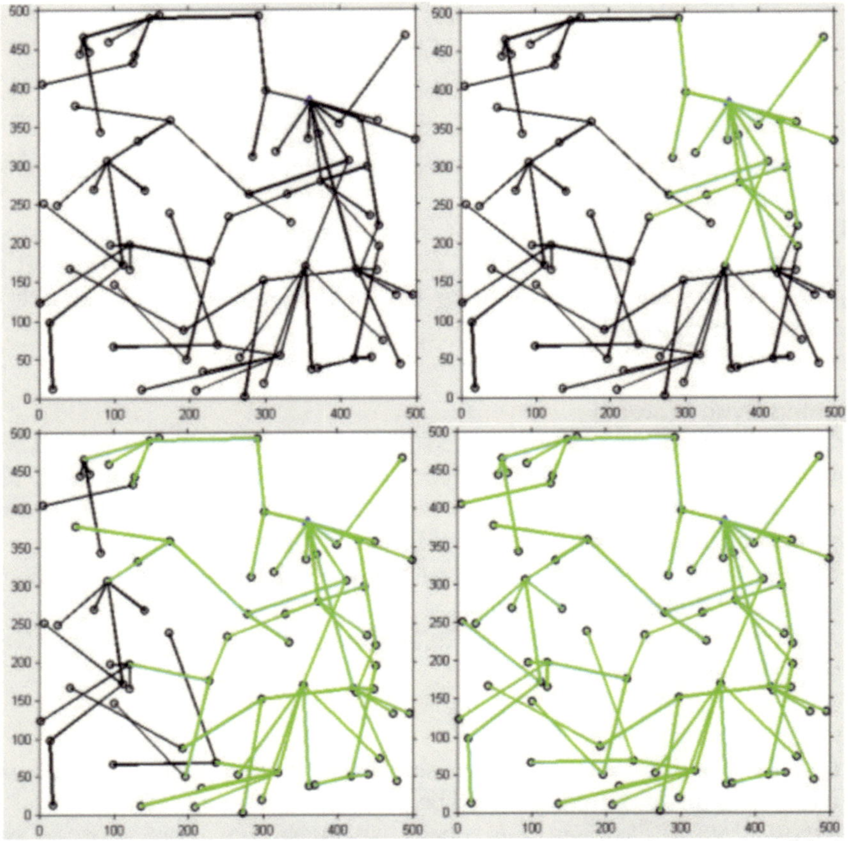

Fig. 2. TPSN Synchronization of a Wireless Sensor Network (The initial solid dark lines represent the network's topology after flooding; the subsequent light lines represent successful transmitter-to-receiver synchronizations).

TPSN is a great improvement over RBS in terms of accuracy since it employs a 2-way handshake, which reduces uncertainty to half since the average of the time differences is used. However, the main drawback TPSN faces is that it consumes energy in sparse networks; a 2-way handshake requires each node to receive a packet and to send one in response. In addition, TPSN shares the same problem with RBS with respect to lost network coverage when nodes begin losing power. A dead transmitter node will drop all of its receivers from the network, lowering the WSN's coverage area. Network restructuring is not included in the TPSN algorithm.

RBS and TPSN are some of the first efforts in creating synchronization algorithms tailored towards low-power sensor networks. They both have unique strengths when dealing with energy consumption. RBS is most effective in networks where transmitting sensors have few receivers, while TPSN excels when transmitters have many receivers.

2.3 Energy-Aware Time Sychronization

A new hybrid algorithm is proposed in this section.

2.3.1 Hybrid Flooding

Before the sensors can be synchronized, a network topology must be created. Table 1 shows the algorithm for the hybrid flooding algorithm that is used by each sensor node to efficiently flood the network.

Algorithm 1: Hybrid Flooding Algorithm

Accept *flood_packets*
Set *receiver_threshold* to *low_power*
Set *num_receivers* to 0
If *current_node* is root node
    Broadcast *flood_packet*
Else If *current_node* receives *flood_packet* and is accepting them
    Set parent of *current_node* to source of broadcast
    Set *current_node* level to parent's node level + 1
    Rebroadcast flood request with *current_node* ID and level
    Broadcast *ack_packet* with *current_node* ID
    Ignore subsequent *flood_packets*
Else If *current_node* receives *ack_packet*
    Increment *num_receivers*

Table 1. The Hybrid Flooding algorithm

Each sensor is initially set to accept *flood_packets*, but will ignore subsequent ones in order not to be continuously reassigned as the flood broadcast propagates. The *num_receivers* variable keeps track of the node's receivers and is used in the synchronization algorithm.

2.3.2 Hybrid Synchronization

Once the network flooding has been completed, the network can be synchronized using the determined hierarchy. In networks where the sensors are dispersed at random, there will be patches of high density node distribution interspersed with lower density regions. A transmitter in a high density area will usually have a large number of receivers, while another transmitter in a lower density section will usually have 1 or 2 receivers at most. As discussed in the previous sections, RBS excels when the transmitter has few receivers and TPSN excels with many receivers connected to each transmitter.

The hybrid algorithm minimizes power regardless of the network's topology by choosing the best synchronization technique depending on the number of children connected to the transmitter. Since the energy required for reception usually differs from that of a transmission, the ratio of the reception power to the transmission power is needed in order to find the optimal point at which to switch from receiver-receiver synchronization to transmitter-receiver synchronization. In order to find the ratio of reception-to-transmission power, α, we combine equations (1), (2), (3), and (4):

$$\alpha = \frac{TX_{TPSN} - TX_{RBS}}{RX_{RBS} - RX_{TPSN}} = \frac{2n}{-(n^2 - 3n)} \tag{5}$$

In general, the following equation can be used to determine the *receiver_threshold* by solving equation (5) for *n*:

$$n^2 - 3n - \frac{2}{\alpha} = 0 \tag{6}$$

Table 2 shows the algorithm for the hybrid Synchronization algorithm.

Algorithm 2: Hybrid Synchronization Algorithm

Set *receiver_threshold* to *high_power*
If *num_receivers* < *receiver_threshold* // *Use RBS algorithm*
    Transmitter broadcasts *sync_request*
    For each receiver
        Record local time of reception for *sync_request*
        Broadcast *observation_packet*
        Receive *observation_packet* from other receivers
Else // *Use TPSN algorithm*
    Transmitter broadcasts *sync_request*
    For each receiver
        Record local time of reception for *sync_request*
        Broadcast *ack_packet* to transmitter with local time

Table 2. The Hybrid Synchronization Algorithm

2.3.3 Energy Depletion

Another issue that the hybrid algorithm addresses when synchronizing a sensor network is the effect that a depleted sensor has on the topology. Once the battery is exhausted, the node will be dropped from the network, but so will all of the receivers depending on it. This loss of connectivity cascades through each receiver, so a drastic restructuring can occur when a high-level sensor is drained. The hybrid algorithm keeps track of the number of powered nodes. Once this number decreases below another user-defined threshold, the network is re-flooded using the flooding algorithm described earlier in Table 2. Should the source node lose power, a new source node is chosen from the original one's receivers. These receivers communicate their power levels with each other and the one with the most remaining energy is elected as the new root node, as show in Table 3.

Algorithm 3: Root Node Election Algorithm
If *cur_node_level* == 1 and *cur_node_power* allows 1 more TX
Broadcast *elect_packet* with *cur_node_ID*
If *cur_node_level* == 2
Broadcast *elect_packet* with *cur_node_ID, cur_node_power*
If *cur_node* receives *elect_packet* and *elect_packet_power* >= *cur_node_power*
Set *elect_packet_ID* to root node

Table 3. The Root Node Election Algorithm

In addition, receivers will only analyze the *sync_request* packets from their respective transmitters when using the TPSN-style synchronization. This saves additional battery power since the receivers do not have to analyze packets they overhear from other broadcasting transmitters. Lastly, the dropped packets are monitored. This is a useful statistic since it keeps track of algorithm efficiency and wasted energy. Dropped packets also allow us to compare various network topologies and determine which ones allow for the most energy conservation.

3. Results and Analysis

3.1 Hybrid Algorithm Validation

Several simulations were run to compare the power consumption of the TPSN, the RBS, and our hybrid algorithm discussed in the previous section. A transmitting sensor can dynamically switch between RBS and TPSN by simply comparing the number of connected receivers to the reception/transmission power ratio. This ratio is changed in order to observe how each of the algorithms is affected. All other parameters are kept constant. Our simulations are run on a 1000m x 1000m area, which is randomly populated with 500 sensors, and the path loss coefficient is set to 3.5. In each simulation, the *receiver_threshold* value is changed from 1 to the largest number of receivers connected to a sensor. The hybrid synchronization algorithm is executed for each of these *receiver_threshold* values and the energy consumption is stored and compared to the consumption of TPSN, RBS, and the optimal hybrid synchronization algorithm. Each of the data points is plotted, along with a line representing the average from all of the simulations. For the MICA2Dot platform, a reception uses approximately 24 mW of power, while a transmission requires 75 mW at -5 dBm (MICA2DOT, 2005). Solving for a and n in equations (5) and (6), we get a= 0.32 and n= 4.42, respectively.

The hybrid algorithm will use the least amount of energy when the *receiver_threshold* is set to 4.42. This means that transmitters with 4 or fewer sensors will use RBS for synchronization while those with 5 or more receivers will use TPSN. Fig. 3 illustrates how changes in the *receiver_threshold* value affect the hybrid algorithm.

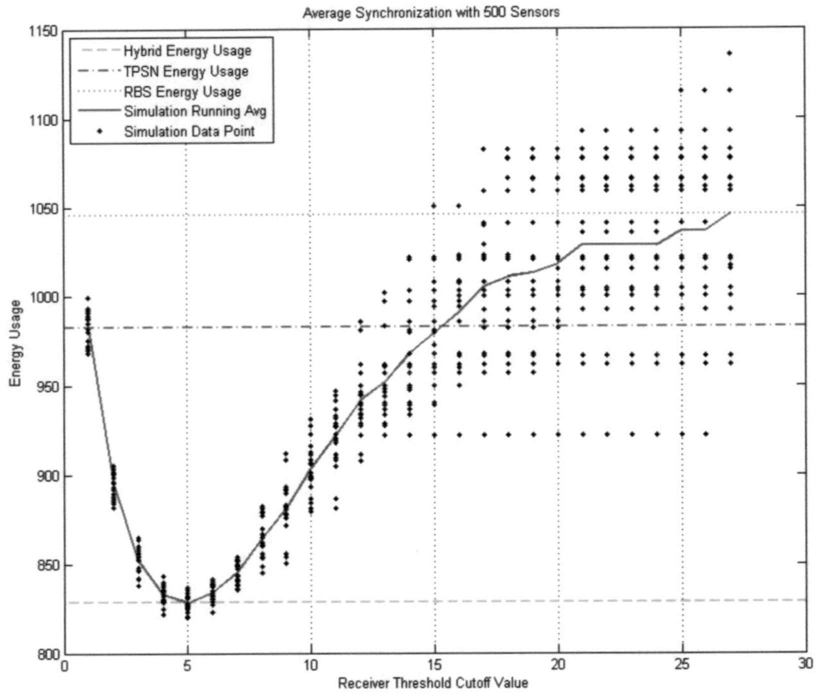

Fig. 3. Mica2DOT Synchronization Comparison

The energy consumption from the hybrid algorithm when using the optimal *receiver_threshold* value is lower than both TPSN and RBS. The minimum value is found between values of 4 and 5. Lastly, the spread amongst data points increases dramatically as the receiver threshold increases beyond 13.

More importantly, setting the *receiver_threshold* value to 1 will force a transmitter to use TPSN. The hybrid algorithm in this case will have the same energy consumption as TPSN. On the other hand, a *receiver_threshold* set to the largest number of receivers connected to a transmitter will force a transmitter to use RBS.

The hybrid synchronization algorithm is very dynamic and will adapt itself to multiple equipment specifications. The power requirements for the MicaZ sensor platform are drastically different from the Mica2DOT platform; MicaZ uses 59.1 mW for a reception, but only uses 42 mW for each transmission at -5 dBm (MICAz, 2005). Similarly, solving for a and n in equations (5) and (6), we get a= 1.407 and n= 3.42, respectively. When using MicaZ, the optimal *receiver_threshold* value is 3.42. This property is reflected in Fig. 4.,where the local minimum has shifted further to the left when compared to the Mica2DOT platform.

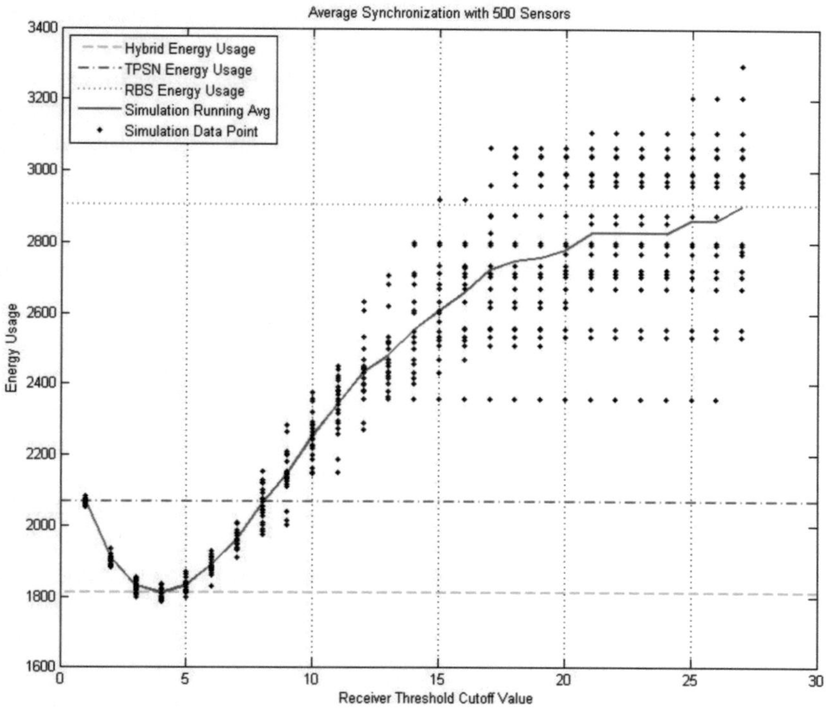

Fig. 4. MicaZ Synchronization Comparison

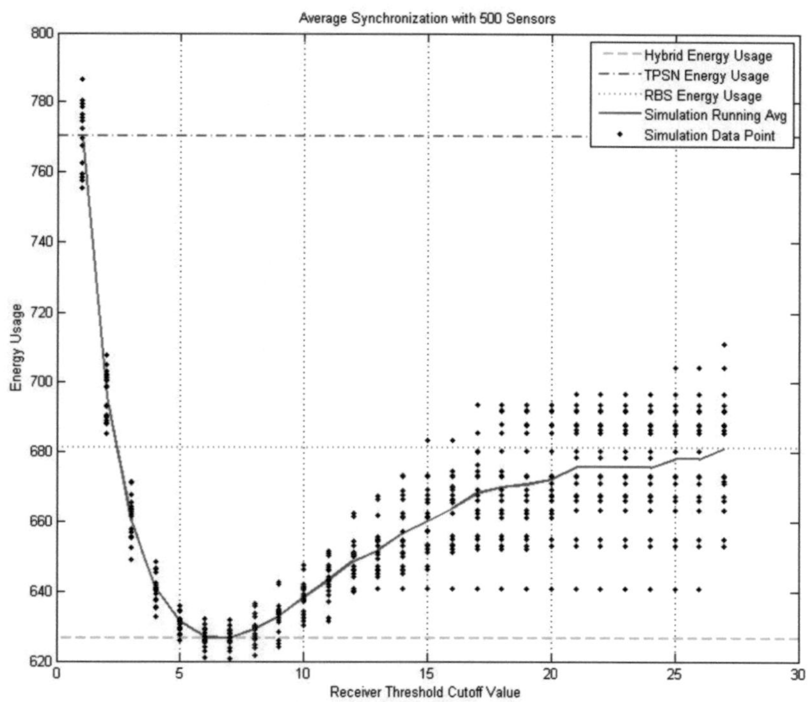

Fig. 5. Synchronization Comparison for Architecture with n=6

Despite the differences in architecture, both of the above examples yield relatively similar values for the optimal *receiver_threshold*. Assume that there is an improvement in the Mica2DOT platform which allows for much lower power in receiving mode. Each transmission still requires 75 mW at -5 dBm, but only 8 mW is needed for a reception. Then, a and n from equations (5) and (6) are 0.107 and 6, respectively. Fig. 5 illustrates the energy usage when the *receiver_threshold* changes.

In this particular example, the hybrid algorithm produces a local minimum when using the optimal *receiver_threshold*, as was expected. It is also interesting to note that now, RBS becomes more energy efficient than TPSN.

3.2 Power Consumption

The next set of simulations demonstrates the algorithm's reduction in power consumption in several network sizes. The number of sensors was changed from 250 up to 1500, in increments of 250. Just as before, 20 simulations were run over a 1000m x 1000m area which was randomly populated with 500 sensors, and the path loss coefficient was set to 3.5. The Mica2DOT platform was used and the ratio of reception/transmission power remained fixed. The *receiver_threshold* value is once again changed from 1 to the largest number of receivers connected to a sensor. The hybrid synchronization algorithm is executed for each of these *receiver_threshold* values and the energy consumption is stored and compared to the consumption of TPSN, RBS, and the optimal hybrid synchronization algorithm. Each of the data points is plotted, along with a line representing the average from all of the simulations as show in Fig. 6, Fig. 7, and Fig. 8.

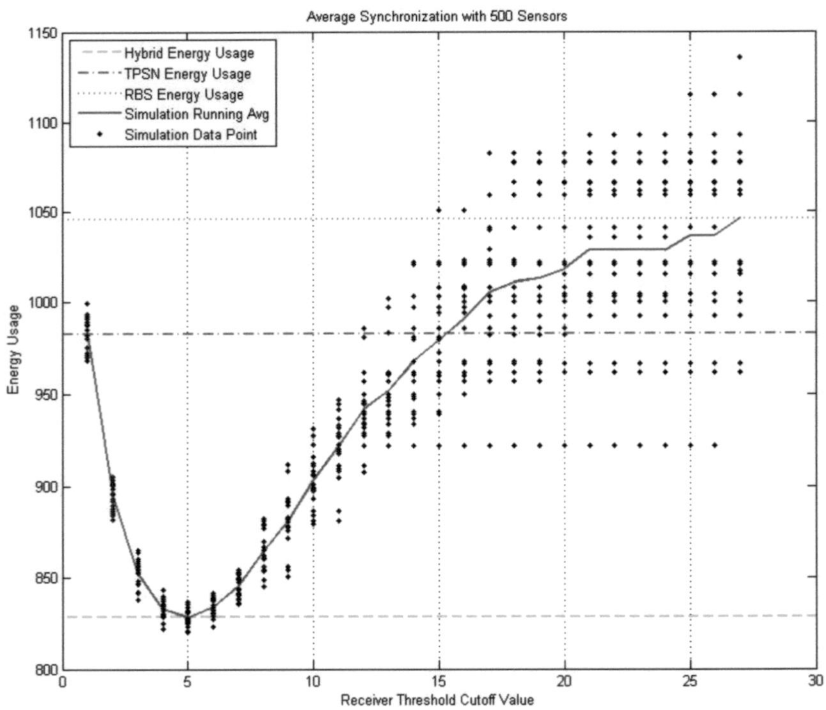

Fig. 6. Energy usage consumption for 500 sensors between RBS, TPSN, and our Hybrid algorithm for different values of *receiver_threshold* values using Mica2Dot platform. Energy usage is in mW.

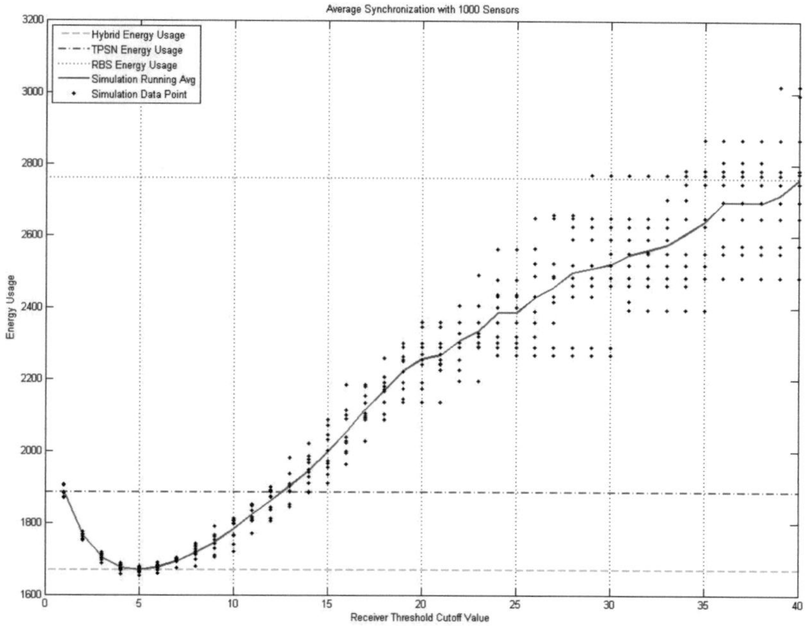

Fig. 7.Energy usage consumption for 1000 sensors between RBS, TPSN, and our Hybrid algorithm for different values of *receiver_threshold* values using Mica2Dot platform. Energy usage is in mW.

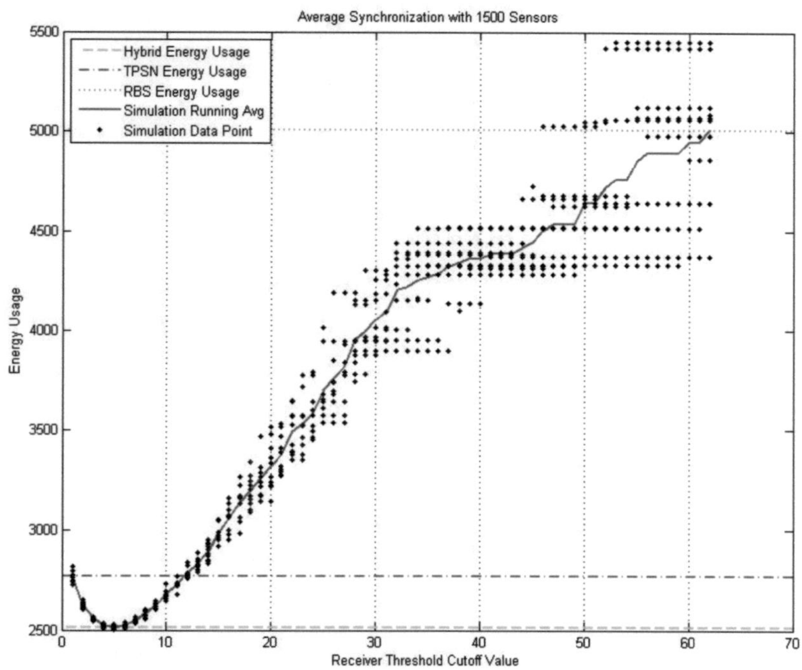

Fig. 8. Energy usage consumption for 1500 sensors between RBS, TPSN, and our Hybrid algorithm for different values of *receiver_threshold* values using Mica2Dot platform. Energy usage is in mW.

As more sensors are introduced into the network, RBS becomes dramatically less feasible for a wireless sensor network. As shown in Table 4, the hybrid algorithm's energy savings over RBS increases from 58% with 750 sensors to over 74% when the network uses 1500 sensors.

Sensors	250	500	750	1000	1250	1500
RBS	615	1709	3421	5510	7833	11128
TPSN	498	998	1498	1998	2498	2998
Hybrid	447	924	1415	1898	2386	2879
RBS Savings	27.44 %	45.94 %	58.65 %	65.55 %	69.54 %	74.13 %
TPSN Savings	10.27 %	7.43 %	5.57 %	4.99 %	4.47 %	3.97 %

Table 4. Average Number of Receptions

In contrast, as the network becomes large, the hybrid algorithm mimics TPSN's behavior, but uses less energy. The difference is 5.57% with 750 sensors and 3.97% with 1500 sensors. Although the number of receptions when using TPSN increases linearly with network size, this number increases much more quickly when using RBS. The hybrid algorithm greatly reduces the number of receptions when compared to RBS; for small networks, the advantage is 27%, but it increases to over 74% in networks with a large number of sensors. Therefore, the hybrid algorithm has a large advantage over TPSN in small networks, but that advantage decreases as more sensors are added.

Table 5 shows the standard deviation in the number of receptions for each of the synchronization algorithms. These results help to determine how sensitive an algorithm is to modifications in the network's topology and sensor density.

Sensors	250	500	750	1000	1250	1500
RBS	54.71 8.89 %	150.09 8.78 %	365.43 10.68 %	524.32 9.52 %	614.26 7.84 %	1129.50 10.15 %
TPSN	0.73 0.15 %	0.00 0.00 %	0.00 0.00 %	0.00 0.00 %	0.00 0.00 %	0.00 0.00 %
Hybrid	11.73 2.63 %	13.16 1.42 %	15.89 1.12 %	14.75 0.78 %	15.99 0.67 %	16.77 0.58 %

Table 5. Standard Deviation for Receptions

The table above shows that there is very large variation in the number of receptions for RBS, meaning that the number of receptions when using RBS is highly dependent on the topology of the network. The table also shows that the deviation in receptions when using TPSN is usually 0, with the exception once again in the 250 sensor network. This exception is due to orphaned nodes which do not participate in the synchronization. The hybrid algorithm has a relatively low deviation, which decreases further with large numbers of sensors. This behavior is attributed to the hybrid algorithm behaving similarly to TPSN when the network is large.

Another simulation results are shown in Table 6 and Table 7. These results show that RBS's energy consumption is more dependent on the density of sensors in a given area. In contrast, TPSN and the hybrid algorithm are less affected by the size of the network.

Sensors	250	500	750	1000	1250	1500
RBS	446	1046	1844	2762	3756	5060
TPSN	511	983	1434	1885	2331	2770
Hybrid	404	828	1253	1672	2095	2514
RBS Savings	9.29%	20.79%	32.04%	39.46%	44.22%	50.31%
TPSN Savings	20.80%	15.73%	12.65%	11.28%	10.11%	9.23%

Table 6. Average Energy Consumption in mW

Sensors	250	500	750	1000	1250	1500
RBS	17.38 3.90%	48.03 4.59%	116.94 6.34%	167.78 6.07%	196.56 5.23%	361.44 7.14%
TPSN	7.67 1.50%	8.88 0.90%	14.31 1.00%	14.48 0.77%	18.22 0.78%	22.09 0.80%
Hybrid	4.00 0.99%	4.72 0.57%	5.23 0.42%	6.85 0.41%	6.33 0.30%	6.84 0.27%

Table 7. Standard Deviation of Energy Consumption

When the network size increases from 250 sensors to 500 sensors (for the same area of 1 km^2), RBS becomes less energy efficient than TPSN. The hybrid algorithm outperforms TPSN by 15.7%, while outperforming RBS by 20.8%. Once the network increases to 750 sensors, RBS clearly becomes less efficient than TPSN. The hybrid algorithm still outperforms TPSN by 12.7%. Since RBS consumes more energy, the hybrid algorithm now outperforms it by 32%. As more sensors are introduced into the network, RBS becomes dramatically less feasible for a wireless sensor network. As shown in Table I, the hybrid algorithm's energy savings over RBS increases from 39% with 1000 sensors to over 50% when the network uses 1500 sensors. In contrast, as the network becomes large, the hybrid algorithm mimics TPSN's behavior, but uses less energy. The energy savings over TPSN are 11% with 1000 sensors and 9% with 1500 sensors. For extremely large networks (10,000+ sensors) TPSN has the same efficiency as our proposed algorithm.

4. Conclusion and Future Work

Wireless sensor networks have tremendous advantages for monitoring object movement and environmental properties but require some degree of synchronization to achieve the best results. The hybrid synchronization algorithm was designed to switch between Timing-sync Protocol for Sensor Networks (TPSN) and the Reference Broadcast Synchronization algorithm (RBS). These two algorithms allow all the sensors in a network to synchronize themselves within a few microseconds of each other, while at the same time using the least amount of energy possible. The savings in energy varies upon the density of the sensors as well as the reception-to-transmission ratio of energy usage; networks, which are saturated with sensors, for example 1500 sensors in a 1 km^2 area, will favor TPSN over RBS. TPSN also becomes more favorable as receptions consume more power. The hybrid algorithm compromises between both of these previous algorithms. The energy savings over RBS can range from 9.3% in small networks of 250 sensors, to over 50% for large networks using 1500 sensors. In contrast, the hybrid algorithm's savings over TPSN range from 20.8% in the same small networks down to 9% in the large networks. Furthermore, analysis of the standard deviation for each of the algorithms shows RBS's energy consumption can vary dramatically, from nearly 4% to over 7%, generally increasing for larger networks. In contrast, the standard deviation for TPSN's energy usage decreases from 1.5% to less than 1%, generally decreasing for larger networks. The hybrid algorithm's deviation is always less than 1% and continuously decreases to 0.3% as more sensors are used.

5. References

Akl, R. & Saravanos, S. Hybrid Energy-Aware Synchronization Algorithm in Wireless Sensor Networks, *18th IEEE International Symposium on Personal, Indoor, and Mobile Radio Communications, PIMRC'07,* pp. 1-5, September 2007, Athens.

Crossbow MICAz Wireless Measurement System, Document Part Number 6020-0060-03 Rev A, http://www.xbow.com/Products/Product_pdf_files/Wireless_pdf/MICAz_Datasheet.pdf

Elson, J.; Girod, L. & Estrin, D. Fine-Grained Network Time Synchronization using Reference Broadcasts, *Proceedings of the Fifth Symposium on Operating Systems Design and Implementation (OSDI 2002),* December 2002, Boston.

Ganeriwal, S.; Kumar, R. & Srivastava, M. Timing Sync Protocol for Sensor Networks, *ACM SenSys '03*, pp. 138-149, November 2003, Los Angeles.

Ganeriwal, S. & Srivastava, M. Timing-sync Protocol for Sensor Networks (TPSN) on Berkeley Motes, *NESL*, 2003.

He, L. & Kuo, G. A Novel Time Synchronization Scheme in Wireless Sensor Networks, *IEEE 63rd Vehicular Technology Conference, VTC 200,*. pp. 568-572, May 2006, Melbourne.

Langendoen, K. & Reijers, N. (2003). Distributed Localization in Wireless Sensor Networks: A Quantitative Comparison. *The International Journal of Computer and Telecommunication Networking*, Vol. 43, No. 4, 2003, pp. 499-518.

Lee, H.; Yu, W. & Kwon, Y. Efficient RBS in Sensor Networks, *3rd International Conference on Information Technology: New Generations, ITNG*, pp. 279-284, April 2006, Las Vegas.

Mica2Dot Wireless Microsensor Mote Document Part Number: 6020-0043-05 Rev A. 2005, [Online], available: http://www.xbow.com/Products/Product_pdf_files/Wireless_pdf/MICA2DOT_Datasheet.pdf.

MICAz Wireless Measurement System Document Part Number: 6020-0060-03 Rev A. 2005, [Online], available: http://www.xbow.com/Products/Product_pdf_files/Wireless_pdf/MICAz_Datasheet.pdf

NTP: The Network Time Protocol, (October 29, 2009), http://www.ntp.org/(visited January 23, 2010).

Palchaudhuri, S.; Saha, A. & Johnsin, D. Adaptive Clock Synchronization in Sensor Networks, *3rd International Symposium on Information Processing in Sensor Networks, IPSN*, pp. 340-348, April 2004, Berkeley.

Patwari, N.; Ash, J.N.; Kyperountas, S.; Hero, A.O.; Moses, R. L. & Correal, N.S. (2005). Locating the nodes: Cooperative Localization in Wireless Sensor Networks. *IEEE Signal Processing Magazine*, Vol. 22, No. 4, July 2005, pp. 54-69.

Qun, L. & Rus, D. Global clock synchronization in sensor networks. *IEEE Trans. On Computers*, Vol. 55, No. 2, Feb. 2006, pp. 214-226.

Sivirkaya, F. & Yener, B. Time Synchronization in Sensor Networks: A Survey. *IEEE Network*, Vol. 18, No. 4, Jul-Aug. 2004, pp. 45-50.

Sun, K.; Ning, P. & Wang, C. Secure and resilient clock synchronization in wireless sensor networks. *IEEE Journal on Selected Areas in Communications*, Vol. 24, No. 2, Feb. 2006, pp.395-408.

Tubaishat, M.; Qi, Q.; Shang, Y. & Shi, H. (2008). Wireless Sensor-Based Traffic Light Control, *5th IEEE Consumer Communications and Networking Conference,CCNC'08*, pp. 702-706, January 2008, Las Vegas.

Wenjie, C.; Lifeng, C.; Zhanglong, C. & Shiliang, T. (2005). A Realtime Dynamic Traffic Control System based on Wireless Sensor Network, *International Conference Workshops on Parallel Processing ICPP'05*, pp. 258-264, June 2005, Oslo.

19

Maximizing Lifetime of Data Gathering Wireless Sensor Network

Ryo Katsuma*, Yoshihiro Murata†, Naoki Shibata‡,
Keiichi Yasumoto* and Minoru Ito*
** Nara Institute of Science and Technology,*
†Hiroshima City University, ‡Shiga University
Japan

1. Introduction

Wireless Sensor Networks (WSNs) are networks consisting of many small sensor nodes capable of wireless communication, and they are used for environmental monitoring, border guards, and so on. Among many types of WSNs, data gathering WSN periodically collects to a *sink node* environmental information such as temperature and amount of sunlight at each point in a wide agricultural area or forest. Some data gathering WSN applications need sufficient sensing quality and robustness of the system, and such systems may require *k*-coverage[1] of the target sensing field. Data gathering WSNs that require *k*-coverage of the field should also operate for a long term. Thus, many research efforts have been devoted to the *k*-coverage problem and the WSN lifetime extension problem.

In order to make such a WSN operate for a long term, Tang et al. reduced power consumption by regulating communication frequency among sensor nodes [Tang et al. (2006)]. Heinzelman et al. reduced total data transmission by merging the data received from multiple sensor nodes [Heinzelman et al. (2000)]. However, since the above existing approaches degrade sensing quality with respect to collected data amount and sensing frequency, some applications that always need sufficient sensing quality may not accept such a quality degradation.

Cao, et al. proposed a sleep scheduling method which lets nodes sleep when they need not communicate, in order to save the overall power consumption in WSN [Cao et al. (2005)]. Keshavarzian, et al. proposed a method to minimize active nodes and guarantee that the event information sensed by sensor nodes arrives to the sink node in a specified time [Keshavarzian et al. (2006)]. In these methods, sleeping nodes consume small power, but do not communicate with other nodes, and become active after specified time interval. These existing methods target applications collecting events occurring rarely and do not consider the field *k*-coverage.

In order to *k*-cover the field, Poduri et al. used mobile sensor nodes to *k*-cover the target sensing field in short time under the constraint that for each sensor node, *k* other sensor nodes always exist in its proximity [Poduri et al. (2004)]. They also discussed about the optimal locations of sensor nodes for *k*-covering the field. This method does not consider maintaining *k*-coverage of the field for a long time though it makes *k*-coverage in short time.

---

[1] Any point in the target area is covered by at least *k* sensor nodes.

In this chapter, we propose two methods to maximize the operation time of the data gathering WSN during which the whole target field is k-covered (we call the time *k-coverage lifetime*, hereafter). The first method uses mobile sensor nodes [Katsuma et al. (2009)]. The second method uses more-than-enough number of static sensor nodes [Katsuma et al. (2010)].

First, we define a k-coverage lifetime maximization problem for WSNs consisting of both static and mobile sensor nodes sparsely deployed in the field. The target problem is to decide a moving schedule of mobile nodes (when and to which direction each mobile node should move at each time during WSN operation time) and a tree spanning all sensor nodes for data collection (we use a tree as data communication paths). This problem is NP-hard. So, we propose a genetic algorithm (GA) based scheme to find a near optimal solution in practical time. In order to speed up the calculation, we devised a method to check a sufficient condition of k-coverage of the field. To mitigate the problem that nodes near the sink node consume a lot of energy for forwarding the data from farther nodes, we construct a tree where the amount of communication traffic is balanced among all nodes, and add this tree to the initial candidate solutions of our GA-based algorithm. Through the simulations, we confirmed that the k-coverage lifetime of our method is about 140% to 190% longer than the other conventional methods for 100 to 300 nodes WSNs.

Next, to maximize k-coverage lifetime of a WSN with static sensor nodes deployed in high density, we define a problem to decide a sleep schedule of all nodes and a data collection tree. In this problem, we assume that each sensor node has three operation modes: sensing, relaying, and sleeping. Each sensing node senses environmental data and sends/relays the data to the sink node via multi-hop wireless communication. Each relaying node just forwards the data received from its uplink node to its downlink nodes. Each sleeping node does nothing and keeps its battery. We propose a method to solve this problem by making the minimal number of the nodes required for k-coverage active, and replacing the node that exhausted battery by another one. This method chooses active nodes one by one in the order of the impact degree the selected node has for k-coverage of the field. In order to evaluate the effectiveness of our algorithms in terms of k-coverage lifetime, we compared our method with methods in which some of the proposed features are disabled. Through simulation-based comparison, we confirmed that the proposed methods achieve 1.1 to 1.7 times longer k-coverage lifetime regardless of k and the number of nodes, than the other methods in which some of the proposed features are disabled for 100 to 500 node WSNs.

2. Assumptions of WSN

In this section, we present the common WSN model, assumptions, and common definitions used by each proposed method. Assumptions and definitions specific for each method will be described later.

2.1 Assumptions on Target WSN

We suppose a WSN in which a massive number of small battery-driven sensor nodes are deployed in a *target field*. Sensor nodes periodically sense such environmental information as temperature, humidity, sunlight, or moving object, and send it by multi-hop communication to a base station called a *sink node*. We denote the target field, the sink node, and the sensing frequency as $Field$, Bs, and I, respectively. We denote the set of sensor nodes by $S = \{s_1, ..., s_l\}$. Sensor nodes have three operation modes: *sensing*, *relaying*, and *sleeping*. A node whose operation mode is sensing, relaying, or sleeping is called *sensing node*, *relaying node*, or *sleeping node*. We denote the sets of sensing, relaying, sleeping nodes by $U = \{u_1, u_2, ...\}$, $V = \{v_1, v_2, ...\}$,

$W = \{w_1, w_2, ...\}$, respectively, where $U \cup V \cup W = S$. Once a node changes its mode to the sleeping mode, it does not wake up until the specified sleeping time elapses. Sleeping nodes can change their modes upon wakeup. Sensing nodes and relaying nodes can change their modes instantly.

A sensing node collects environmental data from a disk with radius R centered at the node. We denote the covered range of sensing node $s \in U$ by $s.range$. Each sensing node obtains data by sensing. We assume that the data size is fixed and the data are sent to the sink node without compression or unification along a multi-hop path to the sink node. We use a tree connecting all sensing and relaying nodes to the sink node as communication paths (we call *data collection tree*). We denote the sensing data size by D.

Each sensing/relaying node has a wireless communication capability and its radio transmission range is a disk with a specified radius centered on it. Each node can change its transmission power to change the communication distance. Since there is little influence on radio interference when sensing frequency I is small enough, we assume that there is no packet collision between nodes. A transmitted packet is always successfully received if the destination node (sensing/relaying node) is within the radio transmission range, and always fails if outside of the range. We assume that each node uses only one-hop unicast communication by designating a destination node.

We assume that each sensor node knows its position and sink node Bs is informed of positions of all nodes at their deployment time (e.g., with single-hop or multi-hop communication from each node to Bs). For each sensor node s, we denote its location by $s.pos$. Similarly, we denote the location of the sink node by $Bs.pos$. The sink node conducts the centralized calculation and informs the solution to all nodes by single-hop or multi-hop flooding.

2.2 Assumptions for Power Consumption

Each sensor node s has a battery, where the initial energy amount and the remaining energy amount at time t are denoted by e_{init} and $s.energy[t]$, respectively. Each node consumes energy for data transmission, data reception, and sensing data, and even during idle time and sleeping time.

Powers $Trans(x, d)$ and $Recep(x)$ required to transmit x[bit] for d[m] and receive x[bit] conform to formulas (1) and (2), respectively [Heinzelman et al. (2000)].

$$Trans(x, d) = E_{elec} \times x + \epsilon_{amp} \times x \times d^n \tag{1}$$

$$Recep(x) = E_{elec} \times x \tag{2}$$

Here, E_{elec} and ϵ_{amp} are constants representing the power required by information processing and the power for amplification, respectively. The value of $n(\geq 0)$ is defined by the antenna properties.

Powers $Sens()$, $Listen(y)$, and $Sleep(y)$ required to sense the information which is D[bit] data, listen to whether radio messages come or not for y [s], and sleep for y [s] conform to the following formulas (3), (4), and (5), respectively.

$$Sens() = E_{elec} \times D + E_{sens} \tag{3}$$

$$Listen(y) = E_{listen} \times y \tag{4}$$

$$Sleep(y) = E_{sleep} \times y \tag{5}$$

Here, E_{sens}, E_{listen}, and E_{sleep} are constants representing the powers required for sensing data, listening for 1 second, and sleeping for 1 second, respectively.
The energy consumption of sensor node s per unit of time $C(s)$ is as follows:
For each sensing node $s \in U$,

$$C(s) = I \times (Sens() + Recep(D \times s.desc) + Trans(D \times (s.desc + 1), Dist(s, s.send)) + Listen(1) \tag{6}$$

For each relaying node $s \in V$,

$$C(s) = I \times (Recep(D \times s.desc) + Trans(D \times (s.desc), Dist(s, s.send))) + Listen(1) \tag{7}$$

For each sleeping node $s \in W$,

$$C(s) = Sleep(1) \tag{8}$$

where $s.desc$ is the number of sensing nodes except for s in the subtree of the data collection tree rooted on s, $s.send$ is the parent node of s, and $Dist(s_1, s_2)$ is the distance from s_1 to s_2.

2.3 Definition of k-coverage

We define k-coverage as follows:

$$\forall t \in [t_0, t_{end}], \forall pos \in Field, |Cover(pos, t)| \geq k. \tag{9}$$

where

$$Cover(pos, t) \stackrel{def}{=} \{s | pos \in s.range \wedge Mode(s, t) = sensing \wedge s.energy[t] > 0\}. \tag{10}$$

The condition (9) guarantees the k-coverage of the target field. In general, k-coverage can be achieved by a part of all sensor nodes ($U \subseteq S$) whose remaining energy amounts are not exhausted.
We define the k-coverage lifetime t_{life} of WSN as the time from initial deployment to the time when condition (9) cannot be satisfied by the remaining sensor nodes. Our objective is to maximize t_{life}.

3. k-coverage Lifetime Maximization Method Using Mobile Nodes

In this section, we formulate the problem to maximize the k-coverage lifetime by using mobile sensor nodes, propose an algorithm to solve the target problem, and show simulation results to validate the usefulness of our proposed method.

3.1 Assumptions and Problem Definition

In this section, we present assumptions for mobile nodes and formulate the problem of maximizing the k-coverage lifetime of a WSN with mobile sensor nodes.

3.1.1 Assumptions for Mobile Nodes

Both *static nodes* and *mobile nodes* are used as sensor nodes. Static nodes cannot be moved from their originally placed locations, while mobile nodes can move by wheels. We denote the sets of static and mobile sensor nodes by $P = \{p_1, ..., p_l\}$ and $Q = \{q_1, ..., q_m\}$, respectively. We assume that there is no obstacle in *Field*, and a mobile node can move straight to an arbitrary position in *Field*. The sensor nodes is deployed over the field without the excess and deficiency for k-coverage of the field. So, each static and mobile sensor node is always sensing node ($P \cup Q \subset U$).
Mobile nodes consume battery power not only by communication but also by movement. Power $Move(d)$ required to move d[m] conforms to formula (11) [Wang et al. (2005)].

$$Move(d) = E_{move} \times d \tag{11}$$

Here, E_{move} is a constant. Each mobile node can move at V [m/s] where V is a constant value.

3.1.2 Problem Definition

When a WSN operates for a long time, batteries of some sensor nodes will be exhausted and k-coverage will be broken. Then, it is necessary to move mobile nodes one after another. So, we formulate a problem to find the data collection tree and the schedule of moving for all mobile nodes in order to maximize the k-coverage lifetime.
The initial WSN deployment time is denoted by t_0. t_{end} denotes the time when the k-coverage of the WSN cannot be maintained any longer due to battery exhaustion or failures of multiple nodes ($t_{end} \geq t_{life}$). For each $q \in Q$ and each $t \in [t_0, t_{end}]$, the speed (0 or V) and direction of q at time t is denoted by $Run(q, t)$. Then, for each $q \in Q$, the speed-direction schedule for q's movement during time interval $[t_0, t_{end}]$ is denoted as follows.

$$schedule(q, [t_0, t_{end}]) = \bigcup_{t \in [t_0, t_{end}]} \{Run(q, t)\} \tag{12}$$

Given the information on the target field *Field*, a sink node *Bs* and its position *Bs.pos*, *s.pos*, *s.energy*, and *s.range* for each sensor node $s \in P \cup Q$, and constants E_{elec}, ϵ_{amp}, n, E_{sens}, E_{listen}, E_{move}, V, D, and I, our target problem for maximizing k-coverage lifetime denoted by t_{life} is to decide the schedule $schedule(s, [t_0, t_{end}])$ for each node $s \in P \cup Q$ and a data collection tree containing all sensor nodes that satisfies condition (9).

3.1.3 Modified Target Problem

Our target problem formulated in Section 3.1.2 is to decide speed-direction schedule of each mobile sensor node $q \in Q$ during time interval $[t_0, t_{end}]$. Then, we must decide a data collection tree including all sensor nodes whenever the positions of mobile nodes change. Solving the problem is considered to be very difficult because of the wide solution space. Therefore, we adopt a heuristic method to solve this problem stepping on the several stages as the following procedures:

1. Solving the problem to find the positions of mobile nodes and the data collection tree for maximizing *the WSN forecast endtime* (defined later) satisfying condition (9).
2. Whenever the battery of any sensor node is newly exhausted, go to step 1.

In the problem of step 1, its input is the same as the original problem. Its output is the new position of each mobile node $q \in Q$ denoted by $q.newpos$ satisfying condition (13) and the parent node of each sensor node $s \in P \cup Q$ denoted by $s.send$. We have the following constraint on $q.newpos$.

$$|q.pos - q.newpos| < \frac{V}{I} \tag{13}$$

Here, the new position of each mobile node is in the area where each mobile node can move in $\frac{1}{I}$ seconds.
The WSN lifetime t_{life} is the time of WSN termination considering the movement of mobile nodes in the future. It is difficult to calculate t_{life} strictly. So, we define the *WSN forecast endtime* when the battery of some sensor node is newly exhausted instead of t_{life}. Thus, we use the following objective function.

$$\textbf{maximize}\left(t_{now} + \min_{s \in P \cup Q}\left(\frac{s.energy}{C(s)} - \frac{Move(|s.pos - s.newpos|)}{C(s)}\right)\right) \tag{14}$$

where t_{now} is current time, and $C(s)$ is the energy consumption of sensor node s per second. If $s \in P$, $|s.pos - s.newpos| = 0$. So, $\frac{s.energy}{C(s)} - \frac{Move(|s.pos - s.newpos|)}{C(s)}$ means the time from present until the battery of the sensor node $s \in P \cup Q$ is exhausted.

3.1.4 NP hardness

The Minimum Geometric Disk Cover Problem(GDC), which is an NP-hard problem, is defined as follows [Srinivas et al. (2006)].

Problem GDC: Given a set N of RNs (points) distributed in the plane, place the smallest set M of Cover MBNs (disks) such that for every RN $i \in N$, there exists at least one MBN $j \in M$ exists such that $d_{ij} \leq r$.

The instance of GDC is set to (N, m, r), where N is a set of points, m is the number of disks, and r is a radius of each disk. Now, we assume that the WSN is constructed by m mobile nodes (no static nodes), sensing radius is set to r, and the target field is set to N. Existence of solution of GDC and 1-coverage of the field N using m mobile nodes are equivalent. Polynomial-time reduction from GDC to our target problem is possible. Thus, our target problem is NP-hard.

3.2 Algorithm

In this section, we describe the algorithm to solve the problem defined in Section 3.1.3.

3.2.1 Overview

Our algorithm decides the destinations of mobile sensor nodes and a data collection tree. Whenever the battery of any sensor node is newly exhausted, our algorithm is applied, as shown in Section 3.1.3. The proposed GA-based algorithm is supposed to be executed at the initial deployment time and ends when k-coverage of the target field is unable to be maintained.
In the calculation algorithm, each GA chromosome contains the positions for all the mobile nodes and the structure of the data collection tree. As a standard GA, it first generates initial candidate solutions to which it repeatedly applies GA operations. GA performance is largely

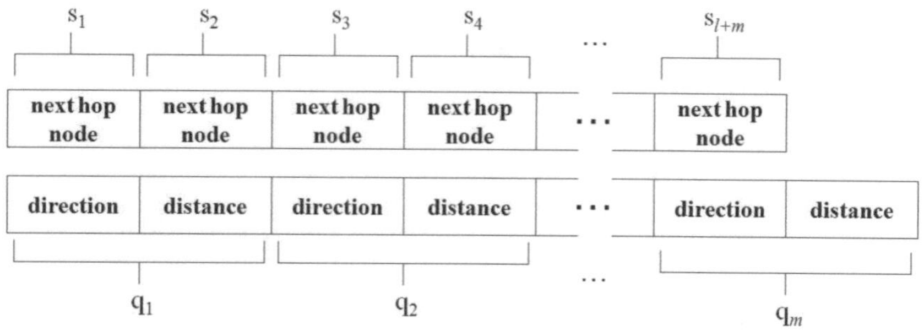

Fig. 1. Encoding of Candidate Solution

influenced by the quality of the initial candidate solutions. To improve its performance, we provide trees for the initial candidate solutions by the *balanced edge selection method*, which is described later in Section 3.2.4. For calculation speed, we developed the *delta-k-coverage* judgment method that decides a sufficient condition for the k-coverage of the field by sensor nodes in Section 3.2.5.

3.2.2 Algorithm details

GA is a well-known meta-heuristic algorithm. The following is its basic procedure.

1. **Generation of initial candidate solutions**: N candidate solutions are randomly generated.
2. **Evaluation**: Objective function for each candidate solution is evaluated to grade each candidate solution.
3. **Selection**: N candidate solutions with better evaluations are selected.
4. **Crossover**: New candidate solutions are generated by mixing two randomly selected candidate solutions.
5. **Mutation**: Part of candidate solutions are randomly mutated.
6. **Check termination**: If the termination condition is met, the candidate solution with the highest evaluation is output as the solution. Otherwise, go to Step 2.

Below, we show our algorithm for each GA operation.

Encoding of candidate solution: To apply a GA, each candidate solution has to be encoded, and the way of encoding sometimes greatly affects the algorithm performance. The coding in the proposed algorithm is shown in Fig. 1. Each candidate solution contains positions for $|Q|$ mobile nodes and the structure of the data collection tree consisting of $|P \cup Q|$ sensor nodes. The positions for the mobile sensor nodes are represented in polar coordinates to avoid generating impossible destinations of mobile sensor nodes. A data collection tree is represented by a set of node IDs.

Generation of initial candidate solutions: Initial candidate solutions are made from random variables. Angles and distances of mobile nodes are uniformly assigned distributed random values between 0 and 2π, and 0 and *Dist*, respectively (here, *Dist* is a constant and typically set to the longest movable distance in the target field). As an initial parent node for each node, a node geographically closer to the sink node is randomly selected. For efficiency, three

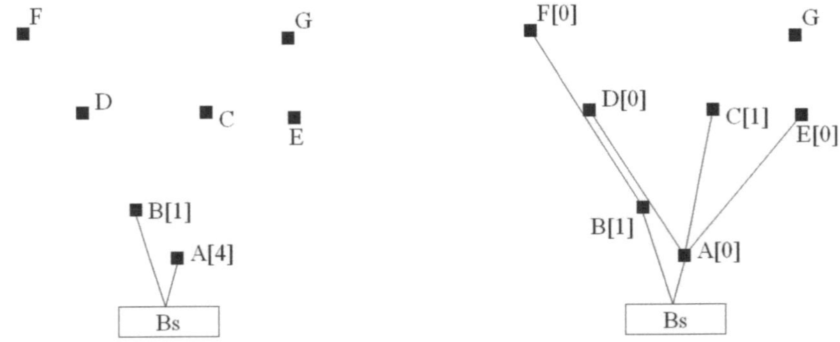

(a) Repeats 1 time (b) Repeats 3 times

Fig. 2. Example of Balanced Edge Selection Algorithm

candidate solutions are added to the initial candidate solutions whose collection trees are made using the minimum cost spanning tree method where an edge cost is the square of the distance, the balanced edge selection method proposed in Section 3.2.4, and a method that directly connects all sensor nodes to the sink node.

Evaluation: The evaluation of each candidate solution verifies how long the target sensing field is k-covered by a simulation of WSN data transmission. The k-coverage lifetime is between the time when all mobile sensor nodes arrive at their new positions and the time when k-coverage cannot be maintained due to battery exhaustion of some nodes. If the decoded data collection network does not form a tree, the resulting evaluation is 0.

Strictly checking the k-coverage of the field is very expensive, and in the proposed algorithm, a sufficient condition for k-coverage is verified as described in Section 3.2.5.

Genetic operators: In our proposed method, we adopted roulette selection, an elite preservation strategy, uniform crossover, and mutation per locus. For uniform crossover, we treated each combination of angle and distance for a mobile sensor node as a gene. For mutation, random value is overwritten to a randomly selected locus.

Termination condition: The algorithm stops after a constant number of generations (one generation corresponds to one iteration of the GA algorithm in Section 3.2.2). In the experiment, we set 20 generations as the constant.

3.2.3 Local search technique

Our method uses the local search technique in addition to GA to improve the quality of solution.

For each mobile node $q \in Q$, we give moving destination randomly in a circle (radius is $1[m]$) centered on q. If WSN lifetime improves when all mobile nodes move to the destination, they move actually and are given new destinations. If it is not improved, this algorithm terminates.

3.2.4 Balanced edge selection method

The nodes near the sink node tend to consume more battery by forwarding the data transmitted from other nodes. In the balanced edge selection method, we first decide the set of nodes called *first-level nodes* directly connecting to the sink node. Next, we connect the remaining nodes to the first-level nodes one by one. The idea to select the first level nodes is as follows.

Step-1: The first level nodes is decided by testing Step-2 for every number of the nodes from 1 to $|P \cup Q|$ so that the maximum power consumption by all the first level nodes is minimized. Here, we select each node in the increasing order of the distance from the node to the sink node.

Step-2: Data sent from the remaining nodes (other than the first-level nodes) must be forwarded through one of the first-level nodes to the sink node. Thus, the remaining nodes are distributed among the first-level nodes so that the power consumption is balanced among the first-level nodes. Here, the power consumption of each first-level node is estimated by the number of assigned nodes and the distance to the sink node.

Next, for each of the first-level nodes and the remaining nodes assigned to the node, we apply the above Step-1 and Step-2, recursively.

We will explain how the algorithm works using an example. Fig. 2(a) depicts the situation just after the first-level nodes A and B have been decided. In the figure, 'A[4]' means that the node A has been assigned 4 remaining nodes. Here, node A is closer to the sink Bs than node B, A has been assigned more remaining nodes. We suppose that A and B have been assigned {C, D, E, G} and {F}, respectively.

Next, the algorithm is recursively applied, and the second-level nodes are decided as shown in Fig. 2(b). Among nodes C, D, and E, D is closest to node A. Then, finally node G is assigned to node C, and the data collection tree completes.

3.2.5 Algorithm for checking k-coverage

Geometrically verifying whether any points of the target sensing field is contained by at least k sensor nodes' sensing ranges is very difficult.

In Wang et al. (2007), Wang et al. proposed a sufficient condition for k-coverage, where the target field is divided into squares whose diagonals have the same lengths as the sensing radius to check if there is at least k sensor nodes in each square.

We propose a looser sufficient condition for the k-coverage of the target sensing field. In our method, we put checkpoints on grid points at intervals of δ in the target sensing field, and only check if each checkpoint is k-covered. However, even if all checkpoints are k-covered, some points between checkpoints may not be k-covered. The smaller δ is, the more the judgement accuracy improves. The judgment accuracy worsens when δ is too large. For $\delta < \sqrt{2}R$, we define delta-k-coverage which is a sufficient condition of k-coverage of the target field.

Definition 1

Checkpoint c is delta-k-covered if a circle whose center and radius are c and $R - \frac{\sqrt{2}}{2}\delta$, respectively, includes at least k nodes.

Fig. 3 shows that a checkpoint is delta-3-covered.

Theorem 1

Given checkpoints on grid points at intervals of δ ($\delta < \sqrt{2}R$) in a given field[2], if each checkpoint is delta-k-covered, then the field is k-covered.

Proof

As shown in Fig. 3, if checkpoint c is delta-k-covered, then any points in the circle with radius $\frac{\sqrt{2}}{2}\delta$ centered at c are k-covered. Thus, as shown in Fig. 4, for neighboring checkpoints c_1, c_2, c_3, and c_4, if all are delta-k-covered, any points in the square formed by those checkpoints are k-covered. Therefore, Theorem 1 holds.

[2] Note that outermost checkpoints must surround the target field.

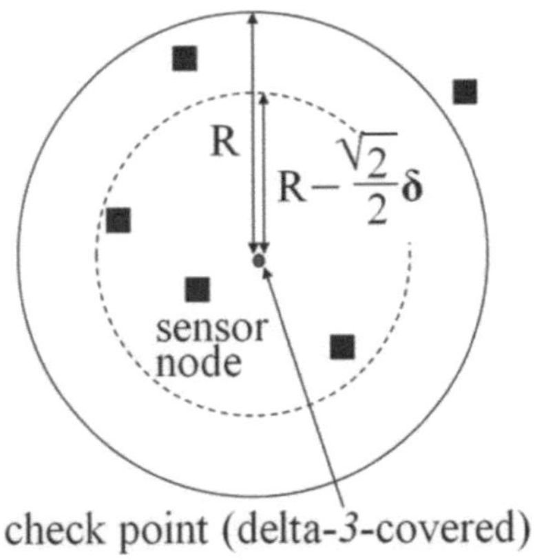

Fig. 3. Condition of delta-k-coverage

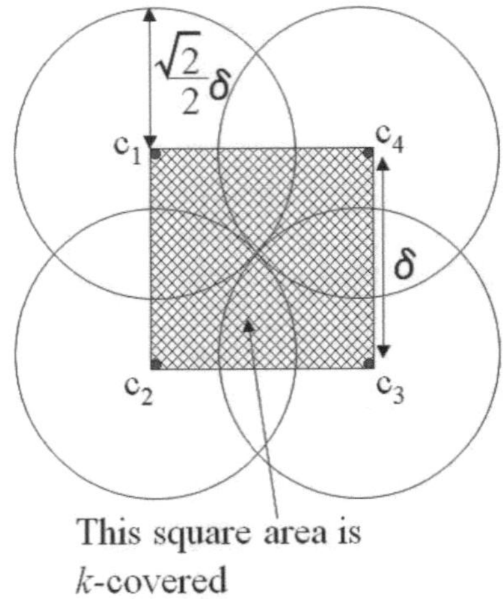

Fig. 4. Checking k-coverage by delta-k-coverage

Theorem 1 only provides a sufficient condition of k-coverage. If we use a smaller value for δ, the condition is closer to the necessary and sufficient condition for k-coverage. However, the smaller value of δ will cause more checkpoints to be checked by delta-k-coverage. In our experiment in Section 3.3, $\frac{\delta}{R} = \frac{1}{10}$ is used.

3.3 Experimental validation

In this section, we show simulation results to validate the usefulness of our proposed method. In order to evaluate the overall performance of our proposed method, we have measured the k-coverage lifetime, and compared it with the performance of other conventional methods including Wang's method [Wang et al. (2007)], for several simulation configurations.

As a common configuration among the simulations, we used the parameter values shown in Table 1. Parameters of GA are determined by preliminary experiments as follows: the number of solution candidates is 20, the number of generations is 20, crossover rate is 1, and mutation rate is 0.01.

3.3.1 k-Coverage lifetime

We have compared k-coverage lifetime of our proposed method with conventional methods named as follows: (i) *Proposed Method* which uses all techniques in Section 3.2; (ii) *No Balancing Method* which randomly generates data collection trees as initial solution candidates in our method; (iii) *Static Method* which prohibits movement of mobile nodes in our method; and (iv) *Wang+Balancing Method* which decides the new positions of the mobile nodes by Wang's Method [Wang et al. (2007)], uses Wang's k-coverage condition, and constructs a data collection tree by the balanced edge selection method and GA. In Wang's k-coverage condition, the field is divided into grids at intervals of $\frac{R}{\sqrt{2}}$, and the number of coverage is the number of the sensor nodes in each grid.

The configuration of this experiment other than Table 1 is provided as follows.

- Field size: 50m $\times$ 50m, 100m $\times$ 50m, and 100m $\times$ 100m
- Position of the sink node : around the south (bottom) end in the field

Parameter	Value
Initial energy amount of each node	$s.energy$ = 32400 J (two AA batteries)
Power consumption exponent	$n = 2$ (by referring to [Wang et al. (2005)])
Power consumption coefficient for data processing	E_{elec} = 50 nJ/bit (by [Wang et al. (2005)])
Power consumption coefficient for signal amplification	ϵ_{amp} = 100 pJ/bit/m^2 (by [Wang et al. (2005)])
Power consumption coefficient for moving	E_{move} = 8.267 J/m (by [Dantu et al. (2005)])
Power consumption coefficient for sensing	E_{sens} = 0.018J/bit (by [Wang et al. (2005)])
Power consumption on idle time	E_{listen} = 0.043 J/s (by [Crossbow (2003)])
Power consumption on sleeping time	E_{sleep} = 0.000054 J/s (by [Crossbow (2003)])
Radius of sensing range of each sensor	R= 20m (by [Ganeriwal et al. (2004)])
Size of data for sensed information	D= 128bit (by [Kamimura et al. (2004)])
Sensing frequency	0.1Hz (by [Kamimura et al. (2004)])
Maximum radio transmission distance	300m (by [Crossbow (2008)])

Table 1. Common configuration for experiments

- Number of sensor nodes: 100, 200, and 300
- Proportion between numbers of static and mobile nodes: 25% and 75%
- Required coverage: $k = 3$

Note that the size of the target field should be appropriately decided so that the field can be sufficiently k-covered by a given number of nodes and coverage degree k. Thus, we used field size 50m × 50m with 100 nodes for the basic case, and enlarged the field size proportionally to the number of nodes. In the experiment, the initial positions of nodes are given by uniform random variables.

We show simulation results in Figs. 11 and 6 for 3-coverage. These results are average of 30 trials.

Fig.11 shows that two Proposed Methods (Balancing and No-Balancing) outperform Static Method to a great extent, independently of the number of nodes. The reason is that finding the appropriate positions of mobile nodes in a wide area greatly affects the performance. Wang+Balancing Method was not so different from Static Method. Initially, the field was k-covered by sensor nodes in all methods. In many cases, however, Wang's k-coverage condition was not satisfied. Then, Wang+Balancing Method moved mobile nodes to the new positions so as to satisfy Wang's k-coverage condition. When a node exhausted its battery, Wang+Balancing Method often could not find the new positions of mobile nodes satisfying Wang's k-coverage condition.

The figure also shows that Proposed Method achieves better performance than No Balancing Method. Thus, our proposed balanced edge selection algorithm is effective in extending the k-coverage lifetime. In the figure, we see that the k-coverage lifetime of all methods decrease as the number of nodes increases. The reason is that the nodes that directly connects sink node Bs have to forward more data transmitted from their upstream nodes as the number of nodes increases, even though mobile nodes move closer to the sink node to help forwarding

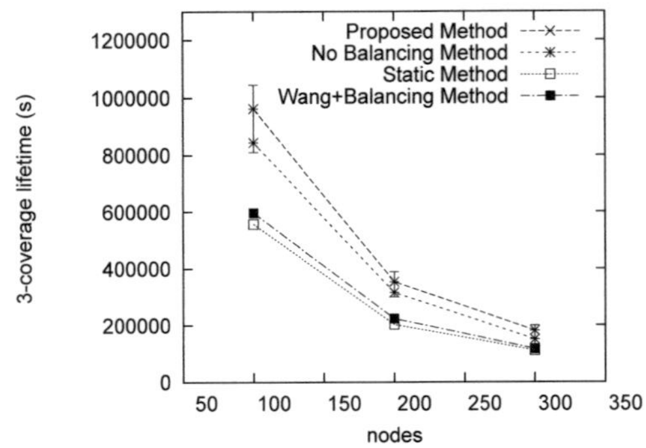

Fig. 5. 3-coverage lifetime

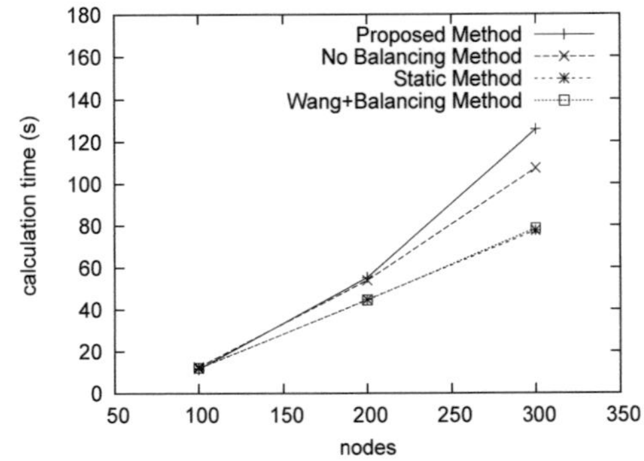

Fig. 6. 3-coverage computation time

	k=1	k=2	k = 3
Wang's Method [Wang et al. (2007)]	93	44	4
Proposed Method ($\delta = 0.5$m)	100	100	100
Proposed Method ($\delta = 1.0$m)	100	100	100
Proposed Method ($\delta = 2.0$m)	100	100	100
Proposed Method ($\delta = 4.0$m)	100	100	100
Proposed Method ($\delta = 8.0$m)	100	100	100
Proposed Method ($\delta = 12.0$m)	100	100	97
Proposed Method ($\delta = 16.0$m)	96	82	48
Proposed Method ($\delta = 20.0$m)	39	1	0

Table 2. The number of occurrences that the field is judged as k-covered (out of 100 simulation runs)

the data. In Fig. 11, the best and worst values of 30 trials by our algorithm were also shown. The difference of k-coverage lifetime of our algorithm was in the range from 84% to 109% compared with the average. We see that our algorithm does not output the solution with extremely bad performance.

Fig. 6 shows the computation time of each method. Proposed Method takes about 120 second in the case of 300 nodes for $k = 3$. This shows that it is possible to operate our method actually.

3.3.2 Efficiency of k-coverage judgment algorithms

We have measured and compared the accuracy and computation time of our delta-k-coverage judgment method and Wang's method [Wang et al. (2007)]. Both methods are based on their own sufficient conditions for checking k-coverage. Our k-coverage condition is described in Section 3.2.5. Wang's k-coverage condition is that the field is divided into grids at intervals of $\frac{R}{\sqrt{2}}$, and the number of coverage is the number of the sensor nodes in each grid (described in Section 3.3.1). Thus, if one of the methods judges affirmatively, then the field is actually k-covered. Conversely, even if both methods judge negatively, it is not always the case that the field is not k-covered. The higher the ratio to judge that the field is k-covered is, the higher the judgment accuracy is.

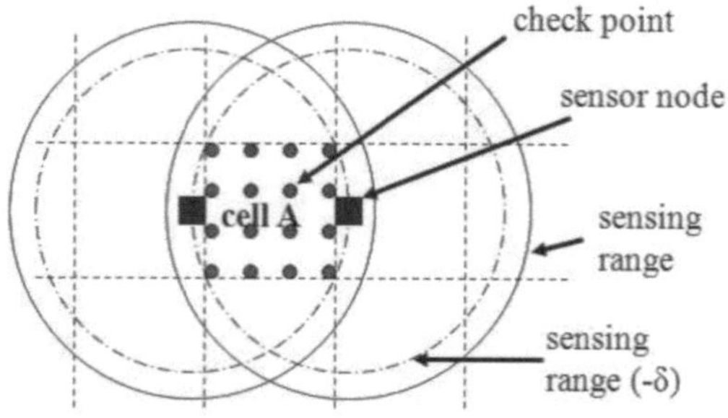

Fig. 7. Example of Misjudge by Wang's Method

In this experiment, 300 static nodes are randomly deployed in the 100m×100m field. In this case, the field is almost always 3-covered. Therefore, it is expected to judge that 1, 2, and 3-coverage of the field are satisfied in all trials. We conducted the above simulation 100 times and measured the number of the occurrences that the field is judged to be k-covered out of the 100 simulations. Note that on some occurrences, the whole field is not actually k-covered since node positions are randomly decided.

We conducted the above simulations by changing the value of δ from 0.5m to 23.5m by 0.5m step for our delta-k-coverage judgment method, while the diagonal length of all squares in Wang's method is fixed to $10\sqrt{2}$m, which is the sensing radius of sensor nodes, and cannot be changed.

The experimental results on measured accuracy is shown in Table 2. Note that Table 2 shows part of the results for some important δ values.

Table 2 suggests that our delta-k-coverage judgment method is better than Wang's method for all numbers for k when δ is no bigger than 16m. The difference becomes bigger as k increases. Especially, when δ is no bigger than 12m, our algorithm almost perfectly judged k-coverage of the field, whereas Wang's method judged that only 4 occurrences out of 100 was 3-covered.

Fig. 7 shows the example of node positions such that the difference of the judgement between our method and Wang's method is extreme. In Fig. 7, cell A is 2-covered actually. Wang's method judges that cell A is not covered because there is no sensor node in cell A. On the other hand, our method judges that cell A is 2-covered, since each check point is delta-2-covered.

Wang's method takes a constant computation time around 0.13ms, while our method takes longer computation time, which is inversely proportional to δ, for example, 159ms for $\delta = 1$m, 2ms for $\delta = 8$m, and 1ms for $\delta = 12$m.

As a result, our algorithm takes longer computation time, however it is much more practical since it is adjustable depending on the required accuracy of k-coverage judgment within allowable computation time.

3.3.3 Influence of mobile nodes ratio for k-coverage lifetime

It is obvious that using n mobile nodes will achieve longer k-coverage lifetime than using n static nodes. However, a mobile node is much more expensive than a static node. In order to investigate the influence of mobile nodes ratio to all nodes, we measured k-coverage lifetime for 100, 200, and 300 sensor nodes, changing the mobile nodes ratio from 0% to 100% by 5% step.

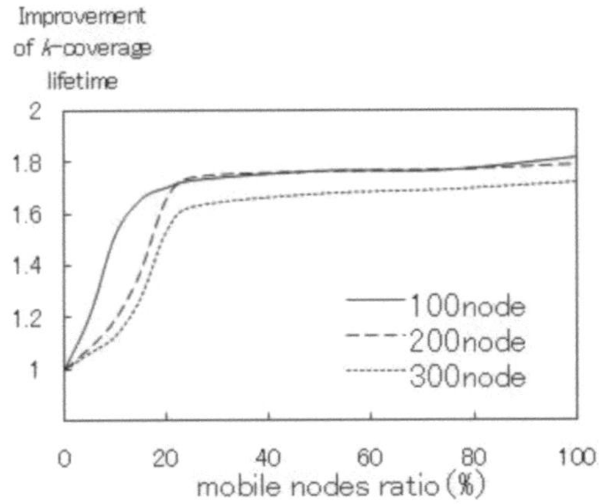

Fig. 8. Improvement of k-coverage Duration for Mobile Nodes Ratio

We show the results in Fig. 8. The results are average values of 30 simulations. Fig. 8 suggests that the k-coverage lifetime increased sharply in the ratio from 0% to 25%, and loosely from 25% to 100%. That means about 25% ratio of mobile nodes will be the best when we consider the deployment cost.

4. k-coverage Lifetime Maximization Method by Sleep Scheduling

In this section, we formulate the problem to maximize the k-coverage lifetime by sleep scheduling, propose an algorithm to solve this problem, and show simulation results to validate the usefulness of our proposed method.

4.1 Problem Definition

In this section, we formulate a problem to maximize the k-coverage lifetime of WSN by scattering more-than-enough number of static sensor nodes at random over the field.

If a particular set of sensing nodes are used for a long time, their batteries will be exhausted. Then, it is necessary to dynamically change the set of sensing nodes. So, we formulate a problem to derive the schedules of when and to which mode each sensor node should change at each time during WSN operation time.

Let t_0 and t_{end} denote the initial WSN deployment time and the time when the k-coverage of the WSN is no longer maintained due to battery exhaustion of some nodes ($t_{end} \geq t_{life}$). For each $s \in S$ and each $t \in [t_0, t_{end}]$, let $Mode(s,t)$ denote the operation mode of s at time t[3] Then, for each $s \in S$, we denote a *schedule* to switch the operation mode of s during time interval $[t_0, t_{end}]$ by the following formula.

$$schedule(s, [t_0, t_{end}]) = \bigcup_{t \in [t_0, t_{end}]} \{Mode(s,t)\}$$

Given the information on the target field *Field*, *s.pos*, *s.energy*, and *s.range* for each sensor node $s \in S$, the position of a sink node *Bs.pos*, and constants $E_{elec}, E_{sens}, E_{listen}, E_{sleep}, \epsilon_{amp}, n,$

[3] We assume that the time domain is discrete.

D, and I, our target problem for maximizing t_{life} is to decide the schedule $schedule(s, [t_0, t_{end}])$ for each node $s \in S$ that satisfies condition (9).

4.2 Modified Target Problem

Our target problem consists of the following three sub-problems.
The first sub-problem is to decide the set of sensing nodes for maximizing t_{life} and satisfying condition (9). Since sensing nodes periodically carry out sensing operation they consume more energy than relaying and sleeping nodes. This problem is presupposed to imply a Dominating Set Problem (DS) that is NP-Complete as a special case [Yang et al. (2006)].
The second sub-problem is to decide the set of relaying nodes for maximizing t_{life}, when the set of sensing nodes are given. Some remaining nodes can reduce critical nodes' transmission distance and transmission data amount so that the overall WSN lifetime is extended.
The third sub-problem is to decide the data collection tree for maximizing t_{life}, when the sets of sensing and relaying nodes are given. It is required to balance the energy consumption among all sensor nodes in the tree. Because a node near the sink node tends to consume more battery by forwarding the data transmitted from other nodes to the sink node.
Since the above problems are dependent on each other in maximizing the WSN lifetime, solving these problems at the same time is considered to be very difficult. Therefore, we adopt a heuristic that solves these problems stepping on the following stages.

(1) Solving the problem to find the minimum set of U satisfying the condition (9).

(2) Solving the problem to find a data collection tree that is rooted on sink node Bs and include all sensing nodes U and some relaying nodes $V \subseteq S - U$ for maximizing *the WSN forecast lifetime.*

(3) Sleeping nodes $W = S - U - V$ are set for a sleeping duration based on the *next battery exhaustion time.*

(4) At next battery exhaustion time, the stages (1), (2), and (3) are executed.

In the above stage (2), the WSN forecast lifetime is the approximated WSN lifetime without considering the changes of the mode of each sensor node in the future. We define the WSN forecast lifetime as follows:

$$t_{now} + \min_{pos \in Field} \left(\frac{\sum_{s \in Cover(pos, t_{now})} (s.energy[t_{now}])}{\sum_{s \in Cover(pos, t_{now})} (C(s))} \right) \tag{15}$$

where, t_{now} is current time, and $C(s)$ is the energy consumption of sensor node s per second, The WSN forecast lifetime is the earliest time when some point in the field is no longer k-covered due to battery exhaustion of some nodes.
Before sleeping nodes sleep, they must be set for the time to wake up. The modes of all sensor nodes are recalculated and informed to them by Bs when the battery of any sensor node is exhausted. When listening to the information of the next mode from Bs, sleeping nodes should be waking up. Therefore, the earliest time when the battery of some sensor node is exhausted (called the *next battery exhaustion time*) is set as the time to wake sleeping nodes up. We define then next battery exhaustion time as follows:

$$t_{now} + \min_{s \in S} \left(\frac{s.energy[t_{now}]}{C(s)} \right) \tag{16}$$

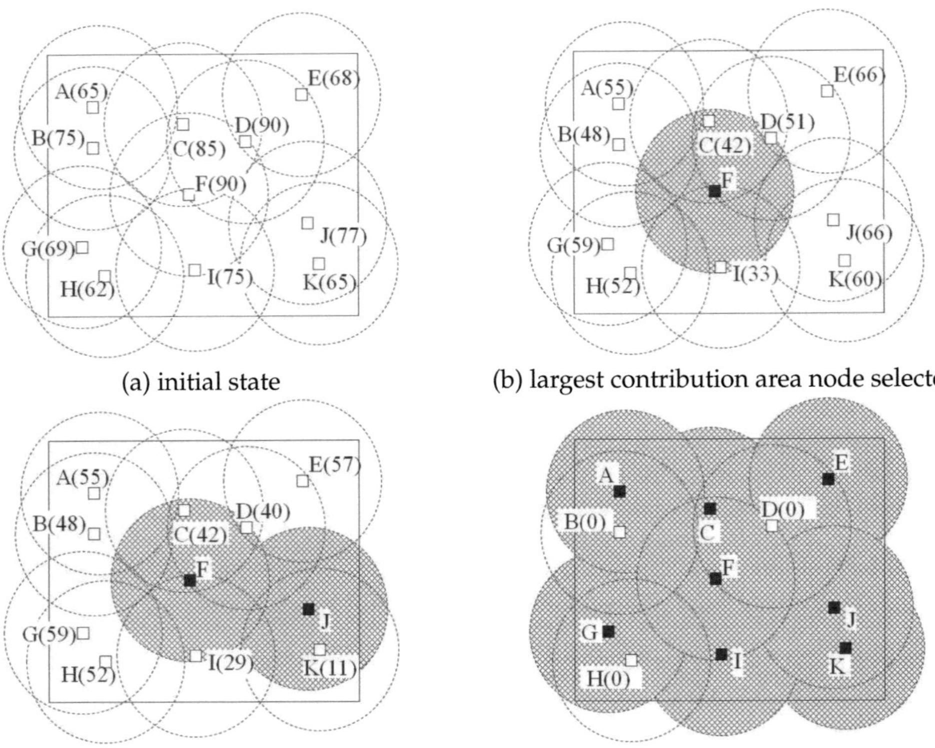

(a) initial state (b) largest contribution area node selected

(c) 2nd largest area node selected (d) resulting state

Fig. 9. Example of Applying Wakeup Method

where $\frac{s.energy[t]}{C(s)}$ is the time duration that the remaining battery amount of sensor node s at time t is exhausted.

4.3 Algorithm

4.3.1 Overview

In this section, we describe an algorithm to solve the problem defined in Section 4.2. Our algorithm finds operation modes for sensor nodes and a data collection tree for each unit time. In our algorithm, we make the minimal number of the nodes required for k-coverage active, and replacing the node that exhausted battery by another one.

The algorithm is supposed to be executed at the initial deployment time and each of the next battery exhaustion time. The lifetime of the whole system ends when there are no sets of sensing nodes that satisfy condition (9).

Our algorithm consists of the following three methods: (1) Wakeup method, (2) Relay selection method, and (3) Mode switching method.

4.3.2 Wakeup Method

Wakeup method finds the minimal number of sensing nodes to k-cover the target field, by letting the more influential nodes to be sensing nodes one by one. We show the algorithm of Wakeup method below. Note that the sink node executes it to just derive the set of sensing nodes, and does not change nodes' actual operation modes.

1. First, all sensor nodes are regarded as sleeping nodes.
2. For each sleeping node, the area called *contribution area* that is not k-covered but included in its sensing range is calculated.
3. Select the node which has the largest contribution area as a sensing node. If there are more than one such nodes, one of those nodes is randomly selected and selected as a sensing node.
4. If there is no sleeping sensor nodes remaining, the algorithm terminates with no solution.
5. If the whole target field is k-covered, the algorithm terminates with the selected set of sensing nodes as a solution. Otherwise, go to Step 2.

We now show an example of finding the nodes to 1-cover the target field. Fig. 9 shows how the sensing nodes are selected by the Wakeup method. In the figure, the squares are sensor nodes, and dotted circles are the sensing ranges of sensor nodes. Each label like 'A(65)' represents the sensor node id 'A' and the contribution area size '65'. Fig. 9(b) shows the result after the first iteration of the algorithm. By selecting sensor node F as a sensing node, the corresponding contribution area has been 1-covered (gray circle in Fig. 9(b)). Then the algorithm is applied to other sensor nodes. Fig. 9(c) shows the result after the second iteration of the algorithm. In this case, nodes E and J have the same largest contribution area size 66, thus node J has been randomly chosen to be a sensing node. Fig. 9(d) is the result after the algorithm terminates with a solution.

4.3.3 Relay Selection Method

The data size and the communication distance have large impact on energy consumption for data communication. We use the *Balanced edge selection method* proposed in Section 3.2.4 to balance transmitted data amount among all nodes. In order to reduce the communication distance, we propose Relay selection method.

In Relay selection method, the tree generated by Balanced edge selection method is modified to improve WSN lifetime by utilizing relay nodes. There are areas with shorter lifetime although the area is k-covered because of non-uniform node density. In some cases, the communication energy can be saved by relaying communication. The proposed relay selection algorithm is shown as follows.

Suppose that there is a link between sensor nodes $s_1 \in U \cup V$ and $s_2 \in U \cup V$. We choose a sleeping or relaying node $s_{relay} \in V \cup W$ such that distance between s_1 and s_{relay} is shorter than that between s_1 and s_2. By making s_{relay} relay the communication between the two nodes, the communication power can be reduced. If this change worsens the value of the objective function, the change is discarded. s_{relay} investigates all sleeping and relaying nodes in the ascending order of distance from s_1. This operation is performed to all links including the new links.

4.3.4 Mode Switching Method

This section describes how and when the operation mode of each sensor node is changed. The algorithm for switching operation modes of all sensor nodes is shown as follows:

1. After the initial deployment of sensor nodes, *Bs* decides the sets of sensing, relaying, and sleeping nodes and the data collection tree by Wakeup method, Balanced edge selection method, and Relay selection method.

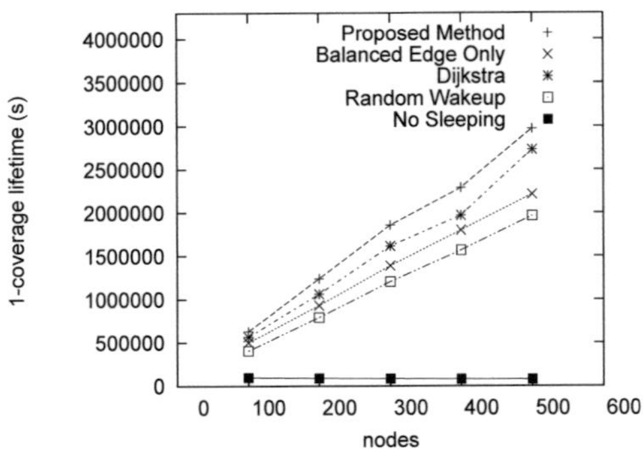

Fig. 10. 1-Coverage Lifetime

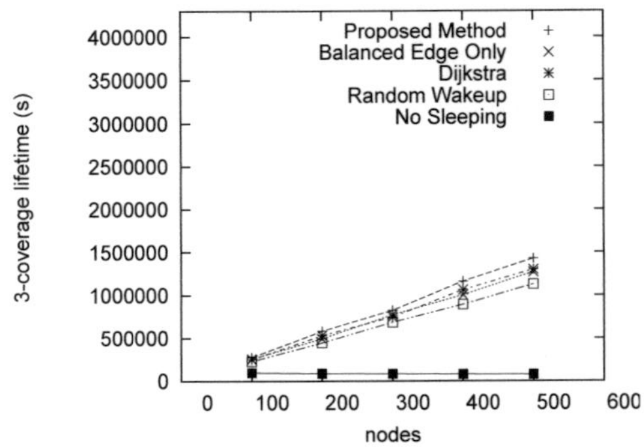

Fig. 11. 3-Coverage Lifetime

2. *Bs* calculates the sleeping time of all sleeping nodes by formula (17).
3. *Bs* informs the information to all sensor nodes by single-hop or multi-hop flooding, that is the mode of each sensor node, the data collection tree, and next battery exhaustion time.
4. Each sensor node switches to the specified mode and sets the destination node..
5. WSN operates, and the energy of each sensor node is reduced as time passes.
6. At next battery exhaustion time, sleeping nodes wake up and prepare for listening the information from *Bs*.
7. The above steps 1 to 6 are repeated during the WSN lifetime.

We define the earliest time when the battery of some sensor node is exhausted (called the *next battery exhaustion time*) as follows:

$$t_{now} + \min_{s \in S} \left(\frac{s.energy[t_{now}]}{C(s)} \right) \tag{17}$$

where, t_{now} is current time and the energy consumption of sensor node s per unit of time ($C(s)$) is calculated by formula (6),(7), or (8).

4.4 Experimental Validation

In order to evaluate the overall performance of our proposed method, we have conducted computer simulations for measuring the k-coverage lifetime, and compared the k-coverage lifetime with other conventional methods, for several experimental configurations.

As a common configuration among the experiments, we used the parameter values shown in Table 1.

We have measured k-coverage lifetime among our proposed method and several other conventional methods named as follows: (i) *Proposed Method* which uses all techniques in Section 4.3; (ii) *Balanced Edge Only* which is the method same as the Proposed Method without Relay selection method; (iii) *Dijkstra* which is the method using a minimum spanning tree instead of a data collection tree generated by Balanced edge selection method in Proposed Method; (iv) *Random Wakeup* which is the method using random selection to find a minimal set of sensing nodes for k-coverage instead of Wakeup Method in Proposed Method; and (v) *No Sleeping* which is the method letting all nodes to be sensing nodes and gathering sensed data from all nodes to the sink node.

For the above conventional algorithm (iii) , we constructed minimum cost spanning trees by Dijkstra method [Dijkstra (1959)] as data collection trees, where cost of each edge is the square of the distance. For the conventional algorithm (iv), we show the detail of Random wakeup method below:

1. First, all sensor nodes are set to sleep mode.
2. A sleeping sensor node is selected randomly, if its sensing range includes the area that is not k-covered, it is set to a sensing node.
3. If there is no sleeping sensor nodes remaining, the algorithm terminates.
4. If the whole target field is k-covered, the algorithm terminates. Otherwise, go to Step 2.

The difference from Wakeup method is the way of node selection in the above step 2. Random wakeup method selects a sleeping node randomly, and if the sensing area of the node includes the area which is not k-covered, its mode is changed to sensing mode. On the other hand, Wakeup method sequentially selects a sleeping node whose sensing area covers the widest area which is not k-covered, and changes its mode to sensing mode.
The configuration of this experiment other than Table 1 is provided as follows.

- Field size: 50m $\times$ 50m
- Position of the sink node: around the south (bottom) end in the field
- Number of sensor nodes: 100, 200, 300, 400, and 500
- Required coverage: k=1 and 3

Note that the size of the target field should be appropriately decided so that the field can be sufficiently k-covered for a given number of nodes and coverage degree k. Thus, we used field size 50m $\times$ 50m, that is, when 100 sensing nodes are randomly deployed in the target field, there will be extremely surplus nodes for k=1, 2, and 3. In the experiment, the initial positions of nodes are given in the target field by uniform random values.
We show experimental results obtained through computer simulations in Fig. 10 for 1-coverage and Fig. 11 for 3-coverage. These results are average of 40 trials.
Figs. 10 and 11 show that Proposed Method, Balanced Edge Only, Dijkstra, and Random Wakeup outperform No Sleeping to a great extent, independently of k and the number of nodes. The reason is that these four methods were able to use the sleep mode well, and reduce the power consumption on idle time of some sensor nodes. The figures also show that Proposed Method achieves better performance than Balanced Edge Only. This is an evidence that our proposed Relay Selection Method is effective to extend the k-coverage lifetime. The figures also show that Proposed Method achieves better performance than Dijkstra. This is an evidence that our proposed balanced edge selection algorithm is effective to extend the k-coverage lifetime. The figures also show that Proposed Method achieves better performance than Random Wakeup Method. This is an evidence that our proposed Wakeup method that greedily selects a node the most effective to the k-coverage guarantees longer k-coverage lifetime than selecting nodes at random.
In these figures, all methods except for No Sleeping extended k-coverage lifetime almost proportionally to the number of surplus nodes. The reason is that until sensing nodes exhaust their battery, surplus nodes are able to keep their battery by sleeping.
In the No Sleeping, we see that the k-coverage lifetime of all methods decrease as the number of nodes increases. The reason is that the nodes that directly connects to the sink node Bs

have to forward more data transmitted from their upstream nodes as the number of nodes increases. We see in the figures that the k-coverage lifetime decreases gradually as k increases. This is because more nodes are required to achieve k-coverage of the field as k increases.
We also confirmed that our proposed algorithm (decision of sensing nodes and construction of a data collection tree) takes reasonably short calculation time. In these experiments, maximum calculation time of the proposed algorithm was 1.2 seconds when the number of nodes is 500.

5. Conclusion

In this chapter, we proposed two methods to maximize k-coverage lifetime of the data gathering WSN.
First, we formulated a k-coverage lifetime maximization problem for a WSN with mobile and static sensor nodes. We proposed a GA-based algorithm to decide the positions of mobile sensor nodes and to construct a data collection tree with balanced power consumption for communication among nodes. We also defined a new sufficient condition for k-coverage based on checkpoints and proposed an algorithm to accurately judge k-coverage in reasonably short time. Through computer simulations, we confirmed that our method improved k-coverage lifetime to about 140% to 190% compared with other conventional methods for 100 to 300 nodes. Also, we confirmed that the best cost-performance is achieved when the mobile nodes ratio is about 25%.
Next, we formulated a k-coverage lifetime maximization problem for a WSN using more-than-enough number of static sensor nodes with sleeping mode. We proposed Wakeup method to decide the modes of sensor nodes, and Relay selection method to modify the data collection tree which includes sensing and relaying nodes. As a result, we confirmed that our method improved k-coverage lifetime to a great extent compared with other conventional methods for several hundreds of sensor nodes.

6. References

Tang, X. & Xu, J. (2006). Extending Network Lifetime for Precision-Constrained Data Aggregation in Wireless Sensor Networks, *Proceedings of The 30th IEEE International Conference on Computer Communications (INFOCOM 2006)*, pp. 1–12, ISBN: 1-4244-0221-2, Apr. 2006, Barcelona, Spain.

Heinzelman, W.R., Chandrakasan, A., & Balakrishnan, H. (2000). Energy-Efficient Communication Protocol for Wireless Microsensor Networks, *Proceedings of the 33rd Annual Hawaii International Conference on System Sciences (HICSS 2000)*, pp. 1–10, Vol. 2, ISBN: 0-7695-0493-0, Jan. 2000, Hawaii.

Cao, Q., Abdelzaher, T., He, T., & Stankovic, J. (2005). Towards Optimal Sleep Scheduling in Sensor Networks for Rare-Event Detection, *Proceedings of The 4th International Symposium on Information Processing in Sensor Networks (IPSN2005)*, pp. 20–27, ISBN: 0-7803-9201-9, Apr. 2005, Los Angeles, California, USA.

Keshavarzian, A., Lee, H., & Venkatraman, L. (2006). Wakeup scheduling in wireless sensor networks, *Proceedings of The 7th ACM International Symposium on Mobile Ad Hoc Networking and Computing (MobiHoc2006)*, pp. 322–333, ISBN: 1-59593-368-9, Apr. 2006, Florence, Italy.

Poduri, S. & Sukhatme, G.S. (2004). Constrained coverage for mobile sensor networks, *Proceedings of International Conference on Robotics and Automation (ICRA2004)*, pp. 165–171, ISBN: 0-7803-8232-3, Apr. 2004, New Orleans, Louisiana, USA.

Wang, G., Cao, G., La Porta, T., & Zhang, W. (2005). Sensor Relocation in Mobile Sensor Networks, *Proceedings of The 29th IEEE International Conference on Computer Communications (INFOCOM 2005)*, pp. 2302–2312, ISBN: 0-7803-8968-9, Mar. 2005, Miami, Florida, USA.

Wang, W., Srinivasan, V., & Chua, K. C. (2007). Trade-offs Between Mobility and Density for Coverage in Wireless Sensor Networks, *Proceedings of The 13th Annual International Conference on Mobile Computing and Networking (MobiCom 2007)*, pp. 39–50, ISBN: 978-1-59593-681-3, Sep. 2007, Montreal, Canada.

Katsuma, R., Murata, Y., Shibata, N., Yasumoto, K., & Ito, M. (2009). Extending *k*-Coverage Lifetime of Wireless Sensor Networks Using Mobile Sensor Nodes, *Proceedings of The 5th IEEE International Conference on Wireless and Mobile Computing, Networking and Communications (WiMob2009)*, pp. 48–54, ISBN: 978-0-7695-3841-9, Oct. 2009, Marrakech, Morocco.

Katsuma, R., Murata, Y., Shibata, N., Yasumoto, K., & Ito, M. (2010). Extending *k*-Coverage Lifetime of Wireless Sensor Networks with Surplus Nodes, *Proceedings of The 5th International Conference on Mobile Computing and Ubiquitous Networking (ICMU 2010)*, pp. 9–16, Apr. 2010, Seattle, Washington, USA.

Srinivas, A., Zussman, G., & Modiano, E. (2006). Mobile Backbone Networks - Construction and Maintenance, *Proceedings of the 7th ACM International Symposium on Mobile Ad Hoc Networking and Computing (MobiHoc 2006)*, pp. 166–177, ISBN: 1-59593-368-9, May 2006, Florence, Italy.

Dantu, K., Rahimi, M., Shah, H., Babel, S., Dhariwal, A., & Sukhatme, G. S. (2005). Robomote: enabling mobility in sensor networks, *Proceedings of The 4th International Symposium Information Processing in Sensor Networks (IPSN 2005)*, pp. 404–409, ISBN: 0-7803-9201-9, Apr. 2005, Los Angeles, California, USA.

Crossbow Technology, Inc. (2003). MICA2: Wireless Measurement System, *http://www.xbow.com/Products/Product_pdf_files/Wireless_pdf/MICA.pdf*.

Ganeriwal, S., Kansal, A., & Srivastava, M. B. (2005). Self aware actuation for fault repair in sensor networks, *Proceedings of International Conference on Robotics and Automation (ICRA2004)*, pp. 5244–5249, ISBN: 0-7803-8232-3, Apr. 2004, New Orleans, Louisiana, USA.

Kamimura, J., Wakamiya, N., & Murata, M. (2004). Energy-Efficient Clustering Method for Data Gathering in Sensor Networks, *Proceedings of the First Workshop on Broadband Advanced Sensor Networks (BaseNets2004)*, pp. 31–36, Oct. 2004, San Jose, California, USA.

Dijkstra, E.W. (1959). A Note on two Problems in Connection with Graphs, *Journal of Numerische Mathematik*, Vol. 1, pp. 269–271.

Crossbow Technology, Inc. (2008). IRIS mote, *http://www.xbow.jp/mprmib.pdf*.

Yang, S., Cardei, M., Wu, J., & Patterson, F. (2006). On Connected Multiple Point Coverage in Wireless Sensor Networks, *Proceedings of International Journal of Wireless Information Networks*, Vol.13, No.4, pp. 289–301.

20

Energy-Efficient Data Aggregation for Wireless Sensor Networks

Rabindra Bista and Jae-Woo Chang
Chonbuk National University
South Korea

1. Introduction

A Wireless Sensor network (WSN) (Heinzelman et al., 2000; Yick et al., 2008) consists of a large number of spatially distributed autonomous resource-constrained tiny sensor devices which are also known as sensor nodes (Horton et al., 2002). WSNs have some unique features, for instance, limited power, ability to withstand harsh environmental conditions, ability to cope with node failures, mobility of nodes, dynamic network topology, communication failures, heterogeneity of nodes, large scale of deployment and unattended operation. Although sensor nodes forming WSNs are resource-constrained, i.e., limited power supply, slow processor and less memory, they are widely used in many civilian application areas, including environment and habitat monitoring, healthcare applications, home automation, traffic control and in military applications such as battlefield surveillance (Pottie & Kaiser, 2000).

Because data from sensor nodes are correlated in terms of time and space, transmitting only the required and partially processed data is more meaningful than sending a large amount of raw data. In general, sending raw data wastes energy because duplicated messages are sent to the same node (implosion) and neighboring nodes receive duplicate messages if two nodes share the same observing region (overlapping). Thus, data aggregation, which combines data from multiple sensor nodes, has been actively researched in recent years. An extension of this approach is in-network aggregation (Considine et al., 2004; Madden et al., 2002; Bista et al., 2009) which aggregates data progressively as it is passed through a network. In-network data aggregation can reduce the data packet size, the number of data transmissions and the number of nodes involved in gathering data from a WSN.

The most dominating factor for consuming precious energy of WSNs is communication, i.e., transmitting and receiving messages. Therefore, reducing generation of unnecessary traffics in WSNs enhances their lifetime. In addition, involving as many sensor nodes as possible during data collections by the sink node can utilize maximum resources of every sensor node. As a result, an adverse scenario will not happen in a WSN in which the sensor nodes closer to the sink run out of energy sooner than other nodes and the network loses its service ability, regardless of a large amount of residual energy of the other sensor nodes.

Since communication is responsible for the bulk of the power consumption, many routing schemes in WSN are carefully designed to provide highly efficient communications among the sensor nodes (Heizelman et al., 1999). Among them, data-centric schemes are very popular where data transmissions are based on their knowledge about the neighboring nodes. Directed Diffusion (DD) (Intanagonwiwat et al., 2002a) and Hierarchical Data Aggregation (HDA) (Zhou et al., 2006) schemes are two representative data-centric schemes. A usual concept of conventional data gathering schemes is that they collect data by a sink node from sensor nodes and transfer data towards the sink node through multi-hop. However, it gives rise to two problems. The first one is the hotspot problem, in which the sensor nodes closer to the sink run out of energy sooner than other nodes. As a result, network loses its service ability regardless of a large amount of residual energy of the other nodes. The second one is that network generates unnecessary traffics during data transmission for choosing a proper path to send data.

Aggregated result of sensor data at the sink node is used for making important decisions. Because WSNs are not always reliable, it cannot be expected that all nodes reply to all request. Therefore, the final aggregated result must be properly derived. For this, the information of the sensor nodes (Node Identifications, IDs) contributing to the final aggregated result must be known by the sink node. And, the communication cost of transmitting IDs of all contributed sensor nodes along with the aggregated data must be minimized. Following are some promising reasons for transmitting IDs of sensor nodes along with their sensed data.

- To know the exact picture of sensors data by identifying which sensor nodes are sending their data for data aggregation.
- Data loss due to collision is inevitable in WSNs. Therefore, IDs of sensor nodes are needed to deal with data loss resiliency and accuracy of the final aggregated result of sensors data at the sink node.
- To know either a sensor node is providing service or not (survivability of a sensor node).
- In end-to-end encryption techniques such as (Girao et al., 2005; Castelluccia et al., 2005) sensor nodes share a common symmetric key with the sink node. Therefore, without knowing the sensor nodes that are contributing data in the aggregated result decryption of the encrypted aggregated result is impossible at the sink node.
- Many privacy preserving data aggregation techniques (Bista et al., 2010; He et al., 2007; Conti et al., 2009; Zhang et al., 2008) use seeds to hide sensor data. The sink node must know the IDs of sensor nodes that are contributing data to the aggregation result so that it can deduce the real aggregated result by subtracting seed values of the sensor nodes which were previously used for data hiding.
- In health care application, to support a common type of query like *"Select the sensor nodes which measure temperature > 98"* for knowing the patients with abnormal temperature.

Hence, a sink node must be aware of node IDs of those sensor nodes which contribute in aggregated value of sensors data in order to derive exact result of the collected data in WSNs. This is possible only when if there exists such a scheme which can transmit IDs of all the participating sensor nodes to the sink node. But, currently existing TinyOS (Hill et al., 2000) - an operating system running on the Berkeley motes (i.e., Mica Motes) (Horton et al., 2002) which has been envisioned as application development platform for WSNs- based privacy preserving data aggregation protocols for WSNs, like (Castelluccia et al., 2005), can not transmit the IDs of those all sensor nodes which contribute to the aggregated value of sensor data to the sink node due to following two reasons. The first is that TinyOS offers limited payload size of 29-byte. The second is that each sensor node ID is transmitted as a plaintext (2-byte) to the sink node. As a result, it restricts sending IDs of all contributed sensor nodes. Handling power is of utmost important. A small size packet is always preferable to WSNs because the communication of even a single bit consumes a significant amount of energy.

For Mica Motes, TinyOS predefined a packet of maximum 36 bytes size. As shown in Fig. 1, out of the 36-byte of the packet, 29-byte are allocated to sensor data (payload) and rest bytes to destination address, Active Message (AM) type, length, group and Cyclic Redundancy Check (CRC) to detect transmission errors. The payload may consist of sampled data, an encryption key/s for security reason and source ID. Since the size of the payload is limited to 29-byte there must be an optimal method in order to adjust IDs of a large number of sensor nodes in a single packet for huge WSNs.

Dest	AM	Len	Grp	Data	CRC
(2)	(1)	(1)	(1)	(0 - 29)	(2)

Fig. 1. TinyOS packet format for Mica Motes. The byte size of each field is indicated below the label. The shaded grey color is data field which can be encrypted.

For these reasons, we, in this chapter, propose a Designated Path (DP) scheme for energy-efficient data aggregation for WSNs. The propose scheme pre-determines a set of paths and runs them in round-robin fashion so that all sensor nodes can participate in the workload of gathering data from WSNs and transmitting the data to the sink node without generating unnecessary traffics during data transmissions. The main idea of our scheme is that each sensor node knows when the sensed/received data has to be sent through which one of its parent nodes for data aggregation before reaching to the sink node by avoiding the communication cost for knowing an appropriate parent node selection in order to aggregate data. In addition, we propose a novel mechanism in which a special set of real numbers are assigned as the IDs to sensor nodes so that a single bit is sufficient to hold an ID of a sensor node while transmitting aggregated data to the sink node. For this, we, first, generate signatures of fixed size for all IDs of respective sensor nodes and then superimpose the signatures of IDs of contributed sensor nodes during data aggregation phase. The analytical and simulation results show that our scheme is more efficient than existing methods in terms of energy dissipation while collecting data from WSNs.

The rest of the chapter is organized as follows. In Section 2, we present related work. In Section 3, we describe how DP scheme works to aggregate data in WSNs and present our signature method to transmit IDs of many sensor nodes to the sink node. In Section 4, we show analytical models for our schemes and the existing schemes. Analytical performance evaluations are shown in Section 5. Section 6 presents simulation results. In Section 7, we conclude this chapter with some future directions.

2. Related Work

In this section, we, first, present a short review of the most related previous work on energy efficient data aggregation for WSNs and then briefly describe the work dealing with sending IDs of sensor nodes to the sink node.

Some researchers have explored in-network aggregation to achieve energy efficiency when propagating data from sensor nodes to the sink node (Madden et al., 2002; Madden et al., 2005; Intanagonwiwat et al., 2002b); Yao & Gehrke, 2003). In-network aggregation approaches are mainly differentiated by their network protocols for routing data. Among them, data-centric routing schemes are very popular where data transmissions are based on their knowledge about the neighboring nodes. Although there are many data-centric approaches (Akkaya & Younis, 2005), DD (Intanagonwiwat et al., 2002a) and HDA (Zhou et al., 2006) are two most related works to our research. In DD scheme, four phases are piggyback with four steps: *interest, exploratory data, reinforcement,* and *data.* A sink node broadcasts an interest describing the desired data to its neighbors. As interests are passed throughout the network, gradients are formed to indicate the direction in which the collected data will flow back. However, DD has two main problems to achieve an energy efficient data aggregation in WSNs. First, even though source nodes are near to the sink node, many other unnecessary nodes in the network are involved to propagate interests and setup gradients to the whole network. Due to this, DD generates unnecessary traffics during data transmissions. Second, DD achieves energy inefficient data aggregation because sources do not know where to forward data for aggregation. In DD, data are aggregated only by chance if the gradients are established as a common path for all sources nodes. As a result, many unnecessary nodes involved to gather data is energy inefficient. On the other hand, HDA overcomes the aforementioned two limitations of DD scheme. For this, HDA proposes a hierarchical structure to constrain exploratory data in a small scope between sink and source nodes. It also proposes parent-select aggregation principle to provide stronger aggregation capability than DD. However, the parent-select aggregation still suffers to achieve energy balanced data aggregation for WSNs. In HDA, there are two types of parent-select aggregation methods to perform data-level aggregation. In the first method, sources choose the parents which have the best attribute, in terms of number of child nodes, to save energy as shown in Fig. 2. Best attributes means the strongest data gathering capacity from as maximum number of sources as possible. This method suffers from hotspot problem and cannot balance energy for WSNs because some core nodes near to the sink, i.e., nodes 2 and 5 in the Fig. 2(a), are frequently used to gather data and run out of energy sooner than other nodes in the network. In the second method, sources choose the parents which have much energy than their siblings. It can balance energy for WSN but cannot guarantee data aggregation frequently as shown in Fig. 2(b & c). Due to this, the number of sensor nodes involved to gather data from the network increases leading to energy inefficiency. Moreover, in HDA, parent-select aggregation is achieved by periodically exchanging exploratory data and reinforcement between sources and the sink node. As a result, it generates unnecessary traffic during data transmissions. In addition, a common problem of both DD and HDA approaches is that they cannot be used for continuous data delivery for event-driven applications (Akyildiz et al., 2002).

On the other hand, CMT (Castelluccia et al., 2005) proposes additively homomorphic scheme to achieve secure data aggregation for WSNs. In the CMT scheme, each sensor node shares a key with the base station (BS) and uses the key to protect data privacy during their aggregation on the way to the BS. Therefore, the BS has to know which sensor has sent the data in order to decrypt the received aggregated data. This process requires transmission of all participated sensor nodes' IDs to the BS. For this, the CMT scheme first divides sensor nodes of a WSN into two groups (a group of data contributing sensor nodes and another group of data not contributing sensor nodes) and then sends IDs of sensor nodes from the group with lower number of sensor nodes as plaintexts (2 bytes of each ID) to the BS. Finally, the BS filters out real aggregated value from the collected data by subtracting proper key stream from the received encrypted aggregated data. However, considering TinyOS based Mica Motes for WSNs, the CMT scheme is not scalable because by using this scheme IDs of just twelve (12) sensor nodes are possible to send along with encrypted aggregated data. For larger size WSNs, it is impossible to decrypt the received data at the BS because of lack of knowledge of participated sensor nodes. In Reference (Zhang et al., 2008), each sensor node adds a seed to hide its data from other sensor nodes for achieving data privacy. Therefore, the knowledge of all source nodes is mandatory for the sink node to compute real aggregated value from the received aggregated data. For this, the work in (Zhang et al., 2008) transmits the IDs of data contributing sensor nodes as plaintexts to the sink node. A WSN is always prone to message-loss due to inevitable data collision property existed in wireless communications. Twin-key approach (Conti et al., 2009) deals with data-loss resiliency while achieving privacy preserving data aggregation by assuring a pair of common key alive for node to node communication. The IDs of those sensor nodes from which data is not getting are sent as plaintexts to the sink node. Like in the work (Castelluccia et al., 2005), both schemes (Conti et al., 2009; Zhang et al., 2008) are not scalable and they need much energy to transmit IDs of sensor nodes.

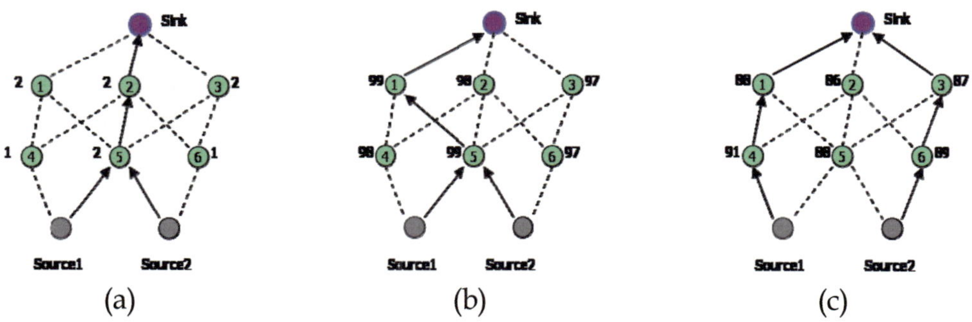

Fig. 2. Parent selection two data aggregation methods in HDA. Best attribute approach (a). Best energy approach with data aggregation (b). Best energy approach without data aggregation (c).

3. Propose Schemes

In this section, we first present our data aggregation scheme and then a scheme for transmitting IDs of a large number of sensor nodes to the sink node which we named signature scheme.

3.1 Our Data Aggregation Scheme

To overcome the shortcomings of DD and HDA schemes, we propose a new energy balanced and efficient approach for data aggregation in wireless sensor networks, called Designated Path (DP) scheme. In DP scheme, a set of paths is pre-determined and run them in round-robin fashion so that all the nodes can participate in the workload of gathering data form the network and transferring the data to the sink node. We use Semantic Routing Tree (SRT) (Madden et al., 2005) for disseminating any kind of aggregation query to get aggregated value such as *MIN, MAX, AVG, SUM* and *COUNT* (Madden et al., 2002).

3.1.1 Network Model

We assume a wireless sensor network model which is appropriate for data gathering applications such as target tracking. The network model has the following properties. First, a sink node without energy constraint is the root of the network topology and located on the top of it. Second, a large number of energy-constrained sensor nodes (e.g., MICA Motes) are deployed uniformly in the network area and they are equipped with power control capabilities to vary their output power. They are arranged in different levels based on the hop-count from the sink node. Third, each sensor node has the capabilities of sensing, aggregating and forwarding data and it can send fixed-length data packets to the sink node periodically. Finally, the sensor nodes can switch into sleep mode or a low power mode to preserve their energy when they do not need to receive or send data (Madden et al., 2005).

Our wireless sensor network model is similar to the structure of HDA scheme which is a multi-parent-multi-child hierarchical structure as shown in Fig. 3. In the multi-parent-multi-child tree structure, one sensor node can have many parent and child nodes and so the sensor node maintains them in two different lists, one for parent nodes and another for child nodes. But, packets are only transmitted between two nodes in neighboring levels. In this structure, all sensor nodes ($M \times N$) are arranged in M levels starting from a sink node. The sink node is the root of the topology and is at level 0; nodes being one hop far from the sink are at level 1; nodes being two hops far from the sink are at level 2 and so on. As a result, lower the level a node is in, the nearer to the sink. Nodes at level i-1 are called 'parents' of nodes at level i, and nodes at level i+1 are called 'children' of nodes at the level i. To have a parent-child relationship between two sensor nodes, they must be within the communication range of each other.

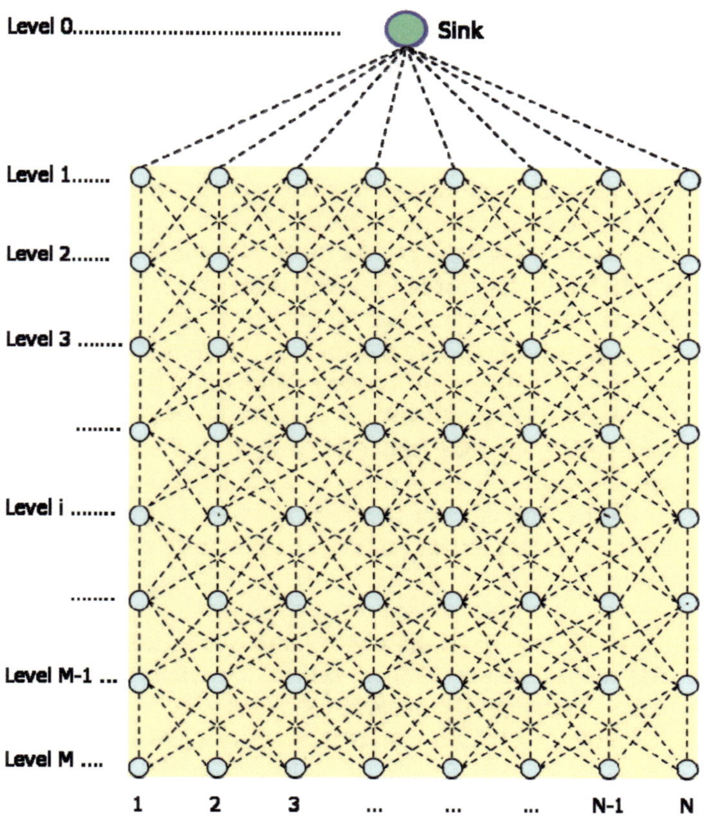

Fig. 3. A general view of network model for our data aggregation scheme.

3.1.2 Designated Path (DP) Scheme

Designated paths are a set of in-built paths, especially, designed for energy balance and efficient data aggregation for WSNs. In the DP scheme, a set of paths is pre-determined and run them in round-robin fashion so that all the nodes can participate in the workload of gathering data form the network and transferring the data to the sink node. In DP scheme, the forwarding behavior of all the nodes is scheduled to balance their burden of aggregation and transmitting network data. By using data aggregation knowledge, each sensor node knows when sensed or received or aggregated data has to send to which one of its parent nodes during data transmissions. In this way, unlike the existing schemes, DP does not generates unnecessary communication traffics to find an appropriate parent node and hence it works in energy efficient way. There are four main phases of DP scheme which are *path construction phase, best node selection phase, knowledge injection phase,* and *paths running phase.*

(a) Path construction phase: After deploying sensor nodes in a field, a multi-parent-multi-child hierarchical tree structure is constructed to provide communication paths for a WSN. In addition, N number of paths (for simplicity, N is equals to the number of columns of the WSN) are constructed for achieving energy-balanced data aggregation in the WSN. Each path is the shortest path from a sensor of level 1 to that of level M. So the first path, P1, consists of the sink and a sequence of the 1st sensor nodes of level 1 to level M, the second

path, P2, consists of the sink and a sequence of the 2nd sensor nodes of level 1 to level M and so on. In this way, we can create N paths for any M×N WSN and store them into a list of paths, PList. Because the paths of the PList will be allocated mainly for data aggregation in WSNs, we termed them as designated paths (DP).

(b) Best node selection phase: Based on the network connectivity, the best node from each path is determined for all of the sensor nodes of the WSN. A sensor node is said to be the best node among other sensor nodes of a path when the sensor node can be reached by any other sensor node of the network in the cost of minimum hop-count. By using Dijkstra's shortest path algorithm (Dijkstra, 1959), we can compute the best nodes for every sensor node of the network. If a sensor node can not reach to a path, then it inserts 'NULL' value and PathID of the path into its routing table. Otherwise, it inserts 'NodeID' of the best node and 'PathID' of the path. In this way, every node maintains the information of the best N nodes from the N number of designated paths, one node from each path in its routing table. The main goal of this phase is to create the routing table in order to use it as data aggregation knowledge for the WSN. Based on the routing table of the best nodes of a sensor node, the sensor node maps the best nodes to its parent sensor nodes so that it doesn't need to store a full path to reach the best node of any path.

(c) Knowledge injection phase: The application knowledge about designated paths and the best nodes is now loaded to each sensor node to achieve an efficient data aggregation in the WSN. By using this knowledge, in DP scheme, each sensor node of the WSN knows where to forward network data during their transmissions without generating unnecessary traffics. On the other hand, most of the existing routing protocols for sensor networks have to decide this task during data transmissions. For this, sensor nodes have to exchange unnecessary messages frequently among each others. It hurts a system in terms of energy efficiency because communication is the bulk of the power consumption and it decreases lifetime of a WSN. It also introduces a delay to the system.

(d) Paths running phase: The N paths from the PList are globally scheduled to all sensor nodes of the WSN so that the sensor nodes can run the paths in round-robin fashion. So, in one round, only one path, for instance P1, of the PList becomes active during data gathering and all the sensor nodes of the network are aware of P1 is active in this round. They send sensed/received/aggregated data to their best nodes from the path (P1) by using the data aggregation knowledge and data is automatically aggregated during their course to the sink node because all the sensor nodes use the same path which is active for the round. In the next round, the next path will be active, for example P2, and all of the sensor nodes send their data through P2 to the sink node. Data is aggregated progressively on their way to the sink node through P2. In the same way, the rests of the paths of PList are active one at a time to collect data from the WSN. The process is repeated after finishing one turn of all paths of the PList. Using designated paths in a round-robin mechanism provides an opportunity to all sensor nodes of the WSN to participate in the workload of gathering data from the network and transferring the data to the sink node. The forwarding behavior of all the nodes is scheduled to balance their burden of aggregating and transmitting the network data to the sink node. In this way, we overcome hotspot problem of the conventional approaches and believe that our DP scheme can achieve energy-efficient data aggregation in WSNs.

Furthermore, as DP scheme does not need to generate unnecessary traffics to select a path during data transmissions, it makes the networks energy efficient. In addition, our DP scheme can support continuous data delivery for event-driven applications.

3.1.3 Data Aggregation Algorithm

To avoid unnecessary communications overheads and achieve energy efficient data aggregation for WSNs, we present an algorithm for data aggregation in WSNs as given below in Fig. 4. The main goal of the propose algorithm is to generate data aggregation application knowledge for sensor nodes and they use it during data transmissions to the sink node.

For example, an 8×6 sensor nodes with a powerful sink are organized in a multi-parent-multi-child hierarchical structure, as shown in Fig. 5, where the total number of levels, M = 8, and the total number of columns, N = 6. In the first step, our algorithm creates six designated paths, P1, P2, P3, P4, P5 and P6 by selecting a sequence of appropriate sensor nodes for each path. The sequence of the nodes for P1, P2, P3, P4, P5 and P6 are < 1, 7, 13, 19, 25, 31, 37, 43 >, < 2, 8, 14, 20, 26, 32, 38, 44 >, < 3, 9, 15, 21, 27, 33, 39, 45 >, < 4, 10, 16, 22, 28, 34, 40, 46 >, < 5, 11, 17, 23, 29, 35, 41, 47 >, and < 6, 12, 18, 24, 30, 36, 42, 48 > respectively, starting from the sink node. All of the six paths are stored into a list of paths, PList. In the second step, the algorithm chooses the nearest nodes (in terms of minimum hop-count, MIN_hopc), called Best_nodes, one for each path for all of the sensor nodes of the network by using Dijkstra's shortest path algorithm (Dijkstra, 1959). If the algorithm can not find the best node from a path for any sensor node, it simply assigns value 'NULL' to the path. The meaning of 'NULL' is that when the path becomes active, the sensor node sends data through its default path (i.e., the path in which a node is situated in the network) because it is not located at the sub-tree of the path. This information is stored into the routing table (RTable) of the network. A sample of RTable to store the information of the best nodes is presented in Table 1. In this table, the first column represents the node identity of a sensor node for which we want to find the best nodes from the designated paths. The second column has entry type <Pi, Nj> where Nj represents the best node from path Pi to the sensor node of the first column. In the third step, the sink node uploads the routing table to all of the sensor nodes and each sensor node updates its original routing table which has already stored such information as a list of parent nodes, a list of child nodes, and its level in the network. The final step of this algorithm is to initialize the WSN. For this, the sink node either receives a SQL like aggregate query from a user or generates itself such type of query. Before propagating the query to the WSN, a query scheduler fetches the time duration of the query and assigns six time slots to the respective paths since the number of designated paths is 6 in this example. Then, it attaches the time schedule to the query and issues it to the WSN by instructing sensor nodes to run them in round-robin mechanism accordingly. When the sensor nodes receive the query, they send the data to the sink node according to the schedule. In this way, all the sensor nodes are synchronized to send the data through the particular active path and data are automatically aggregated during their course to the sink node through the active path. In the example, P3 is active at the moment, so all the source nodes, shown as dark nodes, send their data to their respective best nodes from P3 (for instance, node 15 is the best node for nodes 19 and 20) and data are aggregated before reaching to the sink node.

```
Input: Hierarchical (multi-parent-multi-child) M×N WSN, and
       SQL type aggregation query
Output: Aggregated data from the network

Step1. Create a set of N number of designated paths through
each column of the WSN
       for sensor nodes Nj =1 to N, Pj=1 to N; Nj++, Pj++;
         for level Li =1 to M; Li++
           select LiNj
             insert into NList[ LiNj]   // list of nodes of a path
         Pj = NList
         insert into PList[Pj]

Step2. Select N number of best nodes, one from each path, for
every sensor node
       for sensor nodes LiNj =[1,1] to [M,N], Li++, Nj++;
         for Pj=1 to N, Pj++
           MIN_hopc = infinite value
           Best_node = NULL
           for Li =1 to M; Li++ make shortest hopc Array
        // using Dijkstra's algorithm, it finds hopc for LiNj and Pj
             Arry_hopc = DDistance(LiNj, Pj) ;
               if ( MIN_hopc > Array_hopc[Pj [Li]] )
                 MIN_hopc = Array_hopc[Pj [Li]]
                 Best_node = Li
         insert Pj and Best_node into RTable   // routing table

Step3. Load routing information to the sensor nodes
       for sensor nodes LiNj =[1,1] to [M,N], Li++, Nj++;
         load (RTable);

Step4. Schedule and run the designated paths to collect data
       Initialize ( );  // issuing an aggregation query
       Time_to_run  =T   // life time of a query
       Schedule( T);
         Pj = T/N  // Slotting T into N number of designated paths
       for Pj =1 to N; Pj++
         Round_robin(PList [Pj] ) // running a path for a time slot
           Send_data(value)  //  sending  data through the path
           Aggregate(value);  /*data is aggregated during the
                                  course through the path*/
       return value;
```

Fig. 4. Data aggregation algorithm for our DP scheme.

NodeID	Best Nodes For the Designed Paths
N1	{ <P1, NULL>, <P2, NULL>, <P3, NULL>, <P4, NULL>, <P5, NULL>, <P6, NULL> }
...	....
N8	{ <P1, N1>, <P2, N2>, <P3, N3>, <P4, N4>, <P5, NULL>, <P6, NULL> }
...	....
N18	{ <P1, N1>, <P2, N2>, <P3, N3>, <P4, N4>, <P5, NULL>, <P6, NULL> }
...	....
N29	{ <P1, N13>, <P2, N14>, <P3, N21>, <P4, N22>, <P5, N23>, <P6, N24> }
...	....
N48	{ <P1, N25>, <P2, N32>, <P3, N33>, <P4, N34>, <P5, N41>, <P6, N42> }

Table 1. Routing information of sensor nodes.

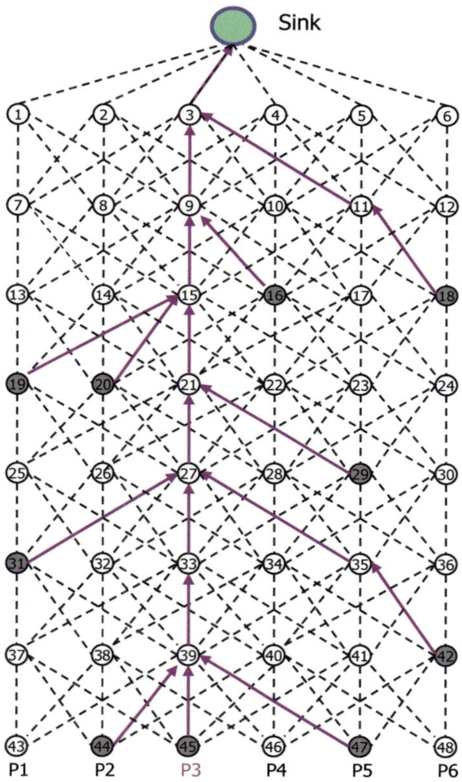

Fig. 5. Data aggregation in our DP scheme where path P3 is being active.

3.1.4 Scheduling

There are two levels of time scheduling in DP scheme. They are *path scheduling* and *communication scheduling*. For path scheduling, DP scheme applies a simple TDMA (Time Division Multiple Access) transmission scheduling mechanism which can be done either using the life time value of WSN or that of a user query (T), depending on the requirement of an application. Its basic idea is to subdivide T into as many number of fixed-length time intervals (slots) as the number of designated paths in a WSN. If the value of the T is very large, like in the case of continuous aggregate query, the path scheduler first divides T into M time slots and each time slot is further divided into the same number of slices as the number of designated paths, N. Fig. 6 shows the path scheduling for DP scheme. The designated paths are run in round-robin mechanism to collect data from the network. For each slice, only the scheduled path becomes active and path synchronization is maintained by all the sensor nodes of the WSN. The communication scheduling is related to how to synchronize the working behavior of all sensor nodes when the sink node collects data from the WSN. During processing of aggregation queries, it is required to coordinate the awaking times of children and parents in such a way that parent nodes can receive data from their child nodes before aggregating. To manage it, we adopt slotted approach (Madden et al., 2005) where the epoch is subdivided into a number of intervals, and assigned the intervals to the sensor nodes based on their position in the routing tree level of the hierarchical structure. It has been shown that the slotted approach can save a significant amount of energy in a hierarchical network structure.

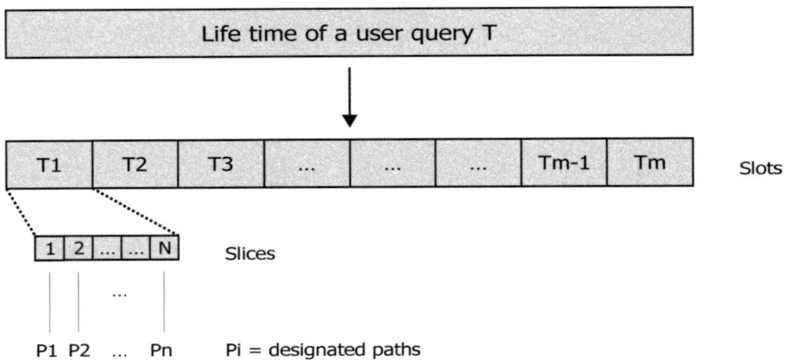

Fig. 6. Time division for designated paths in DP scheme.

3.2 Signature Scheme

To transmit IDs of a large number of sensor nodes in resource-constraint WSNs, we propose a novel approach based on signature of node ID so called signature scheme. There are five (5) steps in our signature scheme which we briefly describe each of them as follows.

(a) Assigning node ID to each sensor node: In this step, we assign a special type of positive integer 2^n (where, n = 0 to $Bn \times 8 - 1$, such that Bn is the number of free bytes available in the payload) to every sensor node as node ID. This is because the binary value of every integer of 2^n type has only one high bit (1). In addition, the position of the high bit for all integers of this type is unique. We termed this node ID as Real ID of a sensor node. The sink node knows a data contributing sensor node through its Real ID.

(b) Generating signatures of each sensor node ID: The Real ID of a sensor node assigned in the previous step is used to generate a signature of a fixed length. A signature is a fixed size bit stream of binary numbers for a given integer. Signature of a senor node ID can be generated by using the technique presented in the work (Zobel et al., 1998). We can determine the length of the signature based on the size of a given WSN. When the size of the WSN increases we can increase the length of the signature up to the Bn bytes. In other words, different size WSNs can have signatures of different lengths.

(c) Transmitting sensor data with signature of sensor ID: In this step, every source sensor node appends its signature as a sensor node ID rather than a plaintext used in the case of the

existing work. After including signature of its nodes ID in the payload, the sensor node forwards its packet to the upper layer sensor node. The sink node is the final destination of all sensor data where they ultimately aggregated.

(d) Data aggregation and superimposing signatures of IDs of sensor nodes: In this step, data aggregators collect data and signatures of the associated sensor nodes to perform following tasks. First of all, they aggregate received data according to the provided aggregation function such as *Average* of sensor data. Next, they superimpose signatures of the sensor nodes by performing bitwise *OR* operation on the bit streams of their Real IDs. Finally, the data aggregators rout aggregated result with the superimposed signatures of Real IDs of contributed sensor nodes to the sink node. Sine this approach needs just one bit to carry an ID of a sensor node it is 16 times scalable than the existing work where plaintexts (2-byte each) are used for carrying IDs of sensor nodes by simply concatenating them.

(e) Computing the final aggregated result and fetching IDs of contributed sensor nodes: When the sink node received partially aggregated data and the superimposed signatures from every sub-tree, it deduces the final aggregated result from the received aggregated data. Since the payload of the partially aggregated data contains signatures of IDs of sensor nodes the sink node can know all the contributed sensor nodes. To know the knowledge of contributed sensor nodes, the sink node separates the high bits (1s) of the superimposed signature of the each sub-tree by performing bitwise *AND* operation with the pre-stored signature files of Real IDs of sensor nodes.

<table>
<tr><th>SN ID</th><th>Real ID</th><th>2-byte Signature</th></tr>
<tr><td>1</td><td>$2^{0} = 1$</td><td>0000000000000001</td></tr>
<tr><td>2</td><td>$2^{1} = 2$</td><td>0000000000000010</td></tr>
<tr><td>3</td><td>$2^{2} = 4$</td><td>0000000000000100</td></tr>
<tr><td>4</td><td>$2^{3} = 8$</td><td>0000000000001000</td></tr>
<tr><td>5</td><td>$2^{4} = 16$</td><td>0000000000010000</td></tr>
<tr><td>6</td><td>$2^{5} = 32$</td><td>0000000000100000</td></tr>
<tr><td>7</td><td>$2^{6} = 64$</td><td>0000000001000000</td></tr>
<tr><td>8</td><td>$2^{7} = 128$</td><td>0000000010000000</td></tr>
<tr><td>9</td><td>$2^{8} = 256$</td><td>0000000100000000</td></tr>
<tr><td>10</td><td>$2^{9} = 512$</td><td>0000001000000000</td></tr>
<tr><td>11</td><td>$2^{10} = 1024$</td><td>0000010000000000</td></tr>
<tr><td>12</td><td>$2^{11} = 2048$</td><td>0000100000000000</td></tr>
<tr><td>13</td><td>$2^{12} = 4096$</td><td>0001000000000000</td></tr>
<tr><td>14</td><td>$2^{13} = 8192$</td><td>0010000000000000</td></tr>
<tr><td>15</td><td>$2^{14} = 16384$</td><td>0100000000000000</td></tr>
<tr><td>16</td><td>$2^{15} = 32768$</td><td>1000000000000000</td></tr>
<tr><td colspan="2">Signature Superimposing by using bitwise OR operator (|)</td><td>1111111111111111</td></tr>
<tr><td colspan="2" rowspan="2">Example: The sink node fetches SN 8 using the signature of Real ID 128 and AND operator (&)</td><td>1111111111111111
& 0000000010000000</td></tr>
<tr><td>= 0000000010000000</td></tr>
</table>

Table 2. Real ID of sensor nodes with signature.

Table 2 illustrates Real ID of 16 sensor nodes (SNs) with 2-byte size signature of each Real ID, signature superimposing process by using bitwise OR operator and an example of fetching a sensor node (SN 8) from the superimposed signature by using the Real ID 128 of SN 8 at the sink node.

3.2.1 Extension to Real ID Assignment and Signature Structure

In the previous section, we described about assigning Real ID to each sensor node using a set of positive integers of type 2^n. Now, we present variants of the integer type 2^n are also applicable to use as Read IDs for sensor nodes. For simple exposition of our idea, we consider three types of integer set: $2^n - 1$, 2^n and $2^n + 1$. For a Real ID of each set, we allocate memory of 2 bytes. Therefore, the total space required to include three Real IDs one for each integer set in the payload is 6 bytes. They can be organized in ascending order, i.e., first an ID of type $2^n - 1$, then ID of type 2^n and finally ID of type $2^n + 1$ occupying continuous 6 bytes space. Fig. 7 shows an algorithm for providing 6-byte signature containing all the three types of Real ID of sensor nodes. The main notion of this algorithm is to make use of the signatures of 2^n type Real IDs for both $2^n - 1$ and $2^n + 1$ types Real IDs and they are distinguished by allocating a particular slot to each type of Real IDs in the memory space of the payload. Every source node transmits its data along with 6-byte bit stream of its Real ID to the immediate parent node. The parent node aggregates sensor data of its child nodes, superimpose their 6-byte size signatures and forwards the packet towards the sink node. When the sink node receives a packet of aggregated data from each sub-tree it executes the algorithm shown in Fig. 8 to identify the contributed source nodes. The sink node first separates the superimposed 6-byte signature into three chunks each of continuous 2-byte size. Next, it generates a list of Real IDs from each chunk as shown in Table 2 and assembles them. By mapping Real IDs to SN IDs, the sink node finally knows all the contributed sensor nodes of the received aggregated data.

```
Input: Real IDs of sensor nodes
Output: Signatures of Real IDs
// Check the types of Real IDs
  if Real ID type = 2^n
    GenSig (Real ID);                //2 bytes
     Padding zeros left and right;   // 2 bytes in each sides
  else if Real ID type =2^n – 1
    GenSig(closest 2^n);
     Padding zeros right;            //4 bytes
  else                               //type = 2^n + 1
    GenSig(closest 2^n);
     Padding zeros left;             //4 bytes
```

Fig. 7. An algorithm to fix spaces for the signatures of Real IDs of types 2^n -1, 2^n and $2^n + 1$ by padding zeros.

```
Input: Superimposed fixed size bit stream (6-bytes)
Output: List of contributed sensor nodes
// Separates the superimposed bit stream from the payload
  split(superimposed bit stream);
    A = 2-byte; B=2-byte; C=2-byte;
     select A;                          // the first 2 bytes
      {fetch_Real_IDs(A);               // as shown in Table 1
  for all Real IDs
     Real ID = Real ID – 1;         // 2^n – 1 type
    List1 = Real ID;}
    select B;                           // middle 2-byte
     {fetch_Real_IDs(B);
        for all Real IDs
                List2= Real ID;}              // 2^n type
    select C;                           // the last 2-byte
     {fetch_Real_IDs(C);
             for all Real IDs
               Real ID = Real ID + 1;       // 2^n + 1 type
                 List3 = Real ID;}
    List =List1 + List2+ List3;  // list of all Real IDs
    List_SN_ID = List;             // using mapping file
    Retrieve List_ SN_ID;
```

Fig. 8. An algorithm to show the process of generating IDs of contributed sensor nodes from the superimposed bit stream of a packet by the sink node.

Table 3 illustrates ID of sensor nodes (SN ID), their respective Real ID with signatures of 6-byte for 32 sensor nodes. First, out of 32 sensor nodes, SNs <3, 6, 9, 12, 15, 18, 21, 24, 27, and 30> have Real IDs of type $2^n - 1$ and they have signatures of the closest 2^n type integers. For instance, SN 6 has Real ID 7 and the Real ID 7 takes the signature of Real ID 8 because latter is the closest 2^n type integer to former. Since every $2^n - 1$ type integer is smaller than respective 2^n type integer former occupies earlier position in the 6-byte space than latter. So, in the signature of every 2^n -1 integer a high bit (1) appears within the first 2-byte of the 6-byte signature and the remaining 4-byte space is padded with zeros. Next, SNs <1, 2, 4, 7, 10, 13, 16, 19, 22, 25, 28 and 31> have Real IDs of 2^n type integers. For instance SN 10 has Real ID 16, and the signature of this type takes the middle position of the 6-byte space having 2-byte zero padding in both left and right sides. Finally, the remaining SNs <5, 8, 11, 14, 17, 20, 23, 26, 29 and 32> have Real ID of type $2^n + 1$ and they have signature of the closest 2^n type integers. For instance, SN 14 has Real ID 33 and it takes the signature of Real ID 32 which is the closest integer of type 2^n. Since every $2^n + 1$ type integer is larger than respective 2^n type integer it occupies the last 2-byte of the 6-byte signature. For instance, SN 17 has Real ID 65 and the Real ID 65 takes the signature of Real ID 64 with 4-byte zero padding in the beginning.

SN ID	Real ID	2-byte Signature	6-byte Signature (Padding 4-byte Zeros)
1	$2^0 = 1$	0000000000000001	000000000000000000000000000000010000000000000000
2	$2^1 = 2$	0000000000000010	000000000000000000000000000000100000000000000000
3	$2^2 -1 = 3$	0000000000000100	000000000000010000000000000000000000000000000000
4	$2^2 = 4$	0000000000000100	000000000000000000000000000001000000000000000000
5	$2^2 +1 = 5$	0000000000000100	000000000000000000000000000000000000000000000100
6	$2^3 -1 = 7$	0000000000001000	000000000000100000000000000000000000000000000000
7	$2^3 = 8$	0000000000001000	000000000000000000000000000010000000000000000000
8	$2^3 +1 = 9$	0000000000001000	000000000000000000000000000000000000000000001000
9	$2^4 -1 = 15$	0000000000010000	000000000001000000000000000000000000000000000000
10	$2^4 = 16$	0000000000010000	000000000000000000000000000100000000000000000000
11	$2^4 +1 = 17$	0000000000010000	000000000000000000000000000000000000000000010000
12	$2^5 -1 = 31$	0000000000100000	000000000010000000000000000000000000000000000000
13	$2^5 = 32$	0000000000100000	000000000000000000000000001000000000000000000000
14	$2^5 +1 = 33$	0000000000100000	000000000000000000000000000000000000000000100000
15	$2^6 -1 = 63$	0000000001000000	000000000100000000000000000000000000000000000000
16	$2^6 = 64$	0000000001000000	000000000000000000000000010000000000000000000000
17	$2^6 +1 = 65$	0000000001000000	000000000000000000000000000000000000000001000000
18	$2^7 -1 = 127$	0000000010000000	000000001000000000000000000000000000000000000000
19	$2^7 = 128$	0000000010000000	000000000000000000000000100000000000000000000000
20	$2^7 +1 = 129$	0000000010000000	000000000000000000000000000000000000000010000000
21	$2^8 -1 = 255$	0000000100000000	000000010000000000000000000000000000000000000000
22	$2^8 = 256$	0000000100000000	000000000000000000000001000000000000000000000000
23	$2^8 +1 = 257$	0000000100000000	000000000000000000000000000000000000000100000000
24	$2^9 -1 = 511$	0000001000000000	000000100000000000000000000000000000000000000000
25	$2^9 = 512$	0000001000000000	000000000000000000000010000000000000000000000000
26	$2^9 +1 = 513$	0000001000000000	000000000000000000000000000000000000001000000000
27	$2^{10} -1 = 1023$	0000010000000000	000001000000000000000000000000000000000000000000
28	$2^{10} = 1024$	0000010000000000	000000000000000000000100000000000000000000000000
29	$2^{10} +1 = 1025$	0000010000000000	000000000000000000000000000000000000010000000000
30	$2^{11} -1 = 2047$	0000100000000000	000010000000000000000000000000000000000000000000
31	$2^{11} = 2048$	0000100000000000	000000000000000000001000000000000000000000000000
32	$2^{11} +1 = 2049$	0000100000000000	000000000000000000000000000000000000100000000000

Table 3. Real ID of thirty-two (32) sensor nodes with 6-byte signature.

In this way, we can assign Real ID to sensor nodes by using small size integers which is convenient to use rather than using big size integers. If necessary, we can easily create further Real ID of types like $2^n -2$, $2^n + 2$ and so on. For this, we have to add just 2 more bytes for every new type in the signature and pad zeros accordingly. Hence, we can assure that our approach is technically feasible for transmitting IDs of very large number of sensor nodes in data aggregation for WSNs.

4. Analytical Models

In this section, first we present analytical model for the data aggregation schemes and then for carrying maximum number of node ID by pre-defined payload of resource-constraint sensor node.

4.1 Analytical Model for Data Aggregation Schemes

Parameters	Descriptions	Parameters	Descriptions
P_{DP} or E_{DP}	Energy consumed by DP Scheme	N_{msg}	Number of message generated by DP per round
P_{HDA} or E_{HDA}	Energy consumed by HDA Scheme	E_{Rx}	Energy consumed by a node to *receive* data
P_{DD} or E_{DD}	Energy consumed by DD	E_{Tx}	Energy consumed by a node to *transmit* data
C	Number of source groups within a WSN	E_{Idle}	Energy consumed to be in idle state for a node
M, M'	Number of rows of WSN, the highest level of a source node	α	Energy dissipation to be in idle state
N	Number of columns of the WSN	β	Energy dissipation to transmit data
A_n	ID of an *Active path*	γ	Energy dissipation to receive data
level	WSN hierarchy level	X	Number of sources
m_j	Number of associated nodes to collect data per level	Y	Number of aggregation nodes
G_i	Source group	Z	Number of routing nodes
n_i	Number of source nodes in a group	r	One side coverage range of a parent
P_α	Communication overhead due to missing data aggregation	$n_{c,}$	Average no. of children per parent (network cardinality)
P_β	Communication overhead due to frequent transmission of parent nodes' energy information	n_p	Average no. of parents per child (network cardinality)
P_γ	Communication overhead for sending gradients from children to their parents	T_{NP}	Total number of parent nodes
$f1$	A ratio of sampling rate to frequency of attributes/parents' energy status sending	T_{NC}	Total number of children nodes
$f2$	A ratio of sampling rate to frequency of gradients set-up	W	Weight that represents excess number of messages than DP generates

Table 4. Parameters used in power consumption cost model.

The energy consumption issue for WSNs is the most important because the lifetime of a sensor node is extremely depends on the available energy of its battery. There are three domains to be considered regarding energy consumption: (i) sensing activity (data collection from the environment), (ii) communication (sending and receiving packets) and (iii) data processing/in-network data aggregation. Although all these activities waste energy, communication is responsible for the bulk of the power consumption which is the main point of attention in many algorithms designed for sensors networks. That is to say, energy saving by reducing the communication activity consequently increases WSN lifetime (Madden et al., 2005). Inspired by this notion, we design a mathematical cost model to compute how much power dissipates by our DP scheme in order to gather data with aggregation in WSN. In addition, we present the cost model in terms of the same metric for DD and HDA schemes. Table 4 lists the parameters used to design the power dissipation cost models.

4.1.1 Power Consumption by DP Scheme

We first divide the source nodes into different groups based on their positions in a WSN. This is done by determining how far they are located, in terms of hop count, from an active designated path. By using following equation we can know the number of groups of source nodes (C) for the given WSN.

$$C = \{\max(N- A_n, A_n-1) +1\}\times 1/r \quad (1)$$

Here, r is the one-side coverage range of a parent node and its value is determined during hierarchical multi-parent multi-child tree construction. For instance, in the Fig. 9, there are 48 sensor nodes (M=8 and N=6), a designated path P3 is active and the value of r equals 2. By substituting the values to parameters, we get

$$C = \{\max(6- 3, 3-1) +1\}\times 1/2 = \{\max(3, 2) +1\}\times 1/2= 2.$$

Therefore, source nodes can be divided into two groups, say group one is G1 (shown in dotted rectangle) and another is G2 (rest part of the network), as shown in Fig. 9. It means that the source nodes of G1 and G2 are located one hop and two hops away from P3, respectively.

The next step is to calculate the number of messages generated during data transmission from all of the source nodes to the sink node. The number of messages *Nmsg* can be calculated by using following expression.

$$N_{msg} = \sum_{i=1}^{C} G_i * n_i + (M' - 1) \quad (2)$$

As we can see in the Fig. 9, G1 and G2 consists of eight and two source nodes out of total ten source nodes (shown as dark colored nodes), respectively. Moreover, data from sources nodes of G1 and G2 need one hop and two hops to reach P3, respectively. If we substitute the values for the parameters, we can get *Nmsg* = (8×1 + 2×2) + (8-1) = 12+7 = 19. It is exactly the same number of messages generated (i.e., 19 solid arrows as shown in the Fig. 9) in the network.

Alternatively, there is another way to compute *Nmsg*. In this method, we simply use the number of all levels of WSN and associated number of sensor nodes in each level involved during data transmissions. Since each of the involved sensor nodes generates one message, the number of messages generated is equivalent to the number of the sensor nodes involved for data transmission. For this, we use following expression.

$$N_{msg} = \sum_{i=1}^{M} level_i \sum_{j=1}^{N} m_j \quad (3)$$

To prove the correctness of this expression, we can substitute the values for its parameters in the Fig.9. In this calculation, we put the value of involved sensor nodes in the decreasing order of level, i.e., starting from level M (in this case M=8) to 1. Then, we can get *Nmsg* = (3+2+3+2+3+3+2+1) = 19. Out of the 19 nodes, 10 nodes are source nodes (X) and 5 nodes are aggregation nodes (Y) which receive more than one message and partially aggregate data. The rest 4 nodes are routing nodes (Z) which just forward the incoming message to their parents. Hence, the number of messages generated in WSN is the sum of the source nodes, aggregation nodes and routing nodes involved during data transmissions.

Mathematically, we can express it as *Nmsg* = X+Y+Z. Since both of the methods result the same number of messages one method verifies the correctness of another and vice-versa.

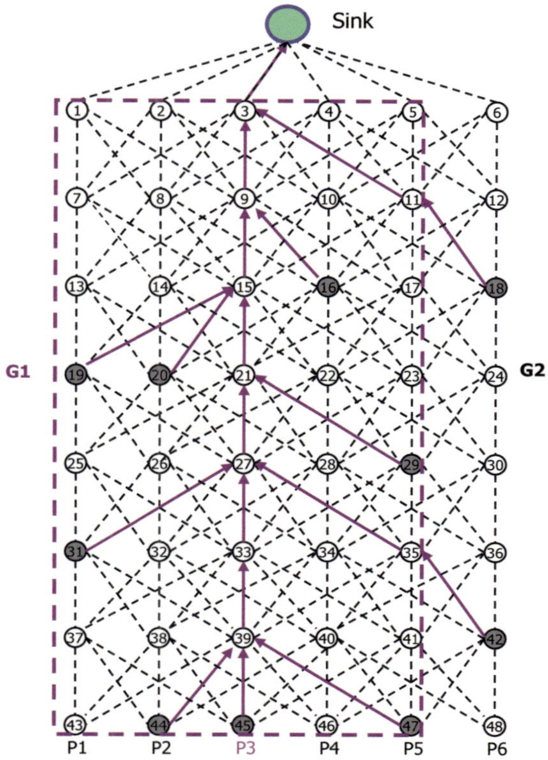

Fig. 9. Two groups of source nodes (G1 and G2).

For a given M×N WSN, the energy dissipation can be defined as the sum of the energy consumed by four types of nodes involved during data transmission to the sink node which are: sensor nodes being in the idle state, source nodes, aggregation nodes and routing nodes, and this can be calculated as below.

$$E_{DP} = (M \times N) \times E_{Idle} + \sum_{s=1}^{numSources} (E_{Tx}) + \left(\sum_{m=1}^{numAgrNodes} \left(\sum_{l=1}^{numSampleRcv} E_{Rx} + E_{Tx} \right) \right) + \sum_{n=1}^{numRouteNodes} (E_{Rx} + E_{Tx}) \qquad (4)$$

The first part of the right hand side of the expression is the energy required for all of the sensor nodes of the M×N WSN which are in the idle state. The second part gives the energy consumed by the sources nodes. The third part measures summation of the energy dissipated by each aggregation node. The second summation notation of the third part counts the number of received messages by an aggregation node. The fourth and the final part gives the energy required to receive and transmit a message for routing nodes. By using the notations of the Table 4, we can deduce the above expression as follow.

$$\begin{aligned} E_{DP} &= (M \times N) \times \alpha + X \times \beta + Y(\gamma + \beta) + Z(\gamma + \beta) \\ &= (M \times N)\alpha + (X + Y + Z)\beta + (Y + Z)\gamma \\ &= (M \times N)\alpha + N_{msg} \times \beta + (N_{msg} - X)\gamma = P_{DP} \end{aligned} \qquad (5)$$

This is the cost model which can compute the power dissipation by our DP scheme while collecting data in WSNs.

4.1.2 Power Consumption by HDA Scheme

HDA requires more power than our DP due two factors. The first one is that HDA frequently misses data aggregation and thus more number of messages is generated, due to the involvement of the many sensor nodes to forward data to the sink node. When we denote this extra communication overhead by weight factor W, in terms of number of messages, the power dissipated by HDA can be given as follow.

$$P_{\alpha} = W \times (E_{Rx} + E_{Tx}) \tag{6}$$

The second factor is that, in HDA, parent nodes have to frequently notify their energy-status/best-attributes/interests to their child nodes so that the child nodes can determine appropriate parent nodes for forwarding data to the sink node. Therefore, each parent node transmits a message to its child nodes and each of the child nodes has to receive the same number of messages as the number of its parent nodes, due to the multi-parent multi-child hierarchy tree structure. But our DP can avoid such type of unnecessary traffic during data transmission because every node has data gathering application knowledge. We can compute this messages overhead of HDA mathematically, as shown below.

Total number of parent nodes: $T_{NP} = (M-1) \times N$
Total number of child nodes: $T_{NC} = N + (M-1) \times N \times n_c$

Hence, the power dissipation to transmit a message by many parents ($P_{\beta 1}$) and that to receive a message by many child nodes ($P_{\beta 2}$) are given below. Here, f_1 is the ratio of sample rate to the frequency of notifying/receiving energy-status/best-attributes.

$P_{\beta 1} = ((M-1) \times N \times E_{TX}) \times 1/f_1$
$P_{\beta 2} = (N + (M-1) \times N \times n_c \times E_{RX}) \times 1/f_1$

By combining above two expressions, we get,

$$\begin{aligned} P_{\beta} = P_{\beta 1} + P_{\beta 2} &= (((M-1) \times N \times E_{TX}) \times 1/f1) + (N + ((M-1) \times N \times n_c \times E_{RX}) \times 1/f_1) \\ &= ((M-1) \times N \times E_{TX} + N + (M-1) \times N \times n_c \times E_{RX}) \times 1/f_1 \\ &= N\,((M-1)\,(E_{TX} + n_c \times E_{RX}) + 1) \times 1/f_1. \end{aligned}$$

As a result, the total power dissipation by HDA for data transmission to the sink node can be computed as below.

$$P_{HDA} = P_{DP} + P_{\alpha} + P_{\beta} \tag{7}$$

4.1.3 Power Consumption by DD Scheme

In the DD scheme, there are three more factors responsible for power consumption than that of DP scheme. Because the first two factors are the same as those of HDA, we just use them here. The third factor is that, in DD, each child node sends gradients to its all parent nodes in the response of frequently received interests from parent nodes. We derive the cost of gradients as below.

Total number of parent nodes: TNP = $(M-1)\times N\times n_p$
Total number of child nodes: $T_{NC} = M\times N$

Hence, the power dissipation to receive a gradient by many parents ($P_{\gamma 1}$) and that to transmit a gradient by many child nodes ($P_{\gamma 2}$) are as follows. Here f_2 is the ratio of sample rate to the frequency of receiving/sending gradients.

$P_{\gamma 1} = ((M-1)\times N \times n_p \times E_{RX}) \times 1/f_2$
$P_{\gamma 2} = (M\times N) \times E_{TX} \times 1/f_2$

By combining above two expressions, we get,

$P_\gamma = P_{\gamma 1} + P_{\gamma 2}$
$= (M-1) \times N \times n_p \times E_{RX} \times 1/f_2 + (M\times N) \times E_{TX} \times 1/f_2$
$= N ((M-1) \times n_p \times E_{RX} + M\times E_{TX}) \times 1/f_2$

As a result, the total power dissipation by DD for data transmission to the sink node can be by using following expression.

$$P_{DD} = P_{DP} + P_\alpha + P_\beta + P_\gamma \tag{8}$$

In summary, above analytical model shows that our DP scheme is an energy efficient scheme to aggregate data in WSN because it can aggregate data efficiently by avoiding unnecessary traffics during data transmissions.

4.2 Analytical Model for Sending ID of Sensor Nodes

As we mentioned earlier, communication is responsible for the bulk of the power consumption in WSNs. The limited power of sensor nodes can be saved by reducing communication overhead so that the lifetime of WSNs can be prolonged. There are many ways to reduce the communication overhead in WSNs. Some of them are: minimizing generation of messages in the network, shortening duty cycling and determining small size packet. Former two processes are applications dependent in WSNs whereas determining small size packet, in the case of low powered sensor nodes (Mica Motes), is controlled by TinyOS, an operating system that runs motes hardware. For Mica Motes, TinyOS predefined a 36-byte packet out of which 29-byte is allocated to the payload. With the commence of in-network data processing for WSNs, aggregation of sensor data became popular because data aggregation can reduce the number of data transmissions to the sink node by combining correlated sensor data . But, in many applications, data aggregation in WSNs needs the sink node to acquire knowledge of the contributed sensor nodes so that the sink node can compute actual result of aggregated data. This requirement creates a problem of sending IDs of participated sensor nodes to the sink node for larger size WSNs because the payload is of limited size. In this section, we present an analytical model for sending IDs of the contributed sensor nodes to the sink node for the existing CMT and our schemes. We assume that N is the total number of sensor nodes of a sub-tree rooted at the sink node in a WSN. We also assume that N_{cl} and N_{ncl} are the lists of contributing nodes and the list of non-contributing nodes of the WSN respectively. Hence, $N=N_{cl}+N_{ncl}$, where $Ncl < N_{ncl}$.

4.2.1 CMT Scheme

In this method, each node ID is considered as a plaintext (2-byte) and all the IDs are concatenated while sending to the sink node. Out of the fixed 29 bytes payload, an encrypted sensor data uses 4 bytes leaving 25 bytes as free space for carrying IDs. Therefore, the number of sensor node IDs can be included in the list of N_{cl} is 12 while sending the aggregated data to the sink node. For the CMT scheme, the value for scalability in terms of carrying IDs is $O(N_{cl})=12$.

4.2.2 Signature Scheme

On the other hand, since we superimpose signatures of sensor node IDs, a single bit is enough to hold ID of a sensor node. Therefore, for the available 25 bytes free space of the payload, our scheme can include $25\times8 = 200$ sensor node IDs in the list of N_{cl} while sending the aggregated data to the sink node. Hence, for our scheme, the value for scalability in terms of carrying IDs is $O(N_{cl}) = 200$.

This analytical model shows that, if necessary, our scheme can transmit around 16 times more number of sensor node IDs than does the CMT scheme. Therefore, our scheme is obviously a scalable one to apply in such data aggregation applications for WSNs that need the information of contributed sensor nodes at the sink node e.g., privacy preserving data aggregation for WSNs.

5. Analytic Performance Evaluation

Based on the previous mathematical models, first we compare the performance of DP scheme with HDA and DD schemes in terms of energy dissipation required to collect data from WSNs and then compare the performance of our signature scheme with CMT scheme in terms of energy efficiency and scalability in order to transmit IDs of sensor nodes to the sink node.

5.1 Analytic Performance Evaluation of DP, HDA and DD Schemes

We consider the scenario where the frequency of attributes/parents-energy-status/gradients sending is once per 50 seconds as in HDA. We use such parameters as idle-time power dissipation of 35 mW, receiving power dissipation of 395 mW, and transmitting power dissipation of 660 mW, as presented in DD. The sampling rate is one sample per second. For this evaluation, we study on the impacts of *network size, the number of source nodes* and *network cardinality* over the energy consumption.

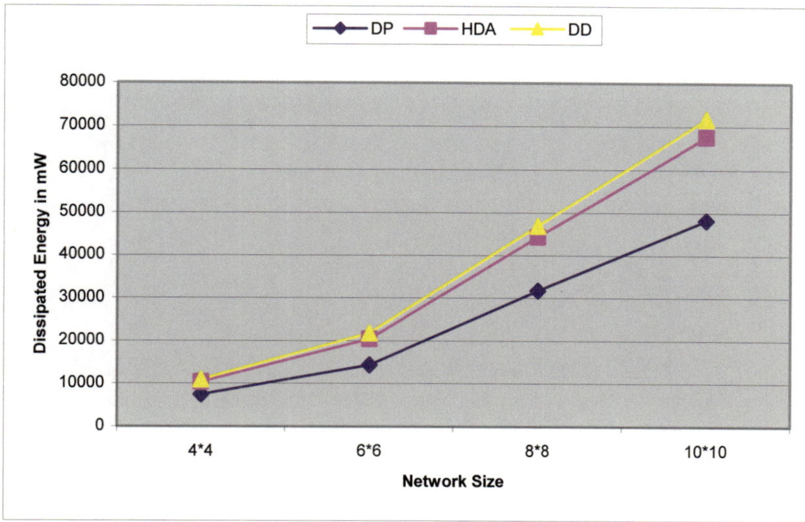

Fig. 10. Energy consumption for varying network size.

(a) Network size: For this, the density of source nodes is fixed to 25% of sensor nodes from different sizes of WSNs. In Fig. 10, it is shown that the performances of all the three schemes DP, HDA and DD are decreased as the size of the network increases from 4×4 to 10×10. This is because as the size of a network increases, the number of source nodes also increases. As a result, the number of generated messages increases during data transmissions in the networks. Consequently, a larger WSN consumes much amount of energy than a smaller one. However, the performance of our DP scheme is always better than both of HDA and DD schemes. It is because DP scheme generates less number of messages in the networks by avoiding unnecessary traffics generation during data transmissions to the sink node. Moreover, as the size of network increases, the performance gap between DP and HDA as well as that between DP and DD get wider. It indicates that data aggregation scalability of our scheme is better than both HDA and DD schemes.

(b) Source nodes: We change the density of the sources nodes from 10 to 50 for a fixed size 10×10 WSN. In Fig. 11, it is shown that as the number of source nodes increases from 10 to 50, the amount of dissipated energy for transmitting data to the sink node also increases for all DP, HDA and DD schemes. The reason is that a larger number of source nodes means that the network generates more number of messages and it needs larger amount of energy to transmit them. However, as the number of source nodes increases, the rate of increase in the amount of the dissipated energy is lower for DP scheme than both HDA and DD schemes. In this way, the performance of the DP scheme improves further for higher number of source nodes in a WSN. It justifies the efficiency of DP scheme to aggregate data in WSNs.

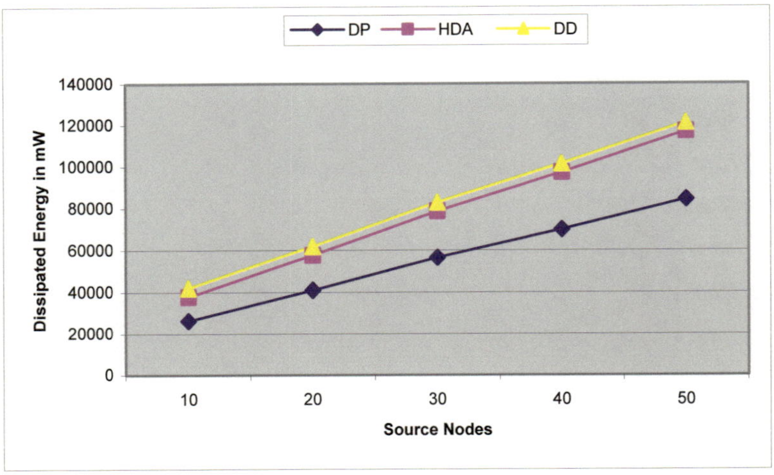

Fig. 11. Energy consumption for varying source nodes.

(c) Network cardinality: The network size and the number of source nodes are fixed to a 10×10 WSN and 15% of sensor nodes respectively. We change network cardinality from 3 to 5 as shown in Fig. 12. The cardinality of a network means an average number of child nodes and parent nodes per sensor node in the WSN and it is determined during the construction of the multi-parent-multi-child hierarchical network structure. The Fig. 12 depicts that our DP scheme has better performance than HDA and DD schemes although the amount of dissipated energy for all the three schemes decreases when the network cardinality increases. This is because the coverage of sensor nodes increases with the increase in the network cardinality. As a result, the number of messages generated in the network is reduced while transmitting data to the sink node.

Above analytical performances show that proposed DP scheme is a more energy efficient scheme to aggregate data in WSNs than HDA and DD schemes.

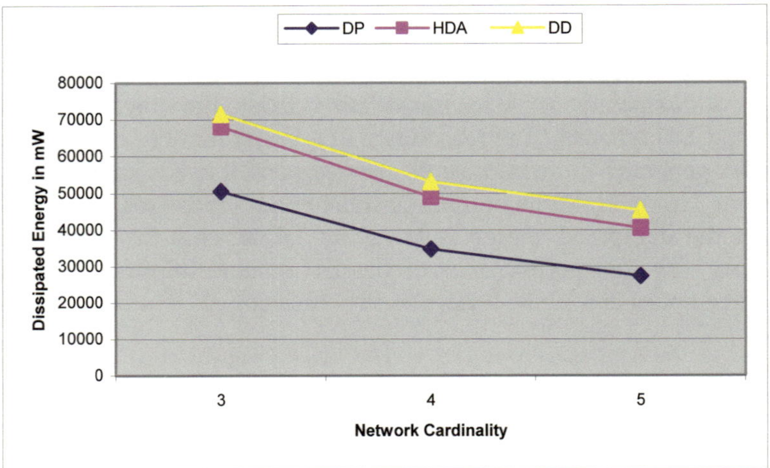

Fig. 12. Energy consumption for varying network cardinality.

5.2 Analytical Performance Evaluation of CMT and Signature Schemes

In this section, we show the efficiency of our scheme by comparing it with the CMT scheme considering transmissions of IDs of contributed sensor nodes along with aggregated data to the sink node. The CMT scheme is the standard work that deals with sending IDs of sensor nodes to the sink node for WSNs. We present the performance results of both schemes in terms of four metrics: *scalability, energy consumption, payload size* and *computation overhead.*

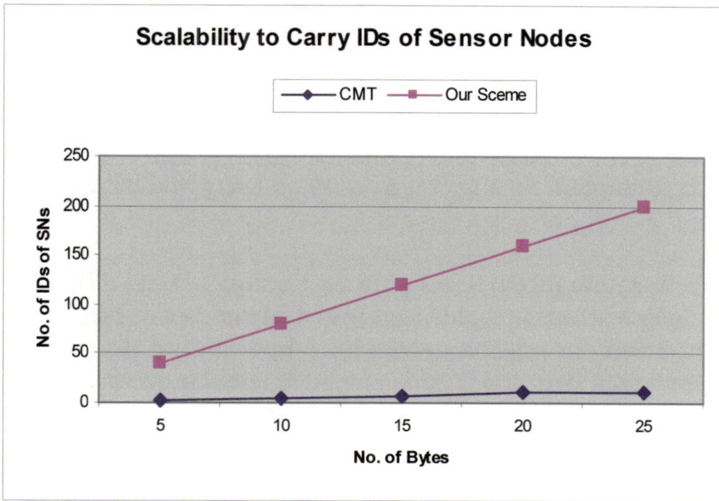

Fig. 13. Carrying IDs of sensor nodes by Our and CMT schemes.

(a) Scalability: For TinyOS based Mica Motes, the maximum payload size is of 29-byte. We assume each of sensor data and a key is of 2-byte. Therefore, the remaining maximum free space of the payload is 25-byte. The scalability measure is given in terms of IDs of how many sensor nodes can be sent by using the available limited free space (25-byte) by both schemes. As shown in Fig. 13, for the given limited 25-byte free space, our scheme can send IDs of up to 200 sensor nodes while transmitting aggregated sensor data to the sink node. On the other hand, the CMT scheme is unable to send IDs of more than 12 sensor nodes. The reason is that our scheme can hold ID of a sensor node just by a single bit whereas the CMT scheme needs 2-byte for the same task. Therefore, it is obvious that our scheme is much more (about 16- time) scalable than the CMT scheme in terms of carrying the number of IDs of sensor nodes in the course of transmitting aggregated value to the sink node in WSNs.

(b) Energy consumption: In this measure, we consider the amount of energy required to transmit and receive a packet by a sensor node. This is calculated as given in (Bi et al., 2007). The total energy (E_{Total}) to communicate a packet is calculated by adding transmission energy (E_{Tx}) and receiving energy (E_{Rx}) as below.

$$E_{Tx} = L \times E_{elec} + L \times \varepsilon \times d^2 \quad (9)$$

$$E_{Rx} = L \times E_{elec} \quad (10)$$

$$E_{Total} = E_{Tx} + E_{Rx} \quad (11)$$

where, L is the length of the packet in bits, E_{elec} is electronic energy (= 1.16 μJ/bit), the parameter ε = 5.46 pJ/bit/m^2, and d is crossover distance (= 40.8 m).

Table 5 illustrates energy efficiency of our scheme over the CMT scheme to communicate a packet which consists of 2-byte sensor data, 2-byte key and IDs of 12 sensor nodes. To achieve this, our scheme dissipates just about 36% of that energy which is required by the CMT scheme. This is because our scheme needs less number of bytes than that of CMT scheme to transmit the packet with aforementioned features. By saving the precious energy of sensor nodes In this way, our signature scheme can enhance the lifetime of WSNs.

Method	Energy Dissipation in mJ	Energy Gain Ratio
CMT	0.670778	63.88%
Our Scheme	0.242225	

Table 5. Energy consumption by a packet to carry an encrypted data along with IDs of 12 sensor nodes.

(c) Payload size: We measure this in terms of bytes required to send different number of IDs of sensor nodes along with an encrypted aggregated sensor data (4-byte) to the sink node. In Fig. 14, it is shown that our scheme needs only 5-byte to send IDs of up to eight sensor nodes with the encrypted data and it adds one more byte for every additional ID of up to eight sensor nodes. On the other hand, the CMT scheme needs 2 more bytes for each additional sensor node ID. Therefore, the size of payload in the CMT scheme is directly proportional to the number of IDs of sensor nodes. For instance, to send IDs of 12 sensor nodes with their encrypted aggregated value, our signature scheme needs just 6-byte (4-byte for encrypted aggregated value and 2-byte for carrying IDs of 12 sensor nodes) payload whereas the CMT scheme needs 28- byte (4-byte for encrypted aggregated value and 24-byte for carrying IDs of 12 sensor nodes). In this way, our signature scheme reduces the size of payload greatly. As a result, the proposed signature scheme not only reduces the packet communication cost but also decreases the message loss rate because the probability of message interference is higher for larger size messages (Muller et al., 2007).

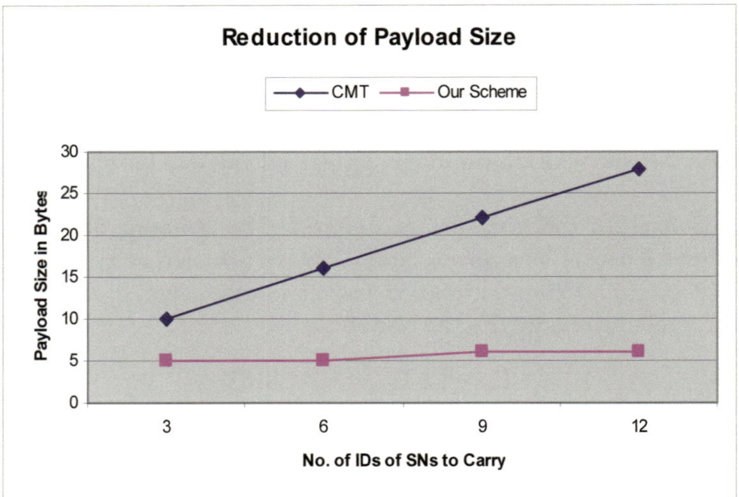

Fig. 14. Variation of payload size with increasing number of node ID.

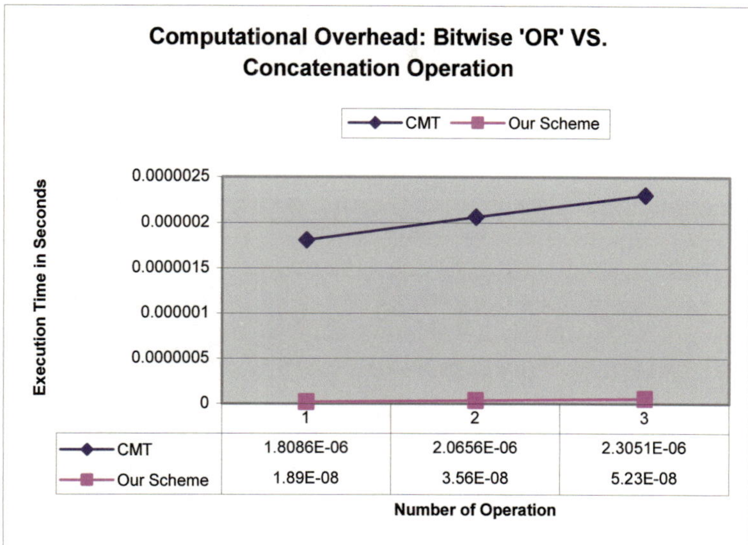

Fig. 15. Computational efficiency of Our scheme over CMT scheme.

(d) Computation overhead: We measure execution time required to: i) concatenate IDs of sensor nodes (plaintexts) in the case of the CMT scheme and ii) superimpose IDs of sensor nodes in our scheme. We use MATLAB® 7.6.0.324 (R14) to compute the execution time. In this experiment, we consider the execution time required for one, two and three concatenation and bitwise *OR* operations to combine IDs of two, three and four sensor nodes (each ID is of 2-byte size, a positive integer type) for the CMT and our scheme respectively. In Fig. 15, it is shown that the execution time of our approach to combine IDs of sensor nodes is always faster than that of the CMT scheme by an order of two-magnitude. The reason is that our scheme uses bitwise *OR* operation to combine signatures of node IDs. Needless to say that the bitwise operation is the fastest one among all available operations for a processor.

6. Performance Evaluation

In this section, by using TOSSIM (Levis et al., 2003) simulator, we evaluate the performances of our DP scheme comparing with HDA and DD schemes, in terms of dissipated energy. We consider the scenario where the frequency of attributes/parents-energy-status/gradients sending is once per 50 seconds as in HDA. We use such parameters as packet receiving, packet transmitting and data aggregation for power dissipation. The sampling rate is one sample per second. We study on the impacts of network size, the number of source nodes and network cardinality over the energy consumption. We consider the same network scenarios for simulations as we did in the previous section for all the three analytic evaluations.

(a) Network size: Similar to the analytic performance, Fig. 16 shows that our DP scheme requires less amount of energy than HDA and DD schemes to collect data from different size WSNs. It is because our DP scheme generates less number of messages in the networks

by avoiding unnecessary traffics generation during data transmissions to the sink node. Moreover, as the size of network increases, the performance gap between DP and HDA schemes as well as that between DP and DD schemes get wider. It indicates that, in of our DP scheme, data aggregation efficiency improves further with the increasing size of the networks.

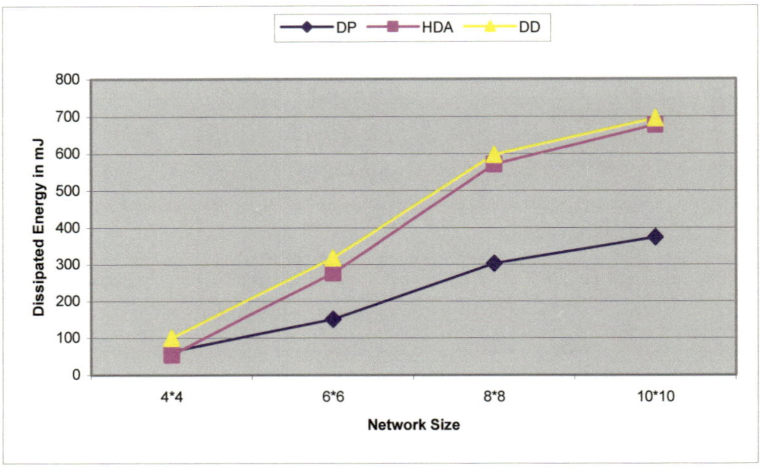

Fig. 16. Energy consumption for varying size of WSN when source nodes are fixed to 25% of the sensor nodes.

(b) Source nodes: Similar to the analytic performance, Fig. 17 shows that our DP scheme always require less amount of energy to aggregate data than HDA and DD schemes when the number of source nodes in a WSN varies. In addition, the rate of increase in the amount of the dissipated energy improves further in DP scheme with the increasing number of source nodes in a WSN. The reason is that, unlike HDA and DD schemes, DP scheme doesn't generate extra traffics and it guarantees data aggregation in WSNs.

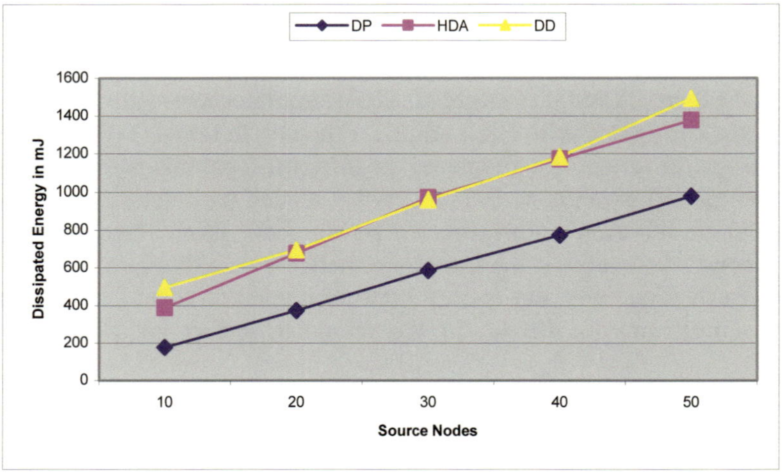

Fig. 17. Energy consumption for varying source nodes in a 10×10 WSN.

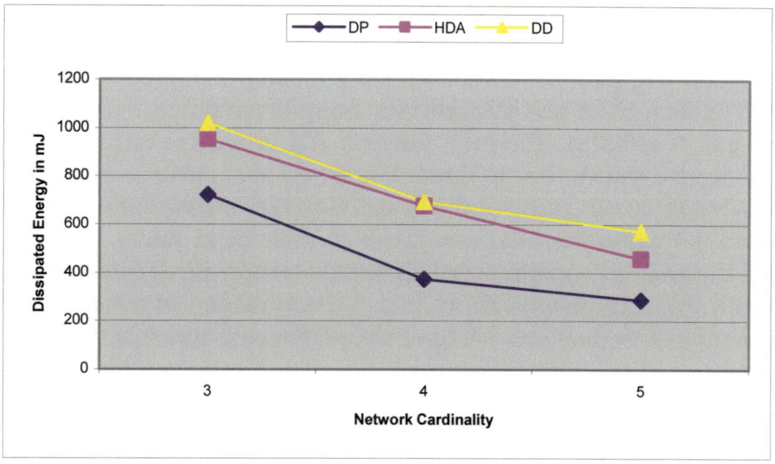

Fig. 18. Energy consumption for varying network cardinality when source nodes are fixed to 15% of sensor nodes in a 10×10 WSN.

(c) Network cardinality: Fig. 18 depicts that when the network cardinality increases the amount of dissipated energy for data transmissions to the sink node decreases for all DP, HDA and DD schemes. This is because with the increase in the network cardinality, the coverage range of each node also increases. As a result, it reduces the total number of messages in the network and so does the dissipated energy. As above analytical performance evaluation, the performance of our DP scheme is always better than those of HDA and DD schemes for varying network cardinality. The reason is that, in DP scheme, all sensor nodes utilize data aggregation application knowledge for when and where to send data during their transmissions to the sink node. However, on the one hand, a larger value for network cardinality gives more energy efficiency to a WSN; but on the other hand, increasing data transmission rage of sensor nodes costs much energy. Therefore, there must be a reasonable trade-off of the network cardinality over the data transmission range. For this time, we would like to keep this issue as our future work.

7. Conclusion and Future Work

In this chapter, we proposed two energy efficient schemes for resource-constraint WSNs. First, we proposed DP scheme as energy efficient data aggregation for WSNs in which a pre-determined set of paths is run in round-robin-fashion in order to tackle the unnecessary traffics and hotspot problem of the conventional data aggregation schemes which always drive data flow towards the sink node/s. In our DP scheme, all sensor nodes participate in gathering all the sensed data and transferring them to the sink node. Because all the nodes in the network are charged for the heavy workload, we believe that the sensor nodes consume their energy almost equally and the hotspot problem can be significantly relieved. In addition, DP scheme avoids unnecessary traffics during data transmissions to the sink node by utilizing data aggregation application knowledge. Moreover, unlike both DD and HDA schemes, DP scheme can be used for continuous data delivery for event-driven applications because unnecessary traffics do not intervene during data collection processes.

The presented analytical performance evaluations and simulation results have similar trends to achieve energy efficiency. Both of them show that DP scheme is more energy efficient for aggregating data in WSNs and hence it can prolong the lifetime of resources-constraints WSNs than HDA and DD schemes. Second, we propose a novel scheme called signature scheme in order to efficiently transmit IDs of a large number of sensor nodes along with aggregated sensor data to the sink node. In our signature scheme, first, the sink node generates a unique signature for the Real ID of every sensor node. Then, parent nodes (data aggregators) superimpose the signatures of their child nodes including their own signatures and transmit the superimposed signatures along with aggregated data to the sink node. For this, a single bit is enough to hold the information of a sensor node. Through analytical performance evaluations, we have shown the efficiencies of the signature scheme over the existing work in terms of scalability, energy consumption, payload size and computation overhead.

Transmitting IDs of contributed sensor nodes along with sensed data is mandatory for many applications designed for WSNs. Therefore, as our future work, first we would like to show simulation results of the signature scheme and then we will mingle DP scheme with signature scheme in order to provide further more energy efficient scheme to collect data in WSNs. In addition, we would like to apply our combined scheme to arbitrary types of WSN and networks with multiple sink nodes.

8. Acknowledgment

This research was financially supported by the Ministry of Education, Science Technology (MEST) and Korea Institute for Advancement of Technology (KIAT) through the Human Resource Training Project for Regional Innovation. This work was also supported by the Korea Science and Engineering Foundation (KOSEF) grant funded by the Korea government (MEST) (No. 2010-0000202).

9. References

Akkaya, K. & Younis, M. (2005). A survey on routing protocols for wireless sensor networks, Ad Hoc Networks 3 (2005) pp. 325-349.

Akyildiz, I.; Su, W.; Sankarasubramaniam, Y. & Cyirci, E. (2002). Wireless sensor networks: a survey, In Computer Networks 38 (4) (2002), 393–422.

Bi, Y.; Li, N. & Sun, L. (2007). DAR: An energy-balanced data-gathering scheme for wireless sensor networks, In Computer Communication 30 (2007) 2812-2825.

Bista, R.; Kim, Y-K. & Chang, J-W. (2009). A New Approach for Energy-Balanced Data Aggregation in Wireless Sensor Networks, In CIT09, cit, vol. 2, pp. 9-15.

Bista R., Chang J-W. Privacy-Preserving Data Aggregation Protocols for Wireless Sensor Networks: A Survey, Sensors 2010, 10(5) : 4577-4601.

Castelluccia, C.; Mykletun, E. & Tsudik, G. (2005). Efficient aggregation of encrypted data in wireless sensor networks, In MobiQuitous, pp. 109–117, 2005.

Considine, J.; Li, F.; Kollios, G. & Byers, J. (2004). Approximate aggregation techniques for sensor databases, In Proceedings of ICDE, pp. 449-460, April, 2004.

Conti, M.; Zhang, L.; Roy, S.; Pietro, R-D.; Jajodia, S. & Mancini, L-V. (2009). Privacy-preserving robust data aggregation in wireless sensor networks, Security and Communication Networks, 2009; 2:195–213.

Dijkstra, E-W. (1959). A Note on Two Problems in Connection with Graphs, Numeriche Mathematik, Vol. 1 (1959) pp. 269-271.

Girao, J.; Westhoff, D. & Schneider, M. (2005). CDA: Concealed Data Aggregation for Reverse Multicast Traffic in Wireless Sensor Networks, In ICC 2005, Vol.5, pp. 3044-3049.

He, W.; Liu, X.; Nguyen, H.; Nahrstedt, K. & Abdelzaher, T. (2007). Pda: Privacy-preserving data aggregation in wireless sensor networks, In Proceeding of INFOCOM, pp. 2045–2053, 2007.

Heinzelman, W-R.; Kulik, J. & Balakrishman, H. (1999). Adaptive protocols for information dissemination in wireless sensor networks, In Proceedings of MOBICOM, pp. 174–185, August, 1999.

Heinzelman, W.; Chandrakasan, A. & Balakrishnan, H. (2000). Energy-efficient communication protocols for wireless microsensor networks, In Proceedings of HICSS, January, 2000.

Hill, J.; Szewczyk, R.; Woo, A.; Hollar, S.; Culler, D-E. & J.Pister, K-S. (2000). System Architecture Directions for Networked Sensors, In ASPLOS, pp. 93–104, 2000. TinyOS is available at http://webs.cs.berkeley.edu.

Horton, M.; Culler, D.; Pister, K.; Hill, J.; Szewczyk, R. & Woo, A. (2002). MICA the commercialization of micro sensor motes, In IEEE Sensors J., April 2002, 19(4): 40-48.

Itanagonwiwat, C.; Govindan, R. & Estrin, D. (2002a). Directed Diffusion: A Scalable and Robust Communication Paradigm for Sensor Networks, In Proceedings of MOBICOM, pp. 56-67, 2002.

Itanagonwiwat, C.; Estrin, D.; Govindan, R. & Heidemann, J. (2002b). Impact of Network Density on Data Aggregation in Wireless Sensor Networks, In Proceedings of the 22nd ICDCS, pp. 457-458, 2002.

Levis, P.; Lee, N.; Welsh, M. & Cullar, D. (2003). TOSSIM: Accurate and scalable simulation of entire TinyOS applications, http://www.cs.berkely.edu/~pal/research/tossim.html.

Madden, S.-R.; Franklin, M.-J.; Hellerstein, J.-M. & Hong, W. (2002). TAG: a tiny aggregation service for ad hoc sensor networks, In Proceedings of the OSDI02, pp. 1-16, December, 2002.

Madden, S.-R.; Franklin, M.-J.& Hellerstein, J.-M. (2005). TinyDB: an acquisitional query processing system for sensor networks, ACM TDS 30 (1) (2005), pp.122–173.

Mueller, R.; Kossmann, D. & Alonso, G. (2007). A Virtual Machine for Sensor Networks, In EuroSys'07, pp. 145-158, March 2007.

Pottie, G-J. & Kaiser, W-J. (2000). Wireless integrated network sensors, Communications of ACM, May 2000.

Yao, Y. & Gehrke, J. (2003). Query processing for sensor networks, In Proceedings of the CIDR 2003.

Yick, J.; Mukherjee, B. & Ghosal, D. (2008). Wireless sensor network survey, In Computer Networks, 2008, 52(12): 2292-2330.

Zhang, W-S.; Wang, C. & Feng, T-M. (2008). GP2S: generic privacy-preservation solutions for approximate aggregation of sensor data, concise contribution, In Proceedings of PerCom, pp.179–184, 2008.

Zhou, B.; Ngoh, L. H.; Lee, B. S. & Fu, C-P. (2006). HDA: A hierarchical Data Aggregation Scheme for Sensor Networks, Computer Communication 29 (2006) 1292-1299.

Zobel, J.; Moffat, A. & Ramamohanarao, K. (1998). Inverted Files versus Signature File for Text Indexing, In ACM TDS, Vol. 23, No. 4, 1998, pp. 453-490.

21

A Chaos-Based Data Gathering Scheme Using Chaotic Oscillator Networks

Hidehiro Nakano, Akihide Utani, Arata Miyauchi and Hisao Yamamoto
Tokyo City University
Japan

1. Introduction

Recently, wireless sensor networks have been studied extensively with a great amount of interest. In wireless sensor networks, many wireless sensor nodes are deployed in an observation area, and monitor status information such as temperature around them. Sensing information is transmitted to and gathered by one or more sink nodes. Each wireless sensor node not only transmits own sensing data but also relays the sensing data from the other wireless sensor nodes. By such a multi-hop wireless communication, the wireless sensor networks are available to observation for large-scale area, and have various applications including natural environmental monitoring. Since wireless sensor nodes generally operate by batteries, efficient data gathering schemes with saving energy consumption of each wireless sensor node are needed for prolonging wireless sensor network lifetime. Ant-based algorithms (Caro et al., 2004; Marwaha et al., 2002; Ohtaki et al., 2006; Subramanian et al., 1998) and cluster-based algorithms (Dasgupta et al., 2003; Heinzelman et al., 2000) have been proposed as routing algorithms. They are more scalable, efficient and robust than the other conventional routing algorithms (Clausen & Jaquet, 2003; Johnson et al., 2003; Ogier et al., 2003; Perkins & Royer, 1999). Sink node allocation schemes based on particle swarm optimization algorithms (Kumamoto et al., 2009; Yoshimura et al., 2009) aim to minimize total hop counts in wireless sensor networks and to reduce energy consumption in each wireless sensor node. Forwarding node set selection schemes (Nagashima et al., 2009; Sasaki et al., 2009) can significantly reduce the number of transmissions of duplicate query messages as compared with original flooding schemes. Secure communication schemes considering energy savings (Li et al., 2009; Wang et al., 2009) have also been proposed. Common purpose of these studies is to prolong wireless sensor network lifetime by saving energy consumption of each wireless sensor node.
Along this line, this study focuses on control schemes for timings of transmissions and receptions of sensing data, proposed as a synchronization-based data gathering scheme (Wakamiya & Murata, 2005). In this scheme, each wireless sensor node has a timer characterized by an integrate-and-fire neuron (Keener et al., 1981). Coupling the timers of wireless sensor nodes which can directly communicate to each other, they construct a pulse-coupled neural network. It is known that pulse-coupled neural networks can exhibit various synchronous and asynchronous phenomena (Catsigeras & Budelli, 1992; Mirollo & Strogatz, 1990). The conventional synchronization-based data gathering scheme is based on the synchronization in pulse-coupled neural networks. As synchronization is achieved, the following control for timings of transmissions and receptions of sensing data is possible: wireless sensor nodes turn

off their power supplies when they do not transmit and receive sensing data. Hence, long-term observation to target area is possible. As a hardware module, a passive wake up scheme for wireless sensor networks has also been proposed (Liang et al, 2008). In the conventional synchronization-based data gathering scheme, it is assumed that wireless sensor nodes do not have any complex routing tables; they transmit and receive sensing data by only referring values of hop counts to the nearest sink node. However, simple pulse-coupled neural networks consisting of integrate-and-fire neurons can exhibit periodic synchronization only. In the conventional synchronization-based data gathering scheme, many duplicate sensing data can be relayed by many wireless sensor nodes. Generally, wireless sensor nodes consume a lot of energy in transmitting sensing data (Heinzelman et al., 2000). Also, in multiple sink wireless sensor networks, multiple sink nodes are allocated on target area, where these are generally distant to each other. If they are not coupled to each other by some communications, it is hard to synchronize all wireless sensor nodes. In order to prolong wireless sensor network lifetime and realize long-term observation, more efficient data gathering schemes are needed.

In the previous works, a chaos-based data gathering scheme has been proposed (Nakano et al., 2009; 2010). In the chaos-based data gathering scheme, each wireless sensor node has a timer characterized by a chaotic spiking oscillator which generates spike-trains with chaotic inter-spike intervals (Nakano & Saito, 2002; 2004). Coupling multiple chaotic spiking oscillators, a chaotic pulse-coupled neural network is constructed. Chaotic pulse-coupled neural networks can exhibit various chaos synchronous phenomena and their breakdown phenomena. The proposed chaos-based data gathering scheme especially applies the breakdown phenomena in chaotic pulse-coupled neural networks. In the phenomena, all chaotic spiking oscillators do not exhibit perfect synchronization. However, partial synchronization on network space and intermittent synchronization on time-domain can be observed depending on parameters. The partial and intermittent synchronization can significantly reduce the redundant transmissions and receptions of sensing data. In the method presented in (Nakano et al., 2009), sensing data is transmitted in the timings when transmitting wireless sensor nodes generate spike signals. In this case, lost sensing data may appear. But, it is confirmed in the numerical experiments that high delivery ratio for sensing data can be kept. In the method presented in (Nakano et al., 2010), sensing data is transmitted in the timings when transmitting wireless sensor nodes accept the spike signals from the other wireless sensor nodes. In this case, it is guaranteed that all sensing data must be transmitted to sink nodes without lost sensing data. Since all chaotic spiking oscillators do not exhibit perfect synchronization, wake up time of each sensor node becomes longer, compared with the conventional synchronization-based data gathering scheme. This method does not aim to reduce energy consumption by turning off power supply of transceivers. However, the partial and intermittent synchronization in the chaos-based data gathering scheme can significantly reduce the total number of transmissions and receptions of sensing data. It can contribute to prolonging wireless sensor network lifetime. Also, the proposed chaos-based data gathering scheme can flexibly adapt not only single sink wireless sensor networks but also multiple sink wireless sensor networks.

This chapter consists of five sections. In Section 2, the conventional synchronization-based data gathering scheme is introduced, and some assumptions for wireless sensor networks in this research is explained. In Section 3, a model of the proposed chaos-based data gathering scheme is explained, and typical phenomena from a simple master-slave network are presented. Then, a basic mechanism of partial and intermittent synchronization in the proposed chaos-based data gathering scheme is discussed. In Section 4, simulation results for two types of wireless sensor networks, a single sink wireless sensor network and a multiple sink wireless sensor network, are presented. Through simulation experiments, effectiveness of the proposed chaos-based data gathering scheme is shown, and its development potential is discussed. In Section 5, the overall conclusions of this chapter are given and future problems are discussed.

2. Synchronization-Based Data Gathering Scheme

First, a synchronization-based data gathering scheme presented in (Wakamiya & Murata, 2005) are explained. A wireless sensor network consisting of M wireless sensor nodes and L sink nodes are considered. Each wireless sensor node S_i $(i = 1, \cdots, M)$ has a timer which controls timing to transmit and receive sensing data. The timer in S_i is characterized by a phase $\phi_i \in [0,1]$, an internal state $x_i \in [0,1]$, a continuous and monotone function f_i, a non-negative integer distance level $l_i > 0$, and an offset time δ_i. If each wireless sensor node does not communicate to each other, dynamics of the timer in S_i is described by the following equation.

$$\frac{d\phi_i(t)}{dt} = \frac{1}{T_i}, \text{ for } \phi_i(t) < 1, \tag{1}$$

$$\phi_i(t^+) = 0, \text{ if } \phi_i(t) = 1, \tag{2}$$

where T_i denotes a period of the timer in S_i. That is, if the phase ϕ_i reaches the threshold 1, S_i is said to fire, and the phase ϕ_i is reset to 0 based on Equation (2), instantaneously. The internal state x_i is determined by the continuous and monotone function $f_i(\phi_i)$ where $f_i(0) = 0$ and $f_i(1) = 1$ are satisfied. The following equation is an example of the function f_i.

$$x_i = f_i(\phi_i) = \frac{1}{b_i} \ln(1 + (e^{b_i} - 1)\phi_i), \tag{3}$$

where $b_i > 0$ is a parameter which controls rapidity to synchronization (Mirollo & Strogatz, 1990). From Equations (1) and (3), increase of the phase ϕ_i causes increase of the internal state x_i. If x_i reaches the threshold 1, x_i is reset to the base state 0, instantaneously.
The couplings between each wireless sensor node are realized by the following manner. Let S_j be one of the neighbor wireless sensor nodes allocated in the radio range of a wireless sensor node S_i. The wireless sensor node S_i has a nonnegative integer distance level l_i characterized by the number of hop counts from the nearest sink node. The wireless sensor node S_i transmits a stimulus signal with the own distance level l_i. If S_j receives the signal from S_i, S_j compares the received distance level l_i with the own distance level l_j. If $l_j > l_i$ is satisfied, S_j is said to be stimulated by S_i, and the phase and internal state of S_j change as follows:

$$x_j(t^+) = B(x_j(t) + \varepsilon_j), \tag{4}$$

$$B(x) = \begin{cases} x, & \text{if } 0 \le x \le 1, \\ 0, & \text{if } x < 0, \\ 1, & \text{if } x > 1, \end{cases} \tag{5}$$

$$\phi_j(t^+) = f_j^{-1}(x_j(t^+)), \tag{6}$$

where ε_j denotes a strength of the stimulus. After S_j is stimulated, S_j does not respond to all stimulus signals from the neighbor wireless sensor nodes during an offset time δ_j. That is, each wireless sensor node has a refractory period corresponding to the offset time.

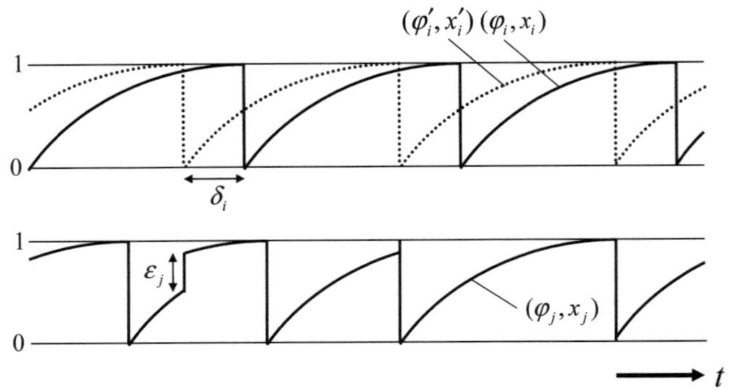

Fig. 1. Time-domain waveforms of internal states x_i and x_j ($l_j > l_i$).

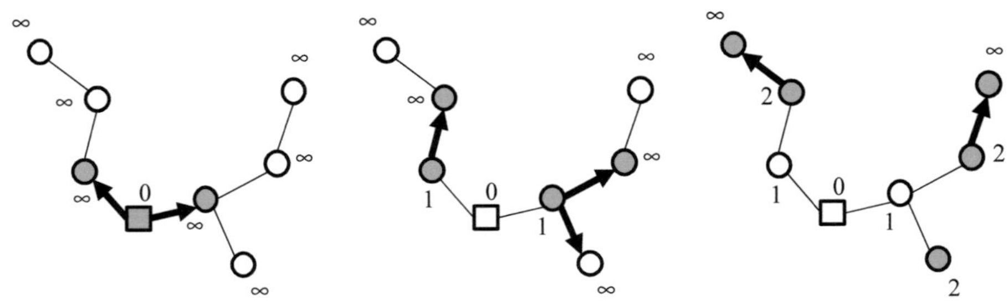

Fig. 2. Propagation of stimulus signals and update of distance levels.

The stimulus signals are transmitted by the following manner. A wireless sensor node S_i broadcasts stimulus signals offset time δ_i earlier than the own firing time. That is, S_i broadcasts the stimulus signals if the following virtual internal state x_i' considered the offset time δ_i reaches the threshold 1.

$$\phi_i' = \phi_i + \delta_i \pmod 1, \tag{7}$$

$$x_i' = f_i(\phi_i'). \tag{8}$$

Fig. 1 shows time-domain waveforms of internal states x_i and x_j, where $l_j > l_i$.
Distance levels of each wireless sensor node are adjusted as shown in Fig. 2. Initially, distance levels of each wireless sensor node are set to sufficiently large values, and that of the sink node is set to 0. A sink node broadcasts "level 0" as a beacon signal. Then, each wireless

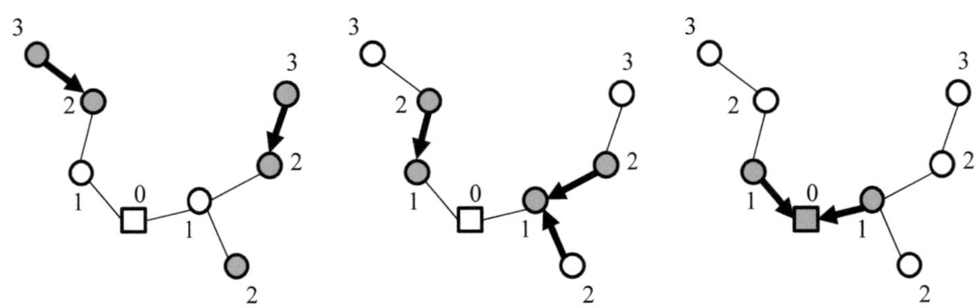

Fig. 3. Transmission of sensing data based on distance levels.

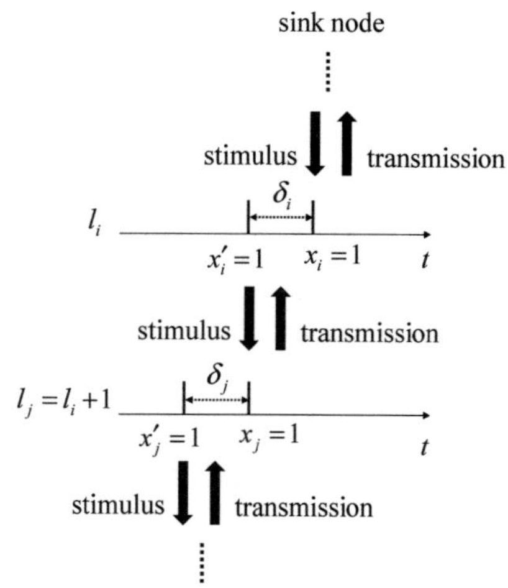

Fig. 4. Relaying sensing data ($l_j > l_i$).

sensor node forwards the beacon signal by using flooding, and adjusts each own distance level as corresponding to hop counts to its nearest sink node. The beacon signal is transmitted when each wireless sensor node transmitts stimulus signals. That is, for a stimulus signal from a wireless sensor node S_i, a wireless sensor node S_j adjusts own distance level l_j as follows:

$$l_j = l_i + 1, \text{ if } x_i'(t) = 1 \text{ and } l_j > l_i \tag{9}$$

As a result, each wireless sensor node has a distance level as corresponding to hop counts to its nearest sink node.
Sensing data is transmitted and received as shown in Fig 3. S_i is assumed to receive sensing data from its neighbor wireless sensor node S_j if $l_j = l_i + 1$ is satisfied. Then, S_i aggregate the received sensing data and own sensing data. After that, S_i transmits the aggregated sensing data. Sensing data is assumed to be transmitted and received in each firing period.
The communications between a wireless sensor node S_i and its neighbor wireless sensor node S_j are summarized as follows (see Fig. 4).

- If $l_j = l_i + 1$, S_i receives sensing data from S_j, and aggregates it with the own sensing data. Then, the aggregated sensing data is transmitted to the other wireless sensor nodes.
- If $l_j > l_i$, S_j is stimulated by S_i, and the internal state x_j is changed based on Equation (4). At the same time, the distance level l_j is updated as $l_j = l_i + 1$. After that, S_j does not respond to all stimulus signals during an offset time δ_j.
- Otherwise, both stimulus signals and sensing data are ingored.

As synchronization is achieved by the above explained manner, wireless sensor nodes having large distance levels can transmit sensing data earlier than those having small distance levels. As the offset time is set to sufficiently large value considered conflictions in MAC layer, the sensing data can be relayed sequentially to sink nodes as shown in Fig. 4.

3. Chaos-Based Data Gathering Scheme

In this section, a chaos-based data gathering scheme using a chaotic pulse-coupled neural network presented in (Nakano et al., 2009; 2010) is explained. As same as synchronization-based data gathering scheme, a wireless sensor network consisting of M wireless sensor nodes and L sink nodes are considered. Each wireless sensor node S_i $(i = 1, \cdots, M)$ has a timer which controls timing to transmit and receive sensing data. The timer in S_i is characterized by an oscillator having two internal state variables x_i and y_i, a non-negative integer distance level l_i, and an offset time δ_i. Basic dynamics of the timer in S_i is described by the following equation.

$$\frac{d}{dt}\begin{bmatrix} x_i(t) \\ y_i(t) \end{bmatrix} = \begin{bmatrix} \Delta_i & \omega_i \\ -\omega_i & \Delta_i \end{bmatrix}\begin{bmatrix} x_i(t) \\ y_i(t) \end{bmatrix}, \text{ for } x_i(t) < 1 \wedge \bigwedge_j \left(x'_j(t) < 1\right) \quad (10)$$

$$\begin{bmatrix} x_i(t^+) \\ y_i(t^+) \end{bmatrix} = \begin{bmatrix} q_i \\ y_i(t) - p_i(x_i(t) - q_i) \end{bmatrix}, \text{ if } x_i(t) = 1 \quad (11)$$

$$\begin{bmatrix} x_i(t^+) \\ y_i(t^+) \end{bmatrix} = \begin{bmatrix} a_i \\ y_i(t) - p_i(x_i(t) - a_i) \end{bmatrix}, \text{ if } \bigvee_j \left(x'_j(t) = 1\right) \quad (12)$$

where Δ_i is a damping, ω_i is a self-running angular frequency, p_i is a slope in firing, q_i is a base sate for *self-firing* and a_i is a base state for *compulsory-firing*. j denotes an index of a neighbor wireless sensor node S_j such that $l_j < l_i$. $x'_j(t)$ is a virtual internal state variable of S_j considered an offset time δ_j such that

$$x'_j(t) = x_j(t + \delta_j) \quad (13)$$

If the internal state variable x_i reaches the threshold 1, S_i exhibits *self-firing*, and the internal state (x_i, y_i) is reset to the base state based on Equation (11). If a virtual internal state variable x'_j reaches the threshold 1, S_i exhibits *compulsory-firing*, and the internal state (x_i, y_i) is reset to the base state based on Equation (12). After S_i exhibits *compulsory-firing*, S_i does not exhibit the next *compulsory-firing* during an offset time δ_i. That is, each wireless sensor node has a refractory period corresponding to the offset time. It should be noted that the unit oscillator presented in Section 2 has one internal state variable, and can exhibit periodic phenomena only. The unit oscillator of the proposed chaos-based data gathering scheme has two internal state variables x_i and y_i, and can exhibit various chaotic and bifurcating phenomena (Nakano & Saito, 2002; 2004). Also, it can generate chaotic spike-trains such that series of interspike intervals is chaotic.

Fig. 5 shows a typical chaotic attractor from a unit oscillator without couplings. As $\Delta_i > 0$, the trajectory rotates divergently around the origin. If the trajectory reaches the threshold, it is reset to the base state based on Equation (11). Repeating in this manner, this oscillator exhibits chaotic attractors. Fig. 6 shows typical phenomena from a simple master-slave network consisting of two oscillators, where $M = 2$ and $l_1 < l_2$. As shown in the figure, the first (master) oscillator exhibits chaotic attractors for both $q_i = -0.2$ and $q_i = -0.6$. The second (slave) oscillator is synchronized to the first oscillator for $q_i = -0.2$. That is, the network exhibits master-slave synchronization of chaos. On the other hand, the second oscillator is not perfectly synchronized but intermittently synchronized to the first oscillator for $q_i = -0.6$. These phenomena can be explained by error expansion ratio between the master and slave trajectories (Nakano & Saito, 2002). The case $a_i = 1$ is considered. Let t_n be the n-th *compulsory-firing*

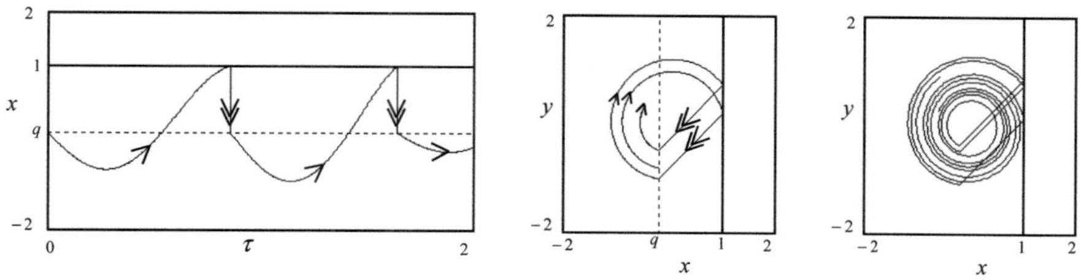

Fig. 5. A typical chaotic attractor from a unit oscillator without couplings. $\Delta_i = 0.25, \omega_i = 5, p_i = 1, q_i = -0.2$.

time of the slave oscillator, let the slave trajectory starts from $(q_i, y_2(t_n^+))$, and let the virtual master trajectory starts from $(q_i, y_1'(t_n^+))$. Let us consider that the $(n+1)$-th *compulsory-firing* of the slave oscillator occurs at $t = t_{n+1}$ and that each trajectory is reset to each base state. Then, the following average error expansion ratio is defined.

$$\overline{\alpha} \equiv \frac{1}{N}\sum_{n=1}^{N} \ln \alpha_n, \quad \alpha_n \equiv \left| \frac{y_1'(t_{n+1}^+) - y_2(t_{n+1}^+)}{y_1'(t_n^+) - y_2(t_n^+)} \right| \tag{14}$$

If the average error expansion ratio is negative for $N \to \infty$, the slave oscillator is synchronized to the master oscillator as shown in Fig. 6(a). Otherwise, the slave oscillator is not synchronized to the master oscillator. However, depending on sequence $\{\alpha_n\}$, the slave oscillator can be intermittently synchronized to the master oscillator as shown in Fig. 6(b). Such intermittent synchronization plays an important role for effective data gathering by the chaos-based data gathering scheme. Basically, the sequence $\{\alpha_n\}$ is determined by the parameters and initial states of the master and slave oscillators.

Distance levels of each wireless sensor node are adjusted as the the same manners explained in Section 2. Each sink node broadcasts "level 0" as a beacon signal. As each wireless sensor node forwards the beacon signal and adjusts each own distance level, each wireless sensor node has a distance level as corresponding to hop counts to its nearest sink node.

Also, sensing data is transmitted and received as the same manners explained in Section 2. By comparing received distance level with own distance level, sensing data is relayed sequentially to sink nodes. However, chaos-based data gathering scheme can exhibit not only synchronization but also intermittent synchronization. Hence, an assumption as shown in Fig. 7 is additionally introduced. In the figure, stimulus signal is transmitted at $t = t_i'$ from S_i and is received by S_j. Then, S_j broadcasts own sensing data at $t = t_j$. This sensing data can be received by S_i if $t_i' \le t_j \le t_i$ and $l_i = l_j - 1$ are satisfied. Each wireless sensor node transmits sensing data to the nearest sink node when stimulus signals are received. Therefore, at least one neighbor wireless sensor node can receive the sensing data even if the chaos-based data gathering scheme exhibits intermittent synchronization.

In wireless sensor networks, energy consumption of transceivers in transmitting sensing data is a dominant factor (Heinzelman et al., 2000). The intermittent synchronization can reduce redundant relays such that the same sensing data is relayed to sink nodes, and can reduce the total number of transmissions in wireless sensor networks. It can contribute to prolonging wireless sensor network lifetime. Also, for effective data gathering, multiple sink nodes should be allocated in an observation area where they are distant from each other (Kumamoto et al., 2009; Yoshimura et al., 2009). If all sink nodes are not coupled to each other via some

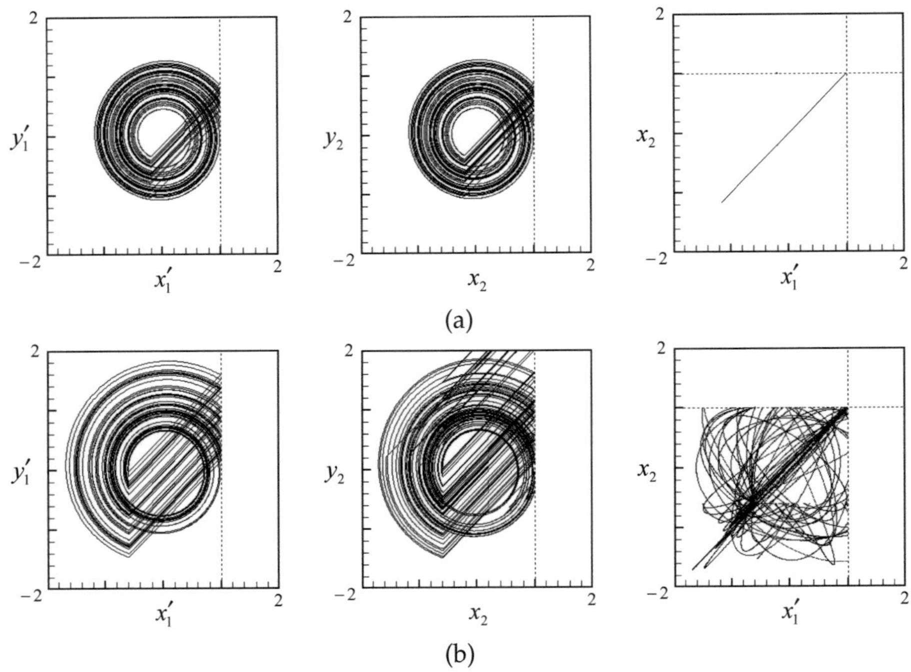

Fig. 6. Typical phenomena from a master-slave chaotic pulse-coupled neural network. Left: Master attractors. Center: Slave attractors. Right: Phase relationships. $\Delta_i = 0.25$, $\omega_i = 5$, $p_i = 1$, $a_i = 1$, $\delta_i = 0$ $(i = 1,2)$. (a) Synchronization of chaos: $q_i = -0.2$ $(i = 1,2)$. (b) Intermittent synchronization: $q_i = -0.6$ $(i = 1,2)$.

communications, it is hard to synchronize all wireless sensor nodes. Because, oscillators without couplings never synchronize to each other. The intermittent synchronization can flexibly adapt various wireless sensor networks not only with a single sink node but also with multiple sink nodes. These advantages can be confirmed by the simulation experiments in the next section.

The chaos-based data gathering scheme is based on the conventional synchronization-based data gathering scheme, and does not use any complex protocols using routing tables. Therefore, this method can easily control transmitting and receiving wireless sensor nodes and can flexibly adapt dynamical changes of network topologies. In the conventional synchronization-based data gathering scheme, power supply of transceivers can be turned off when wireless sensor nodes do not transmit or relay sensing data. However, many wireless sensor nodes can relay the same sensing data. The chaos-based data gathering scheme does not aim to reduce energy consumption by turning off power supply of transceivers. However, partial and intermittent synchronization in the chaos-based data gathering scheme can significantly reduce the number of transmitting and receiving sensing data. In addition, this method can guarantee that sensing data from all wireless sensor nodes must be transmitted to sink nodes without loss.

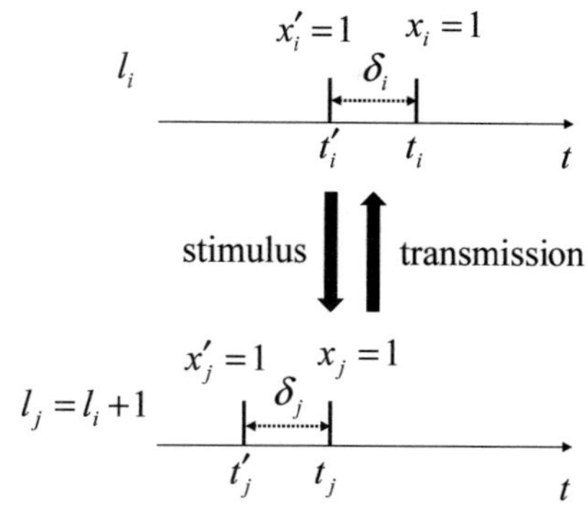

Fig. 7. Relaying sensing data in a chaos-based data gathering scheme ($l_j > l_i$).

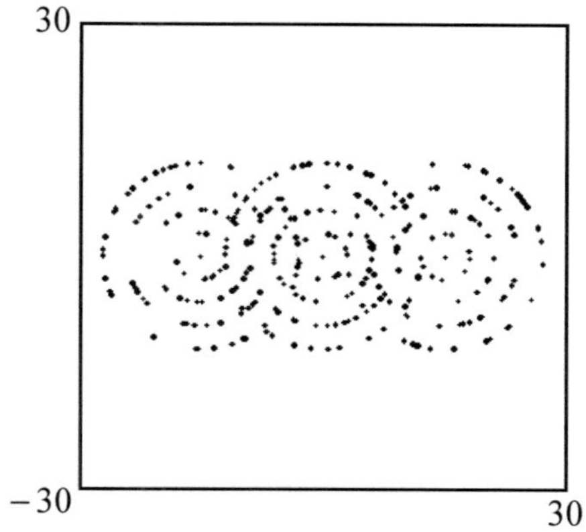

Fig. 8. A model of a wireless sensor network.

4. Numerical Simulations

In order to confirm effectivity of the chaos-based data gathering scheme, numerical simulations are performed. Fig. 8 shows a wireless sensor network model for the simulations. In the figure, 300 wireless sensor nodes are deployed at random locations on 12 concentric circles whose centers are $(-15, 0)$, $(0, 0)$ or $(15, 0)$, and 3 sink nodes are allocated on each center, which is called 3-sink wireless sensor network. On the other hand, in the simulations for 1-sink wireless sensor network, let only a node at $(0, 0)$ be a sink node and let nodes at $(-15, 0)$ and $(15, 0)$ be wireless sensor nodes. The radio range of each wireless sensor node and each sink node is set to 5. The radii of the concentric circles are set to 3, 6, 9 and 12, respectively. $10n$ wireless sensor nodes are set on the n-th concentric circle from each center. Initial values of internal states in each wireless sensor node are set to random values. In the chaos-based data gathering scheme, the parameters are fixed as follows.

$$\forall i, \quad \Delta_i = 0.25, \ \omega_i = 5, \ p_i = 1, \delta_i = 0.2, a_i = 1.$$

Typical simulation results for q_i as a control parameter are shown.

Figs. 9 and 10 show firing time of each wireless sensor node in 1-sink wireless sensor network and 3-sink wireless sensor network, respectively. In the figures, horizontal axis denotes time,

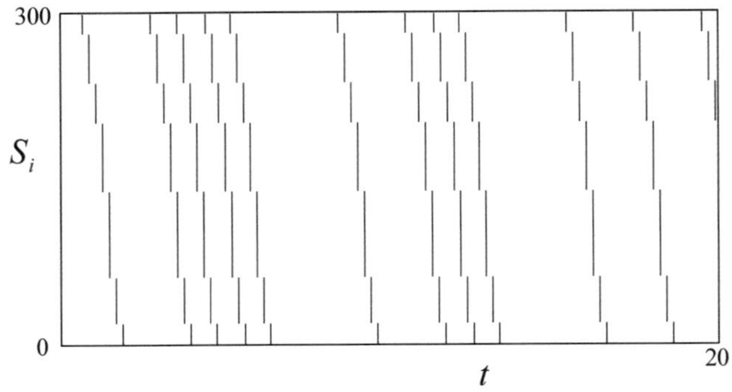

(a)

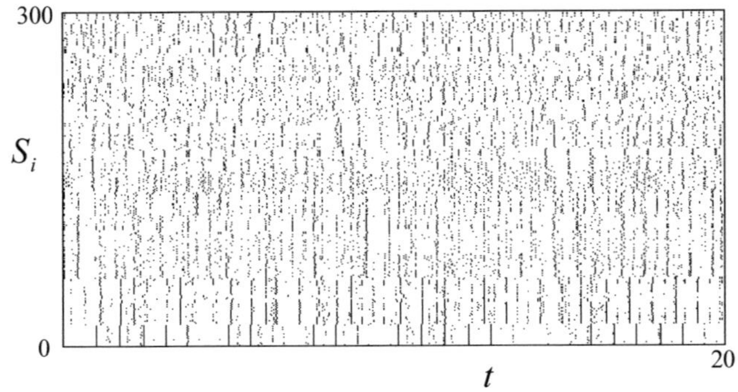

(b)

Fig. 9. Firing time of each sensor node in 1-sink wireless sensor network. (a) $q_i = -0.2$. (b) $q_i = -0.6$.

and vertical axis denotes the indexes of each wireless sensor node, where the indexes are sorted by each distance level.

Fig. 9(a) show the results for 1-sink wireless sensor network in $q_i = -0.2$. All internal states are synchronized to each other with time difference depending on their own distance levels. It can also be found that the sequence of the firing time is chaotic. Fig. 9(b) shows the results for 1-sink wireless sensor network in $q_i = -0.6$. All internal states are not synchronized to each other. However, some regularity of firings can be found. Fig. 10(a) shows the results for 3-sink wireless sensor network in $q_i = -0.2$. As compared with Fig. 9(a), chaos synchronization is broken down. It should be noted that it is also hard for the periodic synchronization-based data gathering scheme to synchronize all wireless sensor nodes in the case of multiple sink nodes. Because, frequency and/or phase of each sink node is not synchronized unless each sink node is coupled to each other. Fig. 10(b) shows the results for 3-sink wireless sensor network in $q_i = -0.6$. As compared with Fig. 9(b), significant differences between the cases in a single sink node and in multiple sink nodes can not be found.

Here, wireless sensor nodes which relay sensing data to sink nodes are considered. If all the wireless sensor nodes are synchronized to each other, all sensing data must be relayed to the sink nodes without lost sensing data. However, it is considered that many wireless sensor nodes relay the same sensing data. This problem becomes more serious if density of wireless sensor nodes increases, and the number of wireless sensor nodes and sink nodes increases.

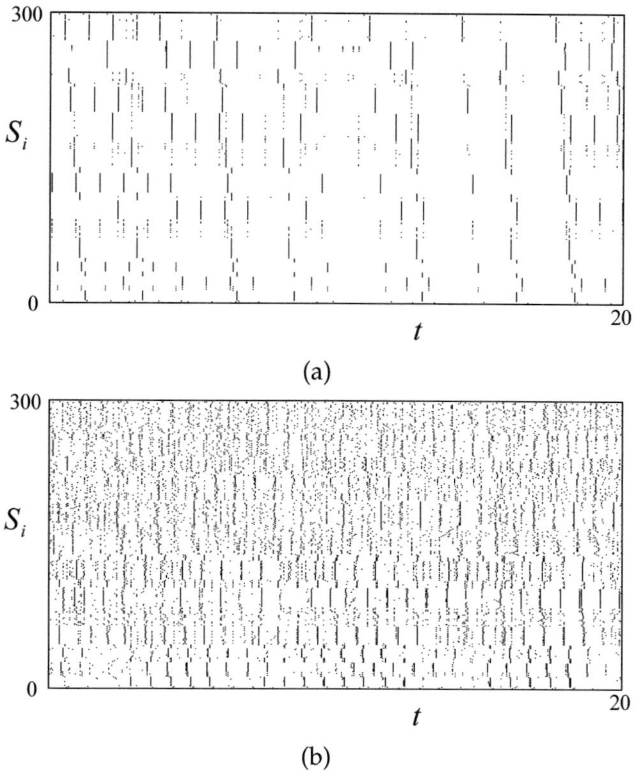

Fig. 10. Firing time of each sensor node in 3-sink wireless sensor network. (a) $q_i = -0.2$. (b) $q_i = -0.6$.

However, it should be noted that sensing data can be relayed to at least one sink node if at least one active path to the sink node exists, although a part of broken paths due to asynchronous firings of transmitting and receiving wireless sensor nodes exists.

In order to evaluate transmission efficiency in more detail, the total number of relays for sensing data from a wireless sensor node to sink nodes are evaluated. 40 wireless sensor nodes S_k ($k = 1, \cdots, 40$) are selected, which are allocated on the most outside of the center concentric circles shown in Fig. 8. S_k transmits sensing data n times. Each sensing data is transmitted in each *compulsory-firing* timing of S_k. It is assumed that only one wireless sensor node in S_k transmits sensing data and the other wireless sensor nodes do not transmit own sensing data. Then, total number of relays for $n = 100$ is calculated.

Figs. 11 and 12 show total number of relays for sensing data in 1-sink wireless sensor network and 3-sink wireless sensor network, respectively. The horizontal axis denotes sorted indexes of the transmitting wireless sensor nodes S_k. The vertical axis denotes the total number of relays, where each value is averaged for the number of transmissions ($n = 100$). The number of relays changes depending on the transmitting wireless sensor nodes. This is due to differences of the number of relay wireless sensor nodes to the sink nodes and/or the number of transmission paths to the sink nodes. That is, this is due to network topology. In the case of 1-sink wireless sensor network and $q_i = -0.2$, all the wireless sensor nodes are synchronized to each other as shown in Fig. 9(a). Then, all sensing data must be transmitted to the sink node without

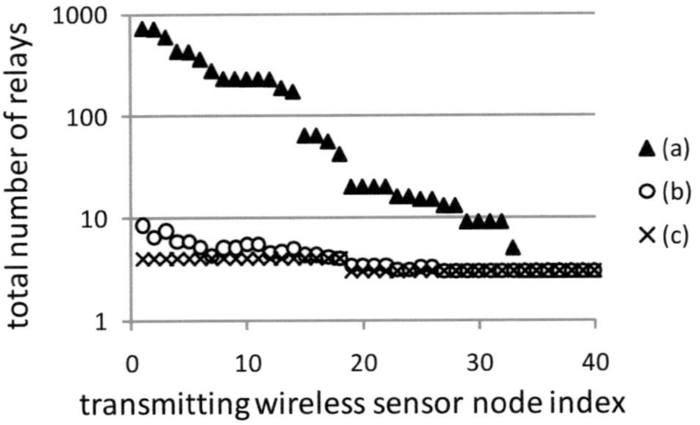

Fig. 11. Total number of relay wireless sensor nodes in 1-sink wireless sensor network. (a) $q_i = -0.2$. (b) $q_i = -0.6$. (c) distance level.

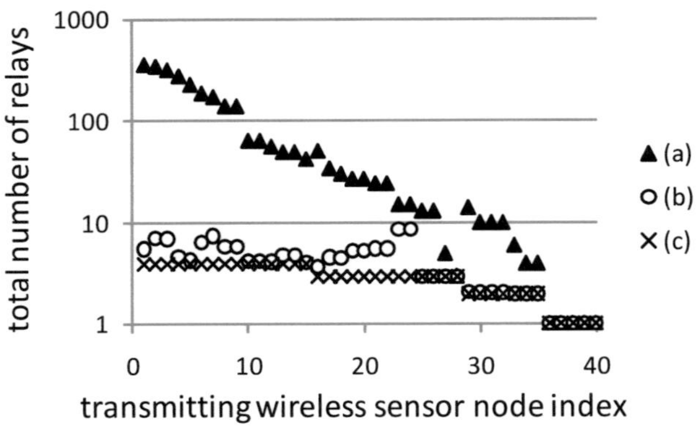

Fig. 12. Total number of relay wireless sensor nodes in 3-sink wireless sensor network. (a) $q_i = -0.2$. (b) $q_i = -0.6$. (c) distance level.

lost sensing data, but the sensing data is relayed by many wireless sensor nodes as shown in Fig. 11(a). In the case of 3-sink wireless sensor network and $q_i = -0.2$, each wireless sensor node is synchronized partially and intermittently to each other as shown in Fig. 10(a). Then, the number of relays for each transmitting wireless sensor node deceases as shown in Fig. 12(a), compared with the case of 1-sink wireless sensor network as shown in Fig. 11(a). In the case of 1-sink wireless sensor network and $q_i = -0.6$, each wireless sensor node is synchronized partially and intermittently as shown in Fig. 9(b). This result is the same also in the case of 3-sink wireless sensor network and $q_i = -0.6$ as shown in Fig. 10(b). Then, the number of relay wireless sensor nodes can be significantly reduced as shown in Figs. 11(b) and 12(b). It can contribute to saving energy consumption of each sensor node. Table 1 shows statistics values of the number of relays for 40 transmitting wireless sensor nodes. These results show that partial and intermittent synchronization can reduce the number of relays. Sensing data can be relayed to a sink node if at least one active path to the sink node exists, although a part of broken paths due to asynchronous firings exists. By the intermittent synchronization in chaos-based data gathering scheme, the number of relays can be significantly reduced. It can contribute to prolonging wireless sensor network lifetime.

	$q_i = -0.2$		$q_i = -0.6$	
	1-sink	3-sink	1-sink	3-sink
average	137.5	71.1	4.1	4.1
maximum	725.0	360.0	8.5	8.6
minimum	3.0	1.0	3.0	1.0

Table 1. Statistics values of number of relays for 40 transmitting wireless sensor nodes.

5. Conclusions

This chapter has analyzed transmission efficiency of a chaos-based data gathering scheme using chaotic pulse-coupled neural networks. Through numerical simulations, it has been shown that this scheme can reduce the total number of wireless sensor nodes which relay the same sensing data, without lost sensing data. For prolonging the lifetime of wireless sensor networks, it is important that the number of transmissions is reduced. In addition, this scheme can be easily applied to wireless sensor networks with multiple sink nodes and shows great performances in the viewpoints of prolonging the lifetime of wireless sensor networks.

Future problems include evaluation of energy consumption and comparison with periodic synchronization-based data gathering schemes in more detail.

6. References

Caro, G.D.; Ducatelle, F. & Gambardella, L.M. (2004). AntHocNet: An ant-based hybrid routing algorithm for mobile ad hoc networks, *Proceedings of 8th International Conference on Parallel Problem Solving from Nature*, 461–470.

Catsigeras, E. & Budelli, R. (1992). Limit cycles of a bineuronal network model, *Physica D*, Vol. 56, 235–252

Clausen, T. & Jaquet, P. (2003). Optimized link state routing protocol, *Request for Comments 3626*

Dasgupta, K.; Kalpakis, K. & Namjoshi, P. (2003). An efficient clustering-based heuristic for data gathering and aggregation in sensor networks, *Proceedings of IEEE Wireless Communications and Networking Conference*, 16–20

Heinzelman, W.R.; Chandrakasan, A. & Balakrishnan, H. (2000). Energy-efficient communication protocol for wireless microsensor networks, *Proceedings of Hawaii International Conference on System Sciences*, 3005–3014

Johnson, D.B.; Maltz, D.A.; Hu, Y.C. & Jetcheva, J.G. (2003). The dynamic source routing protocol for mobile ad hoc networks, *IETF Internet Draft*, draft-ietf-manet-dsr-09.txt

Keener, J.P.; Hoppensteadt, F.C. & Rinzel, J. (1981). Integrate-and-fire models of nerve membrane response to oscillatory input, *SIAM J. Appl. Math.*, Vol. 41, 503–517

Kumamoto, A.; Utani, A. & Yamamoto, H. (2009). Advanced Particle Swarm Optimization for Computing Plural Acceptable Solutions, *International Journal of Innovative Computing, Information and Control*, Vol. 5, No. 11(B), 4383–4392

Li, C.; Hwang, M. & Chu, Y. (2009). An Efficient Sensor-to-sensor Authenticated Path-key Establishment Scheme for Secure Communications in Wireless Sensor Networks, *International Journal of Innovative Computing, Information and Control*, Vol. 5, No. 8, 2107–2124

Liang, S.; Tang, Y. & Zhu, Q. (2008). Passive Wake-up Scheme for Wireless Sensor Networks, *ICIC Express Letters*, Vol. 2, No. 2, 149–154

Marwaha, S.; Tham, C.K. & Srinivasan, D. (2002). A novel routing protocol using mobile agents and reactive route discovery for ad hoc wireless networks, *Proceedings of IEEE International Conference on Networks*, 311–316

Mirollo, R.E. & Strogatz, S.H. (1990). Synchronization of pulse-coupled biological oscillators, *SIAM J. Appl. Math.*, Vol. 50, 1645–1662

Nagashima, J.; Utani, A. & Yamamoto, H. (2009). Efficient Flooding Method Using Discrete Particle Swarm Optimization for Long-Term Operation of Sensor Networks, *ICIC Express Letters*, Vol. 3, No. 3(B), 833–840

Nakano, H. & Saito, T. (2002). Basic dynamics from a pulse-coupled network of autonomous integrate-and-fire chaotic circuits, *IEEE Transactions on Neural Networks*, Vol. 13, No. 1, 92–100

Nakano, H. & Saito, T. (2004). Grouping Synchronization in a Pulse-Coupled Network of Chaotic Spiking Oscillators, *IEEE Transactions on Neural Networks*, Vol. 15, No. 5, 1018–1026

Nakano, H.; Utani, A.; Miyauchi, A. & Yamamoto, H. (2009). Data Gathering Scheme Using Chaotic Pulse-Coupled Neural Networks for Wireless Sensor Networks. *IEICE Transactions on Fundamentals*, Vol. E92-A, No. 2, 459–466

Nakano, H.; Utani, A.; Miyauchi, A. & Yamamoto, H. (2010). Prolonging Lifetime of Multiple-Sink Wireless Sensor Networks Using Chaos-Based Data Gathering Scheme, *Proceedings of ICUFN*, 12–16

Ogier, R.; Lewis, M. & Templin, F. (2003). Topology dissemination based on reverse-path forwarding (TBRPF), *IETF Internet Draft*, draft-ietf-manet-tbrpf-10.txt

Ohtaki, Y.; Wakamiya, N.; Murata, M. & Imase, M. (2006). Scalable and efficient ant-based routing algorithm for ad-hoc networks, *IEICE Transactions on Communications*, Vol. E89-B, No. 4, 1231–1238

Perkins, C.E. & Royer, E.M. (1999). Ad hoc on-demand distance vector routing, *Proceedings of 2nd IEEE Workshop on Mobile Computing Systems and Applications*, 90–100

Sasaki, T.; Nakano, H.; Utani, A.; Miyauchi, A. & Yamamoto, H. (2009). An Efficient Flooding Scheme Using Chaotic Neural Networks in Wireless Sensor Networks, *Proceedings of NOLTA*, 523–526

Subramanian, D.; Druschel, P. & Chen, J. (1998). Ants and reinforcement learning: A case study in routing in dynamic networks, *Technical Report TR96-259*, Rice University

Wakamiya, N. & Murata, M. (2005). synchronization-based data gathering scheme for sensor networks, *IEICE Trans Commun*, Vol. E88-B, No. 3, 873–881

Wang, C.; Hong, T.; Horng, G. & Wang, W. (2009). A GA-based Key-management Scheme in Hierarchical Wireless Sensor Networks, *International Journal of Innovative Computing, Information and Control*, Vol. 5, No. 12(A), 4693–4702

Yoshimura, M.; Nakano, H.; Utani, A.; Miyauchi A. & Yamamoto, H. (2009). An Effective Allocation Scheme for Sink Nodes in Wireless Sensor Networks Using Suppression PSO, *ICIC Express Letters*, Vol. 3, No. 3(A), 519–524

Part 4

Systems Implementation

Energy-efficient Reprogramming of Heterogeneous Wireless Sensor Networks

Seán Harte[1,2], Emanuel M. Popovici[1,2], Stefano Rollo[1] and Brendan O'Flynn[1]
[1] Tyndall National Institute, Cork, Ireland
[2] University College Cork, Cork, Ireland

1. Introduction

In order to build wireless sensor network (WSN) applications, there are many challenges. WSNs are distributed networks with a potentially high number of nodes and unreliable inter node communications, and energy constraints due to the limited power. Much research is ongoing into efficient communication protocols, device level software for energy-efficient control of hardware, and higher level software for network control. The challenge that this chapter is concerned with is efficiently reprogramming WSNs after they have been deployed. This can be due to bugs in the original software, or if parameters in the current application need to be changed, or the nodes are being re-tasked.

Microcontrollers are typically programmed by a wired connection to a PC. This can be done by the software developer or can be done as part of the node manufacture process if the application is already developed. However, after deployment it is not practical to physically connect to each node to upload new code to its microcontroller. There are a number of reasons for this: in a large network it can be too costly to go to each node; some nodes may not be accessible if they are in remote areas, or inside industrial machinery; or it may be required to update many nodes. If the node supports a method to receive data and reprogram itself with this data, then it can be reprogrammed wirelessly.

However programs can be quite large. This requires a lot of energy to send, and may cause communication problems due to flooding the network. If we consider a node which is sending 8 bytes of sensor data every 15 minutes, and has a battery long enough to last one year, then sending a 15 kByte program would shorten the lifespan by 20 days (if the energy cost for receiving and transmitting are similar). If the entire network is being reprogrammed, then the effect would be far more dramatic on nodes that have to forward code to other nodes. It is for this reason that two more energy-aware solutions are looked at in this chapter. The first is delta encoding, which is used to analyse the binary program images for two applications to find similarities between them. This information can be used to send a set of update commands, instead of sending the full new application. The second technique presented is data compression, based on the Lempel-Ziv-Welch (LZW) algorithm

1.1 Heterogeneous WSNs

Before looking at the solutions, we first introduce the idea of heterogeneity in WSNs. Each application presents different requirements and constraints and for some applications, it can be advantageous to have many different types of nodes with different functions that together create a heterogeneous network. This can be because nodes have different components depending on what type of sensors are being used. Another reason is that, to keep costs to a minimum, each node should only have the minimum hardware required to perform its task. For example, if a node is required to only take a reading every 10 minutes and then transmit it, a very low-powered processor is sufficient. Conversely, if a node is required to do relatively complex tasks such as forward error correction, encryption, signal processing, or routing in large networks, a more powerful processor is required, as these tasks are not possible on a very low-powered microcontroller. In this chapter, we focus on a network with two different types of node that construct a two-tiered heterogeneous network as in Fig. 1. One node has a small form factor with less processing and memory capability and can be used for sensor interfacing. The small size also opens up new application possibilities where the node can be embedded easily in existing objects or clothing (Foster-Miller, 2010), or for medical applications (Marinkovic et al., 2009). A cluster of these small nodes can be supported by a larger node. These larger nodes provide the backbone of the network, and are capable of more complicated tasks.

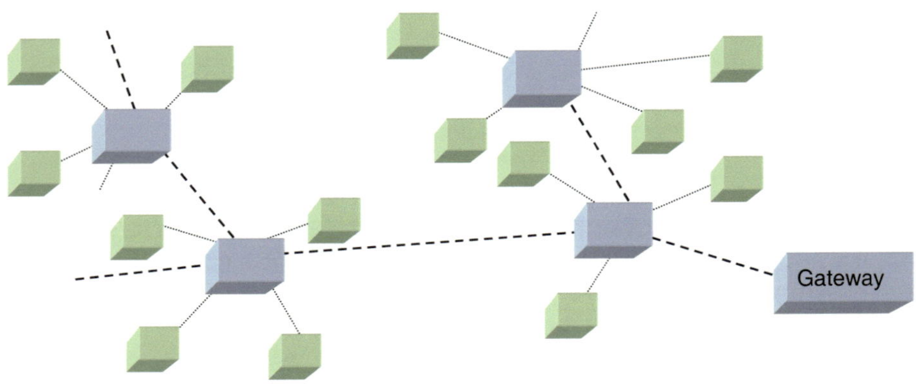

Fig. 1. Two-tiered heterogeneous network

The two nodes were developed at the Tyndall National Institute, and are called the Tyndall 25mm node (Bellis et al., 2005), and the Tyndall 10mm node (Harte et al., 2007). Both nodes are designed to be made from a number of different layers that are connected together. This provides a very high level of modularity, and allows application specific nodes with the desired sensing capabilities to be built quickly, by attaching layers together. The larger node has an ATmega128L (Atmel, 2009) microcontroller with 128 kBytes of program memory, and 4 kBytes of RAM. A number of different radios are available, but in this work a Nordic nRF905 radio operating in the 433 MHz band with 50 kbps data rate is used. The smaller, 10mm node uses a Nordic nRF9E5 chip (Nordic Semiconductor, 2008). This chip has an 8051 derivative microcontroller with 4 kBytes of program memory and 256 bytes of RAM. The chip also includes a radio which can communicate with the Nordic nRF905. Its processing power is very limited compared to the 25mm node. However its smaller size and lower energy requirements give it advantages. Fig. 2 shows the two nodes, and Table 1 shows the energy usage of the nodes in different modes.

Mode	10mm Node	25mm Node
Sleeping, with wakeup timer	20.0 μW	52.9 μW
Processing	9.73 mW	29.3 mW
Accessing memory	13.3 mW	31.0 mW
Radio receiving/listening	55.1 mW	75.1 mW
Radio transmitting at -10 dBm	42.2 mW	62.5 mW
Radio transmitting at +10 dBm	109 mW	128 mW

Table 1. Power used by Tyndall nodes from a 3.7 V Li-ion battery

Fig. 2. Tyndall 10mm node and 25mm node

2. Related Work

One of the big problems with network reprogramming is how to efficiently propagate the updates through the network. The simplest case for reprogramming is when each node in a network has the same application and they need to be updated. The new program can be sent across the entire network using a flooding protocol, where each node forwards the updated program to every node within its RF range. This helps ensure that every node receives the update, but it is also wasteful as some nodes receive the update more than once. To help improve data dissemination, the Trickle (Levis et al., 2004) algorithm was developed. Using Trickle, nodes regularly broadcast which version of data they currently have. If a neighbouring node detects has a different version, then the transfer of the update can begin. This algorithm requires far less power to propagate the update across the network, and scales to larger networks.

TinyOS (Berkeley, 2010) which is one of the most popular operating systems used in WSNs uses a Trickle based algorithm called Deluge (Hui and Culler, 2004) to support wireless reprogramming. Deluge modifies Trickle to support sending very large amounts of data.

The program update can be broken up into a number of pages. When a node has received a page, it can then start sending that page to other nodes that request it. Therefore it does not have to wait for the complete program update, before it can begin propagating the update.

A big limitation of Deluge is that it assumes that every node in the network is running the same code. Aqueduct (Phillips, 2005) extends Deluge to support heterogeneous networks. This is done by adding an identifier to each program update. A node only updates itself if its current identifier matches the identifier of the incoming update. However nodes must still cache updates and forward them to other nodes even if the identifiers do not match, to ensure that every node can receive updated code. This greatly increases memory requirements.

A big problem with the above solutions is that the entire updated program needs to be sent, even if only a small fraction of the code has changed. One solution to this is to a have an interpreter running on the nodes. An interpreter called Maté (Levis and Culler, 2002) has been developed using TinyOS. It can receive a script which describes the functions for the node to perform in a very condensed format. This means that far less data needs to be sent to update the node. However, the application is limited by what functions are possible in the scripting language and also requires the programmer to become familiar with the scripting language.

The concept of mobile agents is another method for making easily reprogrammable wireless sensor networks (Georgoulas and Blow, 2008). In this approach a virtual machine is running on each node. This virtual machine supports "agents" which can move from node to node to carry out their desired task. Each agent contains code that executes on the virtual machine and data that can be modified by the code. For example a tracking agent can follow an event of interest by sending itself to the node it believes to be closest to the event. New agents can be inserted into the network, which is ideal when it is expected that the function of a network will require many changes over its lifetime. However, the agent approach requires sending the agent from node to node, which is wasteful of radio transmission energy when a smaller packet could be sent, and more complicated logic on each node to interpret the packet.

A different approach is taken in the Contiki operating system (Dunkels, 2010). This operating system has core code that runs on the node constantly. This kernel supports loading and unloading of modules which are developed in C. This means that modules can be updated without having the reprogram the entire memory. The modules can either linked with each other at compile time, if the addresses of functions are known, or can be linked dynamically at run-time. However there is still a problem if the kernel needs to be changed due to newer versions becoming available or bugs. A similar approach supporting dynamic linking of modules at run-time in TinyOS is implemented by FlexCup (Marrón et al., 2006). In FlexCup an extra step is done after compiling to generate meta-data describing how to integrate individual components.

The above systems were based on operating systems with very low footprints. However, these operating systems may still not be suitable for very resource constrained systems. The overheads required for scheduling, and the demands placed on the stack by context switching etc., limit the complexity of possible applications. Applications can be developed that manage their own scheduling, and carefully limit the amount of context switching caused by interrupts. Such an optimized program rules out the use of an interpreter, or loadable modules. So another way to limit the amount of data that has to be sent is to only send the parts of the application that have changed. This is called delta encoding. A bug that is found might require just changing a single value in the source code of an application. However this single change can cause many changes in the binary code. The addresses of instructions could change and therefore all JMP instructions will need different operands etc. In this case, the minimum data that could be sent is a description of what changed in the source code. However this would require the application to be able to decompile its code, make the change and recompile. This is too complex for the typical hardware of wireless sensor nodes.

The UNIX tool Rsync (Tridgell, 1999) was developed for synchronizing data efficiently over a network connection. Assuming the receiver has first detected that the sender has a newer version of code, the receiver splits its data up into chunks of n bytes, and calculates a hash value for each chunk. The sender calculates a hash value for every chunk of n bytes. The hash values can then be compared to find out which sections of the data need to be updated. A compact list of commands can then be sent to the receiver telling it how to construct the new file, from a combination of its existing data, and new data. (Jeong and Culler, 2009) analyses a wireless network reprogramming technique based on the Rsync algorithm.
The Rsync algorithm can work for any type of data; however there are more efficient algorithms for executable code. (Reijers and Langendoen, 2003) presents a method for efficient code updating. It is based on analyzing the op-codes to find the minimum amount if data that needs to be sent in order to update the current code. To do this, it relies on knowing the structure of the op-codes, and is thus tied to be used for nodes using a Texas Instruments MSP430 type microcontroller. (Panta, 2009) modifies the compiler to introduce a function indirection table. Function calls are replaced to a jump to a specific location within a function table. This location then contains the call to the real function. This allows functions to be moved easily without requiring all addresses to be changed. However it requires an extra compiler step which will be difficult in a heterogeneous network where multiple processor architectures are being used.

A more general algorithm, called Bsdiff, for finding the difference between executable files is presented in (Percival, 2006). This algorithm begins by calculating which sections are the same with similar methods as Rsync. The difference is that sections which almost match are also noted. This can be done extending the matching areas until a limit of mismatched bytes is reached. This decreases the size of the list of commands that needs to be sent, as in binary program files, there are often sections that almost match, but just have different addresses in the instructions. This means it performs much better than Rsync for executable code and small changes in source code do not introduce large changes in the compiled program file, as they can with Rsync. This is shown by the comparison in (Motta et al., 2007). As this tool is not dependent on a specific instruction set, it is advantageous in a heterogeneous network such as the one presented in this work.

3. Self Programming Methods

Before examining further how to minimise to data that needs to be sent, we will now look at the methods used to allow the nodes to update their own code. The two nodes that we use, the 25mm node, and the 10mm node, have different microcontrollers and memory structures so two different update mechanisms have been developed. First, we will look at the 10mm node with its 8051-based microcontroller, and then consider the case of the 25mm node with its Atmel AVR based microcontroller.

3.1 Tyndall 10mm node

The 8051-derivative microcontroller in the nRF9E5 chip has a Harvard architecture with different memory address spaces for instructions and data. For node programming, only the memory containing instructions (program memory) is relevant. Fig. 3 shows how this program memory is arranged in the 10mm node. There is a RAM and a ROM within the nRF9E5, and an external EEPROM, which is communicated with using the SPI protocol. The EEPROM provides persistent storage of the code, but the actual code is run from the internal RAM.

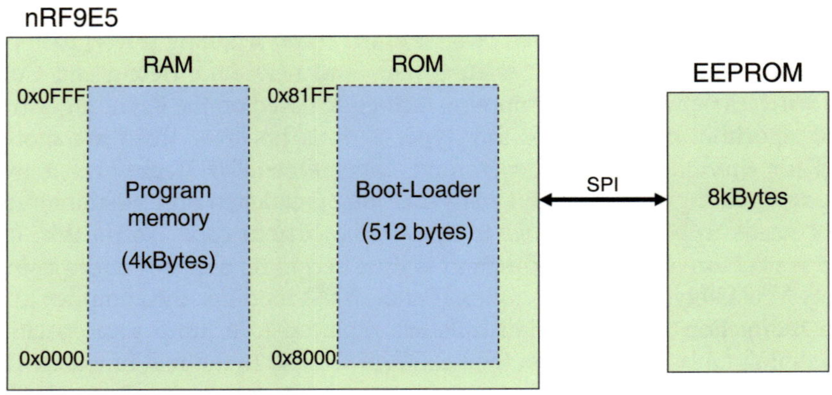

Fig. 3. nRF9E5 program memory structure

When the node is first powered up, it starts executing at address 0x8000, which is located in the internal ROM. This ROM contains boot-loader code that copies the lower 4 kBytes of data from the external EEPROM to internal RAM. Then the node program counter jumps to address 0x0000, and starts executing the application. In order to reprogram the node it is necessary to change the lower 4 kBytes of the EEPROM. When the update is complete the node can then restart itself and start executing the new application. However, there is still a potential problem with this method. It is likely that reprogramming would take a relatively long time, due to receiving commands over the radio, and allowing the current application to send other application data still. If the node should inadvertently restart itself (due to power problems, or a watchdog timer timeout) it is likely that a partially updated program would not function correctly. It is for this reason that an 8 kByte external EEPROM is used. This allows the updated program to be first written to the upper half of the EEPROM. When the entire program is fully written the top half of memory is copied to the bottom half, and the node is restarted. This greatly reduces the potential for a corrupted application due to unexpected restarts.

Fig. 4 shows how the program code is stored in the EEPROM. The first 3 bytes are used by the boot-loader to know where the actual code starts, and how much of the memory is used by the program code. This means it is possible to insert some extra data into the EEPROM. Four bytes are added: two bytes are a count of bytes in the actual program code; and two bytes contain a CRC checksum of the program code. The upper 4 kBytes of memory has the same contents as the lower 4 kBytes.

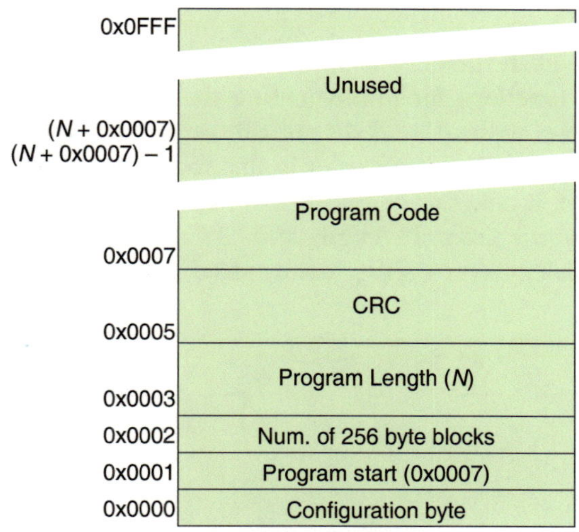

Fig. 4. nRF9E5 EEPROM memory format (lower 4 kBytes)

When all updates have been received, the current application uses the program length to calculate a CRC of the program code. This is then compared with the CRC stored in the EEPROM, and only if they match is the code copied to the lower half of memory, and the node reset (by forcing a watchdog timer timeout). If the CRC values do not match, then the node has to request the program to be fully retransmitted.

3.2 Tyndall 25mm node

The ATmega128L microcontroller used on the 25mm node also has a Harvard architecture. Its program memory is in an internal 128 kByte flash. This provides persistent storage, and the microcontroller can execute instructions directly from the flash memory. The ATmega128L provides support for reprogramming using the SPM instruction. However, this instruction only works when executed from the bootloader section of flash, which is the top 8 kBytes. This means that two approaches for reprogramming are possible. The first is that the bootloader section can be entirely self-contained. When the application detects an update is available, it can jump to the bootloader section. The bootloader can then handle receiving the data over RF, and creating the new application. When the application is fully updated, the bootloader can jump back to the application section. The second option is to split the memory in half, and write the new application to the upper half, as with the 10mm node. With this option the application handles receiving the data. It can call a function in the bootloader section of memory that modifies the version of the application in the upper half of the memory. When the program has been completely updated, the application calls a function that runs in the bootloader section, and copies the code from the upper half of memory to the lower half. This is the only function that writes to the lower half of memory.

The first of these options has the advantage that a much larger area is available for the application, which would allow applications that are more complicated. However, it means that the application cannot run while the program is being updated. As our current applications can comfortably fit within half of the available memory, we chose to implement the second option.

Fig. 5 shows how the flash memory is split into different regions. Within the bootloader section, there are the functions for implementing the program update mechanism. These functions are fully self-contained, and do not call or jump to any code in the application section, to avoid corruption. The bootloader is able to fit into 1 kByte, this leaves 63 kBytes free for the application. It also means that the top 1 kByte in the lower half is available to store information about the program length, and CRC. As with the 10mm node, these bytes are used by the bootloader code to verify that updated application is complete.

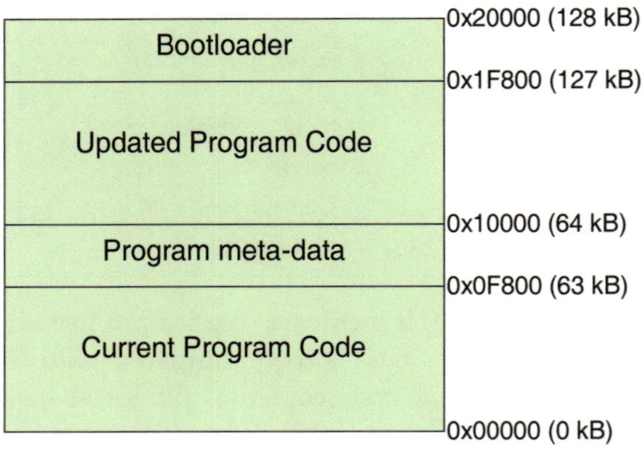

Fig. 5. ATmega128 program memory structure (byte addresses)

4. Delta Encoding

After looking at the mechanism the nodes use to reprogram themselves, we now look at how to reduce the amount of data that has to be sent in order to reprogram the nodes, thus saving energy. As discussed in section 2 of this chapter, delta encoding algorithms exist that can take two files and generate a set of commands to turn the first file into the second file. If the files are similar, then the set of commands can be smaller than the second file.

In a WSN, the node has one version of a program, and it is desired to update this program to a newer version. In our case, a PC has access to the network, and has both versions of the program. It is the PC that does the delta encoding, so the computation costs of this are not important. It can determine a set of commands that turn the old file into the new file. The commands are able to copy current sections of the code to any location, and able to write new data to any location. Although it requires some processing and extra memory reads to implement the handling of these commands, it is advantageous over just sending the new file, as less data is transmitted. In WSNs it has been shown that processing data uses much less energy per bit than transmission and reception (Raghunathan et al., 2002). Therefore, the savings from less radio usage will be greater than the extra processing required.

4.1 Bsdiff Algorithm

To generate the commands, our work uses the Bsdiff algorithm. This algorithm analyses two files, and finds sections that partially match. It outputs data that are arranged in three sections. The third section (*extra* section) contains new data that is written directly. The second section (*difference* section) contains a list of values that are added byte-wise to the current data. As there are many similarities, most values in this section have the value 0, and it is therefore very compressible. The first section (*control* section) is an array of 3-tuples (X, Y, Z). X is the number of bytes that are copied from the old data to the new data, adding byte-wise X bytes from the difference section. Y is the number of bytes from the extra section that are written. A pointer to the last offset read in the new file is moved Z bytes before starting the next operation.

The three sections output by the Bsdiff algorithm are actual larger than the file itself. In the freely available Bsdiff application (Percival, 2010) the bzip2 compression algorithm is used to compress all the sections. The data in the *difference* section is very compressible, and if the compared data is similar there will be far more data in this section than in the *extra* section. This is how the overall data size is greatly reduced, achieving a average compression ratio of 8.33% for program updates in the tests carried out in (Motta et al., 2007). As the nodes do not have processors powerful enough to decompress bzip2 data, it is not used here. Alternatives to work around this limitation are presented in the next section.

4.2 Adapting Bsdiff for use in WSNs

Besides being unable to use bzip2, another potential problem is that we do not want to wait for the node to receive all the Bsdiff output sections before starting to create the new program code. This would require too much buffering of data. To solve this, the *difference* and *extra* sections are broken up, and attached to the relevant 3-tuple from the *control* section. We will refer to this new structure as a command. In each command, the first three values (X, Y, Z), are the control 3-tuple. Then there is a value, P, which specifies how many bytes within X bytes of the diff section are non-zero. After this, there is array of P pairs. The first element of the pair says where to add this byte, and the second element is the byte to add. At the end, there are Y bytes taken from the *extra* section. Each command is structured as shown in Fig. 6.

In the case where commands are still too large, there might not be enough memory available to buffer the commands. For this reason, commands sent to 10mm nodes are limited to 28 bytes, and for 25mm nodes, a size of 112 bytes is used. The value for the 10mm node was picked as it is the size of the data payload that is sent in each radio packet and the 10mm node has very limited memory for buffering. The 25mm node has more buffering space available, so the effect of a command size limit against compression ratio was measured.

The commands for converting between the two applications were generated with different maximum command sizes, and the compression ratio recorded. The results are shown in Fig. 7. 112 bytes was chosen because increasing the size further has very little effect on the compression ratio, and it is a multiple of 28. As the node has to remember the location that it last read from in the current code, and the location in the new code that it last wrote to, it is also necessary to handle the commands in the correct sequence.

```
typedef struct {
    uint8_t index;        /* Where to add this byte */
    uint8_t value;        /* Byte to add to original data */
} pair_t

typedef struct {
    uint16_t copy;        /* How many bytes to copy (adding to diff section) */
    uint8_t write;        /* How many bytes to write from extra section */
    int16_t seek;         /* How many places to move pointer */
    uint8_t numPairs;     /* How many pairs in the diff section */
    pair_t  diff();       /* Array that is 'numPairs' long */
    uint8_t extra();      /* Array that is 'write' bytes in size */
} command_t;
```

Fig. 6. Reprogramming command structure and examples

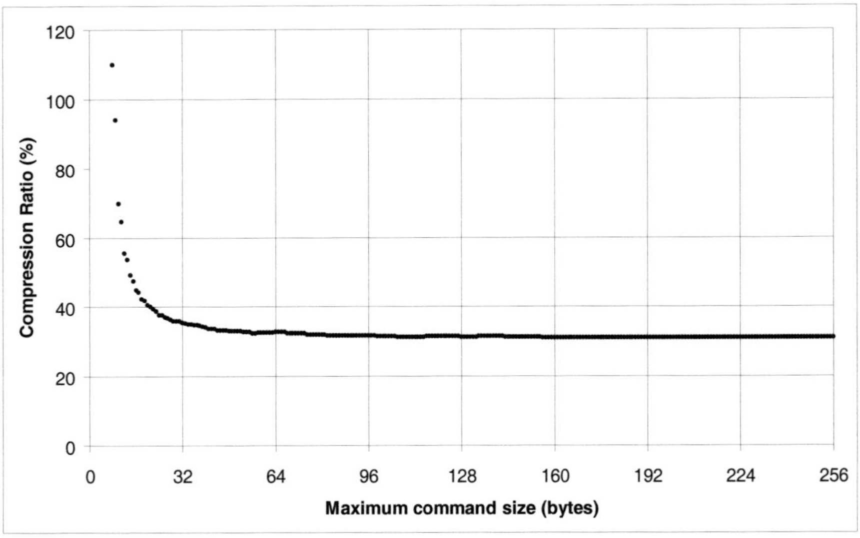

Fig. 7. Effect of command size on compression ratio

4.3 Analysis of delta encoding

To analyse the benefit of delta encoding, we compare the amount of data that would be sent if the complete new program were transmitted, and the amount of data that is sent with delta encoding. This is done using a real WSN application where nodes are arranged in a tree. Each node takes a sensor reading regularly and transmits to its parent node, and it also forwards sensor readings it receives from its children. The effects of changing the sampling frequency; replacing an framelet based (Roedig et al., 2006) MAC algorithm with a very simple form of CSMA (Carrier Sense Multiple Access); changing the sensor used from a Sensirion SHT11 temperature/humidity sensor, to an Analog Devices AD7998 ADC; and changing the application completely, to an application for implementing the Modbus protocol over wireless links are measured. Table 2 and Table 3 show the compression ratio achieved using delta encoding in each of these cases on the 10mm node, and 25mm node code, respectively.

Change	Full Size	Delta-encoded size	Details of commands	Compression Ratio
Changing sampling frequency	2896 bytes	14 bytes	2 command 1 diff pair 0 extra bytes	0.48%
Enabling CSMA	2922 bytes	208 bytes	13 commands 52 diff pairs 26 extra bytes	7.12%
Changing sensor	2744 bytes	919 bytes	36 commands 164 diff pairs 375 extra bytes	33.49%
Different application	2548 bytes	2228 bytes	83 commands 95 diff pairs 1540 extra bytes	87.48%

Table 2. Effects of changing application on 10mm node

Change	Full Size	Delta-encoded size	Details of commands	Compression Ratio
Changing sampling frequency	3407 bytes	14 bytes	2 commands 1 diff pairs 0 extra bytes	0.41%
Enabling CSMA	3419 bytes	78 bytes	6 commands 15 diff pairs 12 extra bytes	2.28%
Changing sensor	3365 bytes	1054 bytes	22 commands 194 diff pairs 534 extra bytes	31.32%
Different application	4238 bytes	3323 bytes	54 commands 94 diff pairs 2811 extra bytes	78.41%

Table 3. Effects of changing application on 25mm node

The tables show that our implementation of Bsdiff reduces greatly the data that needs to be sent to update a node, especially when only small changes are made. In a homogeneous network, the overall savings will be as above, as the same set of commands need to be sent to each node. Limiting the size of reprogramming commands on the 10mm node increases the compression ratio compared to the 25mm node, as more commands must be sent. The tables also show how as the amount of change in the program files increases, more of the sent data is in the extra section, and not the difference section.

In our current network, nodes are arranged in a fixed pre-defined tree. In the tree, nodes can transmit to their parent node, to one of their child nodes, or to all of their child nodes with a multicast transmission. To expand our Bsdiff technique to a heterogeneous network, with multiple different types of nodes, and multiple different node functions, the simplest approach is to generate the commands needed to update each node individually. However, if we consider a heterogeneous network where some nodes have almost the same program, it may be better to first reprogram all nodes so that they have the same application. Then perform the update using multicast transmissions, and then make the changes to each node so that they are unique again. To illustrate the usefulness of this method, we can use data in the above tables. If there are a number of nodes which differ only in sampling frequency and it is desired to change the sensors on each node, then the size of the commands needed to change the sensor compared to the size of commands needed to change the sampling frequency means that the simple approach of sending a single set of commands to each node may be far from optimal.

To decide which method is better we need to calculate the energy cost of each approach. In the tables above, the compression ration is used as the metric to examine the effectiveness of our Bsdiff implementation. This is valid, as when programming a single node, the number of bytes transmitted will be directly related to the energy used. However, the use of multicast transmissions in a heterogeneous network complicates this, as the energy per bit will change depending on how many nodes receive the message. For this reason, a new metric is required to analyse the use of Bsdiff in a heterogeneous network. The radio we use is capable of sending a 32 byte payload, with a 6 byte header, and 10 bit preamble, added by the radio. From this 32 byte payload, 4 bytes are used for routing control, packetisation, and a message type identifier, leaving 28 bytes for use. This means that a full packets is 314 bits long, of which 90 bits are overhead. The radio sends data at a rate of 50 kpbs, and has a 650 μs start-up time. Therefore, for a message with *len* bytes, the time to send it, *T*, can be calculated:

$$T(len) = \frac{314(\lceil len/28 \rceil - 1) + 8(len \bmod 28) + 90}{50000} + 0.00065\lceil len/28 \rceil \quad (1)$$

For a message to be sent to a particular node, or set of nodes, *S*, the message will have to be sent S_{TX} times, received by 25mm nodes S_{RX25} times, and by 10mm nodes S_{RX10} times. In out network the 10mm nodes only act as leaf nodes, so they are never required to transmit the commands. Using values for transmission P_{TX} and reception P_{RX25} and P_{RX10} from Table 1, the energy required to send the message can be calculated:

$$E(len,S) = P_{TX}S_{TX}T(len) + P_{RX25}S_{RX25}T(len) + P_{RX10}S_{RX10}T(len) \quad (2)$$

This value is not fully accurate due to ACKs, and other network management costs, however these costs will affect every message similarly, so it is still a valid metric for comparing the cost of send a message.

This metric can be used to help reduce the energy cost of reprogramming a heterogeneous network. In the network, there are nodes 0, 1, ... , *n*, and applications i_α and i_β refer to different versions of an application that run on node *i*. $B(i_\alpha, i_\beta)$ is the sum of the number of bytes in the commands that are needed to convert a node from running application i_α to running application i_β.

$$c_{separate} = \sum_{i=0}^{n} E(B(i_\alpha, i_\beta), \{i\}) \tag{3}$$

$$c_{combined} = \sum_{i=1}^{n} E(B(i_\alpha, 0_\alpha), \{i\}) + E(B(0_\alpha, 0_\beta), \{0,1,\ldots,n\}) + \sum_{i=1}^{n} E(B(0_\beta, i_\beta), \{i\}) \tag{4}$$

If each node were updated separately, the cost of update in terms of bytes transmitted would be $c_{separate}$. If we take the approach of converting every node to have the same application then the cost will be $c_{combined}$. Depending on the current state of the nodes, and the desired changes, either approach could require less data to be transmitted.

This idea can be expanded further. Instead of reprogramming the entire network to have the same application, the technique is restricted to sub sections, which have very similar applications. For example, a large network carrying out environmental monitoring could have different types of sensors in different areas. In this case, if we want to update the network with a new communication protocol, it might be best to convert all the nodes with the same sensors to run the same application, and the reprogram them all using multicast transmissions.

```
Define Function should_be_grouped(set1, set2):
e1 = energy required to program set1
e2 = energy required to program set2
e3 = energy required to program set1 and set2 with same update commands
return e3 < (e1 + e2)

For each node i that is not in a set
create a set s_i := {i}
joinSiblings := True
For each node j that is a sibling of i
if not should_be_grouped(s_i, {j}): joinSiblings := False
If joinSiblings is True
For each node j that is a sibling of i
Add j to s_i

For each set k
For each set l
If should_be_grouped(k, l): Join k and l
```

Fig. 8. Pseudo code for grouping nodes for efficient reprogramming

This leads to the problem of how to determine which sections of the network should be grouped together. We want to create a number of sets, S_a, S_b, S_c, ..., where all the nodes in a set are reprogrammed together. Initially there are n sets with one node in each set. The cost of reprogramming will be the same as c_{sep} above. To try reducing the cost, the number of sets is reduced. As multicast transmissions can be used to address a group of siblings, we first try to group nodes based on this. Each group of siblings is analysed to see if it is more efficient to update them together or separately. If it is more efficient to update then together then the sets are joined. After doing this, a second iteration is performed over each set, to check if it would reduce costs to join it with any other sets. Sets that have 10mm nodes are not compared with sets that have 25mm nodes, as they cannot execute each other's code. This algorithm is defined in the pseudo-code in Fig. 8.

The amount of data saved is heavily dependent on the current application and on the desired changes in the network, but below we present savings from a simple yet realistic scenario. In Fig. 9 there is a network with five 10mm nodes, and five 25mm nodes. Three of the nodes (1, 5, and 6) have a SHT71 temperature/humidity sensor and the rest are using an AD7998 ADC. They have different sampling frequencies. Table 4 shows the size of the new application, the number of bytes to convert from the old application to the new, the parent of each node, and the number of hops to the gateway node. The table embedded in Fig. 9 shows the number of bytes needed to convert an application to another application that is currently running.

Node	0	1	2	3	4	5	6	7	8	9
Size	9061	8787	8737	8737	8737	3465	3465	3300	3300	3300
Update	273	248	203	203	203	755	755	814	814	814
Parent	-	0	0	2	2	1	1	3	3	4
Hops	0	1	1	2	2	2	2	3	3	3
Type	25mm	25mm	25mm	25mm	25mm	10mm	10mm	10mm	10mm	10mm

Table 4. Update sizes for each node (bytes)

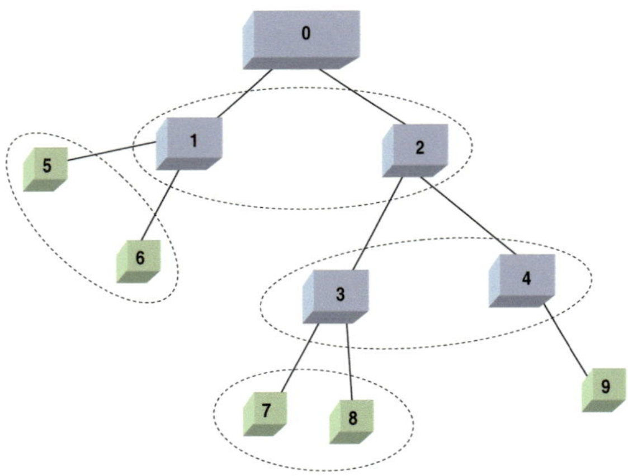

	0	1	2	3	4	5	6	7	8	9
0		1158	588	591	591					
1	1156		1126	1127	1127					
2	613	1147		9	9			Original		
3	616	1148	9		0					
4	616	1148	9	0						
5							0	1276	1276	1276
6						0		1276	1276	1276
7			New			1273	1273		0	9
8						1273	1273	0		9
9						1273	1273	9	9	

Fig. 9. Heterogeneous WSN topology and node application conversion costs

After using the algorithm in Fig. 8, we are left with five sets of nodes. These sets are shown in Table 5.

Set	**Energy Cost**
$S_a = \{0\}$	0 J
$S_b = \{1\}$	0.0436 J
$S_c = \{2, 3, 4\}$	0.1179 J
$S_d = \{5, 6\}$	0.2844 J
$S_e = \{7, 8, 9\}$	0.6412 J

Table 5. Heterogeneous network update costs for each set of nodes

In Table 6, the energy cost for reprogramming the entire network is given. For this particular scenario the energy cost has been reduced to 6.57% of the energy cost of sending the full application program data. Taking advantage of the similarities between nodes in a heterogeneous network reduces the energy cost to 55.15% the cost of sending program update commands to each node separately.

Method	Energy Cost	Energy cost compared to uncompressed
Uncompressed	16.54 J	100%
All nodes separate	1.971 J	11.91%
Grouping nodes into sets	1.087 J	6.57%

Table 6. Comparison of reprogramming methods

5. LZW Compression

As mentioned in Section 4.1, the Bsdiff algorithm usually uses the bzip2 algorithm. As bzip2 decompression could not be performed on our nodes, we were not able to use it. In this section of the chapter, we examine the potential usefulness of a compression algorithm that can be implemented on our nodes. We use sensor-LZW (S-LZW), a variant of the Lempel-Ziv-Welch algorithm. S-LZW was developed specifically for low powered wireless sensor nodes and was shown to use far less memory and instruction cycles for performing compression when compared to other commonly used algorithms such as LZO and bzip2 (Sadler and Martonosi, 2006). However, due to the severely limited memory on the 10mm nodes, it has not been possible to implement it on the 10mm nodes. LZW is a dictionary based compression algorithm, where strings are replaced by a fixed-length code that references an entry in a dictionary. When a new string is found in the data stream, it can be encoded based on previous strings. Such compression works well for repetitive data. S-LZW adds a mini-cache to improve performance for recently accessed strings in the dictionary.

Our data is not as repetitive as the sensor data examined in (Sadler and Martonosi, 2006). This is due to the very primitive form of compression performed when converting the output of the Bsdiff algorithm into reprogramming commands with a set maximum size. To examine this we compared two large applications implementing the ZigBee protocol on a version of the Tyndall 25mm node with a ZigBee compatible Ember EM2420 transceiver. The effects of compressing the Bsdiff output, and the output after it has been converted into reprogramming commands is shown in Table 7.

	Algorithm	Compressed size	Output file compression ratio	Overall compression ratio
Bsdiff output (25968 bytes)	PPM	7859 bytes	30.26%	31.65%
	LZMA	8086 bytes	31.14%	32.56%
	Deflate	8748 bytes	33.69%	35.23%
	Bzip2	9048 bytes	34.84%	36.43%
	S-LZW	1,0476 bytes	40.34%	42.18%
Reprogramming Command size (12801 bytes)	PPM	9548 bytes	74.59%	38.45%
	LZMA	9616 bytes	75.12%	38.72%
	Deflate	9868 bytes	77.09%	39.74%
	Bzip2	1,0298 bytes	80.45%	41.47%
	S-LZW	1,1379 bytes	88.89%	45.82%

Table 7. Compressing Bsdiff output and reprogramming commands

The other algorithms are PPM (Prediction by Partial Matching), LZMA (Lempel Ziv Markov-chain Algorithm), Deflate (as used in Zip files), and BZip2 (Huffman based encoding). These algorithms were performed by the 7-Zip application with default parameters (Pavlov, 2010). The table shows that converting the Bsdiff output into standalone commands, as we did in Section 4.2, leads to a larger end file size in each case. However, this is necessary due to the limited memory available for buffering. Table 7 also shows that S-LZW is not as effective as other compression algorithms, which was expected due to its speed and low memory usage.

S-LZW has a number of parameters that affect the compression ratio: the dictionary size; the mini-cache size; and the block size. LZW can compress streams of data of any length, so here block size refers to the size of chunks that the data stream is split into. This is necessary because of limited memory on the sensor nodes. These parameters can have positive effects by increasing the compression ratio, and negative effects by increasing the time taken to decode, or the memory required. Another method to increase the compression ratio is to use the Burrows Wheeler Transform (BWT) (Burrows and Wheeler, 1994). This algorithm can sort the data into an order that should compress better. It is a reversible transform so the original data can be regenerated.

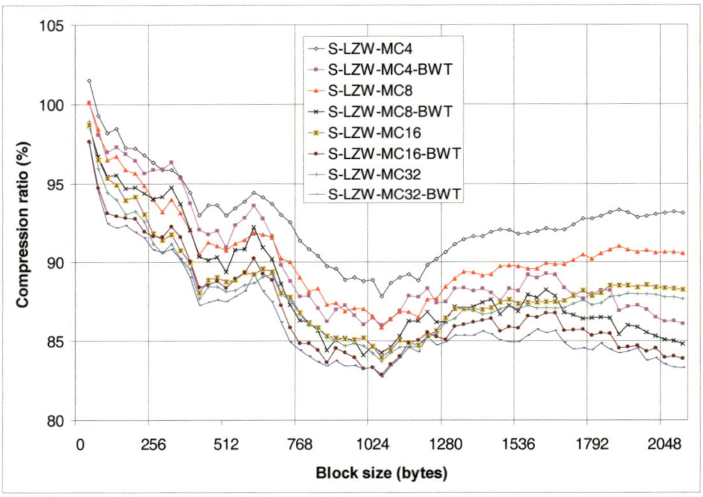

Fig. 10. Effects of mini-cache, block size, and BWT on compression ratio

The effect of the changing the dictionary size was found to be very small, and so was set at 512 entries. Fig. 10 shows the effects of the mini-cache size, how big a block is compressed, and BWT on compressing a set of commands 2,082 bytes in size (this is actually all the commands that are sent to node 0, in the network in Fig. 9). It can be seen that BWT has a positive effect on the compression ratio, and that an increased mini-cache size leads to increased compression too. Fig. 10 shows only the effect on compression ratio. However, the effect on energy consumption is more important. For this, it is necessary to analyse the processing costs of decompressing the data. The data compression is done on a PC, so it is not considered here, as the data sets used here are very small compared to the available processing power of a PC.

To analyse the cost of decompressing the code we measure the time taken to decompress a single block of data. The results of this are shown in Table 8 along with memory requirements in Flash (program memory) and RAM (data memory) for implementing S-LZW on the 25mm node. The memory used by BWT is minimised by sharing buffers with S-LZW. The results show that the mini-cache size has a negligible effect on processing time, and only a small effect on RAM size. For this reason, a 32 byte mini-cache is optimal, as it has a better compression ratio. The results also show that the time to decompress a single byte is not dependent on the block size that was compressed. The block-size should therefore be chosen based on the size that gives the best compression ratio, and still fits within the memory requirements (less than 4096 bytes). From Fig. 10 it can be seen that a block size that is a power of 2 is not always optimal. The PC that is compressing the commands can use a range of block sizes and chose the option that gives the best compression ratio.

Block size	Algorithm	Compressed Size (bytes)	Flash (bytes)	RAM (bytes)	Time (ms)	Time/byte (μs)
512 bytes	S-LZW-MC4	438	1768	3348	12.27	28.01
	S-LZW-MC8	426	1744	3356	13.02	30.57
	S-LZW-MC16	415	1744	3372	12.21	29.42
	S-LZW-MC32	417	1744	3404	11.87	28.46
	S-LZW-MC4-BWT	420	2116	3604	21.28	50.66
	S-LZW-MC8-BWT	415	2092	3612	20.77	50.05
	S-LZW-MC16-BWT	417	2092	3628	20.97	50.28
	S-LZW-MC32-BWT	416	2092	3660	20.71	49.79
256 bytes	S-LZW-MC32	213	1482	2892	6.39	29.98
	S-LZW-MC32-BWT	215	1826	3148	11.59	53.90
128 bytes	S-LZW-MC32	98	1354	2636	3.2	32.78
	S-LZW-MC32-BWT	88	1698	2892	6.5	74.14

Table 8. Memory usage and time for decompression

Our implementation of BWT requires memory that is twice the block size, however we have minimised the impact of this by using the same buffer that S-LZW uses for storing its dictionary. BWT however has a large impact on processing time, and still has some impact on memory usage. Whether or not it should be used depends on the increased compression ratio it offers. From Fig. 10, we see that BWT has very little advantage at the range of block sizes that can be decompressed (less than 512 bytes). If more memory were available, it would be more useful. To consider the energy savings by compression, the energy to send and receive the data and the energy required for decompression must be determined. The 2,082 byte file above can be compressed to 1,826 bytes using S-LZW-MC32 with a block size of 416 bytes. Using the power consumption values from Table 1, we can calculate the energy required with and without compression. The time to decompress a byte is from the table above, for S-LZW-MC32.

$$\begin{aligned} E_{uncompressed} &= P_{TX}T(2082) + P_{RX}T(2082) \\ &= 0.128 \times 1.7977 + 0.075 \times 1.7977 \\ &= 0.36\ \text{J} \end{aligned} \quad (5)$$

$$\begin{aligned} E_{compressed} &= P_{TX}T(1826) + P_{RX}T(1826) + P_{CPU}(1826 \times 28.46 \times 10^{-6}) \\ &= 0.128 \times 1.577 + 0.075 \times 1.577 + 0.0293 \times 0.052 \\ &= 0.32\ \text{J} \end{aligned} \quad (6)$$

The transceivers throughput rate of 50 kbps is very slow compared to the Atmega128L processor running at 8 MHz, so the time taken for decompressing the data is minimal compared to the time taken for transmitting the data. Therefore even for very modest compression ratios, it is worthwhile to use S-LZW.

6. Conclusions

We presented efficient methods for reducing the energy cost of reprogramming wireless sensor networks, by using delta encoding and LZW based compression. We have modified the Bsdiff delta encoding algorithm to make suitable it for use in WSNs, and also tuned the S-LZW algorithm for energy efficiency. In our example heterogeneous network with two different hardware nodes, and two different sensor types we reduced the cost of updating the communication protocol to 6.57 % of an approach that requires sending the full application program. The use of S-LZW gives a further reduction to about 90% of this value.

The solutions we provided can be applied to any type of reprogramming. The Bsdiff algorithm is not dependent on knowledge of instruction sets, and does not require any special compilation methods to keep functions at the same addresses. Very limited support is needed in the existing program. Support could be added on top of existing operating systems such as TinyOS or Contiki. This work has been implemented on a two-tiered heterogeneous network, but can be extended for multi-tier networks. The techniques presented are useful for simpler homogeneous networks.

The work presented in this chapter is already of great use in reducing the energy costs to reprogram a wireless node or network. However, in ad-hoc networks where the topology is not centrally managed, algorithms such as MSP (Kulkarni and Wang, 2009) or Freshet (Krasniewski et al., 2008) are suitable for managing the propagation of commands, and would complement the techniques presented in this chapter.

7. Acknowledgments

This work was supported by Science Foundation Ireland under grant 07/CE/I1147.

8. References

Atmel (2009). *Atmega128L datasheet*, rev. S. http://www.atmel.com

Bellis, S. J.; Delaney, K.; O'Flynn, B.; Barton, J.; Razeeb, K. M. & Ó Mathúna, S. C. (2005). Development of field programmable modular wireless sensor network nodes for ambient systems. *Computer Communications*, 28, 13, Aug. 2005, pp. 1531-1544, ISSN:01403664

Berkeley (2010), University of California. *Tinyos Community Forum*, http://www.tinyos.net

Burrows, M. & Wheeler, D. J. (1994). A block-sorting lossless data compression algorithm. *Digitial SRC Research Report 124*

Dunkels, A. (2010) *The Contiki Operating System*, http://www.sics.se/contiki/

Foster-Miller (2010). *Electrotextiles*, http://www.foster-miller.com/m_m_electrotextiles.htm

Georgoulas, D. & Blow, K. (2008). Intelligent Mobile Agent Middleware for Wireless Sensor Networks: A Real Time Application Case Study. *Proceedings of 4th Advanced Int. Conf. Telecommunications*, pp. 95-100, ISBN:9780769531625, Athens, Greece, Jun. 2008, IEEE Computer Society, Washington DC

Harte, S.; O'Flynn, B.; Martínez-Catalá, R. V. & Popovici, E. M. (2007). Design and implementation of a miniaturised, low power wireless sensor node. *Proceedings 18th Euro. Conf. Circuit Theory and Design*, pp. 894-897, ISBN:9781424413416, Seville, Spain, Aug. 2007, IEEE Press, New Jersey

Hui, J. W. & Culler, D. (2004). The dynamic behavior of a data dissemination protocol for network programming at scale. *Proceedings of 2nd Int. Conf. Embedded Networked Sensor Systems*, pp. 81-94. ISBN:1581138792, Baltimore, USA, Nov. 2004, ACM, New York

Jeong, J. & Culler, D. (2009). Incremental network programming for wireless sensors. *International Journal of Communications, Network and System Sciences*. 2, 5, Aug. 2009, pp. 433-452, ISSN:17543924

Krasniewski, M. D.; Panta, R. K.; Bagchi, S.; Yang, C. & Chappell, W. J. (2008). Energy-efficient on-demand reprogramming of large-scale sensor networks. *ACM Trans. Sensor Networks*, 4, 1, Jan. 2008, pp 1-38, ISSN:15504859

Kulkarni, S. & Wang, L. (2009). Energy-efficient multihop reprogramming for sensor networks. *ACM Trans. Sensor Networks*, 5, 2, Mar. 2009, pp. 1-40, ISSN:15504859

Levis, P. & Culler, D. (2002). Maté: a tiny virtual machine for sensor networks. *Proceedings of 10th Int. Conf. Architectural Support for Programming Languages and Operating Systems*, pp. 85-95, ISBN:1581135742, San Jose, USA, Oct. 2002, ACM, New York

Levis, P.; Patel, N.; Culler, D. & Shenker, S. (2004). Trickle: a self-regulating algorithm for code propagation and maintenance in wireless sensor networks. *Proceedings of 1st Conf. on Networked Systems Design and Implementation*, pp. 15-18, San Francisco, USA, Mar. 2004

Marinkovic, S.; Spagnol, C. & Popovici, E. M. (2009). Energy-efficient TDMA-based MAC protocol for wireless body area networks. *Proceedings of 3rd Int. Conf. Sensor Technologies and Applications*, pp. 604-609, ISBN:9780769536699, Athens, Greece, June 2009, IEEE Computer Society, Washington DC

Marrón P. J.; Gauger, M.; Lachenmann, A.; Minder, D.; Saukh, O. & Rothermel, K. (2006). FlexCup: A Flexible and Efficient Code Update Mechanism for Sensor Networks. *Proceedings of 3rd Euro. Workshop on Wireless Sensor Networks*, pp. 212-227, ISBN:3540321586, Zurich, Switzerland, Feb. 2006, Springer, Berlin

Motta, G.; Gustafson, J. & Chen, S. (2007). Differential compression of executable code. *Proceedings of Data Compression Conf.*, pp. 103-112, ISBN:0769527914, Snowbird, USA, Mar 2007, IEEE Computer Society, Washington DC

Nordic Semiconductor (2008). *nRF9E5 datasheet*, rev. 1.5. http://www.nordicsemi.com

Pavlov, I. (2010). *7-zip file archiver*. http://www.7-zip.org

Percival, C. (2006). *Matching with Mismatches and Assorted Applications*, Ph.D. Dissertation. University of Oxford

Percival, C. (2010). *Binary diff/patch utility*. http://www.daemonology.net/bsdiff

Phillips, L. A. (2005). *Aqueduct: robust and efficient code propagation in heterogeneous wireless sensor networks*. Master's thesis, University of Nebraska

Raghunathan, V.; Schurgers, C.; Park, S. & Srivastava, M. B., 2002. Energy-aware wireless microsensor networks. *IEEE Signal Processing Magazine*, 19, 2, Aug. 2002, pp. 40-50, ISSN:10535888

Roedig, U.; Barroso, A. & Sreenan, C. J. (2006). f-MAC: a deterministic media access control protocol without time synchronization. *Proceedings of 3rd Euro. Workshop on Wireless Sensor Networks*, pp. 276-291, ISBN:3540321586, Zurich, Switzerland, Feb. 2006, Springer, Berlin

Reijers, N. & Langendoen, K. (2003). Efficient code distribution in wireless sensor networks. *Proceedings of 2nd ACM Int. Conf. Wireless sensor networks and applications*, pp. 60-67, ISBN:1581137648, San Diego, USA, Sept. 2003, ACM, New York

Sadler, C. M. & Martonosi, M. (2006). Data compression algorithms for energy-constrained devices in delay tolerant networks. *Proceedings of 4th Int. Conf. Embedded Networked Sensor Systems*, pp. 265-278, ISBN:1595933433, Boulder USA, Nov. 2006, ACM, New York

Tridgell, A. (1999). *Efficient algorithms for sorting and synchronization*. Ph.D. Dissertation, Australian National University

23

Programming a Sensor Network in a layered middleware architecture

Michele Albano
Instituto de Telecomunicações - Aveiro
Portugal

Stefano Chessa
University of Pisa
and
ISTI-CNR
Italy

Abstract

Wireless Sensor Networks (WSNs) have become a mature technology aimed at performing environmental monitoring and data collection. Nonetheless, harnessing the power of a WSN presents a number of research challenges. WSN application developers have to deal both with the business logic of the application and with WSN's issues, such as those related to networking (routing), storage, and transport. A middleware can cope with this emerging complexity, and can provide the necessary abstractions for the definition, creation and maintenance of applications. This work discusses the problems related to the development of such a middleware and surveys the state of the art on the field.

1. Introduction

In the last few years, hardware and software innovations have been leading Wireless Sensor Networks (WSNs) from the research labs to deployments in real contexts. A WSN application is a distributed application that is built on a large number of low-cost, low-power, battery-powered sensors (Akyildiz et Al., 2002; Baronti et Al., 2007; Chessa, 2009). Sensors are spread in an environment (*sensor field*) without any predetermined infrastructure and cooperate to execute common monitoring tasks which usually consist in sensing environmental data and monitoring a variable set of objects. The sensed data are collected by a sink (a node that can communicate with both the WSN and an external network), when the sink is connected to the network. The sink, which could be either static or mobile, is in turn accessed by the external operators to retrieve the information gathered by the network.
Distilling a given high-level behavior from a set of sensors is a challenging problem, since the application has to deal with its own business logic, and with the issues that naturally arise when WSNs are taken into account, such as network formation, data transport and data management, security and energy saving. Dealing with these issues can be done either explicitly,

thus adding complexity to the applications, or implicitly by means of a middleware, that is a software layer that abstracts from common issues of the WSNs.
A middleware would let the developers to focus on the applications' business logic. On the other hand, there is no unique way to define which issues belong to the WSN, and which ones are part of the business logic. In general, this depends on the politics/mechanisms dichotomy, and hence on the level of abstraction that is provided by the middleware. Even a minimal middleware can provide benefits to the application developer, nonetheless it presents research challenges.
The level of abstraction provided by a WSN middleware inherently depends on the mechanisms that are used by the middleware to implement the high-level primitives. The analysis of state-of-the-art mechanisms presented in this work is developed into a structured vision of the mechanisms, that were organized into three layers (Programming Abstraction layer, Data Management layer, and Network layer) whose mechanisms can interact with each other or can be used directly by the application, plus a set of Dependability mechanisms that is orthogonal to the layers and that comprises mechanisms that are used by all the layers and by the user applications alike.

2. Organizing middleware mechanisms into layers

Current research papers agree that one of the critical points to leverage on the potential usefulness of WSNs is the possibility of abstracting common WSNs problems by means of convenient middleware, but literature is not coherent when defining what a middleware is (Albano et Al., 2007-1; Bai et Al., 2009; Chu & Buyya, 2007; Hadim & Mohamed, 2006; Mottola & Picco, 2008; Rahman, 2009; Romer et Al., 2002; Rubio et Al., 2007; Sugihara et Al., 2008; Tong, 2009; Wang et Al., 2008).
To this purpose we focus on the goals that a middleware has to achieve. The main purpose of middleware is to support the development, maintenance, deployment and execution of applications, filling in the gap between the application layer and the hardware, operating system and network stacks layers. In the case of WSNs, this can include mechanisms for formulating high-level sensing tasks, communicating this task to the WSN, merging/aggregating the sensor readings, and reporting them. The actual analysis of state-of-the-art middleware presented a variety of different techniques and approaches used to address the aforementioned goals. In fact, WSN systems offer functionalities that can be collectively called "middleware" but that are very different from each other. For example, different middleware offer:

- low-level mechanisms that operate at the datum granularity, such as data centric storage (Albano et Al., 2010; Bruck et Al., 2005)
- abstract mechanisms that hide the single datum, like database-like systems (Amato et Al., 2010; Madden et Al., 2003)
- service oriented architectures, for example the ZigBee standard (Baronti et Al., 2007)
- platforms for mobile agents, for example AFME (Muldoon et Al., 2006) or Agilla (Fok et Al., 2009)

This analysis of the state-of-the-art mechanisms identifies three layers (Programming Abstraction layer, Data Management layer, and Network layer), and the mechanisms are categorized in terms of this structure. The mechanisms can interact with each other or can be used directly by the application. A set of mechanisms for Dependability is also identified, and it can be

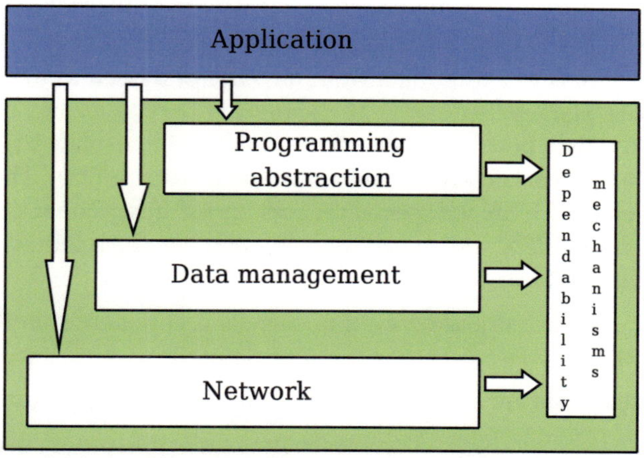

Fig. 1. Middleware is structured as a set of layers between the hardware and the application.

used at different levels for different purposes. Figure 1 provides a graphical representation of this structure.

Programming Abstraction layer comprises the mechanisms that model the behavior of the whole WSN, and that provide the user application with techniques to abstract the WSN details. For example, a user application can perceive a WSN as a relational database (Amato et Al., 2010; Madden et Al., 2003), as a publish/subscribe system (Albano & Chessa, 2009-1), as a service oriented architecture (Baronti et Al., 2007), or as a platform for mobile agents (Muldoon et Al., 2006).

The Data Management layer lets the user perform *store* and *retrieve* operations on data, either storing them on a specific sensor or set of sensors, or selecting the set of sensors from a metadatum of the data to be stored or retrieved, as in Data Centric Storage systems (Albano et Al., 2010; Ee et Al., 2006; Ratnasamy et Al., 2003).

The Network layer features explicit send and receive operations, and lets the user application control the transmissions performed at a finer grain. This layer nonetheless performs some abstraction, since it hides the routing protocol used, hence it lets the invoker specify the goal of the routing process, be it a sensor or a coordinate in the sensor field. This work does not provide details on Network layers, since it focuses on the high-level issues of WSN middleware.

The Dependability mechanisms regard the techniques used to create reliable primitives on the layers, and are used by all the layers alike.

A user application can rely on one of these layers, depending on the level of abstraction required, by addressing the whole WSN using the abstraction provided by the Programming Abstraction layer, or by performing explicit Data Management specifying which data are stored and retrieved, or it can be based on network level send/receive operations.

As an example of interaction of these layers, the Publish/Subscribe system described in (Albano & Chessa, 2009-1) is based on a Data Centric Storage layer provided in (Albano et Al., 2010). The DCS mechanisms such as (Ratnasamy et Al., 2003) used geographical routing such as (Karp & Kung, 2000) (Network layer) to transport data and queries to proper sensors. Erasure coding (Dependability mechanisms) was used to guarantee data availability in (Albano & Chessa, 2009-2), and to optimize access to data in (Dimakis et Al., 2006).

3. Data Management layer

The final goal of a WSN is to gather data from the environment and to route it to data consumers, and the Data Management layer is responsible for controlling dataflows and managing the exchange of data between nodes, towards the data consumers, with the option of caching the data into the WSN before transferring them out of the WSN. The current paradigm considers that data exit the WSN via special sensors, called *sinks*. The sinks are gateways that are connected to both the WSN and an external network, like the internet.
Data can reach the sink in three ways:

1. **local storage**: data are stored on a set of sensors that depends on the sensor that produced the data,
2. **external storage**: data are sent to the sink as soon as they are produced,
3. **in-network storage**: produced data are sent to a set of the sensors that depends on some characteristics of the data.

3.1 Local and External Storage

Local storage is a Data Management paradigm that prescribes data to be stored on a set of sensors that depends on the sensor producing them. The most common implementation of this paradigm stores data on the sensor producing them. When the sink wants to access the data, it must send a request to the sensors that stored them. This approach presents some limitations. The first is that, if some sensors detect a lot of events, their related sets of sensors become burdened by storing more data and hence they deplete their resources earlier (battery, memory, etc). A second problem is that the sink does not usually know in advance which sensor is producing data it is interested into, and hence it is necessary to broadcast a request to contact every node when retrieving data.
External storage is another approach where data are sent to the sink as soon as they are produced. Main drawbacks of this approach are that data can not be pre-processed and aggregated with other data. Moreover, if there are more than one sinks, data must be duplicated and sent to each sink. Finally, the sink must be always connected to the WSN, or the sink would miss data produced while it is away.

3.1.1 Local storage

A number of proposals for WSN middleware are based on the local storage paradigm, since it is the most obvious paradigm to cope with discontinuous connection of the sink to the WSN. The work describing tinyDSM (Piotrowski et Al., 2009), presented substantially a local storage Data Management layer. This middleware allows a node to ask its neighbors to replicate data, hence realizing a high-availability local storage system. When a sink queries the data, data replication assures the information to be available even if some nodes are exhausted or in sleep mode.
The middleware presented in (Dimakis et Al., 2006) proposes a local storage solution that uses a kind of encoding that enables a fully distributed computation of the code. The technique refers to a model featuring a set V_1 of k source sensors, each producing a single data packet, and a set V_2 of n storage sensors, storing the data. The encoding is based on a bipartite graph that connects each source to $O(\log k)$ randomly selected storage sensors. The encoding is performed by each storage sensor using a linear combination of all incoming data, where each incoming data are multiplied by a randomly selected factor. Each storage sensor then

stores the random factors associated to each incoming datum and the result of the linear combination. The authors show that the sink can reconstruct all k packets querying any k storage nodes with high probability. This fully distributed encoding results in a memory overhead that can be ignored only if the data to be encoded are much larger than the random factors.
Another local storage technique for fully distributed coding is based on the Growth Codes (Kamra et Al., 2006), and it implements linear coding using XOR operations. In this model the sensors give to the sink codewords obtained as the XOR of different data, and the sink performs the decoding. The goal of growth codes is coping with the "coupon collection phenomena" with random data collection, since in a pure erasure coding approach, the last few data symbols are the most difficult to get. This coding algorithm implies that for the first data to be encoded, only the original data are stored, and only after some time the encoding composes a number of data to construct the codewords to be stored. As long as the sink receives enough codewords obtained from a single datum, it is able to obtain the different data from the codewords.

3.1.2 External storage

Data Management layers implementing an external storage solution, cope with data management by sending data to the sinks as soon as they are produced. In this paradigm, data are stored and analyzed outside the WSN, hence the WSN's role is limited to data acquisition. Data Management layers of this kind are usually more resource expensive of the other Data Management layers, since they perform a large number of data transmission operations. In this kind of Data Management layer, data can be subject to a filter that decides if it has to be sent to the sink, but the filter must be loose enough not to throw away any data that can become interesting for the user application during the WSN's lifetime.
A refinement of this paradigm is the routing tree, that is used in Directed Diffusion (Intanagonwiwat et Al., 2000), that is a middleware that implements an External storage system that is reprogrammable on-the-fly using interest propagation. Data are named by meta-data that describe them, then the data consumer (a sink) disperses a message into the WSN by a broadcast to instruct nodes to send him data by a multi-hop routing tree that is set up by the interest dispersal: every node takes note about the node he received the interest from, and it sets it up as the next step in a routing process towards the data consumer. At the same time, every node starts considering data pertaining to a certain category as "interesting data" and, instead of discarding them, they send it along the gradient created by the interest dispersal towards the sink. Some Programming Abstraction layers, such as (Amato et Al., 2010; Madden et Al., 2003), employ a refined version of External storage, where data are sent towards the sink as soon as it is collected, but where it is processed while it is moving up the routing tree.
Another solution of the external storage kind is publish/subscribe, where nodes are instructed about sending data concerning interesting data when they collect them. An example is the Data Management solution adopted in Mires (Souto et Al., 2004), that sets up data collection by means of a publish/subscribe service. The main difference with routing trees, is that publish/subscribe is initiated by a node that advertises the data it can produce, and then the node is explicitly instructed to send the data to some data consumer. In contrast, routing trees are about flooding the WSN with an interest, that instructs all the nodes to send data pertaining a meta-datum to the broadcast initiator.

3.2 Data Centric Storage

Data Centric Storage (DCS) is a family of in-network storage techniques, using functions that relate different meta-data describing data to different sets of sensors. Since in WSNs the content of the data is generally more important than the identity of the sensors that gathered the data, a node producing a datum d associates a meta-datum k to d, computes a set of sensors applying a function f to the meta-datum k, and then the node sends d to the set of sensors $f(k)$ for storage. At the high-level, a sink directs a retrieve request towards a meta-datum k. This operation is realized applying the same function f to the meta-datum k to identify the set of sensors $f(k)$ that stored the data. DSWare (Li et Al., 2003-1) is an example of Programming Abstraction layer that relies on DCS to cache data into the WSN before providing them to the user application.

The seminal work described in (Ratnasamy et Al., 2003) introduced DCS, and also compared it with local and external storage. Comparing this approach to the external storage approach, the authors observed that DCS contributes to save sensors' energy and to improve network lifetime. Since sensors have limited memory capacity, the storage of all the data sensed by the WSN may result impractical, however with DCS it is possible to pre-process and aggregate data and thus reduce their size.

A number of different DCS techniques have been proposed, and they differ in the way

1. the datum is assigned a meta-datum;
2. the nodes that store the datum of a meta-datum are selected;
3. the datum is routed to/from the nodes that store it.

The assignment of a meta-datum to the datum is inherently application-dependent, and it will not be discussed in this survey. On the other hand, different DCS architectures can use different functions f_i from meta-datum k to a subset $f_i(k)$ of the sensors, and they can access different Network layers to implement routing from/to these subsets of sensors, and the rest of this section describes the state-of-the-art of DCS architectures based on these two characteristics.

3.2.1 Geographic Hash Table

The reference model of DCS in WSNs is the *Geographic Hash Table* (DCS-GHT) (Ratnasamy et Al., 2003), that constitutes the first proposal of DCS. In DCS-GHT, it is assumed that the geographic coordinates of each sensor are known, and that each datum is described by a *meta-datum* (or *name*). The set of sensors selected to store a datum is computed by means of a hash function applied to the corresponding meta-datum. This function returns a pair of geographic coordinates fitting in the area where the sensor network is deployed.

DCS-GHT exploits the primitives `store` for data storage, and `retrieve` for data retrieval. The `store` primitive takes in input a datum d and its meta-datum k. By hashing k, it produces a pair of coordinates (x, y) and uses the GPSR routing protocol (Karp & Kung, 2000) to route the pair (d, k) towards (x, y). The GPSR routing protocol is able to deliver the data to the sensor closest to the point (x, y) (this sensor is called *home node*). In principle the home node could be a sensor located exactly on the point (x, y), however the chance for this to happen is negligible. As a side effect, GPSR also identifies the inner perimeter of sensors (called *home perimeter*) enclosing (x, y) (the reader is referred to the work of (Karp & Kung, 2000) for more details). Once the home node receives the pair (k, d), it stores the pair in its memory, and, to enforce data persistence against sensors' faults, it also requests the sensors in the home perimeter to store a copy of (k, d). The `retrieve` primitive hashes the input parameter k (the meta-datum) to obtain the coordinate (x, y), then, by means of GPSR, it sends a request for the data with meta-datum k to the point (x, y). When this request reaches the sensors in the perimeter around (x, y), they send back the data they store that correspond to k. See Figure 2 for an example of `store` and `retrieve` execution, where the data producer A stores into the WSN a datum regarding a meta-datum k, and the data consumer (the sink) asks the WSN for data regarding the same meta-datum k. Both A and the sink hash k to the same location (x, y) (represented in the figure by D), then they route their requests towards that location. In the case of A, the `store` primitive semantics involve storing a copy of its datum on all the nodes in the home perimeter around D. In the case of the sink, a `retrieve` response is generated as soon as the query reaches one of the nodes belonging to the perimeter around D.

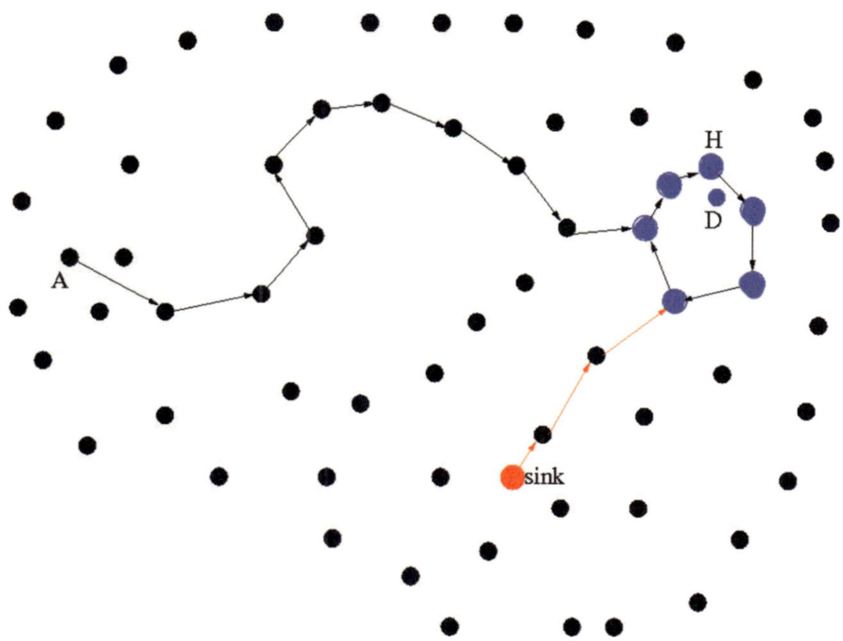

Fig. 2. A stores data, the sink retrieves them, using DCS-GHT

Although innovative, DCS-GHT presents a number of limitations when deployed on real WSNs. It assumes a uniform distribution of sensors and uniformly hashes meta-data on them. Moreover, if WSN produces a large amount of data associated to the same meta-datum, all such data will be stored by the DCS-GHT within the same home perimeter, thus overloading sensors on that perimeter. To avoid this problem DCS-GHT employs **structured replication**, that is a technique that augments event names with a hierarchy depth d and uses a hierarchical decomposition of the key space. Let us consider a single meta-datum that is hashed into a location r, and let us call r the root location for the meta-datum, and d the hierarchy depth. Let us consider the sensing area as the 0-level cell, and given an i-level cell, let us divide recursively it into 4 smaller cells, splitting each coordinate span in half. Given a hierarchy depth d, there are 4^d cells, and $4^d - 1$ mirror images of r, replicating the position of r in its d-level cell into the other cells of d hierarchy level.

For example, Figure 3 shows a $d = 2$ decomposition, and the mirror images of the root point $(3, 3)$ at every level. A node that detects an event, now stores the event at the mirror closest to its location, which is easily computable. Thus, structured replication reduces the storage cost

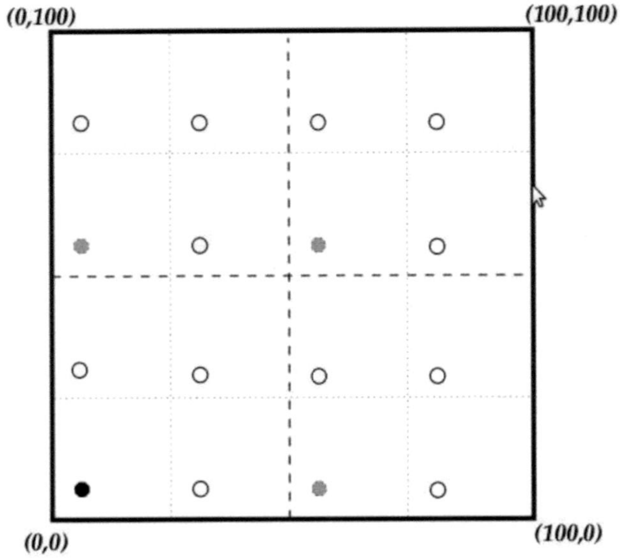

Fig. 3. Example of structured replication with a 2-level decomposition

at one node for one key with n detected events from $O(\sqrt{n})$ to $O(\sqrt{n}/2^d)$. DCS-GHT must route queries to all mirror nodes, to find all data stored into the 4^d mirrors of r.

Structured replication is efficient in limiting the quantity of data stored around a single home node, but this is not sufficient by itself to ensure load balancing, in fact the storage load can become unbalanced even if there is not an unbalance in the meta-data.

Resilient Data Centric Storage (R-DCS) (Ghose et Al., 2003) is an extension of DCS-GHT that addresses the issue of having all data of the same type stored on the same set of nodes. It divides the sensing area into zones, and each sensor can either be a *monitor node*, a *replica node*, or a *normal node*, with respect to a given event type. A normal node generates events and forwards packets, but it does not store data pertaining the given event type. Each zone has one sensor that is on *monitor mode* for each event type; the monitor node does not store data, but knows the location of replica nodes for their event type and fowards data to them. *Replica nodes*, finally, are nodes that can store data regarding a given event type.

Bottom line, R-DCS adds to the efficiency of the DCS system limiting the data transmission from the sensor producing a datum to the monitor node of its zone, and then to the closest replica node. Moreover, resiliency to failures is improved since data are not replicated locally, but instead they are located on replica nodes that are far away from each other, and hence a disaster, that would destroy all sensors close to it, can not exterminate all replica nodes for a given meta-datum.

Another variant of DCS-GHT is Rendezvous Regions (RR) (Seada et Helmy, 2004), that has mechanisms similar to DCS-GHT but, instead of directing queries towards a home node, it makes use of regions in the sensing area, and of all the sensors located into those regions. In RR the network topology is divided into geographical regions, where each region is responsible for a set of keys, with keys representing meta-data of sensed data, or services offered by sensors. The service or data provider stores information in the corresponding region, and the service or data user associates directly its query to the region. The obvious distinction

between RR and DCS-GHT is in using a rendezvous region instead of a rendezvous point. Moreover, RR is also targeted to designing geographic systems that need only approximate location information for the nodes.
Other works explore different routing mechanisms in DCS, based on multiple trees (Ee et Al., 2006), on Geographic Hash Tables over clusters of nodes (Araujo et Al., 2005), on the recursive subdivision of the WSN using K-D trees (Aly et Al., 2006; Li et Al., 2003-2), or on double rulings (Sarkar et Al., 2006).
Let us now consider the storage load on the nodes, and let us define Quality of Service (QoS) for Data Management in the WSN as the capability of the Data Management layer to guarantee that a given datum is stored on a given number of sensors, in order to provide the desired resilience to sensor faults for the datum.
The state-of-the-art mechanisms discussed so far were designed for different goals, but despite their merits, they did not take into account QoS. For example, in DCS-GHT the number of sensors storing a datum depends on the number of sensors in a perimeter, and in RR it depends on the population of a region. On the other hand, the work described in (Albano et Al., 2007-2) and (Albano et Al., 2010) presented DCS systems that do not rely on WSN topology to decide the level of redundancy of the datum. Rather, the systems enable the user application to select the number of nodes that are required to store the datum. Moreover, they select the destination coordinate for the datum using a biased hash function, to adapt the process to the sensor distribution: more meta-data are hashed into more populated regions of the WSN, and the result is that the storage load is balanced between all the sensors.

4. Programming Abstraction layer

Most applications do not need low-level access to a WSN, and a high-level perspective can hide the WSN under a traditional computer science appearance, like a database (Amato et Al., 2010; Madden et Al., 2003), or a publish/subscribe system (Albano & Chessa, 2009-1), or an agent-based platform (Muldoon et Al., 2006).
A thorough analysis of research papers showed that a coherent taxonomy is hard to build, since the approaches applied to WSN middleware design are very different and focus on different abstraction levels. For example, the Programming Abstraction layer comprises both the database approach of TinyDB (Madden et Al., 2003), and the process based approach of the virtual machine Maté (Levis & Culler, 2002). Moreover, current literature has produced taxonomies that do not agree on categories. For example, Maté has been defined as a process based approach, see Wang et Al. (2008), or as a virtual machine approach, see Rubio et Al. (2007). In this last case, the category does not really describes the way the system is programmed, but instead focuses on the underlying structure of the layer. In this survey, Programming Abstraction layers are first classified into **global entity** and **local entities** layers, then the **local entities** layers are further divided into **static local entities** and **mobile local entities**, depending on what is addressed by the user application.
The first category is **global entity**. This category is inspired by the survey of Wang et al (Wang et Al., 2008), that called it *system level abstraction*. The middleware that belong to the **global entity** category "abstract the WSN as a single virtual system and allow the programmer to express a single centralized program (global behavior)" (Wang et Al., 2008), and the WSN is considered a single virtual machine that processes the high-level program. This approach leaves "only a small set of programming primitives for the programmer while making transparent the low-level concerns such as the distributed code generation, remote data access and management, and inter-node program flow coordination" (Wang et Al., 2008). Examples of **global**

entity middleware are the database approach (Amato et Al., 2010; Madden et Al., 2003), that accesses the WSN like a single database management system, and the service oriented interface offered by the "Domain layer" of MiSense (Khedo & Subramanian, 2009), that considers the WSN like a single server that offers a set of services to the application programmer.
The second and third categories let the programmer address a number of entities interacting in the WSN, hence they are called "local approaches". Actually, these two categories differ by the identity of the entities that are considered. Category **static local entities** features programmable entities that do not change over time. Most of these approaches consider the single node as the entity that is executing the program, but this category also comprises cluster-based approaches, where the WSN is composed by a number of clusters of sensors.An example of this approach is the event-driven rule-based middleware of FACTS (Tergloth et Al., 2006), where the same application is deployed over all the sensors, and all information is represented by facts. Rules consist of a predicate over these facts and an action, and the action is triggered by the rule engine whenever the predicate becomes true. Another example of the **static local entities** approach is the virtual machine Maté (Levis & Culler, 2002), that organizes programs into small code capsules and presents a process based interface to the user application. The application code is processed on the local node, and the state of the application can not migrate on different nodes. On the other hand, the virtual machine can execute new programs that are received from the network.
The third category, **mobile local entities**, is based on programmable entities being not in a static relation with a set of real sensors. Approaches of this category consider soft entities that can migrate from sensor to sensor, moving their state with themselves. This category mainly features *mobile agent* middleware, like Agilla (Fok et Al., 2009). This middleware is based on a set of agents, that own a state and have a program flow. The agents, while executing their code, can clone and migrate to other nodes. In particular, clone and migrate operations can have a strong or weak semantics. Weak semantics means that only the code is transferred or cloned to the new node, while strong semantics means that the code and the application's state are migrated/cloned. Thus, the agent execution resumes from where it left off. This taxonomy does not consider Impala (Liu & Martonosi, 2003) as a **mobile local entities** approach, since it only enables code updates but not state migration. The middleware Envirotrack (Abdelzaher et Al., 2004), on the other hand, fits in this category of **mobile local entities**. The goal of Envirotrack is tracking objects, like a fire or a noise emitter, and the set of sensors that are sensing the event create a group to locate the object. As the object moves, the set of sensors that belong to the group, and that are executing the program, changes to follow the object.
The rest of this section reports state-of-the-art middleware that implement a Programming Abstraction layer, and divides them into the three broad categories that were described in the first part of the section, and that are summarized into Figure 4.

4.1 Global Entity

The middleware belonging to this category offer a view of the WSN as a single, centralized element that allows the developer to abstract the low-level details. The drawback is that the developer has less control on resource usage and on the algorithms used for routing and data management.

Databases

The middleware in this category model the whole WSN as a distributed database system. The user formulates data requests using a SQL-like query language, that includes syntax for

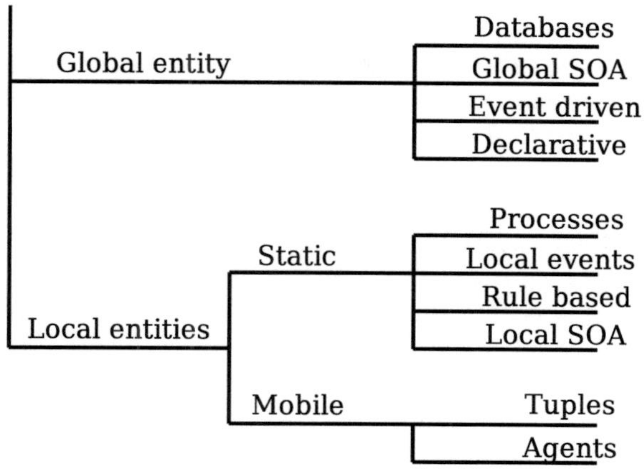

Fig. 4. Programming Abstraction layers.

specifying sample rates as well as query duration. The high level query is translated into a set of data acquisition, data processing and data transfer operations that are carried out by the nodes in the WSN. Query optimization evaluates the various alternatives of task allocation over the sensors, to choose the one that minimizes energy consumption. Examples of the Database approach are TinyDB (Madden et Al., 2003), Cougar (Yao & Gehrke, 2002), MaD-WiSe (Amato et Al., 2010), SINA (Shen et Al., 2001), and Senceive (Hermann & Dargie, 2008).
TinyDB (Madden et Al., 2003), Cougar (Yao & Gehrke, 2002) and MaD-WiSe (Amato et Al., 2010) are all based on a pure database paradigm. They essentially provide a distributed database solution tailored on resource-constrained sensor networks, focusing on efficient query routing and processing. TinyDB (Madden et Al., 2003) uses a SQL-like language with extensions for query duration and sample rates. Queries are expressed over a single `sensors` table that represents all WSN sampled data. Moreover, TinyDB supports spatial aggregation operators and data filtering. Query dissemination is done via Semantic Routing Trees (SRTs), that are routing trees built from the sink.
Cougar (Yao & Gehrke, 2002) shares a number of features with TinyDB. Nodes are modelled as Abstract Data Types (ADTs), and the queries can be addressed toward either single nodes or sets of sensors that satisfying a particular condition, like the physical location of the sensors.
MaD-WiSe (Amato et Al., 2010) is a distributed database system that supports in-network query processing, and query optimization is performed on streams that abstract data sampling, by means of transformation rules based on heuristics that consider query execution plans. Query processing is based on streams that abstract data channels between operators of a query algebra and drive their pipelined behavior (computation and aggregation is carried out on flowing records with almost no need of storage). Operators include selections, projections, spatial aggregates as well as unions and joins. Currently, the ability to perform joins between streams is unique to MaD-WiSe and permits in-network processing of data obtained from different sources.
SINA (Shen et Al., 2001) uses an attribute-based naming scheme in order to facilitate the data-centric characteristics of sensor queries and it allows hierarchical clustering of nodes in order to facilitate scalable operations within sensor networks. The middleware design is based on the creation of clusters of nodes, that cooperate between themselves to orchestrate sensing tasks. The WSN as a whole is considered a collection of logical datasheets. Each cluster of

nodes is related to a datasheet, that is made up of cells, each of them representing a sensor attribute, that can be a single value or a time series. Each cell is unique, and each sensor maintains the whole datasheet. The SQL-like primitives of SINA can be used to issue queries in sensor networks. However, SINA does not provide schemes to hide the faulty nature of both sensor operations and wireless communication, leaving to the application layer the responsibility to provide robustness and reliability for data services.

Senceive (Hermann & Dargie, 2008) is similar to the previous approaches, but based on a graphical interface to define the operations to be performed for data gathering, in terms of SQL-like queries. The WSN is accessed from a special server, that is also a sink to the WSN. Query processing is performed combining all the queries that are active on a sensor. The resulting command is sent to the sensor to retrieve data, and the same command can be sent to a set of nodes using multicast of broadcast routing to save bandwidth and energy. Data are routed towards the data sink using routing trees similar to Directed Diffusion ones (Intanagonwiwat et Al., 2000). Data Storage is realized running a MySQL instance on the server, that is also the access point to the WSN. Data are then stored and processed on the server, and then delivered to the application. The database is also used to store the configuration for the WSN and the middleware.

Global Service Oriented Architectures

Middleware offering Global Service Oriented Architectures model the WSN as a single server that offers a set of services to the application programmers.

In particular, MiSense (Khedo & Subramanian, 2009) is a service-oriented component-based middleware that supports distributed sensor applications with various performance requirements. MiSense copes with application complexity by imposing a structure on top of the component model in the form of composability restrictions and by offering service-specific interfaces to the rest of the system. MiSense breaks up the middleware design into self-contained, interacting components in order to resolve the tension between the optimization requirements for specific scenarios and the need for flexibility and reusability for developing energy efficient WSN applications. The layered approach allows programmer to use different levels of abstraction during application design, and the upper layer of the middleware, the Domain layer, models the WSN as a single server that offers a set of services to the application programmer, and allows the programmer to address the WSN by abstracting the low-level details. The middleware is in charge of taking decisions on communication protocols, network topology organization, sensor operation modes and other functions typical of WSNs, to adapt the middleware to network changes.

Event Driven Global Programming

Abstract Task Graphs (Bakshi et Al., 2005; Mottola et Al., 2007) (ATaG) is a middleware that offers a dataflow programming model with a graphical composition language. It is based on a data-driven program flow and a mixed imperative-declarative specification. It allows developers to declare graphically the data flow and connectivity of virtual tasks, and to specify the functionality of tasks using an imperative language. The application developer produces the declarative part of the ATaG program using a GUI and a description of the target deployment in the form of an annotated network graph (ANG). A code generator analyzes the ATaG program, determines the I/O dependencies between tasks and data objects, and generates code templates for the abstract tasks and data. The programmer populates the code templates with application-specific code. The compiler then interprets the program annotations in the context

of the ANG, and generates configuration files for each node to customize the behavior of that node based on its role in the system. Finally, compile-ready code is generated for each node in the network.
MagnetOS (Barr et Al., 2002) is a distributed, power-aware, adaptive operating system, that targets ad hoc and sensor networks. This middleware implements a virtual machine that considers network-wide energy management as the primary concern in WSNs, and it provides a unified single system image of a Java virtual machine across the nodes that comprise a WSN. MagnetOS optimizes energy consumption, avoids hotspots and increases system longevity by transparently partitioning applications into components and dynamically placing these components on nodes within the WSN. Invocation of methods rely on RMI, and routing is based on the AODV protocol (Perkins & Royer, 1999).
MacroLab (Hnat et Al., 2008) lets the user write a single system-wide program in a C-like language, such that the program manages and processes the data as they are collected from the environment. Then the system generates a number of different versions of the programs, from completely centralized (using an external storage Data Management layer) to completely distributed (using a local storage Data Management layer). A cost analyzer computes which version of the program is the most energy efficient and deploys it to the sensors. During program execution, information about the energy cost of the current deployment is constantly collected. Should the system find out that another version of the program would be more beneficial, it will deploy it to the sensors on the run to enhance the WSN efficiency.

Declarative Systems

The Regiment (Newton et Al., 2007) system consists of a high-level language for WSN programming, and of a compiler that translates a global program into a node-level code. Regiment allows the programmer to look at the WSN as a set of spatially-distributed data streams, that may be defined by topological or geographic relationships between nodes. The middleware provides primitives for in-network data processing and region manipulation. In particular, Regiment calls *deglobalization* the process executed by its compiler, that transforms the network-wide representation of the program into a node-level, event-driven program. The process maps region operation into associated spanning trees that establish region membership and permit in-network data aggregation.
The Smart Messages (Borcea et Al., 2004) middleware addresses high-end sensors equipped with several MBs of memory, and it enables the programmer to reason in terms of Spatial Programming (SP), a space aware programming models, that is used to program an unkown number of volatile embedded systems in order to execute a user-defined application in a certain geographical area. The SP runtime system maintains a mapping between spatial references and the nodes they refer to. The mapping concerning each application is kept into a table, that is persistent during the application execution. Smart Messages (SMs) are actually migratory execution units, with code and data sections, and a lightweight execution stack. The SP program is translated into a SM program, then the nodes cooperate to support the SM execution by providing virtual machines.

4.2 Local Entities

The middleware that feature the **Local Entities** approach, offer to the user application a system composed by a number of interacting entities, and mechanisms to orchestrate their interactions to pursue the application goal.

The advantage of this approach with respect to the **Global Entity** one is that it provides a higher degree of resource control to the developer, and its disadvantage is that the application developer copes with more complexity, since the developer is allowed a glance at the underlying WSN structure.

4.2.1 Static Local Entities

The middleware that offer a Programming Abstraction layer of the **Static Local Entities** kind, adopt the traditional view of considering either a single sensor or a set of sensors as the recipient of the program. The sets are defined at application startup, and the state of the single applications that run on the entities can not migrate to different entities.

Process based

These middleware allow the developers to write applications in separate, small modules. The system injects and distributes the modules throughout the network using tailored algorithms, aiming at minimizing overall energy consumption and resource use.
Solutions in this category include Maté (Levis & Culler, 2002), ASVM (Levis et Al., 2005) and DAViM (Michiels et Al., 2006) that offer explicitly a virtual machine to the user application for the program execution. For example, Maté (Levis & Culler, 2002) is a byte code interpreter that runs on TinyOS. The user code of the application is broken into capsules of 24 byte-long instructions. Each capsule comprises a version number for its code, and the capsules are disseminated throughout the network such that every time a sensor receives a newer version of a capsule, the contained code is saved and then the capsule is forwarded to the sensor's neighbors. Maté does not have to buffer packets nor to store large data because it uses a synchronous model that begins execution in response to an event such as a packet transmission or a timeout. The synchronous model makes application-level programming simpler and less prone to bugs than dealing with asynchronous event notifications, but it limits the expressiveness of the programming model.
Another Process based approach is given by Contiki (Dunkels et Al., 2004), that is a lightweight operating system that supports dynamic loading and replacement of individual programs and services. Contiki is considered a Process based approach since, even though it is built around an event-driven kernel, it also provides preemptive multi-threading.Contiki is implemented in C and it has been ported to a number of micro-controller architectures. This operating system includes mechanisms to reduce energy consumption, and the total size of compiled code fits in 4KB RAM. Contiki has the ability to load and unload individual programs at run-time, and its programs use native code, and can therefore implement low level device drivers without loss of execution efficiency.
Another middleware of the Process based kind is MiLAN (Heinzelman, 2004), that lets programmers to fine-tune the network by setting QoS parameters on the basis of application requirements, that are set through a standard API. The benefits that can be drawn from MiLAN are here considered like a support to the operating system, helping the application to manage low-level mechanisms.
Impala (Liu & Martonosi, 2003) is a middleware designed to be used in the ZebraNet project, that aims at implementing surveillance systems for wildlife environments. Impala novelty relies in its approach of updating at rutime the application that is being executed on the sensors. Applications are modular in order to enable small updates that require little power during transmission. Even though Impala has been defined in a number of surveys as a "Mobile

Agents" approach, the only migration that can happen is about the code being updated with a new program, hence this work considers it a **static local entities** approach.

Event-based programming

Another approach to WSN middleware is based on the notion of events. There, the application specifies interest in certain state changes of the real world (basic events). Upon detecting such an event, the middleware sends a so-called event notification towards interested applications. The application can also specify certain patterns of events (compound events), such that the application is only notified if occurred events match these patterns. In (Yoneki & Bacon, 2005), a reasonably sophisticated set of event operators for describing event patterns in sensor networks has been produced. A crucial limitation of this solution is the complexity that is involved in implementing user applications.

DSWare (Li et Al., 2003-1) provides data-centric and group-based services for sensor networks. A realtime service handles the individual sensor reports, computing correlations among different sensor observations, with the goal of correctly capturing the characteristics of occurred events. The event service supports confidence functions which are designed on data semantics, including relative importance of sub-events and historical patterns. The event service enables partial detection of critical events, and it can also be used to differentiate between the occurrences of events and false alarms. Data are cached into the network using Data Centric Storage (see Subsection 3.2) and an SQL-like script language is used to subscribe events from a set of sensors.

Abstract Regions (Welsh & Mainland, 2004) is a middleware composed by a family of spatial operators that capture local communication within the regions of the network, which may be defined in terms of radio connectivity, geographic location, or other properties of nodes. Abstract Regions provides interfaces for identifying neighboring nodes, sharing data among neighbors, and performing reductions on shared variables. In addition, Abstract Regions exposes the trade-off between the accuracy and resource usage of communication operations. Applications can adapt to changing network conditions by tuning the energy and bandwidth usage of the underlying communication substrate. On the other hand, the group identity is static and set at application start-up and hence this approach can not perform state migration between nodes.

SensorWare (Boulis et Al., 2003) is a middleware that offers good flexibility to the development, at the expense of increased responsibility for the programmer. SensorWare provides a language model to implement distributed algorithms, providing a way to share the resources of a node among many applications and users that might concurrently use the distributed algorithm. The WSN is viewed as executing a set of collaborating programs in a corresponding set of nodes. The sensing, communication, and signal-processing resources of a node are exposed to the control scripts that orchestrate the dataflows to assemble custom protocols and signal processing stacks. SensorWare is also responsible for the dynamic deployment of the distributed algorithms into the WSN, dynamically programming the nodes. The approach of SensorWare is to allow nodes to program their peers, so that the user does not have to worry about deploying the distributed algorithm (because the information on how the algorithm unfolds lies within the algorithm), and the nodes save communication energy because they interact with their immediate neighbors and not with the sink through multi-hop routes.

The Mires middleware (Souto et Al., 2004) is a more pragmatic publish/subscribe solution that has been designed and implemented on top of TinyOS using nesC. It lets the applications specify interests in certain state changes of the real world. Upon detecting such an event, a node sends a so-called event notification towards interested applications. Mires adopts a component-based programming model using active messages to implement its publish/subscribe-based communication infrastructure.

TinyOS (Levis et Al., 2004) is one of the most popular operating systems for networked embedded devices. It is component-based, event-driven and highly configurable, and it does not provide dynamic memory allocation. The programming model of TinyOS is based on the concepts of tasks, events and commands. The task is a low priority code piece that is run while the processor is not requested by event handlers. Components communicate via commands, sent to lower level components, and events, raised to upper level components. Events can preempt tasks and other events. This concurrency mechanism is inherited from the nesC language, used to implement the operating system. A TinyOS application is composed by a component definition file (`.comp`) and a wiring description file (`.desc`). The wiring description file defines dependencies between components through the channel interface connecting components.

Marwis (Wagenknecht, 2008) is a middleware based on Contiki (Dunkels et Al., 2004) (executed on the sensors) and Linux (executed on computers managing the WSN), that aims at managing WSNs composed by different kinds of sensors. Sensors are divided into smaller sensor subnetworks (SSNs), each containing only sensors of one type, then a wireless mesh network (WMN) operates as a backbone for the SSNs and as a gateway to the WSNs. A code updater running on the sensors takes care of code replacement, and some sensors called mesh nodes (MNs) contain a gateway to the WSN and a database of the sensors' status and sensed data, and are used to export the sensed data to the external user applications.

Rule-based systems

A rule-based system considers the application as composed by a program that has to be run on a node, and that is executed whenever a condition is verified. The middleware FACTS (Tergloth et Al., 2006), for example, is both event-driven and rule-based, and it combines these paradigms to perform energy saving. The same application is deployed to all the nodes, and it comprises a set of actions and a set of conditions (rules) for the actions to be executed. All data are defined as "facts", and the rules consist of a combination of a predicate over these facts and an action. The action is triggered by the rule engine whenever the predicate becomes true.

Escape (Russello et Al., 2008) is a framework that is used to simplify application development and deployment, and it promotes the reuse of code. The framework is component-based and it is aimed at the development of sense-and-react applications that combine the use of sensors and actuators. The central component of the framework is a publish/subscribe service broker that manages subscriptions, and that generates data routing towards the data subscriber. The novelty of the approach is the orchestration of two more components, the *service layer* and the *policy manager*. The service layer offers all the services that are orthogonal to the publish/subscribe system, like data collection, routing, and encryption. The policies are enforced on the services offered, to ensure the correct behavior of the middleware. Policies can be used to specify which actions are to be associated with the broker operation, and to coordinate sensors and actuators' operations.

Service Architectures for Static Local Entities

A different approach to WSN middleware is given by the ZigBee standard (Baronti et Al., 2007; ZigBee, 2005). This is a short-range multi-hop wireless protocol constructed over IEEE

802.15.4. At the Network layer it features an inherently node centric behavior, but it offers service-oriented mechanisms at the application level. Since it offers very general services, it does not deal with data management and collection, and it has a high degree of complexity and a big footprint. The ZigBee specification includes mechanisms aimed at limiting the sensors duty cycle, which however are configurable at network creation and that can not be adapted dynamically. ZigBee defines a framework under which the programmers develop applications in terms of Application Objects (APO). Each ZigBee device can host up to 240 APOs, which exploit the services offered by ZigBee which include data transmission, binding, discovery services, and security services. Each APO in the network is uniquely identified by combining its endpoint address and the network address of the hosting device. In the most simple setting, an APO consists of a limited set of attributes which can be accessed from remote APOs using simple *get*, *set*, and *event* transactions. An application profile is the specification in a standard format of the behavior of an application, whose execution may involve several ZigBee devices. An application profile describes a set of devices and clusters and defines the kind of data service. The basic services offered by ZigBee are device and service discovery, binding of devices, network management functions to manage connections/disconnections in a ZigBee network, and security management at network level and device level.
SMEPP Light (Vairo et Al., 2008) is written in NesC and runs on top of TinyOS, and it was inspired by the European Project SMEPP (Albano et Al., 2007-1; SMEPP, 2010), that aimed at creating a secure, service-oriented middleware for embedded peer-to-peer systems. SMEPP Light is a version of SMEPP that is tailored for WSNs, and it supports a subset of SMEPP's primitives. In particular, SMEPP Light does not support full-fledged services, but on the other hand it organizes the sensors into groups and provides eventing mechanisms, based on the directed diffusion paradigm (Intanagonwiwat et Al., 2000), for query dissemination and data collection. A sensor requests events from sensors belonging to the same group and creates a routing tree rooted in the subscriber. Two levels of security are provided. Network security uses two keys that are set at compile time, while group level security uses three keys, and it is based on a masterKey. The masterKey is known in advance by all the sensors that can get into the group, and it is used to restrict the access to the group. For energy management purposes, each group defines a duty cycle that imposes to each node a period of activity followed by a period of inactivity, and each data subscription can add to the activity periods, prescribing to some nodes to sample data from the environment and to relay the data events.
The service approach to WSN was developed also in the direction of web services. Open Sensor Web Architecture (OSWA) (Chu & Buyya, 2007) aims at making various types of web-resident sensors and instruments, discoverable, accessible and controllable via the World Wide Web. The novelty of this approach resides in the efforts that have been made in overcoming the obstacles related to the heterogeneity of the different sensors and instruments that were targeted by sensor web projects.
RESTful (Yazar & Dunkels, 2009) is a web service based middleware, that lets the developer interact with a REST based web service when querying a node. The middleware bases its energy-serving strategies on the X-MAC (Buettner et Al., 2006) protocol, modified to be session-aware. Multi-hop communication is implemented at application level, using a REST call on each communication hop.

4.2.2 Mobile Local Entities

The middleware in the **mobile local entities** category do not focus the programmer's attention on physical nodes or on the whole WSN. Instead, they consider virtual nodes as the target of the programs, so that the actual node that is executing the data collection or the data processing algorithm is changed at runtime. This paradigm is an advanced kind of node coordination, where the identity of the computation is not anymore in the sensor that is performing the task, but it is instead a virtual entity that is associated temporarily with one sensor or a set of sensors, thus it can change over time to follow the application logic or a physical event. This last kind of execution is aimed at applications performing target tracking.

Tuple Spaces

The coordination needed in WSNs has attracted the attention of the Coordination paradigm community (Carriero & Gelernter, 2005). More specifically, different coordination models and middleware based on the Linda abstract model (Gelernter, 1985) have appeared in the area of WSNs. Linda can be considered one of the most representative coordination languages. It is based on a shared memory model where data are represented by elementary data structures called tuples, and the memory is a multiset of tuples and takes the name of *"tuple space"*. Examples of this class of middleware are TinyLime (Curino et Al., 2005) and TeenyLime (Costa et Al., 2006). For example, in TinyLime, a new operational scenario is assumed, with the goal of providing contextual information, not requiring multi-hop communication among sensors, placing reasonable computation and communication demands on the sensors. Sensors are sparsely distributed in the environment, not necessarily able to communicate with each other, and a set of mobile base stations (laptops) move through space accessing the data of sensors nearby. Each base station owns a tuple space and federated tuple spaces can be established in order to communicate and synchronize several base stations.

Tuple Channels

An alternative to tuple spaces is the proposal based on the use of tuple channels (Díaz et Al., 1997) in order to carry out communication and synchronization among the involved WSN nodes. A tuple channel is a FIFO structure that allows one-to-many and many-to-one communication of data structures, represented by tuples. Several advantages can be obtained from the use of channels with respect to shared memory models:

1. Architectural expressiveness: like messaging, using channels to express the communication carried out within a distributed system is more expressive than using shared data spaces, since with a shared data space it is difficult to identify which components exchange data with each other.
2. Channels support data streams in a natural way: the application programmer does not have to deal with head and tail tuples as is necessary in a tuple space based approach to implement information streams.
3. Channel interconnection provides great flexibility for the definition of complex and dynamic interaction protocols: sensor data dissemination can be achieved elegantly, allowing for data redirection, data aggregation and redundant data elimination.

A representative of Tuple Channels middleware is TCMote (Díaz et Al., 2005). This middleware is designed to support an operational setting based on a hierarchical architecture of sensing regions, each one governed by a *region leader* with higher capabilities (power, memory, processing ability) than the rest of the region's sensors. A region leader owns a tuple channel space, which stores tuple channels used to carry out communication and synchronization between the region's sensors and the leader in a single-hop way. Data is consumed when moved through the tuple channels, contributing to dealing with the data-centric characteristics of sensor queries. In addition, tuple channels can be dynamically interconnected through the use of predefined and user-defined connectors, to define new topologies.

Mobile Agents

In the traditional client/server-based computing architecture, data produced by multiple sources is transferred to a destination, whereas in the mobile agent based computing paradigm, a task-specific executable code traverses the relevant sources to gather the data. Mobile agents can be used to reduce the communication cost, by moving the processing function to the data rather than bringing the data to a central node.

Recently, mobile agents have been proposed for efficient data dissemination in WSNs. Some proposals are Agilla (Fok et Al., 2009), MAWSN (Chen et Al., 2006) and actorNet (Kwon et Al., 2006). We discuss the first one as the representative of this category of middleware. Agilla facilitates the rapid deployment of adaptive applications in WSNs by allowing the programmer to create and inject mobile agents, which can migrate across the WSN performing application-specific tasks. Mobile agents can move or clone themselves to desired locations in response to changes in the conditions of the environment. Each node maintains a local tuple space (in fact, can also be considered of the "tuple space" category), and different agents can coordinate through local or remote operations on these tuple spaces. Code allocation is performed using the tuple spaces, allowing an agent to tell Agilla that it is interested in tuples matching a particular template.

Agent Factory Micro Edition (AFME) (Muldoon et Al., 2006) is a middleware featuring the mobile agent approach, that is a version of Agent Factory (O'Hare, 1996) middleware for computationally constrained devices.AFME runs on top of Java 2 Micro Edition (J2ME) and it implements a framework where mobile agents operate under the BDI (Belief-Desire-Intention) paradigm to perform decisions, and where the mobile agents can migrate between devices of different capabilities, for example between personal computers and sensors. Agent design is decoupled into core behaviors, that are constant characteristics of the agent, and platform dependent behaviors, that are changed every time the agent migrates between different devices. Agent communication is agnostic, in the sense that an agent interacts without directly referencing the device of its peer, hence it does not have to know in advance if its peer is running on a personal computer or a sensor. When an agent is created, it is assigned an unique identifier, then communication is addressed by means of the unique identifier, that is resolved to an agent and then to a device, to forward the message appropriately.

Envirotrack (Abdelzaher et Al., 2004) is an object-oriented middleware that aims at providing an interface to the application programmer geared towards tracking the physical environment. Sensors which detect certain user-defined objects in the physical environment form groups, one around each object. A network abstraction layer associates a context label with each such group to represent the corresponding tracked object in the computing system. A context label is the logical address of a virtual host which follows the tracked object in the physical environment. Programs can be attached to context labels to perform context-specific computation. The programs are executed on the sensor group of the context label.

Aware (Gil et Al., 2007) is similar to Envirotrack (Abdelzaher et Al., 2004), since it has the same goal of supporting tracking applications, but it aims also to provide seamless communication between a network of entities with high capabilities (computers and robots, linked by Ethernet and 802.11) and the WSN. Aware's basic premise is to divide sensors into groups that are located around a certain environmental condition, that characterize the physical event to be tracked. The conditions that define the physical event are distributed epidemically in the WSN, then each group of sensors elects a group leader. The system supports the definition of multiple copies of the same physical event (two fires burning at the same time) and it allows the physical event to join (the fires joined into one big fire) and to split (the main fire ignited another fire down a hill) and the sensor group identifiers are managed accordingly to stay consistent with the semantics of the group.

5. Dependability mechanisms

Dependability mechanisms in WSNs are becoming increasingly important, since they are useful for different goals, ranging from network coding (Alon et Al., 2000) to in-network data storage (Kamra et Al., 2006). For example, Data Management layers based on DCS are agnostic to the way that the nodes actually encode the data to perform their storage.
Every kind of dependable data storage must be based on some kind of redundancy on the data that are stored, to be able to reconstruct the data if/when some nodes fail. Two representatives of Dependability mechanisms are the pure replication and the *erasure coding* of data.
Pure replication: Most current approaches adopt pure replication, that implies the replication of the whole data, sometimes in conjunction with the deployment of the copies in regions of the network that are far away (Ratnasamy et Al., 2003), to maximize the lifetime of the data in front of destructive events.
In this kind of scenario, a useful approach to improve the efficiency of data management is the generation of Index Systems (Ganesan et Al., 2005) to manage a small quantity of data at a time.
Erasure coding: Beginning with the work of Shannon (Shannon, 1948) in 1948, a number of redundancy techniques have been designed and employed in very different areas (CD, storage (Alon et Al., 2000), etc).
Erasure coding (Barsi & Maestrini, 1973) consists in encoding a datum into a set of redundant fragments that guarantees the survival of the datum in front of the loss (erasure) of a limited number of fragments. In particular, given a datum d and m keys, the *n out of m* coding of d consists in m fragments (one for each key), with the property that d can be reconstructed from any subset of n fragments, provided the keys used to construct the fragments are known. These codes exploit a set of $m = n + r$ keys to encode a datum d of size L symbols into a set of m fragments of size $\sim \frac{L}{n}$, with the property that d can be reconstructed if up to r fragments are lost and up to $\lfloor \frac{r-e}{2} \rfloor$ fragments are corrupted. Examples of erasure codes are Reed Solomon codes (Plank, 1997), and RNNS (Barsi & Maestrini, 1973).
In (Rodrigues & Liskov, 2005), the authors studied the use of erasure codes in peer-to-peer networks with frequent changes in peers' membership. Previous comparisons (Weatherspoon & Kubiatowicz, 2002) mostly argue that erasure coding is the clear victor, due to huge storage savings for the same availability levels (or conversely, huge availability gains for the same storage levels). The work of Liskov et al, on the other hand, argues that while gains from coding exist, they are highly dependent on the characteristics of the nodes that comprise the overlay. In fact, when a peer leaves the network the fragments it stores are lost, and to reconstruct them (in order to restore the desired level of redundancy) it is necessary first to reconstruct the original data by reading a given number of available fragments. This and the extra complexity can out-weight the benefits of erasure codes in terms of data availability.
Nonetheless, these techniques can lead to improvements in various aspects of WSNs. First of all, they reduce the storage overhead for DCS (Dimakis et Al., 2006), to reduce transport costs for the datum (Albano & Gao, 2010), they increase the robustness of the system because

the system can use the redundancy properties of the erasure coding to recover the datum if a packet gets lost, without having to ask it again to the WSN (Albano & Chessa, 2009-2).

6. Conclusions

This paper considers the issues that arise when designing a middleware for WSNs, and it reviews the state of the art on solutions and basic technologies by means of a layered view. In this work, the solutions representative of the different layers are organized into a taxonomy. More specifically, the upper layer, also called the Programming Abstraction layer, has the most complex structure since it encapsulates many functions aimed at very different goals. At the bottom layer, namely the Data Management layer, this survey presents the low level mechanisms enabling the functionalities of the higher layer. Specifically we placed at this layer the data storage mechanisms. Finally, a discussion of the dependability mechanisms concludes the chapter.
The authors of this survey are positive that this work will be useful to future middleware developers, since decomposing a middleware in a coherent way can help to cope with the complexity that naturally arises when designing a complex system.

7. References

T. Abdelzaher , B. Blum , Q. Cao , Y. Chen , D. Evans , J. George , S. George , L. Gu , T. He , S. Krishnamurthy , L. Luo , S. Son , J. Stankovic , R. Stoleru , A. Wood: EnviroTrack: Towards an Environmental Computing Paradigm for Distributed Sensor Networks. In: Proceedings of the 24th International Conference on Distributed Computing Systems (ICDCS'04), p.582-589, March 24-26, 2004

I.F. Akyildiz et Al.: A survey on sensor networks. In:IEEE Communications Magazine vol. 40, n. 8 (2002)

M. Albano, A. Brogi, R. Popescu, M. Diaz, A. Dianes: Towards Secure Middleware for Embedded Peer-to-Peer Systems: Ob jectives & Requirements. In: RSPSI 2007, Innsbruck, Austria

M. Albano, S. Chessa, F. Nidito, S. Pelagatti: Q-NiGHT: Adding QoS to Data Centric Storage in Non-Uniform Sensor Networks. In: IEEE MDM 2007, Mannheim, Germany (2007)

M. Albano, S. Chessa, F. Nidito, S. Pelagatti: Dealing with Non Uniformity in Data Centric Storage for Wireless Sensor Networks. In: to appear on IEEE Trans. on Parallel and Distributed Systems (2010)

M. Albano, S. Chessa: Publish/subscribe in Wireless Sensor Networks based on Data Centric Storage. In: CAMS 2009, workshop of COMSWARE 2009, Dublin, Ireland, ISBN 978-1-60558-525-3

M. Albano, S. Chessa: Distributed Erasure Coding in Data Centric Storage for Wireless Sensor Networks. In: IEEE ISCC '09, Sousse, Tunisia, pp.6, 5-8 July 2009

M. Albano, J. Gao: In-Network Coding for Resilient Sensor Data Storage and Efficient Data Mule Collection. In: Proceeding of Int. Workshop on Algorithms for Sensor Systems (Algosensors 2010), Bordeaux, July 5th, 2010

N. Alon, H. Kaplan, M. Krivelevich, D. Malkhi, and J. P. Stern: Scalable secure storage when half the system is faulty. In: Automata, Languages and Programming, pages 576-587 (2000)

M. Aly, K. Pruhs, P.K. Chrysanthis: Kddcs: a load-balanced in-network data-centric storage scheme for sensor networks. In: Proceedings of CIKM, p. 317-326 (2006)

G. Amato, S. Chessa, and C. Vairo: MaD-WiSe: A Distributed Stream Management System for Wireless Sensor Networks. In: Software Practice & Experience, 40 (5): 431-451 (2010)

F. Araujo, L. Rodrigues, J. Kaiser, C. Liu, C. Mitidieri: CHR: a Distributed Hash Table for Wireless Ad Hoc Networks. In: Proc. of the 25th IEEE ICDCSW'05 (2005)

L.S. Bai, R.P. Dick, P.A. Dinda: Archetype-based design: Sensor network programming for application experts, not just programming experts. In: ACM IPSN 2009, p. 85-95

A. Bakshi, V.K. Prasanna, J. Reich, D. Larner: The Abstract Task Graph: A Methodology for Architecture-Independent Programming of Networked Sensor Systems. In: Proceedings of International Conference on Mobile Systems, Applications, and Services, p. 19-24, June 2005

P. Baronti et Al.: Wireless Sensor Networks: a Survey on the State of the Art and the 802.15.4 and ZigBee Standards. In: Computer Communications, 30 (7) 1655-1695 (2007)

R. Barr, J.C. Bicket, D.S. Dantas, B. Du, T.W.D. Kim, Danny, B. Zhou, E.G. Sirer: On the need for system-level support for ad hoc and sensor networks. In: ACM SIGOPS Operating Systems Review, vol. 36, n. 2, April 2002

F. Barsi and P. Maestrini: Error Correcting Properties of Redundant Residue Number Systems. In: IEEE Transactions on Computers, Vol. C-22 (3) 307-315 (1973)

C. Borcea, C. Intanagonwiwat, P. Kang, U. Kremer, and L. Iftode: Spatial programming using smart messages: Design and implementation. In: Proceedings the 24th International Conference on Distributed Computing Systems (ICDCS 2004), March 2004

A. Boulis, C.C. Han, and M.B. Srivastava: Design and implementation of a framework for efficient and programmable sensor networks. In: MobiSys '03: Proceedings of the 1st ACM international conference on Mobile systems, applications and services (2003)

J. Bruck, J. Gao, and A. Jiang: MAP: Medial axis based geometric routing in sensor networks. In: ACM/IEEE MobiCom, p. 88-102 (2005)

M. Buettner, V. Yee, E. Anderson, and R. Han: X-MAC: a short preamble MAC protocol for duty-cycled wireless sensor networks. In: SenSys 2006.

N. Carriero, D. Gelernter: Coordination Languages and their Significance. In: Communications of the ACM, 35(2):97–107, 1992. March 2005.

M. Chen, T. Kwon, Y. Yuan, V.C.M Leung: Mobile Agent Based Wireless Sensor Networks. In: Journal of Computers, 1(1):97 (2006)

S. Chessa: Sensor Network Standards. Book chapter in: Wireless Sensor Networks: A Networking Perspective, Wiley-IEEE, ISBN: 978-0-470-16763-2, p. 407-431, September 2009

X. Chu, R. Buyya: Service Oriented Sensor Web. In: Sensor Networks and Configuration: Fundamentals, Standards, Platforms and Applications, Springer-Verlag, pp. 51-74 (2007)

P. Costa, L. Mottola, A.L. Murphy, G.P. Picco: TeenyLIME: Transiently Shared Tuple Space Middleware for Wireless Sensor Networks. In: Proceedings of the 1st International Workshop on Middleware for Sensor Networks (MidSens' 06), Melbourne, Australia, November 2006

C. Curino, M. Giani, M. Giorgetta, A. Giusti, A.L. Murphy, G.P. Picco: Mobile data collection in Sensor Networks: the TinyLime Middleware. In: Pervasive and Mobile Computing, vol 1, n 4, December 2005, p. 446-469 (2005)

M. Díaz, B. Rubio, and J.M. Troya: The tuple channel coordination model. In: Proceedings of ICSE'97 workshop on Software Engineering for Parallel and Distributed Systems, p. 95-106, Boston, May 1997

M. Díaz, B. Rubio, J.M. Troya: A Coordination Middleware for Wireless Sensor Networks. In: Proceedings of the IEEE International Conference on Sensor Networks (SENET' 05), pages 377–382. Montreal, Canada. IEEE Computer Society Press, August 2005

A.G. Dimakis, V. Prabhakaran, and K. Ramchandran: Decentralized erasure codes for distributed networked storage. In: IEEE Transactions on Information Theory, vol. 52, n. 6, p. 2809-2816, June 2006

A. Dunkels, B. Gronvall, T. Voigt: Contiki - a lightweight and flexible operating system for tiny networked sensors. In: Proceedings of the First IEEE Workshop on Embedded Networked Sensors (2004)

C.T. Ee, S. Ratnasamy, and S. Shenker: Practical Data-Centric Storage. In: USENIX NSDI Conference, San Jose, CA, May 2006

C.-L. Fok, G.-C. Roman, C.Lu: Agilla: A mobile agent middleware for self-adaptive wireless sensor networks. In: ACM Trans. Auton. Adapt. Syst., vol. 4, n. 3, p. 1–26 (2009)

D. Ganesan and B. Greenstein and D. Estrin and J. Heidemann and R. Govindan: Multiresolution storage and search in sensor networks. In: Trans. Storage, vol. 1, n. 3, pag. 277-315 (2005)

D. Gelernter: Generative Communication in Linda. In: ACM Transactions on Programming Languages and Systems, 7(1), 80–112, 1985.

A. Ghose, J. Grossklags, and J. Chuang: Resilient data-centric storage in wireless sensor networks. In: IEEE Distributed Systems online, 4(11), November 2003

P. Gil, I. Maza, A. Ollero, P. Marrón: Data Centric Middleware for the Integration of Wireless Sensor Networks and Mobile Robots In: Proceedings of 7th Conference on Mobile Robots and Competitions, ROBOTICA 2007. April 2007

S. Hadim, N. Mohamed: Middleware: Middleware Challenges and Approaches for Wireless Sensor Networks. In: 1st IEEE International Conference on Communication System Software and Middleware (Comsware 2006)

W. Heinzelman, A. Murphy, H. Carvalho and M. Perillo: Middleware to Support Sensor Network Applications. In: IEEE Network Magazine Special Issue, pp. 6–14, January 2004.

C. Hermann and W. Dargie: Senceive: Middleware for a wireless sensor network. In: IEEE 22nd international conference on advanced information networking and applications (AINA 2008), GinoWan, Okinawa, Japan, March 24-28, 2008

T.W. Hnat, T.I. Sookoor, P. Hooimeijer, W. Weimer, and K. Whitehouse: MacroLab: A Vector-based Macroprogramming Framework for Cyber-Physical Systems. In: Proc. of 6th ACM Conf. on Embedded Networked Sensor Systems (SenSys'08), Raleigh, NC, November 2008

C. Intanagonwiwat, R. Govindan, D. Estrin: Directed Diffusion: A Scalable and Robust Communication Paradigm for Sensor Networks. In Proceedings of the Sixth Annual International Conference on Mobile Computing and Networking (MobiCOM '00), p. 56-67, Boston, August 2000.

A. Kamra, J. Feldman, V. Misra, and D. Rubenstein: Growth codes: Maximizing sensor network data persistence. In: Proceedings of ACM SIGCOMM'06, p. 255-266, Pisa, Italy (2006)

B. Karp, H.T. Kung: GPSR: Greedy Perimeter Stateless Routing for Wireless Networks. In: MobiCom 2000, p. 243-254 (2000)

K.K. Khedo, and R.K. Subramanian: A Service-Oriented Component-Based Middleware Architecture for Wireless Sensor Networks. In: IJCSNS International Journal of Computer Science and Network Security, VOL.9 No.3, March 2009

Y. Kwon, S. Sundresh, K. Mechitov, G. Agha: ActorNet: An Actor Platform for Wireless Sensor Networks. In: Proceedings of the IEEE International Joint Conference on Autonomous Agents and Multiagent Systems, Hakodate, Japan, May 2006

P. Levis, S. Madden, J. Polastre, R. Szewczyk, K. Whitehouse, A. Woo, D. Gay, J. Hill, M. Welsh, E. Brewer, and D. Culler: TinyOS: An operating system for wireless sensor networks. In: Ambient Intelligence. Springer-Verlag, 2004.

P. Levis, D. Gay, D. Culler: Active Sensor Networks. In: Proceedings of the 2nh International Symposium on Networked Systems Design and Implementation (NSDI' 05), pages 29-42, San Francisco, CA, USA, March 2005.

P. Levis, D. Culler: Maté: A Tiny Virtual Machine for Sensor Networks. In: 10th Int. Conf. on Architectural Support for Programming Languages and Operating Systems (ASPLOSX' 02), San Jose, USA (2002)

S. Li, S. Son, and J. Stankovic: Event detection services using data service middleware in distributed sensor networks. In: Proceedings of the 2nd International Workshop on Information Processing in Sensor Networks, April 2003.

X. Li, Y. J. Kim, R. Govidan, and W. Hong: Multi-dimensional range queries in sensor networks. In: Proceedings of ACM SenSys (2003)

T. Liu, M. Martonosi: Impala: A Middleware System for Managing Autonomic, Parallel Sensor Systems. In: ACM Symp. on Principles and practice of parallel programming (2003) p. 107-118

S. Madden et Al: The design of an acquisitional query processor for sensor networks. In: SIGMOD 2003, p. 491-502, San Diego, USA.

S. Michiels, W. Horré, W. Joosen, P. Verbaeten: DAViM: a Dynamically Adaptable Virtual Machine for Sensor Networks. In: Proceedings of the 1st International Workshop on Middleware for Sensor Networks (Mid-Sens' 06), co-located with the 7th International Middlewar e Conference (Middleware' 06), Melbourne, Australia, November, 2006.

L. Mottola, A. Pathak, A. Bakshi, V.K. Prasanna, and G.P. Picco: Expressing Sensor Network Interaction Patterns using Data-Driven Macroprogramming. In: Proceedings of the 3rd IEEE International Workshop on Sensor Networks and Systems for Pervasive Computing (PerSens, colocated with IEEE PERCOM), New York (NY, USA), March 2007.

L. Mottola, G. Picco: Programming wireless sensor networks: Fundamentals concepts and state of the art. In: To appear in ACM Computing Surveys (2010)

C. Muldoon, G.M.P. O'Hare, R. Collier, M.J. O'Grady: Agent Factory Micro Edition: A Framework for Ambient Applications. In: Proceedings of Intelligent Agents in Computing Systems (IACS 2) Workshop of International Conference on Computational Science (ICCS), Reading, p. 727-734 (2006)

R. Newton, G. Morrisett, M. Welsh: The regiment macroprogramming system. In: Proceedings of International Symposium on Information Processing in Sensor Networks (IPSN 07), April 2007

G.M.P. O' Hare: Agent Factory: An Environment for the Fabrication of Multi-Agent Systems. In: G. M. P. O'Hare and N. R. Jennings (eds.), Foundations of Distributed Artificial Intelligence , Wiley Interscience, p. 449-484 (1996)

C.E. Perkins and E.M. Royer: Ad hoc On-Demand Distance Vector Routing. In: Proceedings of the 2nd IEEE Workshop on Mobile Computing Systems and Applications, New Orleans, LA, p. 90-100, February 1999

K. Piotrowski, P. Langendoerfer, and S. Peter: tinyDSM: A highly reliable cooperative data storage for Wireless Sensor Networks. In: International Symposium on Collaborative Technologies and Systems, Los Alamitos, USA, p. 225-232, May 2009

J.S. Plank: A Tutorial on Reed-Solomon Coding for Fault-Tolerance in RAID-like Systems. In: Software: Practice and Experience, vol. 27, n. 9, p. 995-1012 (1997)

Md.A. Rahman: Middleware for wireless sensor networks: Challenges and Approaches. In: Seminar on Internetworking, Helsinki University of Technology, Finland, April 27th 2009.

S. Ratnasamy et Al.: Data-centric storage in sensornets with GHT, a geographic hash table. In: Mobile Networking Application (MONET), vol. 8, n. 4, p. 427-442 (2003)

R. Rodrigues and B. Liskov: High availability in DHTs: Erasure coding vs. replication. In: International Workshop on Peer-to-Peer Systems, Ithaca, NY (2005)

K. Romer, O. Kasten, F. Mattern: Middleware Challenges for Wireless Sensor Networks. In: ACM SIGMOBILE Mobile Computing and Communication Review (MC2R), 6(4):59-61, 2002.

B. Rubio, M. Díaz, and J. Troya: Programming Approaches and Challenges for Wireless Sensor Networks. In: IEEE Int. Conf. on Systems and Networks Communications (IC-SNC'07), Cap Esterel, France (2007)

G. Russello, L. Mostarda, and N. Dulay. Escape: A component-based policy frame-work for sense and react applications. In: Proceedings of the 11th International Symposium on Component-Based Software Engineering, pages 212-229, 2008.

R. Sarkar, X. Zhu, J. Gao: Double Rulings for Information Brokerage in Sensor Networks. In: Seventh ACM International Symposium on Mobile Ad Hoc Networking and Computing (MobiCom06), Florence, Italy, May 2006

K. Seada and A. Helmy: Rendezvous Regions: A Scalable Architecture for Service Location and Data-Centric Storage in Large-Scale Wireless Networks. In: IEEE/ACM IPDPS 4th International Workshop on Algorithms for Wireless, Mobile, Ad Hoc and Sensor Networks (WMAN), Santa Fe, New Mexico, April (2004)

C. Shannon: A mathematical theory of communication. In: Bell Sys. Tech. Journal, 27:379-423, 1948

C-C. Shen, C. Srisathapornphat, C. Jaikaeo: Sensor Information Networking Architecture and Applications. In: IEEE Personal Communications:52-59, 2001

Secure Middleware for Embedded Peer-to-peer. In: http://www.smepp.org (final report published in 2010)

E. Souto, G. Guimaraes, G. Vasconcelos, M. Vieira, N. Rosa, C. Ferraz: A Message-Oriented Middleware for Sensor Networks. In: 2nd International Workshop on Middleware for Pervasive and Ad-Hoc Computing (MPAC 2004), pages 127–134, Toronto, Canada, October 2004.

R. Sugihara, R.K. Gupta: Programming Models for Sensor Networks: A Survey. In: ACM Trans. on Sensor Networks, 4(2), March 2008, 1-28

K. Terfloth, G. Wittenburg, J. Schiller: FACTS: A Rule-based Middleware Architecture for Wireless Sensor Networks. In: Int. Conf. on Communication System Software and Middleware, New Delhi, India (2006)

S. Tong: An Evaluation Framework for middleware approaches on Wireless Sensor Networks. In: Seminar on Internetworking, Helsinki University of Technology, Finland, April 27th 2009.

C. Vairo, M. Albano, S. Chessa: A Secure Middleware for Wireless Sensor Networks. In: Proceedings of Middleware for Mobile Embedded Peer-to-Peer Systems (MIMES '08), workshop of Mobiquitous 2008, pp 1-6, Dublin, Ireland

G. Wagenknecht, M. Anwander, T. Braun, T. Staub, J. Matheka, S. Morgenthaler: MARWIS: A Management Architecture for Heterogeneous Wireless Sensor Networks. In: 6th International Conference on Wired/Wireless Internet Communications (WWIC'08), Vol. LCNS, Nr. 5031, p. 177-188, May 28 - 30 2008

M.M. Wang, J.N. Cao, J. Li, S.K. Dasi: Middleware for wireless sensor networks: A survey. In: Journal of Computer Science and Technology 23(3) (May 2008) 305-326

H. Weatherspoon, J.D. Kubiatowicz: Erasure Coding vs. Replication: A Quantitative Comparison. In: LNCS 2429, 1st Int. Workshop on Peer-to-Peer Systems, p. 328 - 338 (2002)

M. Welsh, G. Mainland: Programming Sensor Networks Using Abstract Regions. In: Proceedings of NSDI, 2004

Y. Yao, J. Gehrke: The Cougar Approach to In-Network Query Processing in Sensor Networks. In: SIGMOD Record vol. 31, n.3 (2002)

D. Yazar, A. Dunkels: Efficient Application Integration in IP-based Sensor Networks. In: Proceedings of ACM BuildSys 2009, the First ACM Workshop On Embedded Sensing Systems For Energy-Efficiency In Buildings, Berkeley, CA, USA, November 2009

E. Yoneki, J. Bacon: Unified Semantics for Event Correlation over Time and Space in Hybrid Network Environments. In: IFIP International Conference on Cooperative Information Systems (CoopIS' 05), LNCS vol. 3760, pages 366-384. Springer 2005.

ZigBee Alliance: ZigBee Specification version 1.0, April 2005.

24

Group Key Managements in Wireless Sensor Networks

Ju-Hyung Son* and Seung-Woo Seo
LG Electronics, Seoul National University*
Korea

1. Abstract

A sensor network operating in open environments requires a network-wide group key for confidentiality of exchanged messages between sensor nodes. When a node behaves abnormally due to its malfunction or a compromise attack by adversaries, the central sink node should update the group key of other nodes. The major concern of this group key update procedure will be the multi-hop communication overheads of the rekeying messages due to the energy constraints of sensor nodes. Many researchers have tried to reduce the number of rekeying messages by using the logical key tree. In this chapter, we propose an energy-efficient group key management scheme called Topological Key Hierarchy (TKH). TKH generates a key tree by using the underlying sensor network topology with consideration of subtree-based key tree separation and wireless multicast advantage. Based on our detailed analysis and simulation study, we compare the total rekeying costs of our scheme with the previous logical key tree schemes and demonstrate its energy efficiency.

2. Introduction

Advances in wireless networking, embedded microprocessors, and micro-fabrication technology have enabled a new generation of large-scale sensor networks which target a range of commercial and military applications (Zhao & Guibas, 2004). A typical wireless sensor network is composed of a large number of sensor nodes and one or several sinks collecting data from them (Akyildiz et al., 2002). The sink broadcasts control messages to engage sensor nodes into specific tasks. In response to the control messages, sensor nodes deliver the requested data back to the sink collectively. Usually, a sink has lap-top class computing power while sensor nodes have very limited computing resources. Even though it can vary by specific application scenarios of a sensor network, most sensor nodes are powered by limited batteries. Therefore, energy efficiency should be considered as an important parameter during the design of services or protocols for a sensor network. This also applies to the design of security services. If we want to apply a security service to a sensor network, we should consider the computation & communication efficiencies of the service in addition to its security performances.

From the nature of multi-hop wireless communications, an adversary can easily eavesdrop exchanged messages to gather secret information from a sensor network. Furthermore, it can

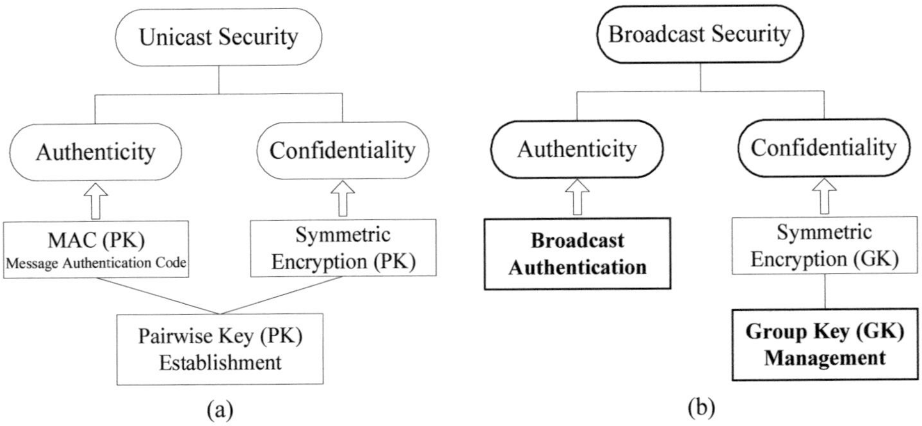

Fig. 1. Classification of security services according to communication patterns.

also inject malicious control messages to disrupt normal operations of sensor nodes. Therefore, in order to protect a sensor network from these attacks, we need *authenticity* and *confidentiality* of exchanged messages.

In general, the authenticity and confidentiality of messages are provided by different security primitives according to their underlying delivery mechanisms (*unicast* and *broadcast*) as depicted in Fig.1. The unicast security provides authentication/encryption of unicast messages between a pair of nodes. For the unicast security, we use MAC (Message Authentication Code) and symmetric encryption for the authenticity and confidentiality respectively while the both schemes use the same pairwise key (PK) established between a pair of nodes. Since there are many standard algorithms for the MAC and encryption such as HMAC (Bellare et al., 1997) and AES (U.S. DoC NIST, 2001) which can be efficiently run even on a sensor node, the problem of the unicast security converted to the pairwise key establishment problem.

The broadcast security provides authentication/encryption of broadcast messages between multiple nodes in a network. For the broadcast authenticity, we need new security primitives called *broadcast authentication*. We can provide the broadcast confidentiality by using the symmetric encryption with the network-wide group key (GK). Unlike the pairwise key, it is hard to secure the group key since it is known to every node in a network. Therefore, we need a new security primitive called *group key management* which updates a group key when a node joins/leaves a network.

While majority of researches on sensor network security have focused on pairwise key management schemes to provide the unicast security, there were not much research efforts to provide the broadcast security. However, the group security is more crucial security service compared to the unicast security for a sensor network from the specific characteristics of a sensor network:

- *Broadcast communications are commonly used in sensor networks*: Unlike the Internet environment where the server-client-based unicast communication type is prevalent, the broadcast communications are commonly used in a sensor network since all sensor nodes are usually controlled by a single administrative domain. A sink should continuously control sensor nodes by using the broadcast control messages, while there are not many scenarios where a pair of sensor nodes communicate each other.

- *Effects of the security breaches of the broadcast communications are much more devastating in sensor networks*: Without proper authentication mechanism, all sensors will react to the false control messages from adversaries which can exhaust energies of all sensor nodes. Also if a network-wide group key is compromised, all communications within a network can be eavesdropped by an adversary. However, revealed pairwise keys from a compromised node will not affect security of other links in a network. Therefore, while the effects of security breaches of the pairwise connections are negligible to other nodes, the compromise of the broadcast security affects the whole network.
- *Existing security primitives for the group security are not applicable to sensor networks*: The existing solutions for the group security such as digital signature (Rivest et al., 1978) and logical key hierarchy (LKH) (Wong et al., 1998) are inadequate to be applied to sensor network environments due to their computation and communication overheads. While the public key cryptography (PKC)-based pairwise key establishment scheme is also computationally heavy, it only occurs once when a pair of sensor nodes first contact. Thus sensors can adopt the PKC-based key establishment without much concerns in energy. However, communication & computation overhead of the broadcast security occurs frequently along with the broadcast messages which can exhaust energies of sensor nodes.

Based on the above motivations for the development of the group security mechanisms for wireless sensor networks, we investigate efficient & secure group key management schemes in this chapter.

2.1 Group Key Managements in Wireless Sensor Networks

In a wireless sensor network (WSN), many sensor nodes collect data from their surroundings, and report them to the central sink node (Akyildiz et al., 2002). The sink broadcasts control messages to sensor nodes to regulate their sensing/reporting operations. From these *many-to-1* and *1-to-many* communication characteristics, typical WSNs utilize a multicast tree topology rooted from the sink. For the design of communication protocol on this topology, the energy efficiency is the most important design principle due to sensor node's energy constraints. This also applies to the design of security services for WSNs. In addition to its security performances, a security service should take into account the energy efficiency of its protocol.

The message *confidentiality* is the imperative security primitive for various security services in a sensor network. Generally, a network-wide group key (GK) is used for message en/decryption for the message confidentiality. The sink should occasionally update GK to prevent a compromised node from decrypting messages. The simplest solution is to separately distribute a new GK to each node after encrypting it by each node's individual key (IK) that is only shared between each node and the sink. However, this will generate $O(N)$ rekeying messages with the network size N.

The Logical Key Hierarchy (LKH) scheme (Wallner et al., 1997) (Wong et al., 1998) reduces the number of rekeying messages to $O(\log N)$ by building a tree of key encryption keys (KEKs). Based on LKH, many researchers tried to further reduce the number of rekeying messages in trade-off of local key computations (Canetti et al., 1999) (Sherman & McGrew, 2003) (Lin et al., 2005). In these schemes, each node requires only several rekeying messages among the total rekeying messages according to its logical position in a key tree. However, in a multi-hop WSN where each node routes messages of other nodes, rekeying messages generated from the logical key tree should be forwarded to many irrelevant nodes before reaching their destinations. In other words, these logical key tree-based schemes can incur heavy communication

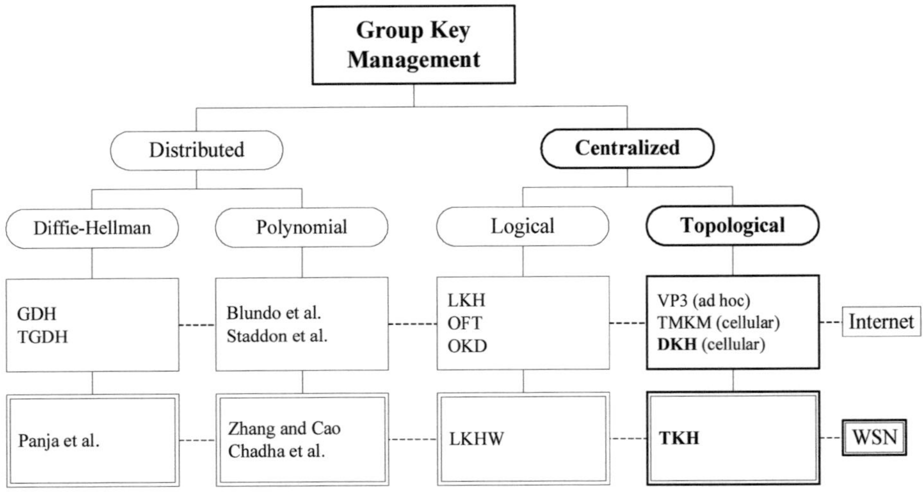

Fig. 2. Taxonomy of group key management schemes.

overheads in multi-hop WSN environments since the key tree structure does not reflect the underlying network topology.

In this chapter, we propse the Topological Key Hierarchy (TKH) scheme which generates a key tree from the sensor network's topology information. The basic principle is to enable topologically adjacent nodes in a network to share the same KEKs so that they can receive the same rekeying messages. Then each rekeying message can be delivered to its designated recipients while minimizing communication costs. While the previous group key management schemes only tried to minimize the number of rekeying messages, our TKH minimizes *the total rekeying cost* which reflects both *the number of rekeying message* and *the communication costs of rekeying messages*. We demonstrate the energy saving of TKH compared to the previous logical key tree-based schemes by using our detailed analysis and simulation study.

2.2 Related Works

A primary method to limit access to information within a group is the message encryption. Along with the message to be encrypted, we need a cryptographic key only shared within a group. Only those who knows the group key are able to decrypt the encrypted message. The most challenging problem of this scenario is to update the group key according to membership changes. We can divide the research literatures for this group key management problem according to their group key establishment style.

In the *Distributed* approaches, members generate a group key in contributory manner by combining their own secret information. Most of group key management schemes for sensor networks mainly focus on the distributed group key management schemes (Zhang & Cao, 2005) (Chadha et al., 2006) (Panja et al., 2006). In these schemes, sensor nodes collaboratively generate and update a group key without the help from a central sink node. However, establishing the group key in a large-scale network by using the distributed manner incurs much overheads. First, these schemes incur computational overheads since they use complex algorithms such as Polynomial (Blundo et al., 1992) (Staddon et al., 2002) or group Diffie-Hellman (Diffie & Hellman, 1976) (Steiner et al., 1996) (Kim et al., 2000) methods between sensor nodes. Also after the collaborative local group key generation procedure between neighbors, each local group key should be merged with other local group keys to generate a single network-wide group key which requires multiple rounds of communications.

On the contrary, in the *Centralized* group key management scheme, a central key distribution center (KDC) randomly generates a new group key and produces related rekeying messages, which eliminates computation overheads of end nodes. Also, the communication costs are the one-time delivery costs of rekeying messages from the KDC to all nodes. Therefore, we think that the centralized group key management scheme is more preferable to a sensor network in terms of rekeying procedure's computation and communication overheads.

After introduction of the logical key hierarchy scheme independently by Wong et al. (Wong et al., 1998) and Wallner et al. (Wallner et al., 1997) in Internet environment, many researchers have tried to further reduce the number of rekeying messages by using the tradeoff between central rekeying and local computation (Sherman & McGrew, 2003) (Lin et al., 2005). In (Pietro et al., 2003), authors combined the directed diffusion data dissemination protocol (Intanagonwiwat et al., 2000) with LKH, and proposed LKHW (LKH for WSN) . LKHW is only compatible with the directed diffusion routing protocol. Our TKH can be applied to any tree-based routing algorithm .

Previously, Sun et al.(Sun et al., 2004) used topological information of a cellular network for efficient group key management. While the wired network part from KDC to each BS (base station) has abundant bandwidth which can easily carry $O(\log N)$ rekeying messages, the wireless network part from BS to each MN (mobile node) suffers from scarce bandwidth. The topological information on the latter part constantly changes due to the mobility of MNs. Therefore, the scheme in (Sun et al., 2004) is superior to the LKH when MN's mobility is not very high. On the contrary, a typical sensor network is a multi-hop wireless network which severely suffers from the limited bandwidth, and each sensor node does not have mobility in most scenarios. Therefore, our TKH can outperform the LKH in most conditions. Also, the hierarchical cellular network topology is quite different from the multi-hop wireless sensor network. Recently, Salido et al. (Salido et al., 2008) proposed VP3 (Vertex-Path, Power-Proximity) scheme for topology-based key management in multi-hop wireless ad hoc network environments. However, they assume dynamic power control capability of each node which is not the case for a sensor network. Also, they assume a subset of nodes belong to a certain group according to application scenarios where all nodes form a group in a sensor network.

3. Logical Key Tree-based Group Key Management

3.1 Logical Key Hierarchy & Related Schemes

The Logical Key Hierarchy (LKH) (Wallner et al., 1997) (Wong et al., 1998) is a centralized group key management scheme which utilizes the logical key tree. A key tree is maintained at the central KDC (Key Distribution Center) and the corresponding rekeying messages are delivered to all nodes when a node joins or leaves a group. A GK (Group Key) which is the root of a key tree is used to encrypt all data traffic within a group. KEKs (Key Encryption Keys) which reside in intermediate edges of a key tree are used to update the root GK and other KEKs. The leaves of a key tree are IKs (Individual Keys) which are individually shared by each node and the KDC. As a result, each node in a group possesses three kinds of keys: its own IK, KEKs (on the path to the root), and a root GK. Figure 3 denotes an example of the logical key tree. By using this example, let us examine the key tree update procedures of both *'Node Join'* and *'Node Leave'* events.

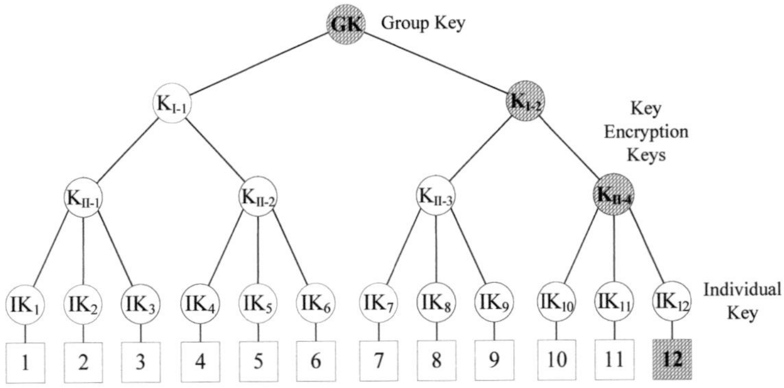

Fig. 3. A logical key tree example consisted of 12 nodes.

3.1.1 Node Join

First, let us assume that there were only eleven nodes initially in Figure 3, then the node 12 newly joins the group. Let $\{K_A\}_{K_B}$ denote key K_A encrypted by key K_B, and K' denote the updated version of key K. The keys that will be possessed by the joining node (GK, K_{I-2}, K_{II-4}) should be updated to prevent the node from decrypting the previously exchanged messages within the group (*Backward Secrecy*) (Kim et al., 2000). After rekeying messages $\{GK'\}_{GK}$, $\{K'_{I-2}\}_{K_{I-2}}$, $\{K'_{II-4}\}_{K_{II-4}}$ are sent to the existing members, the node 12 receives $\{GK', K'_{I-2}, K'_{II-4}\}_{IK_{12}}$. However, the rekeying messages for the existing members can be safely replaced by local key computations (Waldvogel et al., 1999). Each subset of nodes can locally compute keys as $\{1\sim 11\} : GK' = f(GK)$, $\{7\sim 11\} : K'_{I-2} = f(K_{I-2})$, $\{10\sim 11\} : K'_{II-4} = f(K_{II-4})$ with a common one-way function f. It means that the group key update for a node join event only incurs a rekeying message unicast to the joining node.

3.1.2 Node Leave

Second, let us assume that there were initially twelve nodes and the node 12 leaves the group. Then the possessed keys of the leaving node also should be updated to prevent the leaving node from decrypting the future messages (*Forward Secrecy*) (Kim et al., 2000). In this case, however, several current keys cannot be used in the rekeying procedure since the leaving node also knows them. Therefore, more complicated rekeying messages are generated and delivered to the remaining nodes. During the generation of the rekeying messages at KDC, there are two different rekeying strategies in LKH: *group-oriented rekeying* (LKH(g)) and *user-oriented rekeying* (LKH(u)) according to the underlying rekeying message delivery mechanisms (Wong et al., 1998)[1]:

$$\text{LKH}(g) \left\{ \begin{array}{l} m_{\text{KDC}\rightarrow all} : \{GK'\}_{K'_{I-1}} || \{GK'\}_{K'_{I-2}} || \{K'_{I-2}\}_{K_{II-3}} \\ \qquad || \{K'_{I-2}\}_{K'_{II-4}} || \{K'_{II-4}\}_{IK_{10}} || \{K'_{II-4}\}_{IK_{11}} \end{array} \right. \tag{1}$$

$$\text{LKH}(u) \left\{ \begin{array}{l} m_{\text{KDC}\rightarrow\{1\sim 6\}} : \{GK'\}_{K_{I-1}} \\ m_{\text{KDC}\rightarrow\{7\sim 9\}} : \{GK', K'_{I-2}\}_{K_{II-3}} \\ m_{\text{KDC}\rightarrow\{10\}} : \{GK', K'_{I-2}, K'_{II-4}\}_{IK_{10}} \\ m_{\text{KDC}\rightarrow\{11\}} : \{GK', K'_{I-2}, K'_{II-4}\}_{IK_{11}}. \end{array} \right. \tag{2}$$

[1] The *key-oriented rekeying* defined in (Wong et al., 1998) is not considered in this chapter since it equals to the *user-oriented rekeying* in terms of the number of rekeying messages and their delivery mechanism.

In *the group-oriented rekeying,* KDC combines all rekeying messages and broadcasts the whole messages to all nodes. Upon receiving the whole messages, each node selects its messages and decrypts the necessary keys. In *the user-oriented rekeying,* KDC generates rekeying messages for each subset of nodes and multicasts (or unicasts) each rekeying message only to the corresponding subset of nodes. While the group-oriented rekeying generates the smaller number of rekeying messages in total, it incurs more communication overheads in multi-hop WSN since all sensors should receive and forward the whole messages. Even the user-oriented rekeying is more energy-efficient, it requires multicast routing protocol to deliver messages. Without the multicast support in WSNs, rekeying messages for a subset of nodes will be separately delivered to them by unicast.
McGrew and Sherman proposed an improvement over LKH called One-way Function Tree (OFT) (Sherman & McGrew, 2003). OFT reduces the number of rekeying messages from $(2\log_2 N)$ to $(\log_2 N)$ in the binary key tree by using the local key computations (Waldvogel et al., 1999) similar to the node join operation. However, OFT is susceptible to node collusion attacks (Horng, 2002) (Ku & Chen, 2003). There are similar approaches that achieve the same communication overhead as OFT without node collusion vulnerabilities: One-way Function Chain (OFC) (Canetti et al., 1999), and One-way Key Derivation (OKD) (Lin et al., 2005).
In the One-way Key Derivation, KDC reduces the number of rekeying messages by not sending the rekeying messages to nodes that can derive the keys by themselves. Therefore, when node 12 is revoked in Figure 3, the keys can be locally derived in each subset of nodes: $\{1\sim 6\} : \mathrm{GK}' = f(\mathrm{K}_{\mathrm{I\text{-}1}} \oplus \mathrm{GK})$, $\{7\sim 9\} : \mathrm{K}'_{\mathrm{I\text{-}2}} = f(\mathrm{K}'_{\mathrm{II\text{-}3}} \oplus \mathrm{K}_{\mathrm{I\text{-}2}})$, $\{10\} : \mathrm{K}'_{\mathrm{II\text{-}4}} = f(\mathrm{IK}_{10} \oplus \mathrm{K}_{\mathrm{II\text{-}4}})$. Here, f denotes a one-way function and $\oplus$ denotes an exclusive-or computation. After the local key computations, KDC transmits the corresponding rekeying messages to the remaining subset of nodes either by using *group-oriented rekeying* (OKD(g)) or *user-oriented rekeying* (OKD(u)) methods:

$$\mathrm{OKD}(g) \left\{ \; m_{\mathrm{KDC}\rightarrow all} : \{\mathrm{GK}'\}_{\mathrm{K}'_{\mathrm{I\text{-}2}}} || \{\mathrm{K}'_{\mathrm{I\text{-}2}}\}_{\mathrm{K}'_{\mathrm{II\text{-}4}}} || \{\mathrm{K}'_{\mathrm{II\text{-}4}}\}_{\mathrm{IK}_{11}} \right. \tag{3}$$

$$\mathrm{OKD}(u) \left\{ \begin{array}{l} m_{\mathrm{KDC}\rightarrow\{7\sim 9\}} : \{\mathrm{GK}'\}_{\mathrm{K}'_{\mathrm{I\text{-}2}}} \\ m_{\mathrm{KDC}\rightarrow\{10\}} : \{\mathrm{GK}', \mathrm{K}'_{\mathrm{I\text{-}2}}\}_{\mathrm{K}'_{\mathrm{II\text{-}4}}} \\ m_{\mathrm{KDC}\rightarrow\{11\}} : \{\mathrm{GK}', \mathrm{K}'_{\mathrm{I\text{-}2}}, \mathrm{K}'_{\mathrm{II\text{-}4}}\}_{\mathrm{IK}_{11}}. \end{array} \right. \tag{4}$$

Comparing (1)(2) with (3)(4), it is evident that OKD reduces the number of rekeying messages in trade-off of the local key computations.

3.1.3 Total Rekeying Costs

When a group key management scheme properly updates a group key when a node joins or leaves the group as described above, the Backward Secrecy and Forward Secrecy properties are preserved (Kim et al., 2000). Since LKH, OKD, and our TKH are designed to preserve both properties, we argue that they are equal in terms of the security level. However, our TKH achieves the same security level with smaller amount of rekeying cost compared to the logical key tree based schemes including LKH and OKD.
To quantitatively compare the rekeying costs, we define the *Total Rekeying Cost (TRC)* of a group key management scheme as the product of *the number of rekeying messages* and *the communication costs of the rekeying messages.* Previously, most group key management schemes tried to reduce the number of rekeying messages (Rafaeli & Hutchison, 2003). However, it is also important to deliver rekeying messages efficiently to its designated recipients in multi-hop WSN environments. Generally, 1) OKD incurs smaller TRC compared to LKH due to the

reduced number of rekeying messages, and 2) the user-oriented rekeying incurs smaller TRC compared to the group-oriented rekeying since each node receives/forwards the smaller number of messages. However, *OKD's user-oriented rekeying*, currently the most communication-efficient logical key tree-based scheme, is not optimal in multi-hop WSN environments from the following reasons.

First, the multicast routing incurs heavy storage and communication overheads in WSN. Unlike the Internet environment where routers and end-hosts are separated in functionality, each sensor should act as both a router and an end-host in WSNs. Therefore, every sensor should maintain routes to all sensors to support multicast routing. This is infeasible for the resource constrained sensor nodes specifically in large scale networks. Second, even if the multicast routing is supported, it is hard to expect multicast advantage (minimally using the network resources before reaching multiple destinations) with the logical key tree-based schemes. For example, if nodes $\{7,8,9\}$ receiving $\{GK'\}_{K'_{1\text{-}2}}$ in equation (4) are distinctly located in a network, this one multicast session will incur the similar multi-hop communication overheads as three unicast sessions to each of them. To overcome these constraints, we propose Topological Key Hierarchy that does not require multicast routing protocol and utilize multicast advantage by mapping the topological neighbors to the key tree neighbors.

4. Topological Key Hierarchy

In this section, we provide design principles, key tree generation, and key tree update procedures of Topological Key Hierarchy. TKH operates without the multicast routing and minimizes the network usages by using the topology-mapped key tree structure.

4.1 Design Principles

In the key tree-based schemes, the nodes sharing the same KEK mostly receive the same rekeying messages. In order to assign a KEK for a group of topologically adjacent nodes, we use two kinds of tree topology information: *Subtree* and *Sibling* information.

4.1.1 Subtree-based Key Tree Separation (Tree Key)

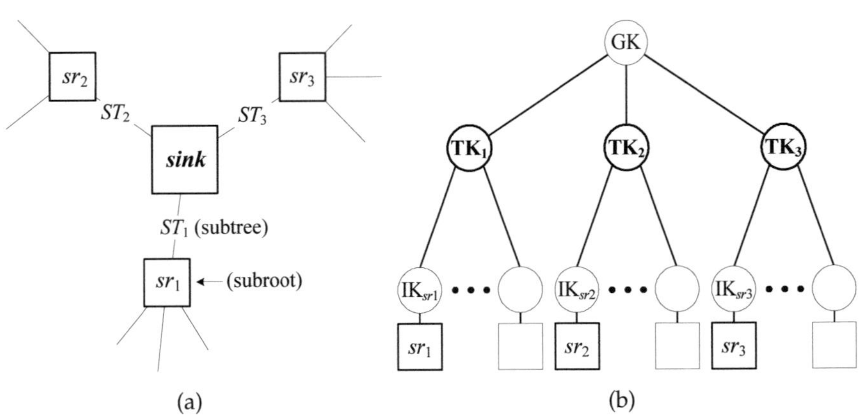

Fig. 4. (a) A sensor network topology and (b) the corresponding TK assignment.

First, we make the nodes in the same subtree share the same KEK called *Tree Key* (TK). The subtree is a tree with nodes below each subroot node, where subroot nodes are direct neighbors of a sink. The sample sensor network topology and its tree key assignment is depicted in Figure 4. From the three subtree branches, three tree keys (TK_1, TK_2, TK_3) are mapped to nodes in each subtree. From this key tree separation, rekeying messages for each subtree will be different from those of other subtrees. It means that TKH separates rekeying messages and delivers each subset only to the corresponding subtree. Nodes in each subtree are required to receive and forward rekeying messages only destined to nodes in their subtree.

4.1.2 Wireless Multicast Advantage Utilization (Sibling Key)

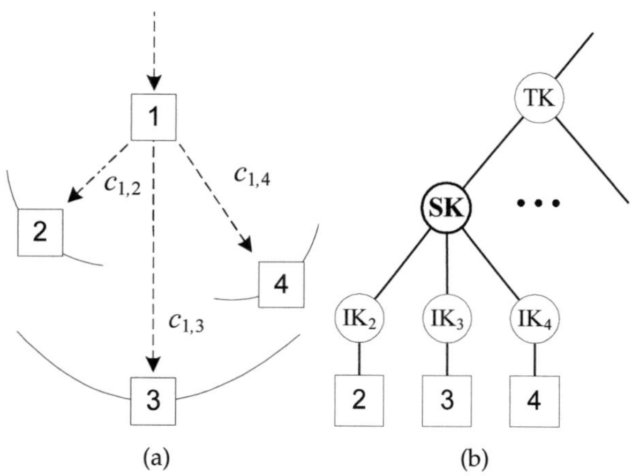

Fig. 5. a) A sensor network topology and (b) the corresponding SK assignment.

Second, we make the nodes sharing the same parent node in a tree topology (*sibling nodes*) to share the same KEK called *Sibling Key* (SK). For a node in a tree, a parent node is a neighbor node that delivers messages from the root sink node. In a wireless medium, since a message transmission can be heard by multiple neighbors, sibling nodes can efficiently receive a message by a single transmission from their parent. For example in Figure 5.(a) where node 1 has three one-hop neighbors $\{2,3,4\}$ in a wireless network, the costs of multicasting a single message to them is $C_{\text{multicast}} = \max(c_{1,2}, c_{1,3}, c_{1,4})$ where $c_{i,j}$ is a unicast cost from node i to j. Therefore, the one-hop multicast in a wireless medium can save energy from the broadcast nature of a wireless medium.
However, the important necessary condition for this wireless multicast advantage is that the message destined to neighbors should be the same. In other words, even if we have n one-hop neighbors which can be heard simultaneously, if the messages destined to them are different from each other, we have no choice but to unicast the messages one-by-one to each recipient. For rekeying messages generated from a key tree, we can make the same message to be destined to specific nodes by locating them under the same KEK. Therefore, we make children nodes of a parent node to share a SK to utilize the wireless multicast advantage.

4.2 Key Tree Generation

Based on the previous design principles, constructing a TKH key tree is composed of three steps: 1) *Routing Tree Construction*, 2) *Routing Tree Learning*, and 3) *Key Tree Generation*. How-

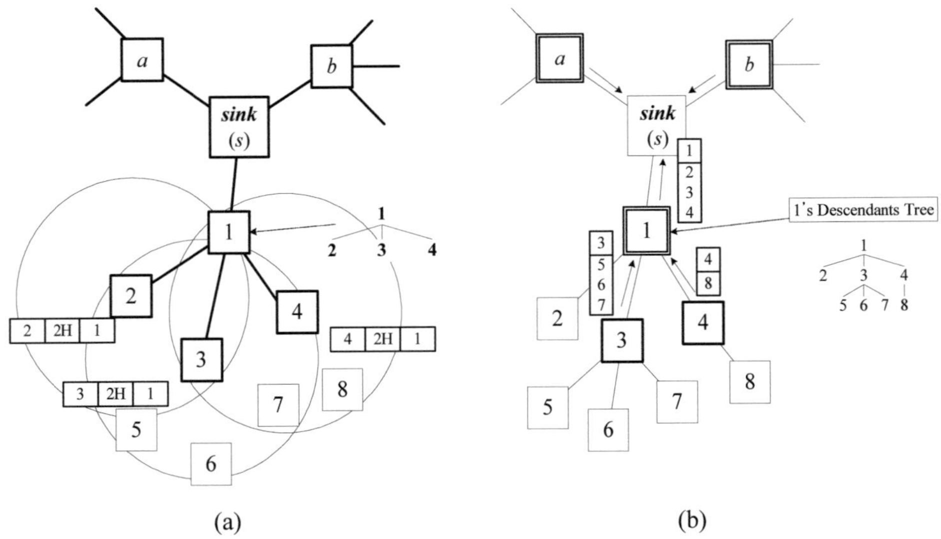

Fig. 6. (a) Routing Tree Construction and (b) Routing Tree Learning procedures.

ever, if a sensor network is already employing the tree-based routing and a central sink knows the topology information, TKH does not require the first two steps. For example, in a ZigBee-based WSN utilizing the tree-based hierarchical routing (ZigBee Alliance, 2006), the central sink can immediately generate the topology-based key tree by using the current topology information. If a WSN does not operate a tree-based routing, TKH needs to setup a sink-based routing tree to generate a topology-mapped key tree. Also the constructed routing tree will be used to deliver rekeying messages afterwards.

4.2.1 Routing Tree Construction

Constructing an efficient multicast source tree has been an active research area both in wired (Diot et al., 1997) and wireless (Wieselthier et al., 2002) networks. Here we introduce a simple routing tree construction method while TKH can generate a key tree from any routing tree construction method. After sensor node deployment, a sink broadcasts *Cost Advertisement (CA)* message to make sensor nodes to setup paths to the sink node. Each CA message contains three information: 1) node ID, 2) hop count to the sink, and 3) parent node ID. For example in Figure 6.(a), the node 3's CA message is '$[3|2H|1]$' since node '3' is '2 Hops' away from the sink through the parent node '1'. After hearing CA messages, a node chooses its parent node which has the minimum hop count to the sink (if multiple CA messages have the same hop count value, a node can choose the CA message received with the highest SNR). After selecting a parent node, each node also broadcasts its own CA message to neighbors. By overhearing CA messages, a parent node can learn the association of its children nodes with itself. In Figure 6.(a), by overhearing CA messages of nodes $\{2,3,4\}$, node 1 learns that it is associated with three children nodes. This routing tree construction procedure continues until it reaches all nodes.

4.2.2 Routing Tree Learning

After construction of a tree topology, every parent node reports *Parent-Child Relationship* (PCR) message to the sink. Each PCR message contains two information: 1) parent node ID and 2)

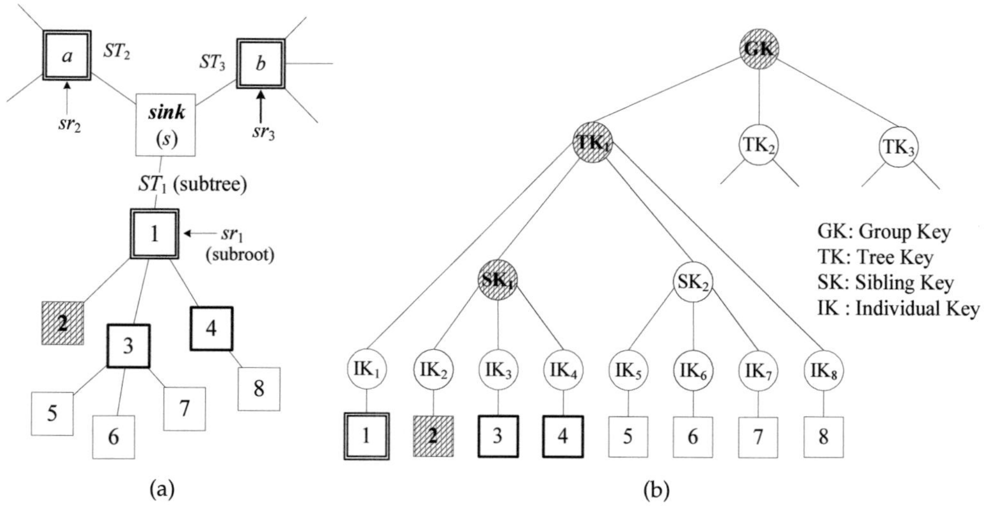

Fig. 7. (a) A sensor network tree topology example and (b) the corresponding TKH key tree structure. We depict the keys that need to be updated as shaded circles when node 2 is revoked.

children node IDs. For example in Figure 6.(b), node 1's PCR message is $[1|2,3,4]$ since it has three children nodes. After collecting all PCR messages, the sink can learn the whole network topology like Figure 6.(b). Also, during the PCR message forwardings, each parent node can learn and save its descendant node IDs in *Descendants Tree*. For example, by overhearing PCR messages from node 3 and 4, node 1 can build its *Descendants Tree* like in Figure 6.(b). By maintaining this tree, each parent can only forward messages destined to its descendants which prevents redundant message forwarding. Therefore, the routing overhead of TKH is only to maintain *Descendants Tree* in each parent node.

4.2.3 Key Tree Generation

Based on the topology information obtained from the previous tree learning procedure, now the sink can build a topology-based key tree. Before describing the key tree generation procedure, we first define several parameters (we show an example of each parameter by using the sample topology of Figure 7.(a)): We describe the key tree generation algorithm of TKH in Figure 8. As an example, Figure 7.(b) depicts the corresponding key tree structure generated from the topology of Figure 7.(a). In addition to GK and IK, Tree Key (TK) is shared by nodes in the same subtree (ST) and Sibling Key (SK) is shared by nodes in the same sibling set (ss). TKH has an advantage that the depth of the key tree is bounded to '4' independent of the network size. Therefore, each sensor is only required to save maximum four keys which are beneficial for storage-limited sensor nodes. In contrast, the logical key tree-based schemes should increase the depth of the key tree according to the network size in order to maintain the optimal tree degree (LKH and OKD achieve the best performance with the tree degree of 4 and 2 respectively (Li et al., 2001) (Lin et al., 2005)). Therefore, they should increase the number of keys in each sensor node as network grows.

parameter	definition
$\mathbb{T}$	a tree topology with a sink at its root and sensors at vertices
N	the total number of sensor nodes in $\mathbb{T}$
l	a number of revoked sensor nodes during a rekeying interval
sr_i	i-th subroot node (e.g. $sr_1 = 1, sr_2 = a, sr_3 = b$ in Figure 7.(a))
ST_i	i-th subtree with sr_i as the subroot
N_i	a set of all nodes in ST_i (e.g. $N_1 = \{1,2,3,4,5,6,7,8\}$)
$ss_{i,j}$	j-th sibling set in ST_i (nodes connected to the same parent) a single child consists a *single-node sibling set* without SK assignment; (e.g. $ss_{1,1} = \{1\}, ss_{1,2} = \{2,3,4\}, ss_{1,3} = \{5,6,7\}, ss_{1,4} = \{8\}$)
rn_i	a set of revoked nodes in ST_i (e.g. $rn_1 = \{2\}$)
rns_i	a set of revoked node's sibling nodes in ST_i (e.g. $rns_1 = \{3,4\}$)
RST	a set of subtrees which have revoked nodes in its vertices (e.g. $RST = \{ST_1\}$)
e_{tx}	energy dissipated during 1-bit transmission by a sensor node
e_{rx}	energy dissipated during 1-bit reception by a sensor node
$cu_{i,j}$	wireless unicast cost delivering 1-bit from node i to j ($cu_{i,j} = e_{tx} + e_{rx}$)
$cm_{i,\{1,\cdots,n\}}$	wireless multicast cost delivering 1-bit from node i to its n neighbors, ($cm_{i,\{1,\cdots,n\}} = e_{tx} + n \cdot e_{rx}$)

Table 1. Parameters for TKH algorithm explanation.

4.3 Key Tree Update

When a sensor node is newly deployed or revoked, a routing tree and the corresponding key tree should also be updated. One may think that the sink does not need to update the group key when a sensor node dies due to energy exhaustion. However, it is secure to update the group key also in this scenario since it is hard to verify by the remote sink whether the non-responding sensor node is pretending to be energy-less due to compromise attack. Therefore, we assume that the revocation of a sensor node take places when it is compromised or it runs out of energy.

Key tree update is composed of three steps: 1) *Routing Tree Repair*, 2) *Routing Tree Re-learning*, and 3) *Key Tree Update*. However, if a sensor network is already employing a tree-based routing or if node join or revocation events do not affect the topology of the remaining nodes, TKH does not require the first two steps.

4.3.1 Routing Tree Repair

When a node joins or leaves a network, a routing tree of the remaining node can be modified according to the node's topological position.

– *Node Join:* A newly deployed sensor node firstly broadcasts *join request* to neighbors. Then each neighbor reply CA messages containing its hop count to the sink. After selecting the parent node, the new node sends its CA message containing the parent ID. Then the selected parent reports a new PCR message to the sink which then locates the new node to the key tree according to its topological position. A joining node can either 1) *create a new single-node sibling set* or 2) *join the existing sibling set*. In both cases, the existing nodes can change the corresponding GK, TK, and SK by using the pre-shared one-way function same as the node

Input: a tree topology $\mathbb{T}$, all nodes' individual keys (IKs)
Output: a key tree

1) generate a group key (GK)
2) **for** (each ST_i) **do**
    if $|N_i| = 1$ **then**
      attach sr_i's IK to GK
    else ($|N_i| \geq 2$)
      generate a new tree key TK_i and attach it to GK
      for each $ss_{i,j}$ in ST_i **do**
        if $\left|ss_{i,j}\right| = 1$ **then**
          attach IK of the node in $ss_{i,j}$ to TK_i
        else $\left(\left|ss_{i,j}\right| \geq 2\right)$
          generate a new sibling key (SK) and attach it to TK_i
          attach all IKs of nodes in $ss_{i,j}$ to SK
        end if
      end for
    end if
  end for
3) return a key tree

Fig. 8. Key Tree Generation Algorithm of TKH.

join procedure in Section 2.1.1. The new node receives the corresponding keys from the sink afterwards. Therefore, we do not consider the node join event since the topology change and the corresponding rekeying cost is negligible.

– *Node Revocation:* We further classify the node revocation event into 1) *leaf node revocation* and 2) *non-leaf node revocation*. The leaf node revocation does not affect the topology of the remaining nodes and the sink can send the rekeying messages based on the current key tree. For example in Figure 7.(a), revocation of the leaf node '2' does not affect the network topology, and rekeying messages can be generated from the current key tree of Figure 7.(b). However, the non-leaf node revocation can disconnect the network topology, and the sink should wait until the orphaned nodes of the revoked parent find new parent nodes. For the routing tree repair, each orphaned node performs the same procedure as the node join case.

4.3.2 Routing Tree Re-learning

If the sink revokes a non-leaf parent node, it waits until it receives new PCR messages from new parents of the orphaned nodes. After receiving PCR messages, the sink modifies the current key hierarchy based on the modified network topology. For example in Figure 9.(a), after revocation of node 3, the sink waits until it receives new PCR messages containing the orphaned nodes $\{5,6,7\}$. Then node 2 and 3, new parents of $\{5,6,7\}$ report their new PCR messages to the sink. Also by overhearing these new PCR messages, other nodes along the path to the sink modifies their *Descendants Tree*. Finally, the sink can send the rekeying messages based on the modified key tree structure.

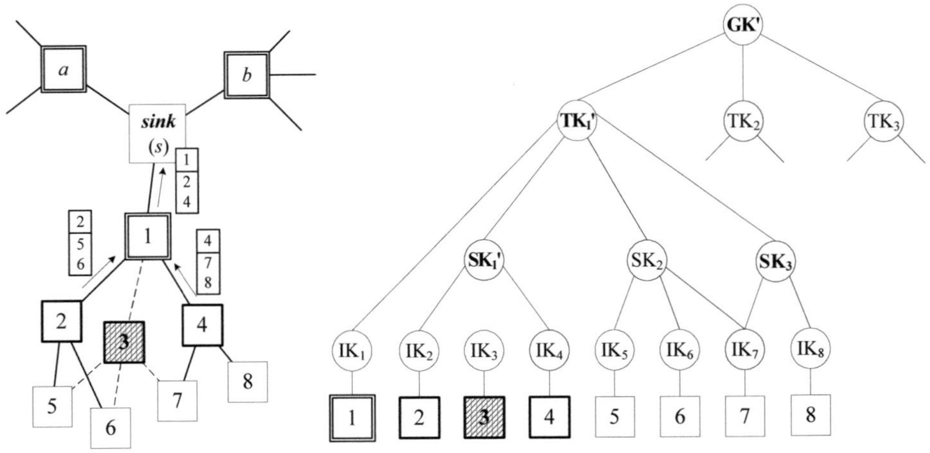

(a) (b)

Fig. 9. After non-leaf node 3 in Figure 7 is revoked, a) the repaired routing tree with the re-learning procedure and b) the modified key tree structure.

4.3.3 Key Tree Update

Based on the modified key tree structure, the sink send the corresponding rekeying messages to each subset of nodes. By using the example of Figure 9, we examine the rekeying message delivery procedures in detail. When the non-leaf node 3 in ST_1 is revoked, rekeying messages (m) and the corresponding communication cost (C) to deliver m from the sink (s) to its recipients are

$$\left\{\begin{array}{l} m_{s\to\{1\}} : \{\mathrm{GK}', \mathrm{TK}'_1\}_{\mathrm{IK}_1} \\ m_{s\to\{2,4\}} : \{\mathrm{GK}', \mathrm{TK}'_1\}_{\mathrm{SK}'_1} \\ m_{s\to\{5,6\}} : \{\mathrm{GK}', \mathrm{TK}'_1\}_{\mathrm{SK}_2} \\ m_{s\to\{7,8\}} : \{\mathrm{GK}', \mathrm{TK}'_1\}_{\mathrm{SK}_3} \\ m_{s\to 2} : \{\mathrm{SK}'_1\}_{\mathrm{IK}_2} \\ m_{s\to 4} : \{\mathrm{SK}'_1\}_{\mathrm{IK}_4} \\ m_{s\to 7} : \{\mathrm{SK}_3\}_{\mathrm{IK}_7} \end{array}\right. \qquad \left\{\begin{array}{l} C_{s\to\{1\}} : e_{tx}+e_{rx} \\ C_{s\to\{2,4\}} : 2e_{tx}+3e_{rx} \\ C_{s\to\{5,6\}} : 3e_{tx}+4e_{rx} \\ C_{s\to\{7,8\}} : 3e_{tx}+4e_{rx} \\ C_{s\to 2} : 2e_{tx}+2e_{rx} \\ C_{s\to 4} : 2e_{tx}+2e_{rx} \\ C_{s\to 7} : 3e_{tx}+3e_{rx}. \end{array}\right.$$

Rekeying messages for ST_2 and ST_3 are $\{\mathrm{GK}'\}_{\mathrm{TK}_2}$ and $\{\mathrm{GK}'\}_{\mathrm{TK}_3}$ respectively. Upon receiving each rekeying message, a node can route it to one of its children nodes based on its *Descendants Tree*. Nodes in the same sibling set $(\{2,4\}, \{5,6\}, \{7,8\})$ will receive the same rekeying messages by using the wireless multicast advantage from their parents.

Comparing Figure 7.(b) and Figure 9.(b), we observe that the sibling sets sharing SK_2 and SK_3 are slightly changed. However, TKH does not update SK_2 and SK_3 since none of the sensors sharing them are revoked. By maintaining the link from node 7 to SK_2 in the key tree, the sink can update both SK_2 and SK_3 later when node 7 is revoked. Finally, the total rekeying cost (TRC) of ST_1 is calculated as

$$\begin{aligned} \mathrm{TRC}_{ST_1} &= 2|m|\times\left(C_{s\to\{1\}}+C_{s\to\{2,4\}}+C_{s\to\{5,6\}}+C_{s\to\{7,8\}}\right) \\ &+ |m|\times(C_{s\to 2}+C_{s\to 4}+C_{s\to 7}+C_{s\to 8}) = |m|\,(25e_{tx}+31e_{rx}). \end{aligned}$$

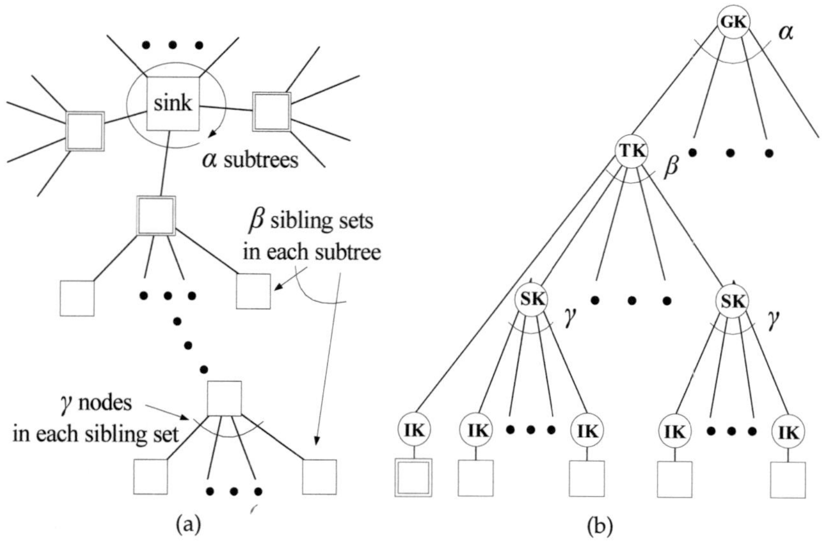

Fig. 10. (a) '$\alpha\beta\gamma$-tree' and (b) the corresponding TKH key tree structure.

where $|m|$ is the size of a unit rekeying message $\{K_A\}_{K_B}$ ($2|m|$ for $\{K_A, K_B\}_{K_C}$). It means that we need 25 transmissions and 31 receptions of a unit rekeying messages to update ST_1 when node 3 is revoked.

5. Analysis of the Total Rekeying Cost

In this section, we analyze and compare the total rekeying costs of LKH, OKD, and TKH in multi-hop WSN environments. For the analysis, we need to derive the average number of rekeying messages and the communication costs. The former is derived in Section 4.3 by employing *the bins-and-balls problem*. To calculate the latter, we model a typical WSN topology as '$\alpha\beta\gamma$-tree' in Section 4.1. Both results are used to derive the total rekeying costs in Section 4.4 while the communication costs of the routing tree maintenance are calculated in Section 4.2.

5.1 '$\alpha\beta\gamma$-tree' Topology Model

For the analysis of the communication cost, we model a sensor network topology by using '$\alpha\beta\gamma$-tree' model. In the $\alpha\beta\gamma$-tree, there are 'α' subtree branches from the sink, and each subtree has 'β' sibling sets, and each sibling set has 'γ' sibling nodes. The resulting topology and the corresponding TKH key tree structure is depicted in Figure 10.(a) and (b) respectively. The total number of sensor nodes excluding a sink is $N = \alpha(\beta\gamma + 1)$ and each subtree has $(\beta\gamma + 1)$ nodes. Among N sensor nodes, $(\alpha\beta)$ nodes are non-leaf parents and the rest $(\alpha(\beta\gamma + 1) - \alpha\beta)$ nodes are leaf children nodes. During the routing tree repair in $\alpha\beta\gamma$-tree, we assume that a revoked non-leaf parent node is replaced by one of its siblings, and a revoked subroot node is replaced by one of its children.

5.2 Cost of Routing Tree Maintenance

When there are N nodes in a network, each node can be identified by using $\lceil \log_2 N \rceil$ bits, and the hop count value ranging from 0 to β can be identified by $\lceil \log_2 \beta \rceil$ bits. Then the size of the CA message ($|m_{CA}|$) and the PCR message ($|m_{PCR}|$) are respectively

$$|m_{CA}| = 2\lceil \log_2 N \rceil + \lceil \log_2 \beta \rceil, \quad |m_{PCR}| = (\gamma+1)\lceil \log_2 N \rceil$$

where $\lceil x \rceil$ denotes the smallest integer equal or greater than x ($\lfloor x \rfloor$ denotes the largest integer equal or smaller than x).

5.2.1 Routing Tree Construction & Learning

The communication cost of the 'Routing Tree Construction & Learning' (C_{CL}) defined in Section 3.2.1 and 3.2.2 is derived as

$$C_{CL} = |m_{CA}| \Big\{ (N+1) \cdot e_{tx} + N\gamma \cdot e_{rx} \Big\} + |m_{PCR}| \Big\{ \alpha\beta \cdot \mathrm{avg}(1,\beta)(e_{tx}+e_{rx}) \Big\} \tag{5}$$

where $\mathrm{avg}(1,n) = \frac{1+2+\cdots+n}{n} = \frac{(n+1)}{2} \left(\mathrm{sum}(1,n) = 1+2+\cdots+n = \frac{n(n+1)}{2} \right)$. We assume that every sensor plus the sink broadcast one CA message and each sensor receives γ CA messages on average. PCR messages are generated by all $\alpha\beta$ parent nodes and they require $\mathrm{avg}(1,\beta)$ hops to reach the sink.

5.2.2 Routing Tree Repair & Re-Learning

Also the communication cost of the 'Routing Tree Repair & Re-learning' (C_{RR}) defined in Section 3.3.1 and 3.3.2 is derived as follows

$$C_{RR} = |m_{PCR}| \sum_{i=0}^{\min(l,\alpha\beta)} \frac{C_i^{\alpha\beta} C_{l-i}^{\alpha(\beta\gamma+1)-\alpha\beta}}{C_l^{\alpha(\beta\gamma+1)}} \times i \times \mathrm{avg}(1,\beta)(e_{tx}+e_{rx}) \tag{6}$$

where C_b^a is the binomial coefficient. Among the total $\alpha(\beta\gamma+1)$ nodes, only revocations of $\alpha\beta$ parent nodes incur new PCR message reports. The corresponding m_{PCR} should be delivered to the sink along $\mathrm{avg}(1,\beta)$ hops.

5.3 Average Number of Rekeying Messages

5.3.1 Basic Functions

When l nodes are revoked, $\overline{\mathbb{B}}(l,v,w)$ calculates *the average number of intermediate KEKs that need to be updated.* v is the total number of intermediate KEKs at a certain key tree level, where each KEK on that level is shared by w nodes. By analogy, $\overline{\mathbb{B}}(l,v,w)$ is equivalent to the average number of *non-full* bins when l balls are randomly picked out from v identical bins each filled with w balls. The picked-out balls represent revoked nodes and the non-full bins represent KEKs need to be updated. The number of non-full bin ($\overline{n}(l,v,w)$) is in the range of $\lceil l/w \rceil \leq \overline{n}(l,v,w) \leq \min(l,v)$. Then, $\overline{\mathbb{B}}(l,v,w)$ is represented as

$$\overline{\mathbb{B}}(l,v,w) \triangleq E[\overline{n}(l,v,w)] = \sum_{i=\lceil l/w \rceil}^{\min(l,v)} \Pr\{\overline{n}(l,v,w)=i\} \times i.$$

In the above equation,

$$\Pr\{\overline{n}(l,v,w)=i\} = C_i^v \cdot N(l,i,w) / C_l^{vw}$$

where $N(l,i,w)$ is the number of ways that there is no full bins when l balls are picked out from i bins containing w balls each. $N(l,i,w)$ is calculated by using the inclusion-exclusion principle (Tucker, 1995, Ch. 3) which results

$$\overline{\mathbb{B}}(l,v,w) = \sum_{i=\lceil l/w\rceil}^{\min(l,v)} \frac{C_i^v \cdot \left(\sum_{j=0}^{i-\lceil l/w\rceil}(-1)^j C_j^i C_l^{w(i-j)}\right)}{C_l^{vw}} \times i. \tag{7}$$

Another function $\underline{\mathbb{B}}(l,v,w)$ calculates *the average number of intermediate KEKs that do not need to be updated* since all the nodes shared the same KEK are revoked. $\underline{\mathbb{B}}(l,v,w)$ is equivalent to the average number of *empty* bins when l balls are randomly picked out from v identical bins each filled with w balls, and calculated as

$$\underline{\mathbb{B}}(l,v,w) = \sum_{i=\max(l-vw+v,0)}^{\lfloor l/w\rfloor} \frac{C_i^v \cdot \left(\sum_{j=0}^{\lfloor l/w\rfloor-i}(-1)^j C_j^{v-i} C_{l-w(i+j)}^{w(v-i-j)}\right)}{C_l^{vw}} \times i. \tag{8}$$

Finally, $\mathbb{B}(l,v,w)$ defined as the difference between (7) and (8) is the actual average number of intermediate KEKs that need to be updated on a certain key tree level when l nodes are revoked

$$\mathbb{B}(l,v,w) = \overline{\mathbb{B}}(l,v,w) - \underline{\mathbb{B}}(l,v,w). \tag{9}$$

While the analysis in this subsection is motivated by the previous results (Sun et al., 2004, Appendix A), we improve them in that 1) we provide non-recursive, closed-form solutions for the bins-and-balls problem and 2) we also analyze the number of KEKs that do not need to be updated by introducing $\underline{\mathbb{B}}(l,v,w)$.

5.3.2 Average Number of Rekeying Messages

We denote a key tree of N nodes as $T(d_1,\cdots,d_h)$ where d_i is the degree of a vertex at the i-th level from the top and h is the height of the tree ($\because d_1 \times\cdots\times d_h = N$). For example, the key tree in Figure 3 is denoted as $T(2,2,3)$. For the simplicity in equations, we assume $d_0 = d_{h+1} = 1$. When l nodes are revoked, the average number of total rekeying messages of LKH and OKD generated by group-oriented rekeying are respectively

$$|M_{\mathrm{LKH}(g)}| = \left\{d_1 + \sum_{i=1}^{h-1} \overline{\mathbb{B}}\left(l, \prod_{j=1}^{i} d_j, \prod_{k=i+1}^{h} d_k\right)\cdot d_{i+1} - l\right\} - \left\{\sum_{i=1}^{h-1} \underline{\mathbb{B}}\left(l, \prod_{j=1}^{i} d_j, \prod_{k=i+1}^{h} d_k\right)\right\} \tag{10}$$

$$|M_{\mathrm{OKD}(g)}| = \left\{(d_1-1) + \sum_{i=1}^{h-1} \overline{\mathbb{B}}\left(l, \prod_{j=1}^{i} d_j, \prod_{k=i+1}^{h} d_k\right)\cdot (d_i-1) - l\right\}. \tag{11}$$

With parameters $N=12, d_1=2, d_2=2, d_3=3, h=3, l=1$ of Figure 3, the number of rekeying messages are calculated as $|M_{\mathrm{LKH}(g)}|=6$ and $|M_{\mathrm{LKH}(g)}|=3$ by using the above (10) and (11), and they are consistent with (1) and (3) respectively.

For the user-oriented rekeying, we assume that each rekeying message is delivered to its recipient by unicast without multicast routing support in WSNs. For example in (13), $m_{\mathrm{KDC}\to\{7\sim9\}}$: $\{\mathrm{GK}'\}_{\mathrm{K}'_{1-2}}$ is calculated as 3 rekeying messages unicast to $(7,8,9)$ independently. When l nodes are revoked, the average number of total rekeying messages of LKH and OKD generated by user-oriented rekeying are respectively

$$|M_{\mathrm{LKH}(u)}| = \sum_{i=1}^{h}\left\{i \times \left(\overline{\mathbb{B}}\left(l, \prod_{j=0}^{i-1} d_j, \prod_{k=i}^{h} d_k\right)\cdot d_i - \overline{\mathbb{B}}\left(l, \prod_{j=1}^{i} d_j, \prod_{k=i+1}^{h+1} d_k\right)\right)\cdot\left(\prod_{k=i+1}^{h+1} d_k\right)\right\} \tag{12}$$

i	m_i	$\lvert m_i\rvert$	tx_i	rx_i (dest+relay)
1	$\{GK\}_{TK}$	$\alpha-\underline{\mathbb{B}}(l,\alpha,\beta\gamma+1)$	$\beta+1$	$(N-l)+0$
2	$\{TK\}_{SK}$	$\mathbb{B}(l,\alpha,\beta\gamma+1)(\beta+1)$	$\mathrm{avg}(1,\beta+1)$	$N_R(l)+\mathbb{B}(l,\alpha,\beta\gamma+1)\cdot\mathrm{sum}(1,\beta)$
3	$\{SK\}_{IK}$	$N_r(l)$	$\mathrm{avg}(2,\beta+1)$	$N_r(l)+N_r(l)\cdot\mathrm{avg}(1,\beta)$

Table 2. For each rekeying message (m_i) in TKH, the number of rekeying messages $(|m_i|)$, the number of transmissions per message (tx_i), and the total number of receptions at destinations and relay nodes (rx_i) are derived.

$$|M_{\mathrm{OKD}(u)}|=\sum_{i=0}^{h-1}\left\{i\times\left(\overline{\mathbb{B}}\left(l,\prod_{j=0}^{i}d_j,\prod_{k=i+1}^{h}d_k\right)\cdot d_{i+1}-\overline{\mathbb{B}}\left(l,\prod_{j=1}^{i+1}d_j,\prod_{k=i+2}^{h+1}d_k\right)\right)+\right. \quad (13)$$
$$\left.\left(\overline{\mathbb{B}}\left(l,\prod_{j=0}^{i}d_j,\prod_{k=i+1}^{h}d_k\right)\cdot(d_{i+1}-1)-\overline{\mathbb{B}}\left(l,\prod_{j=1}^{i+1}d_j,\prod_{k=i+2}^{h+1}d_k\right)+\underline{\mathbb{B}}\left(\left\langle l/\prod_{j=i+2}^{h+1}d_j\right\rangle,\prod_{j=0}^{i}d_j,d_{i+1}\right)\right)^{+}\right\}\left(\prod_{j=i+2}^{h+1}d_j\right)$$

In (13), $(x)^+$ is defined as $\{x \text{ if } x\geq 0, 0 \text{ if } x<0\}$ and $<x>=\lfloor x+0.5\rfloor$. With the parameters of Figure 3, the number of rekeying messages are calculated as $|M_{\mathrm{LKH}(u)}|=18$ and $|M_{\mathrm{OKD}(u)}|=8$ by using the above (12) and (13), and they are consistent with (2) and (4) respectively.
TKH has three kinds of rekeying messages (m_i): $\{GK\}_{TK}$, $\{TK\}_{SK}$, and $\{SK\}_{IK}$. From the key tree structure of Figure 10.(b) generated from the $\alpha\beta\gamma$-tree topology, the average number of rekeying messages are calculated in $|m_i|$ column of Table 2.

5.4 Total Rekeying Costs

By using both the previous results on the average number of rekeying messages and the $\alpha\beta\gamma$-tree model for calculation of the communication costs, we derive the total rekeying costs of LKH, OKD, and TKH as follows

$$\mathrm{TRC}_{\mathrm{LKH}(g)} = |M_{\mathrm{LKH}(g)}|\times\{\alpha(\beta+1)\cdot e_{tx}+(N-l)\cdot e_{rx}\}|m| \quad (14)$$
$$\mathrm{TRC}_{\mathrm{OKD}(g)} = |M_{\mathrm{OKD}(g)}|\times\{\alpha(\beta+1)\cdot e_{tx}+(N-l)\cdot e_{rx}\}|m| \quad (15)$$
$$\mathrm{TRC}_{\mathrm{LKH}(u)} = |M_{\mathrm{LKH}(u)}|\times\left\{\mathrm{avg}(1,\beta)\cdot(e_{tx}+e_{rx})\right\}|m|+C_{\mathrm{CL}}+C_{\mathrm{RR}} \quad (16)$$
$$\mathrm{TRC}_{\mathrm{OKD}(u)} = |M_{\mathrm{OKD}(u)}|\times\left\{\mathrm{avg}(1,\beta)\cdot(e_{tx}+e_{rx})\right\}|m|+C_{\mathrm{CL}}+C_{\mathrm{RR}} \quad (17)$$
$$\mathrm{TRC}_{\mathrm{TKH}} = \sum_{\forall m_i}\left\{(|m_i|\times tx_i)\cdot e_{tx}+(rx_i)\cdot e_{rx}\right\}|m|+C_{\mathrm{CL}}+C_{\mathrm{RR}}. \quad (18)$$

In group-oriented rekeying ((14) and (15)), all rekeying messages are broadcast to all nodes requiring $\alpha(\beta+1)$ transmissions and $(N-l)$ receptions within a network. In user-oriented rekeying ((16) and (17)), each rekeying message is independently unicast to each node requiring $\mathrm{avg}(1,\beta)$ transmissions and receptions on average. While the group-oriented rekeying is independent of the network topology, the user-oriented rekeying and TKH requires topology information to deliver rekeying messages (reflected by $C_{CL}+C_{RR}$ in (16), (17), and (18)). Therefore, LKH(u), OKD(u), and TKH requires additional C_{CL} and C_{RR} costs in the total rekeying cost. In TKH, for each rekeying message (m_i), we calculate the average number of rekeying messages $(|m_i|)$, the number of transmissions per message (tx_i), and the total number of receptions at destinations and relay nodes (rx_i) in Table 2. Here $N_R(l)$ and $N_r(l)$ are defined and derived as follows

$$N_R(l) = \{\text{avg. \# of nodes in revoked subtrees}\} \tag{19}$$
$$= \sum_{\forall ST_i \in RST} |N_i| = \mathbb{B}(l,\alpha,\beta\gamma{+}1)(\beta\gamma{+}1) - \Big(l - \underline{\mathbb{B}}(l,\alpha,\beta\gamma{+}1)(\beta\gamma{+}1)\Big)$$

$$N_r(l) = \{\text{avg. \# of revoked nodes' sibling nodes}\} \tag{20}$$
$$= \sum_{\forall ST_i \in RST} |rns_i| = \sum_{k=\max(0,l-\alpha\beta\gamma)}^{\min(l,\alpha)} \frac{C_k^{\alpha} C_{l-k}^{\alpha\beta\gamma}}{C_l^{\alpha(\beta\gamma+1)}} \Big(\overline{\mathbb{B}}(l-k,\alpha\beta,\gamma)\gamma - (l-k)\Big).$$

5.5 Analysis Results

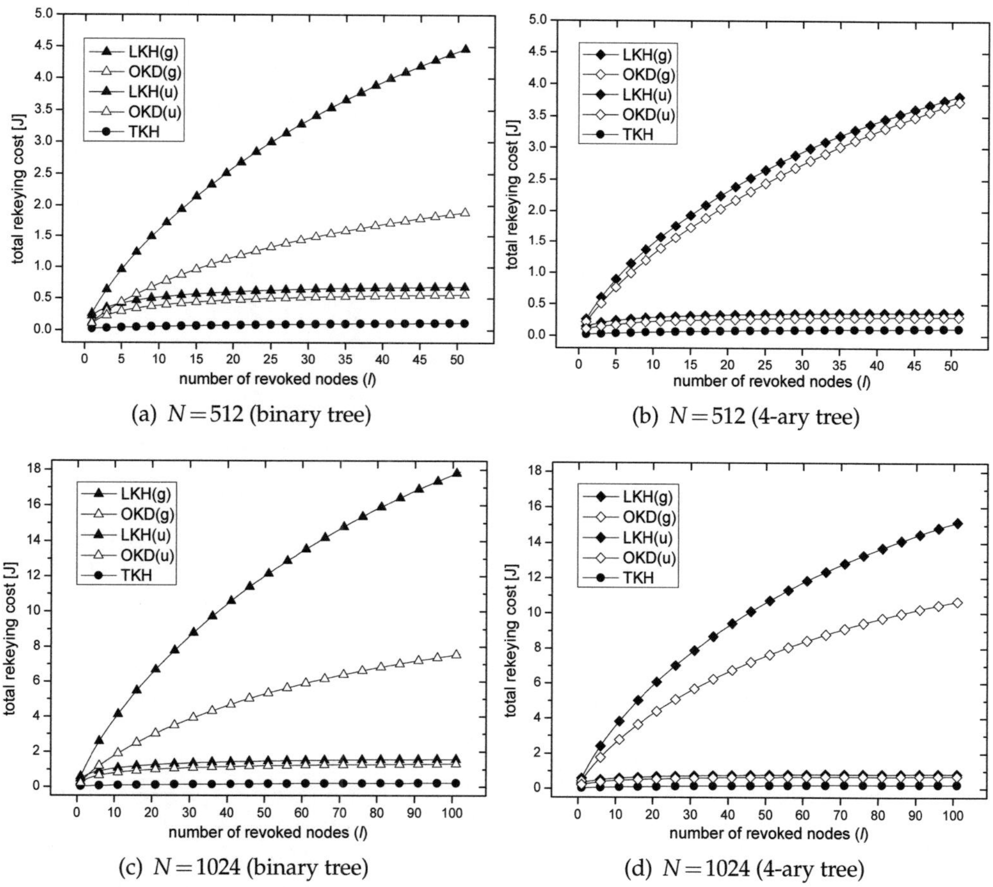

(a) $N=512$ (binary tree)

(b) $N=512$ (4-ary tree)

(c) $N=1024$ (binary tree)

(d) $N=1024$ (4-ary tree)

Fig. 11. Total rekeying costs of LKH, OKD, and TKH.

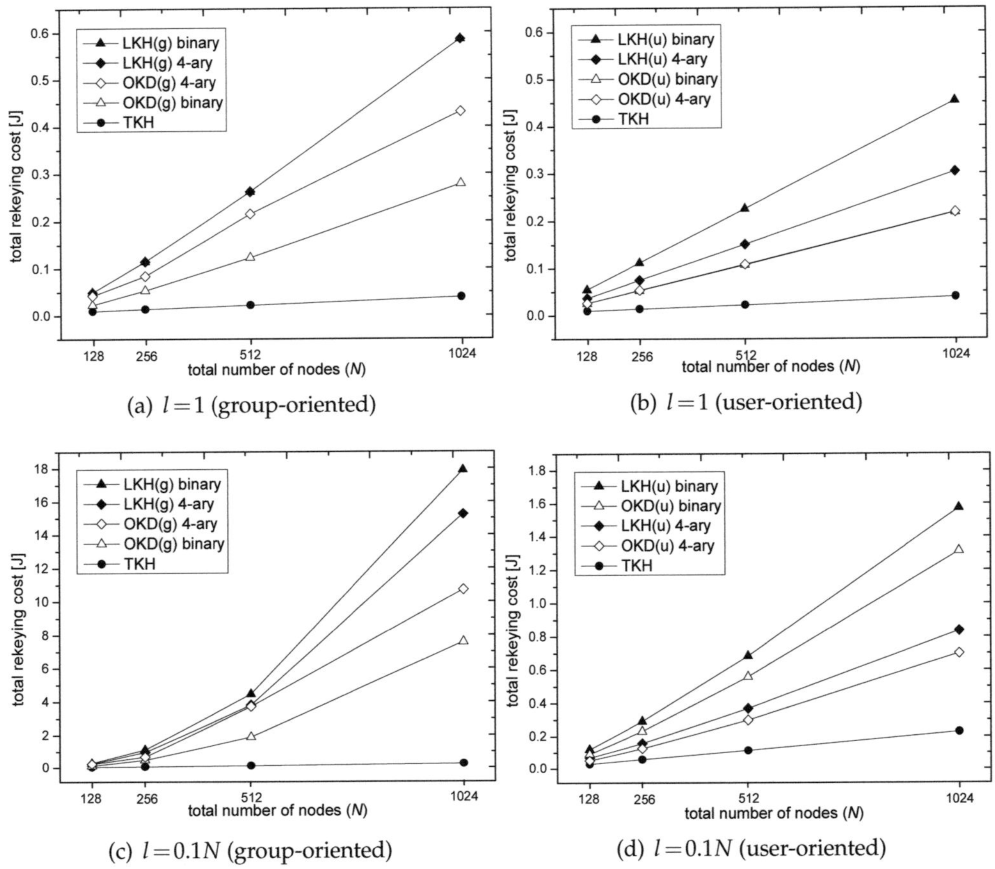

(a) $l=1$ (group-oriented) (b) $l=1$ (user-oriented)

(c) $l=0.1N$ (group-oriented) (d) $l=0.1N$ (user-oriented)

Fig. 12. Total rekeying costs of LKH, OKD, and TKH.

We plot the total rekeying costs (TRC) of LKH, OKD, and TKH in Figure 11 and 12. We vary the total number of nodes (N) as 128, 256, 512, and 1024 by varying (α, β, γ) tuples as (2,7,9), (4,7,9), (8,7,9), and (16,7,9). We consider two logical key trees (binary and 4-ary) for LKH and OKD, while key trees of TKH are directly determined by (α, β, γ) values of each N. The unit rekeying message size is set to $|m| = 128$ bits. The unit communication costs are set to $e_{tx} = 0.209[\mu\text{J}]$ and $e_{rx} = 0.226[\mu\text{J}]$ from the characteristics of the CC2420 transceiver used in the Xbow's MICA-Z and Telos B sensor nodes.

Figure 11 depicts the increasing TRC values according to the increasing number of revoked nodes (l) when $N = 512, 1024$. For various number of the total nodes (N), Figure 12.(a) and (b) depict TRC when one node is revoked ($l=1$), and Figure 12.(c) and (d) depict TRC when 10% of nodes are revoked ($l = 0.1N$). From combinations of three key tree schemes (*LKH, OKD, and TKH*), two rekeying strategies (*User-oriented and Group-oriented*), and two key tree structures (*binary and 4-ary*), we observe the following principles between them in terms of the total rekeying costs.

- TKH is superior to OKD and LKH in all cases.
- OKD is superior to LKH, given the same rekeying strategy and key tree structure.

- User-oriented rekeying is superior to group-oriented rekeying, given the same logical key tree scheme, rekeying strategy, and key tree structure.
- For LKH, 4-ary key tree is superior to binary key tree independent of the rekeying strategies.
- For OKD(g), binary key tree is superior to 4-ary key tree. For OKD(u), 4-ary key tree is superior to binary key tree.

By considering the topological information during the key tree construction, TKH always incurs the lowest rekeying cost compared to the previous logical key tree schemes. Between the logical schemes, OKD is superior to LKH by reducing rekeying messages due to its local key computations. Since rekeying messages are individually delivered to each node in user-oriented rekeying, it is more energy-efficient than group-oriented rekeying which combines-and-broadcasts all rekeying messages. Given the same number of the total nodes, nodes in a 4-ary key tree only stores the half number of keys compared to those in a binary key tree. The reduced number of keys for each node translates into the reduced number of rekeying messages for each node in user-oriented rekeying. Therefore, we observe that LKH(u) and OKD(u) achieve lower rekeying costs when they utilize 4-ary key tree. However, while the 4-ary key tree is also optimal in LKH(g), it is inferior to binary key tree in OKD(g). This is due to the fact that binary key tree is optimal for OKD's local key computations in terms of the number of the total rekeying messages. Our results are consistent with the results of (Li et al., 2001) (Lin et al., 2005) that tried to find the optimal key tree structure for LKH and OKD in terms of the total number of rekeying messages.

In Figure 11.(a),(b),(c),(d) respectively, TKH only requires 17.7%, 32.1%, 14.5%, 26.6% of TRC compared to *OKD(u) with 4-ary* on average, while 13.9%, 25.4%, 11.8%, 21.8% of TRC compared to *LKH(u) with 4-ary* on average. Compared to the best logical key tree scheme: *OKD(u) with 4-ary*, TKH only requires 37.2%, 25.9%, 20.5%, 17.8% of TRC in Figure 12.(b) and 57.2%, 45.5%, 38.1%, 32.6% of TRC in Figure 12.(d) when $N = 128, 256, 512, 1024$ respectively.

5.6 Effects of Wireless Channel Errors

During message delivery between nodes in wireless sensor networks, it is probable that a transmitted message is corrupted due to wireless channel errors. Then the sender should re-transmit the failed message and the receiver should retry to receive it which will consume additional communication costs at both sides. In LKH and OKD, group-oriented rekeying strategy uses multicast communications while user-oriented rekeying uses unicast communications to deliver rekeying messages. Our TKH utilizes the both communication methods according to rekeying message types: $\{GK\}_{TK}$ is delivered by multicast communications, while $\{TK\}_{SK}$ and $\{SK\}_{IK}$ are delivered by unicast communications. Therefore, message retransmissions incurred by wireless channel errors will have different effects on the total rekeying costs of the three schemes.

In unicast communications between a pair of wireless nodes, let p be the probability that a message is not received correctly at a receiver side (correctly received with $1-p$). If we assume the message length is L bits and bit error probability is p_b, p would be $p = 1-(1-p_b)^L$. Then the expected number of transmission attempts required to successfully deliver a message in wireless unicast ($\mathbb{E}(N_U)$) is

$$\mathbb{E}(N_U) = 1\times(1-p) + 2\times p(1-p) + 3\times p^2(1-p) + \cdots = \frac{1}{1-p} = \frac{1}{(1-p_b)^L}$$

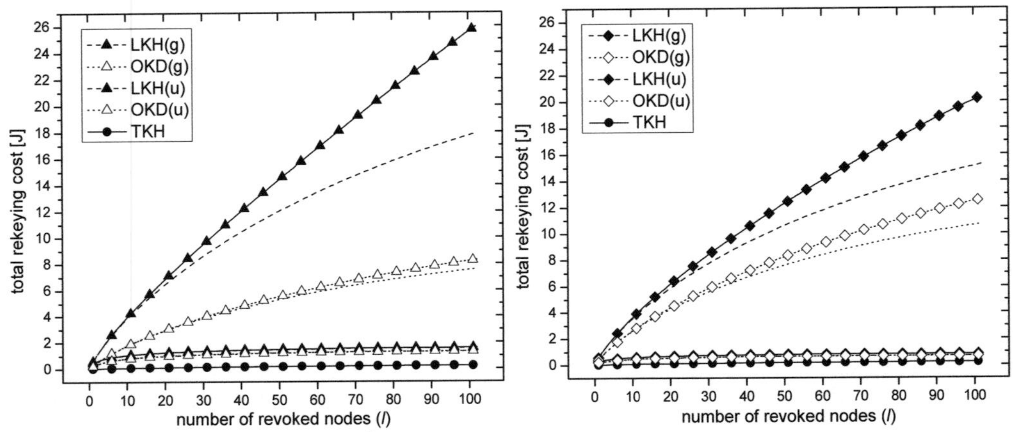

(a) $N = 1024$ (binary tree)

(b) $N = 1024$ (4-ary tree)

Fig. 13. Effects of wireless channel error probability ($p_b = 10^{-5}$) on TRC according to the increasing number of revoked nodes (l) when $N = 1024$.

However, in multicast communications between a group of wireless nodes, the probability of a successful message reception would increase since a receiver can overhear multiple copies of a message not only from a sender but also from its neighbors. Let us assume that each node in multicast communications receives n copies of a message on average. Then the probability that a multicast message is not received correctly at a receiver side is p^n. Similar to (21), the expected number of transmission attempts required to successfully deliver a message in wireless multicast ($\mathbb{E}(N_{\mathrm{M}})$) is

$$\mathbb{E}\left(N_{\mathrm{M}}\right) = \frac{1}{1-p^n} = \frac{1}{1-(1-(1-p_b)^L)^n} \tag{21}$$

If we consider increased communication costs due to wireless channel errors, the total rekeying costs of group-oriented and user-oriented rekeying will be increased by the rates of $\mathbb{E}(N_{\mathrm{M}})$ and $\mathbb{E}(N_{\mathrm{U}})$ respectively, while that of TKH is affected by both. By applying $\mathbb{E}(N_{\mathrm{U}})$ and $\mathbb{E}(N_{\mathrm{M}})$ into the previous total rekeying costs in Section 4.4, we obtain

$$\mathrm{TRC}'_{\mathrm{LKH}(g)} = \mathrm{TRC}_{\mathrm{LKH}(g)} \times \mathbb{E}(N_{\mathrm{M}})_{\mathrm{LKH}(g)} \tag{22}$$

$$\mathrm{TRC}'_{\mathrm{OKD}(g)} = \mathrm{TRC}_{\mathrm{OKD}(g)} \times \mathbb{E}(N_{\mathrm{M}})_{\mathrm{OKD}(g)} \tag{23}$$

$$\mathrm{TRC}'_{\mathrm{LKH}(u)} = \mathrm{TRC}_{\mathrm{LKH}(u)} \times \mathbb{E}(N_{\mathrm{U}})_{\mathrm{LKH}(u)} \tag{24}$$

$$\mathrm{TRC}'_{\mathrm{OKD}(u)} = \mathrm{TRC}_{\mathrm{OKD}(u)} \times \mathbb{E}(N_{\mathrm{U}})_{\mathrm{OKD}(u)} \tag{25}$$

$$\begin{aligned} \mathrm{TRC}'_{\mathrm{TKH}} &= \{(|m_1| \times tx_1) \cdot e_{tx} + (rx_1) \cdot e_{rx}\} \times \mathbb{E}(N_{\mathrm{M}})_{\mathrm{TKH}} \\ &+ \left\{ \sum_{i=2}^{3} \{(|m_i| \times tx_i) \cdot e_{tx} + (rx_i) \cdot e_{rx}\} + C_{\mathrm{CL}} + C_{\mathrm{RR}} \right\} \times \mathbb{E}(N_{\mathrm{U}})_{\mathrm{TKH}}. \end{aligned} \tag{26}$$

To calculate $\mathbb{E}(N_{\mathrm{M}})$ and $\mathbb{E}(N_{\mathrm{U}})$ in the above equations, we input message lengths (L) from (10)$\sim$(13) and $|m_i|$ equations in Table 2.

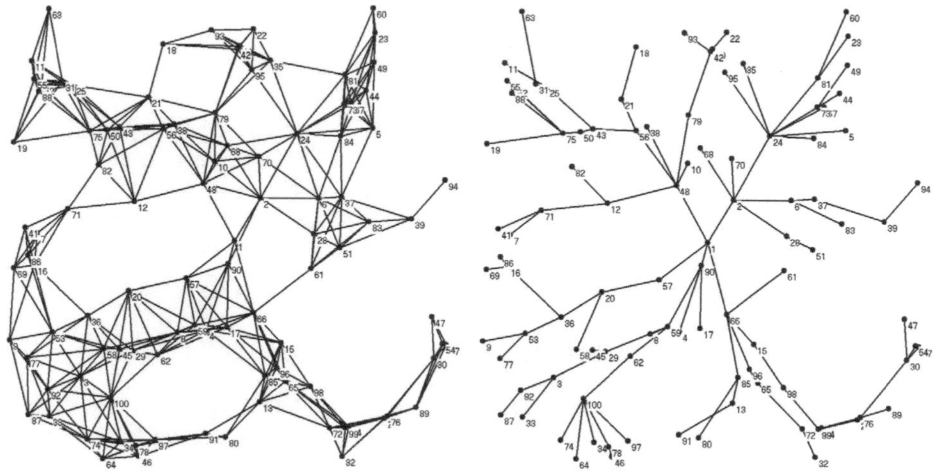

(a) (b)

Fig. 14. (a) A sample sensor network connectivity graph of 100 nodes in an 1×1 unit square area with r=0.171 (P_c=0.99). Sink node numbered as 1 is set to reside at the center of the area. (b) A multicast source tree generated from the topology of Figure 14.(a) by using the DSA heuristic.

Figure 13 depicts the increased TRC values due to the wireless channel error $(p_b = 10^{-5})$ according to the increasing number of revoked nodes when $N = 1024$. We assume that each node in multicast communication can hear two copies of a message on average $(n = 2)$. For comparison purpose, we also plot the original TRC values of LKH, OKD, and TKH as dash, dot, and solid lines respectively. Due to wireless channel errors, LKH(g) and OKD(g) obtain about 20% and 10% increases in their total rekeying costs respectively, while LKH(u) and OKD(u) only obtain about 1% additional rekeying costs. Since group-oriented rekeying combines-and-multicasts all rekeying messages simultaneously, it has a large message size. Therefore, it suffers more from wireless channel errors than user-oriented rekeying which delivers individual small rekeying messages to each node. By combining multicast and unicast communications and exploiting topological information, TKH is resistant to wireless channel errors (only 0.048% TRC increase in Figure 13). TKH's multicast delivery of $\{GK\}_{TK}$ is more error-tolerant than unicast since the message length is always '1' while it can have multicast advantage. Other two unicast rekeying message types ($\{TK\}_{SK}$, $\{SK\}_{IK}$) are also error-tolerant since they also have very small message sizes.

6. Simulation Results

In the previous section, we provided the analysis of the total rekeying costs based on the homogeneous '$\alpha\beta\gamma$-tree' topology model. In this section, we further investigate the rekeying costs of TKH and other schemes in more general and heterogeneous sensor network topology model.

Generating a typical sensor network multicast topology is consisted of two phases: *connectivity graph generation* and *multicast source tree generation*. First, we generate a wireless sensor network connectivity graph by using the *Random Geometric Graph* model (Penrose, 2003). Let us assume that N sensor nodes are randomly deployed in an 1×1 unit square area. Each node has a common communication range of r, and a pair of nodes are connected if they reside within r to each other. The resulting network topology will be a graph (G) consisted of vertices (V) of sensors and edges (E) of wireless connectivity.

Under the given deployment area of a sensor network, increasing the number of nodes (N) or the communication range (r) will respectively increase the number of connections in the network. To obtain the appropriate value of r which connects N sensor nodes with the desired level of connectivity, we utilize the results from (Penrose, 1997). For N points placed uniformly at random on the unit square in the 2-dimensional space, Penrose (Penrose, 1997) found an asymptotic bound on the length of the longest edge (M_n) of MST (Minimum Spanning Tree) as follows

$$\lim_{N\to\infty} \text{Prob}\left[N\pi(M_N)^2 - \log N \le c\right] = \exp(-e^{-c}) \tag{27}$$

with constant c. If we choose the communication range r the same as M_n, we can assure that the graph is almost surely connected with probability of $\exp(-e^{-c})$ because all the nodes have the communication range same as the longest edge of their MST. That is, given the value of N, if we set r as $N\pi r^2 - \log N = c$, "the probability that a given graph is connected" is $\exp(-e^{-c})$. This probability is a "connectivity" of a graph which is denoted as P_c. By setting c according to the desired level of connectivity, we can derive the communication range r. Figure 14.(a) depicts a sample sensor network connectivity graph of 100 nodes in a unit square area with r=0.171 (P_c=0.99).

Second, from the network graph generated by using the previous method, we now transform it into a sink-based multicast source tree which actually delivers the central sink node's multicast messages on it. Among the many source tree generation algorithms (Diot et al., 1997), we use the simple and well-known algorithm: DSA (Dijkstra's Shortest path Algorithm) heuristic. If we overlap all the shortest paths from a source (s) to every nodes obtained from DSA (Cormen et al., 2001), we can build a multicast source tree starting from the central sink. However, our TKH can apply to any multicast source tree structures. We depict the multicast source tree in Figure 14.(b) which is generated from the Figure 14.(a) by using the DSA heuristic.

6.1 Simulation Results

In our simulations, we assume the network area of 1000×1000 size. For $N = 512, 1024$, we randomly placed sensor nodes with the communication range ($r = 82.1, 59.9$) obtained by setting the connectivity (P_c) as 0.99. We set the unit communication costs and the unit rekeying message size same as the analysis settings. For LKH and OKD, binary and 4-ary key trees are generated where each sensor node is randomly assigned in the key trees. TKH's key trees are automatically generated from the generated sensor network multicast topology. After revoking randomly chosen node from a network, we calculated total rekeying costs of the three schemes which occurred during the update of the group key of the remaining nodes. We obtain the total rekeying costs by averaging 1000 independent simulation results for each number of N.

Figure 15 depicts simulation results of TRC according to the increased number of revoked nodes (l). We plot the graphs until 10% of nodes are revoked from $N = 512, 1024$. By comparing Figure 15 with 11, simulation results also possesses similar trends with the analysis results. We also verify that the previous principles in terms of the total rekeying costs obtained in analysis results are still hold in Figure 15. This confirms that our TKH always incurs lower rekeying costs compared to the logical key tree schemes. On average, TKH only requires 18.6%, 33.7%, 15.2%, 27.9% of TRC compared to the most efficient logical key tree scheme (OKD(u) with 4-ary) in Figure 15.(a),(b),(c),(d) respectively.

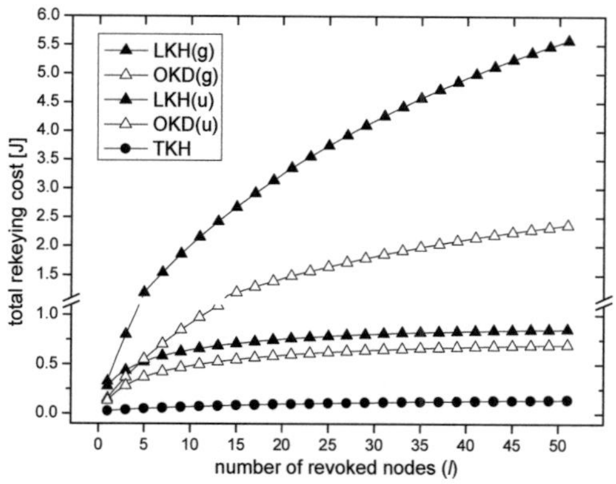

(a) $N=512$ (binary tree)

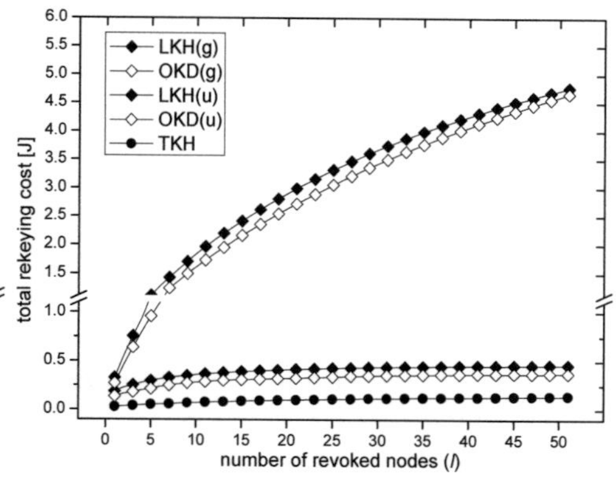

(b) $N=512$ (4-ary tree)

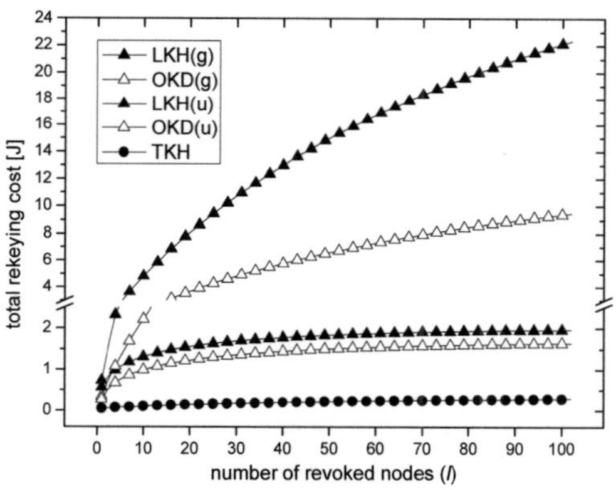

(c) $N=1024$ (binary tree)

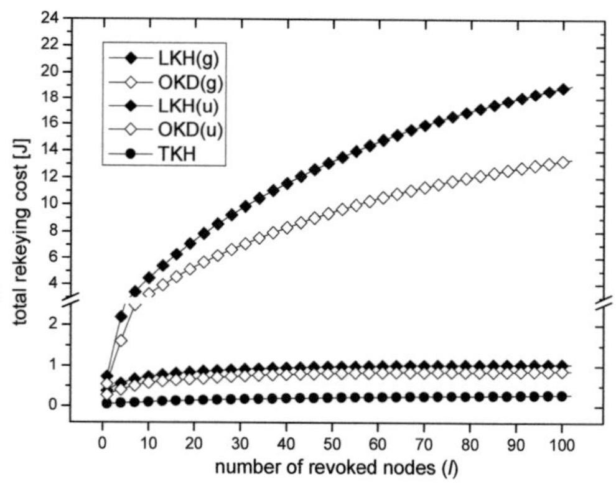

(d) $N=1024$ (4-ary tree)

Fig. 15. Simulation results of totla rekeying costs according to increasing number of revoked nodes (l) when $N=512, 1024$.

Many researchers have proposed methods to construct an efficient multicast tree topology for a multi-hop wireless network (Park & Sahni, 2005) (Wieselthier et al., 2002). These schemes explicitly considers the wireless multicast advantage during multicast tree generation. For example, by applying *the sweep operation* (Wieselthier et al., 2002) after the DSA heuristic will modifies the multicast tree to adopt more sibling nodes in each sibling set. This kind of wireless-optimized topology will further reduce the total rekeying cost of TKH.

7. Conclusions

In this chapter, we proposed energy efficient group key management scheme for a wireless sensor network. By explicitly considering the topological information during a key tree generation, we showed that the Topological Key Hierarchy could greatly reduce the total rekeying costs compared to the previous logical key tree-based schemes. After description of our key tree design principles, we proved performance improvements based on our detailed analysis results. We further compared rekeying costs in realistic simulation environments. TKH only requires about 10 to 30 percentages of rekeying costs compared to the best logical key tree scheme (OKD(u) with 4-ary) in the network of 1024 sensors. We conclude that our TKH can scale to large-scale sensor networks providing small rekeying cost for group key management.

8. References

Akyildiz, I. F., Weilian Su, Y. S. & Cayirci, E. (2002). A survey on sensor networks, *IEEE Communications Magazine* **40**(8): 102–114.

Bellare, M., Canetti, R. & Krawczyk, H. (1997). HMAC: Keyed-hashing for message authentication. IETF RFC 2104.

Blundo, C., Santis, A. D., Herzberg, A., Kutten, S., Vaccaro, U. & Yung, M. (1992). Perfectly-secure key distribution for dynamic conferences, *Advances in Cryptology—CRYPTO '92*, pp. 471–486.

Canetti, R., Garay, J., Itkis, G., Micciancio, D., Naor, M. & Pinkas, B. (1999). Multicast security: A taxonomy and some efficient constructions, *In Proceedings of the 18th IEEE INFOCOM.*

Chadha, A., Liu, Y. & Das, S. K. (2006). Group key distribution via local collaboration in wireless sensor networks, *IEEE International Conference on Sensor and Ad Hoc Communications and Networks (SECON).*

Cormen, T. H., Leiserson, C. E., Rivest, R. L. & Stein, C. (2001). *Introduction to Algorithms*, The MIT Press.

Diffie, W. & Hellman, M. E. (1976). New directions in cryptography, *IEEE Transactions on Information Theory* **22**(5): 644–654.

Diot, C., Dabbous, W. & Crowcroft, J. (1997). Multipoint communication: A survey of protocols, functions, and mechanisms, *IEEE Journal on Selected Areas in Communications* **15**(3): 277–290.

Horng, G. (2002). Cryptanalysis of a key management scheme for secure multicast communications, *IEICE: IEICE Transactions on Communications/Electronics/Information and Systems* **E85-B**(5): 1050–1051.

Intanagonwiwat, C., Govindan, R. & Estrin, D. (2000). Directed diffusion: a scalable and robust communication paradigm for sensor networks, *6th Annual ACM Internation Conference on Mobile Computing and Networking (MobiCom '00)*, pp. 56–67.

Kim, Y., Perrig, A. & Tsudik, G. (2000). Simple and fault-tolerant key agreement for dynamic collaborative groups, *7th ACM Conference on Computer and Communications Security (CCS).*

Ku, W.-C. & Chen, S.-M. (2003). An improved key management scheme for large dynamic groups using one-way function trees, *International Conference on Parallel Processing Workshops.*

Li, X. S., Yang, Y. R., Gouda, M. G. & Lam, S. S. (2001). Batch rekeying for secure group communications, *WWW10.*

Lin, J.-C., Lai, F. & Lee, H.-C. (2005). Efficient group key management protocol with one-way key derivation, *IEEE Conference on Local Computer Networks (LCN)*, pp. 336–343.

Panja, B., Madria, S. K. & K.Bhargava, B. (2006). Energy and communication efficient group key management protocol for hierarchical sensor networks, *IEEE International Conference on Sensor Networks, Ubiquitous, and Trustworthy Computing (SUTC).*

Park, J. & Sahni, S. (2005). Maximum lifetime broadcasting in wireless networks, *IEEE Tran. on Computers* **54**: 1081–1090.

Penrose, M. (2003). *Random Geometric Graphs*, Oxford Studies in Probability, Oxford University Press.

Penrose, M. D. (1997). The longest edge of the random minimal spanning tree, *The Annals of Applied Probability* **7**(2): 340–361.

Pietro, R. D., Mancini, L. V., Law, Y. W., Etalle, S. & Havinga, P. (2003). Lkhw: A directed diffusion-based secure multicast scheme for wireless sensor networks, *Proceedings of the 2003 International Conference on Parallel Processing Workshops*.

Rafaeli, S. & Hutchison, D. (2003). A survey of key management for secure group communication, *ACM Computing Surveys* **35**(3): 303–329.

Rivest, R. L., Shamir, A. & Adleman, L. A. (1978). A method for obtaining digital signatures and public-key cryptosystems, *Communications of the ACM* **21**(2): 120–126.

Salido, J., Lazos, L. & Poovendran, R. (2008). Energy and bandwidth-efficient key distribution in wireless ad-hoc networks: A cross-layer approach, *IEEE/ACM Transactions on Networking (to appear)* **15**(6): 1527–1540.

Sherman, A. T. & McGrew, D. A. (2003). Key establishment in large dynamic groups using one-way function trees, *IEEE Transactions on Software Engineering* **29**(5): 444–458.

Staddon, J., Miner, S., Franklin, M., Balfanz, D., Malkin, M. & Dean, D. (2002). Self-healing key distribution with revocation.

Steiner, M., Tsudik, G. & Waidner, M. (1996). Diffie-hellman key distribution extended to groups, *Third ACM Conference on Computer and Communications Security (CCS)*, pp. 31–37.

Sun, Y., Trappe, W. & Liu, K. J. R. (2004). A scalable multicast key management scheme for heterogeneous wireless networks, *IEEE/ACM Transactions on Networking* **12**(4): 653–666.

U.S. DoC NIST (2001). Advanced encryption standard. FIPS Publication 197.

ZigBee Alliance (2006). Zigbee specifications (version 1.0, r13).

Tucker, A. (1995). *Applied Combinatorics, Ch. 3*, John Wiley & Sons.

Waldvogel, M., Caronni, G., Sun, D., Weiler, N. & Plattner, B. (1999). The VersaKey framework: Versatile group key management, *IEEE Journal on Selected Areas in Communications* **17**(9): 1614–1631.

Wallner, D. M., Harder, E. J., & Agee, R. C. (1997). Key management for multicast: issues and architectures. IETF RFC 2627.

Wieselthier, J. E., Nguyen, G. D. & Ephremides, A. (2002). Energy-efficient broadcast and multicast trees in wireless networks, *Mobile Networks and Applications* **7**(6): 481–492.

Wong, C. K., Gouda, M. G. & Lam, S. S. (1998). Secure group communications using key graphs, *ACM SIGCOMM*.

Zhang, W. & Cao, G. (2005). Group rekeying for filtering false data in sensor networks: a predistribution and local collaboration-based approach, *IEEE INFOCOM*, pp. 503–514.

Zhao, F. & Guibas, L. J. (2004). *Wireless Sensor Networks: An Information Processing Approach*, Elsevier.